Fundamentals of Ecology

FIFTH EDITION

Fundamentals of Ecology

FIFTH EDITION

Eugene P. Odum, Ph.D.
Late of University of Georgia Institute of Ecology

Gary W. Barrett, Ph.D.
Odum Professor of Ecology, University of Georgia Institute of Ecology

Cengage

Australia • Brazil • Canada • India • Mexico • Singapore • United Kingdom • United States

Cengage

Fundamentals of Ecology, Fifth Edition

Eugene P. Odum and
Gary W. Barrett

For product information and technology assistance, contact us at
asia.infoindia@cengage.com

For permission to use material from this text or product, submit all requests online at
www.cengage.com/permissions

ISBN-13: 978-81-315-0020-0
ISBN-10: 81-315-0020-9

Cengage Learning India Pvt. Ltd.
T-303, 3rd Floor
DAV Complex, Plot 3
Mayur Vihar Phase 1
Delhi-110091

Cengage is a leading provider of customized learning solutions with employees residing in nearly 40 different countries and sales in more than 125 countries around the world. Find your local representative at **www.cengage.com**

Cengage is committed to creating inclusive and diverse products, including perspectives that represent the diversity of people across all backgrounds, identities, opinions, and experiences. If you have any feedback related to this, please contact us at **support.cengage.com**

To learn more about Cengage products, platforms and services, visit at **www.cengage.co.in**

Disclaimer:
This book contains link to resources which are a part of the international edition. The same may not be available with the Indian edition. The buyer needs to purchase a copy of the international edition to have full access to those resources. In no event the publisher shall be liable for any limitation or performance due to unavailability of these resources.

Printed by:
Rajiv Book Binding House
D-1518, DSIIDC Industrial Complex,
Narela, New Delhi 110040

Printed in India
Print Number: 27 Print Year: 2026

To Terry Lynn

With respect and gratitude

Brief Contents

Foreword xiii
Preface xv
Acknowledgments xvii

1 The Scope of Ecology 1
2 The Ecosystem 17
3 Energy in Ecological Systems 77
4 Biogeochemical Cycles 140
5 Limiting and Regulatory Factors 177
6 Population Ecology 224
7 Community Ecology 282
8 Ecosystem Development 336
9 Landscape Ecology 374
10 Regional Ecology: Major Ecosystem Types and Biomes 412
11 Global Ecology 459
12 Statistical Thinking for Students of Ecology 479
***Guest Author* R. Cary Tuckfield**

Glossary 511
References 535
Credits 583
Index 585

Contents

Foreword *xiii*
Preface *xv*
Acknowledgments *xvii*

1 The Scope of Ecology 1

1. Ecology: History and Relevance to Humankind 2
2. Levels-of-Organization Hierarchy 4
3. The Emergent Property Principle 7
4. Transcending Functions and Control Processes 9
5. Ecological Interfacing 10
6. About Models 10
7. Disciplinary Reductionism to Transdisciplinary Holism 15

2 The Ecosystem 17

1. Concept of the Ecosystem and Ecosystem Management 18
2. Trophic Structure of the Ecosystem 21
3. Gradients and Ecotones 24
4. Examples of Ecosystems 26
5. Ecosystem Diversity 37
6. Study of Ecosystems 42
7. Biological Control of the Geochemical Environment: The Gaia Hypothesis 43
8. Global Production and Decomposition 46
9. Microcosms, Mesocosms, and Macrocosms 60
10. Ecosystem Cybernetics 67
11. Technoecosystems 71
12. Concept of the Ecological Footprint 74
13. Classification of Ecosystems 75

3 Energy in Ecological Systems 77

1. Fundamental Concepts Related to Energy: The Laws of Thermodynamics 78
2. Solar Radiation and the Energy Environment 82
3. Concept of Productivity 86
4. Energy Partitioning in Food Chains and Food Webs 108
5. Energy Quality: eMergy 121
6. Metabolism and Size of Individuals: The 3/4 Power Principle 124
7. Complexity Theory, Energetics of Scale, and the Law of Diminishing Returns 126
8. Concepts of Carrying Capacity and Sustainability 127
9. Net Energy Concept 132
10. An Energy-Based Classification of Ecosystems 132
11. Energy Futures 135
12. Energy and Money 137

4 Biogeochemical Cycles 140

1. Basic Types of Biogeochemical Cycles 141
2. Cycling of Nitrogen 143
3. Cycling of Phosphorus 149
4. Cycling of Sulfur 151
5. Cycling of Carbon 153
6. The Hydrologic Cycle 156
7. Turnover and Residence Times 162
8. Watershed Biogeochemistry 163
9. Cycling of Nonessential Elements 168
10. Nutrient Cycling in the Tropics 170
11. Recycling Pathways: The Cycling Index 172
12. Global Climate Change 175

5 Limiting and Regulatory Factors 177

1. Concept of Limiting Factors: The Liebig Law of the Minimum 178
2. Factor Compensation and Ecotypes 183
3. Conditions of Existence as Regulatory Factors 185
4. Soil: Organizing Component for Terrestrial Ecosystems 187
5. Fire Ecology 194
6. Review of Other Physical Limiting Factors 199
7. Biological Magnification of Toxic Substances 215
8. Anthropogenic Stress as a Limiting Factor for Industrial Societies 219

6 Population Ecology 224

1. Properties of the Population 225
2. Basic Concepts of Rate 236
3. Intrinsic Rate of Natural Increase 238
4. Concept of Carrying Capacity 241

5. Population Fluctuations and Cyclic Oscillations 246
6. Density-Independent and Density-Dependent Mechanisms of Population Regulation 255
7. Patterns of Dispersion 258
8. The Allee Principle of Aggregation and Refuging 260
9. Home Range and Territoriality 263
10. Metapopulation Dynamics 267
11. Energy Partitioning and Optimization: *r*- and *K*-Selection 268
12. Population Genetics 275
13. Life History Traits and Tactics 280

7 Community Ecology 282

1. Types of Interaction Between Two Species 283
2. Coevolution 286
3. Evolution of Cooperation: Group Selection 288
4. Interspecific Competition and Coexistence 289
5. Positive/Negative Interactions: Predation, Herbivory, Parasitism, and Allelopathy 296
6. Positive Interactions: Commensalism, Cooperation, and Mutualism 304
7. Concepts of Habitat, Ecological Niche, and Guild 311
8. Biodiversity 316
9. Paleoecology: Community Structure in Past Ages 327
10. From Populations and Communities to Ecosystems and Landscapes 329

8 Ecosystem Development 336

1. Strategy of Ecosystem Development 337
2. Concept of the Climax 356
3. Evolution of the Biosphere 360
4. Microevolution Compared with Macroevolution, Artificial Selection, and Genetic Engineering 364
5. Relevance of Ecosystem Development to Human Ecology 368

9 Landscape Ecology 374

1. Landscape Ecology: Definition and Relation to Levels-of-Organization Concept 375
2. Landscape Elements 377
3. Biodiversity at the Community and Landscape Levels 386
4. Island Biogeography 389
5. Neutral Theory 392
6. Temporal and Spatial Scale 396
7. Landscape Geometry 399
8. Concept of Landscape Sustainability 404
9. Domesticated Landscapes 404

10 Regional Ecology: Major Ecosystem Types and Biomes 412

1. Marine Ecosystems 414
2. Freshwater Ecosystems 424
3. Terrestrial Biomes 432
4. Human-Designed and Managed Systems 457

11 Global Ecology 459

1. The Transition from Youth to Maturity: Toward Sustainable Civilizations 460
2. Ecological-Societal Gaps 465
3. Global Sustainability 467
4. Scenarios 472
5. Long-Term Transitions 477

12 Statistical Thinking for Students of Ecology 479

***Guest Author* R. Cary Tuckfield**

1. Ecosystems and Scale 480
2. Theory, Knowledge, and Research Design 482
3. The Ecological Study Unit 484
4. Inference Methods and Reliability 486
5. Experimental versus Observational Method in Ecology 489
6. Statistical Thinking in Ecology 490
7. The Nature of Evidence 492
8. Evidence and Hypothesis Testing 494
9. Formulating the Right Problem 496
10. Publish or "Parish"? 496
11. The Evidence-Oriented Alternative 498
12. The Two Ways of Discovery 506
13. The Weight of Evidence Paradigm 508

Glossary 511

References 535

Credits 583

Index 585

Foreword

Fundamentals of Ecology is an icon among biology textbooks—the most influential such work as measured by the number of students recruited into the field as researchers and teachers. The rebirth here of this classic, in a much modified fifth edition, but under the original title, is welcome.

There has always been a sense of inevitability about ecology, even in the early twentieth century, when it ranked as little more than a pastiche of natural history and schools of thought. Ecology was and remains the discipline that addresses the highest and most complex levels of biological organization. It was and remains a study of holism and emergence, of the properties of life taken from the top down. Even hard-nosed laboratory scientists, who were focused on the less complex (and more accessible) levels of molecules and cells, knew in their hearts that in time, biologists must eventually arrive at this discipline. To understand ecology thoroughly would be to understand *all* of biology, and to be a complete biologist is to be an ecologist. But ecology at the time of the first edition of *Fundamentals of Ecology* was still the most distant subject, enveloped in an intellectual haze, hard to picture except as scattered fragments. Odum's book was a map by which we could take a bearing. We need him still to learn the boundaries and principal features of ecology. The effectiveness of early editions of *Fundamentals of Ecology* is illustrated in a 2002 survey of the American Institute of Biological Sciences (Barrett and Mabry 2002) ranking it as the book that has recruited the most professionals into organismic and environmental biology.

The fifth edition, in comparison with the first edition, shows how far we have advanced in substance and in experimental studies linked to sophisticated theory and models. Subjects that were rudiments at the beginning—ecosystems analysis, energy and materials cycles, population dynamics, competition, biodiversity, and others—have grown to the rank of subdisciplines. They have been increasingly linked to one another and to the biology of organisms.

Furthermore, ecology is now seen as not just a biological but a human science. The future of our species depends on how well we understand that extension and employ it in the wise management of our natural resources. We live both by a market economy—necessary for our welfare on a day-to-day basis—and by a natural economy, necessary for our welfare (indeed, our very existence) in the long term. It is equally true that the pursuit of public health is largely an application of ecology. None of this should be surprising. We are, after all, a species in an ecosystem, exactly adapted to the conditions peculiar to the surface of this planet, and subject to the same principles of ecology as all other species.

The present edition provides a balanced approach among higher levels of biological organization. It can serve as a basic ecology text for college majors—not only in ecology and general biology, but also in the emerging disciplines of conservation biology and natural resource management. Moreover, it takes a futuristic look at such important topics as sustainability, environmental problem solving, and the relationship between market and natural capital.

Edward O. Wilson

Preface

This fifth edition of *Fundamentals of Ecology* retains the classic holistic approach to ecological science found in earlier versions of the text, but with more emphasis on a multilevel approach based on hierarchical theory and more attention to the applications of ecological principles to human predicaments, such as population growth, resource management, and environmental contamination. There is an emphasis on functions that transcend all levels of organization (Barrett et al. 1997), but attention has also been given to the unique emergent properties of individual levels.

The fifth edition also retains the text's original emphasis on the rich history of ecology and environmental science (Chapter 1), and on an understanding of the ecosystem concept and approach (Chapter 2). Chapters 3–5 focus on the major functional components of ecosystem/landscape dynamics, namely system energetics (Chapter 3), biogeochemical cycles (Chapter 4), and regulatory factors and processes (Chapter 5).

In keeping with a greater temporal/spatial approach to ecology, Chapters 6–11 scale levels of organization, including processes that transcend all levels, moving through the population (Chapter 6), community (Chapter 7), ecosystem (Chapter 8), landscape (Chapter 9), regional/biome (Chapter 10), and global (Chapter 11) levels. A final chapter entitled "Statistical Thinking for Students of Ecology" provides a quantitative synthesis to the field of ecology. Our purpose throughout the book is to wed theory with application, to present holistic and reductionist approaches, and to integrate systems ecology with evolutionary biology. The fifth edition text is accompanied by a set of electronic figures and photographs available for download at the book's companion Web site, http://www.brookscole.com/biology.

Although Sir Arthur G. Tansley first proposed the term "ecosystem" in 1935, and Raymond L. Lindeman in 1942 focused attention on the trophic-dynamic relationships of ecosystem structure to function, it was Eugene P. Odum who began the education of generations of ecologists throughout the world when he published the first edition of *Fundamentals of Ecology* in 1953. The clarity and enthusiasm of his holistic approach to both aquatic and terrestrial ecosystems in the second edition (Odum 1959), written in collaboration with his brother Howard T. Odum, were wonderfully compelling (Barrett and Likens 2002). In fact, a survey of the members of the American Institute of Biological Sciences (AIBS) found that *Fundamentals of Ecology* ranked first as the book that had most influenced career training in the biological sciences (Barrett and Mabry 2002).

Since 1970 ecology has fully emerged from its roots in the biological sciences to become a separate discipline, one that integrates organisms, the physical environment, and humans, in keeping with the Greek root of the word ecology, *oikos,* meaning "household." Ecology as the study of Earth as home has, in our opinion, matured enough to be considered the basic and integrative science of the environment as a whole, contributing to C. P. Snow's "third culture," namely, a much needed bridge between science and society (Snow 1963).

The weekly science journal *Nature* runs an occasional feature called "Concepts," one-page commentaries written by noted scientists. In a 2001 commentary entitled "Macroevolution: The Big Picture," Sean B. Carroll noted that "Many geneticists assert that macroevolution is a product of microevolution writ large, but some paleontologists believe that processes operating at higher levels also shape evolutionary trends." Tamas Vicsek extended this notion in a 2002 commentary entitled "Complexity: The Bigger Picture": "The laws that describe the behavior of a complex system are qualitatively different from those that govern its units." In this fifth edition of *Fundamentals of Ecology* we place special emphasis on macroevolution as an extension of traditional evolutionary theory and on self-regulation theory in the development and regulation of complex systems.

Too often, textbooks in successive editions grow ever larger, gradually expanding into encyclopedic tomes containing too much material to be covered in a single term. When the third edition of *Fundamentals of Ecology* was completed in 1971, it was decided that the next edition would be shorter and have a different title. Thus, *Basic Ecology* was born in 1983, as the fourth edition. Now with the fifth edition we return to the original title, *Fundamentals of Ecology.*

As with earlier editions this one is very much a product of students and colleagues associated with the University of Georgia Institute of Ecology. And we are especially indebted to the late Howard T. Odum, whose many footprints appear across the pages that follow.

Eugene P. Odum and Gary W. Barrett

Acknowledgments

I thank Dr. Eugene P. Odum for inviting me to coauthor the fifth edition of *Fundamentals of Ecology.* Dr. Odum was a lifelong mentor and a longtime friend. I have been honored to hold the Odum Professorship of Ecology at the University of Georgia over the past ten years. Dr. Odum and I submitted a draft of this book for publication shortly before his death, at the age of 88 on 10 August 2002. Thus the changes made following extensive review by the publisher have been my responsibility.

Special thanks is extended to R. Cary Tuckfield, Savannah River National Laboratory, for his contribution of Chapter 12, "Statistical Thinking for Students of Ecology," and to Edward O. Wilson, Harvard University, for providing the foreword to this book. Dr. Odum and I are indebted to Terry L. Barrett for transcription, editing, and suggestions regarding all aspects of this book. I thank Lawrence R. Pomeroy for reading and making helpful suggestions regarding Chapter 4 and Mark D. Hunter for his editorial comments regarding Chapter 6. Special thanks are extended to Krysia Haag, Computer Graphics Artist II, University of Georgia, for contributions to the graphics in this book. Others who contributed information or materials pertaining to this fifth edition include Walter P. Carson, University of Pittsburg; Steven J. Harper, Savannah River Ecology Laboratory; Sue Hilliard, Joseph W. Jones Ecological Research Center; Stephen P. Hubbell, University of Georgia; Donald W. Kaufman, Kansas State University; and Michael J. Vanni, Miami University of Ohio. I am also indebted to those who reviewed the text, including Dr. David M. Armstrong, University of Colorado at Boulder; Dr. David L. Hicks, Whitworth College; Dr. Thomas R. Wentworth, North Carolina State University; and Dr. Matt R. Whiles, Southern Illinois University.

Sincere thanks are extended to Peter Marshall, publisher, Brooks/Cole, Cengage Learning; Elizabeth Howe, development editor, Brooks/Cole, Cengage Learning; and Jennifer Risden, editorial production project manager, Brooks/Cole, Cengage Learning, regarding all aspects of the publication process. I thank Gretchen Otto, production coordinator, G&S Book Services; Jan Six, copy editor; and Terri Wright, photo researcher, for individual excellence in their respective fields.

I would also be remiss if I did not thank all those colleagues and graduate students of Dr. Odum who contributed to the earlier editions of this book. Having taught ecology for nearly four decades, twenty-six of those years on the faculty at Miami University of Ohio, I thank all those students of whom I have had the privilege to interact with regarding their ecological education. I trust this fifth edition will help future generations to appreciate the significance of understanding ecological theory, concepts, mechanisms, and natural laws regarding their decision-making processes, just as the second edition had a profound influence on my career and ecological understanding.

Gary W. Barrett

1

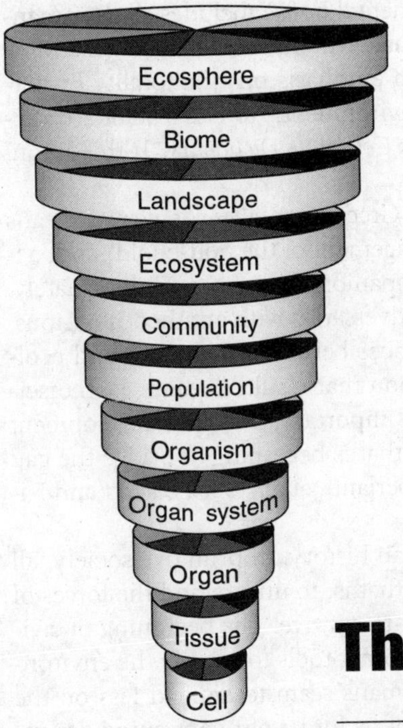

The Scope of Ecology

1 **Ecology: History and Relevance to Humankind**
2 **Levels-of-Organization Hierarchy**
3 **The Emergent Property Principle**
4 **Transcending Functions and Control Processes**
5 **Ecological Interfacing**
6 **About Models**
7 **Disciplinary Reductionism to Transdisciplinary Holism**

1 Ecology: History and Relevance to Humankind

The word *ecology* is derived from the Greek *oikos,* meaning "household," and *logos,* meaning "study." Thus, the study of the environmental house includes all the organisms in it and all the functional processes that make the house habitable. Literally, then, **ecology** is the study of "life at home" with emphasis on "the totality or pattern of relations between organisms and their environment," to cite a standard dictionary definition of the word (*Merriam-Webster's Collegiate Dictionary,* 10th edition, s.v. "ecology").

The word *economics* is also derived from the Greek root *oikos.* As *nomics* means "management," economics translates as "the management of the household" and, accordingly, ecology and economics should be companion disciplines. Unfortunately, many people view ecologists and economists as adversaries with antithetical visions. Table 1-1 attempts to illustrate perceived differences between economics and ecology. Later, this book will consider the confrontation that results because each discipline takes a narrow view of its subject and, more important, the rapid development of a new interface discipline, *ecological economics,* that is beginning to bridge the gap between ecology and economics (Costanza, Cumberland, et al. 1997; Barrett and Farina 2000; L. R. Brown 2001).

Ecology was of practical interest early in human history. In primitive society, all individuals needed to know their environment—that is, to understand the forces of nature and the plants and animals around them—to survive. The beginning of civilization, in fact, coincided with the use of fire and other tools to modify the environment. Because of technological achievements, humans seem to depend less on the natural environment for their daily needs; many of us forget our continuing dependence on nature for air, water, and indirectly, food, not to mention waste assimilation, recreation, and many other services supplied by nature. Also, economic systems, of whatever political ideology, value things made by human beings that primarily benefit the individual, but they place little monetary value on the goods and services of nature that benefit us as a society. Until there is a crisis, humans tend to take nat-

Table 1-1

A summary of perceived differences between economics and ecology

Attribute	*Economics*	*Ecology*
School of thought	Cornucopian	Neo-Malthusian
Currency	Money	Energy
Growth form	J-shaped	S-shaped
Selection pressure	*r*-selected	*K*-selected
Technological approach	High technology	Appropriate technology
System services	Services provided by economic capital	Services provided by natural capital
Resource use	Linear (disposal)	Circular (recycling)
System regulation	Exponential expansion	Carrying capacity
Futuristic goal	Exploration and expansion	Sustainability and stability

Figure 1-1. Earthscape as viewed from Apollo 17 traveling toward the Moon. View of the ecosphere from "outside the box."

© NASA

ural goods and services for granted; we assume they are unlimited or somehow replaceable by technological innovations, even though we know that life necessities such as oxygen and water may be recyclable but not replaceable. As long as the life-support services are considered free, they have no value in current market systems (see H. T. Odum and E. P. Odum 2000).

Like all phases of learning, the science of ecology has had a gradual if spasmodic development during recorded history. The writings of Hippocrates, Aristotle, and other philosophers of ancient Greece clearly contain references to ecological topics. However, the Greeks did not have a word for ecology. The word *ecology* is of recent origin, having been first proposed by the German biologist Ernst Haeckel in 1869. Haeckel defined *ecology* as "the study of the natural environment including the relations of organisms to one another and to their surroundings" (Haeckel 1869). Before this, during a biological renaissance in the eighteenth and nineteenth centuries, many scholars had contributed to the subject, even though the word *ecology* was not in use. For example, in the early 1700s, Antoni van Leeuwenhoek, best known as a premier microscopist, also pioneered the study of food chains and population regulation, and the writings of the English botanist Richard Bradley revealed his understanding of biological productivity. All three of these subjects are important areas of modern ecology.

As a recognized, distinct field of science, ecology dates from about 1900, but only in the past few decades has the word become part of the general vocabulary. At first, the field was rather sharply divided along taxonomic lines (such as plant ecology and animal ecology), but the biotic community concept of Frederick E. Clements and Victor E. Shelford, the food chain and material cycling concepts of Raymond Lindeman and G. Evelyn Hutchinson, and the whole lake studies of Edward A. Birge and Chauncy Juday, among others, helped establish basic theory for a unified field of general ecology. The work of these pioneers will be cited often in subsequent chapters.

What can best be described as a worldwide environmental awareness movement burst upon the scene during two years, 1968 to 1970, as astronauts took the first photographs of Earth as seen from outer space. For the first time in human history, we were able to see Earth as a whole and to realize how alone and fragile Earth hovers in space (Fig. 1-1). Suddenly, during the 1970s, almost everyone became concerned about pollution, natural areas, population growth, food and energy consumption, and biotic diversity, as indicated by the wide coverage of environmental concerns in the popular press. The 1970s were frequently referred to as the "decade of the environment," initiated by the first "Earth Day" on 22 April 1970. Then, in the 1980s and 1990s, environmental issues were pushed into the political background by concerns for human relations—problems such as crime, the cold war, government budgets, and welfare. As we enter the early stages of the twenty-first century, environmental concerns are again coming to the forefront because human abuse of Earth continues to escalate. We hope that this time, to use a medical analogy, our emphasis will be on prevention rather than on treatment, and ecology as outlined in this book, can contribute a great deal to prevention technology and ecosystem health (Barrett 2001).

The increase in public attention had a profound effect on academic ecology. Before the 1970s, ecology was viewed largely as a subdiscipline of biology. Ecologists were staffed in biology departments, and ecology courses were generally found only in the biological science curricula. Although ecology remains strongly rooted in biology, it has emerged from biology as an essentially new, integrative discipline that links physical and biological processes and forms a bridge between the natural sciences and the social sciences (E. P. Odum 1977). Most colleges now offer campus-wide courses and have separate majors, departments, schools, centers, or institutes of ecology. While the scope of ecology is expanding, the study of how individual organisms and species interface and use resources intensifies. The multilevel approach, as outlined in the next section, brings together "evolutionary" and "systems" thinking, two approaches that have tended to divide the field in recent years.

2 Levels-of-Organization Hierarchy

Perhaps the best way to delimit modern ecology is to consider the concept of **levels of organization**, visualized as an ecological spectrum (Fig. 1-2) and as an extended ecological hierarchy (Fig. 1-3). **Hierarchy** means "an arrangement into a graded series" (*Merriam-Webster's Collegiate Dictionary,* 10th edition, s.v. "hierarchy"). Interaction with the physical environment (energy and matter) at each level produces characteristic functional systems. A **system**, according to a standard definition, consists of "regularly interacting and interdependent components forming a unified

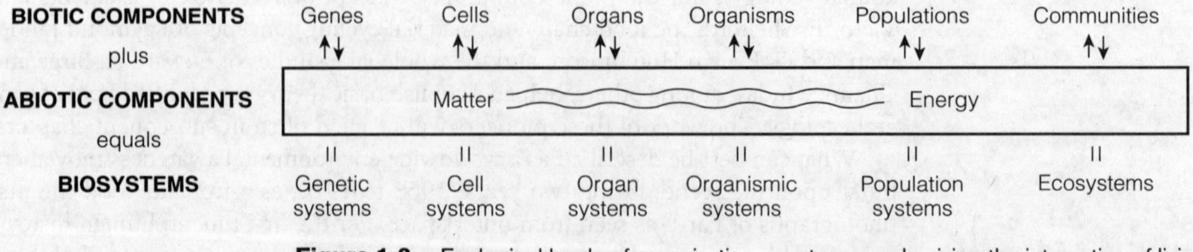

Figure 1-2. Ecological levels-of-organization spectrum emphasizing the interaction of living (biotic) and nonliving (abiotic) components.

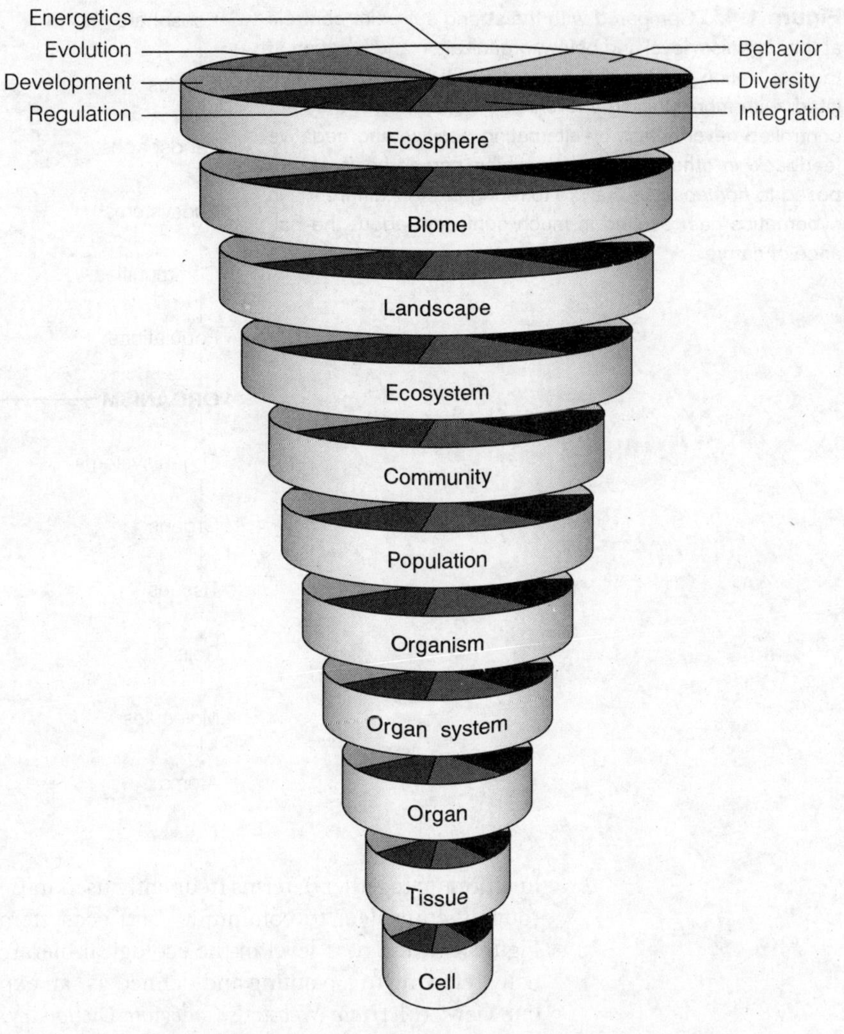

Figure 1-3. Ecological levels-of-organization hierarchy; seven transcending processes or functions are depicted as vertical components of eleven integrative levels of organization (after Barrett et al. 1997).

whole" (*Merriam-Webster's Collegiate Dictionary,* 10th edition, s.v. "system"). Systems containing living (biotic) and nonliving (abiotic) components constitute *biosystems,* ranging from genetic systems to ecological systems (Fig. 1-2). This spectrum may be conceived of or studied at any level, as illustrated in Figure 1-2, or at any intermediate position convenient or practical for analysis. For example, host-parasite systems or a two-species system of mutually linked organisms (such as the fungi-algae partnership that constitutes the lichen) are intermediate levels between population and community.

Ecology is largely, but not entirely, concerned with the system levels beyond that of the organism (Figs. 1-3 and 1-4). In ecology, the term **population,** originally coined to denote a group of people, is broadened to include groups of individuals of any one kind of organism. Likewise, **community,** in the ecological sense (sometimes designated as "biotic community"), includes all the populations occupying a given area. The community and the nonliving environment function together as an ecological system or **ecosystem.** *Biocoenosis* and *biogeocoenosis* (literally, "life and Earth

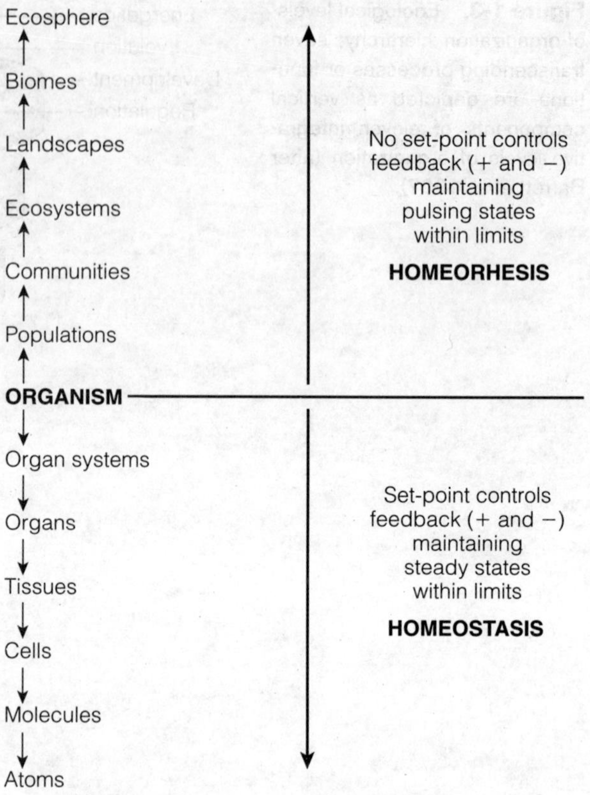

Figure 1-4. Compared with the strong set-point controls at the organism level and below, organization and function at the population level and above are much less tightly regulated, with more pulsing and chaotic behavior, but they are controlled nevertheless by alternating positive and negative feedback–in other words, they exhibit *homeorhesis* as opposed to *homeostasis.* Failure to recognize this difference in cybernetics has resulted in much confusion about the balance of nature.

functioning together"), terms frequently used in European and Russian literature, are roughly equivalent to community and ecosystem, respectively. Referring again to Figure 1-3, the next level in the ecological hierarchy is the landscape, a term originally referring to a painting and defined as "an expanse of scenery seen by the eye as one view" (*Merriam-Webster's Collegiate Dictionary,* 10th edition, s.v. "landscape"). In ecology, **landscape** is defined as a "heterogenous area composed of a cluster of interacting ecosystems that are repeated in a similar manner throughout" (Forman and Godron 1986). A *watershed* is a convenient landscape-level unit for large-scale study and management because it usually has identifiable natural boundaries. *Biome* is a term in wide use for a large regional or subcontinental system characterized by a major vegetation type or other identifying landscape aspect, as, for example, the Temperate Deciduous Forest biome or the Continental Shelf Ocean biome. A *region* is a large geological or political area that may contain more than one biome—for example, the regions of the Midwest, the Appalachian Mountains, or the Pacific Coast. The largest and most nearly self-sufficient biological system is often designated as the **ecosphere,** which includes all the living organisms of Earth interacting with the physical environment as a whole to maintain a self-adjusting, loosely controlled pulsing state (more about the concept of "pulsing state" later in this chapter).

Hierarchical theory provides a convenient framework for subdividing and examining complex situations or extensive gradients, but it is more than just a useful rank-order classification. It is a holistic approach to understanding and dealing with complex situations, and is an alternative to the reductionist approach of seeking answers by reducing problems to lower-level analysis (Ahl and Allen 1996).

More than 50 years ago, Novikoff (1945) pointed out that there is both continuity and discontinuity in the evolution of the universe. Development may be viewed as continuous because it involves never-ending change, but it is also discontinuous because it passes through a series of different levels of organization. As we shall discuss in Chapter 3, the organized state of life is maintained by a continuous but stepwise flow of energy. Thus, dividing a graded series, or hierarchy, into components is in many cases arbitrary, but sometimes subdivisions can be based on natural discontinuities. Because each level in the levels-of-organization spectrum is "integrated" or interdependent with other levels, there can be no sharp lines or breaks in a functional sense, not even between organism and population. The individual organism, for example, cannot survive for long without its population, any more than the organ would be able to survive for long as a self-perpetuating unit without its organism. Similarly, the community cannot exist without the cycling of materials and the flow of energy in the ecosystem. This argument is applicable to the previously discussed mistaken notion that human civilization can exist separately from the natural world.

It is very important to emphasize that hierarchies in nature are *nested*—that is, each level is made up of groups of lower-level units (populations are composed of groups of organisms, for example). In sharp contrast, human-organized hierarchies in governments, cooperations, universities, or the military are *nonnested* (sergeants are not composed of groups of privates, for example). Accordingly, human-organized hierarchies tend to be more rigid and more sharply separated as compared to natural levels of organization. For more on hierarchical theory, see T. F. H. Allen and Starr (1982), O'Neill et al. (1986), and Ahl and Allen (1996).

3 The Emergent Property Principle

An important consequence of hierarchical organization is that as components, or subsets, are combined to produce larger functional wholes, new properties emerge that were not present at the level below. Accordingly, an **emergent property** of an ecological level or unit cannot be predicted from the study of the components of that level or unit. Another way to express the same concept is **nonreducible property**—that is, a property of the whole not reducible to the sum of the properties of the parts. Though findings at any one level aid in the study of the next level, they never completely explain the phenomena occurring at the next level, which must itself be studied to complete the picture.

Two examples, one from the physical realm and one from the ecological realm, will suffice to illustrate emergent properties. When hydrogen and oxygen are combined in a certain molecular configuration, water is formed—a liquid with properties utterly different from those of its gaseous components. When certain algae and coelenterate animals evolve together to produce a coral, an efficient nutrient cycling mechanism is created that enables the combined system to maintain a high rate of productivity in waters with a very low nutrient content. Thus, the fabulous productivity and diversity of coral reefs are emergent properties only at the level of the reef community.

Salt (1979) suggested that a distinction be made between emergent properties, as defined previously, and **collective properties,** which are summations of the behavior of components. Both are properties of the whole, but the collective properties do not involve new or unique characteristics resulting from the functioning of the whole unit. *Birth rate* is an example of a population level collective property, as it is merely a sum of the individual births in a designated time period, expressed as a fraction or percent of the total number of individuals in the population. New properties emerge because the components interact, not because the basic nature of the components is changed. Parts are not "melted down," as it were, but integrated to produce unique new properties. It can be demonstrated mathematically that integrative hierarchies evolve more rapidly from their constituents than nonhierarchical systems with the same number of elements; they are also more resilient in response to disturbance. Theoretically, when hierarchies are decomposed to their various levels of subsystems, the latter can still interact and reorganize to achieve a higher level of complexity.

Some attributes, obviously, become more complex and variable as one proceeds to higher levels of organization, but often other attributes become less complex and less variable as one goes from the smaller to the larger unit. Because feedback mechanisms (checks and balances, forces and counterforces) operate throughout, the amplitude of oscillations tends to be reduced as smaller units function within larger units. Statistically, the variance of the whole-system level property is less than the sum of the variance of the parts. For example, the rate of photosynthesis of a forest community is less variable than that of individual leaves or trees within the community, because when one component slows down, another component may speed up to compensate. When one considers both the emergent properties and the increasing homeostasis that develop at each level, not all component parts must be known before the whole can be understood. This is an important point, because some contend that it is useless to try to work on complex populations and communities when the smaller units are not yet fully understood. Quite the contrary, one may begin study at any point in the spectrum, provided that adjacent levels, as well as the level in question, are considered, because, as already noted, some attributes are predictable from parts (collective properties), but others are not (emergent properties). Ideally, a system-level study is itself a threefold hierarchy: system, subsystem (next level below), and suprasystem (next level above). For more on emergent properties, see T. F. H. Allen and Starr (1982), T. F. H. Allen and Hoekstra (1992), and Ahl and Allen (1996).

Each biosystem level has emergent properties and reduced variance as well as a summation of attributes of its subsystem components. The folk wisdom about the forest being more than just a collection of trees is, indeed, a first working principle of ecology. Although the philosophy of science has always been holistic in seeking to understand phenomena as a whole, in recent years the practice of science has become increasingly reductionist in seeking to understand phenomena by detailed study of smaller and smaller components. Laszlo and Margenau (1972) described within the history of science an alternation of reductionist and holistic thinking (*reductionism-constructionism* and *atomism-holism* are other pairs of words used to contrast these philosophical approaches). The law of diminishing returns may very well be involved here, as excessive effort in any one direction eventually necessitates taking the other (or another) direction.

The reductionist approach that has dominated science and technology since Isaac

Newton has made major contributions. For example, research at the cellular and molecular levels has established a firm basis for the future cure and prevention of cancers at the level of the organism. However, cell-level science will contribute very little to the well-being or survival of human civilization if we understand the higher levels of organization so inadequately that we can find no solutions to population overgrowth, pollution, and other forms of societal and environmental disorders. Both holism and reductionism must be accorded equal value—and simultaneously, not alternatively (E. P. Odum 1977; Barrett 1994). Ecology seeks synthesis, not separation. The revival of the holistic disciplines may be due at least partly to citizen dissatisfaction with the specialized scientist who cannot respond to the large-scale problems that need urgent attention. (Historian Lynn White's 1980 essay "The Ecology of Our Science" is recommended reading on this viewpoint.) Accordingly, we shall discuss ecological principles at the ecosystem level, with appropriate attention to organism, population, and community subsets and to landscape, biome, and ecosphere suprasets. This is the philosophical basis for the organization of the chapters in this book.

Fortunately, in the past 10 years, technological advances have allowed humans to deal quantitatively with large, complex systems such as ecosystems and landscapes. Tracer methodology, mass chemistry (spectrometry, colorimetry, chromatography), remote sensing, automatic monitoring, mathematic modeling, geographical information systems (GIS), and computer technology are providing the tools. Technology is, of course, a double-edged sword; it can be the means of understanding the wholeness of humans and nature or of destroying it.

4 Transcending Functions and Control Processes

Whereas each level in the ecological hierarchy can be expected to have unique emergent and collective properties, there are basic functions that operate at all levels. Examples of such **transcending functions** are behavior, development, diversity, energetics, evolution, integration, and regulation (see Fig. 1-3 for details). Some of these (energetics, for example) operate the same throughout the hierarchy, but others differ in *modus operandi* at different levels. Natural selection evolution, for example, involves mutations and other direct genetic interactions at the organism level but indirect coevolutionary and group selection processes at higher levels.

It is especially important to emphasize that although positive and negative feedback controls are universal, from the organism down, control is *set point,* in that it involves very exacting genetic, hormonal, and neural controls on growth and development, leading to what is often called **homeostasis.** As noted on the right-hand side of Figure 1-4, there are no set-point controls above the organism level (no chemostats or thermostats in nature). Accordingly, feedback control is much looser, resulting in pulsing rather than steady states. The term **homeorhesis,** from the Greek meaning "maintaining the flow," has been suggested for this pulsing control. In other words, there are no equilibriums at the ecosystem and ecosphere levels, but there are *pulsing balances,* such as between production and respiration or between oxygen and carbon dioxide in the atmosphere. Failure to recognize this difference in *cybernetics* (the science dealing with mechanisms of control or regulation) has resulted in much confusion about the realities of the so-called "balance of nature."

5 Ecological Interfacing

Because ecology is a broad, multilevel discipline, it interfaces well with traditional disciplines that tend to have more narrow focus. During the past decade, there has been a rapid rise of interface fields of study accompanied by new societies, journals, symposium volumes, books—and new careers. Ecological economics, one of the most important, was mentioned in the first section in this chapter. Others that are receiving a great deal of attention, especially in resource management, are agroecology, biodiversity, conservation ecology, ecological engineering, ecosystem health, ecotoxicology, environmental ethics, and restoration ecology.

In the beginning, an interface effort enriches the disciplines being interfaced. Lines of communication are established, and the expertise of narrowly trained "experts" in each field is expanded. However, for an interface field to become a new discipline, something new has to emerge, such as a new concept or technology. The concept of nonmarket goods and services, for example, was a new concept that emerged in ecological economics, but that initially neither traditional ecologists nor economists would put in their textbooks (Daily 1997; Mooney and Ehrlich 1997).

Throughout this book, we will refer to natural capital and economic capital. **Natural capital** is defined as the benefits and services supplied to human societies by natural ecosystems, or provided "free of cost" by unmanaged natural systems. These benefits and services include purification of water and air by natural processes, decomposition of wastes, maintenance of biodiversity, control of insect pests, pollination of crops, mitigation of floods, and provision of natural beauty and recreation, among others (Daily 1997).

Economic capital is defined as the goods and services provided by humankind, or the human workforce, typically expressed as the gross national product (GNP). **Gross national product** is the total monetary value of all goods and services provided in a country during one year. Natural capital is typically quantified and expressed in units of energy, whereas economic capital is expressed in monetary units (Table 1-1). Only in recent years has there been an attempt to value the world's ecosystem services and natural capital in monetary terms. Costanza, d'Arge, et al. (1997) estimated this value to be in the range of 16 to 54 trillion U.S. dollars per year for the entire biosphere, with an average of *33 trillion* U.S. dollars per year. Thus it is wise to protect natural ecosystems, both ecologically and economically, because of the benefits and services they provide to human societies, as will be illustrated in the chapters that follow.

6 About Models

If ecology is to be discussed at the ecosystem level, for reasons already indicated, how can this complex and formidable system level be dealt with? We begin by describing simplified versions that encompass only the most important, or basic, properties and functions. Because, in science, simplified versions of the real world are called *models,* it is appropriate now to introduce this concept.

A **model** (by definition) is a formulation that mimics a real-world phenomenon and by which predictions can be made. In their simplest form, models may be verbal or graphic (*informal*). Ultimately, however, models must be statistical and mathematical (*formal*) if their quantitative predictions are to be reasonably good. For example, a mathematical formulation that mimics numerical changes in a population of insects and that predicts the numbers in the population at some time would be considered a biologically useful model. If the insect population in question is a pest species, the model could have an economically important application.

Computer-simulated models permit one to predict probable outcomes as parameters in the model are changed, as new parameters are added, or as old ones are removed. Thus, a mathematical formulation can often be "tuned" or refined by computer operations to improve the "fit" to the real-world phenomenon. Above all, models summarize what is understood about the situation modeled and thereby delimit aspects needing new or better data, or new principles. When a model does not work—when it poorly mimics the real world—computer operations can often provide clues to the refinements or changes needed. Once a model proves to be a useful mimic, opportunities for experimentation are unlimited, because one can introduce new factors or perturbations and see how they would affect the system. Even when a model inadequately mimics the real world, which is often the case in its early stages of development, it remains an exceedingly useful teaching and research tool if it reveals key components and interactions that merit special attention.

Contrary to the feeling of many who are skeptical about modeling the complexity of nature, information about only a relatively small number of variables is often a sufficient basis for effective models because key factors, or emergent and other integrative properties, as discussed in Sections 2 and 3, often dominate or control a large percentage of the action. Watt (1963), for example, stated, "We do not need a tremendous amount of information about a great many variables to build revealing mathematical models." Though the mathematical aspects of modeling are a subject for advanced texts, we should review the first steps in model building.

Modeling usually begins with the construction of a diagram, or "graphic model," which is often a box or compartment diagram, as illustrated in Figure 1-5. Shown are two properties, P_1 and P_2, that interact, I, to produce or affect a third property, P_3, when the system is driven by an energy source, E. Five flow pathways, F, are shown, with F_1 representing the input and F_6 the output for the system as a whole. Thus, at a minimum, there are five ingredients or components for a working model of an ecological situation, namely, (1) an energy source or other outside **forcing function,** E; (2) properties called **state variables,** $P_1, P_2, \ldots P_n$; (3) **flow pathways,** $F_1, F_2, \ldots F_n$, showing where energy flows or material transfers connect properties with each other and with forces; (4) **interaction functions,** I, where forces and properties interact to modify, amplify, or control flows or create new "emergent" properties; and (5) **feedback loops,** L.

Figure 1-5 could serve as a model for the production of photochemical smog in the air over Los Angeles. In this case, P_1 could represent hydrocarbons and P_2 nitrogen oxides, two products of automobile exhaust emission. Under the driving force of sunlight energy, E, these interact to produce photochemical smog, P_3. In this case, the interaction function, I, is a synergistic or augmentative one, in that P_3 is a more serious pollutant for humans than is P_1 or P_2 acting alone.

Alternatively, Figure 1-5 could depict a grassland ecosystem in which P_1 represents the green plants that convert the energy of the Sun, E, to food. P_2 might represent a herbivorous animal that eats plants, and P_3 an omnivorous animal that can eat

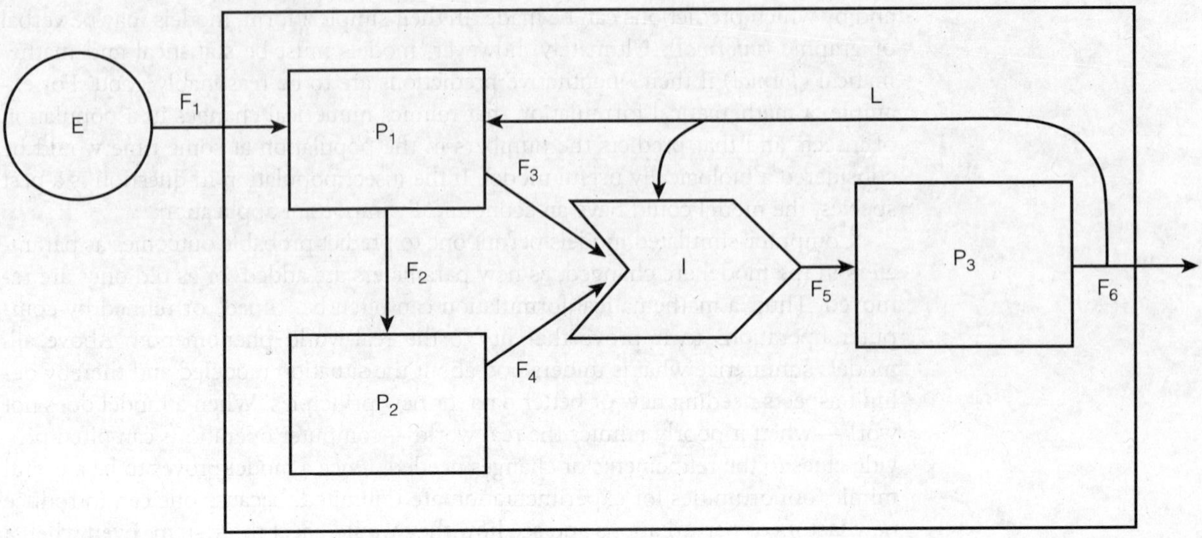

Figure 1-5. Compartment diagram showing the five basic components of primary interest in modeling ecological systems. E = energy source (forcing function); P_1, P_2, P_3 = state variables; F_1–F_6 = flow pathways; I = interaction function; L = feedback loop.

either the herbivores or the plants. In this case, the interaction function, I, could represent several possibilities. It could be a no-preference switch if observation in the real world showed that the omnivore P_3 eats either P_1 or P_2, according to availability. Or I could be specified to be a constant percentage value if it was found that the diet of P_3 was composed of, say, 80 percent plant and 20 percent animal matter, irrespective of the state of P_1 or P_2. Or I could be a seasonal switch if P_3 feeds on plants during one part of the year and on animals during another season. Or I could be a threshold switch if P_3 greatly prefers animal food and switches to plants only when P_2 is reduced to a low level.

Feedback loops are important features of ecological models because they represent control mechanisms. Figure 1-6 is a simplified diagram of a system that features a feedback loop in which "downstream" output, or some part of it, is fed back or recycled to affect or perhaps control "upstream" components. For example, the feedback loop could represent predation by "downstream" organisms, C, that reduce and thereby tend to control the growth of "upstream" herbivores or plants B and A in the food chain. Often, such a feedback actually promotes the growth or survival of a downstream component, such as a grazer enhancing the growth of plants (a "reward feedback," as it were).

Figure 1-6. Compartment model with a feedback or control loop that transforms a linear system into a partially cyclical one.

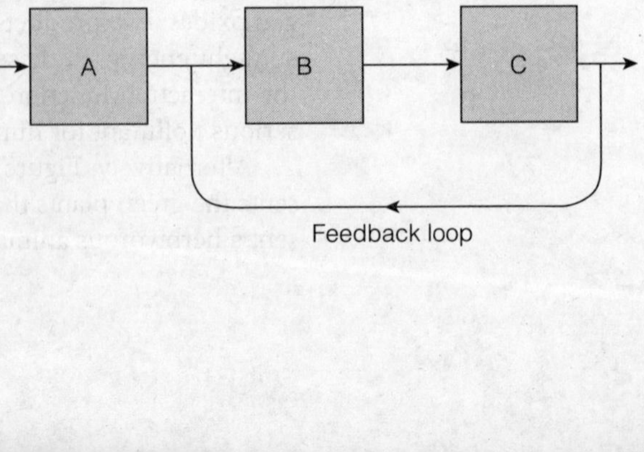

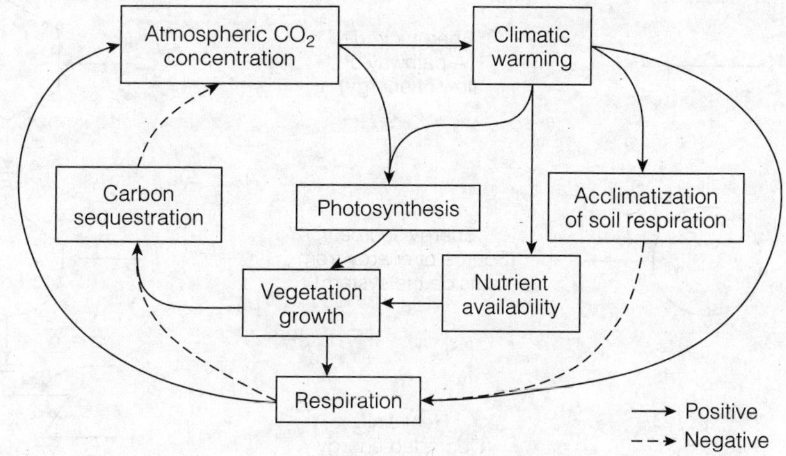

Figure 1-7. Interaction of positive and negative feedbacks in the relationships of atmospheric CO_2, climate warming, soil respiration, and carbon sequestration (modified after Luo et al. 2001).

Figure 1-6 could also represent a desirable economic system in which resources, A, are converted into useful goods and services, B, with the production of wastes, C, that are recycled and used again in the conversion process ($A \rightarrow B$), thus reducing the waste output of the system. By and large, natural ecosystems have a circular or loop design rather than a linear structure. Feedback and cybernetics, the science of controls, are discussed in detail in Chapter 2.

Figure 1-7 illustrates how positive and negative feedback can interact in the relationship between atmospheric CO_2 concentration and climatic warming. An increase in CO_2 has a positive greenhouse effect on global warming and on plant growth. However, the soil system acclimates to the warming, so soil respiration does not continue to increase with warming. This acclimation results in a negative feedback on carbon sequestration in the soil, thus reducing emission of CO_2 to the atmosphere, according to a study by Luo et al. (2001).

Compartment models are greatly enhanced by making the shape of the "boxes" indicate the general function of the unit. In Figure 1-8, some of the symbols from the H. T. Odum energy language (H. T. Odum and E. P. Odum 1982; H. T. Odum 1996) are depicted as used in this book. In Figure 1-9, these symbols are used in a model of a pine forest located in Florida. Also, in this diagram estimates of the amount of energy flow through the units are shown as indicators of the relative importance of unit functions.

In summary, good model definition should include three dimensions: (1) the space to be considered (how the system is bounded); (2) the subsystems (components) judged to be important in overall function; and (3) the time interval to be considered. Once an ecosystem, ecological situation, or problem has been properly defined and bounded, a testable hypothesis or series of hypotheses is developed that can be rejected or accepted, at least tentatively, pending further experimentation or analysis. For more on ecological modeling, see Patten and Jørgensen (1995), H. T. Odum and E. C. Odum (2000), and Gunderson and Holling (2002).

In the following chapters, the paragraphs headed by the word **statement** are, in effect, "word" models of the ecological principle in question. In many cases, graphic models are also presented, and in some cases, simplified mathematical formulations are included. Most of all, this book attempts to provide the principles, concepts,

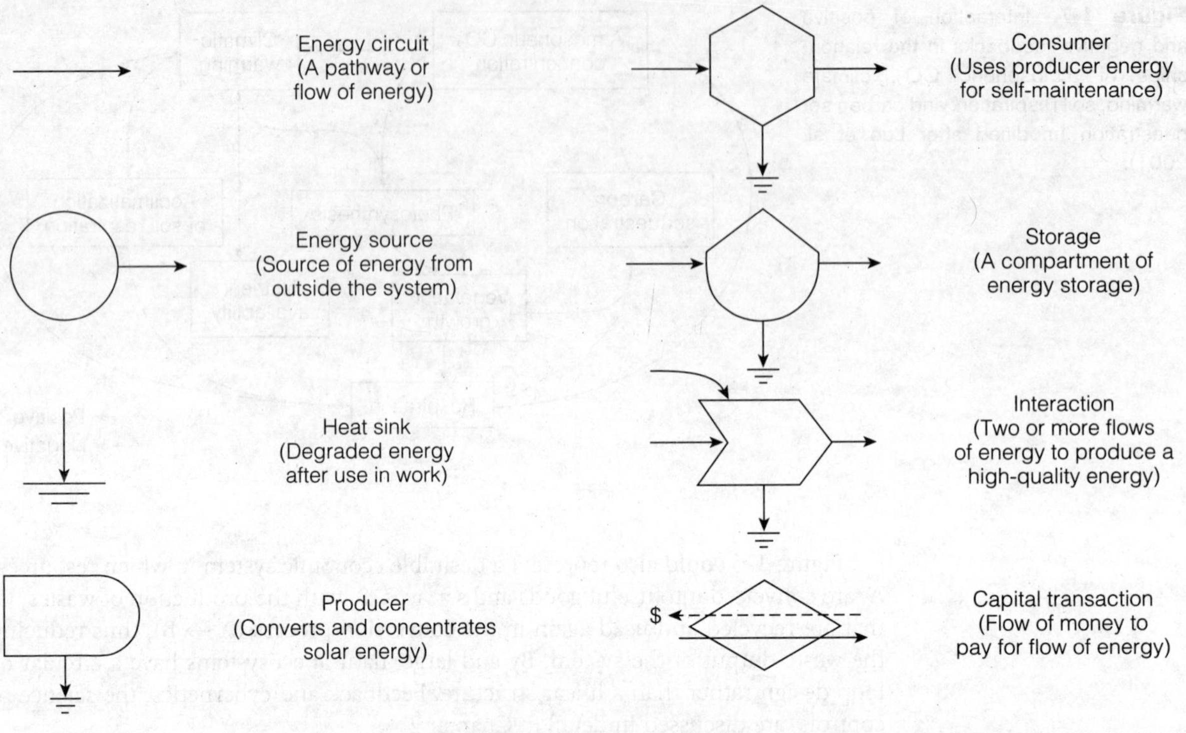

Figure 1-8. The H. T. Odum energy language symbols used in model diagrams in this book.

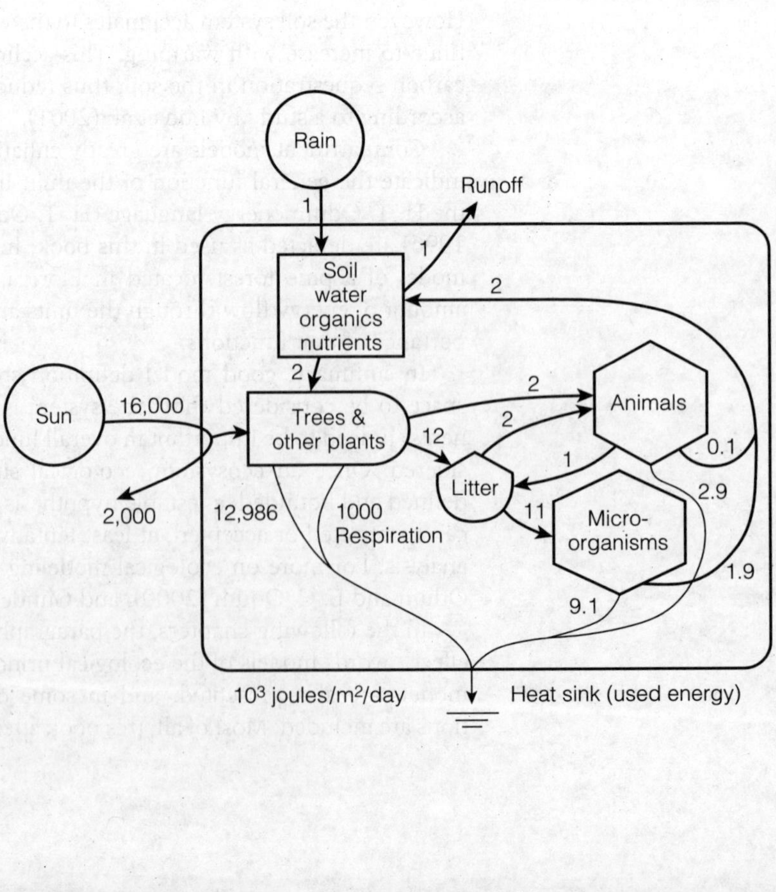

Figure 1-9. Ecosystem model using energy language symbols and including estimated rates of energy flow for a Florida pine forest (courtesy of H. T. Odum).

simplifications, and abstractions that one must deduce from the real world before one can understand and deal with situations and problems or construct mathematical models of them.

7 Disciplinary Reductionism to Transdisciplinary Holism

In a paper entitled "The Emergence of Ecology as a New Integrative Discipline," E. P. Odum (1977) noted that ecology had become a new holistic discipline, having roots in the biological, physical, and social sciences, rather than just a subdiscipline of biology. Thus, a goal of ecology is to link the natural and social sciences. It should be noted that most disciplines and **disciplinary** approaches are based on increased specialization in isolation (Fig. 1-10). The early evolution and development of ecol-

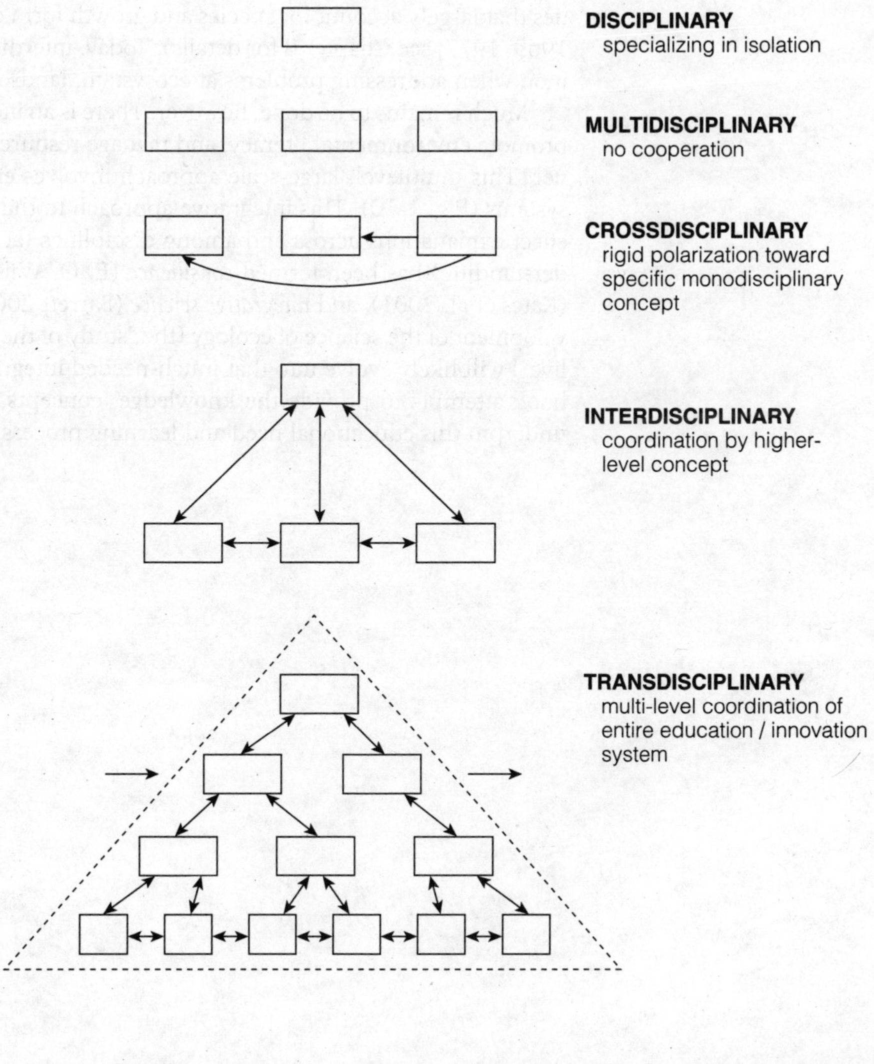

Figure 1-10. Progression of relations among disciplines from disciplinary reductionism to transdisciplinary holism (after Jantsch 1972).

ogy was frequently based on **multidisciplinary** approaches (multi = "many"), especially during the 1960s and 1970s. Unfortunately, the multidisciplinary approaches lacked cooperation or focus. To achieve cooperation and define goals, institutes or centers were established on campuses throughout the world, such as the Institute of Ecology located on the campus of the University of Georgia. These **crossdisciplinary** approaches (cross = "traverse"; Fig. 1-10) frequently resulted in polarization toward a specific monodisciplinary concept, a poorly funded administrative unit, or a narrow mission. A crossdisciplinary approach also frequently resulted in polarized faculty reward systems. Institutions of higher learning, traditionally built on disciplinary structures, have difficulties in administering programs and addressing environmental problems as well as taking advantage of opportunities at greater temporal and spatial scales.

To address the dilemma, **interdisciplinary** approaches (inter = "among") were employed, resulting in cooperation on a higher-level concept, problem, or question. For example, the process and study of natural ecological succession provided a higher-level concept resulting in the success of the Savannah River Ecological Laboratory (SREL) during its conception. Researchers theorized that new system properties emerge during the course of ecosystem development and that it is these properties that largely account for species and growth form changes that occur (E. P. Odum 1969, 1977; see Chapter 8 for details). Today, interdisciplinary approaches are common when addressing problems at ecosystem, landscape, and global levels.

Much remains to be done, however. There is an increased need to solve problems, promote environmental literacy, and manage resources in a **transdisciplinary** manner. This multilevel, large-scale approach involves entire education and innovation systems (Fig. 1-10). This integrative approach to the need for unlocking cause-and-effect explanations across and among disciplines (achieving a transdisciplinary understanding) has been termed *consilience* (E. O. Wilson 1998), *sustainability science* (Kates et al. 2001), and *integrative science* (Barrett 2001). Actually, the continued development of the science of ecology (the "study of the household" or "place where we live") will likely evolve into that much-needed integrative science of the future. This book attempts to provide the knowledge, concepts, principles, and approaches to underpin this educational need and learning process.

2

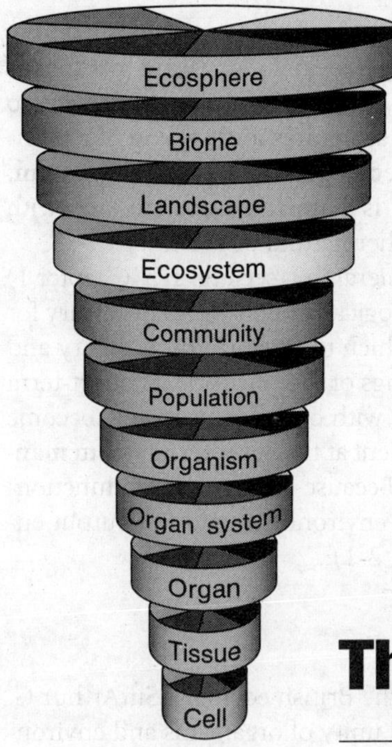

The Ecosystem

1 Concept of the Ecosystem and Ecosystem Management
2 Trophic Structure of the Ecosystem
3 Gradients and Ecotones
4 Examples of Ecosystems
5 Ecosystem Diversity
6 Study of Ecosystems
7 Biological Control of the Geochemical Environment: The Gaia Hypothesis
8 Global Production and Decomposition
9 Microcosms, Mesocosms, and Macrocosms
10 Ecosystem Cybernetics
11 Technoecosystems
12 Concept of the Ecological Footprint
13 Classification of Ecosystems

1 Concept of the Ecosystem and Ecosystem Management

Statement

Living (biotic) organisms and their nonliving (abiotic) environment are inseparably interrelated and interact with each other. Any unit that includes all the organisms (the *biotic community*) in a given area interacting with the physical environment so that a flow of energy leads to clearly defined biotic structures and cycling of materials between living and nonliving components is an **ecological system** or **ecosystem.** It is more than a geographical unit (or *ecoregion*); it is a functional system unit, with inputs and outputs, and boundaries that can be either natural or arbitrary.

The ecosystem is the first unit in the ecological hierarchy (see Fig. 1-3, Chapter 1) that is complete—that has all the components (biological and physical) necessary for survival. Accordingly, it is the basic unit around which to organize both theory and practice in ecology. Furthermore, as the shortcomings of the "piecemeal," short-term technological and economic approaches to dealing with complex problems become ever more evident with each passing year, management at this level (**ecosystem management**) emerges as the challenge for the future. Because ecosystems are functionally open systems, consideration of both the input environment and the output environment is an important part of the concept (Fig. 2-1).

Explanation

The term *ecosystem* was first proposed in 1935 by the British ecologist Sir Arthur G. Tansley (Tansley 1935). Allusions to the idea of the unity of organisms and environment (and the oneness of humans and nature) can be found as far back in written history as one might care to look. Not until the late 1800s, however, did formal statements begin to appear, interesting enough, in a parallel manner in the American, European, and Russian ecological literature. Thus, Karl Möbius in 1877 wrote (in German) about the community of organisms in an oyster reef as a "biocoenosis," and in 1887 S. A. Forbes, an American, wrote his classic essay "The Lake as a Microcosm." The pioneering Russian, V. V. Dokuchaev (1846–1903), and his chief disciple, G. F. Morozov, emphasized the concept of the "biocoenosis," a term later expanded by Russian ecologists to "geobiocoenosis" (Sukachev 1944).

Not only biologists but also physical scientists and social scientists began to consider the idea that both nature and human societies function as systems. In 1925, physical chemist A. J. Lotka wrote in a book entitled *Elements of Physical Biology* that the organic and inorganic worlds function as a single system to such an extent that it is impossible to understand either part without understanding the whole. It is significant that a biologist (Tansley) and a physical scientist (Lotka) independently and at about the same time came up with the idea of the ecological system. Because Tansley coined the word *ecosystem* and it caught on, he received most of the credit, which perhaps should be shared with Lotka.

In the 1930s, social scientists developed the holistic concept of regionalism, especially Howard W. Odum, who used social indicators to compare the southern region of the United States with other regions (H. W. Odum 1936; H. W. Odum and Moore 1938). More recently, Machlis et al. (1997) and Force and Machlis (1997) have promoted the idea of the human ecosystem, combining biological ecology and social theories as a basis for practical ecosystem management.

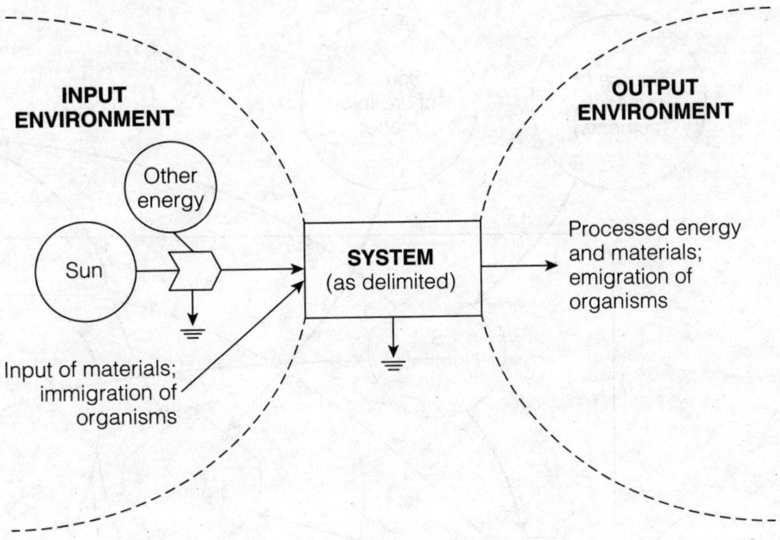

Figure 2-1. Ecosystem model, emphasizing the external environment, which must be considered an integral part of the ecosystem concept (first suggested by Patten 1978).

It was not until a general systems theory was developed in the mid-twentieth century by Bertalanffy (1950, 1968) and others that ecologists, notably E. P. Odum (1953), E. C. Evans (1956), Margalef (1958), Watt (1966), Patten (1966, 1971), Van Dyne (1969), and H. T. Odum (1971), began to develop the definitive, quantitative field of *ecosystem ecology*. The extent to which ecosystems actually operate as general systems and the extent to which they are self-organizing are matters of continuing research and debate, as will be noted later in this chapter. The utility of the ecosystem or systems approach in solving real-world environmental problems is now receiving serious attention.

Some other terms that have been used to express the holistic viewpoint, but that are not necessarily synonymous with *ecosystem*, include *biosystem* (Thienemann 1939); *noösystem* (Vernadskij 1945); and *holon* (Koestler 1969). As is the case for all kinds and levels of biosystems (biological systems), ecosystems are open systems—that is, things are constantly entering and leaving, even though the general appearance and basic function may remain constant for long periods of time. As shown in Figure 2-1, a graphic model of an ecosystem can consist of a box that we can label the system, which represents the area we are interested in, and two large funnels that we can label **input environment** and **output environment.** The boundary for the system can be arbitrary (whatever is convenient or of interest), delineating an area such as a block of forest or a section of beach; or it can be natural, such as the shore of a lake, where the whole lake is to be the system, or ridges as boundaries of a watershed.

Energy is a necessary input. The Sun is the ultimate energy source for the ecosphere and directly supports most natural ecosystems within the biosphere. But there are other energy sources that may be important for many ecosystems, for example, wind, rain, water flow, or fossil fuel (the major source for the modern city). Energy also flows out of the system in the form of heat and in other transformed or processed forms, such as organic matter (food and waste products) and pollutants. Water, air, and nutrients necessary for life, along with all kinds of other materials, constantly enter and leave the ecosystem. And, of course, organisms and their propagules (seeds or spores) and other reproductive stages enter (immigrate) or leave (emigrate).

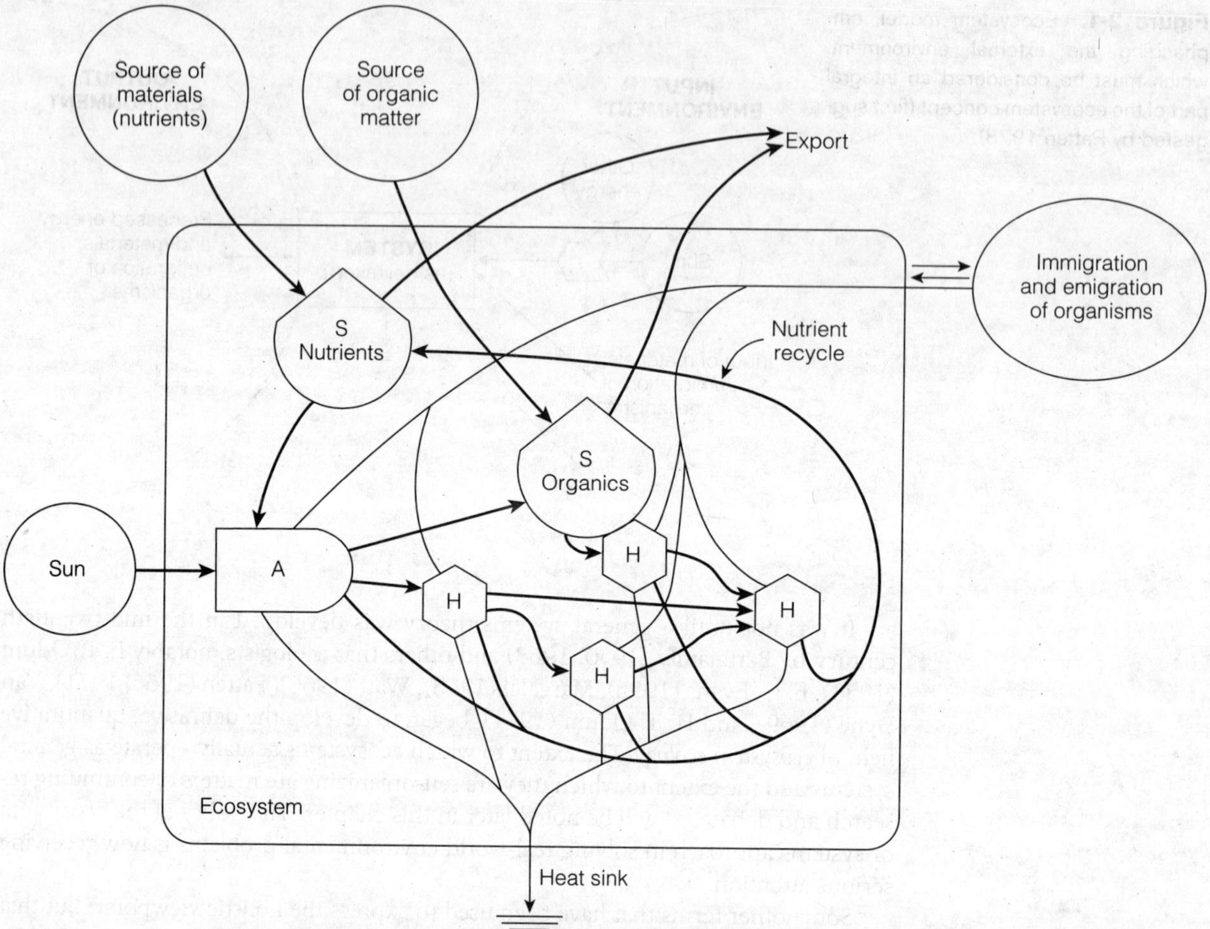

Figure 2-2. Functional diagram of an ecosystem, with emphasis on internal dynamics involving energy flow and material cycles. S = storage; A = autotrophs; H = heterotrophs.

In Figure 2-1, the system part of the ecosystem is shown as a *black box*, which is defined by modelers as a unit whose general role or function can be evaluated without specifying its internal contents. However, we want to look inside this black box to see how it is organized internally and find out what happens to all those inputs. Figure 2-2 shows the contents, as it were, of an ecosystem in model form.

The interactions of the three basic components—namely, (1) the community, (2) the flow of energy, and (3) the cycling of materials—are diagrammed as a simplified compartment model with the general features discussed in the preceding chapter. Energy flow is one-way; some of the incoming solar energy is transformed and upgraded in quality (that is, converted into organic matter, a higher-quality form of energy than sunlight) by the community, but most input energy is degraded and passes through and out of the system as low-quality heat energy (heat sink). Energy can be stored, then "fed back," or exported, as shown in the diagram, but it cannot be reused. The physical laws governing the behavior of energy are considered in detail in Chapter 3. In contrast with energy, materials, including the nutrients necessary for life (such as carbon, nitrogen, and phosphorus) and water, can be used over and over again. The efficiency of recycling and the magnitude of imports and exports of nutrients vary widely with the type of ecosystem.

Each "box" in the diagram (Fig. 2-2) is given a distinctive shape that indicates its general function according to an "energy language," as introduced in Chapter 1 (Fig. 1-8). The community is depicted as a *food web* of autotrophs, A, and heterotrophs, H, linked together with appropriate energy flows, nutrient cycles, and storages, S. Food webs will be discussed in Chapter 3.

Both graphic models (Figs. 2-1 and 2-2) emphasize that a conceptually complete ecosystem includes an input environment (IE) and an output environment (OE) along with the system (S) as delimited, or *ecosystem* = IE + S + OE. This scheme solves the problem of where to draw lines around an entity that one wishes to consider, because it does not matter very much how the "box" portion of the ecosystem is delimited. Often, natural boundaries, such as a lakeshore or forest edge, or political ones, such as province or city limits, make convenient boundaries, but limits can just as well be arbitrary so long as they can be accurately designated in a geometric sense. The box is not all there is to the ecosystem, because if the box were an impervious container, its living contents (lake or city) would not long survive such isolation. A functional or real-world ecosystem must have an input lifeline and, in most cases, a means of exporting processed energy and materials.

The extent of the input and output environment varies extremely and depends on several variables, for example, (1) size of the system (the larger, the less dependent on externals); (2) metabolic intensity (the higher the rate, the greater the input and output); (3) autotrophic-heterotrophic balance (the greater the imbalance, the more externals to balance); and (4) stage of development (young systems differ from mature systems, as detailed in Chapter 8). Thus, a large, forested mountain range has much smaller input-output environments than does a small stream or a city. These contrasts are brought out in the discussion of examples of ecosystems (see Section 4 of this chapter).

Before the agricultural and industrial revolutions, humans were largely hunters and gatherers living on whatever they could kill or harvest from natural systems. Early humans fit into the ecosystem model of Figure 2-2 as the terminal H (top predator and omnivore). Modern urban-industrial society no longer just affects and modifies natural systems but has created a completely new arrangement that we term the *human-dominated technoecosystem,* as we will explain and model in Section 11 of this chapter. For historical reviews of the ecosystem concept, see Hagen (1992) and Golley (1993).

2 Trophic Structure of the Ecosystem

Statement

From the standpoint of **trophic structure** (from *trophe* = "nourishment"), an ecosystem is two-layered: It has (1) an upper, **autotrophic** ("self-nourishing") **stratum** or "green belt" of chlorophyll-containing plants in which the fixation of light energy, the use of simple inorganic substances, and the buildup of complex organic substances predominate; and (2) a lower, **heterotrophic** ("other-nourished") **stratum** or "brown belt" of soils and sediments, decaying matter, roots, and so on, in which the use, rearrangement, and decomposition of complex materials predominate. It is convenient to recognize the following components as constituting the ecosystem: (1) **inorganic substances** (C, N, CO_2, H_2O, and others) involved in material cycles; (2) **organic compounds** (proteins, carbohydrates, lipids, humic substances, and so on) that link biotic and abiotic components; (3) **air, water,** and **substrate environment,** including the **climate regime** and other physical factors; (4) **producers** (autotrophic organisms), mostly green plants that can manufacture food from simple inorganic substances; (5) **phagotrophs** (from *phago* = "to eat"), heterotrophic organisms, chiefly animals, that ingest other organisms or particulate organic matter; and (6) **saprotrophs** (from *sapro* = "to decompose") or decomposers, also heterotrophic organisms, chiefly bacteria and fungi, that obtain their energy either by breaking down dead tissues or by absorbing dissolved organic matter exuded by or extracted from plants or other organisms. **Saprophages** are organisms that feed on dead organic matter. The decomposing activities of saprotrophs release inorganic nutrients that are usable by the producers; they also provide food for the macroconsumers and often excrete substances that inhibit or stimulate other biotic components of the ecosystem.

Explanation

One of the universal features of all ecosystems—whether terrestrial, freshwater, marine, or human-engineered (for example, agricultural)—is the interaction of the autotrophic and heterotrophic components. The organisms responsible for the processes are partially separated in space; the greatest autotrophic metabolism occurs in the upper "green belt" stratum, where light energy is available. The most intensive heterotrophic metabolism occurs in the lower "brown belt," where organic matter accumulates in soils and sediments. Also, the basic functions are partially separated in time, as there may be a considerable delay in the heterotrophic use of the products of autotrophic organisms. For example, photosynthesis predominates in the canopy of a forest ecosystem. Only a part, often only a small part, of the photosynthate is immediately and directly used by the plant and by herbivores and parasites that feed on foliage and other actively growing plant tissue. Much of the synthesized material (leaves, wood, and stored food in seeds and roots) escapes immediate consumption and eventually reaches the litter and soil (or the equivalent sediments in aquatic ecosystems), which together constitute a well-defined heterotrophic system. Weeks, months, or years (or many millennia, in the case of the fossil fuels now being rapidly consumed by human societies) may pass before all the accumulated organic matter is used.

The term **organic detritus** (product of disintegration, from the Latin *deterere,* "to wear away") is borrowed from geology, in which it is traditionally used to designate the products of rock disintegration. As used in this book, **detritus** refers to all the organic matter involved in the decomposition of dead organisms. Detritus seems the most suitable of many terms that have been suggested to designate this important link between the living and the inorganic world. Environmental chemists use a shorthand designation for two physically different products of decomposition as follows: POM is *particulate organic matter;* DOM is *dissolved organic matter.* The role of POM and DOM in food chains is reviewed in detail in Chapter 3. We can also add *volatile or-*

ganic matter (VOM), which mostly functions as "signals"—for example, the fragrance of flowers that attracts pollinators.

Abiotic components that limit and control organisms are discussed in Chapter 5; the role of organisms in controlling the abiotic environment is considered later in this chapter. As a general principle, from the operational standpoint, the living and nonliving parts of ecosystems are so interwoven into the fabric of nature that it is difficult to separate them; hence, operational or functional classifications do not sharply distinguish between biotic and abiotic.

Most of the vital elements (such as carbon, nitrogen, and phosphorus) and organic compounds (such as carbohydrates, proteins, and lipids) are not only found inside and outside of living organisms but are also in a constant state of flux or turnover between living and nonliving states. Some substances, however, appear to be unique to one or the other state. The high-energy storage compound ATP (*adenosine triphosphate*), for example, is found only inside living cells (or at least its existence outside is very transitory), whereas **humic substances**, which are resistant end products of decomposition, are never found inside cells, yet they are a major and characteristic component of all ecosystems. Other key biotic complexes, such as DNA (*deoxyribonucleic acid*) and the chlorophylls, occur both inside and outside organisms but become nonfunctional when outside the cell.

The ecological classification (producers, phagotrophs, decomposers) is one of function rather than of species as such. Some species occupy intermediate positions, and others can shift their mode of nutrition according to environmental circumstances. The separation of heterotrophs into large and small consumers is arbitrary but justified in practice because of the very different methods of study required. The heterotrophic *microconsumers* (bacteria, fungi, and others) are relatively immobile (usually embedded in the medium being decomposed), are very small, and have high rates of metabolism and turnover. Their functional specialization is more evident biochemically than morphologically; consequently, one cannot usually determine their role in the ecosystem by such direct methods as visual observation or counting their numbers. Organisms designated as *macroconsumers* obtain their energy by heterotrophic ingestion of particulate organic matter. These are largely "animals" in the broad sense. These higher forms tend to be morphologically adapted for active food seeking or herbivory, with the development of complex sensory-neuromotor, digestive, respiratory, and circulatory systems in the higher forms. The microconsumers, or saprotrophs, have been typically designated as decomposers. However, it seems preferable not to designate any particular organisms as decomposers but to consider **decomposition** as a process involving all of the biota and abiotic processes as well.

We recommend that students of ecology read Aldo Leopold's "The Land Ethic" (first published in 1933, and in 1949 included in his best-seller, *A Sand County Almanac: And Sketches Here and There*), an eloquent, often quoted and reprinted essay on environmental ethics that has special relevance to the ecosystem concept (more recent critiques of "The Land Ethic" are provided by Callicott and Freyfogle 1999; A. C. Leopold 2004). We also recommend reading *Man and Nature* by the Vermont prophet George Perkins Marsh, who analyzed the causes of the decline of ancient civilizations and forecast a similar doom for modern ones unless an "ecosystematic" view of the world is taken. B. L. Turner (1990) edited a book that reiterates this theme in a review of Earth as transformed by human action over the past 300 years. From another viewpoint, Goldsmith (1996) argued the need for a major paradigm shift from reductionist science and consumption economics to an ecosystem worldview that would provide a more holistic and long-term approach to dealing with increasingly endangered Earth. Especially recommended are Flader and Callicott's (1991) and Callicott and Freyfogle's (1999) reviews of the Leopold philosophy.

3 Gradients and Ecotones

Statement

The biosphere is characterized by a series of gradients, or *zonation,* of physical factors. Examples are temperature gradients from the Arctic or Antarctic to the Tropics and from mountaintop to valley; moisture gradients from wet to dry along major weather systems; and depth gradients from shore to bottom in bodies of water. Environmental conditions, including the organisms adapted to these conditions, change gradually along a gradient, but often there are points of abrupt change, known as *ecotones.* An **ecotone** is created by the juxtaposition of different *habitats,* or *ecosystem types.* The concept assumes the existence of active interaction between two or more ecosystems (or patches within ecosystems), which results in the ecotone having properties that do not exist in either of the adjacent ecosystems (Naiman and Décamps 1990).

Explanation and Examples

Four examples of physical factor zonation as related to biotic communities are shown in Figure 2-3. On land biomes, zonation can often be identified and mapped by indigenous vegetation that is more or less in equilibrium with the regional climate (Fig. 2-3A). In large bodies of water (lakes, oceans) where green plants are small and not a conspicuous visual presence, zonation is best based on physical or geomorphological features (Fig. 2-3B). Zonation based on productivity and respiration or thermal stratification of a pond (Figs. 2-3C and D) will be discussed later in this chapter.

An example of an ecotone as an interface zone with unique properties and species is a sea beach, where alternate flooding and drying tidal action is a unique feature and where there are numerous kinds of organisms that are not found either on land or in the open sea. Estuaries landward to the beaches are other examples, as are prairie-forest zones. In addition to external processes such as tides causing discontinuities in gradients, internal processes such as sediment traps, root mats, special soil-water conditions, inhibitory chemicals, or animal activity (for instance, dam building by beavers) may maintain an ecotone distinct from bordering communities. In addition to unique species, terrestrial ecotones sometimes are populated by more species (*increased biotic diversity*) than can be found in the interior of the adjoining, more homogenous communities. When it comes to game animals and birds, wildlife managers speak of this as the **edge effect** and often recommend special plantings between field and forest, for example, to increase the numbers of these animals. Species that inhabit these edge habitats are frequently referred to as **edge species.** However, a *sharp edge,* such as that between a clear-cut and an uncut forest, may be a poor habitat,

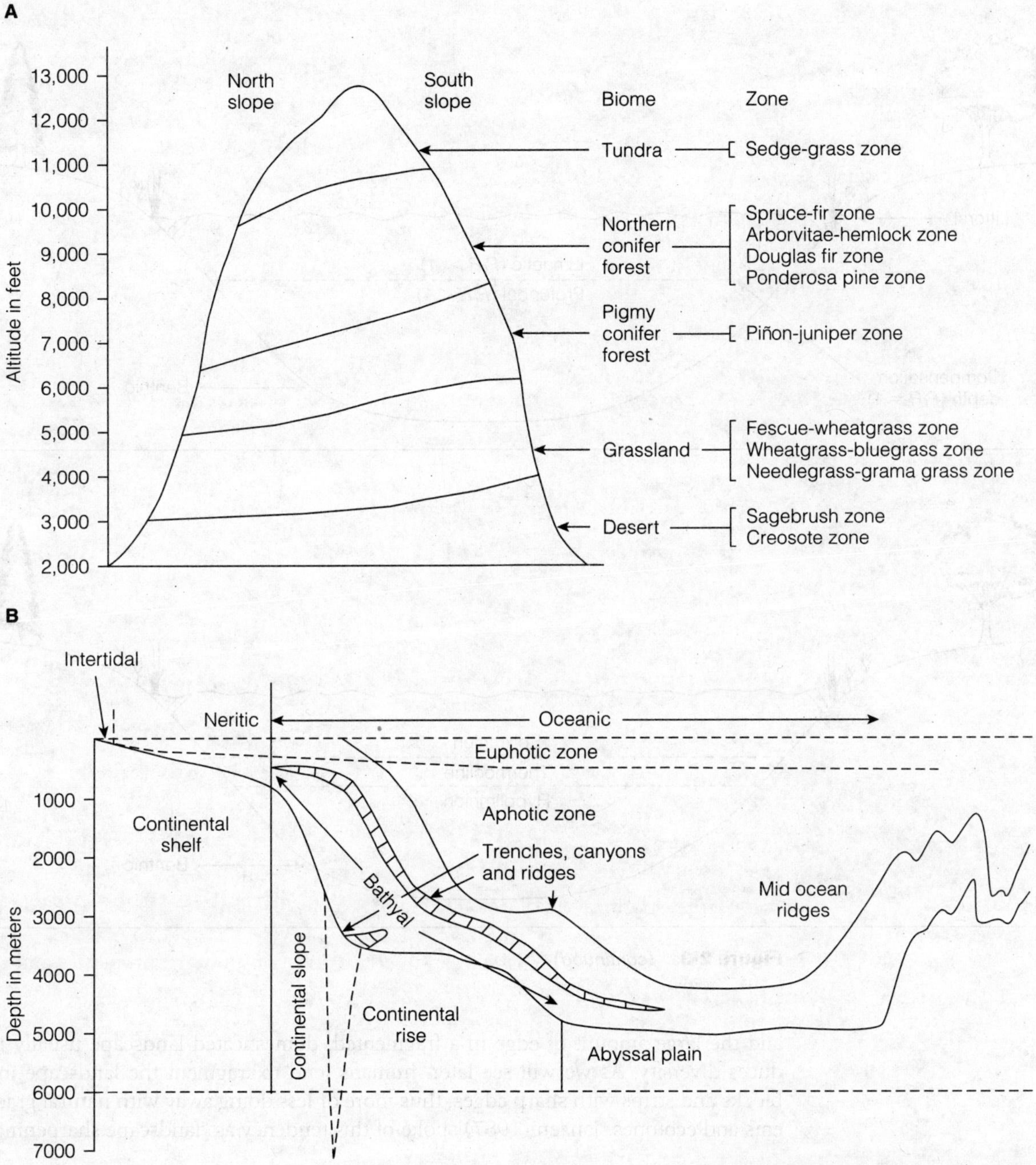

Figure 2-3. (A) Vegetation-based zonation in the mountains of western North America (zone information from Daubenmire 1966). (B) Horizontal and vertical zonation in the sea (diagram based on Heezen et al. 1959). (C) Metabolic zonation of a pond based on productivity (*P*) and respiration (*R;* community maintenance). (D) Zonation of a pond based on thermal (heat) stratification during the summer months in the midwestern United States. (*continued*)

C

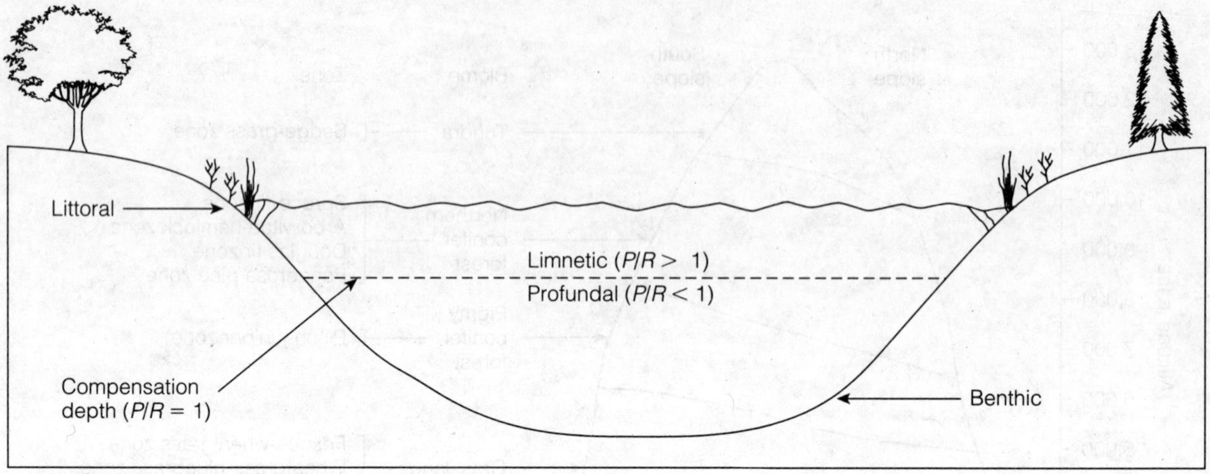

D

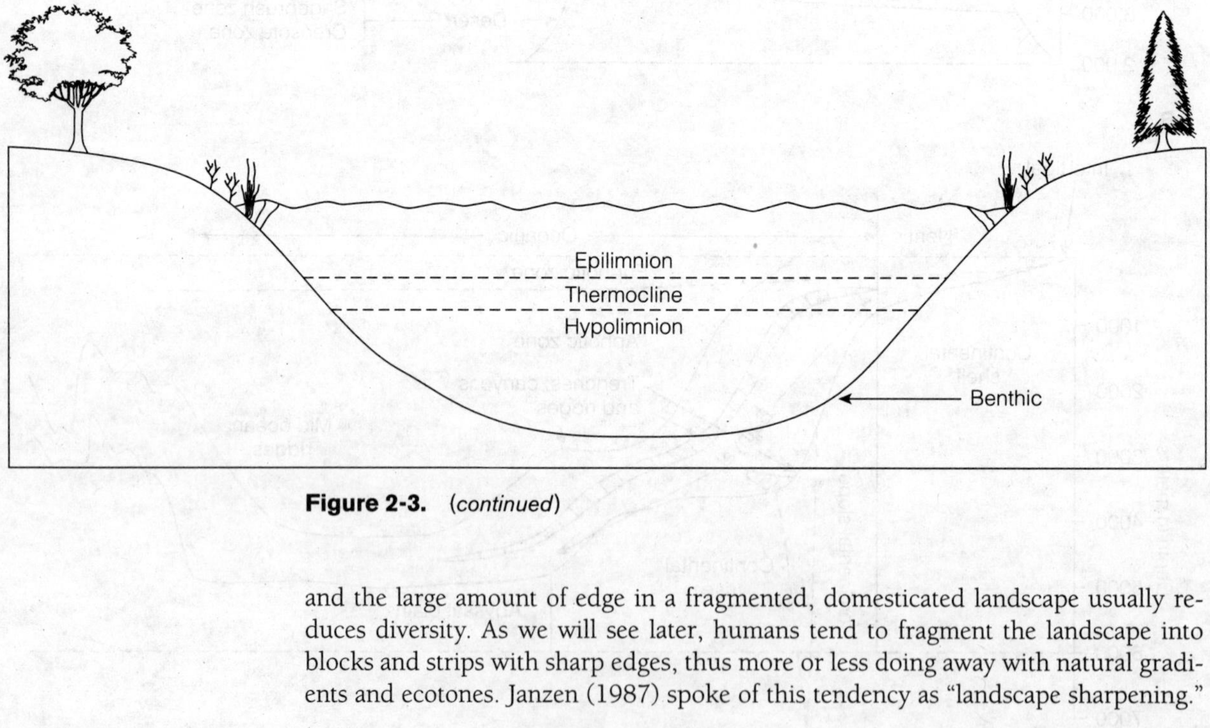

Figure 2-3. *(continued)*

and the large amount of edge in a fragmented, domesticated landscape usually reduces diversity. As we will see later, humans tend to fragment the landscape into blocks and strips with sharp edges, thus more or less doing away with natural gradients and ecotones. Janzen (1987) spoke of this tendency as "landscape sharpening."

4 Examples of Ecosystems

One way to begin studying ecology is to study a small pond and an old field, where the basic features of ecosystems can be conveniently examined, and the nature of aquatic and terrestrial ecosystems can be contrasted. A small pond or an old field is ideal to organize an ecosystem-level field study for an undergraduate class laboratory/

field exercise. In this section, we consider four examples: a pond, an old field, a watershed, and an agroecosystem. In Section 11, we consider the city as a technoecosystem.

A Pond and an Old Field

The inseparability of living organisms and the nonliving environment is at once apparent with the first sample collected. Plants, animals, and microorganisms not only live in the pond and the old field (or grassland), but they also modify the chemical nature of the water, soil, and air that compose the physical environment. Thus, a bottle of pond water or a scoopful of bottom mud or meadow soil is a mixture of living organisms—both plant and animal—and inorganic and organic compounds. Some of the larger animals and plants can be separated from the sample for study by counting, but it would be difficult to completely separate the myriad of small living things from the nonliving matrix without changing the character of the water or soil. True, one could autoclave the sample of water, bottom mud, or soil so that only nonliving material remained, but this residue would then no longer be pond water or old-field soil; it would have entirely different appearances, characteristics, and functions.

The basic components of an aquatic and a terrestrial ecosystem are discussed next.

Abiotic Substances

Abiotic substances include inorganic and organic compounds, such as water, carbon dioxide, oxygen, calcium, nitrogen, sulfur, and phosphorus salts, amino and humic acids, and others. A small portion of the vital nutrients is in solution and immediately available to organisms, but a much larger portion is held in reserve (the "storage" S shown in the functional diagram of Figure 2-2) in particulate matter as well as in the organisms themselves. In a New Hampshire forest, for example, about 90 percent of nitrogen is stored in soil organic matter, 9.5 percent is in biomass (wood, roots, leaves), and only about 0.5 percent is in a soluble, quickly available form in soil water (Bormann et al. 1977).

The rate of release of nutrients from the solids, the solar input, and changes in temperature, day length, and other climatic conditions are the most important processes that regulate the rate of function of the entire ecosystem on a daily basis.

To fully assess the chemistry of the environment, extensive laboratory analysis of samples is necessary. For example, the relative acidity or alkalinity, as indicated by pH or hydrogen ion concentration, often determines what kinds of organisms are present. Acidic soils and waters (pH less than 7) are usually characteristic of regions underlain with igneous and metamorphic rocks; "hard" or alkaline waters and soils occur most frequently in regions with limestone and related substrates.

Producer Organisms

In a pond, the producers may be of two main types: (1) rooted or large floating plants (**macrophytes**) generally growing in shallow water; and (2) minute floating plants, usually algae or green bacteria or protozoa, called **phytoplankton** (from *phyto* = "plant"; *plankton* = "floating"), distributed throughout the pond as deep as light penetrates. An abundance of phytoplankton gives the water a greenish color; otherwise, these producers are not visible, and their presence is not suspected by the casual observer. Yet in large, deep ponds and lakes (and in the oceans), phytoplankton is much more important than rooted vegetation in the production of basic food for the ecosystem.

In the old field or grassland, and in terrestrial communities in general, the reverse is the case; large, rooted plants dominate, but small photosynthetic organisms such as algae, mosses, and lichens also occur on soil, rocks, and stems of plants. Where these substrates are moist and exposed to light, these microproducers may contribute substantially to organic production.

Consumer Organisms

The primary macroconsumers or **herbivores** feed directly on living plants or plant parts. Hereafter, these herbivores will also be termed **primary** (first-order) **consumers.** In the pond, there are two types of microconsumers, **zooplankton** (animal plankton) and **benthos** (bottom forms), paralleling the two types of producers. Herbivores in the grassland or old field also come in two sizes, the small, plant-feeding insects and other invertebrates and the large, grazing rodents and hoofed mammals. The **secondary** (second-order) **consumers** or **carnivores,** such as predaceous insects and game fish (**nekton;** that is, free-swimming aquatic organisms that are able to move about at will through the water) in the pond, and predatory insects, spiders, birds, and mammals in the grassland feed on the primary consumers or on other secondary consumers (thus making them **tertiary consumers**). Another important type of consumer is the **detritivore,** which subsists on the "rain" of organic detritus from autotrophic layers above and, along with herbivores, provides food for carnivores. Many detritivorous animals (such as earthworms) obtain much of their food energy by digesting the microorganisms that colonize detritus particles.

Decomposer Organisms

The nongreen bacteria, flagellates, and fungi are distributed throughout the ecosystem, but they are especially abundant in the mud-water interface of the pond and in the litter-soil junction of the grassland or old-field ecosystem. Although a few of the bacteria and fungi are pathogenic, in that they will attack living organisms and cause disease, the majority attack only after the organism dies. Important groups of microorganisms also form mutually beneficial associations with plants, even to the extent of becoming an integral part of roots and other plant structures (see Chapter 7). When temperature and moisture conditions are favorable, the first stages of decomposition occur rapidly. Dead organisms do not retain their integrity for very long but are soon broken up by the combined action of detritus-feeding microorganisms and physical processes. Some of their nutrients are released for reuse. The resistant fraction of detritus, such as cellulose, lignin (wood), and humus, endures and imparts a spongy texture to soil and sediments that contributes to a quality habitat for plant roots and many tiny invertebrates. Some of the latter convert atmospheric nitrogen to forms usable by plants (nitrogen fixation; see Chapter 4) or perform other processes for their own benefit but also for the enhancement of the whole ecosystem.

Measuring Community Metabolism

The partial stratification into an upper production zone and a lower decomposition/nutrient regeneration zone can be illustrated by measuring diurnal oxygen changes in the water column of a pond. A "light-and-dark bottle" technique may be used to measure the metabolism of the whole aquatic community. Samples of water from different depths are placed in paired bottles; one (the dark bottle) is covered with black tape or aluminum foil to exclude all light. Before the string of paired bottles is lowered into place in the water column, the original oxygen concentration of the water at the selected depths is determined, either by a chemical method or, much more easily, with the use of an electronic oxygen probe. After 24 hours, the string of bottles is removed, and the oxygen concentration in each bottle is determined and compared with the original concentration. The decline of oxygen in the dark bottle indicates the amount of respiration by producers and consumers (the total community) in the water, whereas the change of oxygen in the light bottle reflects the net result of oxygen consumed by respiration and oxygen produced by photosynthesis. Adding dark bottle respiration, *R*, and light bottle net production, *P*, together gives an estimate of the total or gross photosynthesis (gross primary production) for the time period, providing that both bottles had the same oxygen concentration to begin with.

With a light-and-dark bottle experiment in a shallow, fertile pond on a warm, sunny day, one might expect an excess of photosynthesis over respiration in the top 2 or 3 meters, as indicated by a rise in the oxygen concentration in the light bottles. This top area of a pond, where production is greater than respiration ($P/R > 1$), is termed the **limnetic zone** (Fig. 2-3C). Below 3 meters, the light intensity in a fertile pond is usually too low for photosynthesis, so only respiration occurs in the bottom waters. This bottom area of a pond, where respiration is greater than production ($P/R < 1$), is termed the **profundal zone.** The point in a light gradient at which plants can just balance food production and use (zero change in light bottle) is called the **compensation depth** and marks a convenient functional boundary between the autotrophic stratum and the heterotrophic stratum where $P/R = 1$ (Fig. 2-3C).

A daily production of 5 to 10 g O_2/m^2 (5–10 ppm) of excess oxygen production over respiration would indicate a healthy condition for the ecosystem, because excess food is being produced that becomes available to bottom organisms, and to all the organisms when light and temperature are not so favorable. If the hypothetical pond is polluted with organic matter, O_2 consumption (respiration) would greatly exceed O_2 production, resulting in oxygen depletion. Should the imbalance continue, eventually **anaerobic** (without oxygen) conditions would prevail, which would eliminate fish and most other animals. For most free-swimming organisms (*nekton*), such as fish, concentrations of oxygen less than 4 ppm cause detrimental effects on their health. In assaying the "health" of a body of water, we need not only to measure the oxygen concentration as a condition for existence but also to determine the rates of change and the balance between production and use in the diurnal and annual cycle. Monitoring oxygen concentrations, then, conveniently allows one to "feel the pulse" of the aquatic ecosystem. Measuring the *biochemical oxygen demand (BOD)* of water samples incubated in the laboratory is another standard method of pollution assay, but it does not measure the community rate of metabolism.

The community metabolism of flowing waters, such as a stream, can be estimated from diurnal upstream-downstream oxygen changes in much the same way as with the light-and-dark-bottle method. Nighttime change would be equivalent to the "dark" bottle, whereas 24-hour change would correspond to the "light" bottle. Other methods for measuring the metabolism of ecosystems are discussed in Chapter 3.

Although aquatic and terrestrial ecosystems have the same basic structure and similar functions, the biotic composition and size of their trophic components differ, as summarized in Table 2-1. The most striking contrast, as already noted, is in the

Table 2-1
Comparison of density (numbers/m²) and biomass (grams dry weight/m²) of organisms in aquatic and terrestrial ecosystems of comparable, moderate productivity

	Pond			*Meadow or old field*		
Ecological component	*Assemblage*	*No./m²*	*g dry wt./m²*	*Assemblage*	*No./m²*	*g dry wt./m²*
Producers	Phytoplanktonic algae	10^8–10^{10}	5.0	Herbaceous angiosperms (grasses and forbs)	10^2–10^3	500.0
Consumers in the autotrophic layer	Zooplanktonic crustaceans and rotifers	10^5–10^7	0.5	Insects and spiders	10^2–10^3	1.0
Consumers in the heterotrophic layer	Benthic insects, mollusks, and crustaceans*	10^5–10^6	4.0	Soil arthropods, annelids, and nematodes†	10^5–10^6	4.0
Large consumers	Fish	0.1–0.5	15.0	Birds and mammals	0.01–0.03	0.3‡–15.0§
Microorganism consumers (saprophages)	Bacteria and fungi	10^{13}–10^{14}	1–10**	Bacteria and fungi	10^{14}–10^{15}	10–100.0**

*Including animals down to the size of ostracods.
†Including animals down to the size of small nematodes and soil mites.
‡Including only small birds (passerines) and small mammals (rodents, shrews).
§Including two or three large herbivorous mammals per hectare.
**Biomass based on the approximation of 10^{13} bacteria = 1 gram dry weight.

size of the green plants. The autotrophs of land tend to be fewer but very much bigger, both as individuals and in biomass per unit area (Table 2-1). The contrast is especially impressive when one compares the open ocean, where phytoplankton are even smaller than in a pond, and the rain forest with its huge trees. Shallow-water communities (edges of ponds, lakes, oceans, and marshes), grasslands, and deserts are intermediate between these extremes.

Terrestrial autotrophs (producers) must invest a large part of their productive energy in supporting tissue, because the density (and hence the supporting capacity) of air is much lower than that of water. This supporting tissue has a high content of cellulose (a polysaccharide) and lignin (wood) and requires little energy for maintenance because it is resistant to most consumers. Accordingly, plants on land contribute more to the structural matrix of the ecosystem than do plants in water, and the rate of metabolism per unit volume or weight of land plants is correspondingly much lower; for this reason, the rate of replacement or *turnover* differs.

Turnover may be broadly defined as the ratio of throughput to content. Turnover can be conveniently expressed either as a rate fraction or as a "turnover time," which is the reciprocal of the rate fraction. Consider the productive energy flow as the throughput, and the standing crop biomass (grams dry weight/m² in Table 2-1) as the content. If we assume that the pond and the meadow have a comparable gross photosynthetic rate of 5 $g \cdot m^{-2} \cdot day^{-1}$, the turnover rate for the pond would be 5/5, or 1, and the turnover time would be one day. In contrast, the turnover rate for the meadow would be 5/500, or 0.01, and the turnover time would be 100 days. Thus, the tiny plants in the pond may replace themselves in a day when the pond metabolism is at its peak, whereas land plants are much longer-lived and turn over much more slowly (perhaps 100 years for a large forest). In Chapter 4, the concept of turnover will be especially useful for calculating the exchange of nutrients between organisms and environment.

In both land and aquatic ecosystems, a large part of the solar energy is dissipated in the evaporation of water, and only a small part, generally less than 1 percent, is fixed by photosynthesis on an annual basis. However, the role of this evaporation in moving nutrients and in maintaining temperatures differs between terrestrial and aquatic ecosystems. For every gram of CO_2 fixed in a grassland or forest ecosystem, as much as 100 grams of water must be moved from the soil, through the plant tissues, and transpired (evaporated from plant surfaces). No such massive use of water is associated with the production of phytoplankton or other submerged plants.

The Watershed Concept

Although the biological components of the pond and the meadow seem self-contained, they are actually open systems that are parts of larger watershed systems. Their function and relative stability over the years are very much determined by the rate of inflow or outflow of water, materials, and organisms from other parts of the watershed. A net inflow of materials often occurs when bodies of water are small, when outflow is restricted, or when sewage or industrial wastes are added. In such a case, the pond fills up and becomes a swamp, which may be maintained by periodic drawdowns or fires that remove some of the accumulated organic matter. Otherwise, the body of water becomes a terrestrial environment.

The phrase **cultural eutrophication** ("cultural enrichment") is used to denote organic pollution resulting from human activities. Not only do soil erosion and loss of nutrients from a disturbed forest or poorly managed cultivated field impoverish these ecosystems, but such outflows will likely have "downstream" eutrophic or other impacts. Therefore, *the whole drainage basin—not just the body of water or patch of vegetation—must be considered as the minimum ecosystem unit* when it comes to human understanding and resource management. The ecosystem unit for practical management must then include for every square meter or hectare (= 2.471 acres) of water at least 20 times its area of terrestrial watershed. Naturally, the ratio of water surface to watershed area varies widely and depends on rainfall, geological structure of underlying rocks, and topography. In other words, fields, forests, bodies of water, and towns linked together by a stream or river system, or in limestone country by an underground drainage network, interact as an integrative unit for both study and management. This integrative unit, or catchment basin, termed a **watershed,** is also defined as the area of the terrestrial environment that is drained by a particular stream or river. Likens and Bormann (1995) explain the development of the **small watershed technique** to measure the input and output of chemicals from individual catchment areas in the landscape (more will be discussed regarding the cycling and retention of these chemicals in Chapter 4). For a picture of a watershed manipulated and monitored for experimental study, see Figure 2-4.

The watershed concept helps put many of our problems and conflicts in perspective. For example, the cause of and the solutions for water pollution are not to be

A

B

Courtesy of Wayne T. Swank

Courtesy of the U.S. Forest Service

Figure 2-4. (A) Experimental clear-cut watershed at the Coweeta Hydrologic Laboratory in the mountains near Otto, North Carolina–the site of a National Science Foundation–sponsored Long-Term Experimental Research (LTER) program. All trees were cleared from the watershed in 1977 (center of the photograph) and recovery processes have been monitored over the last 27 years. (B) Photograph showing the V-notch weir and recording equipment used to measure the amount of water flowing out of each watershed.

found by looking only at the water; usually, incompetent management in the watershed (such as the conventional agricultural practices that result in fertilizer runoff) is what destroys water resources. The entire drainage or catchment basin must be considered as the unit of management. The Everglades National Park, located in South Florida, provides an example of this need to consider the whole drainage basin. Although large in area, the park does not now include the source of freshwater that must drain southward into the park if it is to retain its unique ecology. In other words, the park does not include the whole watershed. Recent efforts to restore the Everglades are focusing on restoring and cleaning up the freshwater inflow that has been diverted to agriculture and to the urbanized Gold Coast of Florida (Lodge 1994). For more on the coupling of land and water ecosystems, see Likens and Bormann (1974b, 1995) and Likens (2001a).

Agroecosystems

Agroecosystems (short for *agricultural ecosystems*) differ from natural or seminatural solar-powered ecosystems, such as lakes and forests, in three basic ways: (1) the *auxiliary energy* that augments or subsidizes the solar energy input is under the control of humankind and consists of human and animal labor, fertilizers, pesticides, irrigation water, fuel-powered machinery, and so on; (2) the *diversity* of organisms and crops is greatly reduced (again by human management) to maximize yield of specific food crops or other products; and (3) the dominant plants and animals are under *artificial selection* rather than natural selection. In other words, agroecosystems are designed and managed to channel as much conversion of solar energy and energy subsidies as possible into edible or other marketable products by a twofold process: (1) by employing auxiliary energy to do maintenance work that in natural systems would be accomplished by solar energy, thus allowing more solar energy to be converted directly into food; and (2) by genetic selection of food plants and domestic animals to optimize yield in the specialized, energy-subsidized environment. As in all intensive and specialized land use, there are costs as well as benefits, including soil erosion, pollution from pesticide and fertilizer runoff, high cost of fuel subsidies, reduced biodiversity, and increased vulnerability to weather changes and pests.

Approximately 10 percent of the world's ice-free land area is cropland, converted mostly from natural grasslands and forests, but also from deserts and wetlands. Another 20 percent of the land area is pasture, designed for animal rather than plant production. Thus, about 30 percent of the terrestrial surface is devoted to agriculture in the broadest sense. Recent comprehensive analyses of the world food situation emphasize that all the best land (that is, the land most easily farmed by existing technology) is now in use. Recent agricultural practice has switched from conventional agricultural practice, with emphasis on crop yield based on increased subsidies, to alternative (or sustainable) agriculture, with emphasis on low-input, sustainable agriculture, decreased soil erosion, and increased biodiversity (see National Research Council [NRC] 1989).

At the risk of oversimplifying, one can divide agroecosystems into three broad types (see Fig. 2-5 for models):

1. **Pre-industrial agriculture**—self-sufficient and labor intensive (human and animal labor provide the energy subsidy)—provides food for farmer and family and for sale or barter in local markets, but does not produce a large surplus for export (Fig. 2-5A).
2. Intensive mechanized, fuel-subsidized agriculture, termed **conventional** or **industrial agriculture** (machines and chemicals provide the energy subsidy), produces food exceeding local needs for export and trade, thus making food a commodity and a major market force in the economy rather than providing only life-support goods and services (Fig. 2-5B).
3. Lower-input sustainable agriculture (LISA), frequently termed **alternative agriculture** (NRC 1989), places emphasis on sustaining crop yields and profits while reducing inputs of fossil fuels, pesticides, and fertilizer subsidies (Fig. 2-5C).

About 60 percent of the world's croplands are in the pre-industrial category, a large proportion of them in the less developed countries of Asia, Africa, and South America that have large human populations. A great variety of types have been adapted to soil, water, and climate conditions, but for the purposes of general discussion, three

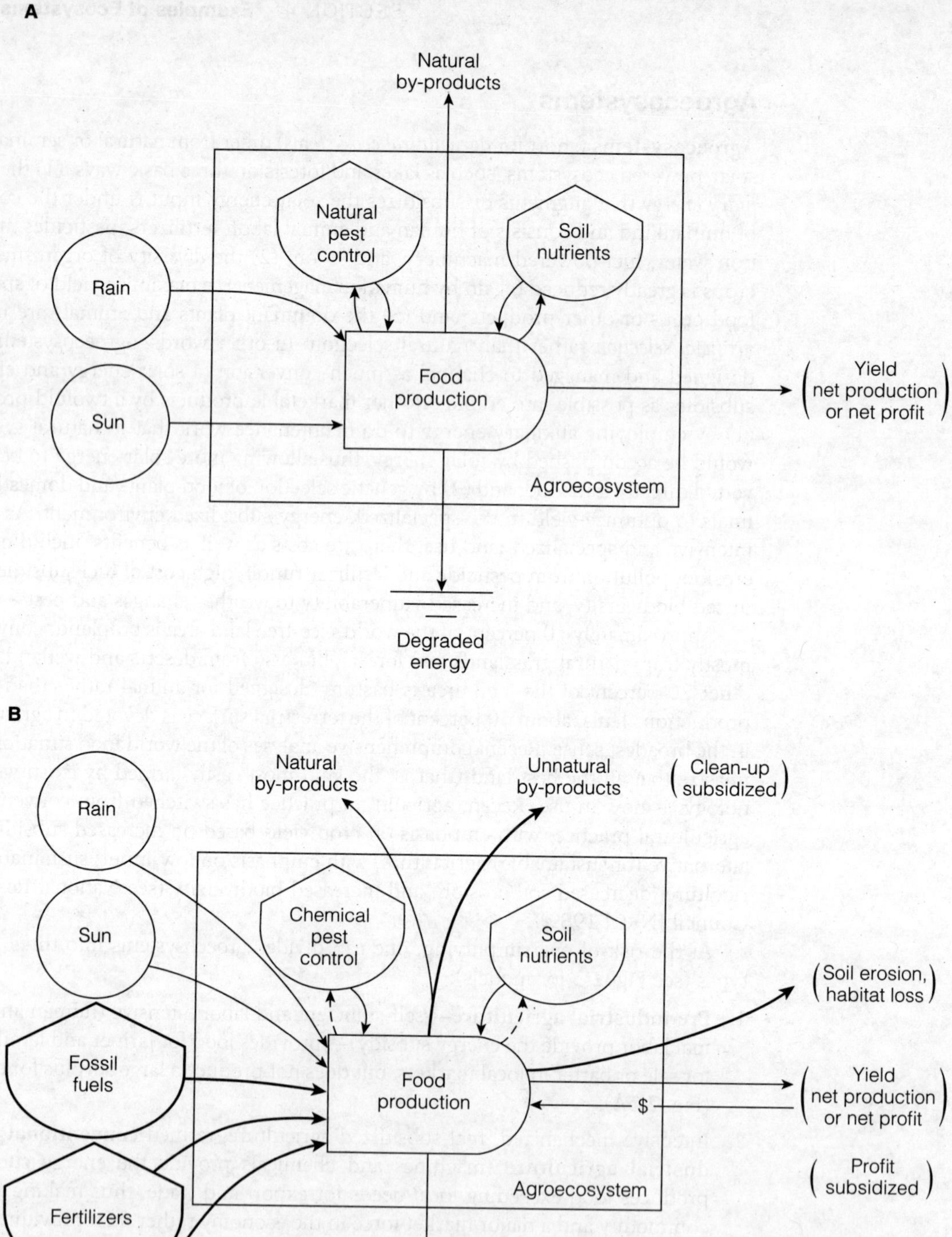

Figure 2-5A–B. (A) Diagram depicting a natural, unsubsidized, solar-powered pre-industrial agroecosystem (after Barrett et al. 1990). (B) Diagram depicting a human-subsidized, solar-powered industrial agroecosystem (after Barrett et al. 1990).

C

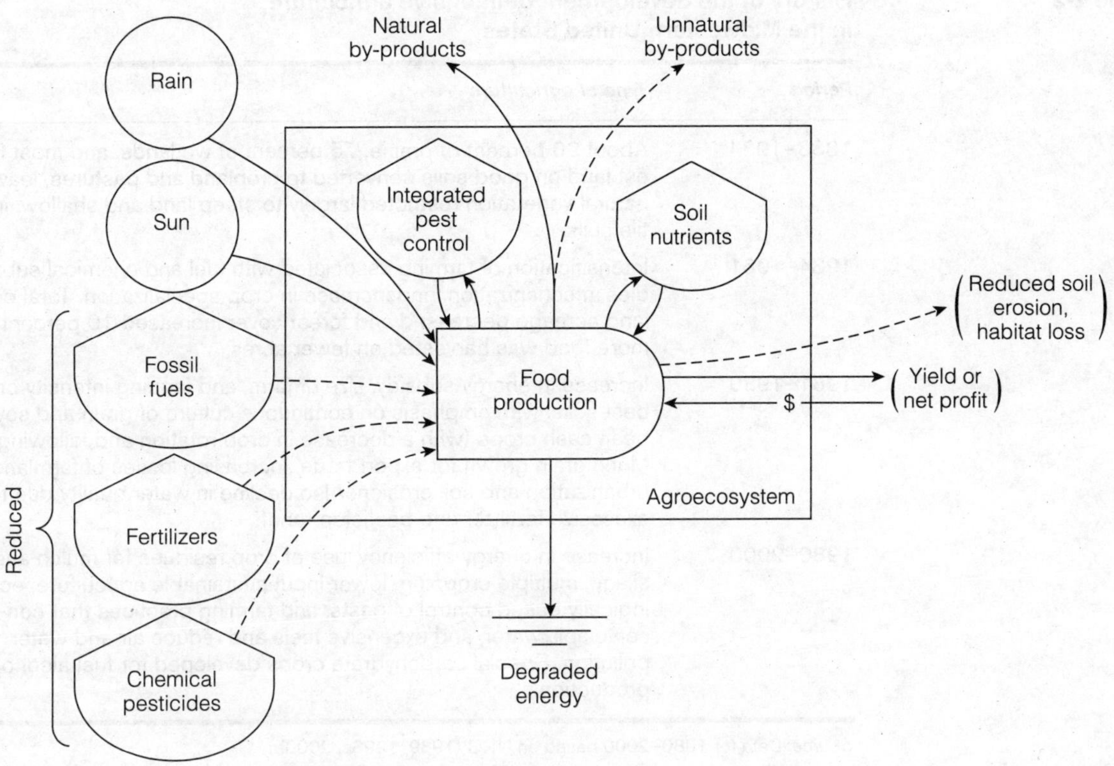

D

Figure 2-5C–E. (C) Example of a low-input, sustainable agroecosystem. (D) Example of conservation tillage in Iowa. Soybeans have been no-till planted in the mulched resi-due of last year's corn crop without the need for plowing.

Table 2-2 **History of the development of intensive agriculture in the Midwestern United States**

Period	*Type of agriculture*
1833–1934	About 90 percent of prairie, 75 percent of wetlands, and most forest land on good soils converted to cropland and pastures, leaving natural vegetation restricted largely to steep land and shallow, infertile soils.
1934–1961	Intensification of farming associated with fuel and chemical subsidies, mechanization, and increase in crop specialization. Total cropland acreage decreased and forest cover increased 10 percent as more food was harvested on fewer acres.
1961–1980	Increase in energy subsidy, size of farm, and farming intensity on best soils, with emphasis on continuous culture of grain and soybean cash crops (with a decrease in crop rotation and fallowing). Much grain grown for export trade. Increasing losses of farmland to urbanization and soil erosion. Also decline in water quality due to excessive fertilizer and pesticide runoff.
1980–2000	Increase in energy efficiency, use of crop residues for mulch and silage, multiple cropping, lower-input sustainable agriculture, ecologically based control of pests, and farming practices that conserve soil, water, and expensive fuels and reduce air and water pollution. Special carbohydrate crops developed for fuel alcohol production.

Source: Data for 1980–2000 based on NRC (1989, 1996a, 2000b).

types predominate: (1) *pastoral* systems; (2) *shifting* or swidden (slash-and-burn) agriculture; and (3) *flood-irrigated* and other nonmechanized systems.

Pastoralism involves herding cattle or other domestic animals in arid and semiarid regions (especially the savanna and grassland regions of Africa), with people subsisting on livestock products such as milk, meat, and hides. **Shifting agriculture,** once practiced throughout the world, is still widely practiced in forested areas of the Tropics. After patches of forest have been cut and the debris burned (or sometimes left lying on the ground as mulch), crops are cultivated for a few years until the nutrients are used up and leached out of the soil. Then the site is abandoned, to be rejuvenated by natural regrowth of the forest. Permanent, nonmechanized agriculture has persisted for centuries in Southeast Asia and elsewhere, feeding millions of people. The most productive of the agroecosystems are subsidized by **flood irrigation,** either naturally by seasonal floods along rivers and in fertile deltas or by artificially controlled flooding, as in the ancient, canal-irrigated paddy rice culture.

Pre-industrial systems, however—even well-adapted, permanent, and energy-efficient ones—do not produce enough surplus food to feed huge cities. In the past, mechanized agriculture capitalized on the availability of relatively inexpensive fuel, fertilizers, and agricultural chemicals (both of which require large amounts of fuel to make), and, of course, advanced technology, not only on the farm but in genetics, food processing, and marketing. In an incredibly short time (relative to the long sweep of agricultural history), farming in the United States (Table 2-2) and other industrialized countries changed from small farms, with a large percentage of people making their living in rural areas, to only 3 percent of the population farming ever larger tracts and producing more food on less land. The yield from fuel-subsidized agroecosystems on about 40 percent of the world's croplands provided a respite from the desperate race between human population growth and food production. The race, however, threatens to become grim, as the cost of the subsidies rises and as more and more countries, no longer able to feed themselves, are forced to import from countries, such as the United States, that have a surplus to export. Figure 2-5D shows a field located in Iowa being cultivated by plow, with large equipment typical of conventional agricultural practice. This field will also be cultivated by disk harrow several times before planting, resulting in extensive loss of soil due to wind and rainfall erosion. Figure 2-5E, on the other hand, illustrates conservation tillage, which greatly reduces soil erosion, promotes increased soil biotic diversity, and enhances nutrient (detritus) recycling.

Whereas the number of farmers in the developed world has declined dramatically, the number of farm animals has not, and the intensity of production of animal products has increased, paralleling that of crops. Thus, grain-fed beef has replaced grass-fed beef, and chickens are bred and managed as egg- and meat-producing machines, encased in wire cages under artificial light, and plied with growth-promoting food mixtures and drugs. In the United States, most of the corn crop (and other grains and soybeans) is fed to domestic animals, which in turn feed—and perhaps overfeed—the affluent of the world (see special issue of *Science,* 7 February 2003, entitled "Obesity," for details).

When thinking about population pressure (now more than 6 billion humans) on the environment and resources, one should not forget that not only are there many more domestic animals than people worldwide, but these animals also consume about five times more calories than do people. That is, the biomass of domestic animals (bovinity, equinity, swine, sheep, poultry) is about five times that of humans in **population equivalents,** a quantity based on the number of calories consumed by the average human. This does not include pets, which also consume a large amount of food.

In summary, industrial agriculture has greatly increased the yield of food and fiber per unit of land. This is the bright side of technology, but there are two dark sides: (1) many small farms go out of business worldwide, and these families gravitate toward cities, where they become consumers rather than producers of food; and (2) industrial agriculture has greatly increased nonpoint pollution and soil loss. To counteract the latter, a new technology called **low-input sustainable agriculture** (LISA) is coming into increasing use (Figs. 2-5C and E). For more on agroecosystem ecology, see Edwards et al. (1990), Barrett (1992), Soule and Piper (1992), W. W. Collins and Qualset (1999), Ekbom et al. (2000), NRC (2000a), Gliessman (2001), and Ryszkowski (2002).

5 Ecosystem Diversity

Statement

Ecosystem diversity may be defined as the genetic diversity, species diversity, habitat diversity, and the diversity of functional processes that maintain complex systems. It is useful to recognize two components of diversity: (1) the **richness** or **vari-**

ety component, which can be expressed as the number of "kinds" of components (such as species, genetic varieties, land-use categories, and biochemical processes) per unit of space; and (2) the **relative abundance** or **apportionment component** of individual units among the different kinds. The maintenance of moderate to high diversity is important not only to ensure that all key functional niches are operating, but especially to maintain redundancy and resilience in the ecosystem—in other words, to hedge against stressful times (such as storms, fires, diseases, or temperature changes) that occur sooner or later.

Explanation

The reason it is important to consider the relative abundance as well as the richness component is that two ecosystems can have the same richness but be very different because the apportionment of the kinds is different. For example, the communities in two different ecosystems might each have 10 species, but one community might have about the same number of individuals (say, 10 individuals) in each species (high *evenness*), whereas most of the individuals in the other community might belong to just one dominant species (low *evenness*). Most natural landscapes have a moderate evenness, with a few common (**dominant**) species for each trophic level or taxonomic group, and numerous rare species. In general, human activities directly or indirectly increase dominance and reduce evenness and variety.

Hanski (1982) called attention to the association between distribution and abundance. He proposed the **core-satellite species hypothesis** to explain this relationship, noting that **core** species are common and widespread in distribution, whereas **satellite** species are rare and local in distribution. According to this hypothesis, the frequency distribution of range sizes should have one peak for core species occupying large areas and a second peak for the satellite species occupying small ranges. Some data do show such a bimodal distribution of range sizes (Gotelli and Simberloff 1987), but most data are not consistent with the core-satellite hypothesis (Nee et al. 1991).

Diversity can be quantified and compared statistically in two basic ways: (1) by calculating diversity indices based on the ratio of parts to the whole, or n_i/N, where n_i is the number or percentage of *importance values* (such as numbers, biomass, basal area, productivity) and N is the total of all importance values; and (2) by plotting semi-log graphic profiles, called **dominance-diversity curves,** in which the number or percentage of each component is plotted in sequence, from most abundant to least abundant. The steeper the curve, the lower the diversity.

Examples

Species or biotic diversity may be divided into the richness and apportionment components (Fig. 2-6A). The total *number of species per unit area* (m^2 or hectare) and the *Margalef diversity index* are two simple equations used to compute species richness. The *Shannon index,* $\bar{H}$ (Shannon and Weaver 1949), and the *Pielou evenness index, e* (Pielou 1966), are two indices frequently used to compute species apportionment.

Species structure and use of diversity indices are illustrated by the comparison of an ungrazed tallgrass prairie and a cultivated, fertilized, but not herbicided millet field (Tables 2-3 and 2-4). The pattern of a few common and many relatively rare spe-

Table 2-3 **Species structure comparison of the vegetation of a natural grassland (A) and a cultivated grain field (B)**

(A) Species structure of the vegetation of an ungrazed tall-grass prairie in Oklahoma

Species	***Percentage of stand****
Sorghastrum nutans (Indian grass)	24
Panicum virgatum (Switch grass)	12
Andropogon gerardii (Big bluestem)	9
Silphium laciniatum (Compass plant)	9
Desmanthus illinoensis (Prickleweed)	6
Bouteloua curtipendula (Side-oats grama)	6
Andropogon scoparius (Little bluestem)	6
Helianthus maximiliana (Wild sunflower)	6
Schrankia nuttallii (Sensitive plant)	6
20 additional species (average 0.8 percent each)	16
Total	100

Source: Rice 1952, based on forty 1 m^2 quadrat samples.

(B) Species structure of the vegetation of a cultivated millet field in Georgia

Species	***Percentage of stand†***
Panicum ramosum (Brown-topped millet)	93
Cyperus rotundus (Nut sedge)	5
Amaranthus hybridus (Pigweed)	1
Digitaria sanguinalis (Crabgrass)	0.5
Cassia fasciculata (Sicklepod)	0.2
6 additional species (average 0.05 percent each)	0.3
Total	100.0

Source: Barrett 1968, based on twenty 0.25 m^2 quadrat samples taken in late July.

*In terms of percentage cover of the total of 34 percent area coverage of soil surface by the vegetation.

†In terms of percentage dry weight of aboveground plant biomass

cies in the tallgrass prairie (Table 2-3A) is typical of most natural plant communities. Such a high diversity is in sharp contrast to the dominance of one species in the cultivated crop field. If no herbicides or other weed control procedures are used, there will likely be some weedy species in the stand (Table 2-3B).

Two widely used apportionment indices are calculated for these two ecosystems (Table 2-4). The Simpson index involves summing the square of each n_i/N probabil-

Table 2-4 **Comparison of species variety, dominance, and diversity in the vegetation of a natural prairie and a cultivated millet field**

Habitat	*Number of species*	*Dominance (Simpson index)*	*Diversity (Shannon index)*
Natural prairie	29	0.13	0.83
Cultivated millet field	11	0.89	0.06

Source: Based on data from Table 2-3.

ity ratio. The Simpson index ranges from 0 to 1, with high values indicating strong dominance and low diversity. The Shannon index, $\overline{H}$, involves log transformations as follows:

$$\overline{H} = -\sum P_i \log P_i$$

where P_i is the proportion of the individuals belonging to the ith species.

With this index, the higher the value, the greater the diversity. The Shannon index is derived from information theory and represents a type of formulation widely used in assessing the complexity and information content of all kinds of systems. As shown in Table 2-4, the prairie grassland is a low-dominance, high-diversity ecosystem compared with the crop field.

Once $\overline{H}$ is calculated, the evenness, e, can be calculated by dividing the log of the number of species by $\overline{H}$. The Shannon index is also reasonably independent of sample size and is normally distributed, provided that the N values are integers (Hutcheson 1970), so routine statistical methods can be used to test for the significance of differences between the means. Biomass or productivity, which is often more ecologically appropriate, can be used if the numbers of individuals per species are not known.

The use of diversity profiles is illustrated in Figure 2-6B, which compares tree diversity in four forests in a gradient from temperate mountain to tropical regions. The relative importance of each species is plotted in sequence, from most abundant to least abundant. The flatter the curve, the greater the diversity. This graphic method reveals both the richness and the relative abundance components. Note that the number of kinds of trees ranges from less than 10 in the high mountain forest to more than 200 in the moist tropical rain forest. Most of the species in the more diverse forests are rare—less than 1 percent of the total of importance percentage values (biomass or production).

Between 1940 and 1982, the diversity of crops (one of several land-use types) declined in rural Central Ohio, even though the number of land-use types remained the same (Barrett et al. 1990). As the variety of crops grown declined, crop fields became larger, especially for corn and soybean crops, thereby increasing dominance and reducing overall diversity as measured by the Shannon index (Table 2-5). This is an example where the diversity component changed but the variety component did not change at the landscape scale.

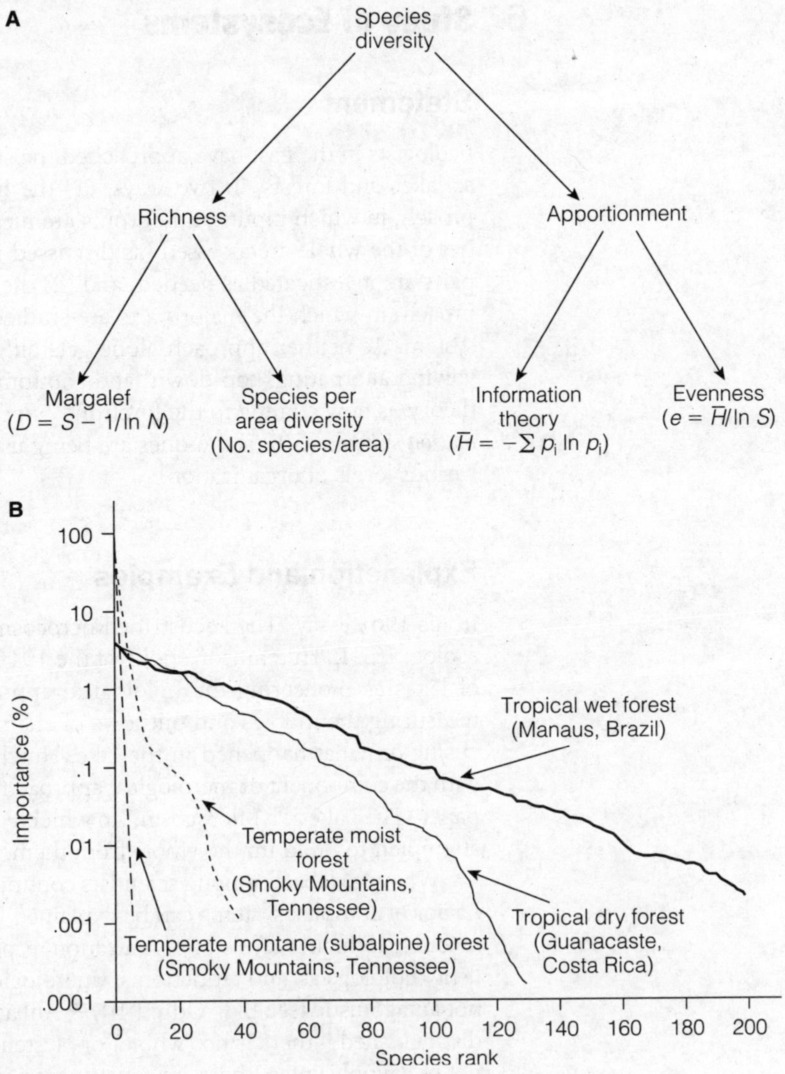

Figure 2-6. (A) Diagram depicting equations for measuring species richness and species apportionment. (B) Comparison of dominance-diversity curves for two tropical forests and two temperate forests. Importance values for the temperate forests are based on annual net primary production; importance values for the dry forest in Costa Rica are from cross-sectional basal area values of all stems of a given species; importance values in the Amazonian forest are based on above-ground biomass (after Hubbell 1979).

Table 2-5

Landscape diversity for rural Central Ohio

	Variety (Number, kinds)		*Diversity (Shannon index)*	
Component	***1940***	***1982***	***1940***	***1982***
Crops	18	9	0.80	0.60
Landscapes	6	6	0.61	0.48

Source: Data from Barrett et al. 1990.

6 Study of Ecosystems

Statement

Ecologists in the past have approached the study of large, complex ecosystems, such as lakes and forests, in two ways: (1) the **holological** (from *holos* = "whole") approach, in which inputs and outputs are measured, collective and emergent properties of the whole are assessed (as discussed in Chapter 1), and then the component parts are investigated as needed; and (2) the **merological** (from *meros* = "part") approach, in which the major parts are studied first and then integrated into a whole system. As neither approach alone gets all the answers, a multilevel approach, involving alternating "top-down" and "bottom-up" approaches based on hierarchical theory, is now coming to the forefront. Experimental modeling and *geographic information systems* (GIS) techniques are being used more and more to test hypotheses at various levels of organization.

Explanation and Examples

In his 1964 essay "The Lacustrine Microcosm Reconsidered," the eminent American ecologist G. E. Hutchinson spoke of the 1915 work of E. A. Birge on the heat budget of lakes as pioneering the holological approach. Birge concentrated his study on measuring the inflows and outflows of energy to and from the lake, rather than focusing on what happened in the lake. Hutchinson contrasted this method of study with the component or merological approach of Stephen A. Forbes in his classic 1887 paper "The Lake as a Microcosm," in which Forbes focused on parts of the system and attempted to build up the whole from them.

Whereas debates among scientists continue to center on just how much of the behavior of complex systems can be explained from the behavior of their parts without recourse to higher levels of organization, ecologists contend that the contrasting holistic (holological) and reductionist (merological) approaches are complementary and not antagonistic (see E. P. Odum 1977; Ahl and Allen 1996). Components cannot be distinguished if there is no "whole" or "system" to abstract them from, and there cannot be a whole unless there are constituent parts. When constituents are strongly coupled, emergent properties will likely reveal themselves only at the level of the whole. Such attributes will be missed if only the merological approach is taken. For further discussion of "top-down" and "bottom-up" approaches, see Hunter and Price (1992), Power (1992), S. R. Carpenter and Kitchell (1993), and Vanni and Layne (1997).

Above all, a given organism may behave quite differently in different systems, and this variability has to do with how the organism interacts with other components. Many insects, for example, are destructive pests in an agricultural habitat but not in their natural habitat, where parasites, competitors, predators, or chemical inhibitors keep them under control.

An effective way to gain insight into an ecosystem is to experiment with it (that is, disturb it in some manner, in the hope that the response will clarify hypotheses that one has deduced from observation). In recent years, *stress ecology* or *perturbation* (from *perturbare* = "to disturb") *ecology* has become an important field of research (as discussed later in this chapter). Besides manipulating the real thing, one can also gain insight by manipulating models, as briefly discussed in Chapter 1. As you read this book, watch for examples of all of these approaches.

7 Biological Control of the Geochemical Environment: The Gaia Hypothesis

Statement

Individual organisms not only adapt to the physical environment but, by their concerted action in ecosystems, they also adapt the geochemical environment to their biological needs. The fact that the chemistry of the atmosphere, the strongly buffered physical environment of Earth, and the presence of a diversity of aerobic life are utterly different from conditions on any other planet in this solar system has led to the Gaia hypothesis. The **Gaia hypothesis** holds that organisms, especially microorganisms, have evolved with the physical environment to provide an intricate, self-regulatory control system that keeps conditions favorable for life on Earth (Lovelock 1979).

Explanation

Although most of us know that the abiotic environment (physical factors) controls the activities of organisms, not all of us recognize that organisms influence and control the abiotic environment in many important ways. The physical and chemical nature of inert materials are constantly being changed by organisms, especially bacteria and fungi, which return new compounds and energy sources to the environment. The actions of marine organisms largely determine the content of the sea and of its bottom oozes. Plants growing on a sand dune build up a soil radically different from the original substrate. A South Pacific coral atoll is a striking example of how organisms modify the abiotic environment. From simple raw materials in the sea, whole islands are built through the activities of animals (corals) and plants (Fig. 2-7). Organisms control the very composition of our atmosphere.

The extension of biological control to the global level is the basis for James Love-

Figure 2-7. A South Pacific atoll, a ring-shaped coral reef, located in Bora-Bora, French Polynesia.

© Tim McKenna/CORBIS

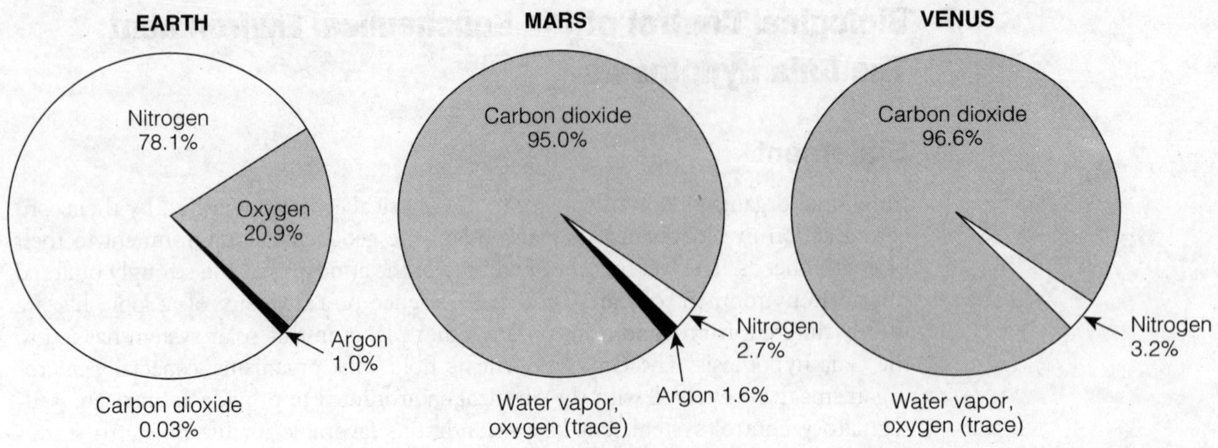

Figure 2-8. Comparison of the major components of the atmosphere on the planets Earth, Mars, and Venus. The percentages represent numbers of molecules (moles), not relative weights. Elements without percentage values are present in trace amounts (after Margulis and Olendzenski 1992).

lock's Gaia hypothesis (from *Gaia*, the Greek name for the Earth goddess). Lovelock, a physical scientist, inventor, and engineer, teamed up with microbiologist Lynn Margulis to explain the Gaia hypothesis in a series of articles and books (Lovelock 1979, 1988; Lovelock and Margulis 1973; Margulis and Lovelock 1974; Lovelock and Epton 1975; Margulis and Sagan 1997). They concluded that the atmosphere of Earth, with its unique high-oxygen/low–carbon dioxide content, and the moderate temperature and pH conditions on the surface of Earth, cannot be accounted for without the critical buffering activities of early life-forms and the continued coordinated activity of plants and microbes that dampen the fluctuations in physical factors that would occur in the absence of well-organized living systems. For example, ammonia produced by microorganisms maintains a pH in soils and sediments that is favorable to a wide variety of life. Without this organismic output, the pH of soils on Earth could become so low that it would preclude all but a very few kinds of organisms.

Figure 2-8 contrasts the atmosphere of Earth with the atmospheres of Mars and Venus where, if there is life, it certainly is not in control. In other words, the atmosphere of Earth did not just develop by a chance interaction of physical forces into a life-sustaining and self-regulating condition, and then life evolved to adapt to this condition. Rather, organisms from the very beginning played a major role in the development and control of a geochemical environment favorable to themselves. Lovelock and Margulis envision the microorganism web of life operating in the "brown belt" as an intricate control system that functions to maintain a pulsing, homeorhetic balance. This control system (Gaia) makes Earth one complex but unified cybernetic system. All of this is very much a hypothesis or a mere metaphor to many skeptical scientists, although most accept a strong biological influence on the atmosphere. Lovelock admits that the "search for Gaia" has been long and difficult, because hundreds of processes would have to be involved in an integrated control mechanism of such magnitude.

Humans, of course, more than any other species, attempt to modify the physical environment to meet their immediate needs, but in doing so we are increasingly shortsighted. Biotic components necessary for our physiological existence are being destroyed, and global balances are beginning to be perturbed and change—a process typically referred to as **global climate change.** Because we are heterotrophs, who thrive best near the end of complex food and energy chains, we depend on the natural environment no matter how sophisticated our technology. Cities can be viewed as "parasites" on the biosphere when we consider what we have already designated as **life-support resources,** namely, air, water, fuel, and food. The larger and the more technologically advanced the cities, the more they demand from the surrounding countryside, and the greater the danger of damaging the natural capital (for the benefits supplied to human societies by natural ecosystems, see Daily et al. 1997).

Examples

One of the classic papers we suggest that students of ecology read is Alfred Redfield's summary essay "The Biological Control of Chemical Factors in the Environment," published in 1958. Redfield marshaled the evidence to show that the oxygen content of the air and the nitrate content in the sea are produced and largely controlled by organic activity and, furthermore, that the quantities of these vital components in the sea are determined by the biocycling of phosphorus. This system is as intricately organized as a fine watch, but unlike a watch, this marine regulator was not built by engineers and is, comparatively speaking, little understood. Lovelock's books extend Redfield's hypothesis to the global level. See also Jantsch's *The Self-Organizing Universe* (1980).

The Copper Basin at Copperhill, Tennessee, and a similar devastated area near Sudbury, Ontario, Canada (Gunn 1995), illustrate on a small scale what land without life would be like. In these areas, sulfuric acid fumes from copper and nickel smelters exterminated all of the rooted plants over areas large enough to be seen from space as scars on the face of Earth. A type of smelting, known as "roasting," involved the igniting of great piles of ore, green wood, and coke. These piles were then smoldered continuously, giving off the acidic fumes. Most of the soil eroded, leaving a spectacular "desert" that looks like a landscape on Mars. Although the mining has stopped, natural recovery has been very slow. At Copperhill, artificial reforestation using heavy fertilization with minerals or sewage sludge has succeeded somewhat. Pine seedlings inoculated with symbiotic root fungi that assist the tree in extracting nutrients from impoverished soils are surviving on their own as the large input of fertilizer is used up (E. P. Odum and Biever 1984). At least, such experiments demonstrate that locally damaged ecosystems can be restored to some degree, but only with great effort and expense.

Mt. Saint Helens demonstrates the impact of a natural disturbance, which has been followed by a rapid recovery or restoration process (Figs. 2-9A and B; see Chapter 8 for additional information regarding ecosystem development). **Restoration ecology** is a relatively new science directed at managing communities, ecosystems, and landscapes that have been damaged by pollution, invasion of exotic species, or human perturbations. *Prevention of pollution, invasive species, or human perturbations is, of course, better than cure.* For this reason, the science of **ecosystem health** needs to be a close companion to restoration (or rehabilitation) ecology, just as preventative

A

Courtesy of Gary W. Barrett

B

Courtesy of Gary W. Barrett

Figure 2-9. (A) Mount Saint Helens, located in the Cascade Mountains of southwestern Washington, following its eruption on 18 May 1980. A luxuriant Douglas fir forest covered this area before the eruption (a natural disturbance) removed all the vegetation. (B) The same area 16 months later. This system has exhibited a rapid rate of ecosystem recovery following this natural disturbance, compared to the slow rate of recovery following a human (toxic) disturbance.

medicine is a close companion to human health and disease control. For additional readings, see also W. R. Jordan et al. (1987) and Rapport et al. (1998).

8 Global Production and Decomposition

Statement

Every year, approximately 10^{17} grams (about 100 billion tons) of organic matter are produced on Earth by photosynthetic organisms. An approximately equivalent amount is oxidized back to CO_2 and H_2O during the same time interval, as a result of the respiratory activity of living organisms. *But the balance is not exact.*

Over most of geological time, since the beginning of the Precambrian period, a very small but significant fraction of the organic matter produced was incompletely decomposed in anaerobic (anoxic) sediments or completely buried and fossilized without being respired or decomposed. This excess organic production probably contributed to a decrease in CO_2 and a buildup of oxygen in the atmosphere to the high levels of recent geological times, although limestone formation, which removes CO_2 from the atmosphere for long periods of time, was likely also important in this regard. In any event, high O_2 and low CO_2 made possible the evolution and continued survival of the higher forms of life.

About 300 million years ago, excess production formed the fossil fuels that made the Industrial Revolution possible. During the past 60 million years, shifts in biotic balances, coupled with variations in volcanic activity, rock weathering, sedimentation, and solar input, have resulted in oscillating CO_2/O_2 atmospheric ratios. Oscillations in atmospheric CO_2 have been associated with and presumably have caused alternate warming and cooling of climates. During the past half century, human agro-industrial activities have significantly raised the CO_2 concentration in the atmosphere. In 1997, heads of state met in Kyoto, Japan, to try to find ways to reduce CO_2 emissions because of the potential for climate alteration that now poses such a serious global problem.

Organic materials produced outside the ecosystem are referred to as **allochthonous** inputs (from the Greek *chthonos,* "of the earth," and *allos,* "other"); photosynthesis that occurs within the ecosystem is referred to as **autochthonous** production. In the subsections that follow, we attempt to give equal attention to the details of organic production and decomposition. *The balance—or lack thereof—between production and respiration (P/R* ratios) *locally, globally, and in ecosystem development sequences* (see Chapter 8), *along with the balance between production and maintenance cost in human technoecosystems, are, perhaps, the most important quantities we need to know if we are to understand and deal with present and future worlds.*

Explanations

Kinds of Photosynthesis and Producer Organisms

Chemically, the photosynthetic process involves the storage of a part of the sunlight energy as potential or "bound" energy in food. The general equation of the oxidation-reduction reaction can be written as follows:

$$CO_2 + 2H_2A \xrightarrow[\text{energy}]{\text{light}} (CH_2O) + 2A,$$

the oxidation being

$$2H_2A \longrightarrow 4H + 2A$$

and the reduction being

$$4H + CO_2 \longrightarrow (CH_2O) + H_2O.$$

For green plants in general (green bacteria, algae, and higher plants), A is oxygen—that is, water is oxidized with release of gaseous oxygen, and the carbon dioxide is reduced to carbohydrate (CH_2O) with release of water. Because it was discovered some years ago, using radioactive tracers, that the oxygen in the following equation comes from the water input, the balanced equation for photosynthesis is as follows:

$$6CO_2 + 12H_2O \xrightarrow[\text{energy}]{\text{light}} C_6H_{12}O_6 + 6O_2 + 6H_2O$$

In some types of bacterial photosynthesis, on the other hand, H_2A (the *reductant*) is not water but either an inorganic sulfur compound, such as hydrogen sulfide (H_2S) in the green and purple sulfur bacteria (Chlorobacteriaceae and Thiorhodaceae, respectively), or an organic compound, as in the purple and brown nonsulfur bacteria (Athiorhodaceae). Consequently, oxygen is *not* released in these types of bacterial processes.

The **photosynthetic bacteria** that do release oxygen as a by-product are largely aquatic (marine and freshwater) **cyanobacteria** that in most situations play a minor role in the production of organic matter. However, they can function under unfavorable conditions, and they do cycle certain minerals in aquatic sediments. The green and purple sulfur bacteria, for example, are important to the sulfur cycle. They are **obligate anaerobes** (able to function only in the absence of oxygen) and occur in the boundary layer between oxidized and reduced zones in sediments or water where the light is of low intensity. Tidal mudflats are good places to observe these bacteria, because they often form distinct pink or purple layers just under the upper green layers of mud algae (in other words, at the very upper edge of the anaerobic or reduced zone, where light but not much oxygen is available). In contrast, nonsulfur photosynthetic bacteria are generally **facultative anaerobes** (able to function with or without oxygen). They can also function as heterotrophs in the absence of light, as can many algae. Bacterial photosynthesis can be helpful in polluted and eutrophic waters and, hence, is being increasingly studied, but it is no substitute for the "regular," oxygen-generating photosynthesis on which the world depends.

Differences in biochemical pathways for carbon dioxide reduction (the reduction portion of the equation) in higher plants have important ecological implications. In most plants, carbon dioxide fixation follows a **C_3 pentose phosphate** or **Calvin cycle,** which for many years was the accepted scheme for photosynthesis. Then, in the 1960s, plant physiologists discovered that certain plants reduce carbon dioxide in a different manner, according to a **C_4 dicarboxylic acid cycle.** Such C_4 plants have large chloroplasts in the bundle sheaths around the leaf veins—a distinctive morphological feature that had been noted a century earlier but had not been suspected as an indicator of a major physiological characteristic. More important, plants using the dicarboxylic acid cycle respond differently to light, temperature, and water. For the purposes of discussing the ecological implications, the two photosynthetic types are designated **C_3 and C_4 plants.**

Figure 2-10 contrasts the response of C_3 and C_4 plants to light and temperature. C_3 plants tend to peak in photosynthetic rate (per unit of leaf surface) at moderate light intensities and temperatures and tend to be inhibited by high temperatures and the full intensity of sunlight. In contrast, C_4 plants are adapted to bright light and high temperatures, greatly exceeding the production of C_3 plants under these conditions. They also use water more efficiently, generally requiring less than 400 grams of water to produce 1 gram of dry matter, compared with the 400- to 1000-gram water requirement of C_3 plants. Moreover, C_4 plants are not inhibited by high oxygen concentrations, unlike C_3 species. One reason why C_4 plants are more efficient at the high end of the light-temperature scales is that they have little photorespiration—that is, the photosynthate of the plant is not respired away as light intensity rises. Some C_4 plants have been reported to be more resistant to grazing by insects (Caswell

Figure 2-10. Comparative photosynthetic response of C_3 and C_4 plants to increasing light intensity and temperature. C_3 plants' photosynthetic rate peaks at moderate light intensities and temperatures, whereas C_4 plants thrive under bright light and high-temperature conditions.

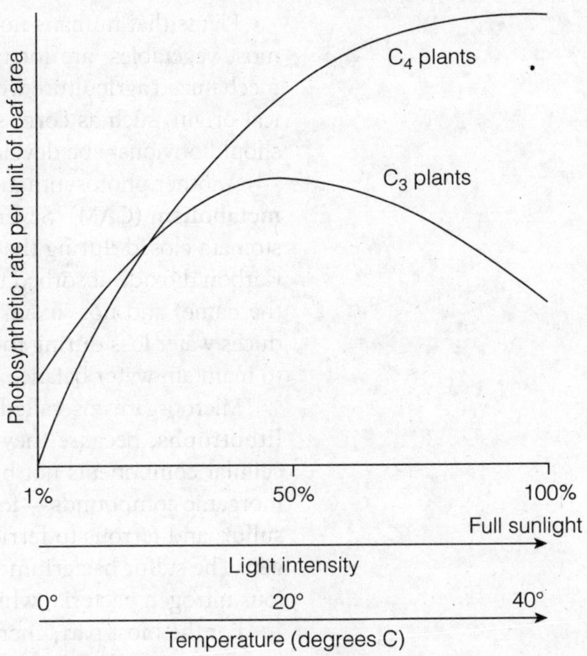

et al. 1973), perhaps because they tend to have a lower protein content. On the other hand, Haines and Hanson (1979) reported that detritus made from C_4 plants was a richer food source for consumers in a salt marsh than was detritus originating from C_3 plants.

Species with a C_4-type photosynthesis are especially numerous in the grass family (Gramineae). As would be expected, C_4 species dominate the vegetation in deserts and grasslands in warm temperate and tropical climates, but they are rare in forests and in the cloudy northern latitudes, where low light intensities and low temperatures predominate. Table 2-6 illustrates how the proportion of C_4 species increases along a gradient from the cool, moist prairies of the midwestern United States to the hot, dry deserts of the Southwest, and also how C_4/C_3 proportions differ with the seasons in temperate deserts. It is not surprising that crabgrass (*Digitaria sanguinalis*), a legendary pest of particular lawns, turns out to be a C_4 species, as are a number of other "weeds" that thrive in human-made warm open spaces.

Despite their lower photosynthetic efficiency at the leaf level, C_3 plants account for most of the photosynthetic production of the world, presumably because they are more competitive in mixed communities, where there are shading effects and where light and temperature are average rather than extreme (observe in Fig. 2-10 that C_3 plants outperform C_4 plants under low light and temperature conditions). This appears to be another good example of the "whole is not the sum of the parts" principle. *Survival in the real world does not always go to the species that are merely physiologically superior under optimal conditions in monoculture, but rather to those species that are superior in multiculture under varying conditions that are not always optimal.* To put it another way, what is efficient in isolation is not necessarily efficient in the ecosystem, where the interaction among species and the abiotic environment is vital to natural selection.

Plants that humans now depend on for food, such as wheat, rice, potatoes, and most vegetables, are for the most part C_3 plants, because most crops suitable for mechanized agriculture were developed in the North Temperate Zone. Crops of tropical origin, such as corn, sorghum, and sugarcane, are C_4 plants. More C_4 varieties should obviously be developed for use in irrigated deserts and in the Tropics.

Another photosynthetic mode especially adapted to deserts is **crassulacean acid metabolism** (CAM). Several desert succulent plants, including the cacti, keep their stomata closed during the hot daytime and open them during the cool of the night. Carbon dioxide absorbed through the leaf openings is stored in organic acids (hence the name) and not "fixed" until the next day. This *delayed photosynthesis* greatly reduces water loss during the day, thereby enhancing the ability of the succulent plant to maintain water balance and water storage.

Microorganisms called **chemosynthetic bacteria** are considered to be **chemolithotrophs,** because they obtain their energy for carbon dioxide assimilation into cellular components not by photosynthesis but by the chemical oxidation of simple inorganic compounds—for example, ammonia to nitrite, nitrite to nitrate, sulfide to sulfur, and ferrous to ferric iron. They can grow in the dark, but most require oxygen. The sulfur bacterium *Thiobacillus,* often abundant in sulfur springs, and the various nitrogen bacteria, which are important in the nitrogen cycle, are examples.

For the most part, chemolithotrophs are involved in carbon recovery rather than in primary production, because their ultimate energy source is organic matter produced by photosynthesis. In 1977, however, unique deep-sea ecosystems were discovered based entirely on chemosynthetic bacteria not dependent on a photosynthate source. These ecosystems are located in totally dark areas where the sea floor is spreading, creating vents from which hot, mineral-rich sulfurous water escapes. Here, a whole food web of mostly **endemic** (that is, species found nowhere else) marine organisms has evolved. The food chain begins with bacteria that obtain their energy to fix carbon and produce organic matter by oxidizing hydrogen sulfide (H_2S) and other chemicals. Snails and other grazers feed on the bacterial mats in the cooler parts of the vent structure; strange, 30-centimeter-long clams and 3-meter-long tube worms have evolved a mutualistic relationship with the chemosynthetic bacteria that live in their tissues; and there are also crab and fish predators. These vent communities are, indeed, an age-old, geothermally powered ecosystem, as the heat of Earth's

Table 2-6 **Percentage of C_4 species in an east-west transect of the United States grasslands and deserts**

Ecosystem	*Percentage of C_4 species*
Tallgrass prairie	50
Mixed-grass grassland	67
Shortgrass grassland	100
Desert–Summer annuals	100
Desert–Winter annuals	0

Source: After E. P. Odum 1983.

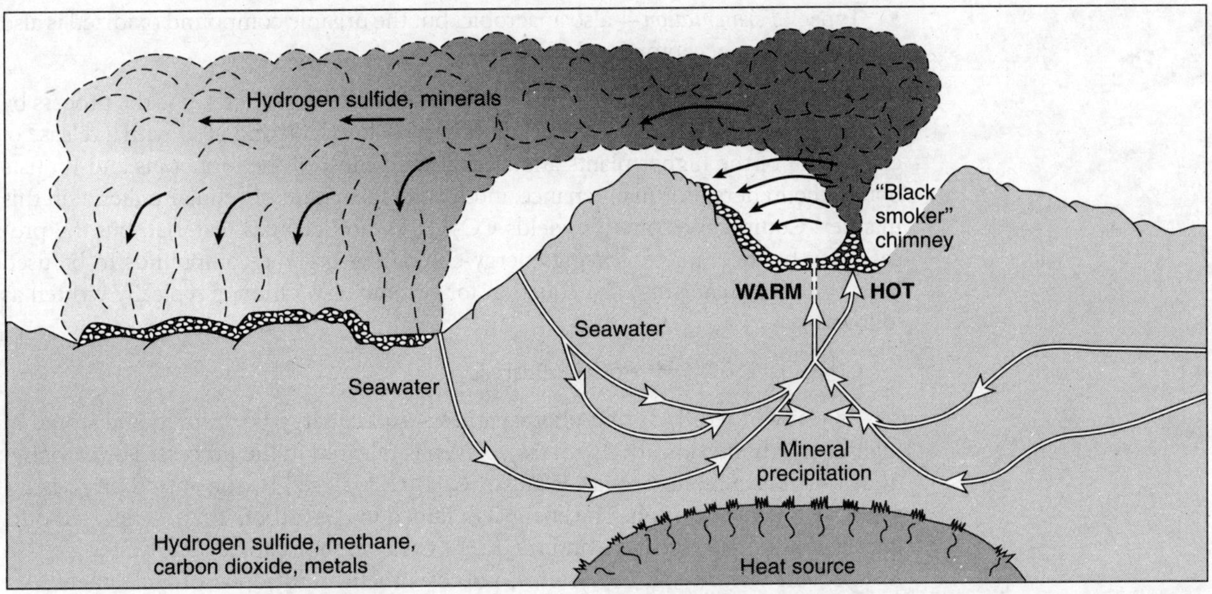

Figure 2-11. Diagram of a deep-sea thermal vent. Volcanic vents in the mid-ocean ranges produce plumes of "black smoke" rich in hydrogen sulfide and minerals. The chimney of the vent is covered by limpets, undersea worms (such as polychaetes), and large tube worms. Crabs and mussels also make their home among the rich deposits (courtesy of V. Tunnicliffe).

core produces the reduced sulfur compounds that are the energy source for this ecosystem. A diagram of the vent structure is shown in Figure 2-11. For reviews and pictures of some of the animals, see Tunnicliffe (1992).

On the global scale, among multicellular life-forms, the distinction between autotrophs and heterotrophs is clear-cut, and gaseous oxygen is essential for the survival of the majority of heterotrophs. But among the microorganisms—the bacteria, fungi, and more primitive algae and protozoa—many species and varieties are not so specialized. Rather, they are adapted to be intermediate or to shift between autotrophy and heterotrophy, with or without oxygen.

Types of Decomposition and Decomposers

In the world at large, the heterotrophic processes of decomposition (*catabolism*) approximately balance the autotrophic metabolism (*anabolism*), as was indicated in this section's statement, but the balance varies widely locally. If decomposition is considered in the broad sense as "any energy-yielding biotic oxidation," then several types of decomposition roughly parallel the types of photosynthesis when oxygen requirements are considered:

- Type 1. *Aerobic respiration*—gaseous (molecular) oxygen is the electron acceptor (*oxidant*).
- Type 2. *Anaerobic respiration*—gaseous oxygen is not involved. An inorganic compound other than oxygen, or an organic compound, is the electron acceptor (*oxidant*).

- Type 3. *Fermentation*—also anaerobic, but the organic compound oxidized is also the electron acceptor (*oxidant*).

Aerobic respiration (Type 1) is the reverse of photosynthesis; it is the process by which organic matter (CH_2O) is decomposed back to CO_2 and H_2O with a release of energy. All of the higher plants and animals and most of the Monerans and Protists obtain their energy for maintenance and for the formation of cellular material in this manner. Complete respiration yields CO_2, H_2O, and cellular material, but the process may be incomplete, leaving energy-containing organic compounds to be used later by other organisms. The equation for aerobic respiration is typically written as follows:

$$C_6H_{12}O_6 + 6O_2 \longrightarrow 6CO_2 + 6H_2O$$

As you will recall, during photosynthesis, solar energy is captured and stored in high-energy bonds in carbohydrates; oxygen is released in the process. The carbohydrate (such as a monosaccharide sugar, $C_6H_{12}O_6$) is used by the autotroph or is ingested by the heterotroph. The energy contained in the carbohydrate is released during respiration via glycolysis and the Krebs cycle; carbon dioxide and water are also released. In virtually all ecosystems, photosynthetic autotrophs provide energy for the total system. Thus, the ultimate source of energy for the system is the Sun.

Respiration without O_2 (**anaerobic respiration**) is largely restricted to the saprophages, such as bacteria, yeasts, molds, and protozoa, although it occurs as a dependent process in certain tissues of higher animals (muscle contraction, for example). The *methane bacteria* are examples of obligate anaerobes that decompose organic compounds with the production of methane (CH_4) through reduction of either organic or mineral (carbonate) carbon (thus employing both types of anaerobic metabolism). In aquatic environments, such as freshwater marshes and swamps, the methane gas, often known as "swamp gas," rises to the surface, where it can be oxidized or, if it catches fire, may be reported as a UFO (unidentified flying object)! The methane bacteria are also involved in the breakdown of forage in the rumen of cattle and other ruminants. As we deplete supplies of natural gas and other fossil fuels, these microbes may be domesticated to produce methane on a large scale from manure or other organic sources.

Desulfovibrio and other varieties of sulfate-reducing bacteria are ecologically important examples of anaerobic respiration (Type 2), because they reduce SO_4 in deep sediments and in anoxic waters, such as the Black Sea, to H_2S gas. The H_2S can rise to shallow sediments or surface waters, where it can be oxidized by other organisms (the photosynthetic sulfur bacteria, for example). Alternatively, H_2S can combine with Fe and Cu and many other minerals. Millions of years ago, the microbial production of minerals may have been responsible for many of our most valuable metal ore deposits. On the negative side, sulfate-reducing bacteria cause billions of dollars damage annually through corrosion of metals by the H_2S they produce. *Yeasts,* of course, are well-known examples of organisms using **fermentation** (Type 3). They are not only commercially important but also abundant in soil, where they help decompose plant residues.

As already indicated, many kinds of bacteria are capable of both aerobic and anaerobic respiration, but the end products of the two reactions will be different, and the amount of energy released will be much less under anaerobic conditions. For example, the same species of bacterium, *Aerobacter,* can be grown under anaerobic and

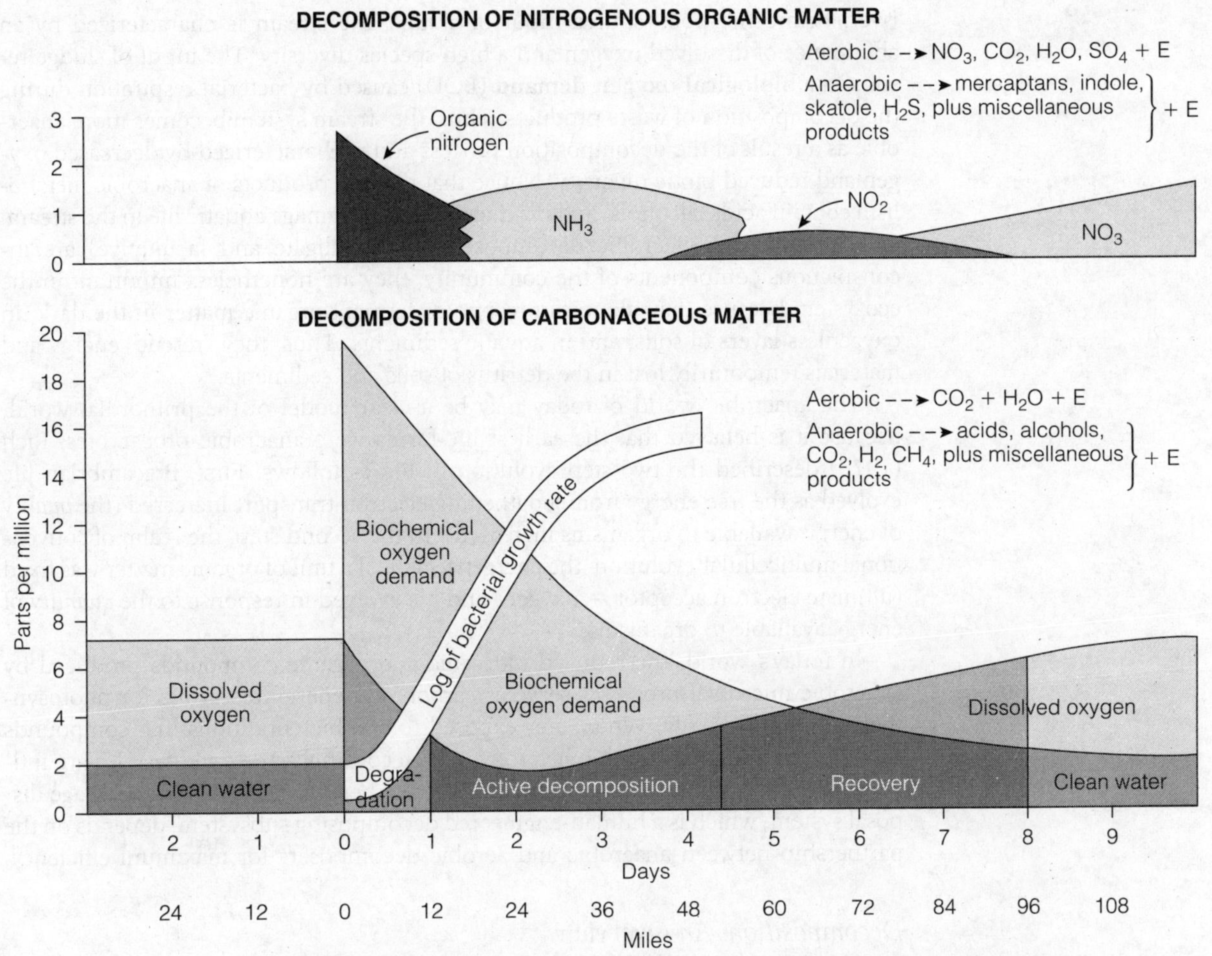

Figure 2-12. Diagram depicting the end products of aerobic and anaerobic metabolism when an acute dosage of sludge containing nitrogenous and carbonaceous matter enters a stream. This input of nutrients quickly causes a biochemical oxygen demand in the system.

aerobic conditions with glucose as the carbon source. When oxygen is present, almost all of the glucose is converted into bacterial biomass and CO_2, but in the absence of oxygen, decomposition is incomplete, a much smaller portion of the glucose ends up as cellular carbon, and a series of organic compounds (such as ethanol, formic acid, acetic acid, and butanediol) is released into the environment. Additional bacterial specialists would be required to oxidize these compounds further and recover additional energy. When the rate of input of organic detritus into soils and sediments is high, bacteria, fungi, protozoa, and other organisms create anaerobic conditions by using up the oxygen faster than it can diffuse into water and soil. Decomposition does not stop then but continues, often at a slower rate, provided an adequate diversity of anaerobic microbial metabolic types is present.

Figure 2-12 illustrates the end products of aerobic and anaerobic metabolism when an input of nutrients (such as untreated municipal sludge) enters a stream or river. Before the point-source input of sludge, the stream is characterized by an abundance of dissolved oxygen and a high species diversity. The input of sludge results in a **biological oxygen demand** (BOD) caused by bacterial respiration during the decomposition of waste products. Thus, the stream system becomes more anaerobic as a result of the decomposition process and is characterized by decreased oxygen and reduced biotic diversity. Notice that the end products of anaerobic metabolism contain acids, alcohols, and products that may damage aquatic life in the stream.

Although the anaerobic decomposers (both obligate and facultative) are inconspicuous components of the community, they are nonetheless important in the ecosystem because they alone can respire or ferment organic matter in the dark, in oxygenless layers of soils, and in aquatic sediments. Thus, they "rescue" energy and materials temporarily lost in the detritus of soils and sediments.

The anaerobic world of today may be a clear model of the primordial world, because it is believed that the earliest life-forms were anaerobic procaryotes. Rich (1978) described the two-step evolution of life as follows: First, Precambrian life evolved as the free energy from lengthening electron transport increased (the *quality* of energy available to organisms increased). In the second step, the realm of conventional multicellular evolution, the energetic value of a unit of organic matter was fixed (ultimate electron acceptor = oxygen) and life evolved in response to the *quantity* of energy available to organisms.

In today's world, the reduced inorganic and organic compounds produced by anaerobic microbial processes serve as carbon and energy reservoirs for photosynthetically fixed energy. When later exposed to aerobic conditions, the compounds serve as substrates for aerobic heterotrophs. Accordingly, the two lifestyles are intimately coupled and function together for mutual benefit. For example, a sewage disposal system, which is a human-engineered decomposing subsystem, depends on the partnership between anaerobic and aerobic decomposers for maximum efficiency.

Decomposition: An Overview

Decomposition results from both abiotic and biotic processes. For example, prairie and forest fires are not only major limiting or controlling factors (as will be discussed later), but they are also "decomposers" of detritus, releasing large quantities of CO_2 and other gases into the atmosphere and minerals into the soil. Fire is an important and even a necessary process in so-called *fire-type* (perturbation-dependent) ecosystems, in which the physical conditions are such that microbial decomposers do not keep up with organic production. The grinding action of freezing and thawing and water flow also break down organic materials, in part by reducing particle size. By and large, however, the heterotrophic microorganisms or saprophages ultimately act on the dead bodies of plants and animals. This kind of decomposition, of course, is the result of the process by which bacteria and fungi obtain food for themselves. Decomposition, therefore, occurs through energy transformations within and between organisms and is an absolutely vital function. If it did not occur, all nutrients would soon be tied up in dead bodies, and no new life could be produced. Within the bacterial cells and the fungal mycelia are enzymes necessary to carry out specific chemical reactions. These enzymes are secreted into dead matter; some of the decomposition products are absorbed into the organism as food, whereas other products remain in the environment or are excreted from the cells. No single species of saprotroph can completely decompose a dead body. However, populations of de-

Figure 2-13. Example of decomposition in a Georgia salt marsh. Rates of decomposition in terms of percentage of dead marsh grass (*Spartina alterniflora*) and fiddler crabs (*Uca pugnax*) remaining in nylon-mesh litter bags placed in the marsh where they would be subjected to daily tidal inundation.

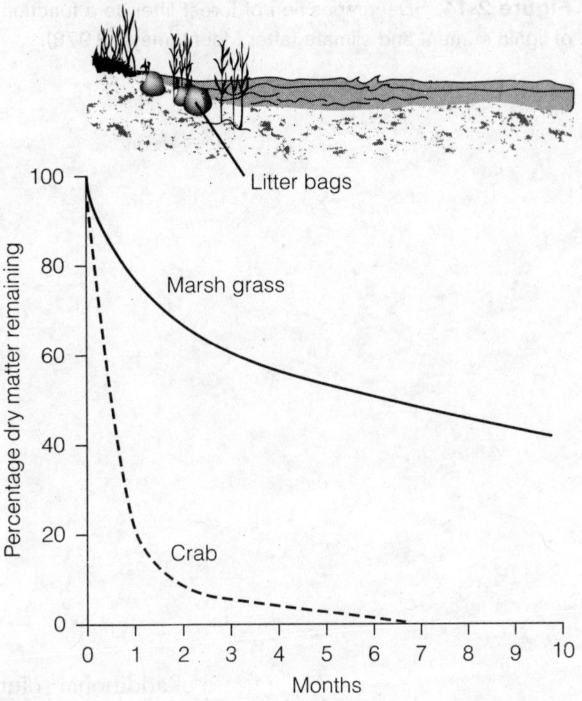

composers prevalent in the biosphere consist of many species that, by their sequential action, can completely decompose a body. Not all parts of the bodies of plants and animals are broken down at the same rate. Fats, sugars, and proteins are decomposed readily, whereas the cellulose of plants, the lignin of wood, the chitin of insects, and the bones of animals are decomposed very slowly. Figure 2-13 shows a comparison of the rate of decomposition of dead marsh grass and fiddler crabs placed in nylon-mesh "litter bags" in a Georgia salt marsh. Note that most of the animal remains and about 25 percent of the dry weight of the marsh grass were decomposed in about two months, but the remaining 75 percent of the grass, largely cellulose, was broken down more slowly. After 10 months, 40 percent of the grass remained, but all of the crab remains had disappeared from the bag. As the detritus becomes finely divided and escapes from the bag, the intense activities of microorganisms often result in nitrogen and protein enrichment, thus providing a more nutritious food for detritus-feeding animals. The graphic model of Figure 2-14 shows that the decomposition of forest litter (leaves and twigs) is very much influenced by lignin (resistant wood polymers) content and climatic conditions. Until a few decades ago, lignin was believed to be decomposed only in the presence of oxygen. However, studies have demonstrated since then that even very resistant compounds can be microbially degraded (although very slowly) under anaerobic conditions (Benner et al. 1984).

The more resistant products of decomposition result in **humus** (or **humic substances**), which is a universal component of ecosystems. It is convenient to recognize four stages of decomposition: (1) initial *leaching,* the loss of soluble sugars and other compounds that are dissolved and carried away by water; (2) the formation of particulate detritus by physical and biological action (*fragmentation*) accompanied by the release of dissolved organic matter; (3) the relatively rapid *production* of humus and

Figure 2-14. Decomposition of forest litter as a function of lignin content and climate (after Meentemeyer 1978).

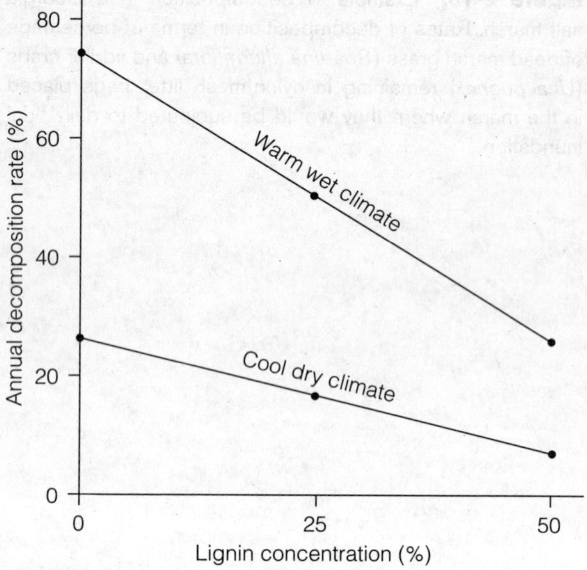

release of additional soluble organics by saprotrophs; and (4) the slower *mineralization* of humus. **Mineralization** is the release of organically bound nutrients into an inorganic form available to plants and microbes.

The slowness of decomposition of humic substances is a factor in the decomposition lag and oxygen accumulation in an ecosystem that has been stressed. In general appearance, humus is a dark, often yellow-brown, amorphous or colloidal substance that is difficult to characterize chemically. No great differences in the physical properties or chemical structure of humus exist between different terrestrial ecosystems, but studies suggest that marine humic materials have a different origin and hence a different structure. This difference is related to the fact that there are no lignin-rich, woody plants in the sea, so humic compounds are derived from less aromatic algal chemicals.

In chemical terms, humic substances are condensations of aromatic compounds (phenols) combined with the decomposition products of proteins and polysaccharides. A model for the molecular structure of humus derived from lignocellulose is shown in Figure 2-15. The phenolic benzene rings and the side-chain bonding make these compounds recalcitrant to microbial decomposition. Ironically, many of the toxic materials that humans are now adding to the environment, such as herbicides, pesticides, and industrial effluents, are derivatives of benzene and are causing serious trouble because of their low degradability and their toxicity.

The overall energy budget of an ecosystem reflects a balance between income and expenditure (that is, a balance between production and decomposition), just as in a bank account. The ecosystem gains energy through the photosynthetic assimilation of light by green plants (autotrophs) and the transport of organic matter into the ecosystem from external sources. More will be said regarding this balance between production, *P*, and decomposition or respiration, *R*, in Chapter 3.

Detritus, humic substances, and other organic matter undergoing decomposition are important for soil fertility. These materials provide a favorable soil texture for plant growth. As gardeners know, adding decaying or decomposed organic matter to most soils greatly increases the ability of the garden plot to produce vegetables and flowers! Furthermore, many organics form complexes with minerals that greatly affect the biological availability of the minerals. For example, **chelation** (from *chele* = "claw," referring to "grasping"), or complex formation with metal ions, keeps the element in solution and nontoxic compared with the inorganic salts of the metal. As industrial wastes are full of toxic metals, it is fortunate that the chelators that are products of the natural decomposition of organic matter work to mitigate toxic effects on organisms. For example, the toxicity of copper to phytoplankton is correlated with the free ion (Cu^{++}) concentration, not the total concentration of copper. Accordingly, a given amount of copper is less toxic in an organically rich, inshore marine environment than in the open sea, where there is less organic matter to complex the metal.

Soils are composed of a variable combination of minerals, organic matter, water, and air. Recent definitions of soil (Coleman and Crossley 1996) include the phrase "with its living organisms." Thus, healthy, fertile soil is alive and is composed of biotic and abiotic components having many interactions (see Chapter 5 for details regarding soil ecology).

Numerous studies have shown that small animals (such as protozoa, soil mites, collembolans, nematodes, ostracods, snails, and earthworms) are very important in decomposition and in maintaining soil fertility. When these microfauna are selectively removed, the breakdown of dead plant material is greatly slowed, as summarized in Figure 2-16. Although many detritus-feeding animals (*detritivores*) cannot actually digest the lignocellulose substrates but obtain their food energy largely from the microflora in the material, they speed up the decomposition of plant litter in a number of indirect ways: (1) by breaking down detritus or **coarse particulate organic matter** (CPOM) into smaller pieces, thus increasing the surface area available for microbial action; (2) by adding proteins or other substances (often in the animals' excretions) that stimulate microbial growth; and (3) by stimulating the growth and metabolic activity of these microbial populations by eating some of the bacteria and fungi. Furthermore, many detritivores are **coprophagic** (from *kopros* = "dung"); that is, they regularly ingest fecal pellets after the pellets have been enriched by fungi and microbial activity in the environment. In the sea, pelagic tunicates feed by straining

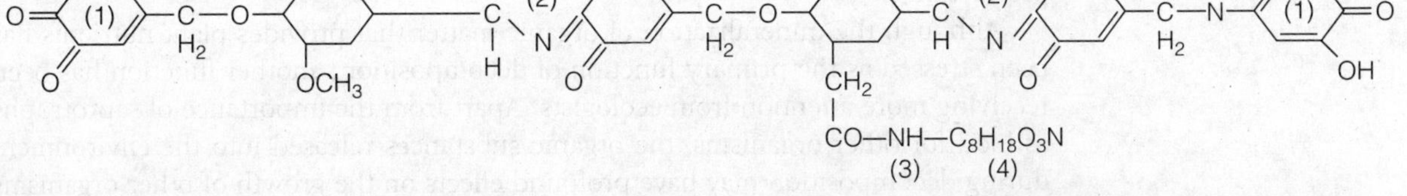

Figure 2-15. Model of a humic acid molecule, illustrating (1) the aromatic or phenolic benzene rings; (2) cyclic nitrogen rings; (3) nitrogen side chains; and (4) carbohydrate residues; all of which make humic substances difficult to decompose.

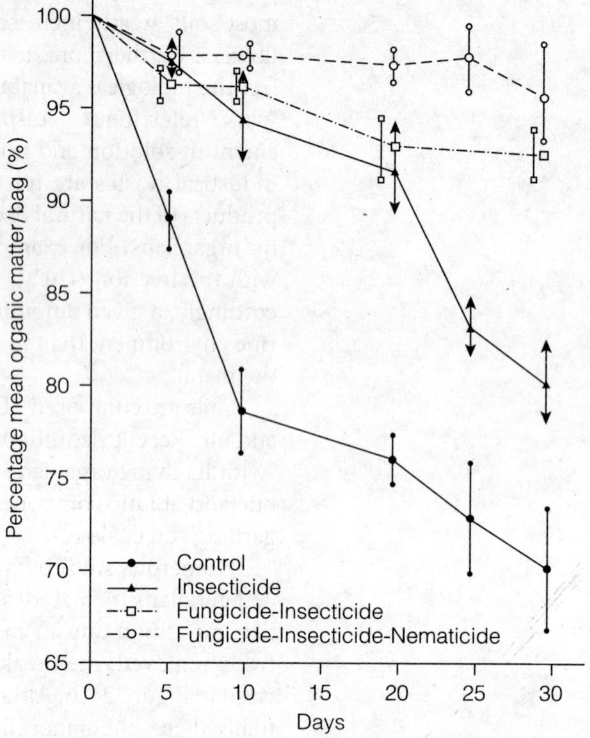

Figure 2-16. Loss of organic matter from buried litter bags in a grassland is greatly slowed when microarthropods, fungi, or nematodes are selectively removed (after Santos et al. 1981). These data demonstrate the importance of small invertebrates and fungi in the decomposition of soil organic matter.

out microflora from the water, producing large fecal pellets that have been shown to provide a major food source for other marine animals, including fish. In grasslands, rabbits also frequently exhibit coprophagic behavior by reingesting their own fecal pellets.

The importance of particle size is revealed in a comparison of till and no-till agriculture. In conventional tillage, involving deep plowing one or more times a year, organic residues are broken up into small pieces, resulting in a bacteria-based detritus food chain. When plowing is reduced or eliminated (limited or no tillage), a fungi-based food chain predominates, because fungi are more efficient at breaking down large particles than are bacteria. Detritivores, such as earthworms, are more abundant when plowing is reduced (Hendrix et al. 1986). Several studies have also indicated that phagotrophs can and do speed up sewage decomposition (see NRC 1996b, "Use of Reclaimed Water and Sludge in Food Crop Production," for details regarding the costs and benefits of using sewage sludge and wastewater in agroecosystems for crop production).

Although the mineralization of organic matter that provides plant nutrients has been stressed as the primary function of decomposition, another function has been receiving more attention from ecologists. Apart from the importance of saprotrophs as food for other organisms, the organic substances released into the environment during decomposition may have profound effects on the growth of other organisms in the ecosystem. Julian Huxley in 1935 suggested the term "external diffusion hormones" for those chemical substances that exert a correlative action (or feedback) on the system via the external medium. However, the terms **secondary metabolites** or

secondary compounds are the labels most often used for the substances excreted by one species that affect others. These substances may be inhibitory, as in the antibiotic penicillin (produced by a fungus), or stimulatory, as in various vitamins and other growth substances (for example, thiamin, vitamin B_{12}, biotin, histidine, uracil, and others), many of which have not been identified chemically.

The direct inhibition of one species by another using noxious or toxic compounds is termed **allelopathy.** These excretions are commonly termed **allelopathic substances** (from *allelon* = "of each other," and *pathy* = "suffering"). Algae release substances that have major effects on the structure and function of aquatic communities. Inhibitory leaf and root excretions of higher plants are also important in this regard. Black walnut trees (*Juglans nigra*), for example, are known for the production of juglone, an allelopathic chemical that interferes with the ability of other plants to establish themselves nearby. It has been demonstrated that allelopathic metabolites interact in a complex way with fire in controlling desert and chaparral vegetation. In dry climates, such excretions tend to accumulate and thus have more effect than under rainy conditions. Whittaker and Feeny (1971), Rice (1974), Harborne (1982), Gopal and Goel (1993), and Seigler (1996) have detailed the role of biochemical excretions in the development and structuring of communities.

In summary, the degradation of organic matter is a long and complex process, involving many species and chemical sequences—*a biodiversity that is extremely important to maintain.* Decomposition controls several important functions in the ecosystem. For example, it (1) recycles nutrients through mineralization of dead organic matter; (2) chelates and complexes mineral nutrients; (3) microbially recovers nutrients and energy; (4) produces food for a sequence of organisms in the detritus food chain; (5) produces secondary metabolites that may be inhibitory or stimulatory and are often regulatory; (6) modifies inert materials on the surface of Earth to produce, for example, the unique earthly complex known as "soil"; and (7) maintains an atmosphere conducive to the life of large-biomass aerobes such as humans.

New Molecular Methods to Study Decomposition

Study of the microbial communities involved in decomposition has been severely limited until recently by our inability to identify or discriminate among the many species of bacteria involved. Except for a few morphologically unusual species, all bacteria appear similar to humans when viewed through the microscope. Cells that look identical, however, may be carrying out vastly different processes. Recently, methods based on the staining of cells with very specific, short DNA probes have made it possible to identify bacteria in samples of soil, sediments, and water, and even to determine whether or not an individual cell possesses a gene for decomposing specific compounds. These molecular techniques also offer the promise of indicating whether or not the gene is "turned on" in the cell. Such techniques will make it possible to describe how decomposer communities function in the same detail that ecologists have been able to apply to communities of higher organisms.

Global Production-Decomposition Balance

Despite the broad spectrum and the great variety of functions in nature, the simple autotroph-heterotroph-decomposer classification is a good working arrangement for describing the ecological structure of a biotic community. *Production, consumption,* and *decomposition* are useful terms for describing overall functions. These and other ecological categories pertain to *functions* and not necessarily to *species* as such, because a particular species population may be involved in more than one basic function. Individual species of bacteria, fungi, protozoa, and algae may be quite specialized metabolically, but collectively, these lower phyla are extremely versatile and can perform numerous biochemical transformations. Human beings and other higher organisms cannot live without what LaMont Cole once called the "friendly microbes" (Cole 1966); microorganisms provide some degree of stability and sustainability in the ecosystem because they can adjust quickly to changing conditions.

The delay in the complete heterotrophic use and decomposition of the products of autotrophic metabolism is one of the most important temporal-spatial features of the ecosphere. Understanding this is especially important to industrialized societies, as fossil fuels have accumulated in the ground and oxygen in the atmosphere because of that delay. It is of immediate concern that human activities are unwittingly, but very rapidly, speeding up decomposition (1) by burning the organic matter stored in fossil fuels; (2) by agricultural practices that increase the decomposition rate of humus; and (3) by worldwide deforestation and burning of wood (still the major energy source for two thirds of people living in the less developed nations of the world). All this activity releases into the air the CO_2 stored in coal and oil and in trees and humus of deep forest soils. Although the amount of CO_2 diffused into the atmosphere by agro-industrial activities is still small compared with the total amount in circulation, the CO_2 concentration in the atmosphere has significantly increased since 1900. Possible consequences for the modification of the climate will be reviewed in Chapter 4.

9 Microcosms, Mesocosms, and Macrocosms

Statement

Little self-contained worlds, or **microcosms**, in bottles or other containers, such as aquaria, can simulate in miniature the nature of ecosystems. Such containers can be considered *microecosystems.* Large experimental tanks or outdoor enclosures, termed **mesocosms** ("midsized worlds"), are more realistic experimental models because they are subjected to naturally pulsing environmental factors, such as light and temperature, and can contain larger organisms with more complex life histories. The planet Earth, large watersheds, or natural landscapes, termed **macrocosms** (the natural or "great" world), are the natural systems used for baseline or "control" measurements.

A self-contained mesocosm including humans, called *Biosphere-2,* was a first attempt to build a bioregenerative enclosure such as might someday be constructed on the Moon or nearby planets. Spacecraft and space stations as now operated are not self-contained and can remain in space only a short time unless they are frequently resupplied from Earth. The characteristics and processes of the naturally evolved biosphere need to be coupled with the human-designed industrial "synthesphere" (Severinghaus et al. 1994) in order to design systems that mimic ecosystem sustainability.

Explanation and Examples

Two basic types of laboratory microcosms may be distinguished: (1) microecosystems derived directly from nature by multiple seeding of culture media with environmental samples; and (2) systems built up by adding species from *pure* or *axenic* (free from other living organisms) cultures until the desired combinations are obtained. The derived systems represent nature "stripped down" or "simplified" to those organisms that can survive and function together for a long time within the limits of the container, the culture medium, and the light-temperature environment imposed by the experimenter. Such systems, therefore, are usually intended to simulate some specific outdoor situation. For example, the microcosm shown in Figure 2-17 is derived from a sewage pond; the example in Figure 2-18 depicts a **standardized aquatic microcosm** (SAM) after Taub (1989, 1993, 1997). One problem with derived microecosystems is that their exact composition—especially the bacteria—is difficult to determine. The ecological use of derived or "multiple-seeded" systems was pioneered by H. T. Odum and his students.

In the axenic approach, defined systems are built up by adding previously isolated and carefully studied components (Fig. 2-17). The resulting cultures are often called **gnotobiotic** because the exact composition, down to the presence or absence of bacteria, is known. Gnotobiotic cultures have been used mostly to study the nutrition, biochemistry, and other aspects of single species or strains or to study the interactions of species, but ecologists have experimented with more complex *polyaxenic* cultures to devise self-contained ecosystems (Taub 1989, 1993, 1997; Taub et al. 1998). These contrasting approaches to the laboratory microecosystem parallel the two long-standing ways (holological and merological) in which ecologists have attempted to study lakes and other large systems in the real world. For a review of the early ecological work with microcosms and a discussion of the balanced aquarium controversy, see Beyers (1964), Giesy (1980), Beyers and Odum (1995), and Taub (1993, 1997).

Figure 2-17. Gnotobiotic microcosm containing three species from axenic ("pure") culture. The tube provides an area in which algae can multiply free from grazing by brine shrimp, to prevent overgrazing (after Nixon 1969).

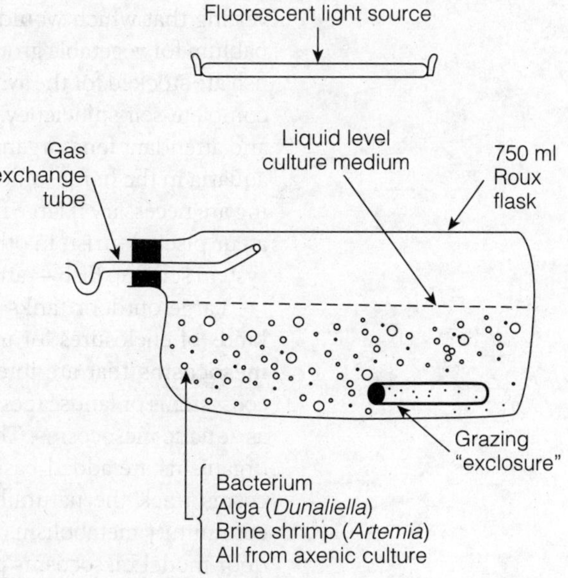

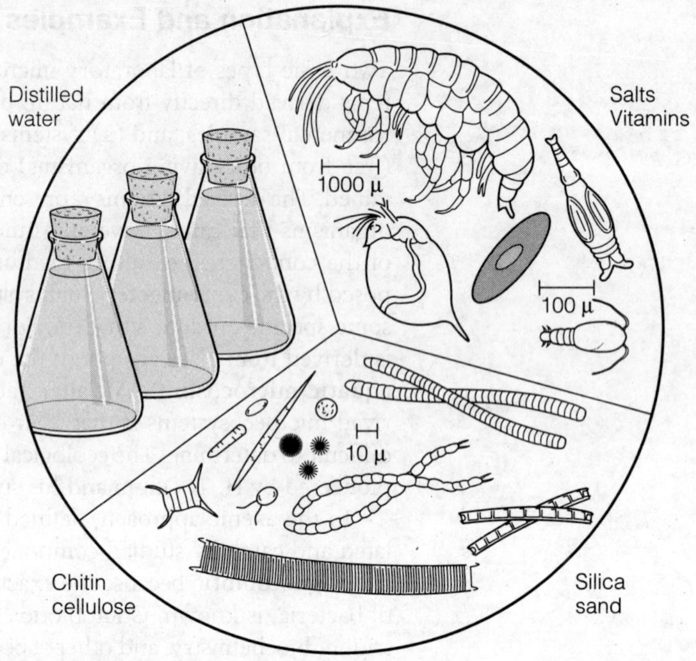

Figure 2-18. Diagram of a standardized aquatic microcosm (SAM) setup. (From Taub, F. B. 1989. Standardized aquatic microcosm-development and testing. In *Aquatic ecotoxicology: Fundamental concepts and methodologies,* vol. 2, ed. A. Boudou and F. Ribeyre, 47–92. Boca Raton, Florida: CRC Press. Used with permission.)

A common misconception exists about the "balanced" fish aquarium. An appropriate balance of gases and food can be achieved in an aquarium if the ratio of fish to water and plants remains small. In 1851, Robert Warington "established that wondrous and admirable balance between the animal and vegetable kingdoms" in a 12-gallon aquarium, using a few goldfish, snails, and lots of eelgrass (*Vallisneria*), as well as a diversity of associated microorganisms (Warington 1851). He not only clearly recognized the reciprocal role of fish and plants but also correctly noted the importance of the snail detritivore "in decomposing vegetation and confevoid mucus," thus "converting that which would otherwise act as a poisonous agent into a rich and fruitful pablum for vegetable growth." Most attempts to balance aquaria fail because too many fish are stocked for the available resources (an elementary case of overpopulation). For complete self-sufficiency, a medium-sized fish requires many cubic meters of water and attendant food organisms. As "fish-watching" is the usual motivation for keeping aquaria in the home, office, or school, subsidies of food, aeration, and periodic cleaning are necessary if large numbers of fish are to be crowded into small spaces. The amateur pisciculturist, in other words, would do well to establish aquaria based on ecosystem science. Fish—and human beings—require more room than one might think!

Large outdoor tanks and flowing water flumes for aquatic systems and various kinds of enclosures for terrestrial systems represent increasingly used experimental **mesocosms** that are intermediate between laboratory culture systems and natural ecosystems or landscapes. Figure 2-19 shows a series of experimental tanks that serve as aquatic mesocosms. The bottom of the tank is filled with sediment, and water and organisms are added based on the research questions to be addressed. These mesocosms track the naturally occurring seasonal changes in organisms' behavior and community metabolism (production and respiration) very well. Both indoor and outdoor model mesocosms provide useful tools for tentatively or preliminarily estimat-

Figure 2-19. Experimental aquatic mesocosms located at the Savannah River Ecology Laboratory (SREL) near Aiken, South Carolina.

Courtesy of J. Whitfield Gibbons

ing the effect of pollutants or experimentally imposed disturbances related to human activity. The role of mesocosm studies in ecological risk analysis has been summarized by Boyle and Fairchild (1997).

Various kinds of fenced terrestrial mesocosms (Fig. 2-20) have also proven useful in assessing the effect of fire, pesticides, and nutrient enrichment on whole ecosystems (Barrett 1968, 1988; Crowner and Barrett 1979; W. P. Carson and Barrett 1988; Hall et al. 1991; Brewer et al. 1994) and in addressing questions and testing hypotheses in agroecology and landscape ecology (Barrett et al. 1995; R. J. Collins and Barrett 1997; Peles et al. 1998). For example, Barrett (1968), while evaluating the effects of an acute insecticide stress on a grassland mesocosm, determined not only that the insecticide application reduced the "target" phytophagous insects in the short term, but also that the treatment decreased the decomposition rate of plant lit-

Figure 2-20. Aerial photograph of 16 terrestrial mesocosms at the Miami University Ecology Research Center, Oxford, Ohio. This figure depicts how the mesocosm approach can be used to investigate habitat restoration (8 subdivided mesocosms on the left), and the effects of habitat (patch) fragmentation on small mammal (meadow vole) population dynamics (8 mesocosms on the right). The mesocosms (enclosures) on the right depict a regular paired experimental design in which enclosures are used to simulate landscape (patch and matrix) components (after R. J. Collins and Barrett 1997).

Courtesy of Gary W. Barrett

ter, delayed the reproduction of small mammals (*Sigmodon hispidus*), and reduced the diversity of predaceous insects in the long term. Thus, this mesocosm approach illustrated how a "recommended" insecticide application affected the dynamics of the system as a whole. For more on the mesocosm concept and approach, see E. P. Odum (1984) and Boyle and Fairchild (1997).

Micro- and mesoecosystem research is also proving useful in the testing of various ecological hypotheses generated from observing nature. For example, the terrestrial mesocosms pictured in Figure 2-20 were designed to evaluate the effects of habitat (patch) fragmentation on meadow vole (*Microtus pennsylvanicus*) population dynamics in experimental landscape patches. The results of this experiment indicated that significantly more female than male voles were found in the fragmented treatment of equal total habitat size, resulting in differences in the social structure of the meadow vole populations (R. J. Collins and Barrett 1997); the findings demonstrated that habitat fragmentation can have both positive and negative effects on particular species. In the next several chapters, the ways in which microcosm and mesocosm research has helped to establish and clarify basic ecological principles will be described.

Spacecraft as an Ecosystem

One way to visualize a model of an ecosystem is to think about space travel. When we leave the biosphere for exploration for a duration of many years, we must take with us a sharply delimited, enclosed environment that can supply all vital needs by using sunlight as the energy input from space. For journeys of a few days or weeks, such as to the Moon and back, we do not need a completely self-sustaining ecosystem, because sufficient oxygen and food can be stored and CO_2 and other waste products can be fixed or detoxified for short periods. However, for long journeys, such as trips to the planets or to establish space colonies, we must design a regenerative spacecraft that includes all vital abiotic substances and the means to recycle them. The vital processes of production, consumption, and decomposition must also be performed in a balanced manner by biotic components or by mechanical substitutes. In a very real sense, the self-contained spacecraft is a *human mesocosm*.

The life-support modules for all crewed spacecraft so far launched have been *storage* types; in some cases, water and atmospheric gases have been partly regenerated by physicochemical means. The possibility of coupling humans and microorganisms, such as algae and hydrogen bacteria, has been considered but found unworkable. Large organisms—especially for food production—considerable diversity, and, above all, large volumes of air and water will be required for a truly regenerative ecosystem that could survive for long periods in space without resupply from Earth (recall our previous comment about the large amount of room needed by a single fish or human). Accordingly, something akin to conventional agricultural and other large plant communities will have to be included.

The critical problem is how the buffering capacity of the atmosphere and the oceans, which stabilizes the biosphere as a whole, will be provided. For every square meter of land surface on Earth, more than 1000 cubic meters of atmosphere and almost 10,000 cubic meters of ocean, plus large volumes of permanent vegetation, are available as sinks, regulators, and recyclers. Obviously, for living in space, some of this buffering work will have to be accomplished mechanically, using solar energy (and perhaps atomic energy). The National Aeronautics and Space Administration

(NASA) concluded, "It is a moot point whether an artificial ecosystem, totally closed to the entry or exit of mass, fully recycling, and completely regulated by its biological components can be constructed" (MacElroy and Averner 1978). However, in 1991–1993, an earthbound prototype mesocosm was built, financed by private funds completely independent of NASA. Here, in brief, is the story of Biosphere-2.

The Biosphere-2 Experiment

To determine what really will be required to support a group of people on the Moon or Mars on a bioregenerative basis, an earthbound capsule called **Biosphere-2** (Biosphere-1 being Earth) was constructed in the Sonoran Desert, 50 kilometers north of Tucson, Arizona. Figure 2-21 shows the 1.27 hectare (3.24 acre), airtight, glass-covered enclosure and its external supporting structures. In the fall of 1991, eight people were sealed into the capsule, where they lived for two years without any material exchange with the outside, although they were provided with abundant energy flow (as would be required for any life-support system) and unrestricted information exchange (such as radio, television, and telephone).

About 80 percent of the space inside Biosphere-2 is occupied by half a dozen natural habitats, ranging from rain forest to desert. These habitat types provide a great deal of biodiversity, as it is expected that some of the species in the enclosure will thrive whereas others may not adapt and will be lost. Most of the remaining area (about 16 percent, or 0.2 hectares), is occupied by crops (the agricultural wing), which feed the humans and a small number of domestic animals (goats, pigs, and chickens) that provide milk and a little bit of meat for the low-cholesterol diet of the human inhabitants. The human habitat, where the eight people have their rooms—the urban area, as it were—is very small, about 4 percent of the space. The allotment of space was divided among the three basic environment types—natural, cultivated, and developed—similar to the proportions of land use in the United States. But in Biosphere-2, there are no automobiles or polluting industries in the "developed area." If there were—or if the population were to increase—a lot more life-supporting environment would be needed. For additional descriptions and pictures, see J. Allen (1991).

In the fall of 1993, the eight "biospherians" emerged from their two-year isolation still speaking to one another and in better health than when they went in. The complex machinery that maintained air and water circulation and recycling, heating and cooling, and so on worked quite well. Incoming solar energy was sufficient to maintain the labor-intensive food gardens, including banana plants in tubs located in sunny spots throughout the enclosure. However, total photosynthesis was not quite enough to maintain the oxygen–carbon dioxide balance; during the last six months, oxygen had to be added to prevent "altitude sickness." Apparently, the reduction of light by the glass, unusually cloudy weather outside, and the rich organic soil, brought in for the agricultural wing, combined to reduce production and increase consumption of oxygen more than anticipated (Severinghaus et al. 1994).

Some scientists criticized the Biosphere-2 experiment as not being "real science" because the crew members were not scientists but people who were selected for their ability to work together, grow all of their own food, and control the devices, and for their willingness to live at subsistence level for two years. The crew, for example, had to spend about 45 percent of their waking hours on growing and preparing food, 25 percent on maintenance and repairs, 20 percent on communication, and 5 percent on small research projects, which left little time (5 percent) for relaxation and

Figure 2-21. Photographs of Biosphere-2, an experimental bioregenerative mesocosm and its supporting structures. (A) The 1.3-hectare, glass-enclosed space combines natural and artificial systems and controls. For two years, 1991–1993, eight people lived in isolation in the capsule with energy inputs (solar and fossil fuels) and information exchanges. Oxygen had to be added during the last six months of the study because total photosynthesis was not enough to maintain the oxygen–carbon dioxide balance (Severinghaus et al. 1994). (B) The greenhouse enclosure contains life-support environments including rain forest, savanna/ocean/marsh, desert, intensive agriculture, and human habitat.

A

Courtesy of Space Biosphere Ventures

B

Courtesy of Space Biosphere Ventures

recreation. Viewed as an exercise in human ecology and environmental engineering, the experiment was a success. Most of all, it demonstrated how difficult and expensive it will be to maintain human life in space without continuous resupply from Earth. This mesocosm approach also demonstrated the benefits of natural ecosystem services supplied to human societies (Daily et al. 1997). Unfortunately, the future of Biosphere-2 as an experimental research facility (mesocosm) remains in doubt (Mervis 2003).

Although we are not yet able to build a human mesocosm—and do not know whether we could afford it even if we knew how—enthusiasts for space colonization predict that during the twenty-first century, millions of people will be living in space colonies supported by a carefully selected biota, free from the pests and other unwanted or unproductive organisms that earthbound people contend with. The successful colonization of the "high frontier" (according to its proponents) would permit the continued growth of human population and affluence long after such growth is no longer possible within the confines of planet Earth. Solar energy and the mineral wealth of moons and asteroids could be exploited to support such growth. However, social, economic, political, and pollution problems within such a mesocosm would be formidable indeed. The extent to which sociopolitical forces shape and limit human life and growth on Earth will be considered later in this book. In any event, it is prudent to care for Earth for the future rather than plan to escape a dying biosphere by moving into space colonies.

10 Ecosystem Cybernetics

Statement

Besides energy flows and material cycles, as briefly described in Section 1 (and in more detail in Chapters 3 and 4), ecosystems are rich in information networks, including physical and chemical communication flows that connect all parts and steer or regulate the system as a whole. Accordingly, ecosystems can be considered *cybernetic* (from *kybernetes* = "pilot" or "governor") in nature, but, as emphasized in Chapter 1, cybernetics above the organism level of organization are very different from those at the level of organisms or mechanical control devices. Control functions in nature are internal and diffuse (no set points) rather than external and specified (by set points), as in human-engineered cybernetic devices. The lack of set-point controls results in a *pulsing state* rather than a *steady state*. The variance, or the degree to which stability is achieved, varies widely, depending on the rigor of the external environment and on the efficiency of internal controls. It is useful to recognize two kinds of stability: *resistance stability* (the ability to remain "steady" in the face of stress) and *resilience stability* (the ability to quickly recover); the two may be inversely related.

Explanation and Examples

The elementary principles of **cybernetics** are modeled in Figure 2-22, which compares A, a goal-seeking automatic control system with specified external control, as in a mechanical device, with B, a nonteleological system with diffuse subsystem reg-

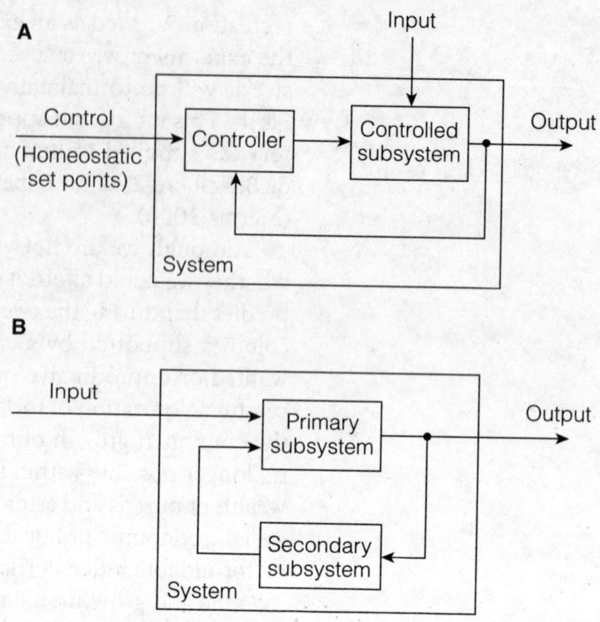

Figure 2-22. Feedback control systems. (A) Model appropriate for human-made automatic control systems and homeostatic, goal-seeking organismic systems. (B) Model appropriate for nonteleological systems, including ecosystems, where control mechanisms are internal and diffuse, involving interactions between primary and secondary subsystems (after Patten and Odum 1981).

ulation. In either case, control depends on **feedback,** which occurs when part of the output feeds back as input. When this feedback input is positive (like compound interest, which is allowed to become part of the principal), the quantity grows. **Positive feedback** accelerates deviation and is, of course, necessary for the growth and survival of organisms. However, to achieve control—for example, to prevent the overheating of a room or the overgrowth of a population—there must also be **negative feedback,** which counteracts deviating input. The energy involved in a negative feedback signal can be extremely small compared with the total energy flow through the system, whether it be a household-controlled heating system, an organism, or an ecosystem. Low-energy components that have much-amplified, high-energy feedback effects are major characteristic features of cybernetic systems.

The science of cybernetics, as founded by Norbert Wiener (1948), embraces both inanimate and animate controls. Mechanical feedback mechanisms are often called **servomechanisms** by engineers, whereas biologists use the phrase **homeostatic mechanisms** to refer to organismic systems. **Homeostasis** (from *homeo* = "same," and *stasis* = "standing") at the organism level is a well-known concept in physiology, as outlined in Walter B. Cannon's classic book, *The Wisdom of the Body* (1932). In servomechanisms and organisms, a distinct mechanical or anatomical "controller" has a specified "set point" (Fig. 2-22A). For example, in the familiar household heating system, the thermostat controls the furnace; in a warm-blooded animal, a specific brain center controls body temperature; and genes tightly control the growth and development of cells, organs, and organisms. There are no thermostats or chemostats in nature; rather, the interplay of material cycles and energy flows, along with subsystem feedbacks in large ecosystems, generates a self-correcting **homeorhesis** (*rhesis* = "flow" or "pulsing"; Fig. 2-22B). Waddington (1975) coined the term *homeorhesis* to denote the evolutionary stability or preservation of a system's flow or pulsing process as a pathway of change through time. The goal of homeorhesis is to keep systems altering in the same manner as they have altered in the past (Naveh and Lieberman

1984). Control mechanisms operating at the ecosystem level include microbial subsystems that regulate the storage and release of nutrients, behavioral mechanisms, and predator-prey subsystems that control population density, to mention just a few examples. (See Engelberg and Boyarsky 1979; Patten and Odum 1981 for contrasting views on cybernetics and the "balance of nature.")

One difficulty in perceiving cybernetic behavior at the ecosystem level is that components at that level are coupled in networks by various physical and chemical messengers that are analogous to but far less visible than the nervous or hormonal systems of organisms. H. A. Simon (1973) pointed out that the "bond energies" that link components become more diffuse and weaker with an increase in system size and temporal scales. At the ecosystem scale, these weak but numerous bonds of energy and chemical information have been called the "invisible wires of nature" (H. T. Odum 1971), and the phenomenon of organisms responding dramatically to low concentrations of substances is more than just a weak analogy to hormonal control. Low-energy causes producing high-energy effects are ubiquitous in ecosystem networks (H. T. Odum 1996); two examples will suffice to illustrate. Tiny insects known as parasitic Hymenoptera account for only a very small portion (often less than 0.1 percent) of the total community metabolism in a grassland ecosystem, yet they can have a very large controlling effect on the total primary energy flow (production) by the impact of their parasitism on herbivorous insects. In a cold spring ecosystem model, Patten and Auble (1981) described a feedback loop in which only 1.4 percent of the energy input to the system is fed back to the detrital substrate of the bacteria. In diagrams of ecological systems (see Figs. 1-5, 1-6, and 1-7), this phenomenon is commonly shown as a reverse loop in which a low quantity of "downstream" energy is fed back to an "upstream" system. This type of amplified control, by virtue of its position in a network, is exceedingly widespread and indicates the intricate global feedback structure of ecosystems. In food chains, herbivores and parasites (downstream components) often enhance or promote the welfare of their hosts (upstream components) by a feedback process known as **reward feedback** (Dyer et al. 1993, 1995). Through evolutionary time, such interactions have stabilized ecosystems by preventing "boom-and-bust" herbivory, catastrophic predator-prey oscillations, and so on. Although, as already noted, the extent of feedback control at the level of the biosphere is a controversial subject, it follows naturally from what is known at the ecosystem level.

In addition to feedback control, *redundancy* in functional components also contributes to stability. For example, if several species of autotrophs are present, each with a different temperature operating range, the rate of photosynthesis of the community as a whole can remain stable despite changes in temperature.

C. S. Holling (1973) and Hurd and Wolf (1974) have suggested that populations and, by inference, ecosystems have more than one equilibrium state and often return to a different equilibrium after a disturbance. CO_2 introduced into the atmosphere by human activities is largely but not completely absorbed by the carbonate system of the sea and other carbon storage, but as the input increases, the new equilibrium levels in the atmosphere are higher. On many occasions, regulatory controls emerge only after a period of evolutionary adjustment. New ecosystems, such as a new type of agriculture or new host-parasite assemblages, tend to oscillate more violently and are more likely to develop overgrowths than mature systems, in which the components have had a chance to adjust jointly to one another.

Part of the difficulty in dealing with the concept of stability is semantic. A dictionary definition of the term **stability** is, for example, "the property of a body that

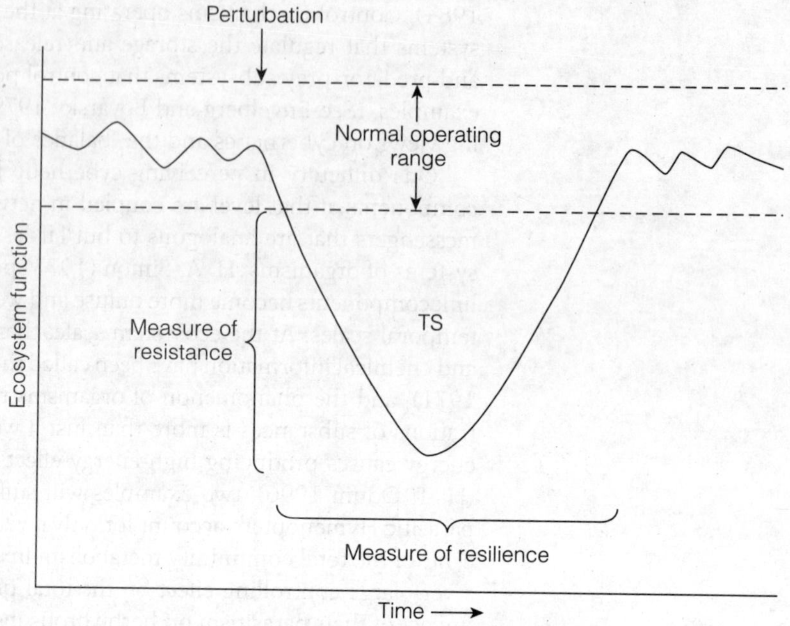

Figure 2-23. Resistance and resilience stability. When a perturbation (disturbance or stress) causes a major ecosystem function to deviate from its normal operating range, the degree of deviation is a measure of relative resistance stability, whereas the time required for recovery is a measure of relative resilience stability. Total stability (TS) may be represented by the area under the curve (after Leffler 1978).

causes it, when disturbed from a condition of equilibrium, to develop forces or moments that restore the original condition" (*Merriam-Webster's Collegiate Dictionary*, 10th edition, s.v. "stability"). This seems straightforward enough, but in practice, *stability* assumes different meanings in different professions (such as engineering, ecology, or economics), especially when one is trying to measure and quantify it. Accordingly, confusion abounds in the literature, and a full discussion of stability theory is beyond the scope of this book. However, from an ecological perspective, two kinds of stability can be contrasted, as shown in Figure 2-23.

Resistance stability indicates the ability of an ecosystem to resist perturbations (disturbances) and to maintain its structure and function intact. **Resilience stability** indicates the ability to recover when the system has been disrupted by a perturbation. Growing evidence suggests that these two kinds of stability may be mutually exclusive; in other words, it is difficult to develop both at the same time. Thus, a California redwood forest is quite resistant to fire (thick bark and other adaptations), but if it does burn, it will recover very slowly or perhaps never. In contrast, California chaparral vegetation is very easily burned (low resistance stability) but recovers quickly in a few years (high resilience stability). In general, ecosystems in benign physical environments can be expected to exhibit more resistance stability and less resilience stability, whereas the opposite is true in uncertain physical environments (see Gunderson 2000 for a review of ecological resilience).

In summary, an ecosystem is not equivalent to an organism; because it is not under direct genetic control, an ecosystem is a supraorganismic level of organization, but it is not a superorganism, nor is it like an industrial complex (such as an atomic power plant). It does have one thing in common with organisms: built-in—although different—cybernetic behavior. Because of the evolution of the central nervous system, *Homo sapiens* has gradually become the most powerful organism, at least as far as the ability to modify the operation of ecosystems is concerned. The human brain requires only an extremely small amount of energy to crank out all sorts of powerful ideas. Much of our short-term thinking has involved positive feedback that promotes the expansion of power, technology, and exploitation of resources. In the long term, however, the quality of human life and the environment will likely be degraded unless adequate negative feedback controls can be established.

In a famous essay, social critic Lewis Mumford (1967) pleaded for "quality in control of quantity," which eloquently states the cybernetic principle of low-energy causes producing high-energy effects. So important is the role of humanity as "a mighty geological agent" becoming that Vernadskij (1945) suggested that we think of the **noösphere** (from Greek *noös* = "mind"), or the world dominated by the human mind, as gradually replacing the biosphere, the naturally evolving world, which has existed for billions of years. Barrett (1985) reviewed the noösystem concept and suggested that the noösystem could serve as a basic unit for the integration of biological, physical, and social components within ecological systems. Although the human brain is a low-quantity–high-quality energy "device" with great control potential, the time probably has not yet come for the noösphere of Vernadskij. When you have finished reading this book, you may agree that we cannot yet manage our life-support system, especially as proven natural processes (natural capital) work so well (and are so inexpensive).

11 Technoecosystems

Statement

Current urban-industrial society not only affects natural life-support ecosystems, but it has created entirely new arrangements, which we term **technoecosystems,** that are competitive with and *parasitic* on natural ecosystems. These new systems involve advanced technology and powerful energy sources. If urban-industrial societies are to survive in a finite world, it is imperative that technoecosystems interface with natural life-support ecosystems in a more positive or mutualistic manner than is the case at present.

Explanation

As noted in Section 1 of this chapter, prior to the Industrial Revolution, humans were a part of—rather than apart from—natural ecosystems. In the ecosystem model of Figure 2-2, humans were top predators and omnivores (the terminal H box in the food web). Early agriculture—as is the case with traditional or pre-industrial agriculture, as still widely practiced in many parts of the world—was compatible with natural systems and often enriched the landscape in addition to providing food. However, with the increasing use of fossil fuels and nuclear fission—energy sources many times more powerful than sunlight—together with the growth of cities and the increase of money-based market economics, the model depicted in Figure 2-2 is no longer adequate. We need to create a model for this new *technoecosystem*—a term suggested by pioneer landscape ecologist Zev Naveh (1982). More recently, Naveh

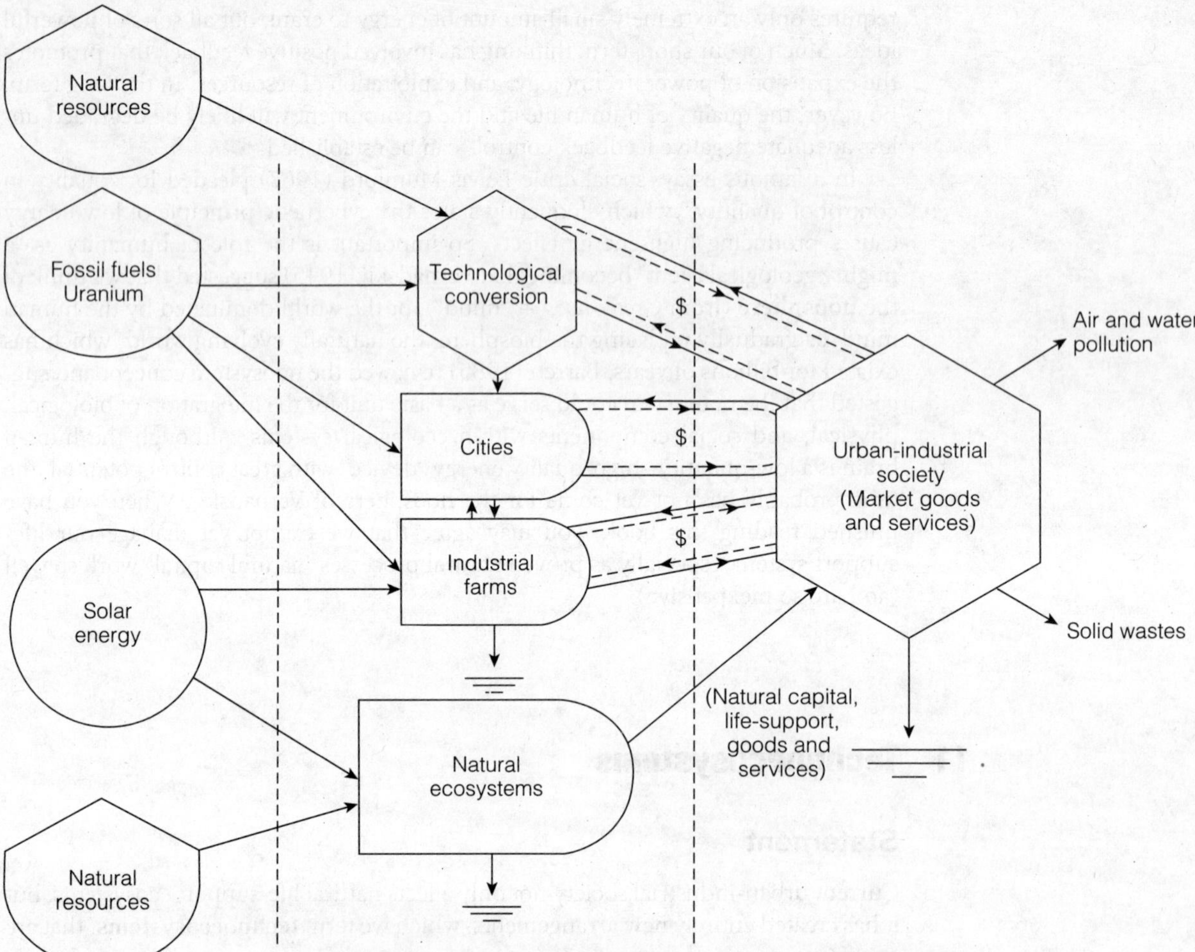

Figure 2-24. Model of the relationships between urban-industrial technoecosystems and natural ecosystems, including money flows. The model depicts how human-dominated techno-ecosystems need to be coupled with the goods and services (natural capital) provided by natural ecosystems in order to enhance landscape sustainability.

has used the term **total human ecosystem** to describe the relationship of the industrial society (the technoecosystem) to the total ecosphere (Naveh 2000).

Figure 2-24 shows our graphic model for these new (in terms of human history) systems. Shown are the inputs of fossil fuels and uranium energy sources and of natural resources, and the increasing outputs of air, water, and solid waste pollution, which are much larger and more toxic than anything that comes out of natural ecosystems! In Figure 2-24, we complete the model with natural ecosystems that provide life-supporting goods and services (breathing, drinking, and eating) and that maintain homeorhetic global balances (sustainability) in the atmosphere, soils, freshwater, and oceans. Note that money circulates as a reverse flow between society and human-made systems, but not natural systems, thereby creating a vast market failure when society fails to pay for ecosystem services.

Examples

A modern city, of course, is the major component of the fabricated technoecosystem—an energetic hot spot that requires a large area of low–energy density natural and seminatural countryside to maintain it. Current cities grow little or no food and generate a large waste stream that affects large areas of downstream rural landscapes and oceans. The city does export money that pays for some natural resources, and the city provides many desirable cultural institutions not available in rural areas, such as museums and symphonies. Figure 2-25 compares a human city with a much lower-powered oyster reef, a natural analog to a city. Note that the energy requirements for the urban technoecosystem are about 70 times greater than for the natural ecosystem.

Essentially, cities can be viewed as parasites on the low-energy countryside. As will be discussed in Chapter 7, parasites and hosts in nature tend to coevolve for coexistence; otherwise, if the parasite takes too much from its host, both die. John Cairns (1997) has expressed hope that somehow natural ecosystems and technoecosystems will coevolve to prevent such a doomsday. Wackernagel and Rees (1996)

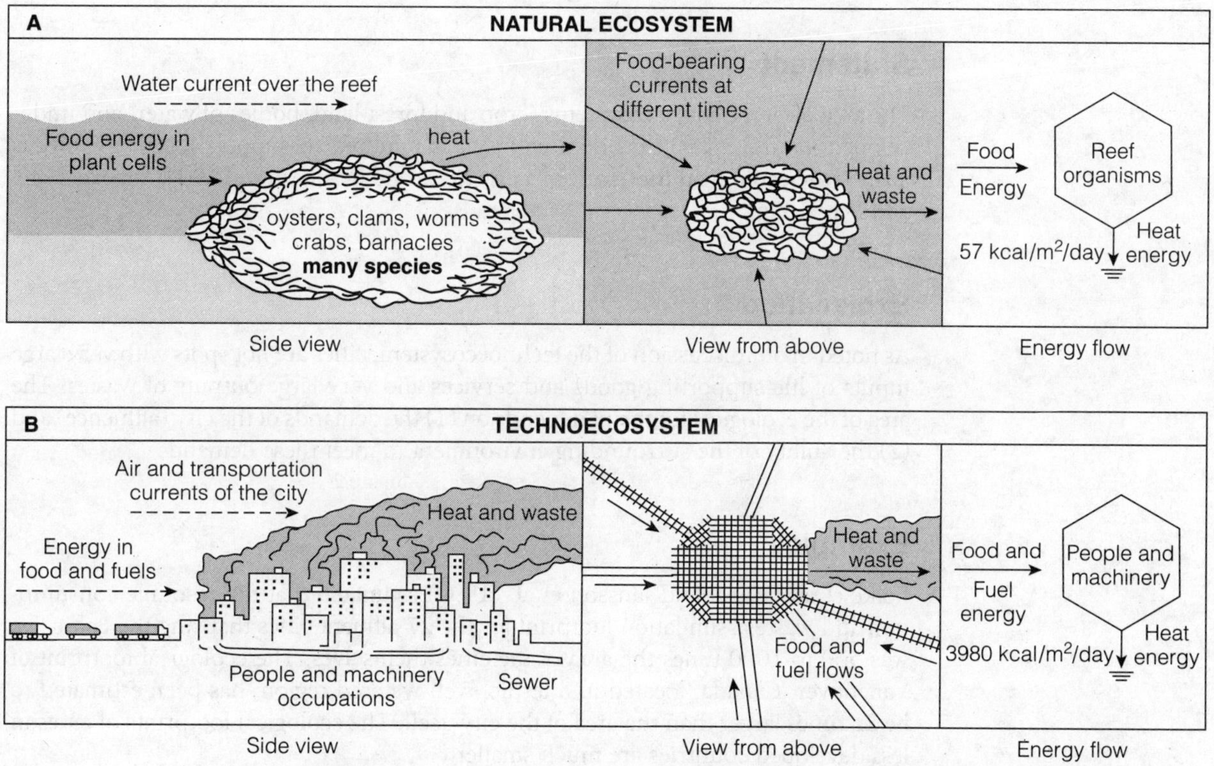

Figure 2-25. Heterotrophic ecosystems. (A) One of the "cities" of nature–an oyster reef that is dependent on the inflow of food energy from a large area of the surrounding environment. (B) A human-built technoecosystem (industrialized city) maintained by a huge inflow of fuel and food, with a correspondingly large outflow of waste and heat. Its energy requirement, on a per m^2 area basis, is about 70 times that of the reef, or about 4000 kcal/m^2/day, which comes to about 1.5 million kcal per year (after H. T. Odum 1971).

used the term *ecological footprint* to describe the impact of and resources needed by a city in order to provide for its citizenry in a sustainable manner.

Especially threatening to the global life-support systems is the explosive growth of *megacities* in the developing nations, caused at least in part by the increasing dominance of another technoecosystem, *industrialized agriculture,* with its often excessive consumption of water and its use of toxic and eutrophicating chemicals. Moreover, these systems not only pollute, but essentially put the small farms out of business worldwide, relocating most of these families into the cities, which are not able to assimilate them at the rate that they are moving into urban areas. This situation reflects what engineer and former president of the Massachusetts Institute of Technology Paul Gray (1992) has written: "A paradox of our time is the mixed blessing of almost every technological development." In other words, technology has its dark as well as its bright sides. Chapter 11 will discuss what can be done about these and other "overshoot problems."

12 Concept of the Ecological Footprint

Statement

The area of productive ecosystems (crop and forest land, bodies of water, and undeveloped natural areas) outside a city that is required to support life in the city is termed the **ecological footprint** of a city (Rees and Wackernagel 1994; Wackernagel and Rees 1996).

Explanation

As noted in our discussion of the technoecosystem, cities are hot spots with very large inputs of life-supporting goods and services and very large outputs of wastes. The area of the ecological footprint depends on (1) the demands of the city (affluence) and (2) the ability of the surrounding environment to meet these demands.

Examples

Folke et al. (1997) and Jansson et al. (1999) estimated that the resource consumption and waste assimilation footprint for the 27 affluent cities that ring the Baltic Sea was 500 to 1000 times the area of the cities themselves. The ecological footprint of Vancouver, Canada, located in a fertile, well-watered region, has been estimated to be 22 times larger than the area of the city itself. The ecological footprints of cities in less developed countries are much smaller.

Luck et al. (2001) provided a good example of the differences in the ability of the matrix environment to provide services in their comparison of the water and food footprints of cities in the United States. The New York and Los Angeles metropolitan areas have about the same human population density, but the water footprint of Los Angeles is twice and the food footprint four times as large as that of New York, which is located in a much wetter area. The water footprint of Phoenix, Arizona, located in the desert, includes half of the neighboring states if irrigation water demand is included.

The footprint concept can also be applied per capita. For example, the ecological footprint of an individual citizen of the United States is estimated to be 5.1 hectares per person; of a citizen of Canada, 4.3 hectares per person; and of a citizen of India, 0.4 hectares per person (Wackernagel and Rees 1996). If highly developed countries would reduce their excessive resource and energy consumption, then international conflicts and terrorists threats would likely be reduced. For example, the United States, with 4.7 percent of the world's human population, consumes 25 percent of the world's energy resources. Schumacher (1973) noted that "small is beautiful"; we suggest that "small ecological footprints" should be perceived as beautiful as well.

13 Classification of Ecosystems

Statement

Ecosystems can be classified by either *structural* or *functional* characteristics. Vegetation and major structural physical features provide the basis for the widely used *biome* classification (a term discussed in detail in Chapter 10). An example of a useful functional scheme is a classification based on the quantity and quality of the energy input "forcing function."

Explanation

Although the classification of ecosystems is not to be considered a discipline in itself, unlike the classification of organisms (*taxonomy*), the human mind seems to require some kind of orderly categorization when it comes to dealing with a large variety of entities, like information in a library. Ecologists have not agreed upon any one classification for ecosystem types, or even on what would be a proper basis for it. However, many approaches serve useful purposes.

Energy provides an excellent basis for a functional classification, as it is a major common denominator for all ecosystems, natural and human-managed alike. Conspicuous, ever-present *structural* macrofeatures are the basis for the widely used biome classification. In terrestrial environments, vegetation usually provides such a macrofeature that "integrates," as it were, the flora and fauna with climate, water, and soil conditions. In aquatic environments, where plants are often inconspicuous, another dominant physical feature, such as "standing water," "running water," "marine continental shelf," and so on, usually provides a basis for recognizing major types of ecosystems.

Examples

An energy-based classification of ecosystems will be discussed in detail after the basic laws of energy behavior are outlined in Chapter 3. A classification based on biomes and global ecosystem types will be illustrated in Chapter 10. These 21 major

Table 2-7 **Major ecosystem types of the biosphere**

Marine ecosystems	Open ocean (pelagic)
	Continental shelf waters (inshore water)
	Upwelling regions (fertile areas with productive fisheries)
	Deep sea (hydrothermal vents)
	Estuaries (coastal bays, sounds, river mouths, salt marshes)
Freshwater ecosystems	Lentic (standing water): lakes and ponds
	Lotic (running water): rivers and streams
	Wetlands: marshes and swamp forests
Terrestrial ecosystems	Tundra: arctic and alpine
	Boreal coniferous forests
	Temperate deciduous forests
	Temperate grassland
	Tropical grassland and savanna
	Chaparral: winter rain–summer drought regions
	Desert: herbaceous and shrub
	Semi-evergreen tropical forest: pronounced wet and dry seasons
	Evergreen tropical rain forest
Domesticated ecosystems	Agroecosystems
	Plantation forest and agroforest systems
	Rural technoecosystems (transportation corridors, small towns, industries)
	Urban-industrial technoecosystems (metropolitan districts)

types of ecosystems represent the matrix in which humans embedded their civilizations (Table 2-7). *Marine* ecosystem types are based on structure and function of marine systems; *terrestrial* ecosystem types are based on natural or native conditions of vegetation; *aquatic* ecosystem types are based on geological and physical structures; and *domestic* ecosystem types depend on the goods and services provided by natural ecosystems.

3

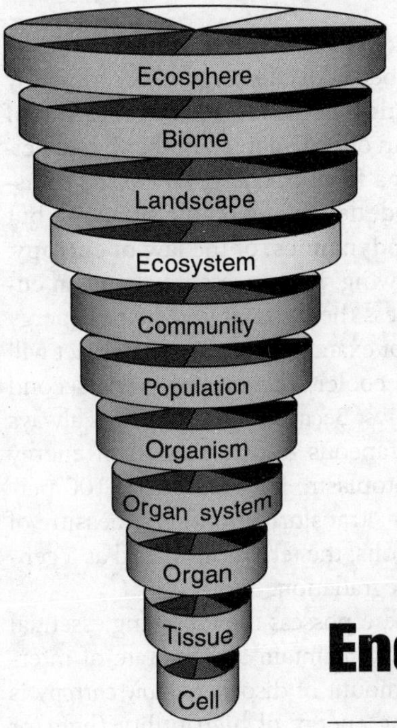

Energy in Ecological Systems

1 Fundamental Concepts Related to Energy: The Laws of Thermodynamics
2 Solar Radiation and the Energy Environment
3 Concept of Productivity
4 Energy Partitioning in Food Chains and Food Webs
5 Energy Quality: eMergy
6 Metabolism and Size of Individuals: The 3/4 Power Principle
7 Complexity Theory, Energetics of Scale, and the Law of Diminishing Returns
8 Concepts of Carrying Capacity and Sustainability
9 Net Energy Concept
10 An Energy-Based Classification of Ecosystems
11 Energy Futures
12 Energy and Money

1 Fundamental Concepts Related to Energy: The Laws of Thermodynamics

Statement

Energy is defined as the ability to do work. The behavior of energy is described by the following laws: The **first law of thermodynamics,** or the **law of conservation of energy,** states that energy may be transformed from one form into another but is neither created nor destroyed. Light, for example, is a form of energy; it can be transformed into work, heat, or potential energy of food, depending on the situation, but none of it is destroyed. The **second law of thermodynamics,** or the **law of entropy,** may be stated in several ways, including the following: No process involving an energy transformation will spontaneously occur unless there is a *degradation* of energy from a concentrated form into a dispersed form. For example, heat in a hot object will spontaneously tend to become dispersed into the cooler surroundings. The second law of thermodynamics may also be stated as follows: Because some energy is always dispersed into unavailable heat energy, no spontaneous transformation of energy (sunlight, for example) into potential energy (protoplasm, for example) is 100 percent efficient. **Entropy** (from *en* = "in" and *trope* = "transformation") is a measure of the unavailable energy resulting from transformations; the term is also used as a general index of the *disorder* associated with energy degradation.

Organisms, ecosystems, and the entire ecosphere possess the following essential thermodynamic characteristic: They can create and maintain a high state of internal order, or a condition of low entropy (a low amount of disorder). *Low entropy* is achieved by continually and efficiently dissipating energy of high utility (light or food, for example) into energy of low utility (heat, for example). In the ecosystem, order in a complex biomass structure is maintained by the total community respiration, which continually "pumps out disorder." Accordingly, ecosystems and organisms are open, non-equilibrium thermodynamic systems that continuously exchange energy and matter with the environment to decrease internal entropy but increase external entropy (thus conforming to the laws of thermodynamics).

Explanation

The fundamental concepts of thermodynamics outlined in the preceding paragraph are the most important of the natural laws that apply to all biological or ecological systems. So far as is known, no exceptions—and no technological innovations—can break these laws of physics. Any system of humankind or nature that does not conform to them is indeed doomed. The two laws of thermodynamics are illustrated by the energy flow through an oak leaf as shown in Figure 3-1.

Various forms of life are all accompanied by energy changes, even though no energy is created or destroyed (first law of thermodynamics). The energy that reaches the surface of Earth as light is balanced by the energy that leaves the surface of Earth as invisible heat radiation. The essence of life is the progression of such changes as growth, self-duplication, and the synthesis of complex combinations of matter. Without the energy transfers that accompany all such changes, there could be no life and no ecological systems. Humankind is just one of the remarkable natural proliferations that depend on the continuous inflow of concentrated energy.

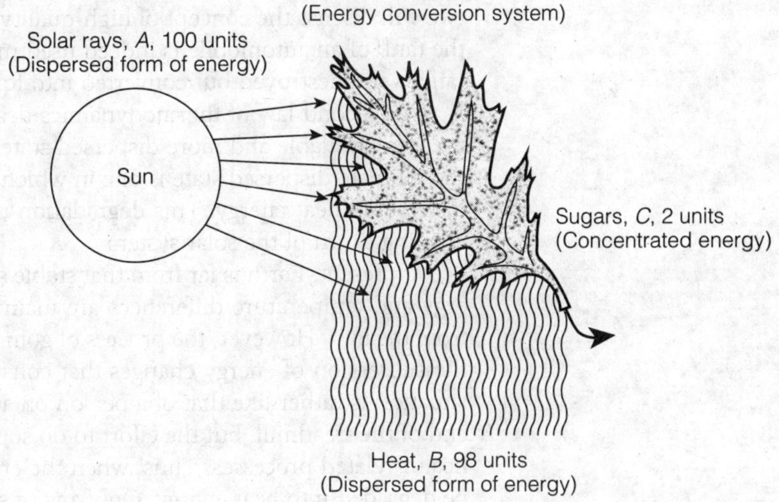

Figure 3-1. Energy flow though an oak leaf, illustrating the two laws of thermodynamics. First law–conversion of energy from the Sun, *A*, to food energy (sugar; *C*) by photosynthesis ($A = B + C$). Second law–*C* is always less than *A* because of heat dissipation, *B*, during the conversion process.

Ecologists understand how light is related to ecological systems and how energy is transformed within the system. The relationships between producer plants and consumer animals, between predators and prey, not to mention the numbers and kinds of organisms in a given environment, are all limited and controlled by the flow of energy from concentrated to dispersed forms. Ecologists and environmental engineers are now using natural ecosystems as models in an attempt to design more energy-efficient human-built systems to transform fossil fuel, atomic energy, and other forms of concentrated energy in industrial and technological societies. The same basic laws that govern nonliving systems, such as automobiles or computers, also govern all types of ecosystems, such as agroecosystems. The difference is that living systems use part of their internally available energy for self-repair and for "pumping out" disorder; machines have to be repaired and replaced by the use of external energy. In their enthusiasm for machines and technology, some forget that a considerable amount of energy resources must be reserved at all times for reducing the entropy created by their operation.

When light is absorbed by some object, which becomes warmer as a result, the light energy has been transformed into another kind of energy: heat energy. *Heat energy* comprises the vibrations and motions of the molecules that make up an object. The differential absorption of the rays from the Sun by land and water causes hot and cold areas, leading to the flow of air, which may drive windmills and perform work, such as the pumping of water against the force of gravity. In this case, light energy changes into heat energy on the land surface of Earth, then into **kinetic energy** of moving air, which accomplishes the work of raising water. The energy is not destroyed by the lifting of the water; instead, it becomes **potential energy,** because the latent energy inherent in having the water elevated can then be transformed into some other type of energy by allowing the water to fall back down to its original level.

As indicated in earlier chapters, food resulting from the photosynthesis of green plants represents potential energy, which changes into other forms of energy when the food is used by organisms. Because the amount of one type of energy is always equivalent in *quantity* (but not in *quality*) to another type into which it is transformed, we can calculate one from the other. Energy that is "consumed" is not actually used up. Rather, it is converted from a state of high-quality to a state of low-quality energy (we will discuss the concept of high-quality energy later in this chapter). Gasoline in the tank of an automobile is indeed used up as gasoline; however, the energy in the tank is not destroyed but converted into forms no longer usable by the automobile.

The second law of thermodynamics deals with the transfer of energy toward an ever less available and more dispersed state. As far as the solar system is concerned, the ultimate dispersed state is one in which all energy ends up in the form of evenly distributed heat energy. This degradation process has often been spoken of as "the running down of the solar system."

At present, Earth is far from that stable state of energy, because vast potential energy and temperature differences are maintained by the continual influx of energy from the Sun. However, the process of going *toward* the stable state is responsible for the succession of energy changes that constitute natural phenomena on Earth. The situation is rather like that of a person on a treadmill; that person never reaches the end of the treadmill, but the effort to do so results in well-defined physiological and health-related processes. Thus, when the energy of the Sun strikes Earth, it tends to be degraded into heat energy. Only a very small portion (less than 1 percent) of the light energy absorbed by green plants is transformed into potential or food energy; most of it goes into heat, which then passes out of the plant, the ecosystem, and the ecosphere. The rest of the biological world obtains its potential chemical energy from the organic substances produced by plant photosynthesis or microorganism chemosynthesis. An animal, for example, takes the chemical potential energy of food and converts a large part of it into heat in order to enable a small part of the energy to be reestablished as the chemical potential energy of new protoplasm. At each step in the transfer of energy from one organism to another, a large part of the energy is degraded into heat. However, entropy is not all negative. As the *quantity* of available energy declines, the *quality* of the remainder may be greatly enhanced.

Over the years, many theorists (Brillouin 1949, for example) were bothered by the fact that the functional order maintained within living systems seemed to defy the second law of thermodynamics. Ilya Prigogine (1962), who won the Noble Prize for his work in non-equilibrium thermodynamics, resolved this apparent contradiction by showing that self-organization and the creation of new structures can and do occur in systems that are far from equilibrium and have well-developed "dissipative structures" that pump out the disorder (see Nicolis and Prigogine 1977). The respiration of the highly ordered biomass is the "dissipative structure" in an ecosystem.

Although entropy in the technical sense relates to energy, the word is also used in a broader sense to refer to the degradation of matter. Freshly made steel represents a low-entropy (high-utility) state of iron; the rusting frame of an automobile represents a high-entropy (low-utility) state. Accordingly, a high-entropy civilization is characterized by degrading energy, such as dilapidating infrastructures (rusting pipes, rotting wood) or eroding soil. Constant repair is one of the costs of high-energy civilizations.

The basic units of energy quantity are presented in Table 3-1. There are two classes of basic units: *potential energy* units, independent of time (Class A), and *power* or *rate* units, with time built into the definition (Class B). Interconversions of power units must take account of the time unit used; thus, 1 watt = 860 cal/h. Of course, Class A units become power units if a time period is included (for example, BTU per hour, day, or year), and power units can be converted back to energy units by "multiplying out" the time unit (as in the case of KWh).

Table 3-1

Units of energy and power and some useful ecological approximations

(A) Units of potential energy

Unit (abbreviation)	*Definition*
calorie or gram-calorie (cal or gcal)	the heat energy required to raise the temperature of 1 cubic centimeter of water by 1 degree Centigrade (at 15° C)
kilocalorie or kilogram-calorie (kcal)	the heat energy needed to raise the temperature of 1 liter of water by 1 degree Centigrade (at 15° C) = 1000 calories
British thermal unit (BTU)	the heat energy needed to raise the temperature of 1 pound of water by 1 degree Fahrenheit
joule (J)	the work energy required to raise 1 kilogram to a height of 10 centimeters (or 1 pound to approximately 9 inches) = 0.1 kilogram-meters
foot-pound	the work energy required to raise 1 pound to a height of 1 foot
kilowatt-hour (KWh)	the amount of electric energy delivered in 1 hour by a constant power of 1,000 watts = 3.6×10^6 joules

(B) Units of power (energy-time units)

Unit (abbreviation)	*Definition*
watt (W)	the standard international unit of power = 1 joule per second = 0.239 cal per second; also the amount of electrical power delivered by a current of 1 ampere across a potential difference of 1 volt
horsepower (hp)	550 foot-pounds per second = 745.7 watts

(C) Reference values (averages or approximations)

Constituent	*Dry weight (kcal/g)*	*Ash-free dry weight (kcal/g)*
Food		
Carbohydrates	4.5	
Proteins	5.5	
Lipids	9.2	
Biomass*		
Terrestrial plants (total)	4.5	4.6
Seeds only	5.2	5.3
Algae	4.9	5.1
Invertebrates (excl. insects)	5.0	5.5
Insects	5.4	5.7
Vertebrates	5.6	6.3

(*continued*)

Table 3-1 ***(continued)***

Species	*Daily food requirement (kcal/g live body weight)*
Human	0.04†
Small bird or mammal	1.0
Insect	0.5

(D) Energy content of fossil fuels (round figures)

Unit of fuel	*Energy content*
1 gram coal	7.0 kcal = 28 BTU
1 pound coal	3200 kcal = 12.8×10^3 BTU
1 gram gasoline	11.5 kcal = 46 BTU
1 gallon gasoline	32,000 kcal = 1.28×10^5 BTU
1 cubic foot natural gas	250 kcal = 1000 BTU = 1 therm
1 barrel crude oil (45 gallons)	1.5×10^6 kcal = 5.8×10^6 BTU

*Because most living organisms are two thirds or more water and minerals, 2 kcal/g live (wet) weight is a very rough approximation for biomass in general.

† = 40 kcal/kg = about 3000 kcal/day for a 70-kg adult

The transfer of energy through the food chain of an ecosystem is termed the **energy flow** because, in accordance with the law of entropy, energy transformations are "one way," in contrast to the cyclic behavior of matter. Later in this chapter, the portion of the total energy flow that passes through the living components of the ecosystem will be analyzed. Furthermore, energy quality, net energy, eMergy, and an energy-based classification of ecosystems will be studied to demonstrate that energy is a common denominator for all kinds of systems, whether natural or designed by humans.

2 Solar Radiation and the Energy Environment

Statement

Organisms at or near the surface of Earth are constantly irradiated by solar radiation and by long-wave thermal radiation from nearby surfaces. Both contribute to the climatic environment (temperature, evaporation of water, movement of air and water). **Solar radiation** reaching the surface of Earth consists of three components: *visible light* and two invisible components, shorter-wave *ultraviolet* and longer-wave *infrared* (Fig. 3-2). Because of its dilute, dispersed nature, only a very small fraction (at most 5 percent) of visible light can be converted by photosynthesis into the much more

Figure 3-2. Spectrum of electromagnetic radiation. Solar radiation reaching Earth's surface is in the middle range of the spectrum, from near infrared to ultraviolet.

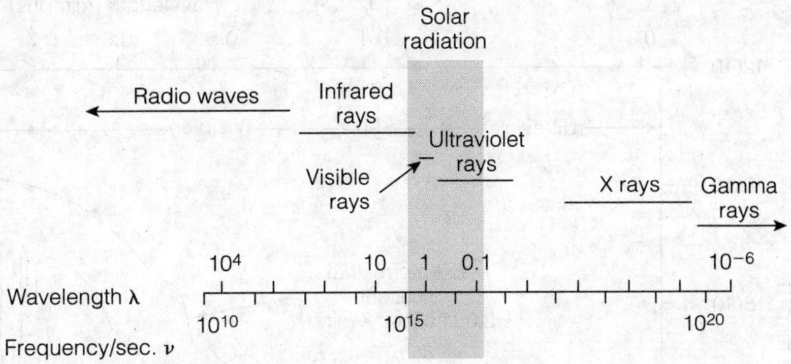

concentrated energy of organic matter for the biotic components of the ecosystem. Extraterrestrial sunlight reaches the ionosphere at a rate of 2 gcal $\cdot$ cm^{-2} $\cdot$ min^{-1} (the **solar constant**) but is attenuated exponentially as it passes through the atmosphere; at most, 67 percent (1.34 gcal$\cdot$cm$^{-2}\cdot$min^{-1}) may reach the surface of Earth at sea level at noon on a clear summer day. Solar radiation is greatly altered as it passes through cloud cover, water, and vegetation. The daily input of sunlight to the autotrophic layer of an ecosystem averages about 300 to 400 gcal/cm^2 (= 3000 to 4000 kcal/m^2) for an area in the North Temperate Zone, such as the United States. The variation in total radiation flux between different strata of the ecosystem, and from one season or site to another on the surface of Earth, is enormous, and the distribution of individual organisms responds accordingly.

Explanation

In Figure 3-3, the spectral distribution of extraterrestrial solar radiation, coming in at a constant rate of 2 gcal $\cdot$ cm^{-2} $\cdot$ min^{-1}, is compared with (1) the solar radiation actually reaching sea level on a clear day; (2) the sunlight penetrating a complete overcast (*cloud light*); and (3) the light transmitted through vegetation. Each curve represents the energy incident on a horizontal surface. In hilly or mountainous country, south-facing slopes receive more solar radiation and north-facing slopes receive less solar radiation than do horizontal surfaces; this results in striking differences in local climates (*microclimates*) and composition of vegetation.

Radiation penetrating the atmosphere is attenuated exponentially by atmospheric gases and dust, but to varying degrees depending on the frequency or wavelength. Short-wave ultraviolet radiation below 0.3 μm is abruptly terminated by the ozone layer in the outer atmosphere (at about 18 miles or 25 km altitude), which is fortunate, because such radiation is lethal to exposed protoplasm. This is why there is great concern regarding the relationship of decreasing ozone (due to chemical degradation by chlorofluorocarbons) and increased risk of skin cancer. Absorption in the atmosphere broadly reduces visible light and irregularly reduces infrared radiation. The radiant energy reaching the surface of Earth on a clear day is about 10 percent ultraviolet, 45 percent visible, and 45 percent infrared. The visible radiation is least attenuated as it passes through dense cloud cover and water, which means that photosynthesis (which is restricted to the visible range) can continue on cloudy days and at some depth in clear water. Vegetation absorbs the blue and red visible wavelengths

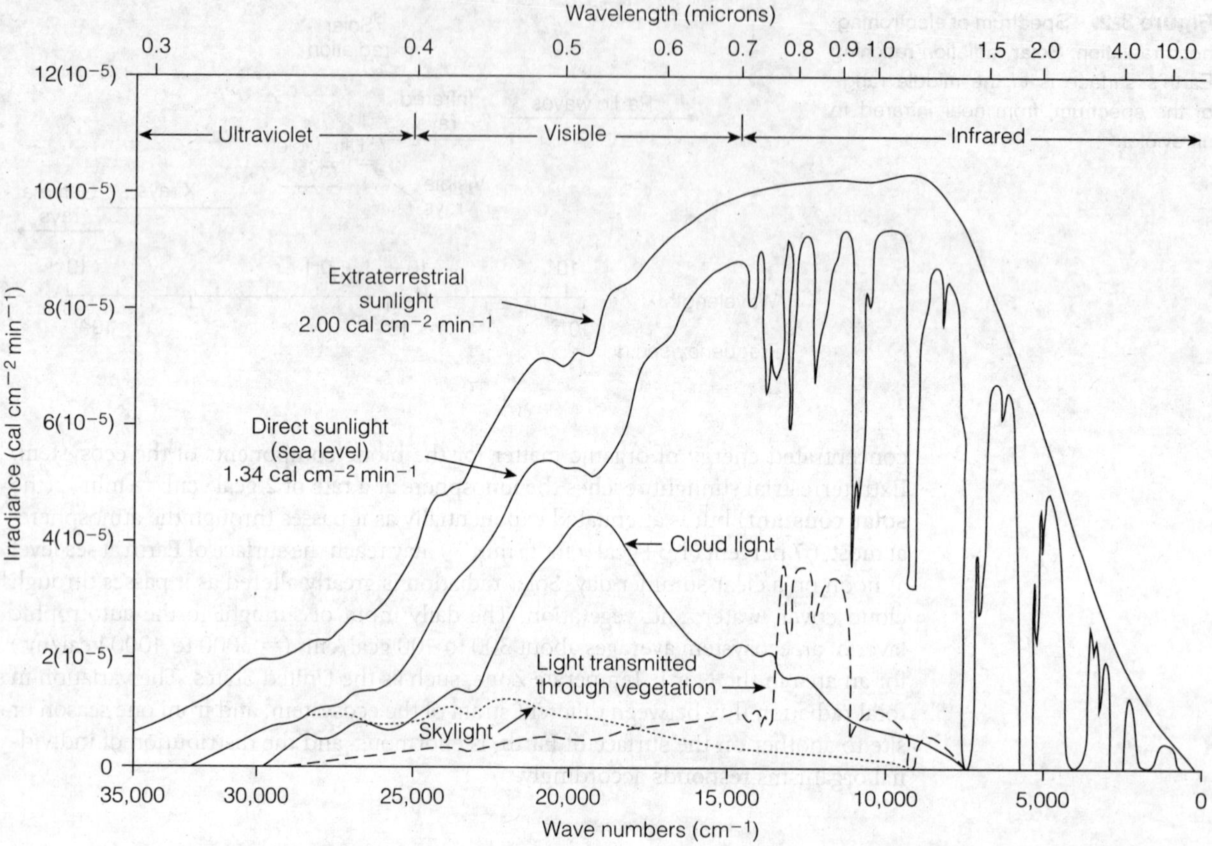

Figure 3-3. Spectral distribution of extraterrestrial solar radiation, of solar radiation at sea level on a clear day, of sunlight from a complete overcast, and of sunlight penetrating a stand of vegetation and diffuse light from the sky scattered by air molecules as distinguished from the direct radiation from the sun. Each curve represents the energy incident on a horizontal surface (from Gates 1965).

and the far infrared strongly and the green less strongly, but the near infrared very weakly. Because the green and near infrared are reflected by vegetation, these spectral bands are used in aerial and satellite remote sensing and photography to reveal the patterns of natural vegetation, condition of crops, presence of diseased plants, and disturbed landscapes.

Thermal radiation, the other component of the energy environment, comes from any surface or object at a temperature above absolute zero. This includes not only soil, water, and vegetation, but also clouds, which contribute a substantial amount of the heat energy radiated downward into ecosystems. For example, temperatures on a cloudy winter night often remain higher than on a clear night. The "greenhouse effect" of re-radiation and heat retention will be considered in greater detail in Chapter 4 in connection with the role of increased CO_2 in global climate change. The long-wave radiation fluxes, of course, are incidental at all times and come from all directions, whereas the solar component is directional (except for the blue and UV light scattered by the atmosphere) and is present only during the daytime. Thermal radiation is absorbed by biomass to a greater degree than is solar radiation. Thus, its daily variation is of greater ecological significance. In places like deserts or alpine Tundra, the daytime flux is much greater than the nighttime flux, whereas in deep water or in the interior of a tropical forest (and, of course, in caves), the total radiation environment may be practically constant throughout the 24-hour period. Water and biomass tend to reduce fluctuations in the energy environment and, thus, make conditions less stressful for life—another example of stress mitigation at the ecosystem level.

Although the total radiation flux determines the conditions of existence for organisms, the integrated direct solar radiation to the *autotrophic stratum*—the energy of the Sun received by green plants over days, months, and the year—is of greater interest for productivity and the cycling of nutrients within the ecosystem. This solar energy input drives all biological and ecological systems. Table 3-2 shows the average daily solar radiation received each month in five regions of the United States. In addition to latitude and season, cloud cover is a major factor, as shown in the comparison between the humid Southeast and the arid Southwest. A range between 100 and 800 $gcal \cdot cm^{-2} \cdot day^{-1}$ would represent most of Earth's surface most of the time, except in the polar regions or in arid tropical regions. There, conditions are so extreme that little biological output is possible. Therefore, for most of the biosphere,

Table 3-2
Solar radiation received regionally over the United States per unit horizontal surface

	Average Langleys (gcal/cm²) per day				
Month	***Northeast***	***Southeast***	***Midwest***	***Northwest***	***Southwest***
January	125	200	200	150	275
February	225	275	275	225	375
March	300	350	375	350	500
April	350	475	450	475	600
May	450	550	525	550	675
June	525	550	575	600	700
July	525	550	600	650	700
August	450	500	525	550	600
September	350	425	425	450	550
October	250	325	325	275	400
November	125	250	225	175	300
December	125	200	175	125	250
Mean $gcal \cdot cm^{-2} \cdot day^{-1}$	317	388	390	381	494
Mean $kcal \cdot m^{-2} \cdot day^{-1}$ (round figures)	3200	3900	3900	3800	4900
Estimated $kcal \cdot m^{-2} \cdot year^{-1}$ (round figures)	1.17×10^6	1.42×10^6	1.42×10^6	1.39×10^6	1.79×10^6

Source: Reifsnyder and Lull 1965.

Table 3-3 **Energy dissipation of solar radiation as percentages of annual input into the biosphere**

Energy	*Percentage*
Reflected	30.0
Direct conversion to heat	46.0
Evaporation precipitation	23.0
Wind, waves, and currents	0.2
Photosynthesis	0.8

Source: Hulbert 1971.

the radiant energy input is on the order of 3000 to 4000 kcal $\cdot$ m^{-2} $\cdot$ day^{-1} and 1.1 to 1.5 million kcal $\cdot$ m^{-2} $\cdot$ $year^{-1}$.

The solar component is usually measured by **solarimeters.** Instruments that measure the total flux of energy at all wavelengths are termed **radiometers.** The **net radiometer** has two surfaces, upward and downward, and records the difference between solar and thermal energy fluxes. Airplanes and satellites equipped with thermal scanners can quantitatively sense heat rising from the surfaces of Earth. Pictures generated from such imagery show the "heat islands" of cities, the location of water bodies, contrasting microclimates (as in north- and south-facing ravines), and many other useful aspects of the energy environment. Cloud cover interferes with this kind of remote sensing much less than it does with visual imagery.

The fate of solar energy coming into the biosphere is summarized in Table 3-3. Although less than 1 percent is converted into food and other biomass, the 70 percent or so that goes into heat, evaporation, precipitation, wind, and so on is not wasted, because these fluxes create a livable temperature and drive the weather systems and water cycles that are necessary for life on Earth. Although energy from tides and the internal heat of Earth may provide useful local sources of energy for humans, little is available globally. An abundance of heat exists deep inside Earth (so-called geothermal energy), but to tap it would require very energy-expensive deep drilling in most parts of the world. However, in the deep sea rifts, there are unique geothermally powered natural ecosystems, as described in Chapter 2.

3 Concept of Productivity

Statement

The **primary productivity** of an ecological system is defined as the rate at which radiant energy is converted by the photosynthetic and chemosynthetic activity of producer organisms (chiefly green plants) to organic substances. It is important to distinguish the four successive steps in the production process as follows:

1. **Gross primary productivity** (*GPP*) is the total rate of photosynthesis, including the organic matter used up in respiration during the period of measurement. This is also known as *total photosynthesis*.
2. **Net primary productivity** (*NPP*) is the rate of storage of organic matter in plant tissues that exceeds the respiratory use, *R*, by the plants during the period of measurement. This is also termed *net assimilation*. In practice, the amount of plant respiration is usually added to measurements of net primary productivity to estimate gross primary productivity ($GPP = NPP + R$).
3. **Net community productivity** is the rate of storage of organic matter not used by heterotrophs (that is, net primary production minus heterotrophic consumption) during the period under consideration, usually the growing season or a year.
4. Finally, the rates of energy storage at consumer levels are referred to as **secondary productivities.** Because consumers use only food materials already produced, with appropriate respiratory losses, and convert this food energy to different tissues by one overall process, secondary productivity should *not* be divided into gross and net amounts. The total energy flow at heterotrophic levels, which is analogous to the gross productivity of autotrophs, should be designated *assimilation* and not *production.*

In all these definitions, the term *productivity* and the phrase *rate of production* may be used interchangeably. Even when the term *production* designates an amount of accumulated organic matter, a time element is always assumed or understood (for instance, a year in agricultural crop production). Thus, to avoid confusion, one should always state the time interval. In accordance with the second law of thermodynamics, the flow of energy decreases at each step due to the heat loss occurring with each transfer of energy from one form to another.

High rates of production, in both natural and cultured ecosystems, occur when physical factors are favorable, especially when *energy subsidies* (such as fertilizers) from outside the system enhance growth or rates of reproduction within the system. Such energy subsidies may also be the work of wind and rain in a forest, tidal energy in an estuary, or the fossil fuel, animal, or human work energy used in cultivating a crop. In evaluating the productivity of an ecosystem, one must consider the nature and magnitude not only of the *energy drains* resulting from climatic, harvest, pollution, and other stresses that divert energy away from the production process but also of the *energy subsidies* that enhance it by reducing the respiratory heat loss (the "disorder pump-out") necessary to maintain the biological structure.

Explanation

The key word in the preceding definitions is *rate*. The **time element**—that is, the amount of energy fixed in a given time—must be considered. Biological productivity thus differs from *yield* in the chemical or industrial sense. In industry, the reaction ends with the production of a given amount of material; in biological communities, the production process is continuous in time, so a time unit must be designated (the amount of food manufactured per day or per year, for example). Although a highly productive community may have more organisms than a less productive community, this is not so if organisms in the productive community are removed or "turn over" rapidly. For example, a fertile pasture being grazed by livestock is likely to have a much smaller standing crop of grass than a less productive pasture not being grazed at the time of measurement. *Biomass or standing crop present at any given time should not be confused with productivity.* Students of ecology often confuse these two quantities. Usually, one cannot determine the primary productivity of a system or the production of a population component simply by counting (or censusing) and weighing the organisms present at any one moment, although net primary productivity can be estimated from standing crop data when living materials accumulate over a period of time (such as a growing season) without being consumed (as in cultivated crops, for example).

Only about half of the total radiant energy of the Sun is absorbed, and at most about 5 percent (10 percent of the energy absorbed) can be converted by gross photosynthesis under the most favorable conditions. Then, plant respiration appreciably reduces—usually by about 20 to 50 percent—the food available for heterotrophs.

During the peak of the growing season, especially during long summer days, as much as 10 percent of the total daily solar input may be converted into gross production, and from 65 up to 80 percent of this may remain as net primary production during a 24-hour period. Even under the most favorable conditions, however, these high daily rates cannot be maintained over the annual cycle, nor can they achieve such high yields over large areas of farmland, as is evident when they are compared with the annual yields actually obtained nationwide and worldwide (see Tables 3-4 and 3-5). Annual primary production varies widely in different kinds of ecosystems as will be detailed in the next section. The relationship between gross and net productivity in natural terrestrial vegetation varies with latitude, as shown in Figure 3-4. The percentage of gross productivity that becomes net primary production is highest at cold latitudes and lowest at hot latitudes, presumably because more respiration is required to maintain the biomass in the Tropics.

The other way in which humans increase food production does not necessarily involve an increase in gross productivity, but rather the genetic selection for an increase in the **food-to-fiber ratio** or **harvest ratio.** For example, a wild rice plant may put 20 percent of its net production into seeds (enough to ensure its survival), whereas a cultivated rice plant is bred to put as much as possible (50 percent or more) into seeds—the edible part. This *grain-to-straw dry weight ratio* has been increased severalfold in most crops. The downside is that the engineered plant does not have much energy available to produce anti-herbivore chemicals (in order to defend itself), so more pesticides have to be used in the cultivation of highly bred varieties.

What has been termed the **Green Revolution** involves the genetic selection for engineered crop varieties with high harvest ratios that are adapted to respond to massive energy, irrigation, and nutrient subsidies. Those who think that developed countries can upgrade the agricultural production of less developed countries by supplying seeds and agricultural recommendations do not realize that the less developed countries cannot afford the necessary energy subsidies. Thus, the Green Revolution so far has benefited the economically rich countries more than the economically poor countries (Shiva 1991). This situation is documented in a dramatic way in Figure 3-5, which compares the trends in agricultural production (since 1950) of those countries with the highest production with those countries having the lowest production of three major food crops: corn, wheat, and rice. The yields have increased two- to threefold in the economically rich countries (United States, France, and Japan) but barely at all in the economically poor countries (India, China, and Brazil).

In the 1960s, plant geneticists developed new varieties of wheat and rice that gave

Table 3-4 Estimates of NPP and plant biomass in major ecosystems

Ecosystem type	Area (10^6 km^2)	Net primary productivity per unit area ($g \cdot m^{-2} \cdot year^{-1}$) Normal range	Mean	Biomass per unit area (kg/m^2) Normal range	Mean
Tropical rain forest	17.0	1000–3500	2200	6–80	45
Tropical seasonal forest	7.5	1000–2500	1600	6–60	35
Temperate evergreen forest	5.0	600–2500	1300	6–200	35
Temperate deciduous forest	7.0	600–2500	1200	6–60	30
Boreal forest	12.0	400–2000	800	6–40	20
Woodland and scrubland	8.5	250–1200	700	2–20	6
Savanna	15.0	200–2000	900	0.2–15	4
Temperate grassland	9.0	200–1500	600	0.2–5	1.6
Tundra and alpine	8.0	10–400	140	0.1–3	0.6
Desert and semidesert shrub	18.0	10–250	90	0.1–4	0.7
Extreme desert, rock, sand, and ice	24.0	0–10	3	0–0.02	0.02
Cultivated land	14.0	100–3500	650	0.4–12	1
Swamp and marsh	2.0	800–3500	2000	3–35	15
Lake and stream	2.0	100–1500	250	0–0.1	0.02
Total continental	149		773		1837
Open ocean	332.0	2–400	125	0–0.005	0.003
Upwelling zones	0.4	400–1000	500	0.005–0.1	0.02
Continental shelf	26.2	200–600	360	0.001–0.04	0.01
Algal beds and reefs	0.6	500–4000	2500	0.04–4	2
Estuaries	1.4	200–3500	1500	0.01–60	1
Total marine	361		152		0.01
Full total	510		333		3.6

Source: Based on Whittaker 1975.

Table 3-5 Two estimates of global net primary production in petagrams of carbon/year (1 Pg = 10^{15} g or 10^9 tons)

Study	*Land*	*Ocean*	*Total*
Whittaker and Likens 1973	57.5	27.5	85.0
Field et al. 1998	56.4	48.5	104.9

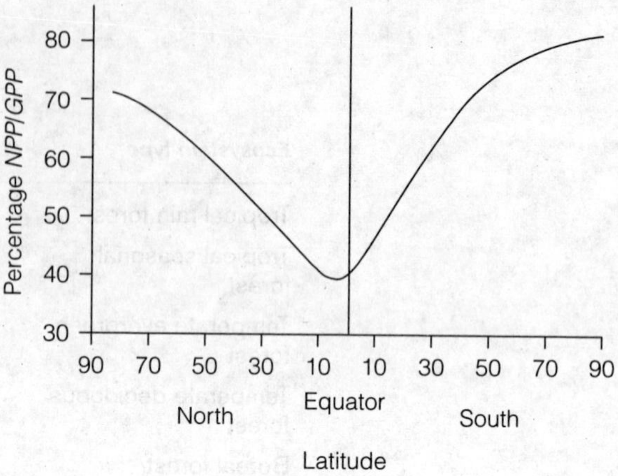

Figure 3-4. Variation by latitude in the percentage of gross primary production (GPP) that ends up as net primary production (NPP) in natural vegetation. The trend is from less than 50 percent in equatorial regions to 60 to 70 percent at high latitudes (graphic model based on data from E. Box 1978).

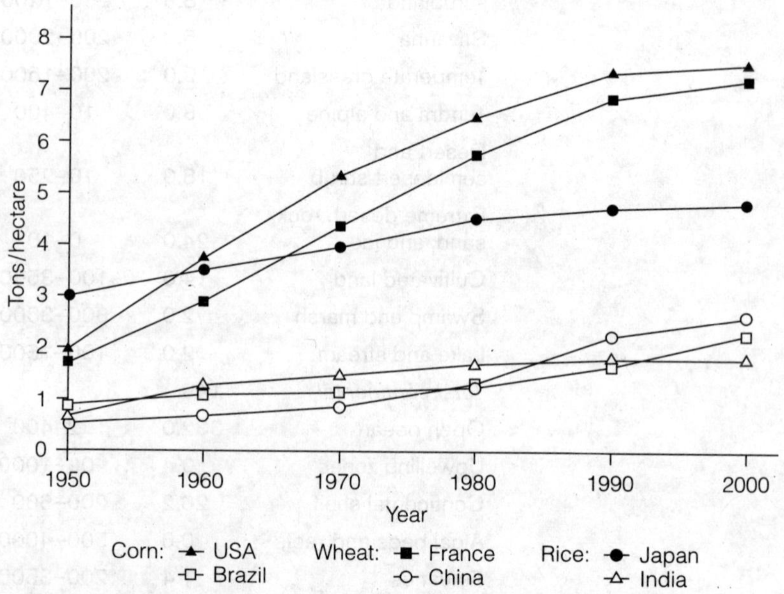

Figure 3-5. Agricultural productivity since 1950 for wheat, rice, and corn in developed countries (United States, France, and Japan) compared to less developed countries (China, India, and Brazil).

yields two to three times those of traditional varieties. Indeed, Norman Borlaug received the Nobel Prize in 1970 for his leadership in the development of these new varieties. This advance in crop breeding was heralded as the beginning of the Green Revolution. It was poorly understood at that time that the need for an increased use of subsidies (such as optimal fertilization and irrigation) that had to accompany these new varieties would negate many of their benefits. This is one of several examples where discoveries judged worthy of Nobel Prizes resulted in unanticipated environmental consequences at a later date.

To take another example, Fritz Haber, a German chemist, received the Nobel Prize in 1918 for his discovery of a catalytic process (termed the **Haber process**) for synthesizing ammonia from nitrogen and hydrogen. Currently, the human alteration of the global nitrogen cycle is one of the major environmental problems confronting society (Vitousek, Aber, et al. 1997). Likewise, the discovery of the pesticide **DDT** during World War II helped to control mosquitoes and thereby greatly reduced the number of human deaths caused by malaria. In fact, the benefits of DDT seemed so tremendous that Paul Muller, a Swiss chemist, was awarded the Nobel Prize in 1948 for its discovery. Proponents of the widespread use of DDT (and other chlorinated hydrocarbon pesticides) failed to understand the long-term ramifications of this discovery (such as the biological magnification of these compounds up the food chain). It was not until 1962 that Rachel Carson's *Silent Spring* (R. Carson 1962) brought attention to and began to document the ecological effects of these large-scale biocide applications. The message is that *what appears to be a breakthrough discovery at one point in time may result in major ecological consequences at a later point in time.*

Concept of Energy Subsidy

High rates of primary production in both natural and cultivated ecosystems occur when physical factors (such as water, nutrients, and climate) are favorable, and especially when auxiliary energy from outside the system reduces maintenance costs (enhances disorder dissipation). Any such secondary or auxiliary energy that supplements the Sun and allows plants to store and pass on more photosynthate is termed an **auxiliary energy flow** or **energy subsidy.** Wind and rain in a rain forest, tidal energy in an estuary, and fossil fuel used in the cultivation of crops are examples of energy subsidies; all of these enhance production by plants and also benefit animals adapted to make use of auxiliary energy. For example, tides do the work of bringing nutrients to marsh grass and food to oysters, as well as taking away waste products, so the organisms do not have to expend energy for these jobs, and can use more of their production for growth (another example of natural capital at work).

High productivity and high net–gross productivity ratios in crops are maintained by large inputs of energy involved in cultivation, irrigation, fertilization, genetic selection, and pest control. The fuel used to power farm machinery is just as much an energy input as sunlight; it can be measured as calories or horsepower diverted to heat in the performance of the work of crop maintenance. In the United States, the energy subsidy input into agriculture increased tenfold between 1900 and the 1980s, from about 1 to 10 calories input per calorie of food harvested (see Steinhart and Steinhart 1974; Tangley 1990; Barrett 1990, 1992). The relationship between inputs of fossil fuels, fertilizers, pesticides, and work energy needed to produce 1 calorie of food energy is shown in Figure 3-6; the doubling of crop yield requires approximately a tenfold increase in all these inputs. Genetic selection for food-to-fiber ratio is the other way in which crop yields have been increased. The ratio of grain-to-straw dry weight for wheat and rice, for example, has been increased from 50 percent to almost 80 percent during the past century.

H. T. Odum was one of the first ecologists to state the vital relationships among energy input, selection, and agricultural productivity. He wrote the following:

> In a real way the energy for potatoes, beef, and plant produce of intense agriculture is coming in large part from the fossil fuels rather than from the Sun. The food we eat is partly made of oil.

High temperatures (and high water stress) generally require a plant to expend more of its gross production energy in respiration. Thus, it costs more to maintain the plant structure in hot climates, although C_4 plants have evolved a photosynthesis

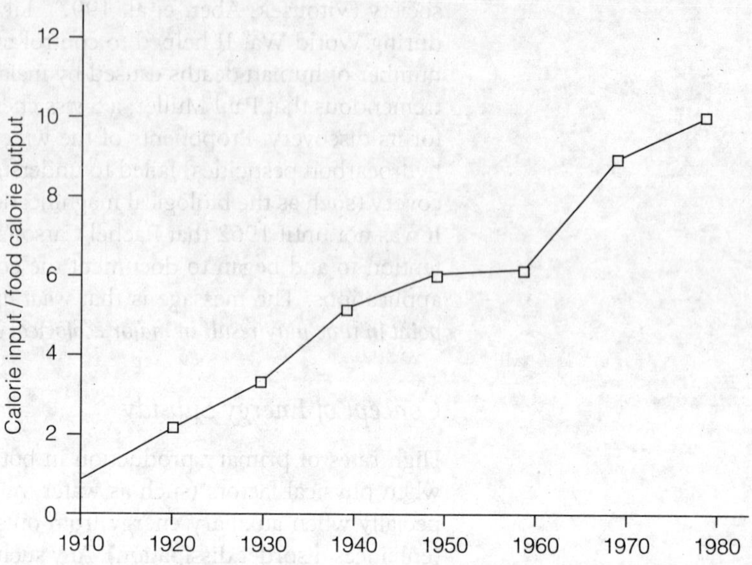

Figure 3-6. Change over time in the amount of energy subsidies used in the food system in the United States to obtain *one* food calorie output (after Steinhart and Steinhart 1974).

cycle that partly circumvents this restraint imposed by hot and dry climates. The general relationship between gross and net production of natural vegetation as a function of latitude was shown in Figure 3-4. These ratios apply to C_3 crops such as rice as well.

Natural communities that benefit from natural energy subsidies (that is, from natural capital; Daily et al. 1997) are those with the highest gross productivity. The role of tides in coastal estuaries and marshes benefiting from an optimal tidal or other water flow subsidy has about the same gross productivity as an intensively farmed Iowa cornfield (see Table 3-4 for comparisons).

As a general principle, the gross productivity of cultivated ecosystems does not exceed that found in nature. We do, of course, increase productivity by supplying water and nutrients in areas where those are limiting (such as deserts and grasslands). Most of all, however, we increase net primary and net community production through energy subsidies that reduce both autotrophic and heterotrophic consumption and thereby increase the harvest.

There is one other important point to be made about the general concept of energy subsidy. A factor under one set of environmental conditions or at a low level of intensity may act as an energy subsidy but under other environmental conditions or at a higher level of input can act as an *energy drain* that reduces productivity. For example, flowing water systems, such as those in Silver Springs, Florida (H. T. Odum 1957), tend to be more fertile than standing water systems, but not if the flow is too abrasive or irregular. The gentle ebb and flow of tides in a salt marsh, a mangrove estuary, or a coral reef contributes tremendously to the high productivity of these communities, but strong tides crashing against a northern rocky shore subjected to ice in winter and heat in summer can be a tremendous drain. Swamps and riverine forests subjected to regular flooding during the winter and early spring dormant period have a much higher production rate than those flooded continually or for long periods in the growing season.

In agriculture, tilling the soil helps in a Temperate Deciduous biome, but not in the Tropics, where the resulting rapid leaching of nutrients and loss of organic matter can severely stress subsequent crops. The trend toward no-till agriculture as a way of reducing these drains was noted in Chapter 2. Finally, certain types of pollution, such as treated sewage, can act as a subsidy or as a stress depending on the rate and periodicity of their input. Treated sewage released into an ecosystem at a steady but moderate rate can increase productivity, but massive, irregular dumping can almost completely destroy the ecosystem as a biological entity.

The Subsidy-Stress Gradient

A factor that under one set of environmental conditions or input level acts as a subsidy can under another set of environmental conditions or at a higher input level act as an **energy drain** or **stress** that reduces productivity. Too much of a good thing (too much fertilizer, too many cars) may be as serious a stress as too little, as humans often belatedly come to realize. The concept of a **subsidy-stress gradient** is illustrated in Figure 3-7. If the input or *perturbation* (from *perturbare* = "to disturb") is poisonous, the response will be negative at any input level. If, however, the input involves usable energy or materials, productivity or other measures of performance may be enhanced, as explained in the previous subsection. As the level of subsidy input increases, the ability of the system to assimilate it can reach saturation; performance will then decline, as shown in the model. For example, a small amount of nitrogen

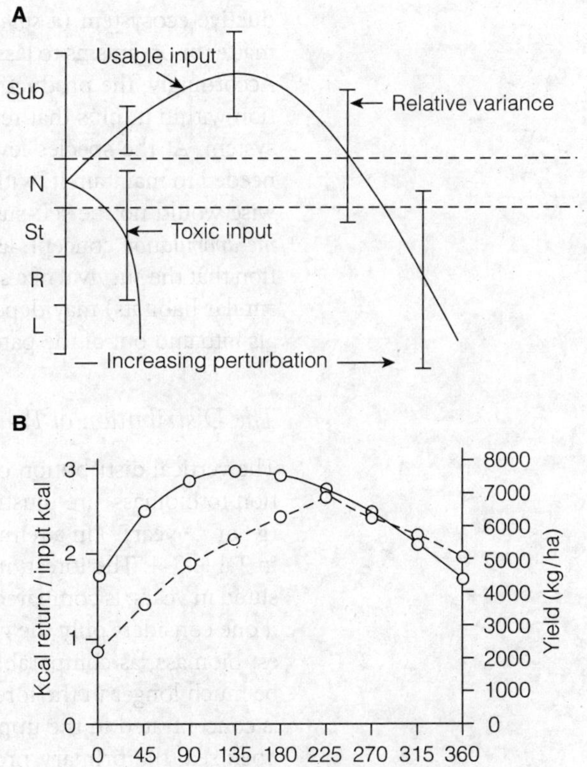

Figure 3-7. Subsidy-stress gradient curves. (A) Generalized curve showing how increasing inputs of energy or materials can bring about a deviation from the normal operating range, N. If the input is usable, basic functions, such as productivity, may be enhanced (subsidy effect; Sub) at moderate levels of input and then depressed with increasing inputs (stress effect; St). If the input is toxic, functions will be depressed, with the likelihood that the community will be either replaced by a more tolerant one or eliminated altogether. R = replacement; L = lethal. (B) Subsidy-stress effects of increasing nitrogen fertilization on a corn crop (phosphorus fertilizer remaining constant). Solid line = efficiency curve, kcal return (harvest) per unit of input. Broken line = yield curve, kg/hectare. Note that the efficiency curve peaks at a lower rate of fertilization than the yield curve (after Pimentel et al. 1973).

fertilizer applied to a lawn will increase growth and improve the health of the lawn; too much nitrogen fertilizer will metabolically "burn up" the lawn or kill the grass (for information on ecological lawn care, see Bormann et al. 2001). As the subsidy begins to turn into stress, the variance increases, as shown by the error bars in Figure 3-7, and the system begins to oscillate out of control until replaced by another system more tolerant of the perturbation or until viable life is no longer possible.

In summary, just about everything that civilization does has a mixed effect on the natural environment and on the quality of human life. Humans can enrich as well as degrade the environment. Very frequently, this is a matter of temporal and spatial scale. We often enhance ecosystem response or quality at low levels of input but degrade both function and quality at high levels of input. A little bit of heat, CO_2, or phosphate may increase the productivity of a body of water if these inputs are limiting under natural conditions, but large amounts of these same inputs may depress basic functions, adversely affecting particular species and reducing water quality for human use.

Humans rarely recognize when increasing returns of scale (which most economists like to talk about) turn to decreasing returns of scale (which most economists do not like to talk about). In other words, *most humans have difficulty determining when enough is enough.*

Concept of Source-Sink Energetics

A corollary to energy subsidy is the concept of **source-sink energetics,** in which excess organic production by one ecosystem (a *source*) is exported to another, less productive ecosystem (a *sink*). For example, a productive estuary may export organic matter or organisms to less productive coastal waters in a process termed **outwelling.** Accordingly, the productivity of an ecosystem is determined by the rate of production within it, plus that received as an import, or minus that exported from a source system. At the species level, one population may produce more offspring than are needed to maintain it, with the surplus moving to an adjacent population that otherwise would not be self-sustaining (Pulliam 1988). Also, at the population level, the *metapopulation* concept, as discussed in Chapters 6 and 9, is based on the observation that the survival of a species in a small landscape patch (semi-isolated from other similar habitats) may depend more on the immigration and emigration of individuals into and out of the patch than on births and deaths within the patch.

The Distribution of Primary Production

The vertical distribution of primary production on land and in the sea, and its relation to biomass, are illustrated in Figure 3-8. The rates of net primary productivity ($g \cdot m^{-2} \cdot year^{-1}$) in a temperate deciduous forest and in the ocean can be compared in Table 3-4. The forest, in which turnover (ratio of biomass to production) is measured in years, is compared with the sea, in which turnover is measured in days. Even if one considers only the green leaves, which compose 1 to 5 percent of the total forest biomass, as comparable to the phytoplankton, the replacement time would still be much longer in the forest. In the more fertile inshore waters, primary production is concentrated in the upper 30 meters or so; in the clearer but poorer waters of the open sea, the primary production zone may extend down to 100 meters or more. This is why coastal waters appear dark greenish and ocean waters blue. In all waters, the peak of photosynthesis tends to occur just under the surface, because the circu-

Figure 3-8. Comparison of vertical distribution of primary production, *P*, and biomass, *B*. (A) shows vertical distribution in a forest. (B) shows vertical distribution in the sea. These data also contrast the radical turnover in the sea (*B*/*P* = 2–4 days) with the slower turnover of the forest (*B*/*P* = 9 years). (Data for young oak-pine forest from Whittaker and Woodwell 1969; data for the sea from Currie 1958 for the northeastern Atlantic ocean.)

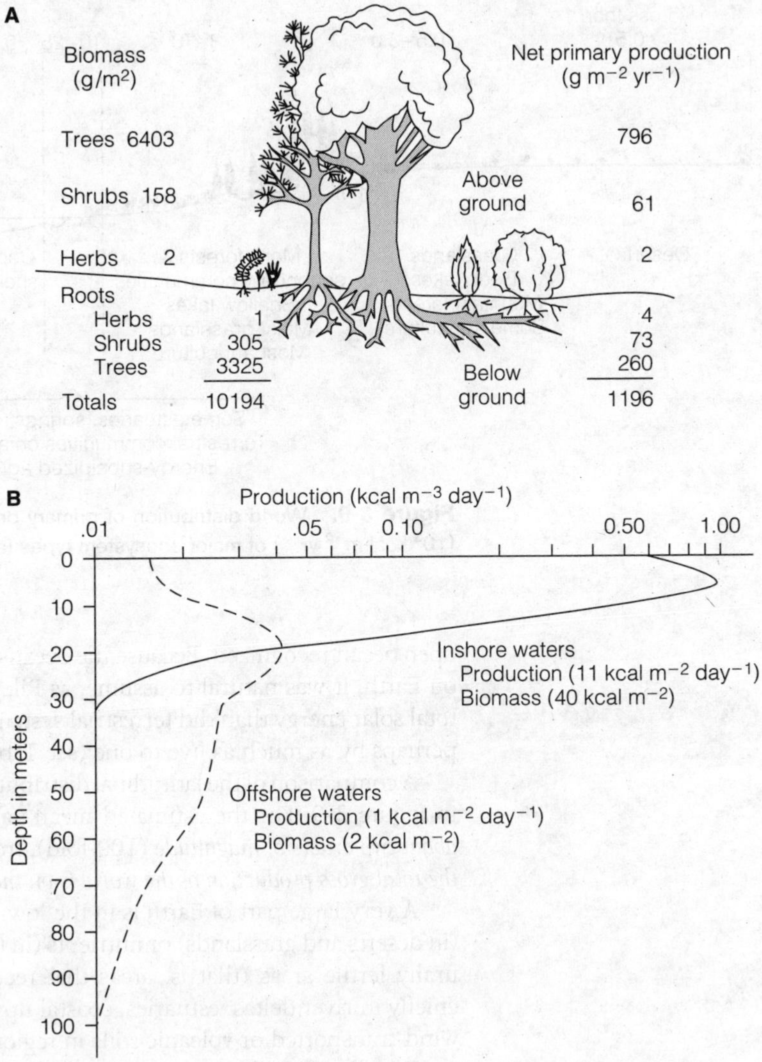

lating phytoplankton are *shade-adapted* and are inhibited by full sunlight. In the forest, where the photosynthetic units (the leaves) are permanently fixed in space, treetop leaves are *sun-adapted*, and understory leaves are shade-adapted (that is, larger and greener).

The attempt to estimate the rate of organic production, or primary productivity, of the world's solar-powered natural systems has an interesting history. In 1862, the pioneer agricultural chemist and plant nutritionist Baron Justus von Liebig based an estimate of the dry-matter production of the global land area on a single sample of a green meadow. Interesting enough, Liebig's estimate of approximately 10^{11} metric tons per year is very close to Lieth and Whittaker's estimate of 118×10^9 tons a year for continental areas (see *Primary Productivity of the Biosphere*, Lieth and Whittaker 1975). Gordon Riley (1944) overestimated ocean productivity by basing his estimate on measurements in fertile inshore waters. Not until the 1960s, after the introduction of the carbon-14 measurement technique, was the low productivity of most of the

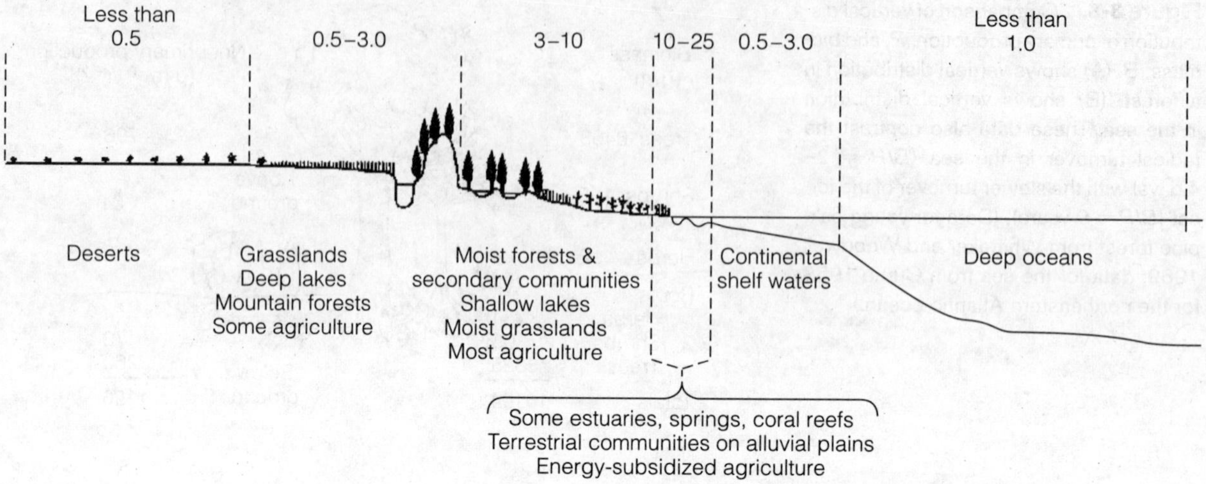

Figure 3-9. World distribution of primary production in terms of annual gross production (10^3 kcal m^{-2} yr^{-1}) of major ecosystem types (after E. P. Odum 1963).

open ocean recognized. Because the oceans cover about 2.5 times the area of the land on Earth, it was natural to assume, as Riley did, that marine ecosystems fixed more total solar energy than did terrestrial systems. Actually, the land outproduces the sea, perhaps by as much as five to one (see Table 3-4).

A comparison of the latitudinal distribution of land and ocean production is shown in Figure 3-9. For the estimated mean values for large areas, *productivity varies by about two orders of magnitude* (100-fold), from 200 to 20,000 kcal · m^{-2} · $year^{-1}$, and *the total gross production of the world is on the order of 10^{18} kcal/year.*

A very large part of Earth is in the low-production category because either water (in deserts and grasslands) or nutrients (in the open ocean) are strongly limiting. Naturally fertile areas (that is, areas that receive natural energy subsidies) are found chiefly in river deltas, estuaries, coastal upwelling areas, areas of rich glacial till, and wind-transported or volcanic soils in regions of adequate rainfall.

For all practical purposes, a level of 50,000 kcal · m^{-2} · $year^{-1}$ can be considered the upper limit for gross photosynthesis. Most agriculture shows low annual productivity, because annual crops are productive for less than half the year. Double cropping (that is, raising crops that produce throughout the year) can approach the gross productivity of the best of natural communities. Recall, however, that net primary production will average about 60 percent of gross productivity and that the "yield to humankind" of crops will be one third or less of the gross productivity. The support system for photosynthesis (cells, leaves, stems, and roots) and respiration of the total biomass, including microbes that decompose organic matter and recycle nutrients, is energy expensive!

Human Use of Primary Production

Vitousek et al. (1986) estimated that although only about 4 percent of terrestrial net production is used directly by humans and domestic animals for food, some 34 percent more is co-opted by humanity, in that it is part of nonedible production (such as lawns) or is destroyed by human activities (such as the clearing of tropical forests).

A similar estimate of 41 percent appropriated by humans has been made by Haberi (1997). Estimates of this sort are difficult to make and are subject to revision, but it would seem that as humankind enters the twenty-first century, at least 50 percent of terrestrial net production and most aquatic net production remains for life-support goods and services (that is, natural capital) and to support all organisms with which we share Earth.

As of the year 2000, the estimated 6.1 billion people in the world each required about 1 million kcal per year, or a total of 6×10^{15} kcal of food energy needed to support the human biomass. The food estimated to be harvested worldwide is inadequate because of poor distribution, waste, and low protein quality. Only about 1 percent of food comes from the sea, and most of that is of animal origin (the small size and rapid turnover of phytoplankton preclude the accumulation of harvestable biomass). Because overfishing has become a worldwide phenomenon, obtaining more food from the sea seems out of the question. Aquaculture is responsible for much of the seafood and fish in today's markets. As already noted, the gap between rich and poor countries in food production has increased during the past half century (see Fig. 3-5), because some countries cannot afford the energy subsidies necessary to support high-yield genetic varieties.

Yields and estimated net primary production (*NPP*) of the major food crops in some developed and less developed countries are compared with world averages in Table 3-6. A *developed country* is defined as a country with a per capita gross national

Table 3-6
Annual yield (net primary production) of edible portion of major food crops at four levels of protein content and two levels of energy subsidy

Harvest weight (kilograms/hectare)

	Developed country (with fuel-subsidized agriculture)				***Underdeveloped country (with little energy subsidy)***				***World average***		
Crops	***Country***	***1970***	***1990***	***1997***	***Country***	***1970***	***1990***	***1997***	***1970***	***1990***	***1997***
Sugars* (< 1 percent protein)	United States	9210	7940	7620	Pakistan	4250	4160	4350	5480	6160	6380
Rice (10 percent protein)	Japan	5630	6330	6420	Bangladesh	1690	2570	2770	2380	3540	3820
Wheat (12 percent protein)	Netherlands	4550	7650	8370	Argentina	1330	1900	2520	1500	2560	2670
Soybeans (30 percent protein)	Canada	2090	2610	2580	India	438	1020	955	1480	1900	2180

Note: Figures are rounded-off averages from the *FAO Production Yearbook* 1997. Yields of basic food crops have leveled off with very little increase (or sometimes decrease) since 1990, but the difference between rich and poor countries remains wide and the world average remains closer to the levels of underdeveloped rather than of developed nations.

*Sugar is estimated as 10 percent of the harvest weight of the sugarcane, as reported in the *Yearbook*.

product (GNP) of more than $8000. About 30 percent of all people live in such countries, which also tend to have a low rate of population growth (1 percent per year or less). In contrast, 65 percent of the world's people live in *underdeveloped countries,* which have a per capita GNP of less than $300, often less than $100, and also have a high population growth rate (more than 2 percent per year). Underdeveloped countries have low food production per hectare because they cannot afford the energy subsidies necessary for high yields. These two masses of humanity are sharply divided—that is, the distribution of per capita income and production per unit area is strongly bimodal.

Figure 3-10 illustrates a natural, unsubsidized, solar-powered ecosystem or landscape including energy inputs (rain and sunlight), natural by-products (carbon dioxide and water runoff), yield (net primary production), and energy loss or respiration (entropy or heat loss from the system). This is a model for natural systems such as forests, grasslands, or organic farming.

Let us now contrast Figure 3-10 with Figure 3-11, which represents a human-subsidized, solar-powered ecosystem or landscape in which, in addition to rain and solar energy inputs, large amounts of fossil fuels, fertilizers, and chemical pesticides enter as inputs. In addition to natural by-products, unnatural by-products "leak" from the system frequently, requiring significant sums of money and energy resources to clean up. Interestingly, even the yield or net primary production of the human-subsidized system is subsidized, in the form of government subsidies or monetary resources. Only the second law of thermodynamics (degraded energy) is not subsidized in this model (Fig. 3-11). Humankind, because of its limited understanding of how natural ecosystems operate, might even try to subsidize the second law, except that it is an impossible task!

Cropland has increased about 15 percent worldwide during recent years, but in

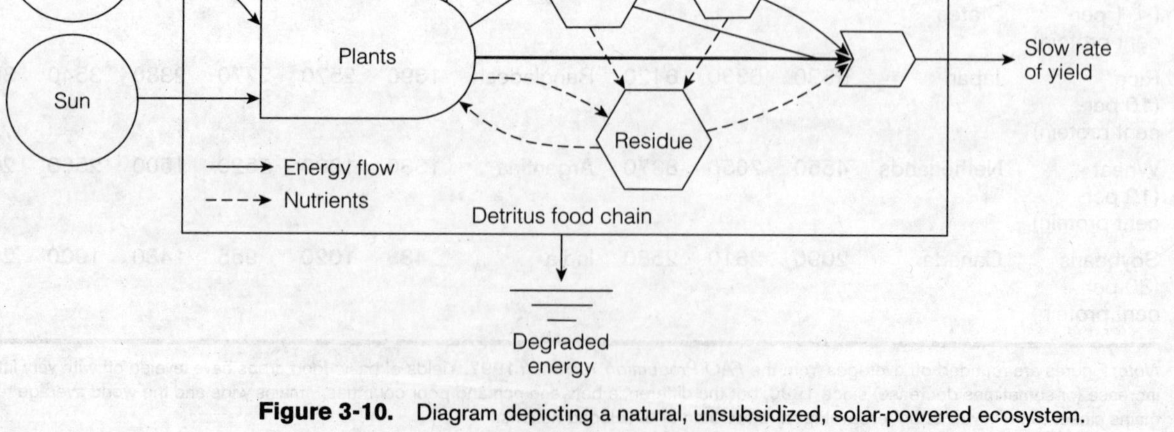

Figure 3-10. Diagram depicting a natural, unsubsidized, solar-powered ecosystem.

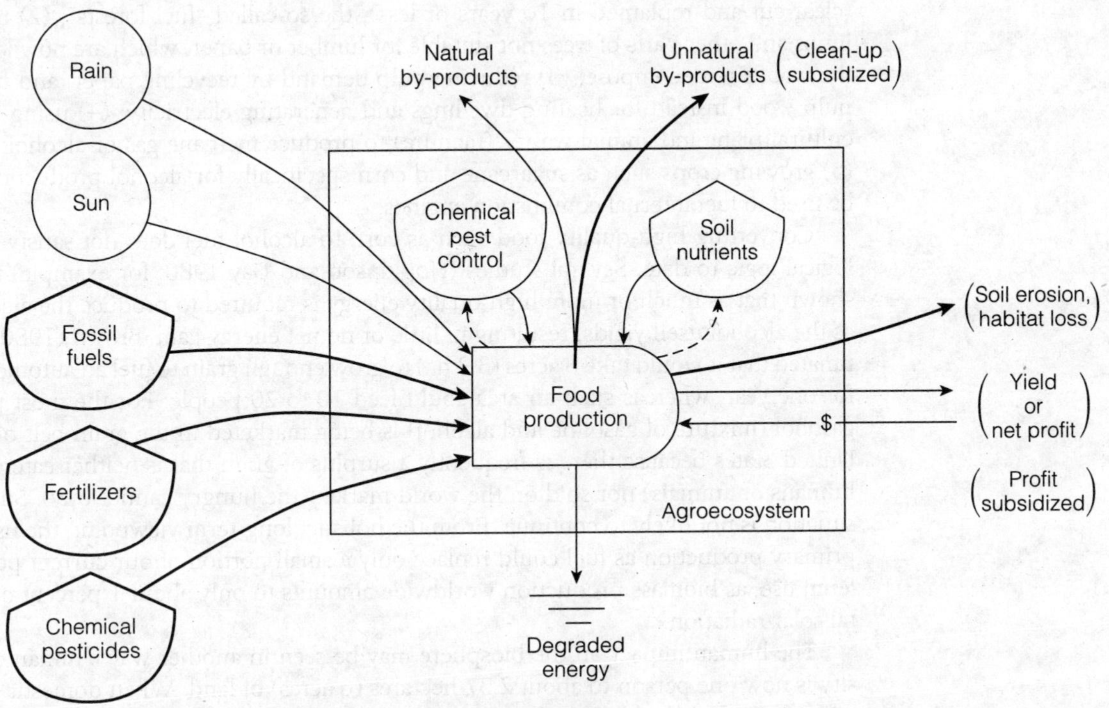

Figure 3-11. Diagram depicting a human-subsidized, solar-powered agricultural system in the Western world, based on solar, fossil fuel, and material subsidy inputs and outputs.

Europe, the United States, and Japan, the harvested area has decreased. Less developed countries have increased their food supply as much by increasing the land area under cultivation as by increasing yields. If such a trend continues, more and more marginal land will be cultivated at increasing cost and risk of environmental degradation. Furthermore, protein content, rather than total calories, tends to limit the diet in the underdeveloped nations. Under equivalent conditions, the yield of a high-protein crop, such as soybean, must always be less (in total calories) than that of a carbohydrate crop, such as sugar (see Table 3-6). It should also be noted that potatoes give a higher yield but have a lower protein content than grains.

There are still other uses of primary production, namely for fiber (such as cotton) and fuel. For more than half the world's population, wood is the chief fuel used for cooking, heating, and light industry. In the poorest countries, wood is burned much faster than it can be grown, so forests are turned into shrublands and then into deserts. The shortage of firewood has been referred to as the "other energy crisis" (oil, of course, is the energy crisis most talked about). In the African countries of Tanzania and Gambia, per capita wood fuel consumption is about 1.5 tons per year, and 99 percent of the population uses wood as fuel.

In North America and other regions with large standing stocks of vegetation, more people have become interested in using biomass, both from forest and agricultural lands, as fuel to supplement or replace dwindling supplies of petroleum, which is a nonrenewable resource. Among the options available are (1) planting fast-growing trees (pines, sycamores, poplars, among others) harvested on short rotation (clear-cut and replanted in 10 years or less), the so-called "fuel forests"; (2) using limbs and other parts of trees not suitable for lumber or paper, which are now left in the woods to decompose; (3) reducing pulp demand by recycling paper, and using pulp wood instead for heating dwellings and generating electricity; (4) using agricultural plant and animal wastes (manure) to produce methane gas or alcohol; and (5) growing crops such as sugarcane and corn specifically for alcohol production to be used to fuel internal combustion engines.

Converting high-quality food such as corn to alcohol fuel does not satisfy ecological logic to date. Several studies (Hopkinson and Day 1980, for example) have shown that as much or more high-quality energy is required to produce the alcohol as the alcohol itself yields, resulting in little or no net energy gain. Brown (1980) estimated that it would take 8 acres (3.2 ha) to grow enough grain to fuel an automobile for one year, whereas such an area could feed 10 to 20 people. For the most part, gasohol (mixture of gasoline and alcohol) is being marketed in the grain belt of the United States because there is frequently a surplus of grain that is neither eaten (by humans or animals) nor sold on the world market (the hungry cannot buy). Such a situation is not likely to continue. From the holistic, long-term viewpoint, the use of primary production as fuel could replace only a small portion of our current petroleum use, as biomass production worldwide amounts to only about 1 percent of total solar radiation.

The human impact on the biosphere may be seen in another way. Human density is now one person to about 2.37 hectares (6 acres) of land. When domestic animals are included, the density is one population equivalent to about 0.65 hectare. This is less than 2 acres for every person and person-equivalent domestic animal consumer. If the population doubles during the twenty-first century, and if humans wish to continue to consume and use animals, there will be only about 1 acre (0.4 ha) to supply all the needs (water, oxygen, minerals, fibers, biomass fuels, living space, and food) of each 50-kilogram consumer; this does not include the pets and wildlife that contribute so much to the quality of human life. Most agroecologists believe that too much emphasis has been placed on the monoculture of annuals. It makes ecological and common sense to consider diversifying crops, establishing multiple cropping systems, adopting limited till procedures (less disturbance of soil structure), and increasing the use of perennial species. Scientists at The Land Institute located near Salina, Kansas, for example, are devoting much thought and research to the feasibility of harvesting native perennial species on the prairie lands of the Midwest. Native prairies are multi-species ecosystems that replenish the fertility of the soils, have deeper roots to build the soil, and are better buffered against nature's vagaries (Pimm 1997). As such natural, unsubsidized systems and landscapes could be turned to "agricultural ends," as scientists such as Wes Jackson believe, they would have great promise to prevail as time-sustainable agriculture. For more on the relationships between energy and food production, see NRC (1989), Barrett (1990, 1992), Soule and Piper (1992), Pimm (1997), Jackson and Jackson (2002), and E. P. Odum and Barrett (2004).

Productivity and Biodiversity: A Two-Way Relationship

In low-nutrient natural environments, an increase in biodiversity seems to enhance productivity, as indicated by experimental research in grasslands (Tilman 1999). However, in high-nutrient or enriched environments, an increase in productivity increases dominance and reduces biodiversity. In other words, a biodiversity increase may increase productivity, but a productivity increase almost always decreases biodiversity—a two-way street. Furthermore, nutrient enrichment (for example, eutrophication, nitrogen fertilization, and runoff) brings on noxious weeds, exotic pests, and dangerous disease organisms, because these kinds of organisms are adapted to and thrive in high-nutrient environments (E. P. Odum and Barrett 2000).

When coral reefs are subjected to human-induced nutrient enrichment, we observe an increase in the dominance of smothering, filamentous algae and the appearance of previously unknown diseases, either of which can quickly destroy these diverse ecosystems that are so beautifully adapted to low-nutrient waters. Another example is the *red tide* that results in periodic massive fish death in Florida estuaries. The red tide microorganism, a dinoflagellate, produces a toxin, presumably as a self-defense against being eaten. At its normal density, not enough toxin is being produced to adversely affect fish, but when estuaries become polluted, the dinoflagellate population sometimes "blooms" (sudden large increase in abundance) resulting in mass dying of fish.

It may be stretching this principle too far, but we can suggest that humans, in their efforts to increase productivity to support increasing numbers of people and domestic animals (which in turn excrete huge amounts of nutrients into the environment) are causing a worldwide eutrophication that is the greatest threat to ecosphere diversity, resilience, and stability—essentially a "too much of a good thing" syndrome. Global warming, which results from CO_2 enrichment of the atmosphere, is one aspect of this overall perturbation, whereas nitrogen enrichment is increasingly responsible for worldwide disorder in both aquatic and terrestrial environments (Vitousek, Aber, et al. 1997). We have here a dilemma or paradox, in which our efforts to feed and produce market goods and services for ever-increasing numbers of people is becoming a major threat to the diversity and quality of our environment.

Chlorophyll and Primary Production

Gessner (1949) observed that the amount of chlorophyll "per square meter" tends to be similar in diverse communities. This finding indicates that the content of the green pigment in whole communities is more uniform than in individual plants or plant parts. The whole is not only different from the parts, but it cannot be explained by them alone. Intact communities containing various plants—young and old, sunlit and shaded—are integrated and adjusted, as fully as local factors allow, to the incoming solar energy, which, of course, impinges on the ecosystem on a "square-meter" basis.

Shade-adapted plants or plant parts tend to have a higher concentration of chlorophyll than light-adapted plants or plant parts; this property enables them to trap and convert as many scarce light photons as possible. Consequently, the use of sunlight is highly efficient in shaded systems, but the photosynthetic yield and the assimilation ratio are low. Algal cultures grown in weak light in the laboratory often become shade-adapted. The high efficiency of such shaded systems has sometimes been mistakenly projected to full-sunlight conditions by those who would feed humankind from mass cultures of algae. However, when the light input is increased to increase yield, the efficiency goes down, as it does in any other kind of plant.

Ecological Pyramids

Relationships between numbers, biomass, and energy flow (metabolism) at the biotic community level can be shown graphically by **ecological pyramids**, in which the first or *producer* trophic level forms the base, and the successive trophic levels form the tiers. Some examples are shown in Figure 3-12. Number pyramids are frequently inverted (base smaller than one or more upper tiers) when individual producer organisms are much larger than the average consumers, as in temperate deciduous forests. Biomass pyramids, on the other hand, tend to be inverted when individual producers are much smaller than the average consumers, as in aquatic communities

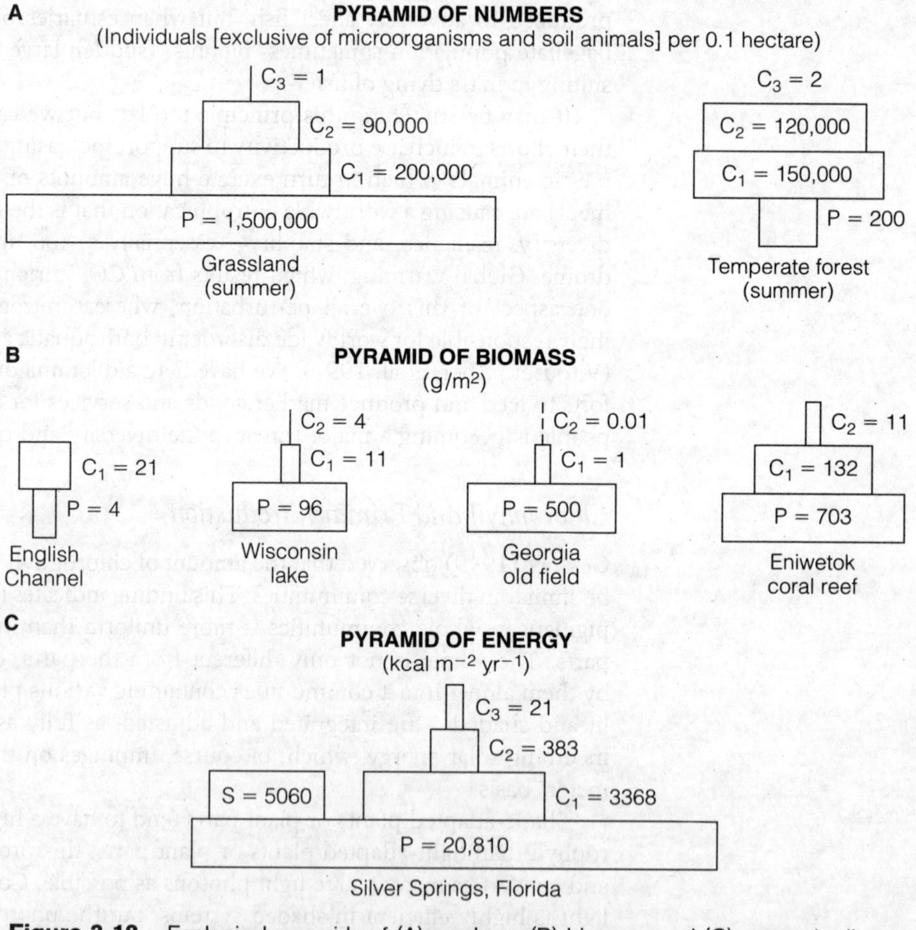

Figure 3-12. Ecological pyramids of (A) numbers, (B) biomass, and (C) energy in diverse ecosystem types. P = producers; C_1 = primary consumers; C_2 = secondary consumers; C_3 = tertiary consumers; and S = saprotrophs (bacteria and fungi). Grassland plant data from F. C. Evans and Cain (1952); temperate forest data from Elton (1966) and Varley (1970); English Channel data from Harvey (1950); Wisconsin Lake data from Juday (1942); Georgia old-field data from E. P. Odum (1957); coral reef data from H. T. Odum and E. P. Odum (1955); and Silver Springs data from H. T. Odum (1957).

Table 3-7 **Density, biomass, and energy flow of six primary consumer populations differing in the size of individuals composing the population**

Population	*Approximate density (no./m²)*	*Biomass (g/m²)*	*Energy flow (kcal · m⁻² · day⁻¹)*
Soil bacteria	10^{12}	0.001	1.0
Marine copepods (*Acartia*)	10^5	2.0	2.5
Intertidal snails (*Littorina*)	200	10.0	1.0
Salt marsh grasshoppers (*Orchelimum*)	10	1.0	0.4
Meadow mice (*Microtus*)	10^{-2}	0.6	0.7
Deer (*Odocoileus*)	10^{-5}	1.1	0.5

Source: E. P. Odum 1968.

dominated by planktonic algae. As dictated by the second law of thermodynamics, however, the energy pyramid must always have a true upright shape—provided all sources of food energy are considered.

Accordingly, energy flow provides a better basis than numbers or biomass for comparing ecosystems and populations with one another. Table 3-7 lists estimates of density, biomass, and energy flow rates for six populations differing widely in size of individual and in habitat. In this series, numbers vary by 17 orders of magnitude (10^{17}), biomass varies by about five orders of magnitude (10^5), whereas energy flow varies only about fivefold. The similarity of energy flows indicates that all six populations are functioning at approximately the same trophic level (primary consumers), even though neither numbers nor biomass indicate this similarity. The ecological rule for this would be as follows: *Numbers overemphasize the importance of small organisms, and biomass overemphasizes the importance of large organisms.* Hence, neither can be used as a reliable criterion for comparing the functional role of populations that differ widely in size-metabolism relationships, although of the two, biomass is generally more reliable than numbers. Thus, *energy flow provides a more suitable index for comparing any and all components of an ecosystem.*

The activities of decomposers and other small organisms may bear very little relation to the total numbers or biomass present at any one moment. For example, a 15-fold increase in dissipated energy resulting from the addition of organic matter to soil may be accompanied by less than a twofold increase in the number of bacteria and fungi. In other words, these small organisms merely turn over faster when they become more active; they do not increase their standing crop biomass proportionally as do large organisms. Trophic structure appears to be a fundamental property that tends to be reconstituted when a particular community is *acutely* perturbed.

When an ecosystem is *continuously* stressed (in contrast to acutely stressed), however, the trophic structure is likely to be altered as the biotic components adapt to the chronic perturbation, as, for example, occurred in the Great Lakes as a result of continuous pollution during the 1960s and 1970s.

Measurement of Primary Production

Before we leave the subject of ecological production, we should comment on the dif-ficulties of measurement. Primary production is best estimated by measuring gaseous exchange— oxygen production or carbon dioxide uptake. This is most easily done in water. In standing water (ponds, lakes, oceans), measuring the diurnal changes in oxygen concentration, as in the light-and-dark bottle experiment described in Chap-ter 2, can be used to estimate both gross and net production. In flowing waters, the upstream-downstream method, involving diurnal measurement of oxygen change at points upstream and downstream, is often effective. Measuring changes in CO_2 with the radioactive isotope ^{14}C is widely used, especially in marine environments.

Measuring gaseous exchanges is much more difficult on land. Measuring the gra-dient in CO_2 concentration of the air from the ground to the top of the vegetation, as was first tried by Transeau (1926) in a cornfield and by Woodwell and Whittaker (1968) in a forest, is widely used today, especially with crops. It is very difficult to measure gross production in large-biomass ecosystems such as forests because it is impractical to try to put the whole forest in a transparent bag or tent (which would have to be cooled, as the air in the bag would heat up rapidly!), although this has been tried on individual trees or limbs. Accordingly, most productivity measurements on land vegetation are estimates of net production obtained by summing annual leaf, trunk, and root growth. For a class exercise, collecting leaf fall with large boxes placed on the forest floor provides a simple method of measuring species diversity and estimating productivity if one takes into consideration latitudinal variations in (1) the ratio of leaves to wood production; and (2) the ratio of gross to net produc-tion. For the North Temperate Zone, the ratio of annual leaf fall to net production is about 1 to 4 g/m^2.

Very close estimates of primary production on large landscapes, regions, and globally are obtained by combining remote sensing of the color of the landscape by satellite with "ground truth" measurements. Thus, a bright green terrestrial landscape or dark green water indicates very productive ecosystems. On land, yellow-green indicates moderate levels and brown very low levels of productivity. Clear blue water indicates low productivity. Quantitative values are obtained by matching color with local quantitative measurements on the surface. Aerial or satellite infrared photography is also often effective. The brighter the infrared, the more productive the landscape. Again, such remote sensing must be calibrated by actual quantitative measurement on the ground.

A Universal Energy Flow Model

What is the basic component of an energy flow model? Figure 3-13 presents what might be termed a **universal model**—one that is applicable to any living component, whether it be plant, animal, microorganism, individual, population, or trophic group. Linked together, such graphic models can depict food chains (Fig. 3-14) or the bioenergetics of an entire ecosystem. In Figure 3-13, the shaded box represents the living, standing crop biomass of the component. Although biomass is usually measured as some kind of weight (living or "wet" weight, dry weight, or ash-free weight), biomass should be expressed in calories, so that the relationships between the rates of energy flow and the instantaneous or average standing crop biomass can be established. The total energy input or *intake* is indicated by *I* in Figure 3-13. For strict autotrophs, this is light, and for strict heterotrophs, it is organic food.

The concept of **trophic level** is not primarily intended for categorizing species. Energy flows stepwise through the community according to the second law of thermodynamics, but a given population of a species may be (and very often is) involved in more than a single trophic level. Therefore, the universal model of energy flow illustrated in Figure 3-14 can be used in two ways. The model can represent a *species population,* in which case the appropriate energy inputs and links with other species would be shown as a conventional, species-oriented food web diagram (see Fig. 3-15),

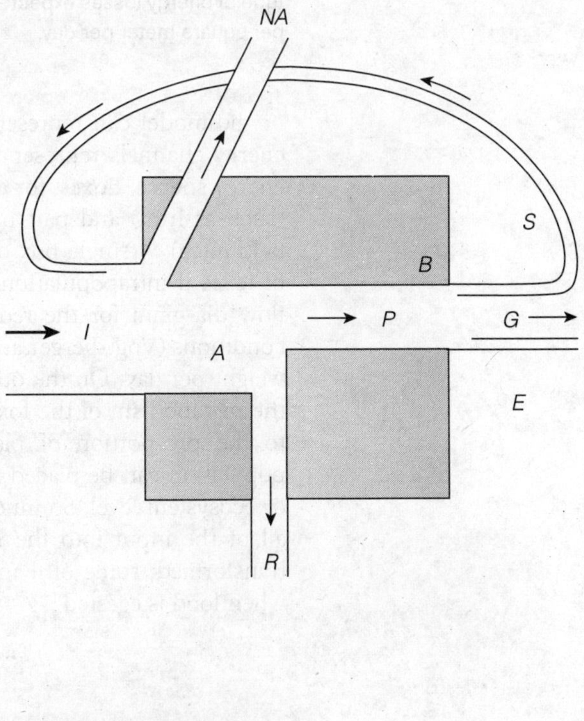

Figure 3-13. Components of a model of ecological energy flow. *I* = input (or ingestion); *NA* = not assimilated (or egestion); *A* = assimilated energy; *P* = production; *R* = respiration; *G* = growth and reproduction; *B* = standing crop biomass; *S* = stored energy; and *E* = excreted energy.

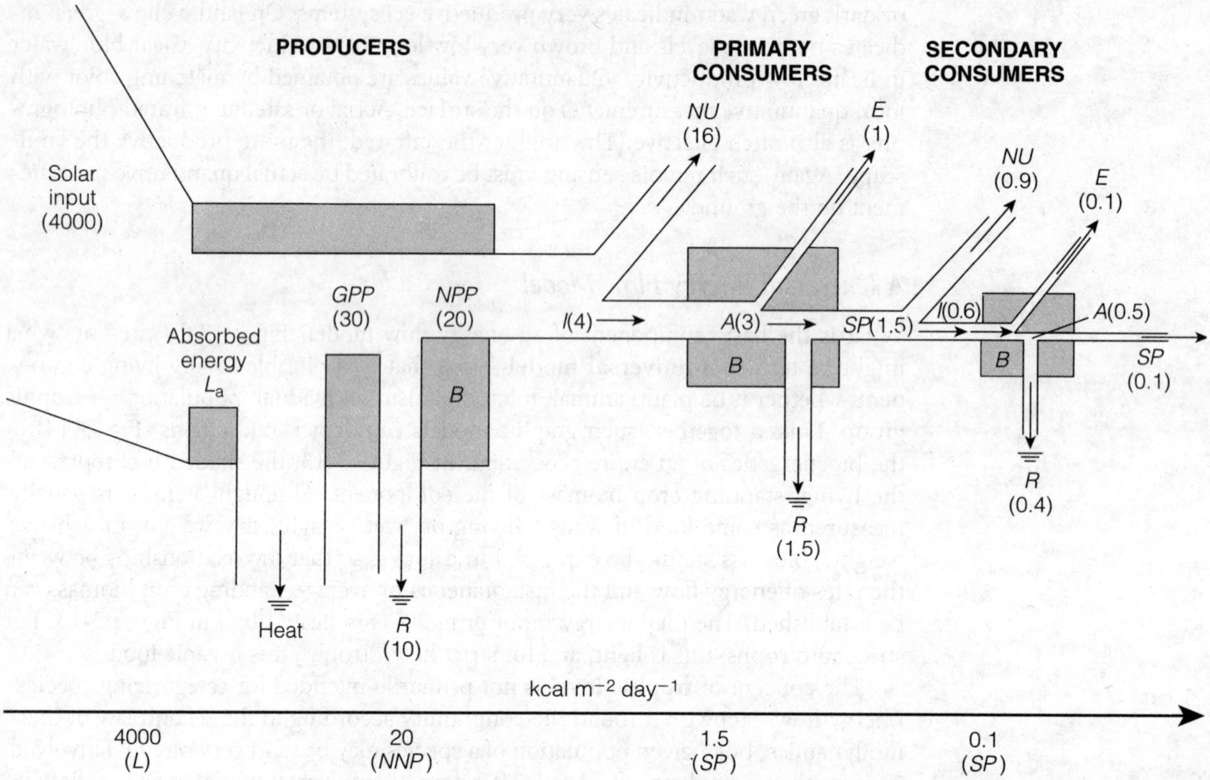

Figure 3-14. Simplified energy flow diagram depicting three trophic levels in a linear food chain. Standard notations for successive energy flows are as follows: L_a = light absorbed by plant cover; *GPP* = gross primary production; *A* = total assimilation; *NPP* = net primary production; *SP* = secondary (consumer) production; *NU* = energy not consumed by next trophic level; *E* = energy not assimilated by consumers (egested); *I* = input (or ingestion); *B* = standing crop biomass; and *R* = respiration. Bottom line in the diagram shows the order of magnitude of energy losses expected at major transfer points, starting with a solar input of 4000 kcal per square meter per day.

or the model can represent a discrete *energy level,* in which case the biomass and energy channels represent all or part of many populations supported by the same energy source. Foxes, for example, usually obtain part of their food by eating plants (such as fruit) and part by eating herbivorous small mammals (such as rabbits or field mice). A single box diagram could be used to represent the whole population of foxes if intrapopulation energetics were to be stressed. Figure 3-15 is an energy flow dia-gram for the red fox (*Vulpes vulpes*) fed a diet of rabbits under penned conditions (Vogtsberger and Barrett 1973). All values are expressed in kcal /kg body weight per day. On the other hand, two or more boxes would be employed should the metabo-lism of the fox population be divided into two trophic levels according to the pro-portion of plant and animal food consumed. In this way, the fox population can be placed into the overall pattern of energy flow at the community or ecosystem level. So much for the question of the source of the energy input. Not all of the input into the biomass of an organism, population, or trophic level is transformed; some of it may simply pass through the biological structure, such as when food is egested from the digestive tract without being metabolized or when light passes through vegetation without being fixed. This energy component is indicated by *NU* (not used) or *NA* (not assimilated; see Fig. 3-13). That portion used or assimilated is indicated by *A* in the diagram. The ratio between *A* and *I* (that is, the efficiency of assimilation) varies widely. It may be very low, as in light fixation by plants or food assimilation by detritus-feeding animals, or very high, as when animals or bacteria consume high-energy food such as sugars or amino acids. In autotrophs, the assimilated energy, *A*, is, of course, the gross primary production or gross photosynthesis. The analogous component (the *A* component) in heterotrophs represents food ingested minus food egested (feces). Therefore, *the term gross primary production should be restricted to auto-trophic production.*

A key feature of the model is the separation of assimilated energy, *A*, into the *P* and *R* components. That part of the fixed energy, *A*, that is burned and lost as heat is designated *respiration, R;* that portion transformed to new or different organic matter is designated *production, P.* Thus, *P* represents net primary production in plants and secondary production in animals. **Secondary production** (*SP*) in consumer individuals is composed of tissue growth and litters of new individuals. The *P* component is energy available to the next trophic level, whereas the *NU* or nonassimilated component (such as feces) enters the detritus food chain (that is, material such as feces becomes available to be broken down by bacteria and fungi).

The ratio between *P* and *R* and between the standing crop biomass, *B,* and *R* varies widely and is ecologically significant. In general, the proportion of energy going into respiration (*maintenance energy*) is large in populations functioning at higher trophic levels and in communities with a large standing crop biomass. *R* increases when a system is stressed (E. P. Odum 1985). Conversely, the *P* component is relatively large in active populations of small organisms, such as bacteria or algae; in youthful, rapidly growing communities; and in systems benefiting from energy subsidies. The relevance of *P/R* ratios to food production for humans will be noted in Chapter 8.

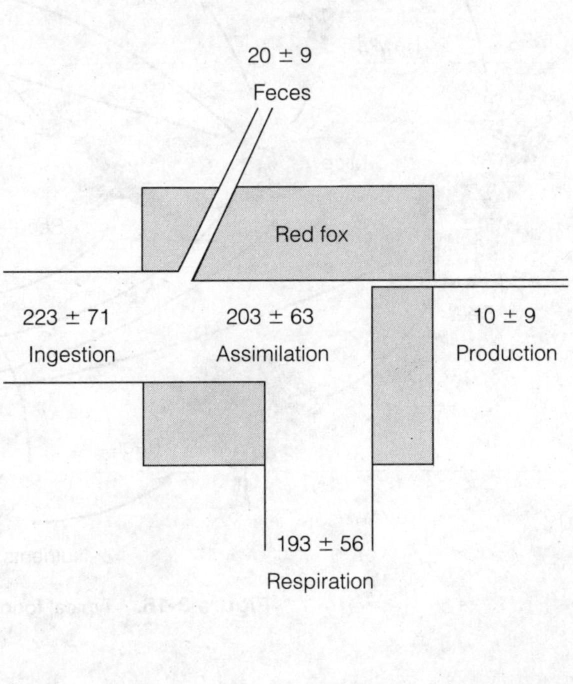

Figure 3-15. Energy flow diagram for the red fox (*Vulpes vulpes*). All values expressed as kcal per kg body weight per day. (From Vogtsberger, L. M., and G. W. Barrett. 1973. Bioenergetics of captive red foxes. *Journal of Wildlife Management* 37:495–500. Reprinted with permission of The Wildlife Society.)

4 Energy Partitioning in Food Chains and Food Webs

Statement

The transfer of food energy from its source in autotrophs (plants) through a series of organisms that consume and are consumed is termed the **food chain.** At each transfer, a proportion (often as high as 80 or 90 percent) of the potential energy is lost as heat. Therefore, the shorter the food chain—or the nearer the organism to the producer trophic level—the greater the energy available to that population. However, *whereas the quantity of energy declines with each transfer, the quality or concentration of the energy that is transferred increases.* Food chains are of two basic types: (1) the **grazing food chain**, which, starting from a green plant base, goes to grazing herbivores (organisms eating living plant cells or tissues) and on to carnivores (animal eaters); and (2) the **detritus food chain**, which goes from nonliving organic matter to microorganisms and then to detritus-feeding organisms (detritivores) and their predators. Food chains are not isolated sequences; they are interconnected. The interlocking pattern is often spoken of as the **food web** (Fig. 3-16). In complex natural commu-

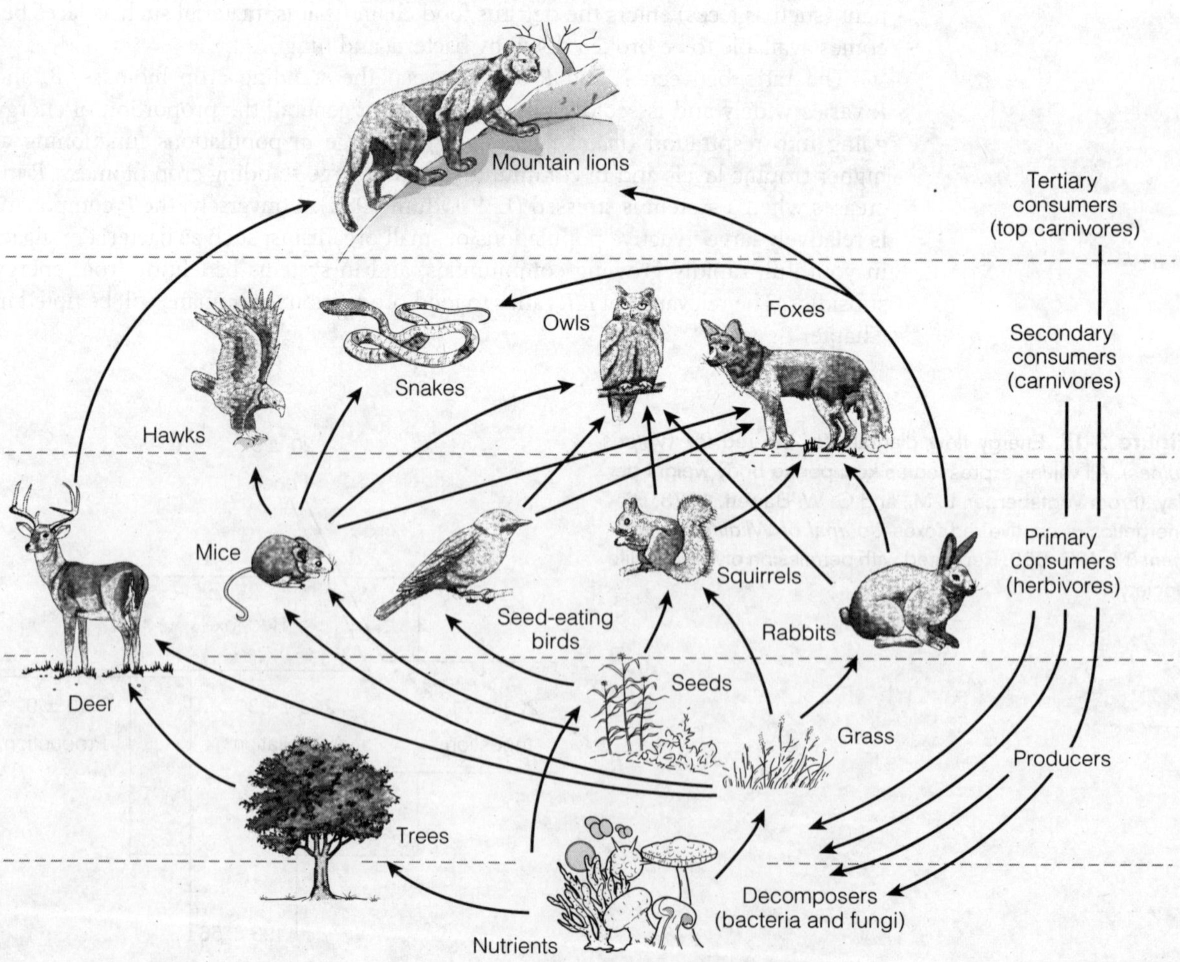

Figure 3-16. Typical food web for a terrestrial ecosystem.

nities, organisms whose nourishment is obtained from the Sun through the same number of steps are said to belong to the same *trophic level.* Thus, green plants occupy the first level (the *producer* trophic level), plant eaters (herbivores) occupy the second level (the *primary consumer* trophic level), primary carnivores occupy the third level (the *secondary consumer* trophic level), and secondary carnivores occupy the fourth level (the *tertiary consumer* trophic level). This trophic classification is one of function and not one of species as such. A given species population may occupy one or more trophic levels according to the source of the energy actually assimilated.

Explanation

Food chains are vaguely familiar to everyone, as we eat the big fish that ate the little fish that ate the zooplankton that ate the phytoplankton that fixed the energy of the Sun; or we may eat the cow that ate the grass that fixed the light energy of the Sun; or we may use a much shorter food chain by directly eating the grain that fixed the energy of the Sun. In the last case, human beings function as primary consumers at the second trophic level. In the grass-cow-human food chain, we function at the third trophic level, as secondary consumers. In general, humans tend to be both primary and secondary consumers as our diet most often comprises a mixture of plant and animal foods. Animals that consume both plant and animal matter are frequently referred to as **omnivores.** Accordingly, the energy flow is divided between two or more trophic levels in proportion to the percentage of plant and animal food eaten.

Potential energy is lost at each food transfer. Only a small portion (typically less than 1 percent) of the available solar energy is fixed by the plant in the first place. Consequently, the number of consumers (such as people) that can be supported by a given output of primary production very much depends on the length of the food chain; each link in our traditional agricultural food chain decreases the available energy by about one order of magnitude (about tenfold). Therefore, fewer people can be supported on a given amount of primary production when large amounts of meat are part of the diet. However, as emphasized in the statement, as the quantity goes down with each transfer, the energy concentration (quality), goes up, a sort of "bad news–good news" story.

Where the nutritional quality of the energy source is high, transfer efficiencies can be much higher than 20 percent. However, because both plants and animals produce a lot of hard-to-digest organic matter (cellulose, lignin, and chitin) together with chemical inhibitors that discourage would-be consumers, typical transfers between whole trophic levels average 20 percent or less.

Table 3-8 provides approximations of the proportion of assimilated energy at each trophic level that is shunted into either production or respiration; the *utilization efficiencies* for each trophic level, illustrating the percentage of available energy from the preceding trophic level that is consumed (used) at that trophic level; and the *assimilation efficiencies* ($I - NA$) for each trophic level. Naturally, these percentages vary depending on food quality, homeothermy versus poikilothermy, and the stage of each species' life history. The approximations, however, do provide estimates regarding the diversity of species that function at each trophic level. The percentages expressed in Table 3-8 can be used to illustrate energy flow through the producer, primary consumer, and secondary consumer trophic levels, as depicted in Figure 3-15.

In Figure 3-18, the grazing and detritus food chains are shown as separate flows in a Y-shaped, or two-channel, energy flow diagram. This model is more realistic than

Table 3-8 **Summary of production, respiration, utilization efficiencies, and assimilation efficiencies expressed as percentages by trophic level**

Trophic level	*Production (percent)*	*Respiration (percent)*
Producers	60–70	30–40
Primary consumers	40–50	50–60
Secondary consumers	5–10	90–95

Trophic level	*Utilization efficiency (percent)*
Producers	<1
Primary consumers	20–25
Secondary consumers	30–40

Trophic level	*Assimilation efficiency (percent)*
Producers	60–70
Primary consumers	70–80
Secondary consumers	90–95

the single-channel model because (1) it conforms to the basic stratified structure of ecosystems; (2) the direct consumption of living plants and the consumption of dead organic matter are usually separated in both time and space; and (3) the macroconsumers (phagotrophic animals) and the microconsumers (saprotrophic bacteria and fungi) differ greatly in size-metabolism relations and in the techniques required for studying them.

The proportion of net production energy that flows down the two pathways varies in different kinds of ecosystems and often varies seasonally or annually in the same ecosystem. In some shallow waters and in heavily grazed pastures or grasslands,

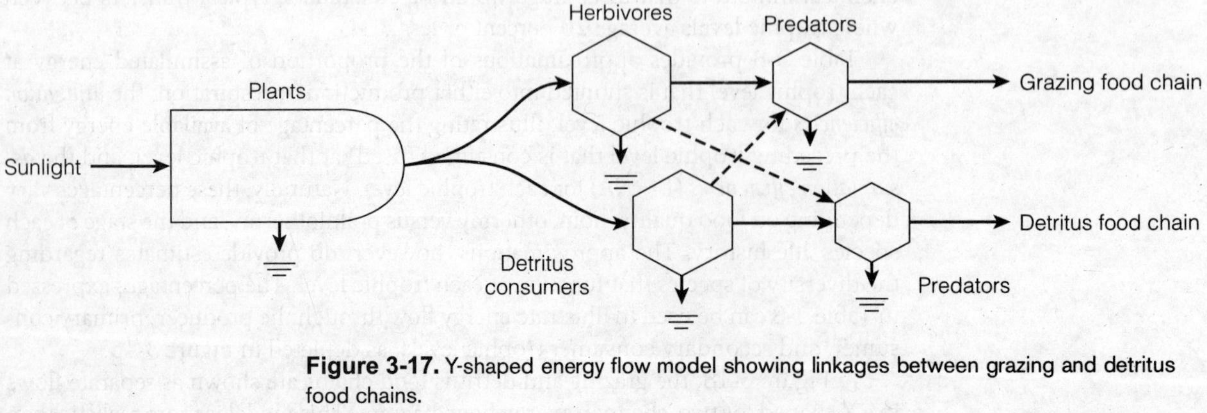

Figure 3-17. Y-shaped energy flow model showing linkages between grazing and detritus food chains.

50 percent or more of the net production may pass down the grazing pathway. In contrast, marshes, oceans, forests, and indeed most natural ecosystems operate as *detrital systems,* in that 90 percent or more of the autotrophic production is not consumed by heterotrophs until the leaves, stems, and other plant parts die and are processed into particulate and dissolved organic matter in water, sediments, and soils. As emphasized in Chapter 2, this delayed consumption increases the structural complexity and biodiversity as well as the storage and buffering capacities of ecosystems. There would be no forests if all tree seedlings were grazed down as soon as they appeared.

In all ecosystems, the grazing and detritus food chains are interconnected, so shifts in flow can occur quickly in response to forcing function inputs from outside the system. Not all food eaten by grazers is actually assimilated; some (such as undigested material in feces) is diverted to the detritus pathway. The impact of the grazer on the community depends on the rate of removal of living plant material, not just on the amount of energy in the food that is assimilated. The direct removal of more than 30 to 50 percent of the annual plant growth by terrestrial grazing animals or by mowing makes the ecosystem less able to resist future stress.

The many mechanisms in nature that control or reduce grazing or herbivory by native species are as impressive as humanity's past ability to control domestic grazing animals is unimpressive. Overgrazing has contributed to the decline of past civilizations. The choice of words is important here. *Overgrazing,* by definition, is detrimental, but what constitutes overgrazing in different kinds of ecosystems is only now being defined in terms of energetics, long-range economics, and ecosystem sustainability.

Undergrazing can also be detrimental. In the complete absence of direct consumption of living plants, detritus would accumulate faster than microorganisms could decompose it, thereby delaying mineral recycling and, perhaps, making the system vulnerable to fires. The positive value of light surface fires in decomposition is discussed in the next chapter, and the *reward feedback* effect of grazers on plants is presented later in this chapter.

Energy flows originating from nonliving organic materials involve several distinct food chain pathways, as shown in Figure 3-18. What was labeled the detritus path-way in Figure 3-17 is subdivided into three flows in Figure 3-18. One flow, often the dominant one, originates with *particulate organic matter* (POM); the other two path-

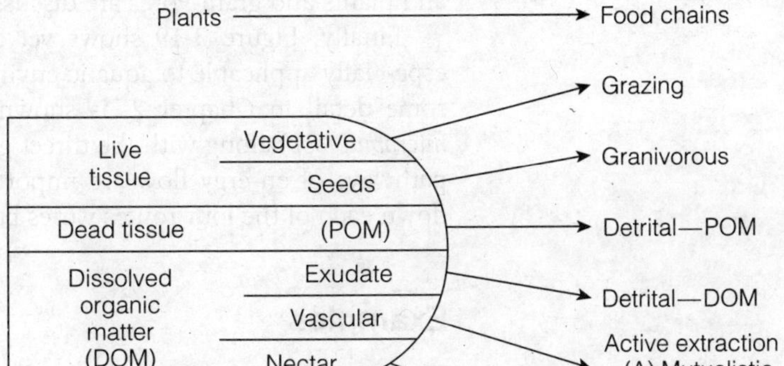

Figure 3-18. Branched food chain model, especially applicable to terrestrial ecosystems. DOM = dissolved organic matter; POM = particulate organic matter.

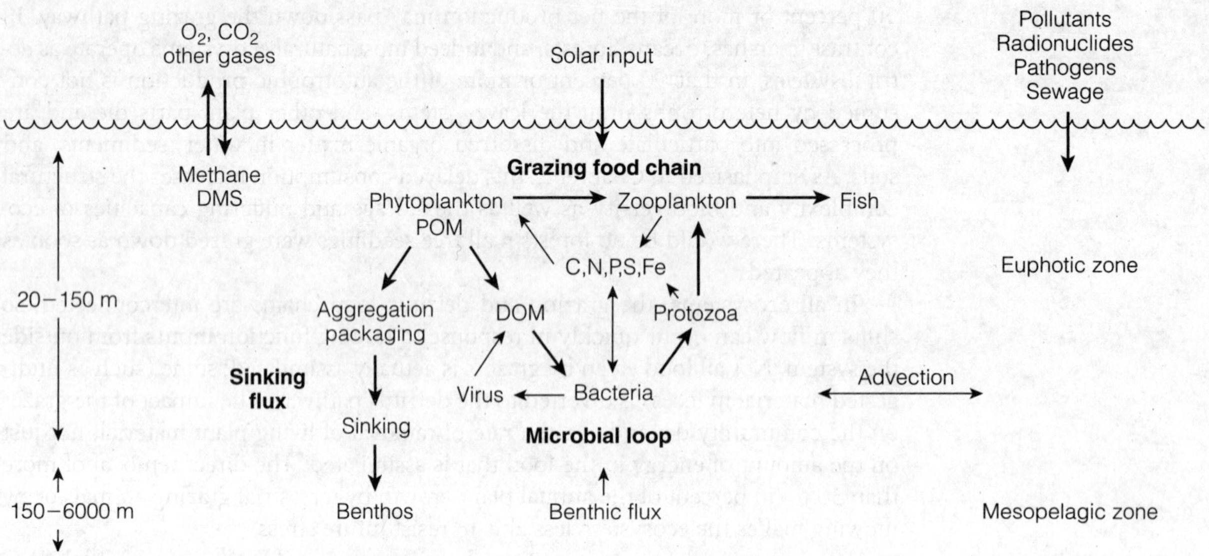

Figure 3-19. Modern view of the pelagic food web, emphasizing the microbial loop as a major pathway for organic matter flux. Competition among the three main flux pathways–grazing food chain, microbial loop, and sinking flux–significantly affects oceanic carbon cycling and productivity. DMS = dimethylsulfide; DOM = dissolved organic matter; POM = particulate organic matter. (From Azam, F. 1998. Microbial control of oceanic carbon flux: The plot thickens. *Science* 280:694–696. Copyright © 1998 AAAS. Reprinted with permission.)

ways start with *dissolved organic matter* (DOM). Symbiotic fungi called mycorrhizae, aphids and other parasites, and pathogens extract photosynthate (DOM) directly from the plant's vascular system or tissues, whereas the great majority of saprotrophic microorganisms (decomposers) consume the DOM most frequently in the form of exudates from cells, roots, and leaves.

Two distinct subsystem food chains are largely restricted to terrestrial or shallow-water ecosystems, as shown in Figure 3-18: the **granivorous food chain**, originating from seeds, high-quality energy sources that are major food items for animals and hu-mans; and the **nectar food chain**, originating from the nectars of flowering plants that depend on insects and other animals for pollination. The intricate, mutualistic relationships that have evolved between plants and pollinators and plants and grani-vores are discussed in Chapter 7.

Finally, Figure 3-19 shows yet another way to picture food chains, this one espe-cially applicable to aquatic environments. The anaerobic pathway, discussed in some detail in Chapter 2, is shown as a separate flow (the *sinking flux* and the *microbial loop*) along with the direct grazing, the DOM, and the POM flows. All four pathways of en-ergy flow are important, but just how much primary energy goes down each of the four routes varies between different kinds of ecosystems.

Examples

Three examples should suffice to illustrate the major features of food chains, food webs, and trophic levels. First, in the far north, in the region known as the Tundra, only relatively few kinds of organisms have become successfully adapted to low temperatures. Food chains and food webs are thus relatively simple. The pioneer British ecologist Charles Elton realized this early and, during the 1920s and 1930s, studied the ecology of Arctic lands. He was one of the first to clarify the principles and concepts relating to food chains (Elton 1927). The plants on the Tundra—reindeer lichens (*Cladonia;* often called "reindeer moss"), grasses, sedges, and dwarf willows—provide food for the caribou of the North American Tundra and for its ecological counterpart, the reindeer of the Old World Tundra. These animals, in turn, are preyed on by wolves and humans. Tundra plants are also eaten by lemmings and by the ptarmigan (*Lagopus lagopus*) or Arctic grouse (*Lagopus mutus*). Throughout the long winter and during the brief summer, the Arctic white fox (*Alopex lagopus*) and snowy owl (*Nyctea scandiaca*) and other raptors prey on the lemmings, voles, and ptarmigans. Any radical change in the numbers of rodents or the density of *Cladonia* affects the trophic levels, because the alternative choices of food are few. This is one reason why the numbers of some groups of Arctic organisms fluctuate greatly, from superabundance to near extinction. The same has often happened to human civilizations that depended on a single or on relatively few local food sources, as for example in the Irish potato famine. We will consider these Arctic cycles in detail in Chapter 6.

Second, a farm pond managed for sport fishing, thousands of which have been built over the years, provides an excellent example of food chains under fairly simplified conditions. Because a fishpond is supposed to provide the maximum number of fish of a particular species and a particular size, management procedures are designed to channel as much of the available energy as possible into the final product by restricting the producers to one group, the floating algae or phytoplankton. Other green plants, such as rooted aquatics and filamentous algae, are discouraged. Figure 3-21 shows a compartment model of a sport-fishing pond in which transfers at

Figure 3-20. Compartment model of the principal food chains in a Georgia pond managed for sport fishing. Estimated energy inputs withrespect to time are in kcal · m^{-2} · $year^{-1}$. The model suggests the interesting possibility that fish production might be increased if the "side food chain" through *Chaoborus* were eliminated, but the possibility that this side chain enhances the stability of the system must be considered (data from Welch 1967).

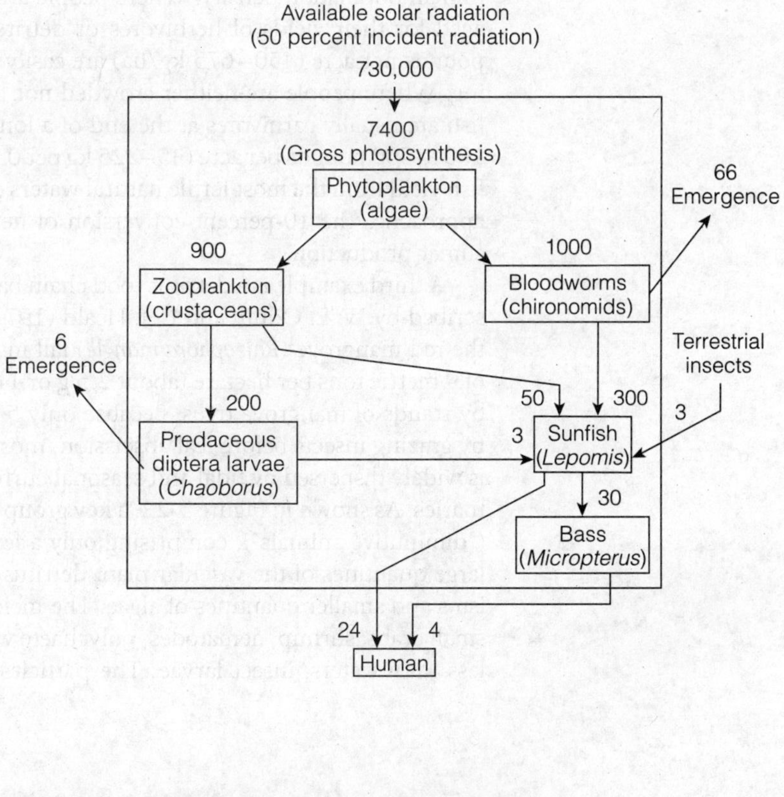

each link in the food chain are quantified in terms of kilocalories per square meter per year. In this model, only the successive inputs of ingested energy are shown; the losses during respiration and assimilation are not shown. The phytoplankton is fed on by the zooplanktonic crustaceans in the water column, and the planktonic detritus is taken in by certain benthic invertebrates, notably bloodworms (chironomids), which are the preferred food of sunfishes; these sunfishes in turn are fed on by bass. The balance between the last two groups in the food chain (sunfish and bass) is very important to the harvesting of fish by humans. A pond with sunfish as the only fish could actually produce a greater total biomass of fish than one with bass and sunfish, but most of the sunfish would remain small because of the high reproduction rate and the competition for available food. Fishing by hook and line would soon be poor. As the sportsperson wants large fish, the final predator (tertiary consumer) is necessary for a good sport-fishing pond.

Fishponds are good places to demonstrate how secondary productivity is related to (1) the length of the food chain; (2) the primary productivity; and (3) the nature and extent of energy imports from outside the pond system. Large lakes and the sea yield fewer fish per hectare, or per square meter, than do small, fertilized, and intensely managed ponds, not only because primary productivity is lower and food chains are longer, but also because only a part of the consumer population (the marketable species) is harvested in large bodies of water. Likewise, yields are several times greater when herbivores, such as carp, are stocked than when carnivores, such as bass, are harvested; the latter, of course, require a longer food chain. High yields are obtained by adding food from outside the ecosystem (by adding subsidies such as plant or animal products that represent energy fixed somewhere else). Actually, such subsidized yields should not be expressed by area unless one adjusts the area to include the land (the *ecological footprint,* as decribed in Chapter 2) from which the supplemental food was obtained. As might be expected, fish culture depends on the human population density. Where people are crowded and hungry, ponds are managed for their yields of herbivores or detritus consumers; yields of 1000 to 1500 pounds per acre (450–675 kg/ha) are easily obtainable without supplemental feeding. Where people are neither crowded nor hungry, game fish are desired. As these fish are usually carnivores at the end of a long food chain, yields are much lower—100 to 500 pounds per acre (45–225 kg per 0.4 ha). Finally, the 300 $kcal \cdot m^{-2} \cdot year^{-1}$ fish yield from the most fertile natural waters or ponds managed for short food chains approaches the 10-percent conversion of net primary production to primary consumer production.

A third example is a detritus food chain based on mangrove leaves, which was described by W. E. Odum and E. J. Heald (1972, 1975). In southern Florida, leaves of the red mangrove (*Rhizophora mangle*) fall into the brackish waters at an annual rate of 9 metric tons per hectare (about 2.5 g or 11 kcal per m^2 per day) in areas occupied by stands of mangrove trees. Because only 5 percent of the leaf material is removed by grazing insects before leaf abscission, most of the annual net primary production is widely dispersed by tidal and seasonal currents over many hectares of bays and estuaries. As shown in Figure 3-22, a key group of small animals, often called *meiofauna* ("diminutive animals"), comprising only a few species but many individuals, ingest large quantities of the vascular plant detritus along with the associated microorganisms and smaller quantities of algae. The meiofauna in estuaries generally comprises small crabs, shrimp, nematodes, polychaete worms, small bivalves and snails, and, in less salty waters, insect larvae. The particles ingested by these detritus consumers

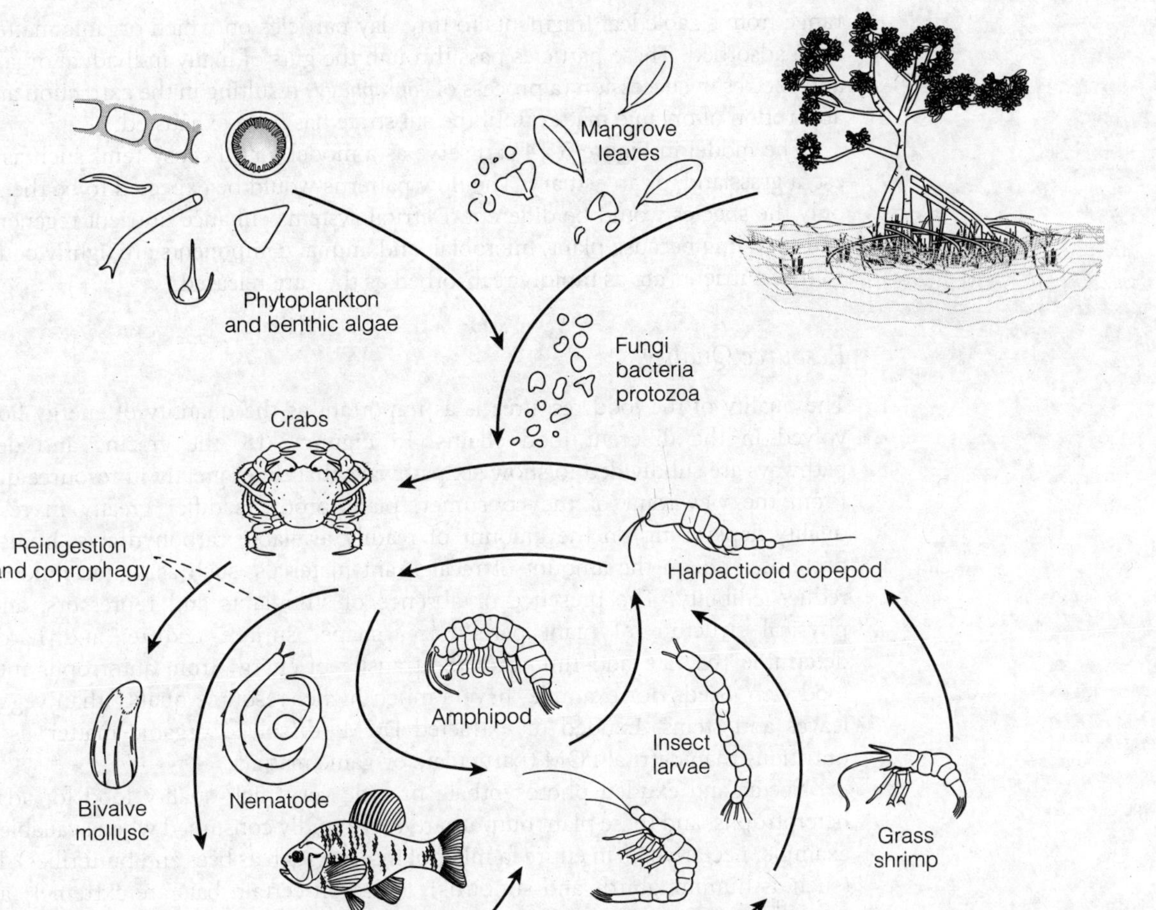

Figure 3-21. Detritus food web based on mangrove leaves that fall into shallow estuarine waters of South Florida. Leaf fragments acted on by saprotrophs and colonized by algae are eaten and re-eaten (coprophagy) by a key group of small detritus consumers, which in turn provide the main food for game fish, herons, storks, and ibis (redrawn from W. E. Odum and Heald 1975).

range from sizable leaf fragments to tiny clay particles on which organic matter has been adsorbed. These particles pass through the guts of many individual organisms and species in succession (a process of *coprophagy*) resulting in the extraction and re-absorption of organic matter until the substrate has been exhausted.

The model in Figure 3-14 can serve as a model for all ecosystems such as a forest, a grassland, or an estuary. The flow patterns would be expected to be the same; only the species would be different. Detrital systems enhance nutrient regeneration and recycling because plant, microbial, and animal components are tightly coupled, so that nutrients are as rapidly reabsorbed as they are released.

Resource Quality

The quality of the food resource is as important as the quantity of energy flow involved in the different food chains. In Figure 3-18, the grazing and detritus pathways are subdivided to show six pathways that differ greatly in resource quality. From the viewpoint of the consumer, plant products differ greatly in resource quality depend-ing on the amount of readily available carbohydrates, lipids, and proteins, and on the amount of recalcitrant materials, such as lignocellulose, that reduce edibility. The presence or absence of stimulants and repressors, and the physical structure of plant parts (size, shape, surface texture, and hardness) determine the rate and timing of the transfer of energy from autotrophs into the food web. Seeds, for example, have a much higher resource quality than vegetative leaves and stems. Exuded or extracted DOM (dissolved organic matter) is more nutritious than detrital POM (particulate or-ganic matter).

Nectar and exuded photosynthate provide very high-quality food for adapted heterotrophs, and these plant outputs are very rapidly consumed when available. For example, nectarivores in entire families of insects (such as bees and butterflies), birds (such as hummingbirds and sunbirds), and even certain bats are extremely active around flowers when nectar is being secreted. Many species have coevolved with flowering plants for mutual benefit (Fig. 3-22). Colwell (1973) documented a case in which nine species of animals in widely different taxonomic groups (insects, mites,

Figure 3-22. A green violet-ear hummingbird feeds on nectar from an elleanthus orchid.

and birds) are closely associated with flowers of four kinds of tropical plants, thus creating a closely knit subcommunity.

Coleman et al. (1998) has estimated that the metabolic "hot spots" of rhizospheres and active decomposition sites in the soil may occupy less than 10 percent of the total soil volume. A similar "halo" of bacteria surrounds living algal cells in the marine environment in what Bell and Mitchell (1972) have termed the **phycosphere.** These bacteria do not make contact with or penetrate the cell membrane and are clearly living off of the exuded organic matter.

Reward Feedbacks and the Role of Heterotrophs in Food Webs

Food webs involve much more than predator-prey relationships, or "who eats whom." There are positive or mutualistic feedbacks as well as negative ones. When a "downstream" organism in the energy flow has a positive effect on its "upstream" food supply, we have **reward feedback,** in the sense that a consumer organism (such as a herbivore or parasite) does something that sustains the survival of its food resource (plant or host).

It has been shown that grazing by the vast herds of antelopes on the East African plains increases the net production of grass; that is, annual vegetative growth is greater with the grazers than without them (McNaughton 1976). The catch is that "timing is everything," in that the herds migrate seasonally over large areas, thereby avoiding overgrazing. Fencing in the animals cancels out the adaptation. A similar effect of bison grazing on North American grasslands has been found in a 10-year study in Kansas (S. L. Collins et al. 1998). Not only was the net primary production increased by moderate grazing, but biodiversity was also increased. It has also been hypothesized that the saliva of grasshoppers and grazing mammals contains growth hormones (complex peptides) that stimulate root growth and the ability of plants to regenerate new leaves, providing a mechanism for this positive feedback effect (Dyer et al. 1993, 1995).

As an aquatic example, fiddler crabs of the genus *Uca,* which feed on surface algae and detritus in coastal marshes, "cultivate" their food plants in several ways. Their burrowing increases water circulation around the roots of marsh grass and brings oxygen and nutrients deep into the anaerobic zone. By constantly reworking the organically rich mud on which they feed, the crabs enhance conditions for the growth of benthic algae. Finally, egested sediment particles and fecal pellets provide substances for the growth of nitrogen-fixing and other bacteria that enrich the system (Montague 1980).

In summary, survival by both the "eater" and the "eaten" on a greater temporal-spatial scale are enhanced when the consumer can promote as well as consume its food supply, just as humans both consume and promote the welfare of domestic plants and animals. The more food webs are studied as a whole, the more partnerships and mutually beneficial relationships between producers and consumers and between different levels of consumers are discovered.

The *length* of food chains is also of interest. The reduction of the energy available to successive trophic levels obviously limits the length of food chains. However, the availability of energy may not be the only factor, as long food chains often occur in infertile systems, such as oligotrophic lakes, and short ones are often found in very productive or eutrophic situations. The rapid production of nutritious plant material may invite heavy grazing, resulting in the concentration of energy flow at the first two or three trophic levels. **Eutrophication** (nutrient enrichment) of lakes also increases energy flow at the first few trophic levels and shifts the planktonic food web from a phytoplankton–large zooplankton–game fish sequence to a microbes-detritus-microzooplankton system not so supportive of recreational fisheries.

Ecological Efficiencies

The ratios between energy flows at different points along the food chain are of considerable ecological interest. Such ratios, when expressed as percentages, are often termed **ecological efficiencies.** Table 3-9 lists some of these ratios and defines them in terms of the energy flow diagram. The ratios have meaning in reference to component populations as well as to whole trophic levels. Because the several types of efficiencies are often confused, defining exactly what relationship is meant is important; the energy flow diagrams (Figs. 3-14 and 3-15) help clarify this definition. We encourage students of ecology to read Raymond L. Lindeman's classic paper "The Trophic-Dynamic Aspect of Ecology" (Lindeman 1942) for an understanding of ecological efficiencies.

Most important, efficiency ratios are meaningful in comparisons only when the numerator and denominator of each ratio are expressed in the same units of measurement. Otherwise, statements about efficiency can be misleading. For example, poultry farmers may speak of a 40 percent efficiency in the conversion of chicken feed to chickens (the P_t/I_t ratio in Table 3-9). But this is actually a ratio of "wet" or live chicken (worth about 2 kcal/g) to dry feed (worth about 4 kcal/g). The true ecological growth efficiency (in kcal/kcal) in this case is more like 20 percent. Thus, wherever possible, ecological efficiencies should be expressed in the same "energy currency" (such as calories to calories).

The general nature of transfer efficiencies between trophic levels has already been

Table 3-9 **Various types of ecological efficiencies**

Ratio	*Designation and explanation*
A. Ratios between trophic levels	
I_t/I_{t-1}	Trophic level energy intake, or Lindeman's efficiency
	For the producer trophic level, this is P_G/L or P_G/L_A
A_t/A_{t-1}	Trophic level assimilation efficiency
P_t/P_{t-1}	Trophic level production efficiency
B. Ratios within trophic levels	
P_t/A_t	Tissue growth or production efficiency
P_t/I_t	Ecological growth efficiency
A_t/I_t	Assimilation efficiency

Symbol key: L = light (total); L_A = absorbed light; P_G = total photosynthesis (gross primary production); P = production of biomass; I = energy intake; A = assimilation; t = present trophic level; t−1 = preceding trophic level.

discussed. The production efficiencies between secondary trophic levels typically approximate 10 to 20 percent. Because the proportion of assimilated energy that must go to respiration is at least 10 times higher in warm-blooded animals (*homeotherms*), which maintain a high body temperature at all times, than in cold-blooded animals (*poikilotherms*), **production efficiency** (*P/A*) must be lower in warm-blooded species. Accordingly, the efficiency of transfer between trophic levels should be higher in an invertebrate than in a mammalian food chain. For example, the energy transfer from moose to wolf on Isle Royale is about 1 percent, compared with a 10 percent transfer in a *Daphnia-Hydra* food chain (Lawton 1981). Herbivores tend to have higher *P/A* but lower *A/I* efficiencies than do carnivores.

To many people, the low primary efficiencies characteristic of intact natural systems are puzzling in view of the relatively high efficiencies apparently obtained in human-designed and other mechanical systems. This has led many to consider ways of increasing nature's efficiency. Actually, the primary efficiencies of long-term, large-scale ecosystems are not directly comparable to those of short-term mechanical systems. For one thing, much fuel is used for repair and maintenance of mechanical systems, and depreciation and repair typically are not included in calculating the fuel efficiencies of engines. In other words, much energy (human or otherwise) other than the fuel consumed in its operation is required to build a machine and keep it running, repaired, and replaced. Mechanical engines and biological systems cannot fairly be compared unless *all* energy costs and subsidies are considered, because biological systems are *self-repairing* and *self-perpetuating*. Moreover, under certain conditions, more rapid growth per unit time probably has greater survival value than maximum efficiency of energy use. By a simple analogy, it might be better to reach a destination quickly at 65 mph than to achieve maximum efficiency in fuel consumption by driving slowly. Engineers should understand that any increase in the efficiency of a biological system will be obtained at the expense of maintenance; thus, a gain from increasing the efficiency will be lost in increased cost—not to mention the danger of increased disorder that may result from stressing the system. As already noted, such a point of diminishing returns may have been reached in industrialized agriculture, because the energy output–energy input ratio has declined as the yield per unit area has increased.

The Interaction of Bottom-Up and Top-Down Control of Energy Flow in Food Webs

Hairston et al. (1960) proposed a "balance of nature" hypothesis that years ago caused much discussion and controversy among ecologists. They argued that because plants by and large accumulate a lot of biomass (the world is green), something must be inhibiting grazing. That something, they theorized, is predators. Accordingly, primary consumers are limited by secondary consumers, and primary producers are thus resource limited rather than grazer limited. Subsequently, much research, especially in aquatic systems, has resulted in "bottom-up" versus "top-down" perspectives in the understanding of food chain dynamics. The **bottom-up hypothesis** holds that production is regulated by upstream factors such as nutrient availability; the **top-down hypothesis** predicts that predators or grazers regulate productivity. There is evidence to support both hypotheses. For example, numerous studies have demonstrated that

Figure 3-23. Diagram depicting the various possible types of control mechanisms under the trophic cascade hypothesis.

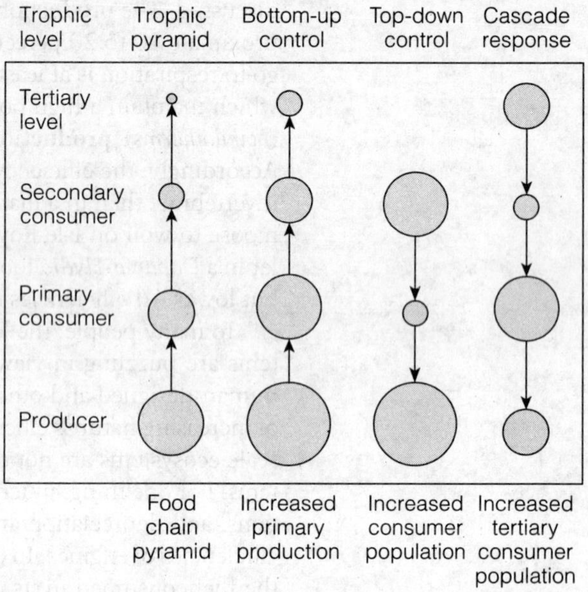

increased inputs of nitrogen fertilizer significantly increase rates of primary productivity in grassland and old-field communities (W. P. Carson and Barrett 1988; Brewer et al. 1994; Wedin and Tilman 1996; Tilman et al. 1996). It should be noted, however, that productivity is related to biodiversity (Tilman and Downing 1994). E. P. Odum (1998b) pointed out that in low-nutrient environments, an increase in biodiversity tends to enhance productivity, but in high-nutrient environments, an increase in productivity increases dominance and reduces diversity.

S. R. Carpenter et al. (1985) proposed that whereas nutrient inputs (**bottom-up controls**) determine the rate of production in a lake, piscivorous and planktivorous fish can cause significant deviations in the rate of primary production. The influences of downstream consumers on ecosystem dynamics are known as **top-down controls.** S. R. Carpenter and Kitchell (1988) proposed that the influence of consumers on primary production propagates through food webs or trophic levels. They termed these effects on trophic levels or ecosystem dynamics *trophic cascades.* The **trophic cascade hypothesis** proposes that feeding by piscivorous and planktivorous consumers affects the rate of primary production in lakes (a top-down influence).

For example, reducing the planktivorous fish population resulted in lower rates of primary production. In the absence of planktivorous minnows, the predaceous invertebrate *Chaoborus* became abundant. Because *Chaoborus* feed on the smaller herbivorous zooplankton, the herbivorous zooplankton shifted in dominance from small to large species. In the presence of large and abundant herbivorous zooplankton, the phytoplankton biomass and the rate of primary production declined. This example illustrates a cascading effect through several trophic levels in an aquatic ecosystem.

It appears that terrestrial consumers can also have important top-down influences on primary productivity (see McNaughton 1985; McNaughton et al. 1997 regarding the effects of large-mammal grazing on primary production and nutrient cycling in the Serengeti grassland ecosystem). Figure 3-23 illustrates both bottom-up and top-down control mechanisms (that is, trophic cascades linking all trophic levels in biotic communities). For more information on bottom-up and top-down control mechanisms, see Hunter and Price (1992); Harrison and Cappuccino (1995); and Price (2003).

There are, of course, mechanisms other than consumers and resources that determine how primary production is used. Allelopathic chemicals produced by plants, such as cellulose, tannins, and lignin, inhibit heterotrophic consumption (see previous sections of this chapter). These mechanisms are all part of the evolutionary arms race between producers and consumers that keeps nature in a pulsing state.

Isotopic Tracers as Aids to the Study of Food Chains

Observation and examination of stomach contents have been the traditional means of determining what food resources heterotrophs consume, but these methods are often not feasible, especially for small or reclusive animals and decomposers (bacteria and fungi). In some cases, isotopic tracers can be used to track food webs in natural ecosystems where many species are interacting. Radioactive tracers have proved useful in determining which insects are feeding on which plants or which predators are feeding on which prey. For example, the use of radioactive phosphorus (^{32}P) to isolate food chains in old-field ecosystems has revealed that food chains in some plants begin with ants attracted by sugars exuded from leaves and stems, whereas other plants support more traditional herbivore-predator food chains. These experiments illustrate two of the three ways in which plants survive herbivory, namely (1) feeding an army of ants for protection; (2) depending on large predators to keep herbivores under control; or (3) producing anti-herbivore chemicals.

With the development of improved detection instrumentation, stable (rather than radioactive) isotopes have come into increased use. The ratios of stable carbon isotopes have proved especially useful in charting material flows in food chains that are otherwise difficult to study. C_3 plants, C_4 plants, and algae have different $^{13}C/^{12}C$ ratios, which are carried over to whatever organisms (animal or microbe) consume that particular plant or plant detritus. For example, in a study of estuarine food chains, Haines and Montague (1979) found that oysters fed largely on phytoplanktonic algae, whereas fiddler crabs used both benthic algae and detritus derived from marsh grass (a C_4 plant) for food.

5 Energy Quality: eMergy

Statement

Energy has quality as well as quantity. Not all calories (or whatever unit of energy quantity is employed) are equal, because equal quantities of different forms of energy vary widely in work potential. Highly concentrated forms of energy, such as fossil fuels, have a much higher quality than do more dispersed forms of energy, such as sunlight. We can express *energy quality* or *concentration* in terms of the amount of one type of energy (such as sunlight) required to develop the same amount of another type (such as oil). The term **eMergy** (spelled with a capital M) has been proposed for

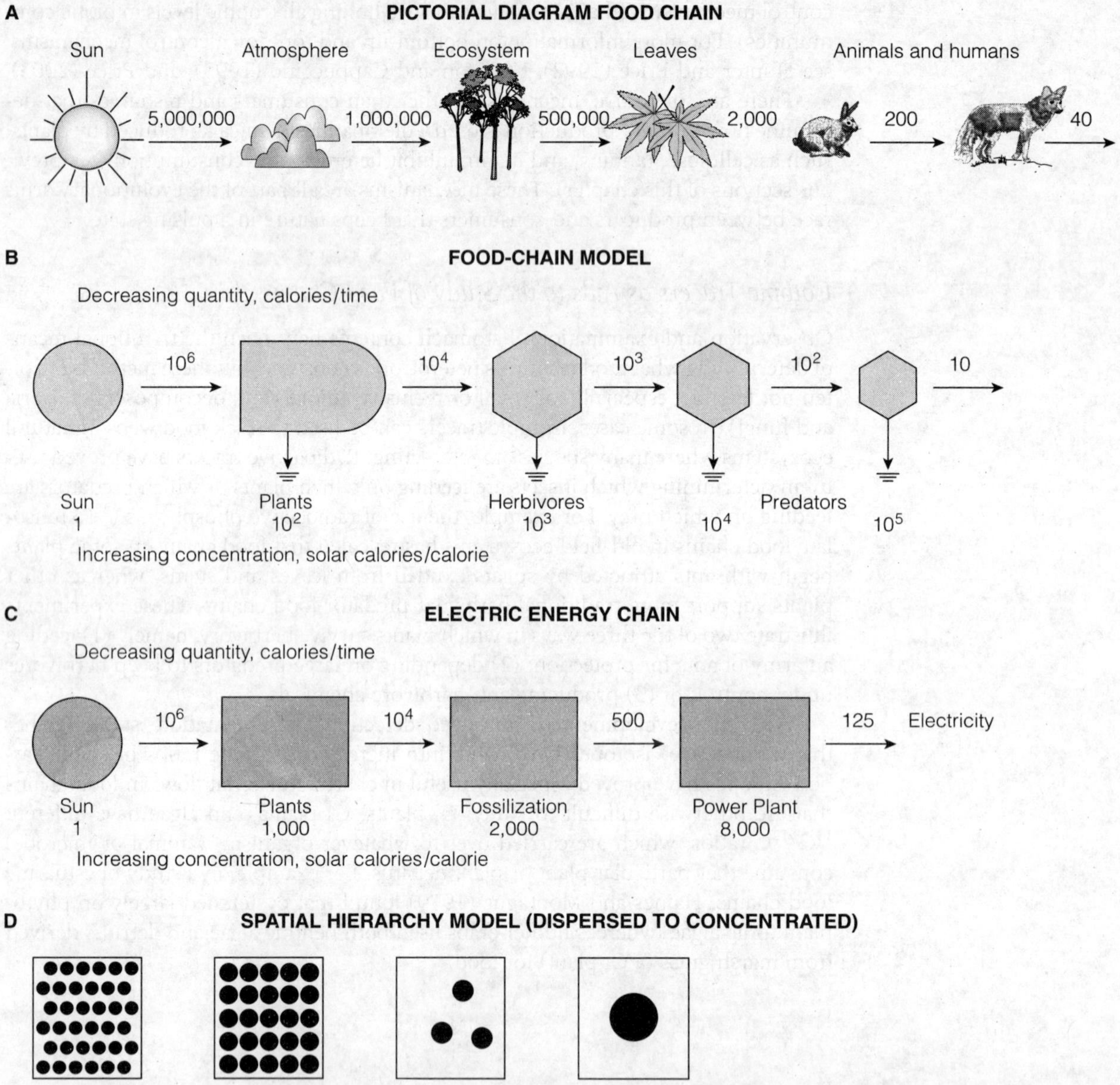

Figure 3-24. Diagram showing how increasing energy concentration (quality) accompanies decreasing energy quantity in (A) food chains, (B) energy flows, (C) electric energy generation, and (D) spatial energy concentration. Energy concentration from input to output involves up to 5 orders of magnitude (10^5). Data are in kcal/m^2 (after H. T. Odum 1983, 1996).

this measure; eMergy can be defined in a general way as the sum of the available energy already used directly or indirectly to create a service or product. In comparing energy sources for direct use by humankind, one should consider the quality as well as the quantity of energy available and, wherever possible, match the quality of the source with the quality of the use (Fig. 3-24).

Explanation and Examples

We have already described in this chapter how quality increases as quantity decreases in food chains and other energy transfer sequences (Fig. 3-15). One reason why most people seem unaware of the importance of the energy concentration or quality factor is that although there are numerous terms for energy quantity (such as calories, joules, and watts), there are no terms for energy quality in general use. In 1971, H. T. Odum proposed the term **embodied energy** as a quality measurement (and renamed it **eMergy** in 1996), defined as all the available energy already used directly or indirectly to create a service or product (H. T. Odum 1996). Thus, if 1000 calories of sunlight are required to produce 1 calorie of food by plants, the transformation (or **transformity**) is 1000 Sun calories to 1 food calorie, and the eMergy of the food is 1000 solar energy calories. EMergy can be thought of as "energy memory," as it is calculated by adding up all the energies transformed to produce the final product or service. For comparative purposes, all contributing energies should be of the same kind and, of course, expressed in the same quantitative units. From another viewpoint, energy quality is measured by the *thermodynamic distance* from the Sun. Whether the upgraded component (food, for example) is available to a consumer depends on the resource quality.

Some eMergy values in round figures are shown in Table 3-10, in terms of both solar and fossil fuel units. As Table 3-10 shows, fossil fuels have a work potential at least 2000 times that of sunlight. For solar power to do the work now being done by coal or oil, it must be upgraded several thousand times. In other words, *sunlight would not run an automobile or a refrigerator unless it is concentrated to the level of gasoline or electricity.* Societies cannot shift from fossil fuels to sunlight as their major energy source unless the dispersed solar energy can be upgraded on a large scale equivalent to electricity or some other kind of concentrated fuel—a conversion that is sure to be expensive.

Solar energy can be used directly, without upgrading, for low-quality jobs such as heating dwellings. Matching the quality of source and use would reduce the current waste of fossil fuels and give societies more time to convert to other possible concentrated energy sources. In other words, oil should be reserved for running ma-

Table 3-10 **Energy quality factors (eMergy): Solar versus fossil fuel units**

Type of energy	*Solar equivalent calories*	*Fossil fuel–equivalent calories*
Sunlight	1.0	0.0005
Plant gross production	100	0.05
Plant net production as wood	1000	0.5
Fossil fuel (delivered for use)	2000	1.0
Energy in elevated water	6000	3
Electricity	8000	4

Source: After H. T. Odum and E. C. Odum 1982.

chinery, not burned in a furnace to heat a house when the Sun can do at least part of that job.

EMergy is an especially useful measure for comparing and interfacing the value of market goods and services with natural (nonmarket) goods and services. See *Environmental Accounting: EMergy and Environmental Decision Making* by H. T. Odum (1996) for a detailed review of the eMergy concept.

6 Metabolism and Size of Individuals: The 3/4 Power Principle

Statement

The standing crop biomass (expressed as the total dry weight or total caloric content of organisms present at any one time) that can be supported by a steady flow of energy in a food chain depends not just on its position in the food web, but also on the size of the individual organisms. Thus, a lower biomass of smaller organisms can be supported at a particular trophic level in the ecosystem. Conversely, the larger the organism, the larger the standing crop biomass. Thus, the biomass of bacteria present at any one time would be very much smaller than the standing crop of fish or mammals, even though the energy use might be the same for both groups. In general, the rate of metabolism in individual animals varies as the 3/4 power of their body weight.

Explanation and Examples

The rate of metabolism per gram of biomass in very small organisms, such as algae, bacteria, and protozoa, is immensely higher than the metabolic rate per unit weight of large organisms, such as trees and vertebrates. In many cases, the metabolically important parts of the community are not the few large, conspicuous organisms, but the many small organisms, including microorganisms that are invisible to the naked eye. Thus, the tiny algae (phytoplankton) weighing only a few kilograms per hectare at any one moment in a lake can have as high a metabolism as a much larger volume of trees in a hectare of forest. Likewise, a few kilograms of small crustaceans (zooplankton) grazing on algae can have a total respiration equal to that of many hundreds of kilograms of buffalo grazing on grasses.

The rate of metabolism of organisms, or groups of organisms, is often estimated by measuring the rate at which oxygen is consumed (or, in the case of photosynthesis, produced). *The metabolic rate of an individual animal tends to increase as the 3/4 power of its weight.* A similar relationship appears to exist in plants, although the structural differences between plants and animals make direct comparisons difficult. The relationships between body weight (or volume) and respiration per individual and per unit weight are shown in Figure 3-26. The lower curve (Fig. 3-26B) is important because it shows how the weight-specific metabolic rate increases as the size of the individual decreases. Various theories about this trend, often called the **3/4 power law,** have focused on diffusion processes—larger organisms have less surface

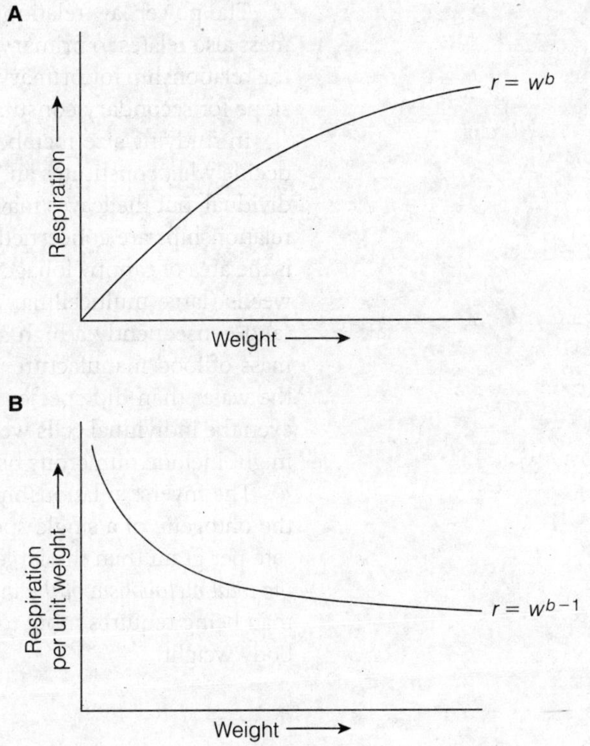

Figure 3-25. Relationships between (A) respiration and body weight per individual and (B) respiration per unit weight and total body mass.

area per unit weight through which diffusion processes might occur. Comparisons should, of course, be made at similar temperatures, because metabolic rates are usually greater at higher temperatures than at lower temperatures (except with temperature adaptation).

West et al. (1999) reviewed allometric scaling relationships in both plants and animals and presented a general model as follows:

$$Y = Y_o M^b,$$

where Y is the rate of metabolism, Y_o is a constant that is characteristic of the kind of organism, M is the mass, and b is the scaling exponent. The scaling exponents often turn out to be multiples of 1/4.

When organisms of the same general size are compared, the relationships shown in Figure 3-25 may not always hold. This is to be expected, because many factors other than size affect the rate of metabolism. For example, warm-blooded vertebrates have a higher rate of respiration than do cold-blooded vertebrates of the same size. However, the difference is actually relatively small compared with the difference between a vertebrate and a bacterium. Thus, given the same amount of available food energy, the standing crop of cold-blooded herbivorous fish in a pond may be of the same order of magnitude as that of warm-blooded herbivorous mammals on land. However, as mentioned in Chapter 2, oxygen is less available in water than it is in air and is, therefore, more likely to be limiting in water. In general, aquatic animals seem to have a lower weight-specific respiratory rate than do terrestrial animals of the same size. Such an adaptation may well affect the trophic structure.

The power law relationship between mammalian population densities and body mass also relates to primary and secondary consumers (Marquet 2000). For example, the relationship for primary consumers (herbivores) has a slope of −3/4, whereas the slope for secondary consumers (carnivores) is much steeper.

In studying size-metabolism relationships in plants, one often finds it difficult to decide what constitutes an "individual." Thus, a large tree can be regarded as one individual, but the leaves may act as "functional individuals" as far as size–surface area relationships are concerned. This relationship is similar to the *leaf-area index,* which is the area of canopy foliage per unit of ground area. A study of various species of seaweeds (large multicellular algae) found that species with thin or narrow branches (and consequently a high surface-to-volume ratio) had a higher rate per gram biomass of food manufacture, respiration, and uptake of radioactive phosphorus from the water than did species with thick branches. Thus, in this case, the branches or even the individual cells were functional individuals, and not the whole plant, which might include numerous branches attached to the substrate by a single holdfast.

The inverse relationship between size and metabolism may also be observed in the ontogeny of a single species. Eggs, for example, show a much higher metabolic rate per gram than the larger adults. Remember, *the weight-specific metabolic rate, not the total metabolism of the individual, decreases with increasing size.* Thus, an adult human being requires more total food than a small child, but less food per kilogram of body weight.

7 Complexity Theory, Energetics of Scale, and the Law of Diminishing Returns

Statement

As the size and complexity of a system increase, the energy cost of maintenance tends to increase proportionally at a higher rate. A doubling in size usually requires more than a doubling in the amount of energy that must be diverted to pump out the increased entropy associated with maintaining the increased structural and functional complexity. There are **increasing returns to scale** or **economies of scale** (which most economists like to talk about) associated with increases in size and complexity, such as increased quality and stability in the face of disturbances; however, there are also **diminishing returns to scale** or **diseconomies of scale** (which most economists do not like to talk about) involved in the increased cost of pumping out the disorder. These diminishing returns are inherent in large and complex systems and can be somewhat reduced by improved design that increases the efficiency of energy transformation. However, they cannot be entirely mitigated. The **law of diminishing returns** applies to all kinds of systems, including electrical power grids in human technoecosystems. As an ecosystem becomes larger and more complex, the proportion of gross primary production that must be respired by the community to sustain itself increases, and the proportion that can go into further growth in size declines. When these inputs and outputs balance, size cannot further increase without overshooting maintenance capacity, resulting in a pulsing "boom and bust" sequence.

Explanation

Engineering experience in dealing with physical networks such as telephone switchboards indicates that as the number of subscribers or calls, C, goes up, the number of switches needed, N, goes up, approaching a square of the number, as follows:

$$C = N\left(\frac{N-1}{2}\right)$$

In 1950, C. E. Shannon of the Bell Telephone Laboratory proved that such a *diseconomy of scale* is an intrinsic feature of networks, and no method of construction, however ingenious, can avoid it. The best that has been achieved in switching networks is a reduction of the diseconomy to something like N to the 1.5 power. See Shannon and Weaver (1949), Pippenger (1978), and Patten and Jørgensen (1995) for a background review of complexity theory.

This sort of diseconomy of scale is also an intrinsic feature of natural ecosystems, but at least some of the increased cost of complexity is balanced by benefits of the kind that economists refer to as *economies of scale*. The metabolism per unit weight decreases as the size of the organism or the biomass (as in a forest) increases, so more structure can be maintained per unit of energy flow (the *B/P efficiency*). Adding functional circuits and feedback loops can also increase the efficiency of energy use, the recycling of materials, and the resistance or resilience to disturbance. EMergy properties and mutualistic relationships between organisms may also develop increased overall efficiency. No matter what the adjustment may be, however, the total entropy that must be dissipated increases rapidly with any increase in size, so that more and more of the total energy flow (gross primary production plus subsidies) must be diverted to respiratory maintenance and less and less is available for new growth. *When maintenance energy costs balance the available energy, the theoretical maximum size or carrying capacity has been reached, beyond which decreasing returns of scale set in.*

Example

Per capita taxes provide an example of the network law of costs rising as a power function of size. Per capita state and local taxes are typically strongly correlated with population density, and especially with the percentage of urbanization within the state. For example, it costs a person about three times more in taxes to live in the state of New York than to live in the state of Mississippi. Citizen "tax revolts" notwithstanding, one cannot avoid higher taxes if one chooses to live in a large city and does not wish to see it become disorderly. For several decades, the difference in per capita taxes between urban and rural areas has increased as urban sprawl has intensified. See Barrett et al. (1999) for a discussion focusing on the need to reconnect rural and urban landscapes.

8 Concepts of Carrying Capacity and Sustainability

Statement

In terms of energetics at the ecosystem level, what is known as the **carrying capacity** is reached when all the available incoming energy is needed to sustain all the basic structures and functions—that is, when P (*production*) equals R (*respiratory maintenance*). The amount of biomass that can be supported under these conditions is known as the **maximum carrying capacity** and is indicated by the capital letter K in theoretical models. This level is not absolute (not a *glass ceiling*), but it is easily overshot when the momentum of growth is strong. Increasing evidence shows that the **optimum carrying capacity** (what is sustainable over long periods in the face of environmental uncertainties) is lower than the maximum carrying capacity (see Barrett and Odum 2000 for details). In terms of individuals and populations, carrying capacity depends not only on number and biomass but also on lifestyle (that is, per capita energy consumption).

Explanation

Simple mathematical models of sigmoid growth, as graphed in Figure 3-27, are discussed in more detail in Chapter 6. For now, two points on the growth curve need to be noted: K, the *upper asymptote,* represents the maximum carrying capacity as defined in the statement, and I, the *point of inflection* where the rate of growth is highest, as shown by the lower diagram (Fig. 3-27B). The I level is often spoken of as the **maximum sustained yield** or **optimum density** by game and fish managers because, theoretically, the harvested biomass would be most rapidly replaced at this point.

The problem with maintaining the maximum level, K, in the fluctuating environment of the world is that overshoots are likely to occur, either because the momentum of growth causes the population size to exceed K or because a periodic reduction in resource availability (such as a drought) temporarily reduces K. When an overshoot occurs and entropy exceeds the capacity of the system to dissipate it, a reduction in

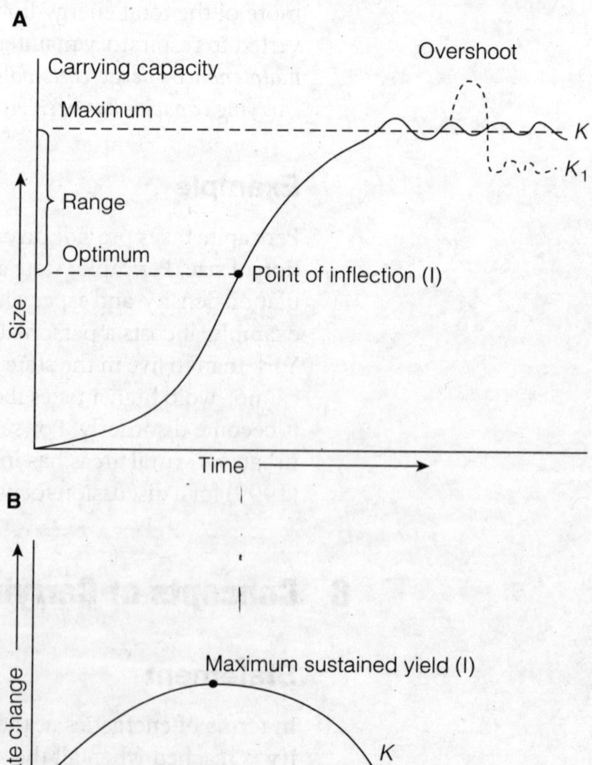

Figure 3-26. Carrying capacity in relation to sigmoid population growth. (A) Growth curve. (B) Change in growth rate depending on population size. K represents the maximum density that can be supported with a given space and resource base. If density overshoots this level, K may be lowered to K_1 at least temporarily. The inflection point (I) represents the population level of highest growth rate and is the theoretical optimum in terms of maximum sustainable yield for a game or fish population.

size or a "crash" must occur. If the productive capacity of the environment is damaged in the crash, *K* itself may be lowered, at least temporarily, to a new level (K_1 in Fig. 3-26A). The global challenge of feeding people is approaching that point where food needs equal maximum production capacity given current technological, political, economic, and distributional constraints. Any widespread perturbation, such as war, drought, disease, or terrorism, that reduces crop yields for even one year means severe malnutrition or starvation for millions of individual humans living on the brink.

The margin of safety at the maximum carrying capacity level is very small. From the viewpoint of long-term safety and stability, some level in the range between *K* and *I* (as shown in Fig. 3-26A) would represent the desirable carrying capacity level that is sustainable over the long term. Many natural populations have evolved mecha-nisms that maintain density at this safe or secure level.

Examples

Concept of Carrying Capacity

A classic study of carrying capacity in the animal world involved a deer herd in Michigan (McCullough 1979). Six deer introduced into a 2-square-mile (5 km^2 or 500 hectare) fenced enclosure in 1928 increased to about 220 by the mid-1930s. When it became evident that this herd was damaging its environment by overbrowsing the vegetation, the population was reduced to about 115 by selective hunting and then maintained at this level for a number of years. McCullough suggested that the ±200 level (2 hectares per deer) represented the *maximum carrying capacity* level, *K*, and that deer populations tend to "track" this maximum level. Left to themselves, the deer will increase right up to this limit of food or other vital resources. The ±100 number (about 4 hectares per deer), accordingly, represents the *carrying capacity of optimum density, I* (see Fig. 3-26), which avoids overshoots, starvation stress, disease, and possible damage to the habitat. In this particular species, predation that keeps the population below the maximum carrying capacity seems to be a function that favors quality over quantity. Other kinds of populations have evolved self-regulatory mechanisms that tend to maintain a carrying capacity level well below the maximum.

A study of energy flow in ant colonies might reveal something about the energetics of scale and carrying capacity that is applicable to *Homo sapiens*. Leaf-cutter ants (*Atta colombica*), which live in wet tropical forests, harvest fresh leaf sections from the vegetation and carry them into underground nests as a substrate for culturing the fungi that provide their food. These fungal gardens are cared for and fertilized (partly by ant excretions) much as a human farmer cultivates his or her food crop. Lugo et al. (1973) estimated the energy expenditures for all the major activities within the colony and concluded that the maximum carrying capacity (the maximum size of the colony) is reached when the input of fuel calories (in the form of harvested leaves) balances the energy cost of the work involved in cutting and transporting leaves, maintaining trails, and cultivating the crops. At any one time in large colonies, approximately 25 percent of the ants were carrying leaves, and 75 percent were maintaining the trails and fungal gardens. When the energy input is balanced by these maintenance costs, the colony stops growing. Reward feedback to other organisms, such as frass deposited by the ants on the forest floor that increases leaf growth, increases efficiency and raises the maximum carrying capacity, *K*.

Estimating the carrying capacity of an agrarian civilization supported by subsistence agriculture is not too difficult, as the input energy comes mostly from local resources and not from distant regions. For example, Mitchell (1979) reported that density in the rural countryside in India is a linear function of rainfall, which determines crop yield in the absence of irrigation or other subsidy. He reported that 10 cm of rainfall supports 2 persons per hectare of harvested land, 100 cm supports 3 persons, 200 cm supports 4.5 persons, and 300 cm supports 6 persons. Another interesting study of agrarian carrying capacity is that of Pollard and Gorenstein (1980), who documented the relation between maize production and human density in an early Mexican (Tarascan) civilization.

Estimating the carrying capacity for urban-industrial societies is much more difficult, because such societies are supported by massive energy subsidies imported from afar and often drawn from storage accumulated before the advent of humankind, such as fossil fuels, underground water, virgin timber, and deep organic soils. All of these resources diminish with intensive use. One thing is certain; humans, like deer, seem to track maximum carrying capacity levels, or *K*, on a landscape or regional scale (*Homo sapiens* is one of the few species that has yet to reach carrying capacity conditions on a global scale). Our population tends to increase right up to or beyond one limit after another (food, fossil fuels, and diseases being the limits of concern at the moment). Reward feedback, or other means of maintaining the optimum rather than the maximum carrying capacity levels, are only weakly developed for several reasons: (1) many people living in developed countries believe that science and technology will continue to find substitutes for declining resources and continue to raise *K* indefinitely; (2) many families living in developing countries often have a strong economic or social reason regarding family size; and (3) traditional belief systems frequently tend to dominate ecological understanding. Thus, the dangerous game of flirting with overshoots continues. There are good ecological reasons to regulate the human population, but the involvement of complex social, economic, and religious issues make such regulation difficult.

The carrying capacity (or *threshold*) concept can also be applied to economics. For example, Max-Neef (1995) compared gross national product (GNP) trends with the Daly and Cobb (1989) *index of sustainable economic welfare* (ISEW). His findings suggested the existence of an optimum economic carrying capacity. Figure 3-28 illustrates trends in the GNP and the ISEW in the United States. The two indexes tracked one another until the mid-1970s and then separated at a time known as the **economic welfare threshold** or **economic carrying capacity** (Barrett and Odum 2000). It seems that during the 1970s, economic growth began to outpace the long-term economic welfare (*quality of life*) trend. This trend suggests that economic growth has already increased well beyond the optimum economic growth carrying capacity for the United States.

Thus, *per capita energy use or lifestyle is as important as numbers in determining human carrying capacity.* For example, a person living in the United States consumes 40 times more high-quality energy than does a person living in India. In other words, 40 times more people with an Indian lifestyle than with an American one can be supported by a given resource base. For more on these two aspects of carrying capacity, see Catton (1980, 1987).

Concept of Sustainability

The concept of sustainability is directly related to the concept of carrying capacity. Dictionary definitions of *to sustain* include "to hold," "to keep in existence," "to support," "to maintain," or "to supply with sustenance or nourishment" (*Merriam-*

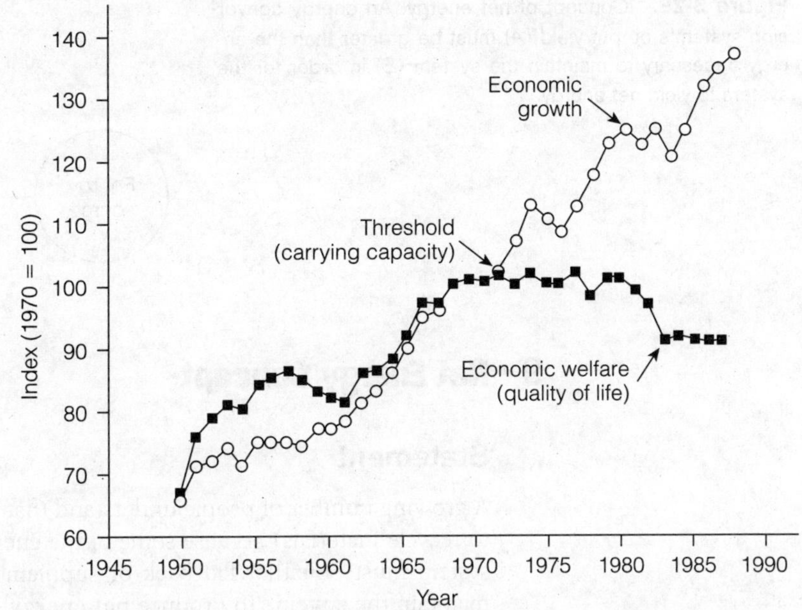

Figure 3-27. Trends in the gross national product (GNP) index and the index of sustainable economic welfare (ISEW) in the United States, with the threshold suggested as the optimum carrying capacity. Open circles = GNP index; solid boxes = ISEW index (modified from Max-Neef 1995; Barrett and Odum 2000).

Webster's Collegiate Dictionary, 10th edition, s.v. "sustain"). In terms of the environment, Goodland (1995) defined **sustainability** as *maintaining natural capital and resources.* The term sustainability is increasingly used as a guide for future development, in view of the fact that much of what humans are now doing in the area of consumption and environmental management is obviously unsustainable. The difficulties of using the concept of *sustainable development* as a goal or policy guide will be discussed in Chapter 11. Barrett and Odum (2000) have suggested that in human affairs in the long term, sustainability may be more effectively understood and dealt with in terms of the *optimum carrying capacity* concept outlined earlier in this section. New strategies and technologies, such as the use of windmills to generate electricity (Fig. 3-29), illustrate how natural capital (in this case, wind) can be used to generate economic capital (electrical energy).

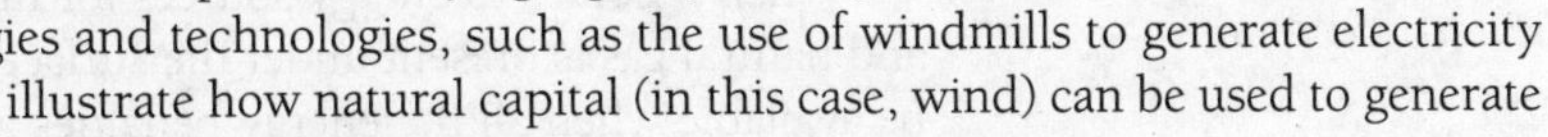

Figure 3-28. Energy-generating windmills in eastern Germany.

Courtesy of Gary W. Barrett

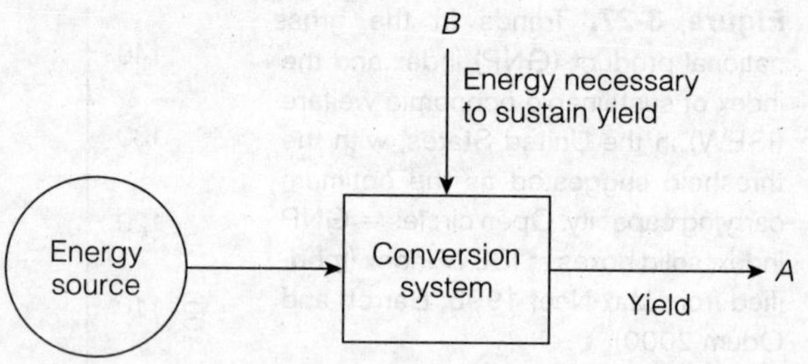

Figure 3-29. Concept of net energy. An energy conversion system's output yield (*A*) must be greater than the energy necessary to maintain the system (*B*) in order for the system to yield net energy.

9 Net Energy Concept

Statement

A growing number of people understand that it takes energy to produce energy (and to recycle materials) because some of the energy produced by any given conversion system must be either fed back or supplemented by additional outside energy to maintain the system. To produce **net energy**, the yield must be greater than the energy cost of sustaining the conversion system.

Explanation and Examples

The concept of net energy is diagrammed in Figure 3-30. To have net energy, the *yield, A,* must be greater than the *energy cost, B,* of maintaining the conversion system. For a power plant to be really worthwhile; for instance, the net energy needs to be at least two times, preferably four times, the energy cost or *energy penalty,* as engineers have termed it.

When it comes to energy sources for humans, the question is not how much oil and natural gas is present under the surface of the land and seas, but how much will be available when all the energy penalties associated with entropy dissipation (such as drilling, protection of human health, and pollution prevention) have been paid. In 1998, when gasoline became very inexpensive in the United States, for example, many offshore drilling rigs, which are expensive to maintain, shut down.

To take another example, in the United States, current uranium-fission power plants are so costly to build and maintain that their net energy is marginal, and various government subsidies (tax dollars)—for example, to pay for waste containment—are necessary to keep the plants running. To date, experiments in nuclear fusion have failed to produce any net energy. The extremely high temperatures and pressures required for fusion to occur make it extremely difficult to "tame" the H-bomb.

10 An Energy-Based Classification of Ecosystems

Statement

The source and quality of available energy determine to a greater or lesser degree the kinds and numbers of organisms, the pattern of functional and developmental processes, and, directly or indirectly, the lifestyle of human beings. As energy is the common denominator and the ultimate forcing function of all ecosystems, whether natural or designed by humans, it provides a logical basis for a "first-order" classification. On this basis, it is convenient to distinguish four classes of ecosystems:

- Natural, unsubsidized solar-powered ecosystems.
- Natural solar-powered ecosystems subsidized by other natural energies.
- Human-subsidized solar-powered ecosystems.
- Fuel-powered, urban-industrial technoecosystems (using energy from fossil fuels or other organic or nuclear fuels).

This classification is based on the input environment (see Fig. 2-1), and it contrasts with and complements the *biome* classification (see Chapter 10), which is based mainly on the major vegetative structure of the ecosystem.

Explanation

The four major types of ecosystems classified according to source, level, and quantity of energy are described in Table 3-11. Ecosystems rely on two dissimilar sources of energy: (1) the Sun; and (2) chemical or nuclear fuels. Accordingly, one can conveniently distinguish between **solar-powered** and **fuel-powered** systems, while recognizing that both energy sources can be used in any given situation. In comparing the major classes of ecosystems (Table 3-11), we can speak of the energy flow per unit area as the **power density** (recall that *power* is energy use or dissipation per unit time). This measure can also be considered to represent the amount of disorder or entropy that must be dissipated if the system is to remain viable.

Table 3-11 **Ecosystems classified according to source and level of energy**

Category	*Example*	*Annual energy flow (Power level) (kcal/m²)* *
1. Unsubsidized natural solar-powered ecosystems†	Open oceans, upland forests, grasslands	1000–10,000 (2000)
2. Naturally subsidized solar-powered ecosystems‡	Tidal estuaries, some rain forests	10,000–40,000 (20,000)
3. Human-subsidized solar-powered ecosystems§	Agriculture, aquaculture	10,000–40,000 (20,000)
4. Fuel-powered urban-industrial systems**	Cities, suburbs, industrial parks	100,000–3,000,000 (2,000,000)

*Numbers in parentheses are estimated round-figure averages.

†These systems constitute the basic life-support module (natural capital) for spaceship Earth.

‡These are the naturally productive systems of nature that not only have high life-support capacity but also produce excess organic matter that may be exported to other systems or stored.

§These are food- and fiber-producing systems supported by auxiliary fuel or other energy subsidies supplied by humans.

**These are our wealth-generating (also pollution-generating) systems, in which fuel replaces the Sun as the chief energy source. They are dependent (parasitic) on the other classes of ecosystems for life support, food, and fuel.

The systems of nature that depend largely or entirely on the direct rays of the Sun can be designated **unsubsidized solar-powered ecosystems** (Category 1 in Table 3-11). They are *unsubsidized* in the sense that there are few, if any, available auxiliary energy sources to enhance or supplement solar radiation. Humans should be encouraged to protect and encompass the benefits of Category 1 in future decision-making models. The open oceans, great tracts of upland forests and grasslands, and large, deep lakes are examples of relatively unsubsidized solar-powered ecosystems. Frequently, they are subjected to other limitations as well, such as a shortage of nutrients or water. Consequently, ecosystems in this broad category vary widely, but they are generally low powered and have a low productivity or capacity to do work. Organisms that populate such systems have evolved remarkable adaptations for living on scarce energy and other resources and use these resources efficiently.

Although the power density of natural ecosystems in this first category is not very impressive, and such ecosystems themselves could not support a high density of people, they are nonetheless extremely important because of their huge extent (the oceans alone cover almost 70 percent of the planet). For humans, the aggregate of unsubsidized solar-powered natural ecosystems can be thought of—and certainly should be highly valued—as the basic life-support module that stabilizes and homeorhetically controls spaceship Earth. In these systems, large volumes of air are purified daily, water is recycled, climates are controlled, weather is moderated, and much other useful work is accomplished. These processes and services are termed **natural capital.** A portion of the food and fiber needs of humans are also produced as a by-product without economic cost or management effort by humans. This evaluation, of course, does not include the priceless aesthetic values inherent in a sweeping view of an ocean, the grandeur of an unmanaged forest, or the cultural desirability of green, open spaces (see Daily 1997; Daily et al. 1997 for an outline of the benefits supplied to human societies by natural ecosystems).

When auxiliary sources of energy can be used to augment solar radiation, the power density can be raised considerably, perhaps by an order of magnitude. Recall that an *energy subsidy* is an auxiliary energy source that reduces the unit cost of self-maintenance of the ecosystem and thereby increases the amount of solar energy that can be converted to organic production. In other words, solar energy is augmented by nonsolar energy, freeing it for organic production. Such subsidies can be either natural or synthetic (or, of course, a combination). For the purpose of simplified classification, **naturally subsidized** and **human-subsidized solar-powered ecosystems** have been listed as Categories 2 and 3, respectively, in Table 3-11.

A coastal estuary is an example of a natural ecosystem subsidized by the auxiliary energy of tides, waves, and currents. Because the ebb and flow of water partly recycles mineral nutrients and transports food and wastes, the organisms in an estuary can concentrate their efforts, so to speak, on more efficient conversion of the energy of the Sun to organic matter. In a very real sense, organisms in the estuary are adapted to use tidal power. Consequently, estuaries tend to be more fertile than, say, an adjacent land area or pond that receives the same solar input but does not have the benefit of the tidal energy subsidy. Subsidies that enhance productivity can take many other forms—for example, wind and rain in a tropical rain forest, the flowing water of a stream, or imported organic matter and nutrients received by a small lake from its watershed.

Human beings learned early how to modify and subsidize nature for their direct benefit, and we have become increasingly skillful not only in raising productivity but more especially in channeling that productivity into food and fiber materials that are easily harvested, processed, and used. Agriculture (land-based culture) and aquaculture (water-based culture) are the prime examples of Category 3 (Table 3-11), the human-subsidized solar-powered ecosystems. High yields of food are maintained by large inputs of fossil fuels (and, in primitive agriculture, human and animal labor) into cultivation, irrigation, fertilization, genetic selection, and pest control. Thus, fuel for tractors (and animal or human labor) is just as much an energy input in agroecosystems as sunlight, and it can be measured as horsepower or calories expended, not only in the field but also in processing and transporting food to the supermarket.

In Table 3-11, the productivity or power levels of natural and human-subsidized solar-powered ecosystems are the same. This evaluation is based on the observation that the most productive natural ecosystems and the most productive agroecosystems are at about the same productivity level; 50,000 kcal $\cdot$ m^{-2} $\cdot$ year^{-1} seems to be about the upper limit for any plant photosynthetic system in terms of continuous, long-term function. The real difference between these two classes of systems is the distribution of the energy flow. People channel as much energy as possible into food they can use immediately, whereas nature tends to distribute the products of photosynthesis among many species and products and to store energy as a "hedge" against bad times, in what will be later discussed as a strategy of *diversification for survival.*

The **fuel-powered ecosystem** (Category 4 in Table 3-11), otherwise known as the urban-industrial system, is a crowning achievement of humanity. The highly concentrated potential energy of fuel replaces—rather than merely supplements—the energy of the Sun. As cities are now managed, solar energy is mostly unused and frequently becomes a costly nuisance by heating up the concrete and contributing to the generation of smog. Food—a product of solar-powered systems—is considered external, as it is largely imported from outside the city. As fuel becomes more expensive, cities will likely become more interested in using solar energy. Perhaps a new class of ecosystem, the Sun-subsidized, fuel-powered city, will increasingly become a new category during the twenty-first century. It may be wise to develop new technologies designed to concentrate solar energy to a level where it might partially replace fuel rather than merely supplement it (see Fig. 3-29 for an example).

11 Energy Futures

Statement

The history of civilization is very closely coupled with the availability of energy sources. Hunters and gatherers lived as part of natural food chains in solar-powered ecosystems, achieving their highest density in naturally subsidized systems at coastal and riverine sites. When agriculture developed about 10,000 years ago, carrying capacity was greatly increased as humans became more skillful at cultivating plants, domesticating animals, and subsidizing edible primary production. For many centuries, wood and other biomass provided the chief energy source. Architectural and agricultural accomplishments were achieved with biomass-fueled muscle power (physical labor, both animal and human). This period can be termed the *age of muscle power.* Next came the *age of fossil fuels,* which provided such a bountiful supply that the global population doubled every half century. Machines powered by oil and electricity have gradually replaced animal and human labor in the developed countries. Until recently, it seemed likely that as fossil fuels became exhausted, the third age of humankind would be the *age of atomic energy*. But pumping out the high disorder associated with this energy source to obtain net energy has so far proved troublesome, so the future of atomic energy is unpredictable. Other options for the future (the *age of technology*) include returning to solar energy and using hydrogen as a fuel.

Explanation

Civilization has progressed through the four ecosystem types outlined in Table 3-11. In the closing decades of the twentieth century and the first decade of the twenty-first century, the part of the world that consumes petroleum and other fossil fuels on a large scale has been operating as a fuel-powered system, whereas the part that by some has been termed the Third World remains essentially dependent on biomass (food and wood) and human labor as its major energy sources. As already noted, the difference in per capita income between high-energy and low-energy countries has created worrisome social, economic, and political conflicts. Despite worldwide efforts to close the gap, happenings such as the terrorist attack on the World Trade Center towers on 11 September 2001 have been increasing rather than decreasing.

At the first International Conference for Peaceful Uses of Atomic Energy, held in Geneva in 1955, the chair of the conference, the late Homi J. Bhabha of India, described three ages of humankind as the age of muscle power, the age of fossil fuels, and the atomic age. Bhabha spoke eloquently of his belief that because of the universal availability of the atom, the coming of the atomic age would close the gap between rich and poor nations. The dream of equal and abundant energy for all from the atom has yet to materialize, because tapping the enormous potential of atomic energy has proved to have a far greater disorder potential than anticipated in 1955. Carroll Wilson, the first general manager of the U.S. Atomic Energy Commission, in an article (1979) entitled "What Went Wrong?" wrote,

> No one appeared to understand that if the whole system does not hang together coherently, none of it might be acceptable.

Until the entire cycle from raw material to waste disposal becomes "coherent," and new, more effective ways to tap energy from nuclear sources are devised, the coming of the atomic age is at least postponed.

In the meantime, other options being considered in the twenty-first century include a return to solar power, coupled with more efficient (less wasteful) use of the remaining fossil fuels to prolong their availability for as many years as possible. As pointed out in the discussion on energy quality (Section 5 of this chapter), concentrating this dispersed but abundant and renewable energy source may be costly and will require new and improved existing technology (such as photovoltaics). Locally, indirect solar energy derived from wind, water flow, and temperature differences in the tropical sea is already being tapped (see Fig. 3-29). The last is especially promising as a renewable source with the use of a technology known as **ocean thermal energy conversion** (OTEC), which taps the temperature differential between warm surface water and deep cold water to drive Rankin-cycle engines and generate electricity. Such power plants using, as it were, stored solar energy can be located on barges anchored in equatorial waters (see Avery and Wu 1994). Geothermal power is another potential renewable source in localities where the heat of the inner Earth is near the surface.

Finally, the use of hydrogen as a substitute for gasoline or natural gas to run automobiles and other machinery is another option. Burning hydrogen instead of carbon-based fossil fuels would reduce the global warming threat, as CO_2, a greenhouse gas, would not be emitted. There is plenty of hydrogen in water, but breaking the bonds in H_2O to release hydrogen requires an abundance of high-quality energy; thus, the question of net energy comes up again. Electricity produced by OTEC or other direct and indirect solar sources could fill this need.

Then there are the cornucopian ("horn of plenty") optimists among us—especially Jesse Ausubel (1996) and his colleagues at Rockefeller University—who suggest that a combination of a hydrogen economy, wasteless industry, and land-reduced agriculture will not only support a large human population, but will make possible the preservation of large areas of the natural environment to provide air and water life support and homes for endangered species.

12 Energy and Money

Statement

Money, as an example of currency, is directly related to energy, because it takes energy to make money, and it costs money to buy energy. Money is a *counterflow* to energy, in that money flows out of cities and farms to pay for the energy and materials that flow in. The trouble with current economic practices is that money tracks *human-made* goods and services, but not the equally important goods and services provided by *natural* systems. At the ecosystem level, money enters the picture only when a natural resource is converted into marketable goods and services, leaving unpriced (and therefore unappreciated) all the work of the natural system that sustains this resource. It is vital that human market capital and natural capital be interfaced and environment quality be maintained if we are to avoid a global boom and bust as the natural capital is unnecessarily depleted to produce ever more market goods and services.

Explanation and Illustrations

Figure 3-31 shows an energy flow model of an estuary that is producing market goods in the form of seafood. Money enters the picture only after the harvest of shrimp and fish, leaving all the work of the estuary unpriced. In this example, the value of the work of the estuary in energy-converted dollars is calculated to be at least 10 times the value of the market goods and services extracted. Figure 3-32 depicts a typical energy support system for humans. Note that money counterflows accompany energy flows from human-made and domesticated ecosystems, but not from natural ecosystems.

Kenneth Boulding was one of the first economists to be elected to the prestigious National Academy of Sciences (NAS). In the 1960s and 1970s, he argued for the development of a more holistic economics that would close the gap between market

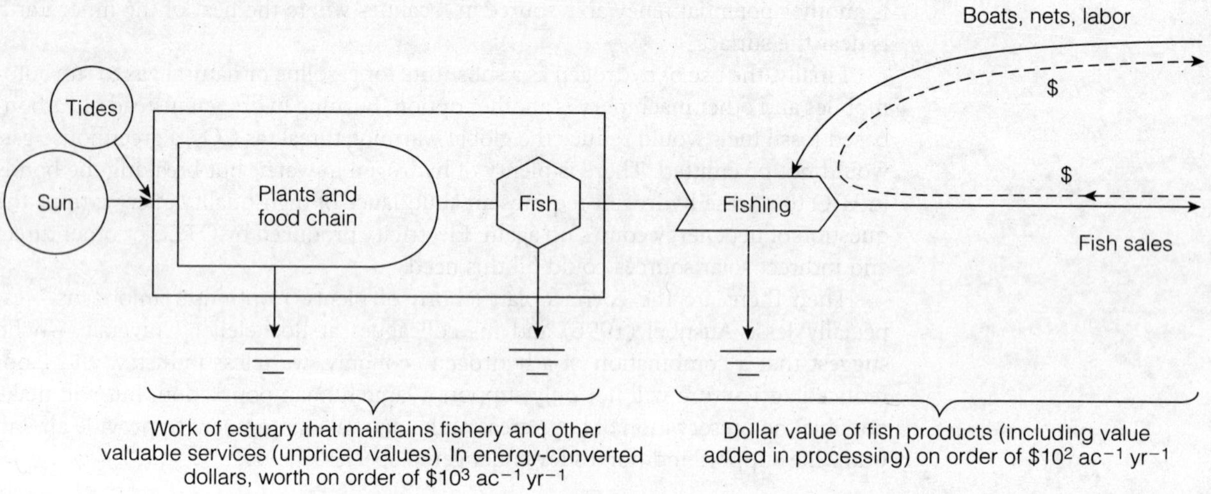

Figure 3-30. Energy flow model of an estuarine ecosystem harvested for seafood, including money flows ($). In conventional economics, money is seldom involved until the fish are caught; the work of the estuary (natural capital) to provide the fish is given no economic value. The total value of the estuary in terms of useful work for humans is at least ten times the value of the harvested product (Gosselink et al. 1974).

Figure 3-31. Conventional energy support system for humans. Money flows ($) accompany energy flows for human-made and domesticated ecosystems, but not for natural ecosystems (courtesy of H. T. Odum).

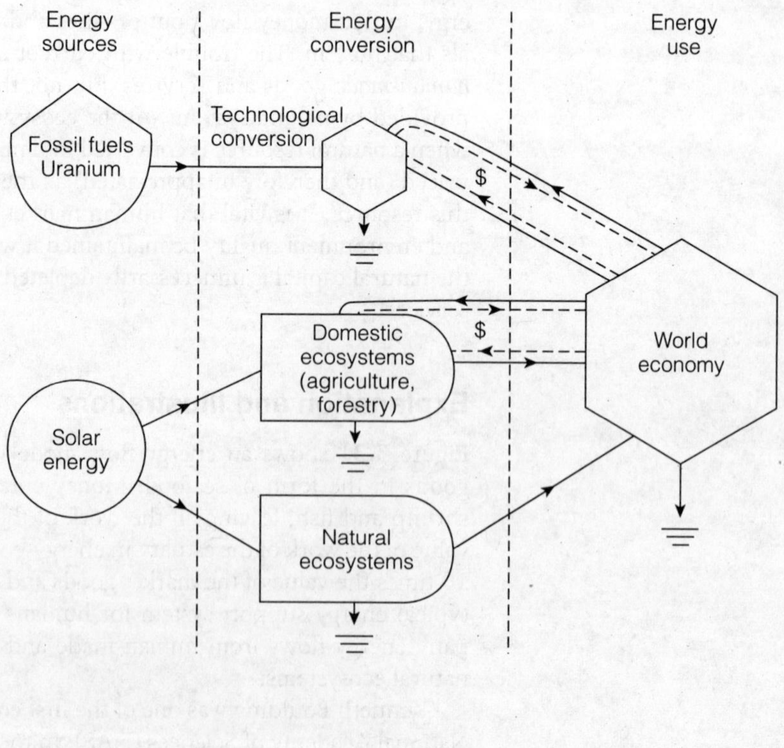

(priced) and nonmarket (unpriced) values. His numerous books and articles have provocative titles, such as *A Reconstruction of Economics* (1962), "The Economics of the Coming Spaceship Earth" (1966), and *Ecodynamics: A New Theory of Societal Evolution* (1978). His writings have been widely cited but have had little effect on the economic practices of his time. However, during the twenty-first century, a serious dialogue between economists and ecologists has begun to create the new interface field of **ecological economics,** with new societies and journals. Daly and Cobb (1989), Costanza (1991), Daly and Townsend (1993), H. T. Odum (1996), Prugh et al. (1995), Costanza, Cumberland, et al. (1997), and Barrett and Farina (2000) are among those leading this dialogue, which is finally getting the attention of global citizenry and political leadership. As is so often the case, timing is of the essence, and the time has come for reforms.

In conclusion, monetary currency is an important invention, and it now provides the basis for decision making at most levels of society. We must remember, however, that monetary systems do not presently take into account all the real costs of living. We must correct this "market failure" and take care that money is not allowed to be the only factor in the decision-making process. When it comes to the quality of human life, money and the consumption of human-made market products on which current economies are based are not the only considerations. An appreciation of natural capital and a commitment to ecosystem and human health deserve higher priorities in the never-ending pursuit of happiness and well-being.

4

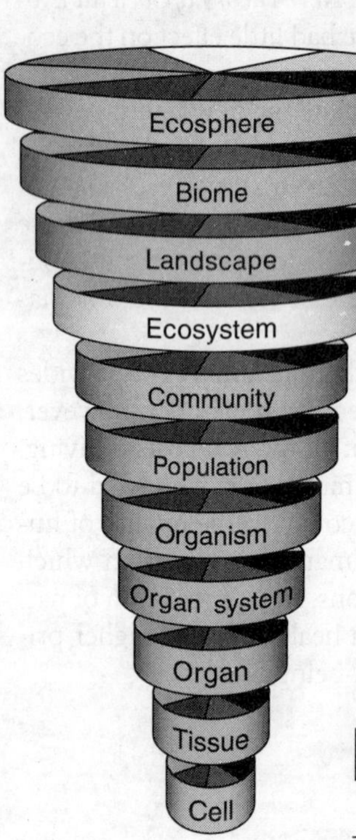

Biogeochemical Cycles

1 Basic Types of Biogeochemical Cycles
2 Cycling of Nitrogen
3 Cycling of Phosphorus
4 Cycling of Sulfur
5 Cycling of Carbon
6 The Hydrologic Cycle
7 Turnover and Residence Times
8 Watershed Biogeochemistry
9 Cycling of Nonessential Elements
10 Nutrient Cycling in the Tropics
11 Recycling Pathways: The Cycling Index
12 Global Climate Change

1 Basic Types of Biogeochemical Cycles

Statement

The chemical elements, including all the essential elements of life, tend to circulate in the biosphere in characteristic pathways from environment to organisms and back to the environment. These more or less circular pathways are known as **biogeochemical cycles.** The movement of these elements and inorganic compounds that are essential to life can be conveniently designated as **nutrient cycling.** Each nutrient cycle can also be conveniently divided into two compartments or *pools:* (1) the **reservoir pool,** the large, slow-moving, generally nonbiological component; and (2) the **labile** or **cycling pool,** a smaller but more active portion that is exchanging (moving back and forth) rapidly between organisms and their immediate environment. Many elements have multiple reservoir pools, and some (such as nitrogen) have multiple labile pools. From the viewpoint of the ecosphere as a whole, biogeochemical cycles fall into two basic groups: (1) **gaseous types,** in which the reservoir is in the atmosphere or the hydrosphere (ocean); and (2) **sedimentary types,** in which the reservoir is in the crust of Earth. The dissipation of energy in some form is always necessary to drive material cycles.

Explanation

As emphasized in Chapter 2, it is essential in ecology to study not only organisms and their environmental relations but also the basic nonliving environment in relation to organisms. We have seen how the two ecosystem divisions—the biotic and the abiotic—coevolve and influence each other's behavior. Of the elements occurring in nature, between 30 and 40 are known to be required by living organisms (*essential elements*). Some elements, such as carbon, hydrogen, oxygen, and nitrogen, are needed in large quantities; others are needed in small, even minute, quantities. Whatever the quantitative need may be, essential elements exhibit definite biogeochemical cycles. The nonessential elements (elements not required for life), though less closely coupled with organisms, also cycle, often flowing along with essential elements either in the water cycle or because of a chemical affinity with them.

Bio- refers to living organisms and *geo-* refers to Earth. **Geochemistry** is concerned with the chemical composition of Earth and with the exchange of elements between different parts of Earth's crust, its atmosphere, and its oceans, rivers, and other bodies of water. The concept of geochemistry is credited to the Russian Polynov (1937) and is defined as the role of chemical elements in the synthesis and decomposition of all kinds of materials, with special emphasis on weathering. **Biogeochemistry,** a science founded in 1926 by the Russian V. I. Vernadskij (1998) and made prominent by the early monographs of G. E. Hutchinson (1944, 1948, 1950), thus involves the study of the exchange of materials between living and nonliving components of the ecosphere. Fortescue (1980) reviewed geochemistry from an ecological and holistic perspective in terms of **landscape geochemistry.** Excerpts from key papers in the development of the field of biogeochemistry are presented by Butcher et al. (1992) and Schlesinger (1997).

In Figure 4-1, a biochemical cycle is superimposed on a simplified energy flow diagram to show how the one-way flow of energy drives the cycle of matter. It is im-

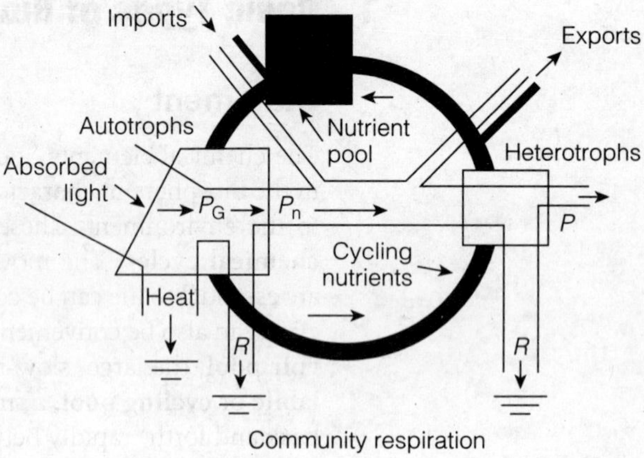

Figure 4-1. Biogeochemical cycle (black circle) superimposed on a simplified energy flow diagram, contrasting the cycling of material with the one-way flow of energy. P_G = gross primary production; P_N = net primary production; P = secondary production; R = respiration (after E. P. Odum 1963).

portant to emphasize that energy in some form usually must be expended to recycle materials—a fact of life to remember when it comes to the increasing need for humans to recycle water, metals, paper, and other materials. Thus, the science of **human ecology**—the study of the impact of humankind on and integration with natural systems—has become a vital component of natural and human-built systems management.

Elements in nature are almost never homogeneously distributed, nor are they present in the same chemical form throughout the ecosystem. In Figure 4-1, the reservoir pool—the portion of the cycle that is chemically or physically remote from organisms—is indicated by the box labeled "nutrient pool," whereas the cycling portion is designated by the circle going from autotrophs to heterotrophs and back again. Sometimes, the reservoir portion is termed the *unavailable pool,* and the active, cycling pool is termed the *available* or *exchangeable pool.* For example, agronomists routinely measure the fertility of the soil by estimating the concentration of *exchangeable nutrients*—that usually small part of the total soil nutrient content that is quickly available to plants. Such designations are permissible provided one clearly understands that the terms are relative. An atom in the reservoir pool is not necessarily permanently unavailable to organisms, because there are slow fluxes between available and unavailable components. The methods used to estimate exchangeable nutrients in soil testing (usually extraction with weak acids and bases) are at best only approximate indicators. The relative size of reservoir pools is important when one is assessing the effect of human activity on biogeochemical cycles. Generally, the smallest reservoir pools will be the first affected by changes in element fluxes. For example, the amount of carbon in the atmosphere (mostly in the form of CO_2) is a very small part of the total carbon in the biosphere, but a small change in this pool has a very large effect on the temperature of Earth.

Human beings are unique in not only requiring the 40 essential elements but also using nearly all the other elements, including the newer, synthetic ones. Humankind has so speeded up the movement of many materials that the self-regulating processes that tend to maintain homeorhesis are overwhelmed and nutrient cycles tend to become imperfect, or *acyclic,* resulting in the paradoxical situation of "too little here and too much there." For example, humans mine and process phosphate rock with such careless abandon that severe local pollution occurs near mines and phosphate mills.

Then, with equally acute myopia, humans increase the input of fertilizers in agricultural systems with little consideration of the inevitable increase in runoff, which severely stresses waterways and reduces water quality.

Pollution has frequently been defined as *resources misplaced*. The aim of the conservation of natural resources in the broadest sense is to make acyclic processes become more cyclic. The concept of *recycling* must increasingly become a major goal for societies. The recycling of water is a good beginning, because if the hydrologic cycle can be maintained and repaired, there is a better chance of controlling the nutrients that move along with the water.

Examples

Five examples will illustrate the principle of cycling. The **nitrogen cycle** is an example of a very complex, well-buffered, gaseous-type cycle; the **phosphorus cycle** is an example of a simpler, less well buffered, regulated sedimentary-type cycle. Both these elements are often very important factors limiting or controlling the abundance of organisms, and, in recent times, overfertilization with both these elements has been creating severe adverse effects on a global scale.

The **sulfur cycle** has been selected to illustrate (1) the links between air, water, and the crust of Earth, as there is active cycling within and between each of these pools; (2) the key role played by microorganisms; and (3) the complications caused by industrial air pollution. The reservoir sizes and turnover times of four biologically active elements are presented in Table 4-1. The **carbon cycle** (Table 4-1) and the **hydrologic cycle** (Table 4-2) are crucial to life and are being increasingly affected by human activities. When discussing biogeochemical cycles, it is also important to distinguish between the boundaries of an ecosystem—both natural boundaries and those set by ecologists for purposes of study and modeling—and what is termed its **footprint** or **region of influence**.

2 Cycling of Nitrogen

Figure 4-2 shows two different ways to picture the complexities of the nitrogen cycle; each illustrates a major overall feature or driving force. Figure 4-2A brings out the circularity of nutrient flows and the kinds of microorganisms required for the basic exchanges between organisms and environment. The nitrogen in protoplasm is broken down from organic to inorganic forms by a series of decomposer bacteria, each specialized for a particular part of the cycle. Some of the nitrogen ends up as ammonium and nitrate, the forms most readily used by green plants. The atmosphere, which is approximately 78 percent nitrogen, is the greatest reservoir and safety valve of the system. Nitrogen is continually entering the atmosphere by the action of denitrifying bacteria and continually returning to the cycle through the action of nitrogen-fixing microorganisms (*biofixation*) and through the action of lighting and other physical fixation.

The steps from proteins down to nitrates provide energy for the organisms that accomplish the breakdown, whereas the return steps require energy from other

Table 4-1 **Reservoir sizes and turnover times of biologically active elements**

Reservoir	*Quantity*	*Turnover time**
Nitrogen (10^{12} g N)		
Atmosphere (N_2)	4×10^9	10^7
Sediments	5×10^8	10^7
Ocean (dissolved N_2)	2.2×10^7	1000
Ocean (inorganic)	6×10^5	
Soil	3×10^5	2000
Terrestrial biomass	1.3×10^4	50
Atmosphere (N_2O)	1.4×10^4	100
Marine biomass	4.7×10^2	
Sulfur (10^{12} g S)		
Lithosphere	2×10^{10}	10^8
Ocean	3×10^9	10^6
Sediments	3×10^8	10^6
Soils	3×10^5	10^3
Lakes	300	3
Marine biota	30	1
Atmosphere	4.8	8–25 days
Phosphorus (10^{12} g P)		
Sediments	4×10^9	2×10^8
Land	2×10^5	2000
Deep ocean	8.7×10^4	1500
Terrestrial biota	3000	~50
Surface ocean	2700	2.6
Atmosphere		0.028 days
Carbon (10^{12} g C)		
Sediments, rocks	77×10^6	$> 10^6$
Deep ocean (DIC)	38,000	2000
Soils	1500	$< 10–10^5$
Surface ocean	1000	decades
Atmosphere	750	5
Deep ocean (DOC)	700	5000
Terrestrial biomass	550–680	50
Surface sediments	150	0.1–1000
Marine biomass	2	0.1–1

*Turnover time in years unless otherwise noted.

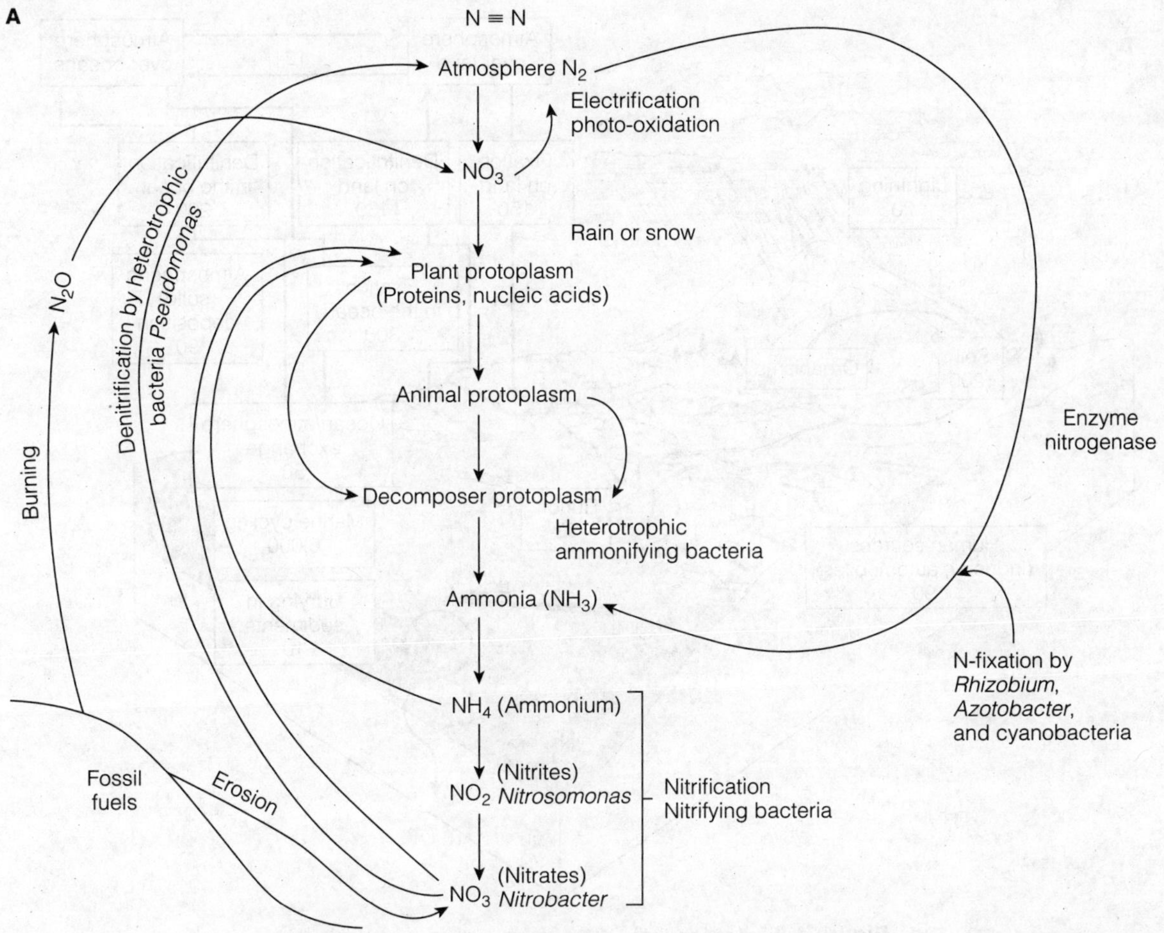

Figure 4-2. Two ways of depicting the biogeochemical cycle of nitrogen, an example of a relatively well-buffered, self-regulating cycle with a large atmospheric reservoir. (A) Circulation of nitrogen between organisms and the environment, showing microorganisms responsible for key steps. (B) Schematic diagram of the nitrogen cycle, depicting global pools and fluxes per year expressed in billions of metric tons (10^{15} g). (Reproduced from Schlesinger 1997; we thank Lawrence Pomeroy for recent corrections in values.) *(continued)*

sources, such as organic matter or sunlight. For example, the chemosynthetic bacteria *Nitrosomonas* (which converts ammonia to nitrite) and *Nitrobacter* (which converts nitrite to nitrate) obtain energy from the breakdown of organic matter, whereas denitrifying and nitrogen-fixing bacteria require energy from other sources to accomplish their respective transformations.

There is also an important short cycle of nitrogen in the living biosphere, in which heterotrophic organisms break down proteins enzymatically and excrete the excess nitrogen as urea, uric acid, or ammonium. Specialized bacteria gain energy for their livelihood by oxidizing the ammonium to nitrite and the nitrite to nitrate. All three—ammonium, nitrite, and nitrate—may be used as basic nitrogen sources by plants. Plants that use nitrate must produce enzymes to convert it back to ammonium, so nitrate is a more energy-expensive source of nitrogen than ammonium from

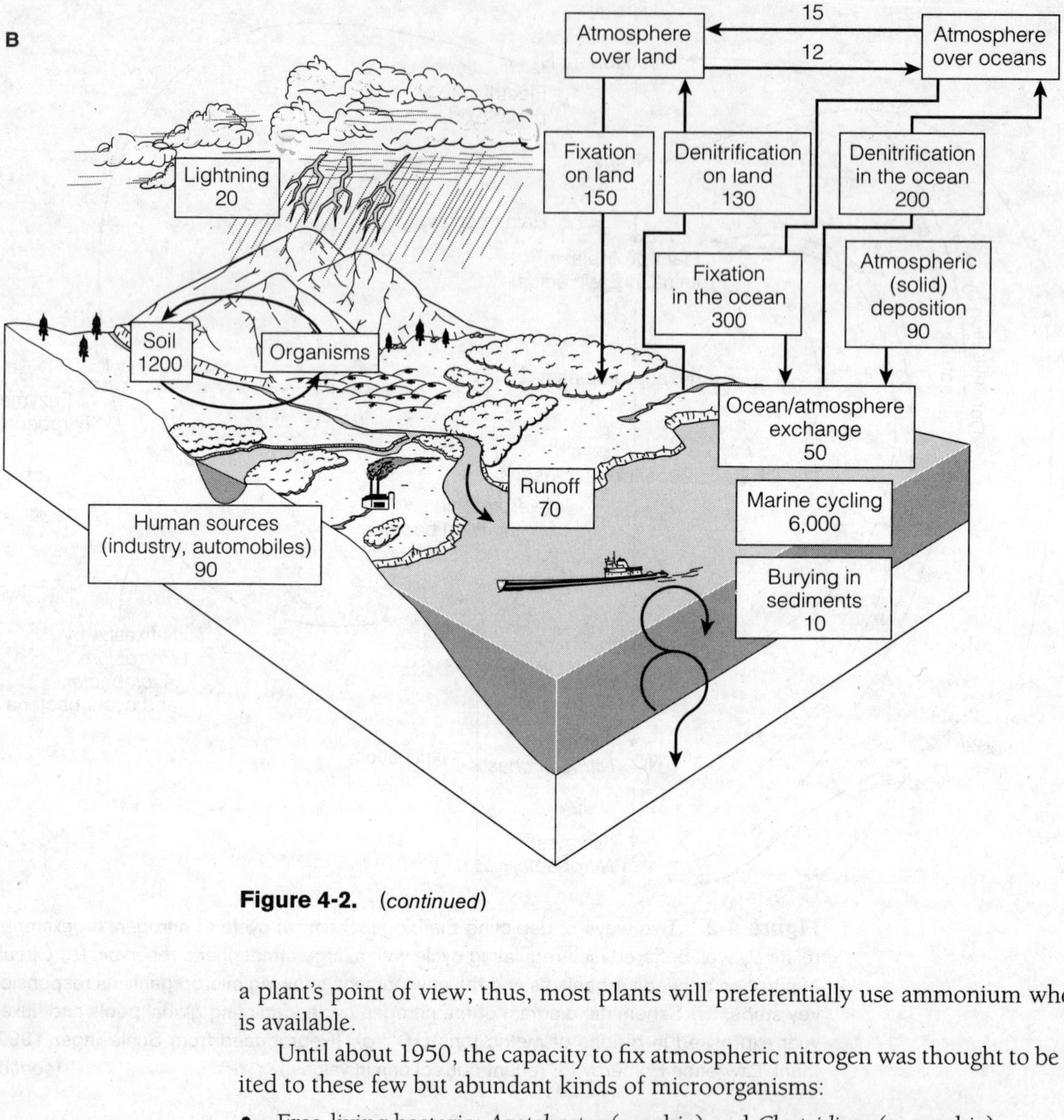

Figure 4-2. *(continued)*

a plant's point of view; thus, most plants will preferentially use ammonium when it is available.

Until about 1950, the capacity to fix atmospheric nitrogen was thought to be limited to these few but abundant kinds of microorganisms:

- Free-living bacteria: *Azotobacter* (aerobic) and *Clostridium* (anaerobic).
- Symbiotic nodule bacteria on legume plants: *Rhizobium*.
- Cyanobacteria: *Anabaena, Nostoc,* and several other genera (formerly termed blue-green algae, but they are green bacteria, not algae).

Then it was discovered that the purple bacterium *Rhodospirillum* and other representatives of the photosynthetic bacteria are also nitrogen fixers and that a variety of *Pseudomonas*-like soil bacteria also have this capacity. Later, it was discovered that actinomycetes (a kind of filamentous bacteria) in the root nodules of alders (*Alnus*) and certain other nonleguminous woody plants fix nitrogen as efficiently as do *Rhizobium* bacteria in legume nodules. Nitrogen fixation also occurs in the ocean. For example, nitrogen fixation by *Trichodesmium* blue-green bacteria in the ocean is limited by iron. As a result, nitrogen fixation is seasonal and controlled by the pattern of dust fallout on the sea from the Gobi and Sahara Deserts and from upwellings or coastal sources. So far, 160 species in eight genera in five families of dicotyledons have been shown to possess actinomycete-induced nodules. Unlike legumes, which are largely tropical in origin, these nitrogen fixers originate in the North Temperate Zone. Most species are adapted to poor sandy or boggy soils where available nitrogen is scarce. Some species, such as alders (*Alnus*), have the potential of increasing forest yields when interplanted with timber species. An example of a nodule nitrogen fixer is shown in Figure 4-3.

Nitrogen fixation by cyanobacteria may take place in free-living forms or in forms symbiotic with fungi, as in certain lichens, or with mosses, ferns, and at least one seed plant. The fronds of the small, floating aquatic fern *Azolla* contain small pores filled with symbiotic *Anabaena* that actively fix nitrogen. For centuries, this fern has played an important role in paddy rice culture in the Orient. Before the rice seedlings are planted, the flooded paddies are covered with the aquatic ferns, which fix enough nitrogen to supply the crop as it matures. This practice, combined with the encouragement of free-living nitrogen-fixing microorganisms, permits rice to be grown season after season in the same paddy without the addition of fertilizer. However, rice paddies are also among the major sites of denitrification and methane production.

The key to biofixation is the enzyme **nitrogenase,** which catalyzes the splitting of N_2 (Fig. 4-2A). This enzyme can also reduce acetylene to ethylene, thereby providing a convenient way to measure nitrogen fixation in nodules, soils, water, or wherever one suspects that fixation is occurring. The acetylene reduction method, together with the use of the isotopic tracer ^{15}N, has revealed that the ability to fix nitrogen is widespread among photosynthetic, chemosynthetic, and heterotrophic microorganisms. There is even evidence that microorganisms growing on leaves and epiphytes in humid tropical forests fix appreciable quantities of atmospheric nitrogen, some of which may be used by the trees themselves. In short, it appears that biological nitrogen fixation goes on in both the autotrophic and the heterotrophic strata of ecosystems and in both aerobic and anaerobic zones of soils and aquatic sediments.

Figure 4-3. Root nodules on a soybean plant (*Glycine max*) depicting the location of mutualistic nitrogen-fixing bacteria (*Rhizobium leguminosarum*). The nodules are the sites where the bacteria fix atmospheric nitrogen and convert it into nitrogenous compounds which are used by the plant in the synthesis of proteins and nucleic acids.

© Wally Eberhart/Visuals Unlimited

Nitrogen fixation is especially energy expensive because much energy is required to break the triple bond of molecular N_2 ($N{\equiv}N$) so that it can be converted (with the addition of hydrogen from water) to two molecules of ammonia (NH_3). For biofixation by legume nodule bacteria, some 10 g of glucose (about 40 kcal) from plant photosynthate is required to fix 1 g of nitrogen (efficiency 10 percent). Free-living nitrogen fixers are less efficient and may require up to 100 g of glucose to fix 1 g of nitrogen (efficiency 1 percent). Similarly, a lot of fossil fuel energy has to be expended in industrial fixation, which is why nitrogen fertilizer is more expensive, pound for pound, than most other fertilizers.

In summary, only procaryotes (primitive microorganisms) can convert biologically useless nitrogen gas into the nitrogen forms required to build and maintain living cells. When these microorganisms form mutually beneficial partnerships with higher plants, nitrogen fixation is greatly enhanced. The plant provides a congenial home (a root nodule or leaf pocket), protects the microbes from too much O_2 (which inhibits N_2 fixation), and furnishes the microbes with the high-quality energy required. In return, the plant gets a readily assimilable supply of fixed nitrogen. This cooperation for mutual benefit—a survival strategy very common in natural systems—could be emulated in human-made systems to greater benefit. Nitrogen fixers work hardest when the nitrogen supply in their environment is low; adding nitrogen fertilizer to a legume crop shuts down biofixation.

It may be possible to genetically engineer nodule formation on corn and other grain crops, reducing the need for and pollution from mineral nitrogen fertilizers, which tend to run off more than organically fixed nitrogen. A number of commercial genomics companies are working on splicing nitrogen-fixing genes into corn. However, there would be a cost in reduced yield, as some of the primary production that would otherwise go to grain production would be diverted to support the nodule, as noted previously.

Detrimental Effects of Too Much Nitrogen

In earlier editions of this book, the primary emphasis was on nitrogen as a major limiting factor. Now, a major concern is the adverse effects of too much nitrogen—a "too much of a good thing" situation. Figure 4-2B provides recent estimates of annual global nitrogen fluxes in teragrams (1 Tg = 10^{12} grams or 1 million metric tons), including estimates for the magnitude of the flows directly related to human activities. Fertilizer production and use, legume crops, and fossil fuel burning worldwide deposit approximately 140 Tg/year of new nitrogen into soil, water, and air—about equal to the estimated nitrogen fixed naturally. Human sewage and domestic animal manure contribute perhaps half again as much. Very few of these inputs are recycled, because they escape into soils or streams, or are mixed with heavy metals or other toxins.

Most natural ecosystems, and most native species, are adapted to low-nutrient environments. Enrichment with nitrogen and other nutrients opens the door for opportunistic "weedy-type" species that are adapted to high-nutrient conditions. For example, in natural grasslands in Minnesota and California that have been nitrogen enriched, almost all native species of plants have been replaced by weedy exotic species, resulting in reduced biodiversity (Tilman 1987, 1988). Based on extensive field evidence, Tilman et al. (1997) predicted that nitrogen deposition is likely to strongly affect ecosystem processes. Annual inputs of nitrogen in fertilizer and municipal sludge applied to old-field communities in Ohio also significantly reduced plant diversity on a long-term basis compared to control plots (Brewer et al. 1994).

As is so often the case, anything that is detrimental to natural ecosystems eventually becomes detrimental to humans. Excess nitrogen compounds in drinking water, in food, and especially in the air, pose threats to human health. Excess nitrate in drinking water can also be caused by exotic legumes; for example, the introduction of the legume *tangan-tangan* (*Leucaena leucocephala*) from the Philippines after World War II has poisoned the groundwater in much of Guam.

In summary, nitrogen enrichment is reducing biodiversity and increasing the number of pests and diseases globally, and is also beginning to adversely affect human health. For more on excess nitrogen as a present and future threat to the environment, see Vitousek, Mooney, et al. (1997).

3 Cycling of Phosphorus

The phosphorus cycle appears somewhat simpler than the nitrogen cycle, because phosphorus occurs in fewer chemical forms. As shown in Figure 4-4, phosphorus, a necessary constituent of protoplasm, tends to circulate with organic compounds in the form of phosphates (PO_4), which are again available to plants. The great reservoir of phosphorus is not the air, however, but in apatite mineral deposits formed in past geological ages (that is, in the lithosphere). Atmospheric dust and aerosols return 5×10^{12} g of phosphorus (not phosphate) to the land yearly, but phosphate continually returns to the sea, where part of it is deposited in the shallow sediments and part of it is lost to the deep sediments.

Contrary to popular belief, seabirds play only a limited role in returning phosphorus to the cycle (witness the guano deposits located on the coast of Peru). This transfer of phosphorus and other materials by birds from the sea to the land is continuing—likely at the same rate at which it occurred in the past—but these guano deposits have been mined out. Although rookeries everywhere produce local concentrations of phosphate and uric acid, their global importance is limited. We currently retrieve phosphate from ancient bone beds located in Florida and Russia.

Unfortunately, human activities appear to hasten the rate of loss of phosphorus and, thus, to make the phosphorus cycle less cyclic. Although an abundance of marine fish are harvested, it is estimated that only about 60,000 tons of phosphorus per year are returned to the cycle in this manner, compared with the one to two million tons of phosphate that is mined and used for fertilizer, much of which is washed away and lost. There is no immediate cause for concern about supplies for human use, as the known reserves of phosphate beds are large. However, the mining and processing of phosphate for fertilizer creates severe local pollution problems, as is evident in the Tampa Bay area of Florida, where there are very large deposits.

Walsh and Steidinger (2001) suggested that phosphate mining is probably part of the cause of the Florida red tides, the other factor being Sahara dust that provides iron for marine N_2 fixation. Walsh and Steidinger's (2001) hypothesis is that dust from the Sahara regularly reaches the Gulf of Mexico, bringing iron, which stimulates a *Trichodesmium* bloom. The nitrogen thus fixed, plus the phosphate from the Florida deposits, stimulate a general bloom of phytoplankton. Zooplankton then eat all

Figure 4-4. (A) Model diagram of the phosphorus cycle. (B) The global phosphorus cycle, depicting reservoir pools and fluxes expressed in units of 10^{12} g phosphorus per year. (From Figure 12.6, p. 397, in Schlesinger, W. H. 1997. *Biogeochemistry: An analysis of global change.* 2nd ed. San Diego: Academic Press. Copyright © 1997 with permission from Elsevier.)

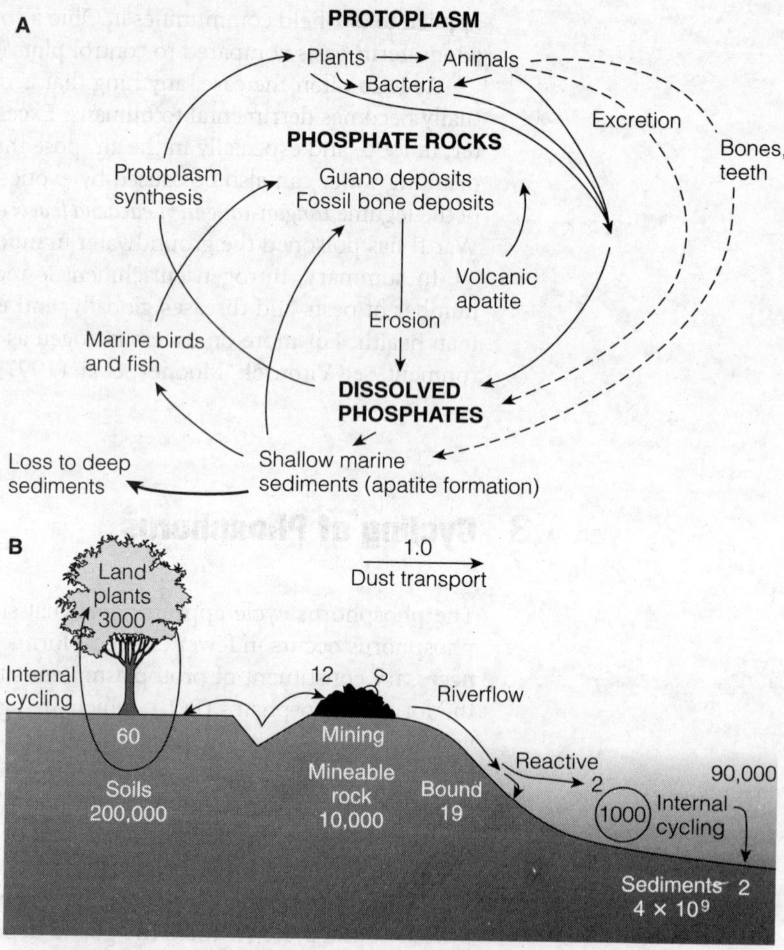

the nontoxic phytoplankton, leaving a residual red tide of toxic *Karenia brevis*. Furthermore, spraying wastewater and sewage on the land is nowadays so common that it has become a new form of pollution. The excess of dissolved phosphate in aquatic systems resulting from this increased input of urban-industrial and agricultural runoff is a concern at present. Ultimately, phosphorus will have to be recycled on a large scale to avoid famine.

One experimental procedure for recycling phosphorus "uphill" involves spraying wastewater on upland or old-field vegetation or passing it through natural or engineered wetlands (marshes and swamps) instead of piping it directly into streams and rivers (see Woodwell 1977; Soon et al. 1980; W. P. Carson and Barrett 1988; Levine et al. 1989). The "poster child" for phosphate pollution is the Everglades, which has a PO_4 concentration in its flowing waters comparable to the concentration in the surface water of the Sargasso Sea. The Everglades has been polluted for decades by agricultural runoff, and the problem is exacerbated by more rapid water movement in canals. Currently, the plan is to create, at great expense, "expendable" wetlands adjacent to agricultural areas to "soak up" the phosphorus and retain it in sediments—likely another "quick fix" with some unplanned payback in the future.

At any rate, as depicted in Figure 4-4, phosphorus will loom large in the future because, of all the macronutrients (vital elements required in large amounts by life), phosphorus is the most scarce in terms of its relative abundance in available pools on the surface of Earth.

The interaction of nitrogen and phosphorus is especially worthy of attention. The N/P ratio in average biomass is about 16 to 1, and in streams and rivers about 28 to 1. Schindler (1977) reported experiments in which fertilizers with different N/P ratios were added to a whole lake. When the N/P ratio was reduced to 5, nitrogen-fixing cyanobacteria dominated the phytoplankton and fixed enough nitrogen to raise the ratio to within the range of many natural lakes. Schindler hypothesized that lake ecosystems have evolved natural mechanisms to compensate for deficits of nitrogen and carbon, but not for deficits of phosphorus, which does not have a gaseous phase. Accordingly, primary production in freshwater systems is very often correlated with available phosphorus.

4 Cycling of Sulfur

Sulfate (SO_4), like nitrate and phosphate, is the principal biologically available form that is reduced by autotrophs and incorporated into proteins, sulfur being an essential constituent of certain amino acids. Not as much sulfur is required by the ecosystem as nitrogen and phosphorus, nor is sulfur as often limiting to the growth of plants and animals. Nevertheless, the sulfur cycle is a key one in the general pattern of production and decomposition. For example, when iron sulfides are formed in sediments, phosphorus is converted from an insoluble to a soluble form, as depicted in Figure 4-4, and thus enters the pool available to living organisms. This is an illustration of how one nutrient cycle regulates another. The recovery of phosphorus as part of the sulfur cycle is most pronounced in the anaerobic sediments of wetlands, which are also important sites for the recycling of nitrogen and carbon.

Figure 4-5A provides estimates of the amount of sulfur in the reservoir pools (lithosphere, atmosphere, and oceans) and the annual fluxes in and out of these pools, including inputs and outputs directly related to human activities. Figure 4-5B emphasizes the key role played by specialized sulfur bacteria, which function like a "relay team" in the cycling of sulfur in soils, freshwaters, and wetlands. The microbially driven process in deep anaerobic zones in soils and sediments results from the upward movement of gaseous hydrogen sulfide (H_2S) in land and wetland ecosystems. The decomposition of proteins also leads to the production of hydrogen sulfide. Once in the atmosphere, this gaseous phase is converted to other forms, principally sulfur dioxide (SO_2), sulfate (SO_4) and sulfur aerosols (very fine floating particles of SO_4). Sulfur aerosols, unlike CO_2, reflect sunlight back into the sky, thereby contributing to global cooling and to acid rain.

Effect of Air Pollution

Both the nitrogen and the sulfur cycles are increasingly being affected by industrial air pollution. The gaseous oxides of nitrogen (N_2O and NO_2) and sulfur (SO_2), unlike nitrates and sulfates, are toxic to varying degrees. Normally, they are only transitory steps in their respective cycles; in most environments, they are present in very low concentrations. The combustion of fossil fuels, however, has greatly increased

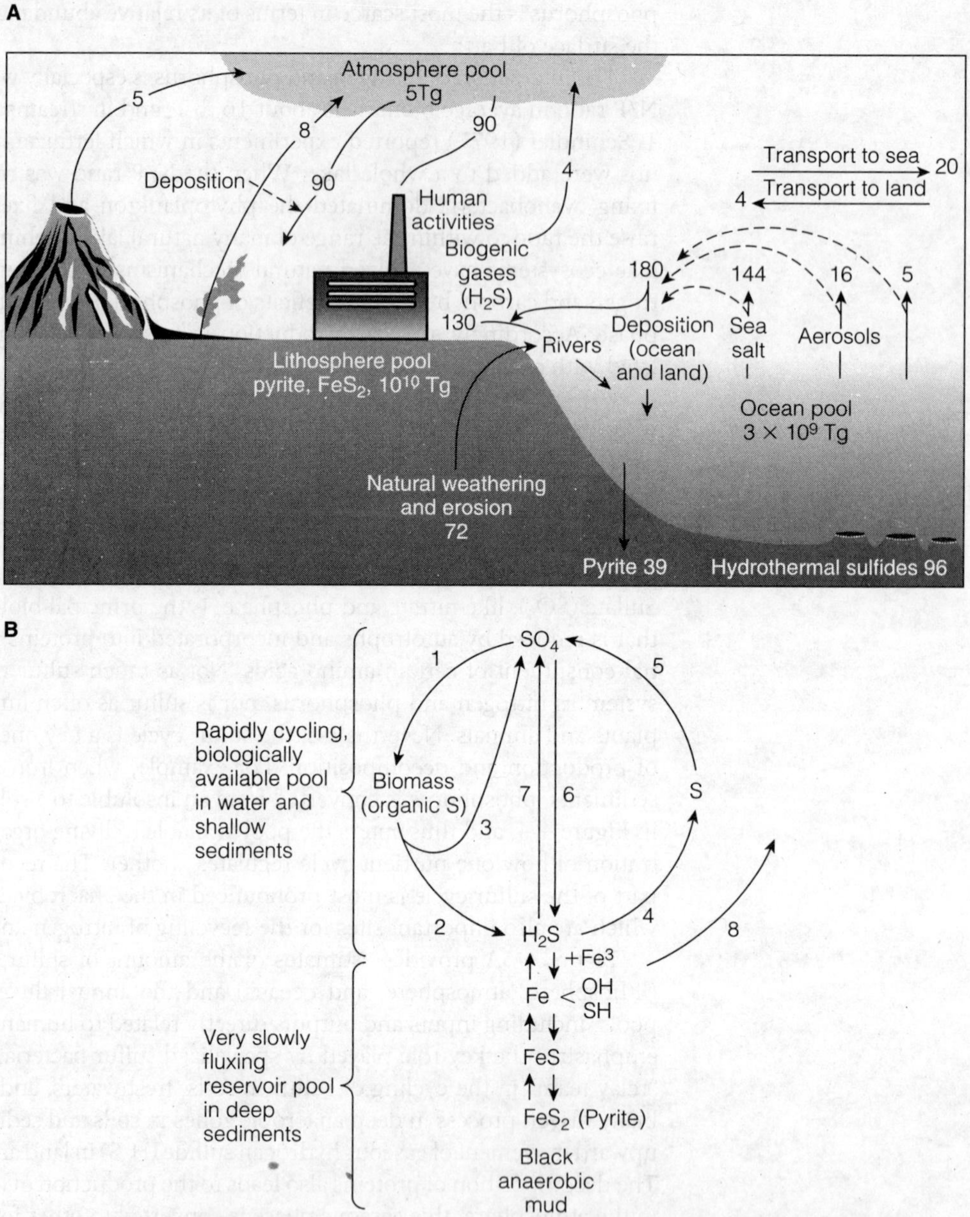

Figure 4-5. (A) The global sulfur cycle, with emphasis on the relationship between reservoir pools and fluxing components. Values expressed in teragrams (Tg) of sulfur per year (modified after Schlesinger 1997). (B) The sulfur cycle in aquatic environments, with emphasis on the role of microorganisms. Step 1 is primary production by plants. Other organisms, most of them specialized microorganisms, carry out steps 2–7: 2 = decomposition by heterotrophic microorganisms; 3 = animal excretion; 4 and 5 = colorless, green, and purple sulfur bacteria; 6 = anaerobic sulfur-reducing bacteria, *Desulfovibrio;* and 7 = aerobic sulfide-oxidizing bacteria, *Thiobacillus.* Step 8 represents the conversion of phosphorus from an unavailable to an available form when iron sulfides are formed, illustrating how the cycling of one vital element can affect another.

the concentrations of these volatile oxides in the air, especially in urban areas and in the vicinity of power plants, to the point where they adversely affect important biotic components and processes of ecosystems. When plants, fish, birds, or microbes are poisoned, humans eventually are also adversely affected. These oxides constitute about one third of the industrial air pollutants discharged into the air over the United States. The passage of the Clean Air Act (1970, amended 1990), which tightened emission standards, has only slightly reduced the volume.

Coal-burning emissions and automobile exhaust are major sources of SO_2 and SO_4 production and, along with other industrial combustion, a major source of poisonous forms of nitrogen. Sulfur dioxide is damaging to photosynthesis, as was discovered in the early 1950s when leafy vegetables, fruit trees, and forests showed signs of stress in the Los Angeles Basin. The destruction of vegetation around copper smelters is largely caused by SO_2. Furthermore, both sulfur and nitric oxides interact with water vapor to produce droplets of dilute sulfuric and nitric acid (H_2SO_4 and H_2NO_3) that fall to Earth as **acid rain**, a truly alarming development (see Likens and Bormann 1974a; Likens et al. 1996; Likens 2001a for details). Acid rain has the greatest impact on soft-water lakes or streams and already acidic soils that lack pH buffers (such as carbonates, calcium, salts, and other bases). The increase in acidity (decrease in pH) in some Adirondack lakes has rendered them incapable of supporting fish. Acid rain has also become a major problem in Scandinavia and other parts of northern Europe. In many ways, the building of tall smokestacks for coal-burning power plants (to reduce local air pollution) has aggravated the problem, because the longer the oxides remain in cloud layers, the more acid is formed. This is a typical example of a short-term "quick fix" that produces a more severe long-term problem (local fallout is extended to regional fallout). The long-term solution is to gasify or liquefy the coal, thereby eliminating the emissions entirely.

The oxides of nitrogen are also directly threatening the quality of human life. They irritate the respiratory membranes of higher animals and humans. Furthermore, chemical reactions with other pollutants produce a *synergism* (the total effect of the interaction exceeds the sum of the effects of each individual substance) that increases the danger. For example, in the presence of ultraviolet radiation in sunlight, NO_2 reacts with unburned hydrocarbons (produced in large quantities by automobiles) to produce a **photochemical smog**, which not only makes eyes tear but may cause lung damage. For more on the nitrogen, phosphorus, and sulfur cycles see Butcher et al. (1992) and Schlesinger (1997).

5 Cycling of Carbon

Statement

At the global level, the carbon cycle and the water cycle are very important biogeochemical cycles, as carbon is the basic element of life and water is essential for all life. Both cycles are characterized by small but very active atmospheric reservoir pools that are vulnerable to human-made perturbations, which, in turn, can change weather and climate in ways that strongly affect life on Earth. Indeed, during the latter half of the twentieth century, the CO_2 concentration in the atmosphere has been significantly increasing, along with other greenhouse gases that reflect solar heat back down to Earth.

Explanation

The global carbon cycle is shown in Figure 4-6A, with estimates of the amounts in reservoir pools and fluxes, whereas Figure 4-6B plots the rise in atmospheric CO_2 measured at Mauna Loa Observatory in Hawaii from 1958 to 2002. As already noted, the atmospheric carbon pool is very small compared with the amount of carbon in the oceans and in fossil fuels and other storages in the lithosphere. The burning of fossil fuels, along with agriculture and deforestation, is contributing to the continuous increase of CO_2 in the atmosphere. The net loss of CO_2 (the addition of more CO_2 into the atmosphere than is removed) in agriculture may seem surprising, but it occurs because the CO_2 fixed by crops (many of which are active for only a part of the year) does not compensate for the CO_2 released from the soil, especially by frequent plowing. Forest removal, of course, may release the carbon stored in wood, especially from wood that is immediately burned, and this is followed by carbon release through the oxidation of humus, if the land is used for agriculture or urban development. In contrast, young, rapidly growing forests are carbon sinks, so reforestation on a large scale might reduce the rate of global warming associated with the increase in atmospheric CO_2.

Before 1850 (before the Industrial Revolution), the concentration of CO_2 in the atmosphere was on the order of 280 ppm. During the last 150 years, atmospheric CO_2 has increased to more than 370 ppm. This increase has led to concern regarding the greenhouse effect. The **greenhouse effect** is the warming of the climate of Earth attributed to the increased concentration of CO_2 and certain other gaseous pollutants in the atmosphere. These *greenhouse gases* (methane, ozone, nitrous oxide, and chlorofluorocarbons) absorb infrared radiation emitted by the solar-heated Earth and reflect most of that heat energy back to Earth, resulting in potential global warming.

The rapid oxidation of humus and the release of the gaseous CO_2 normally held in the soil has other, more subtle effects, including effects on the cycling of other nutrients. Agronomists now recognize that they must add trace minerals to fertilizers to maintain yields in many areas, because agroecosystems do not regenerate these nutrients as well as do natural systems.

In ecological terms, the flux between the reservoir pool and the exchangeable pool of many elements is being fundamentally altered by the present mismanagement of the landscape. There are practices that humans can use to compensate, for example, by promoting conservation tillage practices in agriculture, which reduce runoff and soil erosion. If humans recognize what has happened and learn to compensate, such changes need not be detrimental. Recall how the atmosphere of Earth came to have its present low CO_2 and high O_2 content (see the section on the Gaia hypothesis in Chapter 2).

The production of carbonates in the sea also forms carbon dioxide as a by-product as follows:

$$Ca + 2HCO_3 \longrightarrow CaCO_3 + H_2O + CO_2$$

Because of the reduction in pH that results from moving this reaction to the right, only 0.6 moles of CO_2 per mole of carbonate are actually released into seawater (and ultimately the atmosphere). Coral reefs and other calcifying organisms are a source, not a sink, of CO_2. The sea plays a major role in the sequestering of carbon. The sea

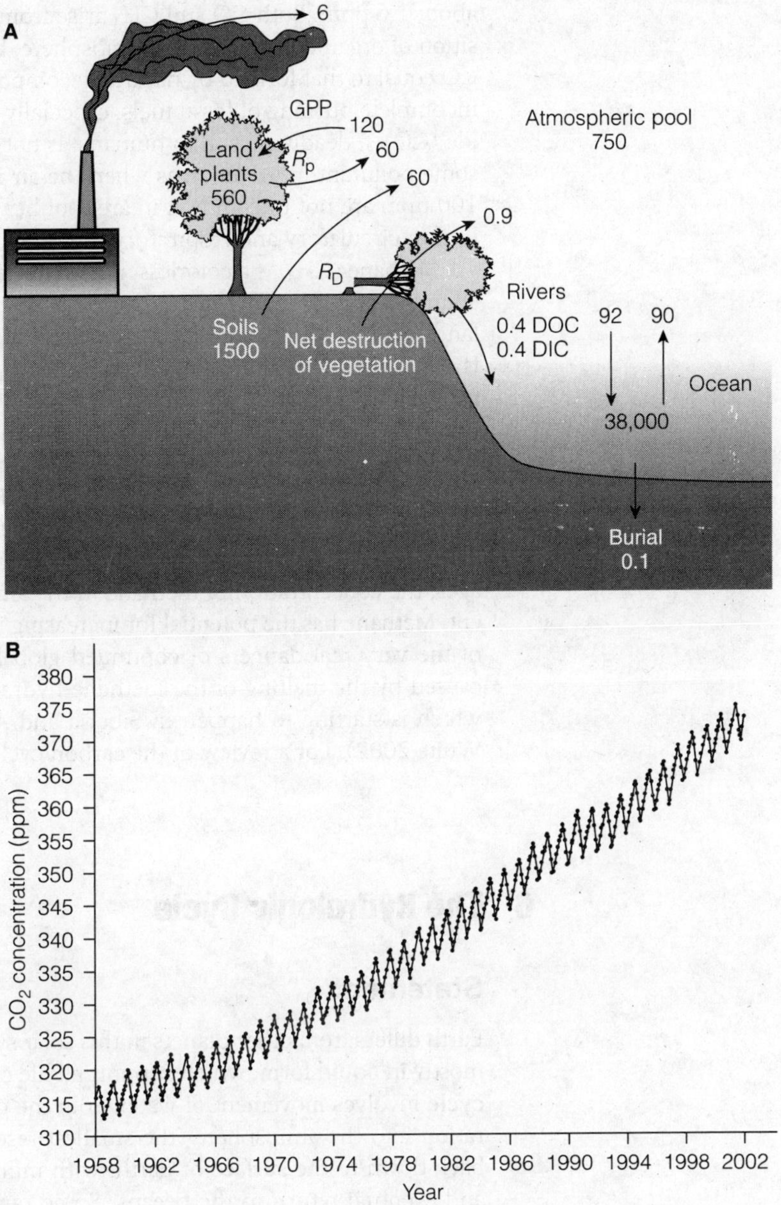

Figure 4-6. (A) The global carbon cycle. Values expressed in 10^{15} g carbon per year (values from Schimel et al. 1995 and Schlesinger 1997). *GPP* = gross primary product; R_p = plant respiration; R_D = detrital respiration; DOC = dissolved organic carbon; DIC = dissolved inorganic carbon. (B) Plot depicting the continuous increase in atmospheric CO_2 from 1958 to 2002 as measured at the Mauna Loa Observatory, Hawaii. Data points represent monthly average CO_2 concentration. (Source: C. D. Keeling and T. P. Whorf, Scripps Institution of Oceanography, UC, La Jolla, California, USA 92093.)

contains 40 atmospheres of carbon as bicarbonate and dissolved organic carbon (DOC), which function as major carbon reservoirs. The sea is, therefore, a very effective buffer of atmospheric CO_2, because the sea and the atmosphere are equilibrating with one another. This is likely the primary control mechanism for atmospheric CO_2. Any large future increase in the burning of fossil fuels, coupled with future decreases in the CO_2 removal capacity of the green belt, is almost certain to result in a continued rise in the CO_2 content of the atmosphere. Recall from the previous discussion that the content of small, active compartments is most affected by changes in fluxes or throughputs.

In addition to CO_2, two other forms of carbon are present in the atmosphere in small amounts: carbon monoxide (CO) at about 0.1 ppm, and methane (CH_4) at about 1.6 ppm. Both CO and CH_4 arise from the incomplete or anaerobic decomposition of organic matter; in the atmosphere, both are oxidized to CO_2. An amount of CO equal to that formed by natural decomposition is now injected into the air by the incomplete burning of fossil fuels, especially in automobile exhaust. Carbon monoxide (CO), a deadly poison to humans, is not a global threat, but it becomes a worrisome pollutant in urban areas when the air is stagnant. CO concentrations of up to 100 ppm are not uncommon in areas of heavy automobile traffic—a stress that can lead to circulatory and respiratory illnesses.

Methane (CH_4) is a colorless, flammable gas that is produced naturally by the decomposition of organic matter by anaerobic bacteria, especially in freshwater wetlands, rice paddies, and digestive tracts of ruminants (such as cattle) and termites. It is also a major component of natural gas, so the geochemical disturbances associated with mining and drilling for fossil fuels result in the release of methane. Although it is now only a very minor constituent of the atmosphere (2 ppm, compared to 370 ppm for CO_2), the methane concentration has doubled during the past century, mostly due to human activities such as landfills and use of fossil fuels. Methane is a greenhouse gas that, molecule for molecule, is 25 times as heat absorbing as CO_2. Its residence time in the atmosphere is about 9 years, compared to 6 years for CO_2. In past ages, the concentration of methane in the atmosphere has been higher than at present. Methane has the potential for increasing its contribution to global warming. One of the very real dangers of continued global warming is another "methane burp," caused by the melting of the methane hydrates in permafrost and on the sea floor, which is starting to happen in Siberia and Alaska (D. J. Thomas et al. 2002; R. V. White 2002). For a review of the carbon cycle, see Schimel (1995).

6 The Hydrologic Cycle

Statement

Earth differs from other planets in this solar system in having large amounts of water, mostly in liquid form, which support all life on Earth. The water cycle or **hydrologic cycle** involves movement of water from the oceans (the largest reservoir) by evaporation into the atmosphere (the smallest reservoir), then by precipitation (rainfall) back down to the surface of Earth, with infiltration and runoff from the continents and eventual return to the oceans. Some rainfall goes directly back into the air by evaporation and transpiration by vegetation. About a third of incident solar energy is involved in driving the water cycle. Although the global amount of water on Earth is about the same now as during the ice ages, the amount frozen has varied widely over geological time. Water movement (fluxing) also varies greatly from place to place and is being increasingly affected by human activities.

Explanation

Two views of the water cycle are shown in Figure 4-7. Figure 4-7A includes estimates of the amount of water and the annual fluxes in and out of the large reservoir pools. In Figure 4-7B, the water cycle is shown in terms of energy, with an "uphill loop"

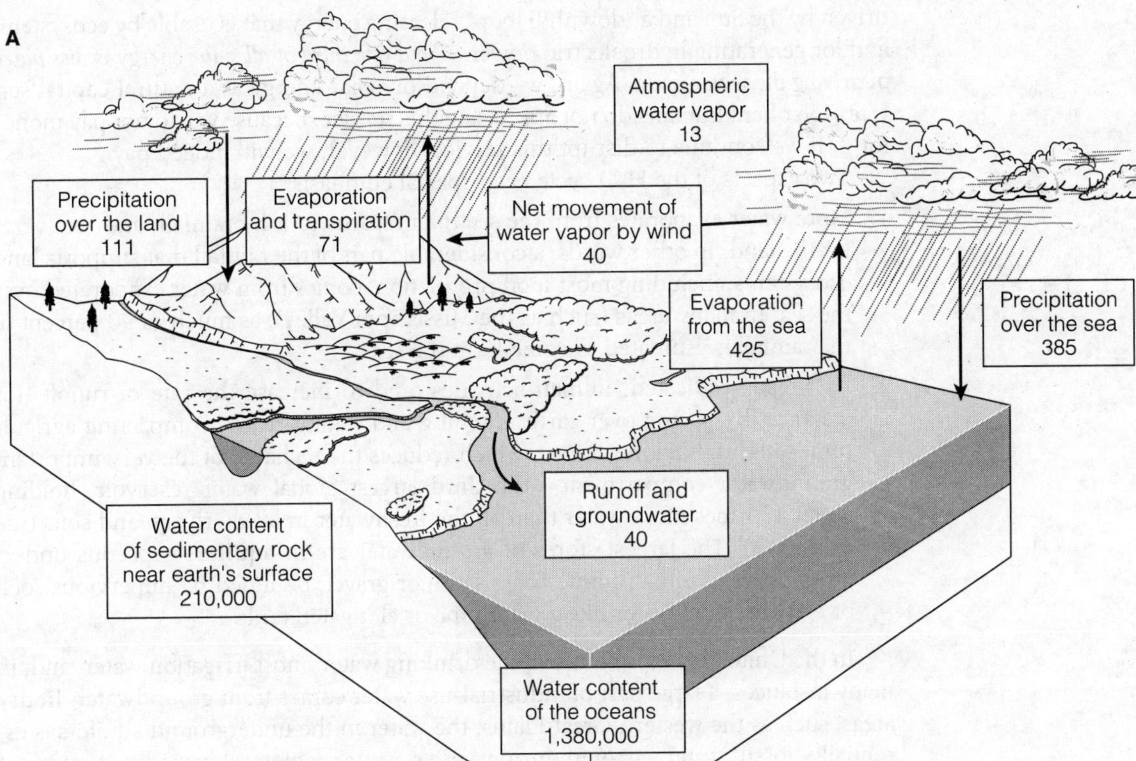

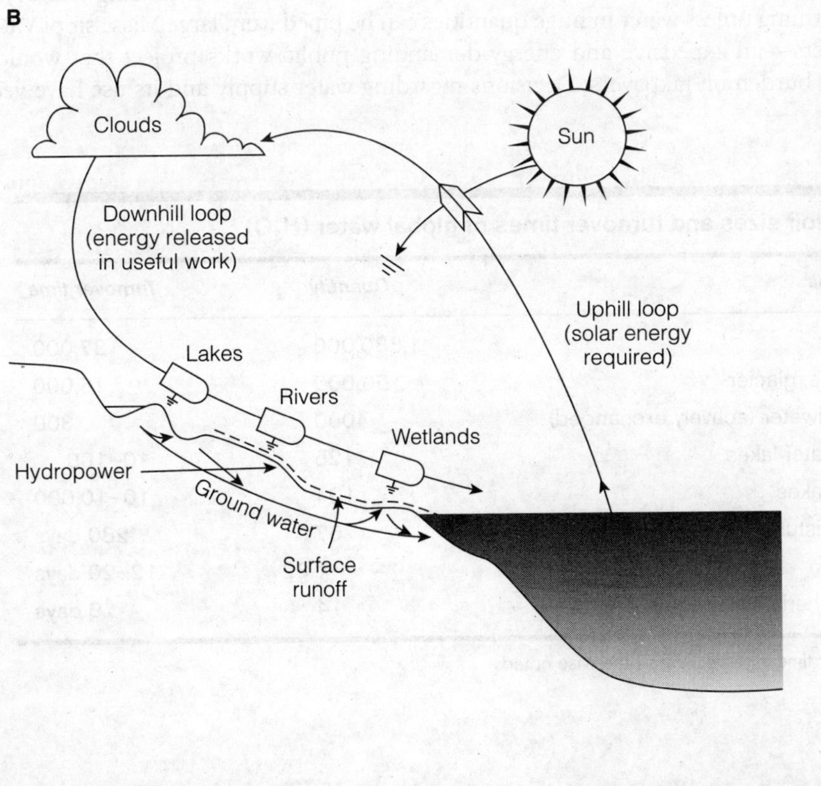

Figure 4-7. (A) Global pools and fluxes of water on Earth. Values are in teratons (10^{18} g) per year (data from Graedel and Crutzen 1995 and Schlesinger 1997). (B) Energetics of the hydrological cycle, viewed as an uphill loop driven by solar energy and a downhill loop that releases energy to lakes, rivers, and wetlands and performs useful work of direct benefit to humans, such as hydroelectric power generation.

driven by the Sun and a "downhill loop" releasing energy that is usable by ecosystems and for generating hydroelectric power. *About one third of all solar energy is dissipated in driving the water cycle.* Again, we depend on solar energy as a natural capital service. Too often, humans do not appreciate this service, because we do not pay money for it. If we continue to disrupt this service, however, we will indeed pay.

Two aspects of the H_2O cycle need special emphasis:

1. More water evaporates from the sea than returns to it by rainfall, and vice versa for the land. In other words, a considerable part of the rainfall that supports land ecosystems, including most food production, comes from water evaporated over the sea. In many areas (such as the Mississippi Valley), as much as 90 percent of the rainfall is estimated to come from the sea.
2. As already indicated, human activities tend to increase the rate of runoff (for instance, by paving over earth, ditching and diking rivers, compacting agricultural soils, and deforestation), which reduces the recharge of the very important groundwater compartment—the third largest global water reservoir, holding about 13 times more water than all the freshwater in lakes, rivers, and soils (see Table 4-2). The largest stores of groundwater are in **aquifers**—porous underground strata, often of limestone, sand, or gravel, bounded by impervious rock or clay that hold water like a giant pipe or elongated tank.

In the United States, about half the drinking water, most irrigation water, and, in many instances, a large part of industrial-use water comes from groundwater. In dry areas, such as the western Great Plains, the water in the underground aquifers is essentially "fossil" water—stored during earlier, wetter geological periods—that is not now being recharged. Consequently, it is a nonrenewable resource, like oil. A case in point is the heavily irrigated grain region of western Nebraska, Oklahoma, Texas, and parts of Kansas, where the principal aquifer, called the Ogallala, will be pumped out by 2030–2040 (Opie 1993). Land use will then have to revert to grazing and dryland farming unless water in huge quantities can be piped from large Mississippi Valley rivers—an expensive and energy-demanding public works project that would place a burden on taxpayers. Decisions regarding water supply and its use have yet

Table 4-2 **Reservoir sizes and turnover times of global water (H_2O)**

Reservoir	*Quantity*	*Turnover time**
Oceans	1,380,000	37,000
Polar ice, glacier	29,000	16,000
Groundwater (actively exchanged)	4000	300
Freshwater lakes	125	10–100
Saline lakes	104	10–10,000
Soil moisture	67	280 days
Rivers	1.32	12–20 days
Atmospheric water vapor	14	9 days

*Turnover time in years unless otherwise noted.

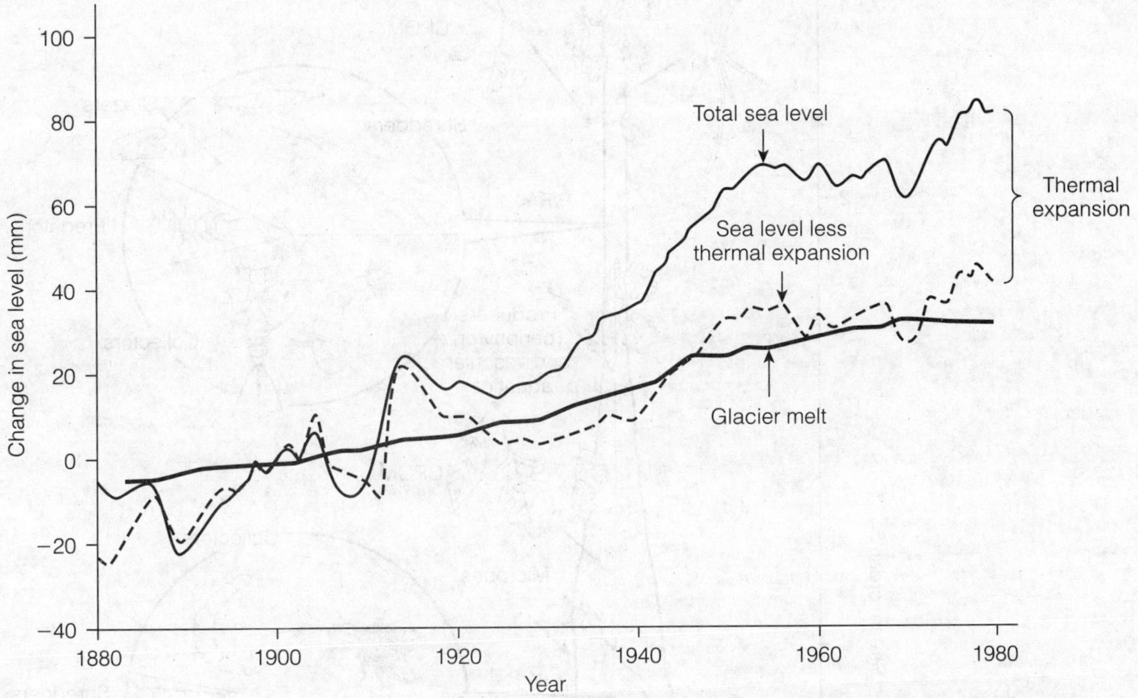

Figure 4-8. Changes in sea level during the last century, indicating the proportion due to thermal expansion of the oceans and that due to melting of glaciers. (From Gornitz, V., S. Lebedeff, and J. Hansen. 1982. Global sea level in the past century. *Science* 215:1611–1614. Copyright © 1984 AAAS.)

to be made, but the political controversy in the future will certainly be bitter, and many people will be caught in the economic collapses that always come when nonrenewable resources are exploited without regard for tomorrow.

As shown in Table 4-2, the polar icecaps and mountain glaciers constitute the second largest water reservoir pool. Due to the melting of the global ice packs, the sea level has been very gradually rising during the past century (Fig. 4-8). About half of this rise is due to thermal expansion, because warm water occupies more space than very cold water or ice. This small but perceptible sea-level rise is the best evidence for a global warming trend.

Figure 4-9A is a model of the downhill loop in the H_2O cycle, showing how biotic communities adjust to the changing conditions in what has been termed the **river continuum concept** (the gradient from small to large streams; Cummins 1977; Vannote et al. 1980). Headwater streams are small and often completely shaded, so that little light is available to the aquatic community. Consumers depend largely on leaf and other organic detritus entering from the watershed. Large *coarse particulate organic matter* (CPOM), such as leaf fragments, predominates, as do aquatic insects and other primary consumers belonging to a class termed *shredders* by stream ecologists. The headwater ecosystem is heterotrophic, with a *P/R* ratio of much less than one.

In contrast, midsections of rivers are wider, no longer shaded, and depend less on imported organic matter from watersheds, because autotrophic algae and aquatic macrophytes provide net primary production. *Fine particulate organic matter* (FPOM)

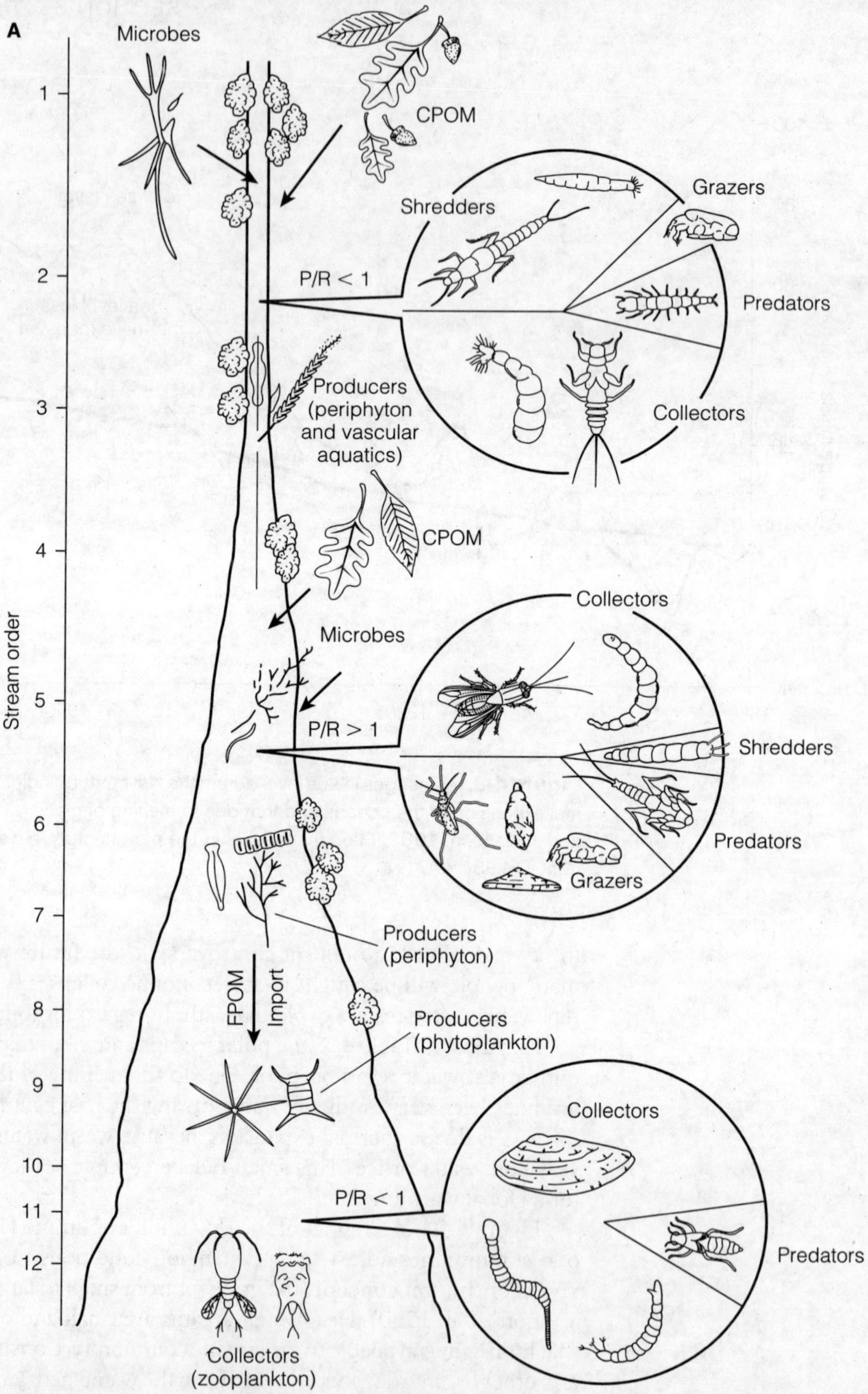

Figure 4-9. (A) Overview of the river continuum, depicting stream order, organisms by feeding type, and changes in particulate matter. CPOM = coarse particulate organic matter; FPOM = fine particulate organic matter (after Cummins 1977). (B) Model of the river continuum, depicting changes in community metabolism (*P*/*R* ratios), diversity, and particle size from headwater streams to large rivers (after Vannote et al. 1980).

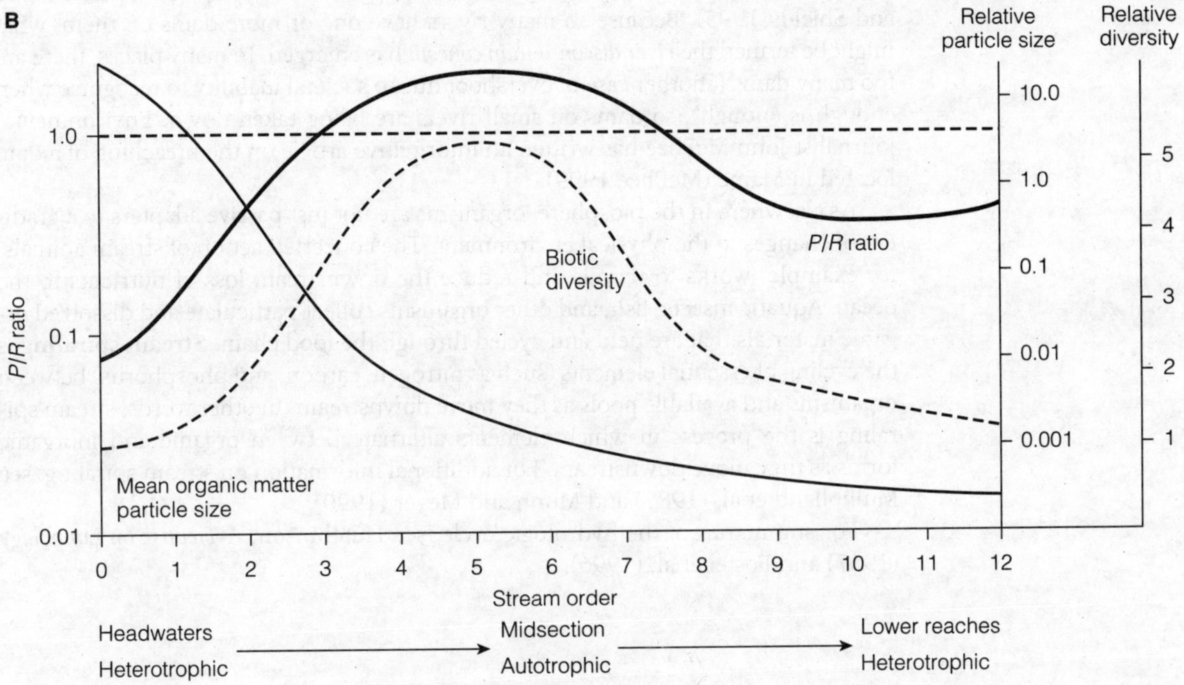

Figure 4-9. *(continued)*

and filter feeders with collector adaptations (equipped with nets and strainers) predominate. Community metabolism in stream midsections is autotrophic, with a *P/R* ratio of one or more (Fig. 4-9B). Species diversity and diurnal temperature range generally peak in the stream midsection. In the lower reaches of large rivers, the current is reduced and the water is usually deeper and muddy, thereby decreasing light penetration and aquatic photosynthesis. The stream then again becomes heterotrophic ($P/R < 1$), with a reduced variety of species at most trophic levels.

Whereas the river continuum concept describes a stream longitudinally, the **flood pulse concept** views the stream both laterally and longitudinally, including both the river itself and its associated riparian floodplain (B. L. Johnson et al. 1995). This concept holds that *periodic flooding* is a natural event to which biological communities are adapted. This annual advance and retreat of floodwaters extends the river onto the floodplain. Thus, the river system includes not only the main channel, but also off-channel streams and the floodplain. The *floodplain* supports a highly productive riparian, bottomland forest, a variety of aquatic habitats, and a gradient of plant species adapted to seasonal degrees of flooding and drying (Junk et al. 1989). During flooding, the floodwaters deposit nutrients and sediments into the riparian system. Floodwaters also bring young fish and aquatic invertebrates into these periodic nursery grounds. Receding waters stimulate rates of decomposition, regrowth of grasses and shrubs, and pulses in the abundance of small mammals.

Channelization of rivers, construction of dams, and increased pollution compromise both the river continuum and flood pulse concepts. An understanding of both concepts is essential for managing and emulating natural hydrologic regimes (Gore and Shields 1995). Because so many rivers have one or more dams on them, what might be termed the *river discontinuum concept* has emerged. In many places, there are too many dams (another case of overshoot due to societal inability to recognize when enough is enough), so dams on small rivers are being taken down. Environmental journalist John McPhee has written an informative article on the breaching of a dam located in Maine (McPhee 1999).

As elsewhere in the biosphere, organisms are not just passive adapters to a gradient of changes in the physical environment. The concerted action of stream animals, for example, works to recycle and reduce the downstream loss of nutrients to the ocean. Aquatic insects, fish, and other organisms collect particulate and dissolved organic materials that are held and cycled through the food chain. **Stream spiraling** is the cycling of essential elements (such as nitrogen, carbon, and phosphorus) between organisms and available pools as they move downstream. In other words, stream spiraling is the process in which elements alternate between organic and inorganic forms as they move downstream. For additional information on stream spiraling, see Mulholland et al. (1985) and Munn and Meyer (1990).

For summaries of the hydrologic cycle, see Hutchinson, *A Treatise on Limnology* (1957) and Postel et al. (1996).

7 Turnover and Residence Times

Statement

The concept of turnover, as first introduced in Chapter 2, is useful for comparing the exchange rates between different compartments of an ecosystem after a pulsing equilibrium has been established. The **turnover rate** is the fraction of the total amount of a substance in a compartment that is released (or that enters) in a given period of time, whereas the **turnover time** is the reciprocal of this—that is, the time required to replace a quantity of a substance equal to its amount in the compartment. For example, if 1000 units are present in the compartment and 10 go out or enter each hour, the turnover rate is 10/1000 (0.01), or 1 percent per hour. The turnover time would then be 1000/10, or 100 hours. **Residence time,** a term widely used in the geochemical literature, is a concept similar to turnover time; it refers to the time a given amount of a substance remains in a designated compartment of a system.

Explanation

As previously emphasized, the flux or rate of movement of nutrients in and out of pools is more important than the amounts within the pools when it comes to understanding how ecosystems function. For example, Pomeroy (1960) commented that "a rapid flux of phosphate is more important than concentration in maintaining high rates of organic production."

Examples

Estimates of reservoir sizes and turnover times in the global cycles discussed in the previous five sections are listed in Tables 4-1 and 4-2. Although turnover time tends to be shorter in the smaller pools, the relationship between pool size and turnover is not linear. Much depends on the location of the reservoir.

Advances in detection technology that made possible the measurement of very small amounts of both radioactive and stable isotopes of all the major biogenic elements have given a tremendous stimulus to cycling studies at the landscape level, because these isotopes can be used as *tracers* or tags to follow the movement of materials. Ponds and lakes are sites especially conducive to tracer studies because their nutrient cycles are relatively self-contained over short periods.

A model of a Georgia salt marsh studied by field observations and experiments using ^{32}P illustrates the importance of filter feeders and detritus complexes in recycling phosphorus in this estuarine system. For example, Kuenzler (1961a) found that a population of filter-feeding mussels (*Modiolus demissus*) alone "recycles" from the water every 2.5 days a quantity of particulate phosphorus equivalent to the amount present in the water (that is, the turnover time for particulate phosphorus is only 2.5 days). Kuenzler (1961b) also measured the energy flow of the population and concluded that the mussel population is more important to the ecosystem as a biogeochemical agent than as a transformer of energy (that is, as a potential source of food for other animals or humans). This example illustrates that a species does not have to be a link in the food chain in order to be valuable to life. Many species are valuable in indirect ways by providing ecosystem services that are not apparent without careful examination.

8 Watershed Biogeochemistry

Statement

Like all ecosystems, bodies of water are open systems, and they need to be considered as parts of larger **drainage basins** or **watersheds.** Insofar as practical management is concerned, the watershed provides a sort of minimum ecosystem or landscape unit. Long-term studies (10 years or more of year-round research) on experimental, instrumented watersheds (outdoor macrocosms)—such as those ongoing at Hubbard Brook Experimental Forest located in New Hampshire; Coweeta Hydrologic Laboratory located in western North Carolina; and Schindler's (1990) long-term studies of a Canadian lake watershed—have greatly advanced understanding of the basic biogeochemical processes as they occur in relatively undisturbed ecosystems. Schindler's work, for example, has demonstrated that phosphorus is frequently in short supply in aquatic systems and often limits freshwater productivity. He also demonstrated the critical role of phosphorus in eutrophication by adding phosphorus to one half of a twin lake system that was divided by a plastic curtain. The system that received phosphorus was covered by a heavy bloom of photosynthetic cyanobacteria within two months. Such studies, in turn, have provided a basis for comparison with agricultural, urban, and other domesticated watersheds, where most people live. These comparisons reveal wasteful human activities and point to means of reducing downhill losses, restoring the cyclic behavior of vital nutrients, and, of course, conserving energy.

Examples

A quantitative model of the calcium cycle for mountainous, forested watersheds of the Hubbard Brook study area located in New Hampshire is presented in Figure 4-10. The data are based on studies of six watersheds ranging in size from 12 to 48 hectares (Bormann and Likens 1967, 1979; Likens et al. 1977, 1996; Likens 2001a). Precipitation, which averaged 123 cm (58 inches) per year, was measured by a network of gauging stations, and the amount of water leaving the watershed in the drainage stream of each watershed unit was measured by a V-notched weir similar to that shown in Figure 2-4B. From the concentration of calcium and other minerals in the input and output water and in the biotic and soil pools, a watershed input-output "budget" can be calculated, as shown in a simplified manner in Figure 4-10A.

Retention by and recycling within the undisturbed but rapidly growing forest proved so effective that the estimated loss from the ecosystem was only 8 kg Ca $\cdot$ ha^{-1} $\cdot$ year^{-1} (and equally small amounts for other nutrients). Because 3 kg of the calcium was replaced in rain, an input of only 5 kg/ha would be needed to achieve a balance. This amount is thought to be easily supplied by the normal rate of weathering from the underlying rock that constitutes the reservoir pool. Experiments with ^{45}Ca to measure turnover in Oak Ridge, Tennessee, watersheds have demonstrated how understory trees, such as dogwood (*Cornus florida*), act as calcium pumps that counter the downward movement in the soil and thus recirculate calcium between organisms and the active upper layers of litter and soil.

Table 4-3 summarizes the mean concentrations of calcium in three plant species sampled at the Coweeta Hydrologic Laboratory located in Franklin, North Carolina. Chestnut oak (*Quercus prinus*) had the largest standing crop of all nutrients due to its large size; the leaves of chestnut oak had the highest standing crop of nitrogen. Rhododendron (*Rhododendron maximum*), an evergreen, had the largest standing crop of leaf biomass of the three species. Because its leaves are evergreen, rhododendron recycled nutrients over a period of seven years instead of the one year typical of the other two species. Flowering dogwood (*Cornus florida;* Fig. 4-11) had the smallest biomass but had the distinction of possessing the highest concentration of calcium in its leaves (Table 4-3). Flowering dogwood concentrated more than three times as much calcium per unit leaf biomass as did chestnut oak. The small dogwood tree recycled 66 percent as much calcium as the chestnut oak and 150 percent more than the rhododendron. Thus, different plant species have pronounced but different influences on nutrient recycling based on their size, life histories, and longevity.

Flowering dogwood is currently under attack by a fatal disease known as dogwood anthracnose (*Discula destructiva*). Caused by a fungus, this disease has spread across nearly 1.6 million hectares (4 million acres), thereby changing the composition and appearance of forests throughout the Appalachian Mountains (Bolen 1998; Rossell et al. 2001). Stiles (1980) categorized the drupes of flowering dogwood as a high-quality fall fruit providing food resources for 40 species of migratory and overwintering birds and for numerous species of mammals. The most significant impact

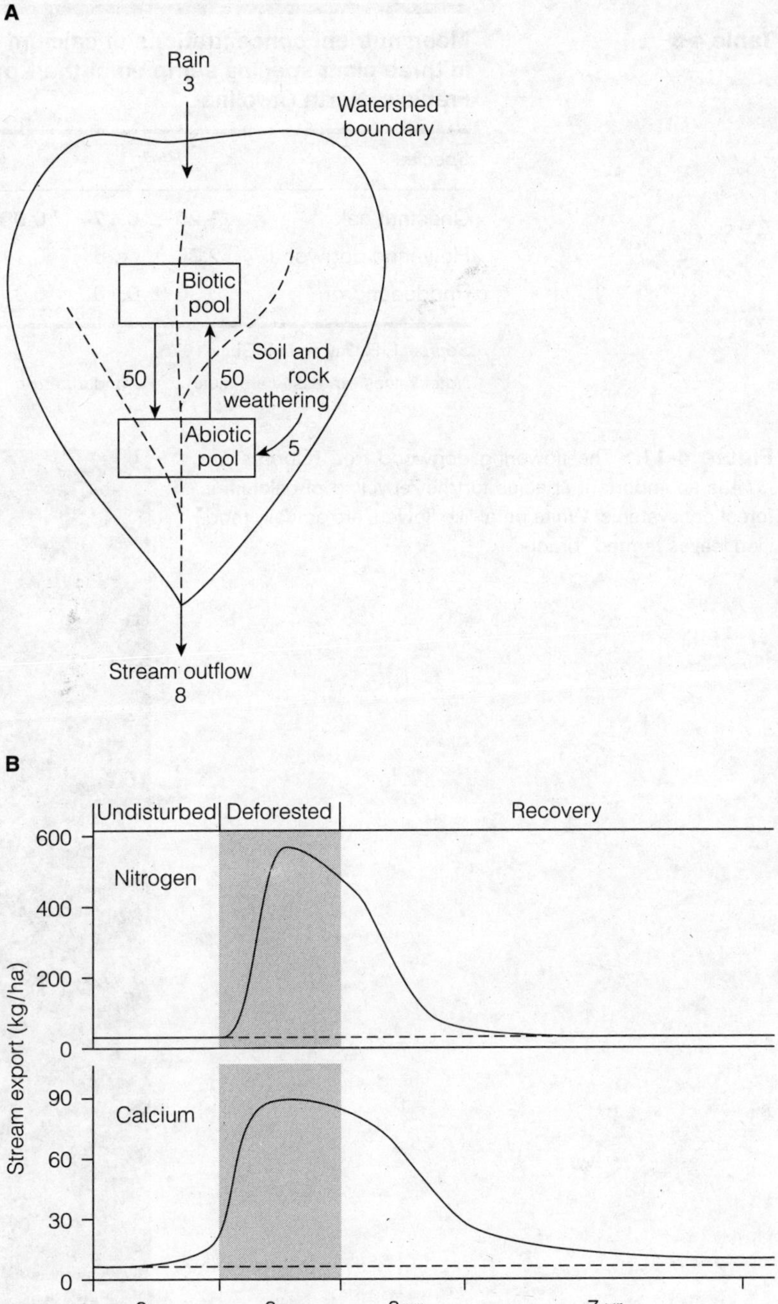

Figure 4-10. (A) Balanced calcium budget of a forested watershed at Hubbard Brook Experimental Forest, New Hampshire. Values are calcium flows in kg per hectare per year. Note that the inputs and outputs are small compared with the exchanges between biotic and abiotic pools within the watershed ecosystem. (B) Effect of deforestation and natural reforestation (recovery) on stream output of nitrogen and calcium (after Bormann and Likens 1979).

of dogwood anthracnose will likely be the loss of flowering dogwoods from the landscape, with the concomitant loss of fruit production. Thus, this plant-fungus relationship at the community (forest) level has direct ramifications at the ecosystem level, because flowering dogwoods are known to be important in the cycling of calcium within forest ecosystems, and likely will lead to changes in productivity and

Table 4-3 **Mean nutrient concentrations of calcium (percentage dry weight) in three plant species sampled at the Coweeta Hydrologic Laboratory, Franklin, North Carolina**

Species	*Bark*	*Wood*	*Twigs*	*Leaves*
Chestnut oak	1.25 ± 0.17	0.09 ± 0.01	0.68 ± 0.06	0.58 ± 0.07
Flowering dogwood	2.36 ± 0.26	0.11 ± 0.01	0.80 ± 0.06	1.85 ± 0.11
Rhododendron	0.30 ± 0.10	0.07 ± 0.31	0.99 ± 0.24	1.20 ± 0.29

Source: F. P. Day and McGinty 1975.

Note: Values expressed are mean ± 1 standard error.

© Farrell Grehan/CORBIS

Figure 4-11. The flowering dogwood tree (*Cornus florida*) is an important species for the recycling of calcium in forest ecosystems. White petal-like flowers are actually modified leaves termed "bracts."

biodiversity at the landscape level. This is an excellent example of how a cause-and-effect relationship at one level of organization can result in cascading effects through several levels of organization.

In one of the Hubbard Brook experimental watersheds, all the vegetation was felled, and any regrowth in the next three seasons was suppressed by the application of herbicides. Even though the soil was little disturbed and no organic matter was removed by this procedure, the loss of mineral nutrients in stream outflow increased 3 to 15 times over the losses from the undisturbed control watersheds. The sixfold increase in loss of calcium and 15-fold increase in nitrogen loss are shown in Figure 4-10B. The increased stream flow out of the razed ecosystem resulted primarily from the elimination of plant transpiration, and it was the additional stream flow that carried out additional minerals. To some extent, outflows are related to what geochemists call *relative mobility.* Potassium and nitrogen are very mobile (easily removed by leaching), for example; calcium, however, is more tightly held in the soil.

When vegetation was allowed to recover (no further application of herbicide), the rate of nutrient loss declined rapidly, with a "balanced budget" being restored in 3 to 5 years, although 10 to 20 years were required for all nutrients to return to the baseline output of an undisturbed forested watershed (Fig. 4-10B). The rapid recovery of nutrient retention—long before the species composition and biomass of the original forest can be restored—is aided by a number of mechanisms, such as what Marks (1974) called the *buried seed strategy.* The seeds of rapidly growing pioneer trees, such as pin cherry (*Prunus pennsylvanica*), remain viable for years when buried in the soil. When the forest is removed, these seeds sprout, and the fast-growing cherry trees quickly form a sort of temporary forest that stabilizes water and nutrient fluxes and reduces soil and nutrient loss from the watershed. Of course, such quick recovery adaptations have evolved in response to natural perturbations, such as storms and fire. In fact, forests (and other ecosystems) subjected to (adapted to) natural periodic disruptions are more resilient and recover more quickly after human disturbances than do forests in benign physical environments that are less subject to severe natural perturbations. Accordingly, *inherent resilience is an ecosystem-level property that needs to be considered when harvest procedures or other management practices are being decided.*

The Coweeta watershed site in the mountainous deciduous forests of North Carolina consists of a series of small tributary headwater streams flowing into a larger stream that runs down the middle of the basin. This basin has been under continuous study since 1934, making it the longest studied of any landscape in North America. Coweeta was one of the first research sites to adopt large-scale experimental approaches to the study of natural landscapes and to set up permanent water flow and measurement weirs (Swank et al. 2001).

Early studies at Coweeta focused on hydrology, especially on water yield downstream as affected by different land uses and forestry practices. Individual head watersheds were left natural, selectively cut, clear-cut, planted in crops, or had hardwoods replaced by pine plantations. In general, these experiments showed that reducing the biomass of the vegetative cover increased water flow downstream but decreased water and soil quality (Swank and Crossley 1988)—an example of the quantity-quality dilemma (seemingly, one cannot maximize for both at the same time).

In recent years, Coweeta, as a Long-Term Ecological Research (LTER) site sponsored by the National Science Foundation, has concentrated research on the biotic elements of the ecosystem—trees, insects, soil biota, stream life, and litter decomposition—and focused on the effects of natural disturbances, such as drought, flooding, storms, and caterpillar defoliations.

Research at Hubbard Brook and Coweeta has demonstrated that forests (and other ecosystems) that are subject to frequent natural perturbations recover quickly from *acute* disturbances but are less resilient when it comes to long-continued, *chronic* disturbances such as eutrophication or toxic chemicals. Whereas nutrient losses from undisturbed forested watersheds along headwater streams are small, and mostly replaced by inputs from rain and weathering, the picture downstream, where human activities are more intense, is quite different. Concentrations of nitrogen and phosphorus in the water of streams and rivers increase sharply as watersheds are increasingly domesticated (that is, as the percentage of watershed area in agricultural and urban use increases). Nutrient concentrations in water flowing out of an urban-agricultural landscape are sevenfold higher than in streams draining a completely forested watershed. Eighty percent of the phosphorus output from agricultural and urban landscapes is inorganic (phosphate), whereas organic phosphorus predominates in runoff from watersheds that are completely occupied by forest or other natural vegetation. Most other nutrients, and many other chemicals (including toxic ones), show a similar pattern of increasing runoff with increasing intensity of land and energy use by humans. The large outputs of nutrients and other chemicals from domesticated and especially from industrialized landscapes are, of course, a more or less direct result of the large inputs of agricultural and industrial chemicals and organic human and domestic animal wastes. Thus, ecosystem processes such as stream eutrophication and biomagnification are increased.

9 Cycling of Nonessential Elements

Statement

Although nonessential elements may have little or no known value to an organism or species, they often pass back and forth between organisms and their environment in the same general manner as do the essential elements. Many of these nonessential elements are involved in the general sedimentary cycle, and some find their way into the atmosphere. Many nonessential elements become concentrated in certain tissues, sometimes because of their chemical similarity to specific vital elements. Chiefly because human activities involve many of the nonessential elements, ecologists have become concerned with the cycling of these elements; indeed, all of us must be concerned with the increasing volume of toxic wastes that are discharged or that inadvertently escape into the environment and contaminate the basic cycles of vital elements.

Explanation

Many marine animals concentrate elements—such as arsenic, a phosphorus analog—that they cannot remove from their environment. They then convert the arsenic into an inert chemical form stored in their tissues. Some elements, such as mercury, are passed up the food chain; thus, large predatory animals tend to accumulate high concentrations of it. This process, termed **biological magnification**, is the reason why a number of fishes, such as swordfish and tuna, contain potentially harmful amounts of mercury. The process of biological magnification will be illustrated and discussed in greater detail in Chapter 5.

Most nonessential elements have little effect at the concentrations normally found in most natural ecosystems, probably because organisms have become adapted to their presence. Therefore, their biogeochemical movement would be of little interest, were it not for the byproducts of the mining, manufacturing, chemical, and agricultural industries that contain high concentrations of heavy metals, toxic organic compounds, and other potentially dangerous materials that all too often find their way into the environment. Consequently, the cycling of every element is important. Even a very rare element can become of biological concern if it takes the form of a highly toxic metallic compound or a radioactive isotope, because a small amount of such material (from the biogeochemical standpoint) can have a marked biological effect.

Examples

Strontium is an example of a previously almost unknown element that must now receive special attention, because radioactive strontium is particularly dangerous to humans and other vertebrates. Strontium behaves like calcium, with the result that radioactive strontium gets into close contact with calcium-rich blood-making tissues in our bones. About 7 percent of the total sedimentary material flowing down rivers is calcium. For every 1000 atoms of calcium, 2.4 atoms of strontium move along with the calcium to the sea. When uranium is fissioned in the preparation and testing of nuclear weapons and in nuclear power plants, it yields radioactive strontium-90 as a waste product—one of a number of fission products that decay very slowly. Strontium-90 is a relatively new material added to the biosphere; it did not exist in nature before the atom was split. Tiny amounts of radioactive strontium, released in fallout from nuclear weapons testing and escaping from nuclear reactors, have now followed calcium from soil and water into vegetation, animals, human food, and human bones. Strontium-90 present in the bones of people can have carcinogenic effects.

Radioactive cesium-137, another dangerous fission product, behaves like potassium and, therefore, cycles rapidly through food chains. The Arctic Tundra is an ecosystem that has been subject to nuclear fallout from weapons testing in the past. The Arctic Tundra received another input of radioactive materials from the explosion in the Chernobyl nuclear power plant in 1986. Large amounts of radioactive fission products are now stored in tanks at atomic energy facilities. The lack of technological knowledge to safely process and store these wastes has limited the peaceful uses of atomic energy. The problem of hazardous wastes will be considered in more detail in Chapter 5.

Mercury is another example of a natural element that had little impact on life before the industrial age because of its low concentration and low mobility. Mining and manufacturing have changed all that, and mercury and other heavy metals (such as cadmium, lead, copper, and zinc) are now severe pollution problems; see Levine et al. (1989), Brewer et al. (1994), and Brewer and Barrett (1995) for a review of a 10-year investigation of heavy-metal concentrations through trophic levels resulting from municipal sludge treatment in an old-field ecosystem.

A number of aquatic plants have the ability to sequester and store large amounts of toxic heavy metals in their tissues without injury to themselves. The feasibility of propagating and bioengineering such plants to clean up industrial spills of mercury, nickel, and lead is now being investigated. For a review of mercury as a global problem, including approaches to its cleanup, see Porcella et al. (1995).

10 Nutrient Cycling in the Tropics

Statement

The pattern of nutrient cycling in the Tropics, especially the Wet Tropics, is different in several important ways from that in the North Temperate Zone. In cold regions, a large portion of the organic matter and available nutrients is located in the soil or sediment at all times; in the Tropics, a much larger percentage is located in the biomass and is recycled rapidly within the organic structure of the system, aided by a number of nutrient-conserving biological adaptations, including mutualistic symbiosis between microorganisms and plants. When this evolved and well-organized biotic structure is removed (by deforestation, for example), nutrients are rapidly lost by leaching under conditions of high temperature and heavy rainfall, especially on sites that are initially poor in nutrients. For this reason, the agricultural strategies of the North Temperate Zone, involving the monoculture of short-lived annual plants, are quite inappropriate for tropical regions. An ecological reevaluation of tropical agriculture and environmental management is urgently needed if past mistakes are to be corrected and if future ecodisasters are to be avoided. At the same time, the rich genetic, species, and habitat diversity of the Tropics must be preserved. Swidden agriculture, invented independently in many parts of the Tropics, is better suited to moist, montane regions.

Explanation

Figure 4-12 compares the distribution of organic matter and nutrients in a northern temperate and a tropical forest. Interesting enough, in this comparison both ecosystems contain about the same amount of organic carbon, but in the temperate forest, more than half is in litter and soil, whereas in the tropical forest, more than three fourths is in vegetation, especially wood biomass.

When a forest in the North Temperate Zone is removed, the soil retains nutrients and structure and can be farmed for many years by *conventional agriculture,* which involves plowing one or more times a year, planting annual species, and applying inorganic fertilizers. During the winter, freezing temperatures help retain nutrients and control pests and parasites. In the humid Tropics, however, forest removal takes away

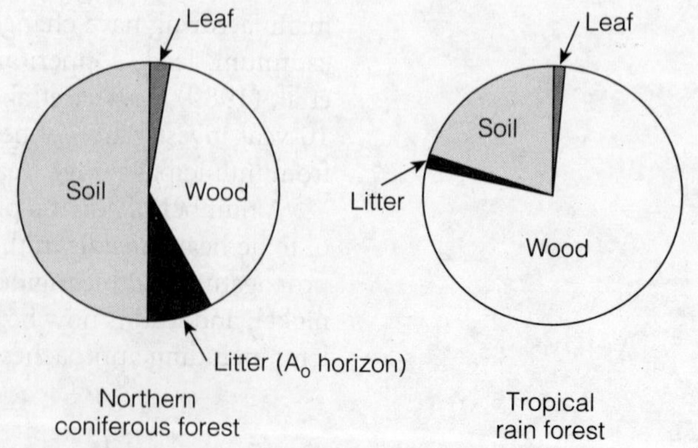

Figure 4-12. Comparison of the distribution of organic carbon accumulated in abiotic (soil, litter) and biotic (wood, leaves) compartments of a northern temperate and a tropical forest ecosystem. Overall quantities are similar (~250 tons/ha), but a much larger percentage of total organic carbon is in the biomass in the tropical forest.

the ability of the land to hold and recycle nutrients (and to combat pests) in the face of high year-round temperatures and long periods of leaching rainfall. Often, crop productivity declines rapidly, and the land is abandoned, creating a pattern of *shifting* or *swidden agriculture.* Community control in general, and nutrient cycling in particular, tend to be more physical in the North Temperate Zone and more biological in the Tropics; in other words, temperate nutrient pools are mostly in the soil and litter, whereas in the Wet Tropics, the nutrient pool is in the biomass.

It should be noted, however, that swidden agriculture can be sustained as long as human population densities are low—as was the case in the past—and long-term rotations are slow and continuous. The problem with swidden agriculture is not with the process itself but with *overpopulation,* which makes more and more clearing necessary and does not allow long enough time intervals between reclearing. Also, not all tropical-latitude agriculture is located in rain forests. For example, people living in Peru, Ecuador, and Papua New Guinea have practiced sustained agriculture for centuries (Rappaport 1968). This brief account, of course, oversimplifies the complexity of the situation, but it reveals the basic ecological reason why subtropic and tropic sites, which support luxurious and highly productive forests or other vegetation, yield so poorly under conventional temperate methods of crop management.

C. F. Jordan and Herrera (1981) pointed out that the degree to which tropical forests "invest," as it were, in nutrient-conserving recycling mechanisms depends on the geology and basic fertility of the site. Large areas of tropical forests (such as in most of the eastern and central Amazon Basin) are on ancient, highly leached Precambrian soils or nutrient-poor sand deposits. These oligotrophic sites nevertheless support forests as luxurious and productive as those found on more eutrophic (fertile) sites, such as in the mountains of Puerto Rico and Costa Rica and in the foothills of the Andean Mountains. Intricate symbiosis between autotrophs and heterotrophs, involving special microorganism intermediaries, is the key to success in these oligotrophic types of ecosystems.

C. F. Jordan and Herrera (1981) listed the following mechanisms that are especially well developed in rain forest ecosystems on oligotrophic sites:

- Root mats consisting of numerous fine feeder roots penetrating the surface litter recover nutrients from leaf fall and rainfall before they can be leached away. Root mats apparently also inhibit the activities of denitrifying bacteria, thus blocking loss of nitrogen to the air.
- Mycorrhizal fungi associated with root systems act as nutrient traps and greatly facilitate the recovery of nutrients and their retention within the biomass. This symbiosis for mutual benefit is widespread on oligotrophic sites in the North Temperate Zone as well.
- Evergreen leaves with thick, waxy cuticles retard the loss of water and nutrients from trees and also resist herbivores and parasites.
- "Drip tips" on leaves (long, pointed leaf tips) drain off rain water, thereby reducing the leaching of leaf nutrients.
- Algae and lichens that cover the surface of leaves scavenge nutrients from rainfall, some of which are immediately available for uptake by the leaves; the lichens also fix nitrogen.
- Thick bark inhibits the diffusion of nutrients out of the phloem and their subsequent loss by stem flow (rain running down tree trunks).

In summary, the nutrient-poor tropical ecosystem is able to maintain high productivity under natural conditions through a variety of nutrient-conserving mechanisms. These evolutionary mechanisms provide more direct cycling from plant back to plant, more or less bypassing the soil. When such forests are cleared for large-scale agriculture or tree plantations, these mechanisms are destroyed, and productivity declines very rapidly, as do crop yields. When the clearings are abandoned, the forest recovers slowly, if at all. In contrast, forests on eutrophic sites are more resilient.

The development and testing of crop plants with well-developed mycorrhizal and nitrogen-fixing root systems and the greater use of perennial plants are ecologically sound goals for high-temperature areas (such as the southeastern United States) and tropical climates (such as the Philippines). Paddy rice culture is successful in the Tropics because of the special nutrient-retention features of this ancient type of agriculture. Rice paddies have been cultivated on the same site for more than 1000 years in the Philippines—a record of success that few conventional agricultural systems in use today can claim. One certainty is apparent: *Industrialized agrotechnology as practiced in the North Temperate Zone cannot be transferred unmodified to tropical regions.*

11 Recycling Pathways: The Cycling Index

Statement

It is instructive to review the subject of biogeochemistry in terms of recycling pathways, because the recycling of water and nutrients is a vital process in ecosystems and is increasingly becoming an important concern for humankind. Five major **recycling pathways** can be distinguished: (1) by microbial decomposition; (2) by animal excretions; (3) by direct recycling from plant to plant through microbial symbionts; (4) by physical means involving direct action of solar energy; and (5) by use of fuel energy, as in the industrial fixation of nitrogen. Recycling requires dissipation of energy from some source such as organic matter, solar radiation, or fossil fuel. The relative amount of recycling in different ecosystems can be compared by calculating a *cycling index* based on the ratio of the sum of the amounts cycling between compartments within the system to total throughflow.

Explanation

It is appropriate to focus on the cycling of nutrients in the biologically active portion of the ecosystem. Recall that the same approach was used for energy in Chapter 3; the total energy environment was considered first, and then attention was focused on the fate of that small energy fraction involved in the food chain. Also, a discussion of biological regeneration is relevant because recycling has increasingly become a major goal for human societies.

A microbial food web consisting of bacteria, fungi, and microorganisms that consume organic detritus is present in somewhat different forms in all soils and all natural waters. Both dissolved and particulate organic matter in soil and water are partially processed by bacteria, some attached to particles and some floating free in the water. The bacteria are eaten by protozoans, which excrete ammonium and phosphate, which in turn can be reused by plants. This food web is often termed the **detritus pathway** or **detritus cycle.** The complex interactions between microbes and small animal detritivores have been described in Chapter 2. Where small plants, such as grass or phytoplankton, are heavily grazed by animals, recycling by way of animal excretions may also be important.

Measurements of turnover rates indicate that the nutrients that protozoa release during their lifetime is many times the amount of soluble nutrients released by microbial decomposition of their bodies after their death (Pomeroy et al. 1963; Azam et al. 1983). These excretions include dissolved inorganic and organic compounds of phosphorus, nitrogen, and CO_2, which are directly usable by producers without any further chemical breakdown by bacteria.

Direct recycling by symbiotic microorganisms, such as dinoflagellates in coral reefs, is expected to be important in nutrient-poor or oligotrophic environments, such as the oceans or the Everglades. Water, as we have seen, is largely recycled by the direct action of solar energy and by the weathering and erosion processes associated with downhill flows of water that bring the sedimentary elements of abiotic reservoirs into biotic cycles. Human beings enter the recycling picture when they expend fuel energy to desalinate water from the sea, produce fertilizers, or recycle aluminum or other metals.

Recycling work accomplished mechanically or physically can provide an energy subsidy for the system as a whole. In the design of disposal systems for human and industrial wastes, it is frequently profitable to provide an input of mechanical energy to pulverize organic matter and thus hasten its rate of decomposition. Physical breakdown by the activities of large free-ranging mammalian grazers is also important in the release of nutrients from resistant pieces of detritus (McNaughton et al. 1997).

Recycling is not a free service; there is almost always an energy cost. When sunlight and organic matter are the energy sources for recycling work, humans do not need to pay for the use of the services provided by natural capital. Without being disrupted or poisoned, natural recycling mechanisms can do most of the work of recycling water and nutrients. Industrial materials (such as heavy metals) involved in manufacturing are quite another matter; their recycling is costly in fuel and money, but there is little choice when supplies become limited or when the wastes endanger human health.

The Cycling Index

Cycling within ecosystems may be defined in terms of the proportion of incoming material that circulates from one compartment to another before exiting from the system. The recycled fraction is the sum of the amounts cycled through each compartment as follows:

$$CI = \frac{TST_c}{TST}$$

where CI is the **cycling index,** TST_c is the portion of the total system throughflow that is recycled, and TST is the total system throughflow. **Throughflow** is defined as the sum of all inputs minus the change in storage within the system if it is negative or, alternatively, all outputs plus the change in storage if it is positive.

Finn (1978) calculated the cycling index for calcium in the Hubbard Brook watersheds to be between 0.76 and 0.80. This means that about 80 percent of the calcium throughflow is recycled. Cycling indices were even higher for potassium and nitrogen. The nutrients in this watershed appear to recycle in the following order of efficiency (from highest to lowest *CI*): K > Na > N > Ca > P > Mg > S. This ordering relates to the input of each element from outside the system, the mobility of the element, and the biological requirements of biota. Cycling indices are generally low for nonessential elements, such as lead, or for essential elements that are required in very small amounts relative to their availability, such as copper. Elements that human societies consider valuable, such as platinum and gold, are 90 percent or more recycled. As would be expected, the cycling index for energy (calorie flow) is zero, because, as emphasized earlier in this book, due to the second law of thermodynamics, energy passes straight through the system and cannot cycle.

Recycling of Paper

Paper provides an excellent example of how recycling develops in urban-industrial systems in a parallel manner to the recycling of important materials in natural systems. As shown in Figure 4-13A, recycling in natural ecosystems, as measured by a cycling index, increases as the biotic components of the ecosystem become larger and more complex, as resources in the input environment become scarce, or as waste products pile up in the output environment to the detriment of the life within the ecosystem.

As long as there were plenty of trees, paper mills, and vacant land for the disposal

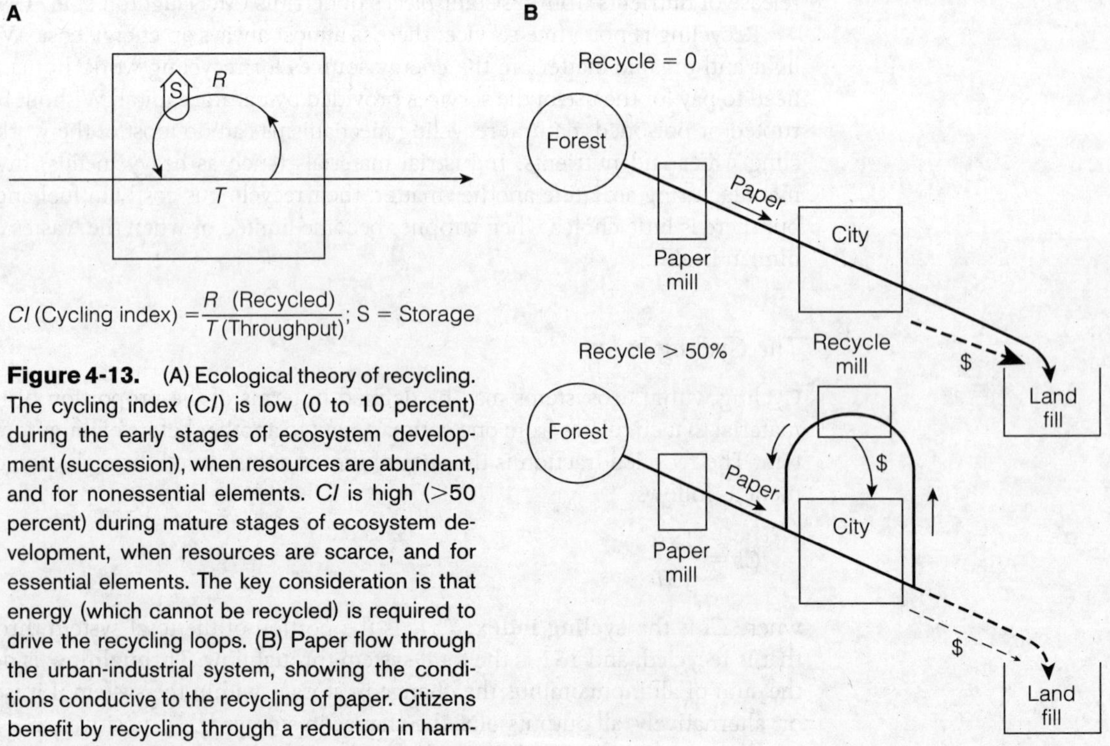

Figure 4-13. (A) Ecological theory of recycling. The cycling index (*CI*) is low (0 to 10 percent) during the early stages of ecosystem development (succession), when resources are abundant, and for nonessential elements. *CI* is high (>50 percent) during mature stages of ecosystem development, when resources are scarce, and for essential elements. The key consideration is that energy (which cannot be recycled) is required to drive the recycling loops. (B) Paper flow through the urban-industrial system, showing the conditions conducive to the recycling of paper. Citizens benefit by recycling through a reduction in harmful environmental impacts (on forests, streams, and land) and taxes for city services.

of waste paper, there was little incentive to invest in facilities and energy to recycle the paper flowing through the urban-industrial system (Fig. 4-13B). As the environs of the city become congested, however, land values rise, and it becomes increasingly difficult and expensive to maintain landfills or disposal sites. Pressure comes from input environments when pulpwood supplies or mill production begin to fall short of demand. In both cases, it "pays" to consider recycling. For paper recycling to be successful, a market (a recycling mill) must exist for used newspaper and cardboard. Such a mill represents an energy-saving recycling mechanism similar to the dissipative structures that are found in natural ecosystems such as forests and coral reefs.

12 Global Climate Change

Statement

As emphasized in Chapter 2, large, complex systems tend to pulse widely in the absence of set-point controls. During the past 150,000 years or so, global climates have fluctuated between two states: warm-moist and cold-dry. Currently, human activities are beginning to affect the forcing functions of global warming.

Explanation

There are at least three ways to study past climate fluctuations: (1) tree-ring dendrology; (2) lake-, bog-, and ocean-bottom cores; and (3) ice cores. Sediment cores reveal that between 130,000 and 115,000 years ago, there was a warm period rather like today's climate, followed by the Wisconsin Ice Age, which ended about 12,000 years ago, when the current, relatively warm Holocene period began (M. B. Davis 1989). Although these large shifts between warm and cool climate occurred over long periods of time, giving organisms time to adapt or shift their ranges geographically, recent studies of polar ice cores reveal that in the past, there have also been periods of rapid climate changes that occurred within less than 50 years. Accordingly, there is a concern that human activities, such as the burning of fossil fuels relating to global increases in CO_2, have triggered such a rapid climate change that humanity will have difficulty dealing with it.

Recall Figure 4-6B, which showed the continuous increase in atmospheric CO_2 concentration from 1958 to 2002, as measured at the Mauna Loa Observatory, Hawaii. The increase in greenhouse gases (especially CO_2 and methane) that tend to warm Earth has been subject to much research and discussion in recent years. Much less is known about the role of dust and aerosols (particles small enough to remain suspended for weeks and months in the troposphere, the lowest layer of the atmosphere, ranging from 10 to 20 kilometers in elevation) or cloud cover, which tend to cool the planet by reflecting solar radiation back into space. There is considerable uncertainty about the indirect cooling effects of aerosols (Andreae 1996; Tegen et al. 1996). Also uncertain are the long-term effects of large volcanic eruptions, such as the eruption of Mount Pinatubo in 1991, which reduced the global mean temperature by about 0.5° C. Likewise, there is uncertainty regarding the effect of reforestation (biomass increase) and sequestration of carbon in soil and ocean sinks, which

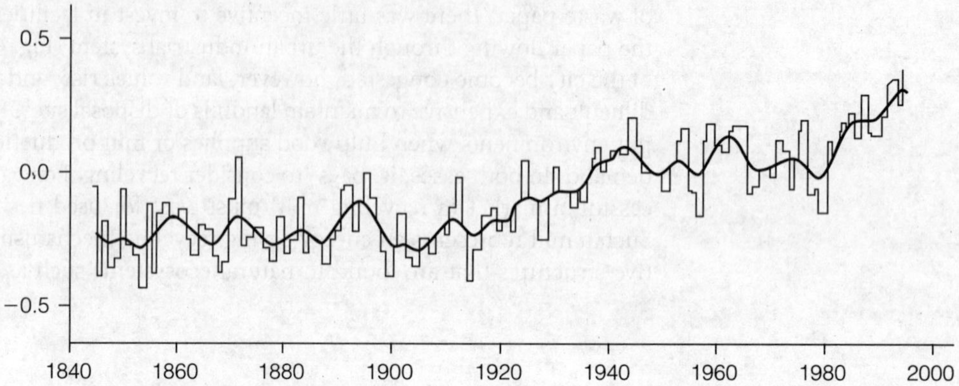

Figure 4-14. Global warming in terms of average mean temperature (1860–2000). Global temperatures pulse strongly, but only since the 1930s has there been significant net warming (courtesy of University of East Anglia, Hadley Centre, Norwich, UK).

could function as a negative feedback to slow global warming (J. L. Sarmiento and Gruber 2002).

During the past several decades, however, the certainty of the rise in CO_2 and other greenhouse gases resulting in increasing global temperatures (Fig. 4-14) has been established. There is now ample evidence to evaluate the ecological impact of recent climate changes on various ecosystem-types, ranging from polar terrestrial to marine environments. There seems to be a pattern of ecological change across systems and levels of organisms (such as changes in assemblages of species in ecological communities, timing of behavioral events, and invasion of exotic species). These responses relate to changes at the population (both flora and fauna), community, ecosystem, landscape, and biome levels of organization (see Walther et al. 2002 for a summary of the ecological responses to recent climate change).

Although it is certain that during the past several decades, an increase in CO_2 (Fig. 4-6B) and other greenhouse gases has resulted in an increase in global temperatures (Fig. 4-14), there is still considerable uncertainty regarding the effects of global warming on rainfall. Grasslands are more vulnerable to changes in rainfall than forests or deserts, as Kaiser (2001) noted. Shrubs and trees will invade grasslands with an increase in rainfall, whereas desert shrubs will invade grasslands with a decrease in rainfall. Changes in grasslands will also depend on the intensity of grazing, as intensive grazing brings on desert shrubs. Only long-term, integrative investigations addressing these global changes will provide the necessary information to manage resources on a worldwide scale.

5

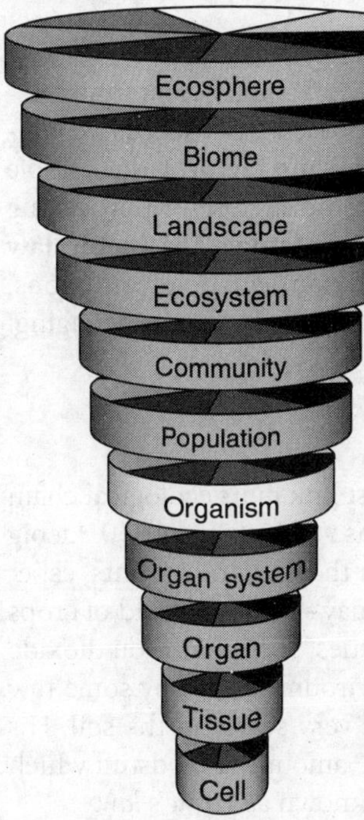

Limiting and Regulatory Factors

1 Concept of Limiting Factors: The Liebig Law of the Minimum
2 Factor Compensation and Ecotypes
3 Conditions of Existence as Regulatory Factors
4 Soil: Organizing Component for Terrestrial Ecosystems
5 Fire Ecology
6 Review of Other Physical Limiting Factors
7 Biological Magnification of Toxic Substances
8 Anthropogenic Stress as a Limiting Factor for Industrial Societies

1 Concept of Limiting Factors: The Liebig Law of the Minimum

Statement

The success of an organism, a group of organisms, or a whole biotic community depends on a complex of conditions. Any condition that approaches or exceeds the limits of tolerance is said to be a **limiting condition** or a **limiting factor.** Under stable conditions, the essential constituent available in amounts most closely approaching the minimum need tends to be the limiting one, a concept termed the **Liebig law of the minimum.** The concept is less applicable under transient-state conditions, when the amounts, and hence the effects, of many constituents are rapidly changing.

Explanation

The idea that an organism is no stronger than the weakest link in its ecological chain of requirements was first clearly expressed by Baron Justus von Liebig in 1840. Liebig was a pioneer in studying the effect of various factors on the growth of plants, especially domestic crops. He found—as do agriculturists today—that the yield of crops was often limited not by nutrients needed in large quantities, such as carbon dioxide and water, because these were often abundant in the environment, but by some raw material (such as zinc) needed in minute quantities but very scarce in the soil. His statement that the "growth of a plant is dependent on the amount of foodstuff which is presented to it in minimum quantity" has come to be known as *Liebig's law.*

Extensive work since the time of Liebig has shown that two subsidiary principles must be added to the concept if it is to be useful in practice. The first is a constraint that *the Liebig law of the minimum is strictly applicable only under relatively stable conditions;* that is, when the average inflows of energy and materials balance the outflows over an annual cycle. To illustrate, suppose that carbon dioxide was the major limiting factor in a lake, and productivity was therefore controlled by the rate of supply of carbon dioxide coming from the decay of organic matter. Assume that light, nitrogen, phosphorus, and other vital elements were available in excess of use (and hence not limiting factors). If a storm brought more carbon dioxide into the lake, the rate of production would change and depend on other factors as well. While the rate is changing, there is less likely to be one minimum constituent. Instead, the reaction depends on the concentration of *all* constituents present, which in this transitional period differs from the usual rate at which the least plentiful constituent is being added. The rate of production would change rapidly as various constituents were used up, until some constituent, perhaps carbon dioxide again, became limiting. The lake system would once again be operating at the rate controlled by the law of the minimum.

The second important consideration is *factor interaction.* Thus, a high concentration or availability of some substance, or the action of some factor other than the minimum constituent may modify the rate of use of the limiting factor. Sometimes organisms can substitute, at least in part, a chemically closely related substance for one that is deficient in the environment. Thus, where strontium is abundant, mollusks can partially substitute strontium for calcium in their shells. Some plants have been shown to require less zinc when growing in the shade than when growing in full sunlight; therefore, a low concentration of zinc in the soil would less likely be limiting to plants in the shade than to plants under the same conditions in full sunlight.

Limits of Tolerance Concept

Not only may too little of something be a limiting factor, as proposed by Liebig (1840), but also too much (as in the case of nitrogen documented in Chapter 4) of such factors as heat, light, and water. Thus, organisms have an ecological minimum and maximum; the range in between represents the *limits of tolerance*. The concept of the limiting effect of maximum as well as minimum constituents was incorporated into the **Shelford law of tolerance** (Shelford 1913). Since then, much work has been done in "stress ecology," so that the limits of tolerance within which various plants and animals can exist are well known. Especially useful are what can be termed *stress tests*, carried out in the laboratory or in the field, in which organisms are subjected to an experimental range of conditions (see Barrett and Rosenberg 1981 for details). Such a physiological approach has helped ecologists to understand the distribution of organisms in nature; however, it is only part of the story. All physical requirements may be well within the limits of tolerance for an organism, but the organism may still fail because of biological interrelations, such as competition or predation (see Chapter 7 for details). Studies in intact ecosystems must accompany experimental laboratory studies, which, of necessity, isolate individuals from their populations and communities.

Some subsidiary principles to the law of tolerance may be stated as follows:

- Organisms may have a wide range of tolerance for one factor and a narrow range for another.
- Organisms with wide ranges of tolerance for limiting factors are likely to be most widely distributed.
- When conditions are not optimal for a species with respect to one ecological factor, the limits of tolerance may be reduced for other ecological factors. For example, when soil nitrogen is limiting, the resistance of grass to drought is reduced (more water is required to prevent wilting at low nitrogen levels than at high levels).
- Frequently, organisms in nature are not actually living at the optimum range (as determined experimentally) of a particular physical factor. In such cases, some other factor or factors are found to have greater importance. For example, cord grass (*Spartina alterniflora*), which dominates East Coast salt marshes, actually grows better in freshwater than in salt water, but in nature it is found only in salt water, apparently because it can extrude the salt from its leaves better than other rooted marsh plants (that is, because this mechanism enables cord grass to outcompete its competitors).
- Reproduction is usually a critical period when environmental factors are most likely to be limiting. The limits of tolerance for reproductive individuals, seeds, eggs, embryos, seedlings, and larvae are usually narrower than for nonreproducing adult plants or animals. Thus, an adult cypress tree will grow continually submerged in water or on dry upland, but it cannot reproduce unless there is moist, unflooded ground for seedling development. Adult blue crabs and many other marine animals can tolerate brackish water or freshwater that has a high chloride

content and, thus, are often found for some distance up rivers. The larvae, however, cannot live in such waters; therefore, the species cannot reproduce in the riverine environment and never becomes established permanently. The geographical range of game birds is often determined by the impact of climate on eggs or young rather than on adults. One could cite hundreds of other examples.

For the relative degree of tolerance, a series of terms have come into general use in ecology that use the prefixes *steno-*, meaning "narrow," and *eury-*, meaning "wide." Thus,

stenothermal-eurythermal	refers to narrow and wide tolerance, respectively, of *temperature*
stenohydric-euryhydric	refers to narrow and wide tolerance, respectively, of *water*
stenohaline-euryhaline	refers to narrow and wide tolerance, respectively, of *salinity*
stenophagic-euryphagic	refers to narrow and wide tolerance, respectively, of *food*
stenoecious-euryecious	refers to narrow and wide tolerance, respectively, of *habitat selection*

These terms apply not only to the organism level but equally well to the community and ecosystem levels. For example, coral reefs are very stenothermal, in that they prosper only within a very narrow range of temperature. A prolonged 2° C temperature drop is stressful, causing "bleaching" or loss of the symbiotic algae that make it possible for corals to prosper in very low-nutrient waters.

The concept of limiting factors is valuable because it gives the ecologist an "entering wedge" into the study of complex ecosystems. Environmental relations of organisms are complex, but fortunately, all possible factors are not equally important in a given situation for a particular organism. Studying a particular situation, the ecologist can usually discover the probable weak links and focus attention, initially at least, on those environmental conditions most likely to be critical or limiting. If an organism has a wide limit of tolerance for a relatively constant factor present in moderate quantity in the environment, that factor is not likely to be limiting. Conversely, if an organism is known to have definite limits of tolerance for a factor that is also variable in the environment, then that factor merits careful study, because it might be limiting. For example, oxygen is so abundant, constant, and readily available in aboveground terrestrial environments that it is rarely limiting to land organisms, except to parasites or organisms living in soil or at high altitudes. On the other hand, oxygen is relatively scarce and often extremely variable in water and, thus, is often an important limiting factor to aquatic organisms, especially animals.

Examples

For an example of limiting factors at the species level, compare the conditions under which brook trout (*Savelinus*) eggs and eggs of the leopard frog (*Rana pipiens*) develop and hatch. Trout eggs develop between 0° and 12° C, with an optimum at about 4° C. Frog eggs develop between 0° and 30° C, with an optimum at about 22° C. Thus, trout eggs are stenothermal and low-temperature tolerant, whereas frog eggs are eu-

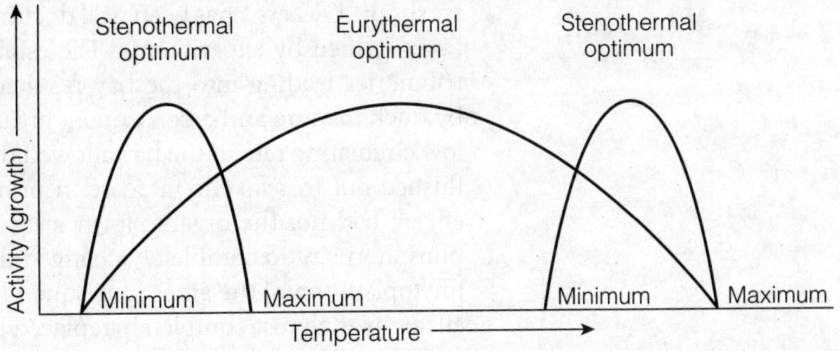

Figure 5-1. Relative limits of tolerance of stenothermal and eurythermal organisms. For a stenothermal species, minimum, optimum, and maximum lie close together, so that a small difference in temperature, which might have little effect on a eurythermal species, is often critical.

rythermal and high-temperature tolerant. Trout—both eggs and adults—are in general relatively stenothermal, but some species are more eurythermal than the brook trout. Likewise, of course, species of frogs differ. These concepts, and the use of terms in regard to temperature, are illustrated in Figure 5-1. In a way, the evolution of narrow limits of tolerance might be considered a form of specialization that contributes to increased diversity in the community or ecosystem as a whole, whereas the evolution of broad limits of tolerance could be considered to promote generalist species that are less susceptible to human perturbations.

An example of limiting factors at the ecosystem level is the finding that two mineral nutrients, iron and silica, limit primary production over very large areas of the world's open oceans. Menzel and Ryther (1961) were among the first to find iron as a limiting factor in their studies of ocean waters off Bermuda. Martin et al. (1991) and Mullineaux (1999) have reviewed the evidence that iron is limiting almost everywhere in the open oceans. Because diatoms require silica for their shells, and this micronutrient is present in very low concentrations in seawater, silica is limiting wherever diatoms are an important part of the phytoplankton (Tréguer and Pondaven 2000).

The following briefly outlined examples further demonstrate the importance of the concept of limiting factors and the limitations of the concept itself:

- Ecosystems developing on unusual geological formations often provide instructive sites for the analysis of limiting factors, as one or more important chemical elements may be unusually scarce or unusually abundant. Such a situation is provided by *serpentine* soils (derived from magnesium-iron-silicate rocks), which are low in major nutrients (calcium, phosphorus, and nitrogen) and high in magnesium, chromium, and nickel, with concentrations of the latter two approaching toxic levels for organisms. Vegetation growing on such soils has a characteristically stunted appearance, which contrasts sharply with adjacent vegetation on nonserpentine soils, and comprises an unusual flora with many *endemic species* (that is, species restricted to certain specialized habitats). For example, bare monkeyflower (*Mimulus nudatus*) thrives only in serpentine soils. Despite the twin limitations of scarce major nutrients and abundant toxic metals, a biotic community has evolved over geological time that can tolerate these conditions, but at a reduced level of community structure and productivity.
- Great South Bay on Long Island, New York, dramatically demonstrated how "too much of a good thing" can completely change an ecosystem, in this case to the detriment of a seafood industry. This story, which might be entitled "The Ducks

versus the Oysters," has been well documented, and the cause-effect relations have been verified by experiments. The establishment of large duck farms along the tributaries leading into the bay resulted in extensive fertilization of the waters by duck manure and a consequent great increase in phytoplankton density. The low circulating rate in the bay allowed the nutrients to accumulate rather than be flushed out to sea. The increase in primary productivity might have been beneficial, had not the organic form of the added nutrients and the low nitrogen-phosphorus ratio completely changed the type of producers. The normal mixed phytoplankton of the area, consisting of diatoms, green flagellates, and dinoflagellates, was almost completely replaced by very small green flagellates of the genera *Nannochloris* and *Stichococcus*. The famous blue-point oysters, which had been thriving for years on a diet of the normal phytoplankton and supporting a profitable industry, could not use the newcomers as food and gradually disappeared. Oysters were found starving to death with their intestines full of undigested green flagellates. Other shellfish were also eliminated, and all attempts to reintroduce them failed. Culture experiments demonstrated that the green flagellates grow well when nitrogen is in the form of urea, uric acid, and ammonia, whereas the diatom *Nitzschia*, a normal phytoplankton producer, requires inorganic nitrogen (nitrate). The invading flagellates could "short-circuit" the nitrogen cycle (they did not have to wait for organic material to be reduced to nitrate). This case provides a good example of how a normally rare specialist in the usual fluctuating environment takes over when unusual conditions are stabilized. This example also demonstrates the frequent experience of laboratory biologists, who find that the common species of unpolluted nature are often difficult to culture in the laboratory under conditions of constant temperature and enriched media, because they have adapted to the opposite conditions (low nutrients and variable temperatures). On the other hand, the "weed" species—normally rare or transitory in nature—are easy to culture because they are stenotrophic and thrive on enriched (that is, polluted) conditions.

- In the 1950s, Andrewartha and Birch (1954) started a lively discussion in the ecological literature when they suggested that distribution and abundance are controlled mainly by physical (abiotic) factors. Accordingly, studies at range margins should be a good way to single out which factors are limiting. However, ecologists now know that both biotic and abiotic factors may limit abundance in the center of ranges and distribution at the margins, especially because population geneticists have reported that individuals in marginal populations may have different gene arrangements from central populations (see discussion of ecotones and ecotypes in the next section). In any event, the biogeographical approach becomes especially interesting when one or more landscape-level environmental factors suddenly or drastically change. Thus, a "natural experiment" is set up that is often superior to a laboratory experiment, because factors other than the one being considered continue to vary normally instead of being controlled in an abnormal, constant manner.
- One experimental approach to determining biotic limiting factors involves adding or removing species populations. The intertidal zone on rocky seashores is a good habitat for such experiments. Extensive work by Paine (1966, 1976, 1984), Dayton (1971, 1975), Connell (1972), and others has shown that intertidal communities tend to have strong dominants (that is, species capable of excluding others

in the same trophic position). With space in the narrow intertidal zones always potentially limiting, the main factors in preventing monopolization by a single species are predation (for animals) and grazing (for plants).

2 Factor Compensation and Ecotypes

Statement

Organisms are not subjugated to the physical environment; they adapt themselves and modify the physical environment to reduce the limiting effects of temperature, light, water, and other physical conditions of existence. Such *factor compensation* is particularly effective at the community level of organization, but it also occurs within species. Species with wide geographical ranges almost always develop locally adapted populations called *ecotypes* that have optima and limits of tolerance adjusted to local conditions. **Ecotypes** are genetically differentiated subspecies that are well adapted to a particular set of environmental conditions. Compensation along gradients of temperature, light, pH, or other factors usually involves genetic changes of ecotypes, but it can occur by physiological adjustments without genetic fixation.

Explanation

Species that range widely along a gradient of temperature or other conditions often differ physiologically and sometimes morphologically in different parts of their range. Usually, genetic changes are involved, but factor compensation can be accomplished without genetic fixation by physiological adjustments in organ functions or by shifts in enzyme-substrate relationships at the cellular level. *Reciprocal transplants* provide a convenient method of determining to what extent genetic fixation is involved in ecotypes. McMillan (1956), for example, found that prairie grasses of the same species (and to all appearances identical) transplanted into experimental gardens from different parts of their range responded quite differently to light. In each case, the timing of growth and reproduction was adapted to the area from which the grasses were transplanted. The importance of genetic fixation in local strains has often been overlooked in applied ecology; restocking or transplanting of plants and animals frequently fail because individuals from remote regions are used instead of locally adapted stock. Transplanting also frequently disrupts local species interactions and regulatory mechanisms.

Examples

Figure 5-2 illustrates temperature compensation for the jellyfish *Aurelia aurita*. Northern jellyfish can swim actively at low temperatures that would completely inhibit individuals from the southern populations. Both populations are adapted to swim at about the same rate, and both function to a remarkable extent independently of the temperature variations in their particular environment.

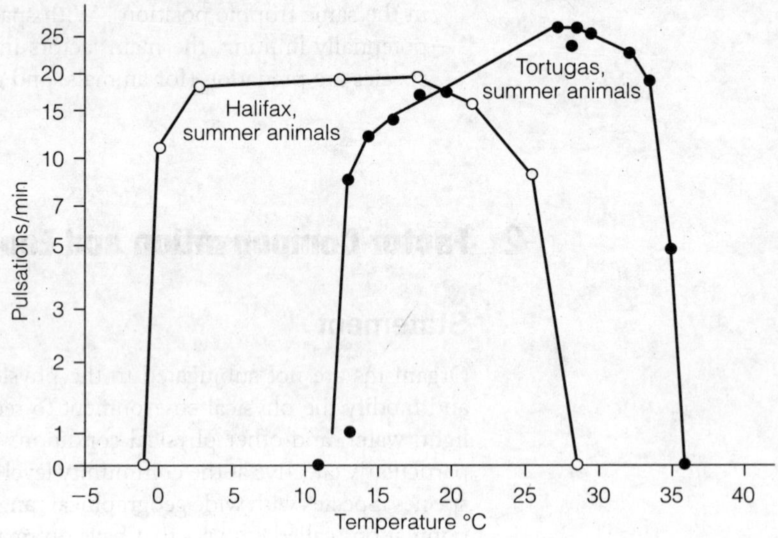

Figure 5-2. Temperature compensation at the species and community levels. The relation of temperature to swimming movement in northern (Halifax) and southern (Tortugas) individuals of the same species of jellyfish, *Aurelia aurita*. The habitat temperatures were 14° C and 29° C, respectively. Note that each population is acclimated to swim at a maximum rate at the temperature of its local environment. The cold-adapted form shows an especially high degree of temperature independence (from Bullock 1955).

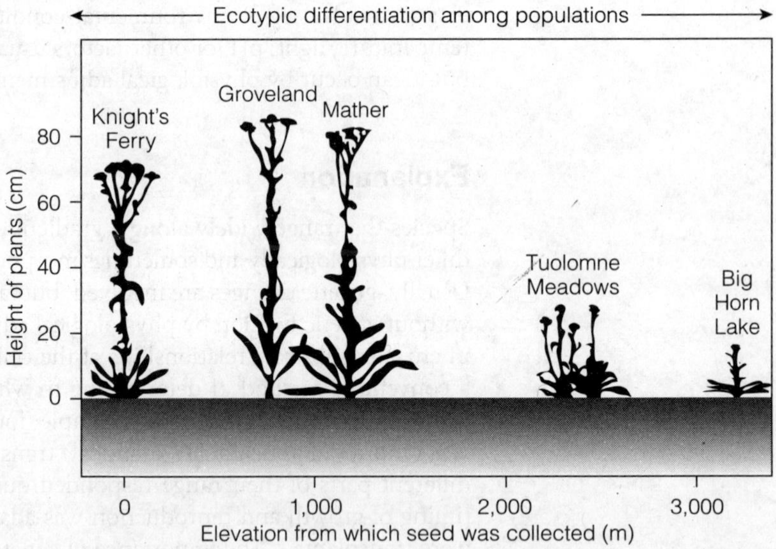

Figure 5-3. Ecotypic differentiation in populations of yarrow, *Achillea millefolium*, as demonstrated by collecting seeds from different elevations and growing these seeds under identical conditions in a sea-level garden. (From Clausen, J., D. D. Keck, and W. M. Hiesey. 1948. *Experimental studies on the nature of species.* Volume 3. *Environmental responses to climactic races of* Achillea. Carnegie Institution of Washington Publication 581:1–129. Reprinted with permission.)

A good example of an experimental approach to determining the extent of genetic fixation in ecotypes is the study of *Achillea millefolium,* a species of yarrow that grows from down in the valleys to high altitudes in the Sierra Mountains. Low-altitude plants are tall, and high-altitude plants are short in structure (Fig. 5-3). When seeds of both varieties were planted in the same garden at sea level, they retained their tall and short statures, indicating that genetic fixation had taken place (see Clausen et al. 1948 for details).

Factor compensation occurs along seasonal as well as geographical gradients. A striking, well-studied example is that of the creosote bush, *Larrea,* which dominates low-altitude, hot deserts of the southwestern United States. Although *Larrea* is a C_3 plant, (using a photosynthetic mode not especially adapted to hot and dry conditions), it can shift its optimal temperature upward from winter to summer by acclimation. High photosynthetic rates are maintained by an additional acclimation to drought stress, as measured by leaf water potential.

In nutrient-poor environments, efficient recycling between autotrophs and heterotrophs often compensates for nutrient scarcity. Coral reefs and rain forests are examples cited previously. Nitrogenous nutrients in the waters of the North Atlantic are so low that they are difficult to detect by standard instruments. Yet phytoplankton photosynthesis occurs at a high rate. The rapid and efficient uptake of nutrients released in zooplankton excretion and bacterial action compensates for the overall scarcity of nitrogen.

3 Conditions of Existence as Regulatory Factors

Statement

Organisms not only adapt to the physical environment in the sense of tolerating it, but also use the natural periodicities in the physical environment to time their activities and to "program" their life histories so they can benefit from favorable conditions. They accomplish this by means of **biological clocks,** physiological mechanisms for measuring time. The most common and perhaps basic manifestation is the **circadian rhythm** (from *circa* = "about" and *dies* = "day"), or the ability to time and repeat functions at about 24-hour intervals even in the absence of conspicuous environmental cues such as daylight. When one adds interactions between organisms and reciprocal natural selection between species (coevolution), the whole community becomes programmed to respond to seasonal and other rhythms.

Explanation

It is our circadian rhythm that gets upset when we suffer "jet lag" after a long airplane trip. The biological clock is set by biological and physical rhythms that enable organisms to anticipate daily, seasonal, tidal, and other periodicities. There is increasing evidence that the actual timing is accomplished by cellular oscillators that operate as a feedback loop involving "clock" genes (see Dunlap 1998). Circadian rhythms and their underlying cellular oscillators are ubiquitous in biological organisms and are used to anticipate the best time to feed, to bloom in the case of plants, to migrate, to hibernate, and so on.

Examples

Fiddler crabs that live in tidal marshes have their clock set to tidal rather than diurnal time. When kept in the lab in the dark and without tide, they become active at the time the tide would be ebbing, when they would normally be emerging from their burrows to feed.

A dependable cue by which organisms time their seasonal activities in the Temperate Zones is the day-length period or **photoperiod.** In contrast to most other seasonal factors, day length is always the same for a given season and locality. The ampli-

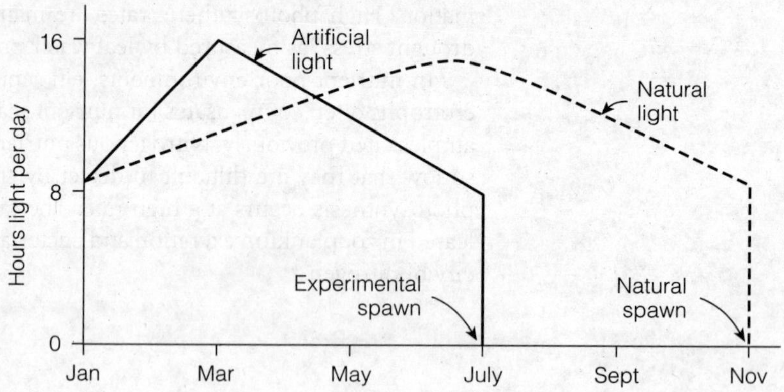

Figure 5-4. Control of the breeding season in the brook trout by artificial manipulation of the photoperiod. Trout, which normally breed in the autumn, spawn in summer when day length is increased artificially in the spring and then decreased in the summer to simulate autumn conditions (redrawn from Hazard and Eddy 1950).

tude in the annual day-length cycle increases with increasing latitude, thus providing latitudinal as well as seasonal cues. In Winnipeg, Manitoba, Canada, the maximum photoperiod is 16.5 hours (in June) and the minimum is 8.0 hours (in late December). In Miami, Florida, the range is only 13.5 to 10.5 hours, respectively. The photoperiod has been shown to be the timer or trigger that sets off physiological sequences that cause the growth and flowering of many plants; the molting, fat deposition, migration, and breeding in birds and mammals; and the onset of hibernation or diapause (resting stage) in insects. *Photoperiodicity* is coupled with the biological clock of the organism to create a timing mechanism of great versatility.

Day length acts through a sensory receptor, such as an eye in animals or a special pigment in the leaves of a plant, which, in turn, activates one or more back-to-back hormone and enzyme systems that bring about the physiological or behavioral response. Although the higher plants and animals are widely divergent in morphology, the physiological linkage with environmental photoperiodicity is similar.

Among the higher plants, some species bloom on increasing day length and are termed *long-day plants;* others that bloom on short days (less than 12 hours) are known as *short-day plants.* Animals likewise may respond to either lengthening or shortening days. In many, but by no means all, photoperiod-sensitive organisms, the timing can be altered by experimental or artificial manipulation of the photoperiod. As depicted in Figure 5-4, an artificially accelerated light regimen can bring brook trout into breeding condition up to four months early. Florists can often force flowers to bloom out of season by altering the photoperiod. In migratory birds, there are several months after the fall migration when the birds are refractory to photoperiod stimulation. The short days of fall are apparently necessary to "reset" the biological clock, as it were, and prepare the endocrine system for a response to long days. Anytime after late December, an artificial increase in day length will bring on the sequence of molting, fat deposition, migratory restlessness, and gonad enlargement that normally occurs in the spring. The physiology of this response in birds was first documented by Farner (1964a, 1964b).

Photoperiodicity in certain insects and annual plant seeds is noteworthy because it provides a sort of birth control. For example, in insects, the long days of late spring and early summer stimulate the "brain" (actually a nerve cord ganglion) to secrete a neurohormone that starts producing a diapause or resting egg that will not hatch until next spring no matter how favorable temperatures, food, and other conditions are.

Thus, population growth is halted before rather than after the food supply becomes critical.

In striking contrast to day length, rainfall in a desert is highly unpredictable, yet desert annuals, which constitute the largest number of species in many desert floras, use this factor as a regulator. These annuals, known as **ephemerals,** persist as seeds during periods of drought, but are ready to sprout, flower, and produce seeds whenever moisture is favorable. The seeds of many such species contain a germination inhibitor that must be washed out by a certain minimum amount of water (for example, half an inch, or 1–2 cm) from a rain shower. This shower provides all the water necessary to complete the life cycle back to seeds again. The young plants grow rapidly in the bright desert sunlight following the rain. They start flowering and setting seed almost immediately. They remain small, with no elaborate stem or root systems, with all energy put into flowering and seed production. If such seeds are placed in moist soil in the greenhouse, they fail to germinate; but they do so quickly when treated with a simulated shower of the necessary magnitude. Seeds may remain viable in the soil for many years, "waiting," as it were, for the adequate shower, which explains why deserts bloom (become quickly covered by flowers) a short time after a heavy rainfall.

4 Soil: Organizing Component for Terrestrial Ecosystems

Statement

It is sometimes convenient to think of the ecosphere as comprising the atmosphere, the hydrosphere, and the *pedosphere,* the latter being the soil. Each is composed of a living and a nonliving component that are more easily separated theoretically than practically. Biotic and abiotic components are especially intimate in **soil,** which by definition consists of a weathered layer of Earth's crust with living organisms intermingled with products of their decay.

Explanation

Because, for the most part, nutrients are regenerated and recycled during decomposition in the soil before they become available to the primary producers (plants), *the soil can be considered a chief organizing center for land ecosystems.* Without life, Earth would have a crust of some sort, but nothing like soil. Thus, soil is not only a factor of the environment to organisms, but it also is produced by them. In general, *soil* is the net result of the action of climate and organisms, especially vegetation and microbes, on the parent material of the surface of Earth. Thus, soil is composed of a parent material—the underlying geological or mineral substrate—and an organic component in which organisms and their products are intermingled with the finely divided and modified parent material. Spaces between the soil particles are filled with gases and water. The texture and porosity of the soil are highly important characteristics that largely determine its fertility.

As in other major parts of the ecosphere, soil activity is concentrated in "hot

Figure 5-5. An area where much of the A horizon (topsoil) is being eroded by rainfall due to lack of a cover crop.

spots" such as root zones (*rhizospheres*) and organic aggregates. **Rhizospheres** are aggregations of microbes around roots, fecal pellets, patches of organic matter, and mucus secretions in soil pore necks (Coleman and Crossley 1996). According to Coleman (1995), approximately 90 percent of metabolic activity may occur in these hot spots, which may occupy as little as 10 percent of the total soil volume. The soil system is the organizing center for terrestrial ecosystems; sediments in aquatic ecosystems may also function in a similar way. Major functions, such as community respiration, *R*, and recycling are controlled by the rate at which nutrients are released by decomposition.

The cut edge of a bank or an eroded field (Fig. 5-5) shows that soil is composed of distinct layers, which often differ in color. These layers are called **soil horizons**, and the sequence of horizons from the surface down is called a *soil profile*. The upper horizon, or **A horizon** (topsoil), is composed of the bodies of plants and animals that are being reduced to finely divided organic material by *humification*. In a mature soil, this horizon is usually subdivided into distinct layers representing progressive stages of humification. These layers (Fig. 5-6) are designated (from the surface downward) as A-0 (litter), A-1 (humus), and A-2 (leached [light-colored] zone). The A-0 layer is sometimes subdivided as A-1 (litter proper), A-2 (duff), and A-3 (leaf mold). The litter, or A-0 horizon, represents the detritus component and can be considered a sort of ecological subsystem in which microorganisms (bacteria and fungi) work in partnership with small arthropods (soil mites and collembolans) to decompose the organic material. These microarthropods are "shredders," in that they break up pieces of particulate detritus into smaller pieces and dissolved organic matter (DOM), which are more readily available to soil microorganisms. When these shredders are removed, the rate of decomposition is markedly reduced (Coleman and Crossley 1996).

The annual input into the litter subsystem from leaf fall in forests increases from arctic to equatorial latitudes (Fig. 5-7). The next major horizon, or **B horizon,** is composed of mineral soil in which the organic compounds have been converted by decomposers into inorganic compounds (*mineralization*) and thoroughly mixed with finely divided parent material. The soluble materials of the B horizon are often formed in the A horizon and deposited or leached by the downward flow of water

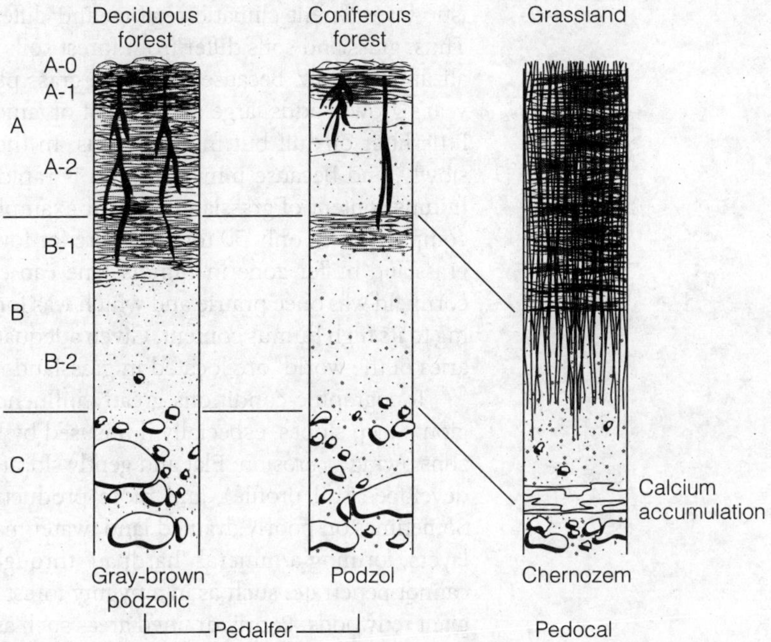

Figure 5-6. Simplified diagrams of three major soil types that are characteristic of three major biomes (deciduous forest, coniferous forest, and grassland).

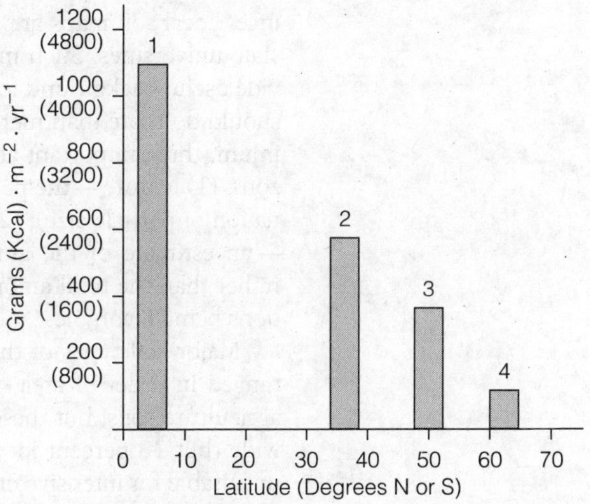

Figure 5-7. Annual litter fall in forests in relation to latitude. (1) Equatorial forests; (2) Warm temperate forests; (3) Cool temperate forests; and (4) Arctic-alpine forests (after Bray and Gorham 1964).

into the B horizon. The dark band in Figure 5-6 represents the upper part of the B horizon where leached materials have accumulated. The third horizon, or **C horizon,** is the more or less unmodified parent material. This parent material may represent the original mineral formation that is disintegrating in place, or it may have been transported to the site by gravity (*colluvial deposit*), water (*alluvial deposit*), glaciers (*glacial deposit*), or wind (*eolian deposit,* or *loess*). Transported soils are often extremely fertile (witness the deep loess soils of Iowa and the rich soils of the deltas of large rivers).

The soil profile and the relative thickness of the horizons are generally characteristic for different climatic regions and different topographical situations (Fig. 5-6). Thus, grassland soils differ from forest soils in that humification is rapid, but mineralization is slow. Because the entire grass plant, including roots, is short-lived, each year's growth adds large amounts of organic material, which decays rapidly, leaving little litter or duff but much humus. In the forest, however, litter and roots decay slowly, and because mineralization is rapid, the humus layer remains narrow. The humus content of grassland soil, for example, may be as much as 600 tons per acre, compared with only 50 tons per acre for forest soils (Daubenmire 1974). In a forest-grassland buffer zone in Illinois, one can easily tell by the color of the soil which cornfield was once prairie and which was forest. The prairie soil is much blacker, owing to its high humus content. Given adequate rainfall, it is no accident that the "granaries of the world" are located in grassland regions.

Topographic conditions greatly influence the soil profile in a given climatic region. Steep slopes, especially if misused by humans, tend to have thin A and B horizons owing to erosion. Flat and gently sloping lands have deeper, more mature (well-developed soil profile), and more productive soils than do steeply sloping lands. Sometimes on poorly drained land, water may leach materials rapidly into the deeper layers, forming a mineral "hardpan" through which plant roots, animals, and water cannot penetrate, such as in a pygmy forest in a region where soils typically support giant redwoods. Poorly drained areas such as bogs also favor the accumulation of humus, because poor aeration slows decay.

The classification of soil types has become a highly empirical subject. The soil scientist may recognize dozens of soil types as occurring within a county, state, or province. Local soil maps are widely available from soil conservation agencies and from state universities. Such maps, and the soil descriptions that accompany them, provide useful background for studies of terrestrial ecosystems. The ecologist, of course, should do more than merely name the soil in his or her study area. At the very minimum, three important attributes should be measured at least in the A and B horizons: (1) texture—the percentage of sand, silt, and clay (or a more detailed determination of particle size); (2) percentage of organic matter; and (3) exchange capacity—an estimate of the amount of exchangeable nutrients. The available minerals, rather than the total amount of materials, determine potential fertility, other conditions being favorable.

Major soil types of the world and the United States are listed in Table 5-1, arranged in order of area occupied worldwide. Alfisols and mollisols make the best agriculture soils, but these constitute only about 24 percent of the land area worldwide (but 38 percent for the United States). Huge areas of the terrestrial world are unsuitable for intensive crop production unless soils are heavily subsidized with fertilizers and water.

Because soil is a product of climate and vegetation, the major soil types of the world form a composite map of climate and vegetation. Given a favorable parent material and not too steep a topography, the action of organisms and climate will tend to build up a soil characteristic of the region. From a broad ecological viewpoint, the soils of a given region may be lumped into two groups: *mature soils,* on level or gently rolling topography, largely controlled by the climate and vegetation of the region; and *immature soils* (in terms of profile development), largely controlled by local conditions of topography, water level, or unusual type of parent material. The degree of soil maturity varies greatly with the region. Wolfanger (1930), for example, estimated that 83 percent of the soils located in Marshall County, Iowa, are mature, compared

Table 5-1 **Distribution of major soil types worldwide and in the United States**

Soil type	*Percentage of land area worldwide*	*Percentage of land area in the United States*
Aridisols (desert soils)	19.2	11.5
Inceptisols (weakly developed soils)	15.8	18.2
Alfisols (moderately weathered forest soils)	14.7	13.4
Entisols (recent soils, profile undeveloped)	12.5	7.9
Oxisols (tropical soils)	9.2	<0.1
Mollisols (grassland soils)	9.0	24.6
Ultisols (highly weathered forest soils)	8.5	12.9
Spodosols (northern conifer forest soils)	5.4	5.1
Vertisols (expandable clay soils)	2.1	1.0
Histosols (organic soils)	0.8	0.5
Miscellaneous soils (steep mountains, for example)	2.8	4.9
Total	100.0	100.0

Source: After E. P. Odum 1997.

with only 15 percent of the soils in Bertie County, North Carolina, which is located on the sandy, geologically young Coastal Plain. For more on the ecology of soil, see Richards (1974), one of the first to consider soil as an ecosystem; Paul and Clark (1989); Killham (1994); and Coleman and Crossley (1996). Effland and Pouyat (1997) suggested we add human-created urban soil or *anthrosol* to the list of soil types. **Anthrosol** is 25 percent "fill," including lots of pulverized concrete, dust, and debris, with more nitrogen and lime runoff than from natural soils.

Soil Displacement: Natural and Human-Accelerated

Soil erosion caused by water and wind occurs naturally at low rates all the time, with periodic large displacements resulting from great floods, glaciers, volcanic eruptions, and other episodic events. Areas that lose soil faster than new soil is formed generally suffer reduced productivity and other detrimental effects. Areas receiving too much soil may also be negatively affected. However, fertility may be enhanced when soils are washed down from hills into river valleys and deltas or are deposited on prairies by wind. As is the case for so many natural processes, humans tend to accelerate soil erosion, often to our long-term detriment.

In the 1930s, the Soil Conservation Service (SCS) was established by the U.S. government to combat the soil erosion that was ruining thousands of hectares of farmland and forest. The Dust Bowl was taking its toll on the western plains at about this time. The SCS program that was developed to save soil is an excellent example of how government should work in the public interest in a democracy. A close linkage was established among the federal government in Washington, D.C., state gov-

Figure 5-8. A dam constructed by the Army Corps of Engineers, located on the Brookville Reservoir in southeastern Indiana. Much of the watershed in this area is in agricultural practice, thus the reservoir is subject to sedimentation that may require dredging in future years.

Courtesy of Gary W. Barrett

ernments, land-grant universities, and counties. Washington provided the money, the universities contributed much of the research, but the decisions were made locally, and county agents worked directly with landowners or with the stakeholders of the region. Terracing, grass waterways, riparian forest buffer strips, crop rotation, and other measures, together with improvements in the economic and educational status of farmers, reversed the tide of soil loss, and a *soil conservation ethic* was generally accepted by farmers and other landowners.

Perhaps partly because of its success, the SCS had so much support in Congress and in the states that it became increasingly bureaucratic (and thus less responsive to relevant needs) and extended its activities into other areas, such as channeling streams and building large dams (Fig. 5-8), that often had questionable value to soil preservation. Then, suddenly, in the 1970s, soil erosion again became an urgent national problem because of two new trends. The first was the *industrialization of farming,* emphasizing cash crops that were treated less as food than as commodities for sale, especially on the overseas market. Unfortunately, when farms are operated strictly as businesses—often by corporations or other absentee owners—short-term crop yield is maximized at the expense of maintaining long-term fertility and productivity. The second trend was *urban sprawl,* as roads and housing developments mushroomed in the rural countryside, with little or no concern about the loss of soil and prime farmland (Forman and Alexander 1998).

The urgent need to counteract the deleterious effects of these two major land-use changes and to reestablish a soil conservation ethic is well documented by governmental reports, for example, the Council on Environmental Quality (CEQ; 1981), and assessments by private conservation foundations. For example, in 1985, the Conservation Reserve Program (CRP) was established, which paid farmers to "retire" 15 million hectares—roughly 10 percent of the cropland located in the United States —converting it back to grassland or forest before it became wasteland. Within five years, U.S. farmers had converted nearly 15 million hectares of cropland to grassland. The CRP reduced excessive soil erosion by some 40 percent, helping to enhance food security on a global basis. The nonmarket (natural capital) benefits from reducing soil erosion and providing habitat quality between 1985 and 2000 are estimated to exceed 1.4 billion dollars (L. R. Brown 2001).

Approximately half of the best farmland in Iowa and Illinois is losing 10 to 20 tons of soil per acre each year, and a quarter of all farmland in the United States is losing soil at a rate greater than the tolerable level. To put this in perspective, consider that an acre (0.4 ha) of topsoil 6 inches (15 cm) deep (about plow depth) weighs about 1000 tons, so 1 acre-inch equals about 167 tons. An annual loss of 10 tons per acre represents a loss of 1 inch (2.54 cm) of topsoil every 17 years—a loss much greater than any known rate of soil formation. Langdale et al. (1979) estimated that for every inch (2.54 cm) of topsoil lost, a crop yield reduction of at least 10 percent occurs. Soil losses from urban and suburban construction, although often of short duration, are even more severe. Losses of 40 tons per acre are not uncommon, and losses of 100 tons per acre have been recorded in extreme cases (E. H. Clark et al. 1985).

Soil erosion resulting from poor land use is, of course, not new. What is new are the accelerated rate and larger scale of soil disturbance due to market pressure, population increase, and use of large, powerful machines; and the toxic agricultural and industrial chemicals that move downhill and downstream with the displaced soil. If the current degradation continues, the needs and demands for more food from fewer hectares cannot possibly be met.

Of course, erosion is not the only problem that threatens the capacity of soil to produce food and fiber for humans. Soil compaction, resulting from intensive cultivation with ever larger and heavier farm machinery, definitely reduces yields. About half of the irrigated lands of the world are damaged to some extent by salinization (salt accumulation) or alkalinization (alkali accumulation). So far, yields have been maintained despite declines in soil quality by pouring on more fertilizer and more water. This method works for as long as these subsidies are relatively inexpensive, which will be less and less the case in the near future.

Soil Quality as an Indicator of Environmental Quality

In the closing years of the twentieth century, scientific attention to and publicity on high-yielding crop varieties diverted attention from the fact that maintaining high yields depends on maintaining soil quality, which, in turn, depends on sustainable tillage and diversity both at the crop and landscape levels. As soil is the chief organizing center for terrestrial and wetland ecosystems, soil quality should be a good indicator of environmental quality in general. In other words, if the quality of the soil is being maintained, then whatever is going on in the landscape, whether natural or managed, should be sustainable.

Soil quality has been defined by the Soil Science Society of America (SSSA; 1994) as "the capacity of a particular kind of soil to function within natural or managed ecosystem boundaries to sustain plant and animal productivity, maintain or enhance water quality, and support human health and habitation." The National Research Council (NRC; 1993) has a shorter definition: "Soil quality is the capacity of the soil to promote growth of plants, protect watersheds, and prevent air and water pollution."

Despite an outpouring of books and papers focusing on sustainable soil management, the Soil and Water Conservation Society (Lal 1991) has noted that there is as yet no agreement on how to measure soil quality. Obviously, measurement must involve multiple indices, including available nutrients; texture; density of organic aggregates; diversity of microorganisms and soil animals, including mycorrhizae, nitrogen fixers and earthworms; and measures of erosion and rates of leaching. For a review of approaches, see Karlen et al. (1997).

Ultimately, the fate of the soil system depends on a society's willingness to intervene in the marketplace to forgo some short-term benefits so that soils are preserved to protect long-term natural capital. The short-term economic costs of soil conservation can be greatly reduced by designing more efficient and more harmonious agroecosystems. The real problem, however, is political and economic, not ecological or technical.

5 Fire Ecology

Statement

Fire is a major factor in shaping the history of vegetation in most of the terrestrial environments of the world. As climate pulses between wet and dry periods, so does fire in the environment. As with most environmental factors, human beings have greatly modified the effect of fire, increasing its influence in many cases and decreasing it in others. Failure to recognize that ecosystems may be "fire adapted" has resulted in a great deal of mismanagement of our natural resources. Properly used, fire can be an ecological tool of great value. It is thus an extremely important limiting factor, if for no other reason than that the control of fire is far more feasible than the control of many other limiting factors.

Explanation

Using a series of global satellite images spread over a 12-month period (1992–1993), Dwyer et al. (1998) were able to present a global picture of fire. On any given day in that year, fires, both natural and human-made, were burning all over the globe. The largest number of fires was in Africa, especially in regions of the Savanna (grasslands with scattered trees or scattered clumps of trees). Most fires that occurred in January were located in equatorial regions and in the South Temperate Zone, whereas in August, there were large numbers of fires located in dry or hot regions of the North Temperate Zone. Although many fires in remote regions are natural, in that they are ignited by lightning, most fires are started by humans, either accidentally or on purpose.

In most parts of the United States, especially in southern and western states, it is difficult to find a sizable area that does not show evidence of fire having occurred there during the last 50 years; witness the great fires that occurred in southern California during October 2003. In many areas, fires are started naturally by lightning. Early humankind (North American Indians, for example) regularly burned woods and prairies for practical reasons. Fire was a factor in natural ecosystems long before modern times. Accordingly, it should be considered an important ecological factor along with such other factors as temperature, rainfall, and soil.

As an ecological factor, fire can be of different types with different effects. Two extreme types are shown in Figure 5-9. *Crown fires* or *wildfires* (that are very intense and out of control) often destroy most of the vegetation and some soil organic matter, whereas *surface fires* have entirely different effects. Figure 5-9A shows the crown fires

Figure 5-9. (A) Major crown fire in Yellowstone National Park in 1988. (B) Controlled (prescribed) burning of a longleaf pine forest in southwestern Georgia. Prescribed burns remove hardwood competition, stimulate growth of legumes, and improve reproduction of valuable pine timber. Burning is done under damp conditions late in the afternoon. Ants, soil insects, and small mammals are not harmed by such light surface fires. (C) Longleaf pine (*Pinus palustris*) seedling with special needles adapted to ground fires. Photograph taken one week following prescribed burning.

© Raymond Gehman/CORBIS

Courtesy of the Joseph W. Jones Ecological Research Center at Ichauway

Courtesy of the Joseph W. Jones Ecological Research Center at Ichauway

in Yellowstone National Park in 1988. **Crown fires** are limiting to most organisms; after a crown fire, the biotic community must start to develop all over again, more or less from scratch, and it may be many years before natural succession restores the area to something that resembles the pre-fire condition. **Surface fires,** on the other hand, exert a much more selective effect; they are more limiting to some organisms than to others and, thus, favor the development of ecosystems with a high tolerance to fire, such as oak forests (McShea and Healy 2002). Figure 5-9B illustrates prescribed (controlled) burning of longleaf pine (*Pinus palustris*) at the Joseph W. Jones Ecological Research Center at Ichauway in southwest Georgia. The longleaf pine forest at Ichauway is a fire-maintained ecosystem.

Light surface fires or prescribed burning supplement bacterial action in breaking down the bodies of plants and in making mineral nutrients more quickly available for new plant growth. Nitrogen-fixing legumes often thrive after a light burn. In regions that are especially subject to fire, regular light surface fires greatly reduce the danger of severe crown fires by keeping the combustible litter (fuel) to a minimum. In examining regions where fire is a factor, the ecologist usually finds some evidence of the past influence of fire. Whether fire should be excluded in the future (assuming that is practical) or should be used as a management tool will depend entirely on the type of community that is desired or seems best from the standpoint of regional land use.

Examples

Several examples taken from well-studied situations illustrate how fire acts as a limiting factor and how fire is not necessarily "bad" from the human standpoint:

1. On the Coastal Plain of the southeastern United States, the longleaf pine (*Pinus palustris*) is more resistant to fire than any other tree species, and pines in general are more resistant to fire than hardwoods. The terminal bud of seedling longleaf pines is well protected by a bunch of long, fire-resistant needles (Fig. 5-9C). Thus, ground fires selectively favor this species. In the complete absence of fire, scrub hardwoods grow rapidly and choke out the longleaf pines. Grasses and legumes are also eliminated, and the bobwhite quail and other animals dependent on legumes do not thrive in the complete absence of fire in forested lands. Ecologists generally agree that the magnificent virgin, open stands of pine of the Coastal Plain, and the abundant game associated with them, are part of a fire-controlled or "fire climax" ecosystem. A place to observe the long-term effects of the intelligent use of fire is the Tall Timbers Research Station in northern Florida and the adjacent plantations of southwestern Georgia, where for many years, more than a million acres have been managed according to principles developed by the late Herbert Stoddard, and E. V. Komarek and R. Komarek, who began studying the relation of fire to the entire ecological complex in the 1930s. H. L. Stoddard (1936) was one of the first to advocate the use of controlled or "prescribed" burning for increasing both timber and game production at a time when most professional foresters believed that all fire was bad. For years, high densities of both quail and wild turkeys have been maintained on land devoted to highly profitable timber crops through the use of a system of "spot" burning, aided by a diversification in the land use.
2. Fire is especially important in grassland and savannas. Under moist conditions (as in the tallgrass prairies of the Midwest), fire favors grass over trees, and un-

der dry conditions (as in the southwestern United States and East Africa), fire is often necessary to maintain grassland against the invasion of desert shrubs. The main growth centers and energy storages of grasses are underground, so they sprout quickly and luxuriously after the dry, aboveground parts burn, which also releases nutrients to the soil surface. A close coupling of fire and grazing has been shown to be the key to maintaining the incredible diversity of antelope and other large herbivores and their predators on the East African Savanna.

3. Perhaps the most studied type of fire ecosystem is the chaparral vegetation of coastal California, the Mediterranean region, and other areas with a wet winter and dry summer climate. Here, fire interacts with plant-produced antibiotics or allelopathics to produce a unique cyclic climax.
4. The use of fire in game management is exemplified on the British heather moors. Extensive experimentation over the years has shown that burning in patches or strips of about 1 hectare each, with about six such patches per square kilometer, results in the highest grouse populations and game yields. The grouse, which are herbivores that feed on buds, require mature (unburned) heather for nesting and protection against enemies, but they find more nutritious food in the regrowth on burned patches. This example of a compromise between maturity and youth in an ecosystem is very relevant to human beings, and will be discussed in Chapter 8.
5. The Konza Prairie is a 3487-hectare native grassland prairie located in the Flint Hills region of northeastern Kansas. The vegetation of the Konza Prairie is predominantly composed of native, perennial, warm-season grasses. Periodic fire (Fig. 5-10A) is one of the main natural processes that regulate and sustain this ecosystem. Figure 5-10B shows how prescribed burning of an experimental prairie at the Ecology Research Center at Miami University of Ohio can be established and used as a learning tool for classes in ecology and resource management.

In the summer of 1988, a very dry year, wildfires, mostly started by lightning, burned out of control over about one half of Yellowstone National Park (about 350,000 hectares). To the casual observer, the scorched earth looked totally devastated, as if no life survived. However, because the fires moved rapidly, killing temperatures penetrated no more than an inch (2.54 cm) into the ground, so no land was made unfit for plant regrowth. The large animals (such as bison and elk) were affected very little. They were able to forage more than expected on sugars in the charred debris—"caramel candy" as the rangers referred to it—and on the herbaceous vegetation that soon covered the burned areas. As a matter of fact, severe winters in the years following the fire had more effect on the herds than did the fire.

In the first summer following the fire, a carpet of fireweed and other herbs covered the burned areas. Fireweed (*Epilobium angustifolium*), a member of the evening primrose family, is a tall plant with a spire of pinkish purple flowers. It is normally found in openings and disturbed places in northern forests from coast to coast and, true to its name, is one of the first plants to follow forest fires; it can turn the scorched earth into a beautiful flower garden! Fireweed also occurs in England where, during World War II, it carpeted the bombed and burned-out areas of London.

In 1998, ten years after the fires, the original dominant conifers, lodgepole pine and Douglas fir, were emerging from the herb-shrub layer. Normally, aspen precedes conifers in ecological succession in the western mountains, but in Yellowstone the succession seems to be going directly to the conifer dominants. Apparently, the large populations of mammalian herbivores are grazing down the aspens.

Figure 5-10. (A) Controlled burning of the Konza Prairie, located in the Flint Hills region of northeastern Kansas. Konza Prairie is dominated by perennial, warm-season grasses such as big bluestem, little bluestem, Indian grass, and switchgrass. Fire is a natural process that helps to regulate and sustain the tallgrass prairie. (B) Prescribed burning of an experimental prairie at the Ecology Research Center, Miami University of Ohio.

A

Courtesy of Donald W. Kaufman

B

Courtesy of Gary W. Barrett

The Yellowstone fires touched off a "firestorm," as it were, over federal fire policy, which since 1972 had been to allow fires to burn unless they threatened people or property. The trouble now, in the first decade of the twenty-first century, is that people are moving from the cities and building homes in the forest, so massive and costly fire-fighting efforts have to be undertaken in the dry years of the normal wet-dry climate oscillations. The wildfires burning in southern California during October 2003, which blackened more than 743,000 acres (297,000 ha) and destroyed nearly 3600 homes, illustrate this point. For more information on the Yellowstone fires and on the recovery a decade later, see Stone (1998) and Baskin (1999).

As might be expected, plants have evolved special adaptations to fire, just as they have for other limiting factors. Fire-dependent and fire-tolerant species can be divided into two basic types: (1) **resprout** species that put more energy into underground storage organs and less into reproductive structures (inconspicuous flowers, little nectar, few seeds) and, thus, can quickly regenerate after fire has killed exposed parts; and (2) **mature-die** species that do just the opposite, producing abundant, resistant seeds ready to germinate just after a fire (fireweed, for example).

The question whether to burn or not to burn can certainly be confusing, as seasonal timing and intensity are so critical to determining the consequences of burning. Human carelessness tends to increase wildfires; therefore, it is necessary to have a strong campaign for preventing human-made fires in national forests and recreation areas. However, one should recognize that the use of fire as a tool by trained persons is part of good land management. Fire is part of the "climate" in many areas, and often beneficial. Recommended reviews of the ecology of fire are H. L. Stoddard (1950), Kozlowski and Ahlgren (1974), Whelan (1995), Knapp et al. (1998), and McShea and Healy (2002).

6 Review of Other Physical Limiting Factors

Statement

The broad concept of limiting factors is not restricted to physical factors, as biological interrelations are just as important in controlling the actual distribution and abundance of organisms in nature. However, biological factors will be considered in Chapters 6 and 7, dealing with populations and communities. This section concludes the brief reviews of the natural, physical, and chemical aspects of the environment. To present all that is known in this field would require several books—especially in physiological ecology or *ecophysiology*—and is beyond the scope of the present outline of ecological principles.

Explanation

Ecophysiology is that part of ecology concerned with the responses of individual organisms or species to abiotic factors such as temperature, light, moisture, atmospheric gases, and other factors in the environment. We focus here on only a few major factors that ecologists need to appreciate in order to understand abiotic and biotic relationships at higher levels of biological-ecological organization.

Temperature

Compared with the range of thousands of degrees known to occur in the universe, life as we know it can exist only within a tiny range of about 300 degrees Celsius—from about −200° to 100° C. Actually, most species and most activities are restricted to an even narrower band of temperatures. Some organisms, especially in a resting stage, can exist at very low temperatures, whereas a few microorganisms, chiefly bacteria and algae, can live and reproduce in hot springs where the temperature is close to the boiling point. The upper temperature tolerance for hot-spring bacteria is about 80° C for cyanobacteria, compared with 50° C for the most tolerant fish and insects. In general, the upper limits are more quickly critical than the lower limits, although many organisms appear to function more efficiently toward the upper limits of their range of tolerance. The range of temperature variation tends to be less in water than on land, and aquatic organisms generally have narrower ranges of tolerance to temperature than do land animals. Temperature, therefore, is universally important as a limiting factor.

Temperature is one of the easiest of environmental factors to measure. The mercury thermometer, one of the first and most widely used precision scientific instruments, has now been replaced by electrical "sensing" devices, such as platinum resistance thermometers, thermocouples, and thermistors, which permit not only measurement in "hard-to-get-at" places but also the continuous and automatic recording of measurements. Furthermore, advances in the technology of radiotelemetry now make it feasible to transmit temperature information from the body of a lizard deep in its burrow or from a migratory bird flying high in the atmosphere.

Variability of temperature is extremely important ecologically. A temperature fluctuating between 10° C and 20° C and averaging 15° C does not necessarily have the same effect on organisms as a constant temperature of 15° C. Organisms that are normally subjected to variable temperatures in nature (as in most temperate regions) tend to be depressed, inhibited, or slowed down by constant temperatures. For example, Shelford (1929), in a pioneer study, found that eggs and larval or pupal stages of the codling moth (*Cydia pomonella*) developed 7 to 8 percent faster under conditions of variable temperature than under a constant temperature having the same mean. Thus, the stimulating effect of variable temperatures, in the Temperate Zones at least, may be accepted as a well-defined ecological principle, especially as the tendency has been to conduct experimental work in the laboratory under conditions of constant temperature.

Light

Light places organisms on the horns of a dilemma: direct exposure of protoplasm to light causes death, yet sunlight is the ultimate source of energy, without which life could not exist. Consequently, many of the structural and behavioral characteristics of organisms are concerned with solving this problem. In fact, as noted in the discussion of the Gaia hypothesis (Chapter 2), the evolution of the biosphere as a whole has chiefly involved the "taming" of incoming solar radiation, so that its useful wavelengths could be exploited and dangerous ones mitigated or shielded out. Light, therefore, is not only a vital factor but also a limiting one, at both the maximum and minimum levels. There is, perhaps, no other factor of greater interest to ecologists.

The total radiation environment, and something of its spectral distribution, was considered in Chapter 3, as was the primary role of solar radiation in ecosystem energetics. Consequently, this chapter discusses light waves over a wide range in wavelength. Two bands of wavelengths readily penetrate the atmosphere of Earth: the *visible band,* together with some parts of adjacent bands, and the *low-frequency radio band,* having wavelengths greater than 1 cm. Whether the long radio waves are ecologically significant is unknown, although some researchers assert positive effects on migrating birds or other organisms. The roles of ultraviolet (below 3900 Å) and infrared light (above 7600 Å) were considered in Chapter 3. The role that high-energy, very short–wave gamma radiation, and other types of ionizing radiation, may play as an ecological limiting factor will be briefly reviewed in the next section.

Ecologically, the *quality* (wavelength or color), the *intensity* (actual energy measured in gram-calories), and the *duration* (length of day) of light are known to be important. Both animals and plants respond to different wavelengths of light. Color vision in animals sporadically occurs in different taxonomic groups, apparently being well developed in certain species of arthropods, fish, birds, and mammals, but not in other species of the same groups (among mammals, for example, color vision is well developed only in primates). The rate of photosynthesis varies somewhat with different wavelengths. In terrestrial ecosystems, the quality of sunlight does not vary enough to have an important differential effect on the rate of photosynthesis, but as light penetrates water, the reds and blues are filtered out by attenuation, and the resultant greenish light is poorly absorbed by chlorophyll. The marine red algae (Rhodophyta), however, have supplementary pigments (*phycoerythrins*) enabling them to use this energy and to live at greater depths than would be possible for the green algae.

The intensity of light (the energy input) impinging on the autotrophic layer controls the entire ecosystem through its influence on primary production. The relationship of light intensity to photosynthesis in both terrestrial and aquatic plants follows the same general pattern of linear increase up to an optimum or *light saturation* level, followed in many instances by a decrease at the high intensities of full sunlight. Plants with the C_4 type of photosynthesis, however, reach light saturation at high intensities and are not inhibited by full sunlight (see Chapter 2).

As would be expected, factor compensation occurs; individual plants and communities adapt to different light intensities by becoming *shade adapted* (reaching saturation at low intensities) or *sun adapted.* Diatoms that live in beach sand or on intertidal mudflats are remarkable in that they reach a maximum rate of photosynthesis when light intensity is less than 5 percent of full sunlight. Yet these diatoms are only slightly inhibited by high intensities. Phytoplankton, in contrast, are shade adapted and are very much inhibited by high intensities, which accounts for the fact that peak production in the sea usually occurs below rather than right at the surface.

Ionizing Radiations

Very high-energy radiations that can remove electrons from atoms and attach them to other atoms, thereby producing positive and negative *ion pairs,* are known as **ionizing radiations.** Light and most other solar radiations do not have this ionizing effect. It is believed that ionization is the chief cause of radiation injury to life and that the damage is proportional to the number of ion pairs produced in the absorbing material. Ionizing radiations are produced by radioactive materials on Earth and are also received from space. Isotopes of elements that emit ionizing radiations are termed *radionuclides* or *radioisotopes.*

Ionizing radiation in the environment has been increased appreciably by human efforts to use atomic energy. Nuclear weapons tests have injected radionuclides into the atmosphere that then return to Earth as global fallout. About 10 percent of the energy of a nuclear weapon is expended in residual radiation. Nuclear power plants (and fuel processing and disposal of wastes at other sites), medical research, and other peaceful uses of atomic energy produce local "hot spots" and wastes that often escape into the environment while being transported or stored. Failure to avoid accidental releases and to solve the radioactive waste problem are the main reasons why atomic energy has not lived up to its potential as an energy source for human societies. Because of the importance of atomic energy in the future, we will review this topic in some detail.

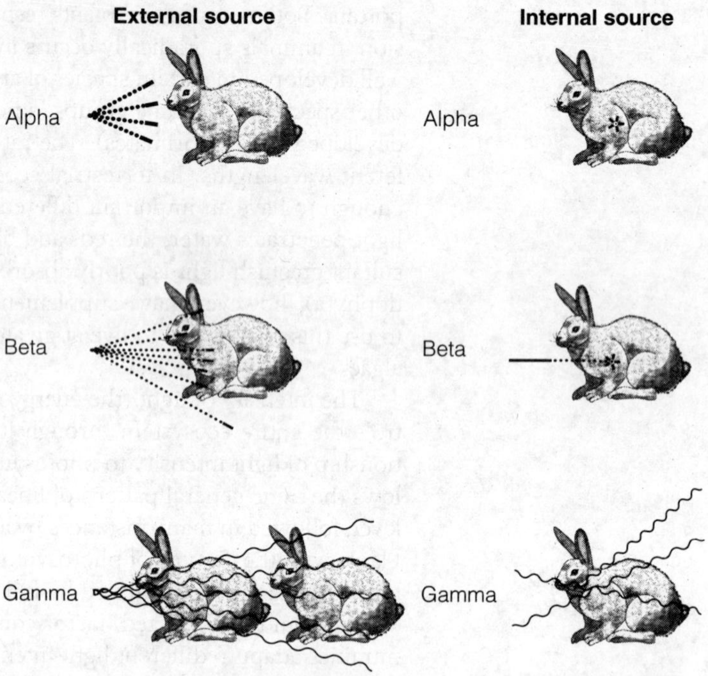

Figure 5-11. Schematic comparison of the three types of ionizing radiations of greatest ecological interest, showing relative penetration and specific ionization effects. The diagram is not intended to be quantitative.

Of the three ionizing radiations of primary ecological concern, two are *corpuscular* (alpha and beta radiations) and one is *electromagnetic* (gamma radiation and the related X-radiation). Corpuscular radiation consists of streams of atomic or subatomic particles that transfer their energy to whatever material they strike. **Alpha particles** are nuclei of helium atoms that travel only a few centimeters in air and may be stopped by a sheet of paper or the epidermis of human skin, but when stopped, they produce a large amount of ionization locally. **Beta particles** are high-speed electrons—much smaller particles that may travel several meters in air or up to a couple of centimeters in tissue and give up their energy over a longer path. Ionizing electromagnetic radiations are of much shorter wavelength than visible light, and they travel great distances and readily penetrate matter, releasing their energy over long paths (the ionization is dispersed). For example, **gamma rays** penetrate biological materials easily; a given gamma ray may go right through an organism without having any effect, or it may produce ionization over a long path. The effect of gamma rays depends on the number and energy of the rays and the distance of the organism from the source, as intensity decreases exponentially with distance. Important biological features of alpha, beta, and gamma radiation are diagrammed in Figure 5-11. The alpha-beta-gamma series is one of increasing penetration but decreasing concentration of ionization and local damage. Therefore, biologists often classify radioactive substances that emit alpha or beta particles as *internal emitters,* because their effect is likely to be greatest when absorbed, ingested, or otherwise deposited in or near living tissue. Conversely, radioactive substances that are primarily gamma emitters are classed as *external emitters,* because they are penetrating and can produce their effect without being ingested.

Other types of radiation that are of interest to the ecologist include **cosmic rays,** which are radiations from outer space that are mixtures of corpuscular and electromagnetic components. The intensity of cosmic rays in the ecosphere is low, but they are a major hazard in space travel. Cosmic rays and ionizing radiation from natural radioactive sources in soil and water produce what is known as **background radiation,** to which the present biota are adapted. In fact, the biota may depend on this background radiation for maintaining genetic fluidity. Background radiation varies three- to fourfold in various parts of the ecosphere; it is lowest at or below the surface of the sea and highest at high altitudes on granitic mountains. Cosmic rays increase in intensity with increasing altitude, and granitic rocks have more naturally occurring radionuclides than do sedimentary rocks.

A study of radiation phenomena requires two types of measurements: (1) a measure of the *number of disintegrations* occurring in a quantity of radioactive substance; and (2) a measure of radiation dose in terms of the *energy absorbed* that can cause ionization and damage. The basic unit of the quantity of a radioactive substance is the *curie* (Ci), defined as the amount of material in which 3.7×10^{10} atoms disintegrate each second, or 2.2×10^{12} disintegrations per minute. The actual weight of material making up a curie is very different for a long-lived, slowly decaying isotope compared with a rapidly decaying one. Because a curie represents a rather large amount of radioactivity from the biological viewpoint, smaller units are widely used: *millicurie* (mCi) $= 10^{-3}$ Ci; *microcurie* (μCi) $= 10^{-6}$ Ci; *nanocurie* (nCi) $= 10^{-9}$ Ci; and *picocurie* (pCi) $= 10^{-12}$ Ci. The curie indicates how many alpha or beta particles or gamma rays are emitted from a radioactive source per unit time, but this information does not tell us how the radiation might affect organisms in the line of fire.

Radiation dose, the other important aspect of radiation, has been measured with several scales. The most convenient unit for all types of radiation is the *rad,* which is defined as an absorbed dose of 100 ergs (10^{-5} joules) of energy per gram of tissue. The *roentgen* (R) is an older unit, which strictly speaking is to be used only for gamma and X-rays. Actually, for effects on living organisms, the rad and the roentgen are nearly the same. The roentgen or rad is a unit of *total dose.* The *dose rate* is the amount received per unit time. Thus, if an organism is receiving 10 mR per hour, the total dose in a 24-hour period would be 240 mR per hour, or 0.240 R/h. The time over which a given dose is received is a very important consideration.

In general, the higher, more complex organisms are more easily damaged or killed by ionizing radiation. Human beings are about the most sensitive of all. The comparative sensitivity of three diverse groups of organisms to single doses of gamma radiation is shown in Figure 5-12. Large, single doses delivered at short time intervals (minutes or hours) are known as *acute doses,* in contrast to *chronic doses* of sublethal radiation that might be experienced continuously over a whole life cycle. The left ends of the bars in Figure 5-12 indicate levels at which severe effects on reproduction (temporary or permanent sterilization, for example) may be expected in the more

Figure 5-12. Comparative radiosensitivity of three groups of organisms to single acute doses of gamma radiation.

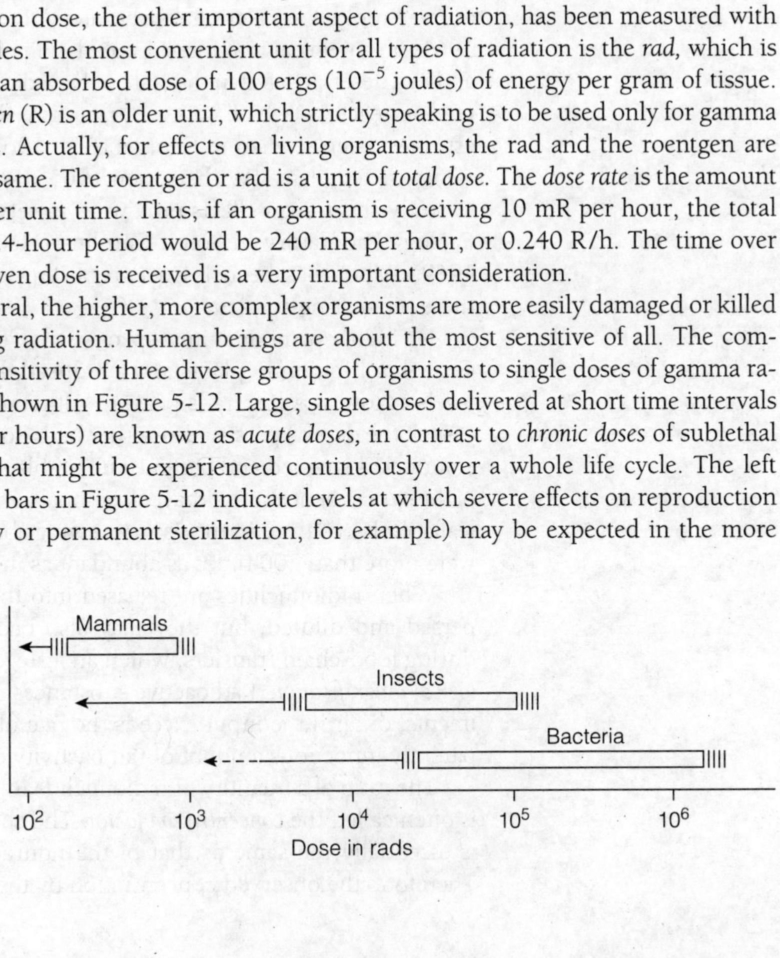

sensitive species of the group, and the right ends of the bars indicate levels at which a large portion (50 percent or more) of the more resistant species would be killed outright. The arrows to the left indicate the lower range of doses that would kill or damage sensitive life-cycle stages, such as embryos. Thus, a dose of 200 rads will kill some insect embryos in the cleavage stage, 5000 rads will sterilize some species of insects, but 100,000 rads may be required to kill all adult individuals of the more resistant insect species. In general, mammals are the most sensitive and microorganisms the most resistant of organisms. Seed plants and lower vertebrates fall somewhere between insects and mammals. Most studies have shown that rapidly dividing cells are most sensitive to radiation (which explains why sensitivity decreases with age). Thus, any component undergoing rapid growth is likely to be affected by comparatively low levels of radiation regardless of taxonomic relationships. The effects of low-level chronic doses are more difficult to measure, because long-term genetic as well as somatic effects may be involved.

In higher plants, sensitivity to ionizing radiation has been shown to be directly proportional to the size of the cell nucleus or, more specifically, to chromosome volume or DNA content. In the field, other considerations, such as the shielding of sensitive growing or regenerating parts (as when underground), would determine relative sensitivity.

Between 1950 and 1970, the effects of gamma radiation on whole communities and ecosystems were studied at several sites. Gamma sources—usually either cobalt-60 or cesium-137—of 10,000 Ci or more were placed in fields and in forests located at the Brookhaven National Laboratory, Long Island, New York (Woodwell 1962, 1965), in a tropical rain forest located in Puerto Rico (H. T. Odum and Pigeon 1970), and in a desert located in the state of Nevada (French 1965). The effects of unshielded nuclear reactors (which emit neutrons as well as gamma radiation) on fields and forests have been studied in Georgia (Platt 1965) and at the Oak Ridge National Laboratory in Tennessee (Witherspoon 1965, 1969). A portable gamma source was used to study short-term effects on a wide variety of communities at the Savannah River Ecology Laboratory in South Carolina (McCormick and Golley 1966; McCormick 1969). Much has been learned regarding ecosystem structure and function from these pioneering studies.

No higher plant or animal species survived when close to these powerful sources. Growth inhibition in plants and a reduced diversity of animal species were noted at levels as low as 2 to 5 rads per day. Although resistant forest trees or shrubs (in the case of the desert) persisted at rather high dose rates (10 to 40 rads per day), the vegetation was stressed and became vulnerable to insects and disease. In the second year of the experiment at Brookhaven National Laboratory, for example, an outbreak of oak leaf aphids occurred in the zone receiving about 10 rads per day. In this zone, aphids were more than 200 times as abundant as in the normal, unradiated oak forest.

When radionuclides are released into the environment, they often become dispersed and diluted, but they may also become concentrated in living organisms during food-chain transfers, which are categorized under the general heading of *biological magnification*. Radioactive substances may also accumulate in water soils, sediments, or air if the input exceeds the rate of natural radioactive decay; thus, an apparently innocuous amount of radioactivity can soon become lethal.

The ratio of a quantity of radionuclide in the organism to that in the environment is often called the *concentration factor*. The chemical behavior of a radioactive isotope is essentially the same as that of the nonradioactive isotope of the same element. Therefore, the observed concentration by the organism is not the result of the radio-

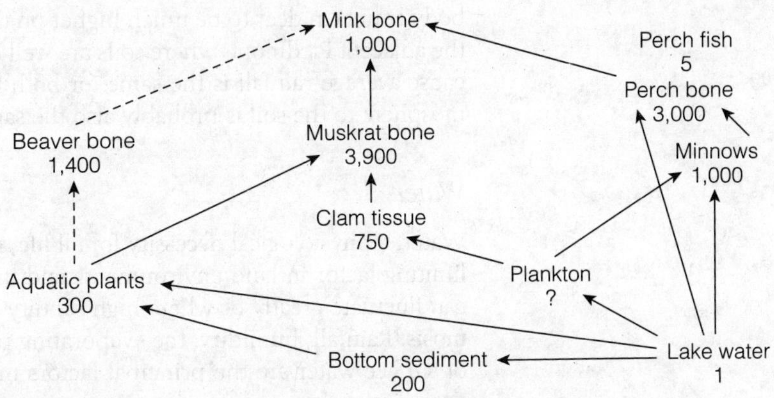

Figure 5-13. Concentration of strontium-90 in various parts of the food web of a small Canadian lake receiving low-level nuclear wastes. Average concentration factors are shown in terms of lake water = 1 (after Ophel 1963; used by permission of Biology and Health Physics Division, Atomic Energy of Canada Limited, Chalk River, Ontario).

activity, but merely demonstrates in a measurable way the difference between the density of the element in the environment and in the organism. Thus, radioactive iodine-131 (^{131}I) concentrates in the thyroid just as nonradioactive iodine does. Also, some of the synthetic radionuclides become concentrated because of their chemical affinity with nutrients that are naturally concentrated by organisms.

Two examples will illustrate the concentrative tendencies of two of the most troublesome, long-lived radionuclides that are by-products of the fission of uranium (hence called *fission products*). Strontium-90 (^{90}Sr) tends to cycle like calcium; cesium-137 (^{137}Cs) behaves like potassium in living tissues. Concentration factors for ^{90}Sr in various parts of a food web in a lake receiving low-level radioactive wastes are diagrammed in Figure 5-13. Because the blood-making bone marrow tissue is especially sensitive to the beta radiation of ^{90}Sr, the 3000- to 4000-fold concentration in perch and muskrat bone is significant. In assessing the release of radioactive material into the environment, one must determine biological concentrations.

Concentration factors are likely to be higher in nutrient-poor than in nutrient-rich soils and water. Concentration is also greater in "thin-cover" vegetation, such as lichen-covered rocks on the Arctic Tundra. A concern is that the Inuit peoples and the Saame of the Lapland region who consume caribou or reindeer meat ingest more fallout radionuclides than do those consuming from a grain-beef food chain.

Table 5-2 shows the concentration of fallout cesium-137 (determined by whole

Table 5-2 **Comparison of the concentration of cesium-137 (resulting from fallout) in white-tailed deer in the Coastal Plain and Piedmont regions of Georgia and South Carolina**

		Cesium-137 (pCi/kg wet weight)	
Region	***Number of deer***	***Mean ± standard error****	***Range***
Lower Coastal Plain	25	18,039 ± 2359	2076–54,818
Piedmont	25	3007 ± 968	250–19,821

Source: Data from Jenkins and Fendley 1968.

*Difference between regions highly significant at the $p < .01$ level of probability.

body count) in deer to be much higher on the sandy, low-lying Coastal Plain than in the adjacent Piedmont, where soils are well drained and have a high clay content. Because average rainfall is the same for both regions, the input of fallout from the atmosphere to the soil is probably also the same.

Water

Water, a physiological necessity for all life, is from the ecological viewpoint chiefly a limiting factor in land environments and in water environments where the amount can fluctuate greatly or where high salinity fosters water loss from organisms by osmosis. Rainfall, humidity, the evaporating power of the air, and the available supply of surface water are the principal factors measured. Each of these aspects is briefly described.

Rainfall is determined largely by geography and by the pattern of large air movements or weather systems. As anyone who accesses a weather report knows, weather systems within the United States move mainly from west to east. As shown in Figure 5-14, moisture-laden winds blowing across the ocean deposit most of their moisture on the ocean-facing slopes; the resulting **rain shadow** produces a desert on the leeward side of the mountains. In general, the higher the mountains, the greater the effect. As the air continues beyond the mountains, some moisture is picked up, and rainfall may again increase somewhat. Thus, deserts are usually found "behind" high mountain ranges. The distribution of rainfall over the year is an extremely important limiting factor for organisms. The situation provided by a 35-inch (89-cm) rainfall evenly distributed over time is entirely different from that provided by 35 inches (89 cm) of rain that falls largely during a restricted part of the year. In the latter case, plants and animals must be able to survive long droughts (and sudden floods). Rainfall generally tends to be unevenly distributed over the seasons in the Tropics and subtropical regions, often resulting in well-defined wet and dry seasons. In the Tropics, this seasonal rhythm of moisture regulates the seasonal activities (especially reproduction) of organisms, much as the seasonal rhythm of temperature and light regulates organisms living in the Temperate Zones. In temperate climates, rainfall tends to be more evenly distributed throughout the year, though with many exceptions. The following tabulation gives a rough approximation of the climax biotic communities (biomes) that may be expected with different annual amounts of rainfall evenly distributed in temperate latitudes:

0–25 centimeters (0–10 inches) per year—desert

25–75 centimeters (10–30 inches) per year—grassland, savanna

75–125 centimeters (30–50 inches) per year—dry forest

>125 centimeters (> 50 inches) per year—wet forest.

Actually, the biotic situation is determined not by rainfall alone but by the balance between rainfall and potential evapotranspiration, which is the loss of water by evaporation from the ecosystem.

Humidity represents the amount of water vapor in the air. **Absolute humidity** is the actual amount of water in the air expressed as weight of water per unit of air (grams per kilograms of air, for example). As the amount of water vapor that air can hold (at saturation) varies with temperature and pressure, **relative humidity** represents the percentage of water vapor actually present compared with the saturation density under existing temperature-pressure conditions. In general, relative humid-

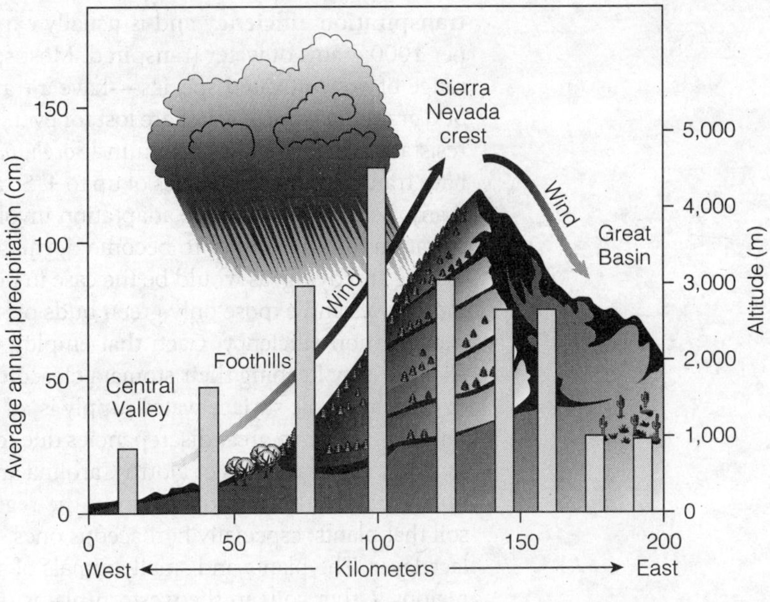

Figure 5-14. Mountain ranges influence patterns of precipitation. In the Sierra Nevada, wind moves from the West Coast of the United States across the central valley of California and then is deflected upward as it reaches the mountain range. This moisture-laden air then cools, and its moisture condenses as it is deflected upward by the mountain. As the air rushes down the eastern slope (the leeward side of the mountain), it warms and creates arid conditions in the Great Basin (data and profile after Pianka 2000).

ity has been the measurement most used in ecological work, although the converse of relative humidity, **vapor pressure deficit** (the difference between the partial pressure of water vapor at saturation and the actual vapor pressure), is often referred to as a measure of moisture relations, because evaporation tends to be proportional to vapor pressure deficit rather than to relative humidity.

Because of the daily rhythm of humidity in nature (high at night, low during the day, for example), as well as vertical and horizontal differences, humidity, along with temperature and light, helps regulate the activities of organisms and limit their distribution. Humidity is especially important in modifying the effects of temperature, as will be noted in one of the following sections.

The evaporative power of the air is ecologically important, especially for land plants. Animals may often regulate their activities to avoid dehydration by moving to protected places or becoming active at night; plants, however, cannot move. Between 97 and 99 percent of the water that enters plants from the soil is lost by evaporation from the leaves. This evaporation, called **evapotranspiration,** is a unique feature of the energetics of terrestrial ecosystems. When water and nutrients are nonlimiting, the growth of land plants is closely proportional to the total energy supply at the ground surface. As most of the energy is heat, and as the fraction providing latent heat for transpiration is nearly constant, growth is also proportional to transpiration.

Despite the many biological and physical complications, total evapotranspiration is broadly correlated with the rate of productivity. For example, Rosenzweig (1968) found that evapotranspiration was a highly significant predictor of the annual aboveground net primary production (P_n) in mature or climax terrestrial communities of all kinds (deserts, tundras, grasslands, and forests); however, the relationship was not reliable in unstable or developmental (nonclimax) vegetation. The poor correlation between assimilated energy and P_n in developmental communities is logical, because such communities have not yet reached equilibrium conditions with their energy and water environment.

The ratio of net primary production to the amount of water transpired is termed

transpiration efficiency and is usually expressed as grams of dry matter produced per 1000 grams of water transpired. Most species of agricultural crops—and a wide range of noncultivated species—have a transpiration efficiency of 2 or less (that is, 500 grams or more of water are lost for every gram of dry matter produced). Drought-resistant crops, such as sorghum (*Sorghum bicolor*) and millet (*Panicum ramosum*), have transpiration efficiencies of up to 4. Strangely enough, desert plants can do little, if any, better. Their unique adaptation involves not the ability to grow without transpiration, but the ability to become dormant when water is not available (instead of wilting and dying, as would be the case in non–desert plants). Desert plants that lose their leaves and expose only green buds or stems during dry periods do show a high transpiration efficiency. Cacti that employ the CAM type of photosynthesis reduce water loss by keeping their stomata closed during the day (see Chapter 2).

The available surface water supply is, of course, related to the rainfall in the area, but there are often great discrepancies due to the nature of the substrate on which the rain falls. The sandhills of North Carolina are frequently referred to as "deserts in the rain," because the abundant rain in the region sinks so quickly through the porous soil that plants, especially herbaceous ones, find very little water available in the surface layer. The plants and small animals of such areas resemble those of much dryer regions. Other soils in the western plains of the United States retain water so tenaciously that crops can be raised without a single drop of rain falling during the growing season (the plants can use the water stored from winter rains).

Artificial impoundment of streams (reservoirs) has helped increase the availability of local water supplies, as well as providing recreation and hydroelectric power. However, these mechanical engineering devices, useful though they often are, should never be regarded as substitutes for sound agricultural and forestry land-use practices, which trap the water at or near its sources for maximum usefulness. The ecological viewpoint—water as a cyclic commodity in the whole ecosystem—is very important. People who think that all floods, erosion, and water-use problems can be solved by building dams, or other mechanical strategies such as channelization, need to acquire an understanding of hydrology and landscape ecology. The Mississippi River, for example, has a long history of flooding and attempts to control this process. Despite millions spent on flood control in constructing dikes and dams and other attempts to "tame" the Mississippi River, the cost of flood damage has increased. The more the river is constricted by dikes and levees and the more the watershed is urbanized, the higher the water rises and the worse the flood is when water does break through or rise over the barriers.

From 1930 to 2000, there has been a dramatic loss of wetlands in the Mississippi Delta of Louisiana, with estimates as high as 100 km^2/year, or a total of 4000 km^2 during this period. According to J. W. Day et al. (2000), canals and dikes designed to expedite the river flow into the Gulf of Mexico as fast as possible have reduced the sediments necessary to maintain the Delta wetlands. For a well documented account of the downside of attempts to control the Mississippi River, see Belt (1975), Sparks et al. (1998), and Jackson and Jackson (2002).

Dew may contribute appreciably and, in areas of low precipitation, vitally to the water supply. Dew and ground fog are especially important in coastal forests and in deserts. Fog on the West Coast may account for as much as two to three times more water than the annual precipitation. Tall trees, such as the coastal redwood tree (*Sequoia sempervirens*), intercept coastal fog as it moves inland and may collect as much as 150 cm of "rainfall" dripping down from the limbs.

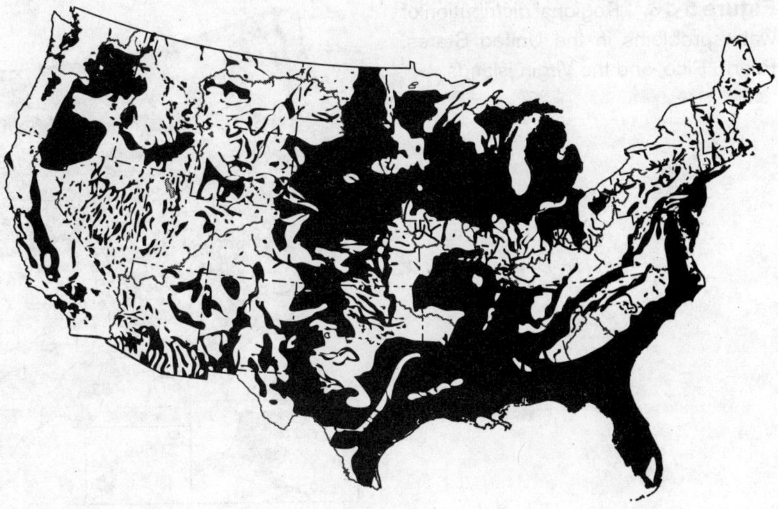

Figure 5-15. Groundwater resources of the United States of America. About half of the country is underlain by aquifers capable of yielding very large volumes of water. Aquifers in the mid-continent and western regions, which are poorly recharged, are being "overdrawn" or "mined" in many areas where withdrawals are the source of irrigation water (courtesy of U.S. Water Resources Council).

Groundwater

For humankind, groundwater is one of the most important resources, because we do have access in many regions to a great deal more water than falls as rain. Cities and irrigated agriculture located in deserts and other dry regions are made possible by this access to groundwater. Unfortunately, much of this underground water was stored in past ages and is either not being replenished at all or is being replenished at a slower rate than it is being pumped out. Arid-region groundwater, like oil, is a nonrenewable resource.

Groundwater provides 25 percent of the freshwater used for all purposes in the United States, and about 50 percent of the drinking water. Irrigation water use in the United States has increased steadily from 1965 to 1980, because irrigation water use is dependent on factors such as precipitation, water availability, energy costs, farm commodity prices, application technologies, and conservation practices. The total amount of water used for irrigation actually decreased from 1980 to 1995, even though the total irrigated area remained consistent at about 23.5 million ha (Pierzynski et al. 2000). These data suggest that factors such as conservation practices, reduced energy usage, and appropriate technologies can significantly reduce the amount of groundwater use for irrigation. As with other abundant natural capital, groundwater tended to be taken for granted and was studied very little until signs of its depletion and pollution showed plainly that limiting factors were involved.

The largest stores of groundwater are in **aquifers,** porous underground strata, often composed of limestone, sand, or gravel, bounded by impervious rock or clay that holds the water in like a giant pipe or elongated tank. Water enters where the permeable strata are close to the surface or otherwise intersect the surface water table; water may leave the aquifer by way of springs or other discharges at or near the surface. Where aquifers slope seaward from higher-ground recharge areas, the water in the deeper aquifer is under pressure and will rise above the surface like a fountain when a well is drilled into it (the so-called *artesian well*). The geographical distribution of aquifers and other substantial stores of groundwater is mapped in Figure 5-15.

The annual input (rain and snowbelt recharge) and output (water returned to the

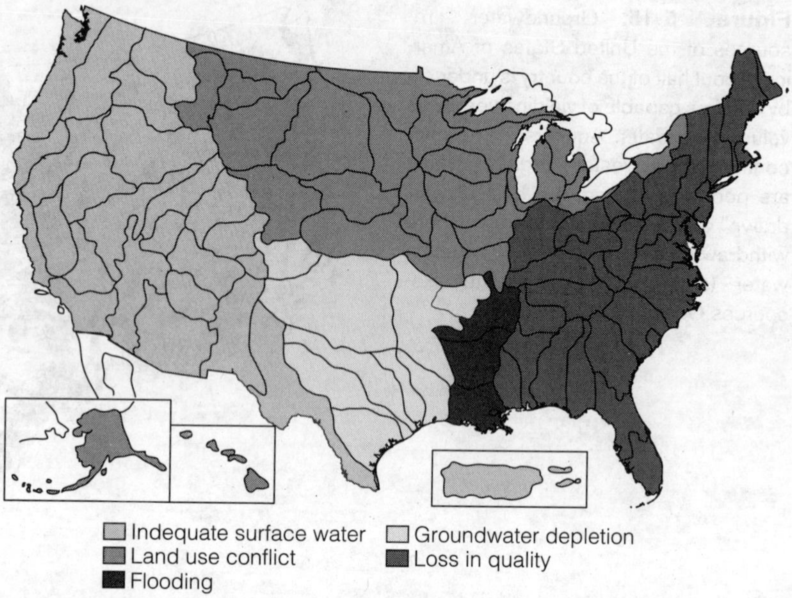

Figure 5-16. Regional distribution of water problems in the United States, Puerto Rico, and the Virgin Islands.

hydrologic cycle of rivers, oceans, and the atmosphere) for this huge reservoir are estimated to be about 1 part in 120 parts of the total volume. Although withdrawals total only about one tenth of the recharge volume, some of the most heavily used aquifers are located in regions of low or no recharge. For example, about one fourth of all aquifer withdrawals are *overdrafts* (exceeding recharge), mostly in agricultural regions of the West. An example is the Ogallala aquifer of the high plains of Texas, Kansas, Oklahoma, Nebraska, and eastern Colorado, where irrigated grain production provides an important part of the export market that the United States counts on to balance its payments for imported oil. Fossil water and fossil fuel (to pump the water) support a billion-dollar economy in this region. It is predicted that within the next couple decades, this aquifer will, for all practical purposes, be "pumped out" (Opie 1993). The fossil water will be gone before the fossil fuel is exhausted, but the latter becomes useless without water. Then, the region will be faced with severe economic depression and depopulation, and the nation will have to find some other place to grow grain—unless, of course, it is judged feasible to build an aqueduct from the Mississippi River system! For more information on groundwater, see the National Geographic Special Edition (1993).

Depletion is not the only threat to groundwater. Contamination with toxic chemicals may be an even greater threat. At least the problem of toxic wastes does have technological solutions, if societies are willing and able to pay the cost to protect a water resource that in the long run is more precious than oil or gold. In fact, one could argue for the proposition that usable freshwater is potentially a greater limiting factor for civilization than is energy. Water problems vary with the region but, as shown in Figure 5-16, no region is without a water problem of some sort. Because water is frequently viewed as a nonmarket commodity, public opinion and political intervention are important to prevent both wasteful allocation and complete depletion of this resource. More important, however, is extending the market economy to include water. If we pay more for water, we will waste and pollute less water. There is no question that quality freshwater is becoming a severe limiting factor for humans on a global scale (see Postel 1992, 1993, 1999; Gleick 2000 for details).

Temperature and Moisture Acting Together

Based on the ecosystem concept, we have avoided creating the impression that environmental factors operate independently of one another. This chapter attempts to show that the consideration of individual factors is a means of approaching complex ecological problems, but it is not the ultimate objective of ecological study, which is to evaluate the relative importance of various factors as they operate together in actual ecosystems. Temperature and moisture are generally so important and so closely interacting in terrestrial environments that they are usually conceded to be the most important aspect of climate.

The interaction of temperature and moisture—as with the interaction of most factors—depends on the relative as well as the absolute values of each factor. Thus, temperature exerts a more severe limiting effect on organisms when there is either abundant or very little moisture than when there are moderate conditions. Likewise, moisture is critical at extremes of temperature. In a sense, this is another aspect of the principle of factor interaction. For example, the boll weevil (*Anthonomus grandis*) can tolerate higher temperatures when the humidity is low or moderate than when it is very high. Hot, dry weather in the cotton belt is a signal for the cotton farmers to look for an increase in the weevil population. Hot, humid weather is less favorable for the weevil but, unfortunately, not so good for the cotton plant.

Large bodies of water greatly moderate land climates because of the high latent heat of evaporation and melting characteristics of water, which is to say that many heat calories are required to melt ice and evaporate water. In fact, there are two basic types of climate: (1) the *continental climates,* characterized by extremes of temperature and moisture; and (2) the *marine climates,* characterized by less extreme temperature and moisture fluctuations because of the moderating effect of large bodies of water (large lakes thus produce local marine climates).

Early classifications of climate were based largely on quantitative measures of temperature and moisture, taking into consideration the effectiveness of precipitation and temperature as determined by seasonal distribution and mean values. The relationship between precipitation and potential evapotranspiration (which depends on temperature) provides a particularly accurate picture of climates. The period of soil moisture use represents the principal period of primary production for the ecosystem as a whole and, thus, determines the supply of food available to the consumers and decomposers for the entire annual cycle. In the Deciduous Forest biome, water is likely to be severely limiting only in late summer, more so in the southern than in the northern portion of the biome. Native vegetation is adapted to withstand periodic summer droughts, but some agricultural crops grown in the region are not. After bitter experiences with many late summer crop failures, farmers in the southern United States are beginning to provide for irrigation in the late summer.

Climographs, or charts in which one major climatic factor is plotted against another, are a useful method of graphically representing temperature and moisture in combination. In temperature-rainfall charts, mean monthly values are plotted with the temperature scale on the vertical axis and either humidity or rainfall on the horizontal axis. The resulting polygon shows the temperature-moisture conditions and makes possible the graphic comparison of one year with another, or the comparison of the climate of one biotic region with that of another. Climographs have been useful in determining the suitability of temperature-moisture combinations for proposed introductions of agricultural plants or game animals. Plots of other pairs of factors, such as temperature and salinity in marine environments, may also be instructive.

Environmental chambers provide another useful approach to the study of combinations of physical factors. They vary from simple *temperature-humidity cabinets* to large controlled *greenhouses* or *phytotrons,* in which any desired combination of temperature, moisture, and light can be maintained. These chambers are often designed to control environmental conditions so that the investigator can study the genetics, physiology, and ecology of cultivated or domesticated species. These chambers can be especially useful for ecological studies when natural rhythms of temperature and humidity can be simulated. Experiments of this sort help single out factors that may be operationally significant, but they can reveal only part of the story, as many significant aspects of the ecosystem cannot be duplicated indoors (in microcosms) but must be experimented with outdoors (in mesocosms). Phytotrons have been used in recent years to determine the effect of increasing CO_2 due to human activities on plants (see next section). Biosphere-2, described in Chapter 2, is the largest "greenhouse" designed to support humans.

Atmospheric Gases

The atmosphere in the major part of the ecosphere is remarkably homeostatic. Interesting enough, the present concentrations of carbon dioxide (0.03 percent by volume) and oxygen (21 percent by volume) are somewhat limiting to many higher plants. It is well known that photosynthesis in many C_3 plants can be increased by moderately increasing the CO_2 concentration, but it is not so well known that decreasing the oxygen concentration experimentally can also increase photosynthesis. Beans, for example, increase their rate of photosynthesis by as much as 50 percent when the oxygen concentration around their leaves is lowered to 5 percent. C_4 plants are not inhibited by high O_2 concentration. Thus, C_4 grasses, including corn and sugarcane, do not show oxygen inhibition. The reason for inhibition in C_3 broad-leaved plants may be that they evolved when the CO_2 concentration was higher and the O_2 concentration lower than they are now.

The situation in aquatic environments differs from that in the atmospheric environment because the amounts of oxygen, carbon dioxide, and other atmospheric gases dissolved in water and thus available to organisms vary greatly from time to time and from place to place. Oxygen is a prime limiting factor, especially in lakes and in waters with a heavy load of organic material. Although oxygen is more soluble in water than is nitrogen, the actual quantity of oxygen that water can hold under the most favorable conditions is much less than that constantly present in the atmosphere. Thus, if 21 percent by volume of a liter of air is oxygen, there will be 210 cm^2 of oxygen per liter. By contrast, the amount of oxygen per liter of water does not exceed 10 cm^2. Temperature and dissolved salts greatly affect the ability of water to hold oxygen; the solubility of oxygen is increased by low temperatures and decreased by high salinities. The oxygen supply in water comes chiefly from two sources: (1) by diffusion from the air; and (2) from photosynthesis by aquatic plants. Oxygen diffuses into water very slowly, unless helped along by wind and water movements; light penetration is an all-important factor in the photosynthetic production of oxygen. Therefore, important daily, seasonal, and spatial variations may be expected in the oxygen concentration of aquatic environments.

Carbon dioxide, like oxygen, may be present in water in highly variable amounts. It is difficult to make general statements about the role of carbon dioxide as a limiting factor in aquatic systems. Although present in low concentrations in the air, carbon dioxide is extremely soluble in water, which also obtains large supplies from respiration, decay, and soil. Thus, the minimum CO_2 concentration is less likely to be important than is the case with oxygen. Furthermore, unlike oxygen, carbon dioxide chemically combines with water to form H_2CO_3 (*carbonic acid*), which in turn reacts with available limestones to form carbonates (CO_3) and bicarbonates (HCO_3). A major reservoir pool of biospheric CO_2 is the carbonate system of the oceans. Carbonate compounds not only provide a source of nutrients, but they also act as *buffers*, helping to keep the hydrogen ion concentration (pH) of aquatic environments near the neutral point. Moderate increases in CO_2 in water seem to speed up photosynthesis and the developmental processes of many organisms. CO_2 enrichment, along with increased nitrogen and phosphorus, may help to explain cultural eutrophication. High CO_2 concentrations may be limiting to animals, especially because such high concentrations of carbon dioxide are associated with low concentrations of oxygen. Fishes respond vigorously to high CO_2 concentrations and may be killed if the water is too heavily charged with unbound CO_2.

Hydrogen ion concentration, or pH, is closely related to the carbon dioxide cycle and has been much studied in natural aquatic environments. Unless pH values are extreme, communities compensate for differences in pH (by mechanisms already described in this chapter) and show a wide tolerance for the naturally occurring range. However, when the total alkalinity is constant, pH change is proportional to CO_2 change and, therefore, is a useful indicator of the rate of total community metabolism (photosynthesis and respiration). Soils and waters of low (acidic) pH are frequently deficient in nutrients and low in productivity.

Macronutrients and Micronutrients

About half of the 92 elements in the periodic table have now been shown to be essential to either plants or animals or, in most cases, both. As already indicated, nitrogen and phosphorus salts are of major importance, and the ecologist may do well to consider these first as a matter of routine (see Chapter 4 for details regarding N/P ratios).

Potassium, calcium, sulfur, and magnesium merit consideration after nitrogen and phosphorus. Mollusks and vertebrates need skeletal calcium in especially large quantities, and magnesium is a necessary constituent of chlorophyll, without which no ecosystem could operate. Elements and their compounds needed in relatively large amounts are known as **macronutrients.**

In recent years, great interest has developed in the study of elements and their compounds that are necessary for the operation of living systems but that are required only in extremely minute quantities, often as components of vital enzymes. These elements are generally termed **trace elements** or **micronutrients.** Because minute requirements seem to be associated with an equal or even greater minuteness in environmental occurrence, these micronutrients are frequently important as limiting factors. The development of modern methods of microchemistry, spectrography, X-ray diffraction, and biological assay has greatly increased our ability to measure even the smallest amounts of micronutrients. Also, the availability of radioisotopes for many trace elements has greatly stimulated experimental studies. *Deficiency diseases* due to the absence of trace elements have been known to exist for a long time.

Pathological symptoms have been observed in laboratory, domestic, and wild plants and animals. Under natural conditions, deficiency symptoms of this sort are sometimes associated with peculiar geological histories and sometimes with a deteriorated environment of some sort—often a direct result of poor habitat or landscape management by humans. An example of a peculiar geological history is found in southern Florida. The potentially productive organic soils of this region did not meet expectations for crops and cattle, until it was discovered that this sedimentary region lacked copper and cobalt, which are present in most other areas.

Ten micronutrients are especially important to plants: iron (Fe), manganese (Mn), copper (Cu), zinc (Zn), boron (B), silicon (Si), molybdenum (Mo), chlorine (Cl), vanadium (V), and cobalt (Co). These elements can be arranged by function into three groups: (1) those required for photosynthesis: Mn, Fe, Cl, Zn, V; (2) those required for nitrogen metabolism: Mo, B, Co, Fe; and (3) those required for other metabolic functions: Mn, B, Co, Cu, Si. All these elements, except boron, are essential for animals, which also may require selenium (Se), chromium (Cr), nickel (Ni), fluorine (F), iodine (I), tin (Sn), and perhaps even arsenic (As; Mertz 1981). Of course, the dividing line between macro- and micronutrients is neither sharp nor the same for all groups of organisms; sodium and chlorine, for example, are needed in larger amounts by vertebrates than by plants. Sodium (Na), in fact, is often added to the preceding list as a micronutrient for plants. Many micronutrients resemble vitamins because they act as catalysts. The trace metals often combine with organic compounds to form *metallo-activators;* cobalt, for example, is a vital constituent of vitamin B_{12}. Goldman (1960) documented a case in which molybdenum was limiting to a whole ecosystem when he found that the addition of 100 parts per billion Mo to the water of a mountain lake increased the rate of photosynthesis. He also found that in this particular lake, the concentration of cobalt was high enough to be inhibitory to the phytoplankton. As with macronutrients, too much of a micronutrient can be as limiting as too little. For an analysis of the pattern of trace elements in a whole watershed, see Riedel et al. (2000).

Wind and Flooding

The atmospheric and hydrospheric media in which organisms live are seldom completely still for any period of time. Currents in water not only greatly influence the concentration of gases and nutrients, but they act directly both as limiting factors at the species level and as energy subsidies that increase productivity at the ecosystem level. Thus, the differences between the species composition of a stream and that of a small pond community are related to the large difference in wind and water currents. Many stream plants and animals are morphologically and physiologically adapted to maintaining their position in the current and are known to have very definite limits of tolerance to this specific factor. On the other hand, water flow that acts as an energy subsidy is a key to the productivity of wetland and tidal ecosystems.

On land, wind exerts a limiting effect on the activities, behavior, and even the distribution of organisms. Birds, for example, remain quiet in protected places on windy days—days on which it is difficult for the ecologist to attempt a bird census. Plants may be modified structurally by the wind, especially when other factors are also limiting, as in alpine regions. Figure 5-17 shows the tree line in Rocky Mountain National Park, where trees are exposed to extreme wind conditions. Years ago, Whitehead (1957) demonstrated experimentally that wind limits the growth of plants in

Figure 5-17. Tree line in Rocky Mountain National Park, Colorado, showing how wind affects tree morphology and physiology.

exposed locations on mountains. When he erected a wall to protect the vegetation from wind, the height of the plants increased.

On the other hand, air movement can enhance productivity in the same manner as water flow, as is apparently the case for certain tropical rain forests. Storms are important, even though they may be only local. Hurricanes transport animals and plants for great distances, and when these storms strike land, the winds may change the composition of the forest communities for many years to come. Oliver and Stephens (1977) reported that the effects in New England forests of two hurricanes that occurred before 1803 could still be seen in the structure of the vegetation. It has been observed that insects spread faster in the direction of the prevailing winds than in other directions to areas that seem to offer equal opportunity for the establishment of the species. In dry regions, wind is an especially important limiting factor for plants, because it increases the rate of water loss by transpiration, but desert plants have developed many special adaptations, such as sunken stomata, to tolerate these limitations.

7 Biological Magnification of Toxic Substances

Statement

The distribution of energy, of course, is not the only quality influenced by food chain phenomena. Some substances become concentrated instead of dispersed with each link in the food chain. Food chain concentration, or **biological magnification,** is dramatically illustrated by the behavior of certain persistent radionuclides, pesticides, and heavy metals.

Explanation

The tendency for certain radionuclide by-products of atomic fission and activation to become increasingly concentrated with each step in the food chain was first discovered at the Atomic Energy Commission's Hanford plant in eastern Washington in the 1950s. Radioactive cesium, strontium, and phosphorus released into the Columbia River were found to have become concentrated in the tissues of fish and birds. A concentration factor (amount in tissue/amount in water) of 2 million times was reported for radioactive phosphorus in the eggs of geese nesting on the islands in the river. Thus, what was considered harmless when released into the water became highly toxic to the downstream components of the food chain.

Rachel Carson, in her famous book *Silent Spring* (R. Carson 1962), called attention to the harmful effects and persistence (remaining active for long periods of time) of chlorinated hydrocarbon insecticides, especially DDT, and their detrimental effects as a biocide on populations, communities, ecosystems, and total landscapes due to the widespread aerial application of these compounds. An example of the buildup of DDT is shown in Table 5-3 and Figure 5-18. To control mosquitoes on Long Island, New York, municipalities sprayed DDT on the marshes for many years. Insect control specialists tried to use spray concentrations that were not directly lethal to fish and other wildlife, but they failed to recognize the negative effect on ecological processes and the long-term toxicity of DDT residues. Instead of being washed out to sea, as some had predicted, the poisonous residues adsorbed on detritus, became concentrated in the tissues of detritus feeders and small fishes, and increasingly concentrated in the top predators, such as fish-eating birds. The concentration factor (ratio of ppm in organism to ppm in water) is about 500,000 times for fish-eaters in the

Table 5-3 **An example of food chain concentration of a persistent pesticide, DDT**

Trophic level	*DDT residues (ppm*)*
Water	0.00005
Plankton	0.04
Silverside minnow	0.23
Sheephead minnow	0.94
Pickerel (predatory fish)	1.33
Needlefish (predatory fish)	2.07
Heron (feeds on small animals)	3.57
Tern (feeds on small animals)	3.91
Herring gull (scavenger)	6.00
Osprey (egg)	13.8
Merganser (fish-eating duck)	22.8
Cormorant (feeds on larger fish)	26.4

Source: Data from Woodwell et al. 1967.

*Parts per million (ppm) of total residues, DDT + DDD + DDE (all of which are toxic), on a wet-weight, whole-organism basis.

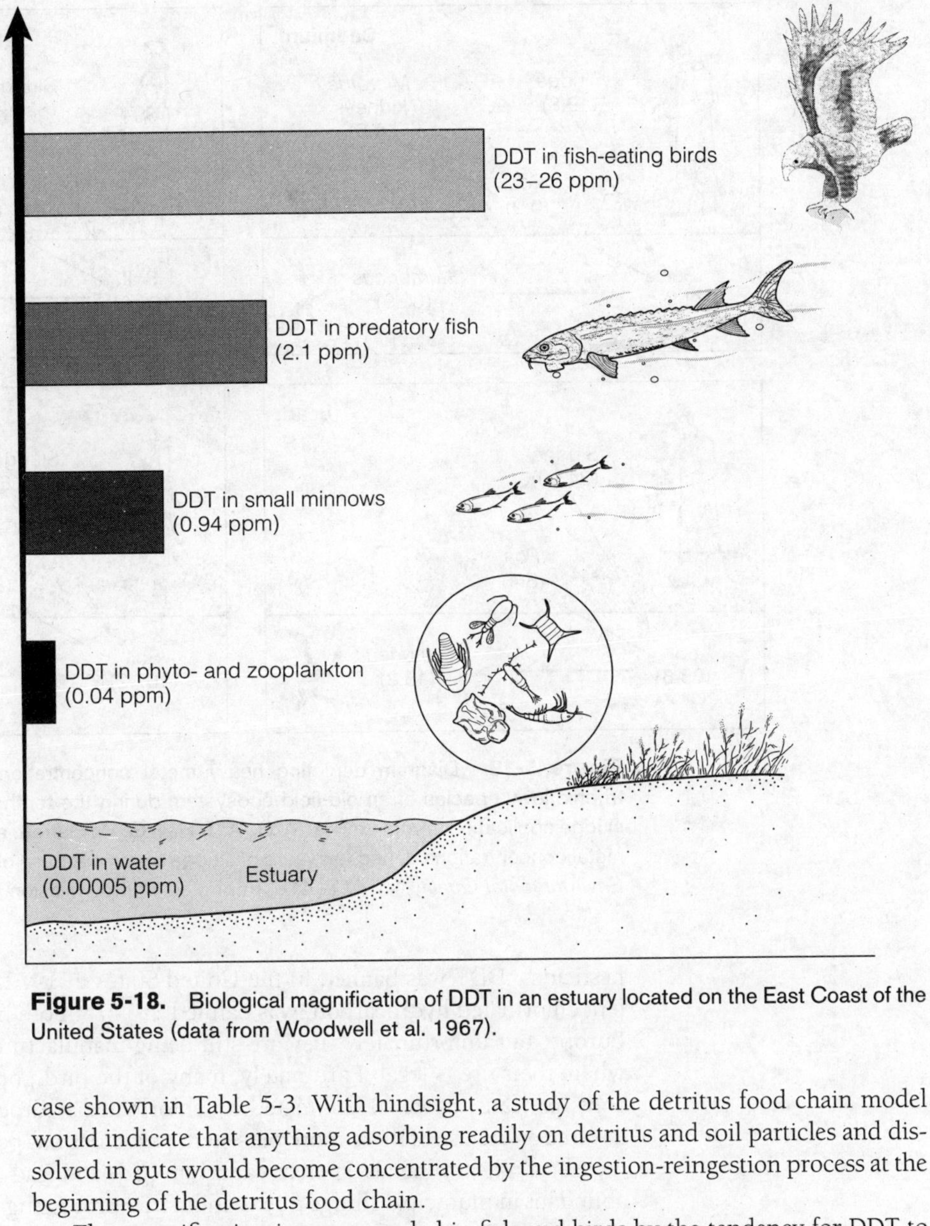

Figure 5-18. Biological magnification of DDT in an estuary located on the East Coast of the United States (data from Woodwell et al. 1967).

case shown in Table 5-3. With hindsight, a study of the detritus food chain model would indicate that anything adsorbing readily on detritus and soil particles and dissolved in guts would become concentrated by the ingestion-reingestion process at the beginning of the detritus food chain.

The magnification is compounded in fish and birds by the tendency for DDT to accumulate in body fat. The widespread use of DDT ultimately wiped out whole populations of predatory birds, such as osprey, peregrine falcons, and pelicans, and detritus feeders such as fiddler crabs. Birds are especially vulnerable to DDT poisoning because DDT (and other chlorinated hydrocarbon insecticides) interferes with the formation of eggshells by causing a breakdown in steroid hormones (Peakall 1967; Hickey and Anderson 1968). These fragile eggs then break before the young can hatch. Thus, very small amounts that are not lethal to the individual can be lethal to the population. Scientific documentation of this sort of frightening buildup (frightening because humans are also partly "top carnivores") and its unanticipated physiological effects finally marshalled public opinion against the use of DDT and similar

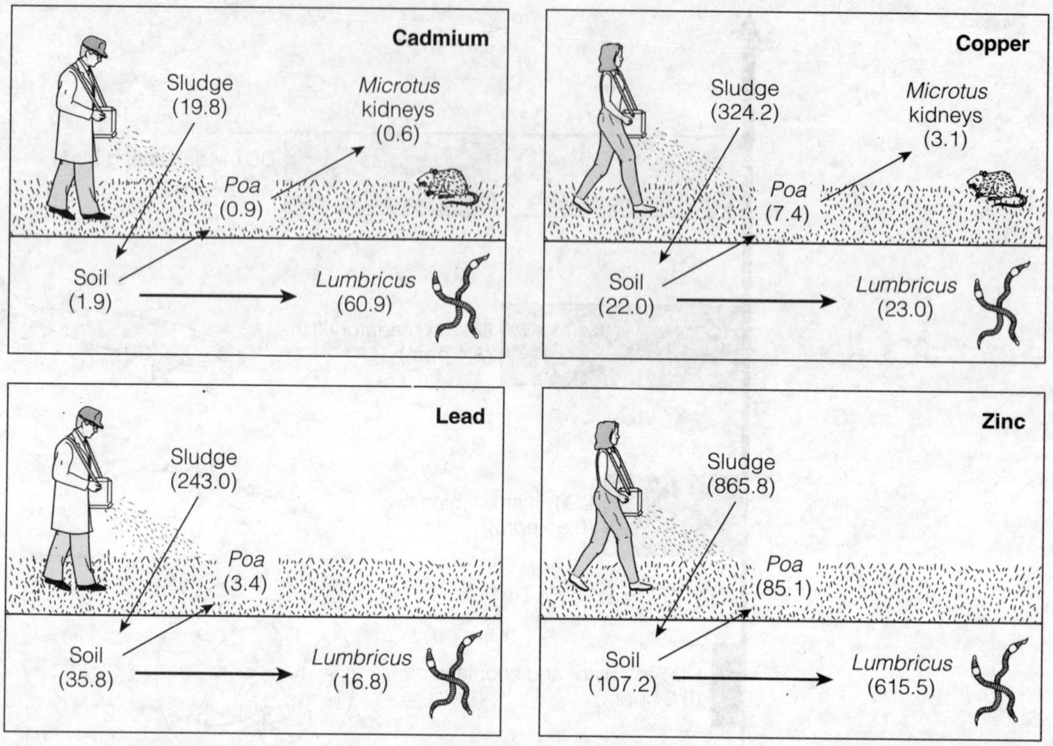

Figure 5-19. Diagram depicting heavy metal concentrations (mg/kg) in representative trophic level species of an old-field ecosystem during the tenth consecutive year of sewage sludge application. (From Levine, M. B., A. T. Hall, G. W. Barrett, and D. H. Taylor. 1989. Heavy metal concentrations during ten years of sludge treatment to an old-field community. *Journal of Environmental Quality* 18:411–418. Reprinted with permission.)

pesticides. DDT was banned in the United States in 1972. Dieldrin, another persistent chlorinated hydrocarbon, was banned in 1975. Both have also been outlawed in Europe, but unfortunately they are still being manufactured for export to countries where their use is legal. Fortunately, many of the bird populations (bald eagles, falcons, pelicans, ospreys) decimated by chlorinated hydrocarbon pesticides have recovered as the use of these more persistent pesticides has been reduced or eliminated.

Certain heavy metals, such as cadmium (Cd) and lead (Pb), which are frequently abundant in municipal sludge due to industrial processing in an urban area or watershed, may also become biologically magnified in the food chain. In a long-term investigation involving soils and three trophic levels in an old-field community treated for 11 consecutive years with sewage sludge, it was determined that the detritivore trophic level bioconcentrated heavy metals more than the producer or primary consumer trophic levels (Fig. 5-19). As illustrated, earthworms (*Lumbricus*) concentrated cadmium more than 30 times over concentration levels found in the soil, more than 60 times over levels found in plants (*Poa*), and more than 100 times over levels in the kidneys of meadow voles (*Microtus*) during the tenth year of sewage sludge application (see W. P. Carson and Barrett 1988; Levine et al. 1989; Brewer et al. 1994 for details). These investigators recommended that earthworms, as a representative or *indicator species* of detritivores, be used for monitoring the effects of sludge disposal on terrestrial communities and landscapes during secondary succession.

8 Anthropogenic Stress as a Limiting Factor for Industrial Societies

Statement

Natural ecosystems exhibit considerable resistance, resilience, or both to periodic severe or *acute* disturbance, probably because through evolutionary time they have naturally adapted to it. Many organisms, in fact, require *stochastic* (random) or periodic disturbances, such as fire or storms, for long-term persistence, as noted in the discussion of fire-adapted vegetation (Section 5). Accordingly, ecosystems may recover rather well from many periodic anthropogenic disturbances, such as harvest removal. *Chronic* (persistent or continued) disturbance, however, may have pronounced and prolonged effects, especially in the case of industrial chemicals that are new to the environment. In such cases, organisms have no evolutionary history of adaptation. Unless the increasing volume of highly toxic wastes, which are the current by-product of high-energy, industrialized societies, is reduced and ultimately eliminated at the source, toxic wastes will increasingly threaten both human and ecosystem health and be a major limiting factor for humankind.

Explanation

Although any classification is somewhat arbitrary, it may be instructive to consider *anthropogenic stress* on ecosystems under two categories: (1) **acute stress,** characterized by sudden onset, sharp rise in intensity, and short duration; and (2) **chronic stress,** involving long duration or frequent recurrence but not high intensity—a "constantly vexing" disturbance. Natural ecosystems exhibit considerable ability to deal with or recover from acute stress. Figure 5-20 depicts an example of an acute stress. An acute dose of municipal sludge input into a stream ecosystem resulted in a fish kill, because bacterial decomposition caused the stream's oxygen content to approach zero. Once the sewage treatment plant whose breakdown had caused the

Figure 5-20. Fish kill in Four Mile Creek near Oxford, Ohio, resulting from an acute input of municipal sludge into the stream ecosystem.

Courtesy of Gary W. Barrett

acute stress was repaired, the stream initiated its recovery process. Another example of recovery following acute stress is the *buried seed strategy,* a quick recovery mechanism that facilitates forest regrowth after clear-cutting (Marks 1974). The effects of chronic stress are more difficult to assess, because responses will not be so dramatic. It may be years before the full effects are known, just as it took many years to understand the link between cancer and smoking or the relationship between cancer and chronic low-level ionizing radiation. Environmental "cancer" (disorderly growth of exotic species at population or community levels) appears to provide an analogous situation regarding ecological systems.

Of special concern to human health are industrial wastes containing potential stressors that are new chemical creations and, hence, are environmental factors to which living organisms and ecosystems have not had a period of evolutionary history for adaptation or accommodation. Chronic exposure to such anthropogenic factors can be expected to result in basic changes in the structure and function of biotic communities, as acclimation and genetic adaptation occur. During the transition or adaptation period, organisms may be especially vulnerable to secondary factors, such as disease, that can have catastrophic results.

The increasing volume of toxic waste that affects human health—either because of direct contact or through contamination of food and drinking water—is approaching crisis proportions. In an issue of *Time* magazine (1980) under the heading "The Poisoning of America," the situation was reviewed as follows:

> Of all humankind's interventions in the natural order, none is accelerating quite so alarmingly as the creation of chemical compounds. Through their genius, modern alchemists brew as many as 1000 new concoctions each year in the United States alone. At last count, nearly 50,000 chemicals were on the market. Many have been an undeniable boon to humankind—but almost 35,000 of these used in the United States are classified by the federal EPA as being either definitely or potentially hazardous to human health.

One of the greatest dangers and potential disasters is the contamination of groundwater in the deep aquifers that provide a large percentage of the water for cities, industry, and agriculture. Unlike surface water, groundwater is almost impossible to purify once it has become polluted, because it is not exposed to sunlight, strong flow, or any of the other natural purification processes that cleanse surface water. Already, cities in the industrial heartlands can no longer use local groundwater for drinking because of contamination; they must pipe in water at great expense (see National Geographic Special Edition 1993, "Water: The Power, Promise, and Turmoil of North America's Fresh Water" for details).

The handling of toxic wastes before 1980 was considered a business "externality" not worthy of serious attention. The unwanted material was just dumped somewhere, until several local disasters came to public attention. The Love Canal disaster in New York, where a residential area built on top of a waste dump had to be abandoned, received wide press coverage, as did the Kepone that poisoned a large section of the James River in Virginia (as well as workers in the plant that made the insecticide). When the plant closed, the river recovered, but some of the workers did not. These and other incidents aroused public concern and government action. However, despite the millions of dollars of Environmental Protection Agency (EPA) Super Fund monies spent on attempts to clean up some of the worst toxic waste dumps, this goal remains elusive.

The clear solution to the toxic waste problem is *source reduction*—that is, eliminating the waste at its source by recycling, detoxification, and seeking less toxic materials in manufacturing (E. P. Odum 1989, 1997). Source reduction can be accomplished by a combination of regulation and incentives.

Examples

It is beyond the scope of this book to discuss or even to list all of the toxic emissions that are potentially limiting to human society. We will be content to comment very briefly on three examples where an ecological problem-solving approach seems especially helpful.

Air Pollution

Air pollution provides the negative feedback signal that may well save industrialized society from extinction because (1) it provides a clear danger signal that is easily perceived by everyone; and (2) almost everyone contributes to it (by driving a car, using electricity, buying a product, and so on) and suffers from it, so it cannot be blamed on a convenient villain or hidden in some remote landfill. A holistic solution must evolve, because piecemeal attempts to deal with any one pollutant (the one problem–one solution approach) is not only ineffective but usually just shifts the problem from one place or environment to another.

Air pollution also provides an example of an *augmentative synergism,* in that combinations of pollutants react in the environment to produce additional pollution, which greatly aggravates the original problem (in other words, the total effect is greater than the sum of the individual effects). For example, two components of automobile exhaust combine in the presence of sunlight to produce new and even more toxic substances, known as *photochemical smog:*

$$\text{nitrogen oxides} + \text{hydrocarbons} \xrightarrow[\text{in sunlight}]{\text{Ultraviolet radiation}} \text{peroxyacetylnitrate (PAN)} + \text{ozone } (O_3)$$

Both secondary substances not only cause eye watering and respiratory distress in humans, but they also are extremely toxic to plants. Ozone increases the respiration of leaves, killing the plant by depleting its food. Peroxyacetylnitrate blocks the "Hill reaction" in photosynthesis, killing the plant by shutting down food production. The tender varieties of cultivated plants become early victims, so that certain types of agriculture and horticulture are no longer possible near large cities. Other photochemical pollutants that go under the general heading of polynuclear aromatic hydrocarbons (PAH) are known carcinogens.

Thermal Pollution

Thermal pollution is becoming a commonplace example of chronic stress, because low-utility heat is a by-product of any conversion of energy from one form to another, as dictated by the second law of thermodynamics. Power plants and other large energy converters release great quantities of heat to both air and water, with nuclear power plants requiring especially large volumes of cooling water. Consequently, a significant amount of water surface is required to disperse heat, something on the order of 1.5 acres/megawatt in a temperate locality, or 4500 acres (1822 ha) for a 3000-megawatt power station.

The use of powered cooling devices, such as cooling towers, can reduce the space and water volume needed, of course, but at a considerable cost, because expensive fuels replace solar energy. Also, cooling towers can cause other environmental impacts if chlorine or other chemicals are used to keep their surfaces free of algae.

In general, the effects of increasing the water temperature in ponds, lakes, or streams follows the subsidy-stress gradient (discussed in Chapter 3) in that both positive and negative responses result. Moderate increases often act as subsidies, in that productivity of the aquatic community and growth of fish may be increased, but in time, or with increasing heat loading, stress effects begin to enter the picture.

The National Environmental Research Park (NERP) located at the Department of Energy Savannah River Plant is a place to observe the long-term effects of thermal pollution. The Savannah River Ecology Laboratory (SREL) located on the site has focused on the study of thermal effects since the establishment of nuclear energy facilities there in the 1950s and has sponsored two large symposia that combine work and ideas from other study sites (Gibbons and Sharitz 1974; Esch and McFarlane 1975). A large artificial lake constructed as a cooling pond is especially interesting because it has a "warm" arm (receiving heated water) and a "cool" arm that receives no heated water and is at ambient temperature (normal for the region). Furthermore, as the reactors are periodically turned off and on, one can observe the effect from one temperature state to the other. Turtles and bass, for example, grow faster and achieve larger sizes when the water temperature is elevated a few degrees, and the active season for alligators is prolonged into the winter months. Thus, the first observed effects were generally subsidies. However, after a few years, definite stress effects began to appear, such as debilitating diseases that shorten the life span and increase mortality. The percentage of bass infected with red-sore disease rises and falls with the seasons, but it is consistently higher in the thermally enriched zones of the lake. Also, after 10 years or more of elevated temperatures, there is evidence of genetic change in populations of fish and also in cattails (*Typha*) that grow along the shore of the warm arm of the lake. For a brief review of all these studies, see Gibbons and Sharitz (1981). These illustrations emphasize the importance of looking for secondary or delayed responses when assessing the effect of a chronic anthropogenic perturbation.

Pesticides

Increasingly heavy applications of insecticides and other pesticides in agriculture have resulted in the contamination of soil and water. This threat to the health of ecosystems and of humans may soon be reduced for the simple reason that the exclusive dependence on chemical poisons fails to achieve long-term control, but rather produces booms and busts in crop yields. Alternative systems of pest control have been developed that may soon reduce the need for massive applications of what are, in reality, very dangerous poisons.

Paradoxically, the resilience and adaptability of nature is the root cause of the failure of broad-spectrum insecticides like the organochlorines (such as DDT) and the organophosphates (such as malathion). All too often, pests develop immunity or become even more abundant after the pesticide has been dissipated or detoxified, because their natural enemies were destroyed by the treatment. Also, a pest species that is successfully exterminated is sometimes replaced by other species that are more resistant, less well known, and, therefore, even more difficult to deal with.

Efforts to control insect pests of cotton provide a clear example of the boom-and-bust syndrome. Cotton was one of the crops most heavily treated with insecticides; prior to 1970, as much as 50 percent of all insecticide used in U.S. agriculture was sprayed on cotton. In the 1950s, massive aerial spraying of chlorinated hydrocarbons in the Canete Valley of Peru, made possible by foreign aid funds from the United States, resulted in a doubling of the yield for about six years. Then followed, however, a complete crop failure, as pests became resistant and other species of insects moved in. The same thing happened in the 1960s in the state of Texas—a major cotton-growing state—as documented in detail by Adkisson et al. (1982). In both cases, yields were restored by the adoption of what is now known as *integrated pest management* (IPM) or *ecologically based pest management* (EBPM; NRC 1996a, 1996b, 2000b), which involves cultural and management practices that discourage pests, promotion of parasites and predators of insect pests and weeds (biological controls), and bioengineering of crop plants to produce their own insecticides, combined with the judicious use of less toxic, short-lived pesticides.

The new control system confirms an age-old, common-sense wisdom that it never pays to put all your eggs in one basket. The diversity and resilience of nature must be met with diverse technological innovations that must be continually updated as conditions change and nature reacts. In other words, the "war" with pests and disease probably can never actually be "won," but involves a continuous effort that is one of the costs of the "pumping out of disorder" necessary to maintain a large and complex civilization. E. P. Odum and Barrett (2000) have reviewed some of the landscape-level management practices such as overfertilization and monoculture vulnerability to pest invasion; mitigating these influences would be helpful in the ecological control of insect pests.

In the 1960s, there was considerable optimism for what Carroll Williams (1967) called the "third generation of pesticides." The first generation, according to the Williams classification, were the botanical pesticides and inorganic salts; the second generation were the broad-spectrum chlorinated hydrocarbons and organophosphates. The third generation are the biochemical pesticides—hormones and pheromones (sexual attractants) that direct behavior and are species-specific—which add to the arsenal available for integrative pest management. For the most part, worldwide, industrial agriculture continues to depend too much on the second generation. For a review of the prospects for ecologically based pest management in the twenty-first century, see the 1996 National Research Council Report "Ecologically Based Pest Management: New Solutions for a New Century."

6

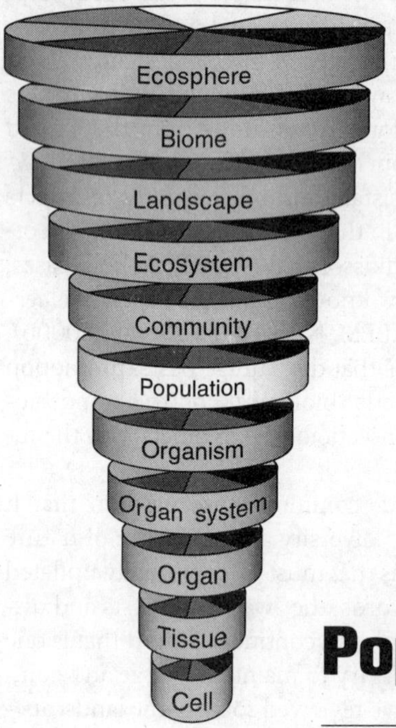

Population Ecology

1 Properties of the Population

2 Basic Concepts of Rate

3 Intrinsic Rate of Natural Increase

4 Concept of Carrying Capacity

5 Population Fluctuations and Cyclic Oscillations

6 Density-Independent and Density-Dependent Mechanisms of Population Regulation

7 Patterns of Dispersion

8 The Allee Principle of Aggregation and Refuging

9 Home Range and Territoriality

10 Metapopulation Dynamics

11 Energy Partitioning and Optimization: *r*- and *K*-Selection

12 Population Genetics

13 Life History Traits and Tactics

In Chapters 3 through 5, the physical and chemical forces that act as primary forcing functions were discussed. Organisms do not just passively adapt to these forces but actively modify, change, and regulate the physical environment within the limits imposed by the natural laws that determine the transformation of energy and the cycling of materials. In other words, human beings are not the only population that modifies and attempts to control the environment. Referring back to the levels-of-organization chart (Figs. 1-2 and 1-3), we see that this chapter and the next one focus on the respective biotic levels of *populations* and *communities*. Interaction at these levels between genetic systems and physical systems affects the course of *natural selection* and, thereby, determines not only how individual organisms survive but also how ecosystems as a whole change over evolutionary time.

1 Properties of the Population

Statement

A **population** is defined as any group of organisms of the same species occupying a particular space and functioning as part of a **biotic community**, which, in turn, is defined as an assemblage of populations that function as an integrative unit through coevolved metabolic transformations in a prescribed area of physical habitat. A population has various properties, which, although best expressed as statistical variables, are the unique possession of the group and are not characteristic of the individuals in the group. Some of these properties are density, natality (birth rate), mortality (death rate), age distribution, biotic potential, dispersion, and *r*- and *K*-selected growth forms. Populations also possess genetic characteristics that are directly related to their ecology, namely, adaptiveness, reproductive success, and persistence (the probability of leaving descendants over long periods of time).

Explanation

As pioneer population ecologist Thomas Park well expressed (in Allee et al. 1949), a population has characteristics or *biological attributes* that it shares with its component organisms, and it also has characteristics or *group attributes* unique to the group or species. Among the biological attributes of population is life history (the population grows, differentiates, and maintains itself as the organism does). Also, the population has a definite structure and function that can be described. By contrast, group attributes, such as birth rate, death rate, age ratio, genetic fitness, and growth form apply only to the population. Thus, an individual is born, dies, and ages, but it does not have a birth rate, a death rate, or an age ratio. These latter attributes are meaningful only at the population level.

Definitions and brief resumés of basic population attributes follow.

Indices of Density

Population density is the size of a population in relation to a definite unit of space. It is generally expressed as the number of individuals or the population biomass per unit area or volume—for example, 200 trees per hectare (1 hectare = 2.471 acres) or five million diatoms per cubic meter of water. Sometimes, it is important to distinguish between **crude density,** the number (or biomass) per unit of *total space,* and **ecological density,** the number (or biomass) per unit of *habitat space* (available area or volume that can actually be colonized by the population). Often, it is more important to know whether a population is changing (increasing or decreasing) than to know its size at any one moment. In such cases, indices of **relative abundance** are useful; these may be time-relative, as, for example, the number of birds seen per hour. Another useful index is the **frequency of occurrence,** as, for example, the percentage of sample plots occupied by a species. In descriptive studies of vegetation, density, dominance, and frequency are often combined to provide an **importance value** for each species.

Density, Biomass, and Trophic Relationships

Figure 6-1 illustrates how densities encountered in populations of mammals are related to trophic level and to the size of individual animals. Although the density of mammals as a class may range over nearly five orders of magnitude, the range for any given species or trophic group is much less. The lower the trophic level, the higher the density, and within a given level, the larger the individuals, the larger the biomass. As large organisms have lower rates of metabolism per unit weight than small organisms, a larger population biomass can be maintained on a given energy base.

When the size and metabolic rate of individuals in the population are relatively uniform, density expressed in terms of the number of individuals is quite satisfactory as a measure, but most often that is not the case. The relative merits of numbers, biomass, and energy flow parameters as indices were discussed in Chapter 3. Recall that numbers overemphasize the importance of small organisms, and biomass overemphasizes the importance of large organisms. Energy flow provides a more suitable index for comparing any and all populations in an ecosystem.

Many special measures and terms apply only to specific populations or groups of

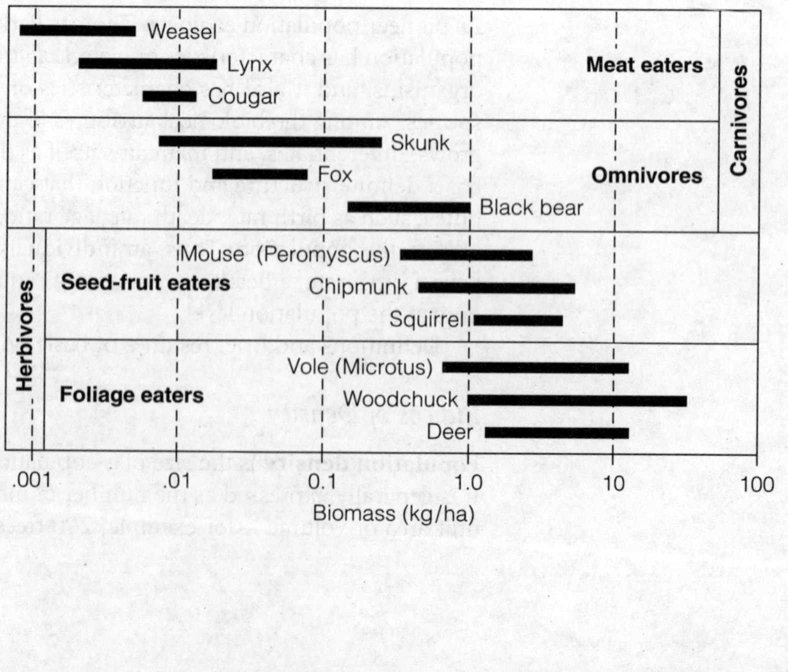

Figure 6-1. The range of population density (biomass per hectare) of various species of mammals as reported for the preferred habitat of the species. Species are arranged according to trophic levels and according to individual size within the four trophic levels to illustrate the limits imposed by trophic level position and individual size of the organism on the expected standing crop.

populations. Forest ecologists, for example, often use "basal area" (total cross-section area of tree trunks) as a measure of tree density. Foresters, however, prefer "board feet per acre" as a measure of the commercially usable part of the tree. These, and many others, are density measures as the concept has been broadly defined, because they all express in some manner the size of the standing crop per unit area.

As might be imagined, *relative abundance* indices are widely used with populations of larger animals and terrestrial plants, where it is imperative that a measure applicable to large areas be obtained without excessive expenditure of time and money. For example, administrators charged with establishing annual hunting regulations for migratory waterfowl must know whether the populations are smaller than, larger than, or the same as in the previous year if they are to adjust the hunting regulations to the best interest of both the birds and the hunters. To do this, these officials must rely on relative abundance indices obtained from field checks, hunter surveys, questionnaires, and nesting censuses. Such information is often summarized in terms of the number observed or killed per unit time. Percentage indices are widely used in the study of vegetation, and specially defined terms have come into general use. For example, *frequency* equals the percentage of sample plots in which the species occurs, *abundance* equals the percentage of individuals in a sample, and *cover* equals the percentage of ground surface covered as determined by the projection of areal parts. One should be careful not to confuse these indices with true density measures, which are always in terms of a *definite* amount of space.

Methods for Estimating Population Densities

The **Lincoln index** is a common mark-recapture method used to estimate the total population density (the number of organisms of a species) in a defined area. This method relies on capturing and marking some fraction of the total population and using this fraction to estimate the total population density.

The following equation is used to obtain this population estimate:

$$\frac{\text{Population estimate } (x)}{\text{Number of animals captured and marked in sample } S_1 \text{ at time } t_1} = \frac{\text{Number of animals captured in sample } S_2 \text{ at time } t_2}{\text{Number of marked animals found in sample } S_2 \text{ at time } t_2}$$

Because we know (for example, by live trapping and marking captured small mammals) three of the four components of this equation, we can estimate the fourth component (the population estimate, x) of the equation.

The validity of this method depends on the following assumptions:

- that the marking technique has no negative effect on the mortality of marked individuals;
- that the marked individuals are released at the original site of capture and allowed to mix with the population based on natural behavior;
- that the marking technique does not affect the probability of being recaptured;
- that the marks (such as ear tags) are not lost or overlooked;
- that there is no significant immigration or emigration of marked or unmarked individuals in the interval between t_1 and t_2; and
- that there is no significant mortality or natality in the interval between t_1 and t_2.

Violation of these assumptions would obviously affect the estimate of population density.

The **minimum known alive** (MKA) method is another mark-recapture method used to estimate population densities over an extended period of time. This method was originally published as the *calendar-of-catches method* (Petrusewicz and Andrzejewski 1962) using a capture history (calendar) kept for each individual, followed by a period of intensive removal to "update the calendar" on termination of the study.

Other methods fall into several broad categories:

1. **Total counts** are sometimes possible with large or conspicuous organisms (for example, bison on open plains or whales in an area of the sea) or with organisms that aggregate into large breeding colonies.
2. **Quadrat or transect sampling** involves counting organisms of a single species in plots or *transects* of appropriate size and number to get an estimate of the density in the area sampled. This method is applicable to a wide variety of terrestrial and aquatic species in environments ranging from forests to the bottom of the sea.
3. **Removal sampling,** in which the number of organisms removed from an area in successive samples is plotted on the *y*-axis of a graph, whereas the total number previously removed is plotted on the *x*-axis. If the probability of capture remains reasonably constant, the points will fall on a straight line that can be extended to the point on the *x*-axis that would indicate the theoretical 100 percent removal from the area (the population density estimate).
4. **Plotless methods** (applicable to sessile organisms such as trees). The *point-quarter method* is based on a series of random points; the distance to the nearest individual is measured in each of four quarters at each point along this series of random points. The density per unit area can be estimated from the mean distance.
5. An **importance percentage value** is the sum of relative density, relative dominance, and relative frequency of a species in a community. **Relative density,** *A*, equals the density for a species divided by the total density for all species × 100. **Relative dominance,** *B*, equals the basal area for a species divided by the total basal area for all species × 100. **Relative frequency,** *C*, is the frequency (occurrence) for a species in a plot divided by the total frequency for all species × 100. Thus, the importance value for each species equals the sum of relative density, relative dominance, and relative frequency: $A + B + C$. Combining density with dominance and frequency of occurrence provides a better index than density alone regarding the importance or function of a species in its habitat. A table or summary of importance values for each tree species (trees greater than 3 inches diameter at breast height) provides a rank order for a particular tree species within a forest community.

Many techniques and methodologies for estimating population density have been tried; and *sampling methodology* is an important field of research in itself. Methods are learned effectively by consulting field manuals or by consulting with an experienced investigator who has reviewed the literature and then modified and improved existing methods to fit a specific field situation. There is no substitute for experience when it comes to a field census.

Natality

Natality is the ability of a population to increase by reproduction. Natality is equivalent to the *birth rate* in the terminology of human population study (demography). In fact, it is a broader term covering the production of new individuals of any organism, whether such new individuals are born, are hatched, germinate, or arise by division. **Maximum** (sometimes called *absolute* or *physiological*) **natality** is the theoretical maximum production of new individuals under ideal conditions (no ecological limiting factors, reproduction being limited only by physiological factors) and is a constant for a given population. **Ecological** or **realized natality** refers to population increase under an actual or specific environmental field condition. It is not a constant for a population but may vary with the size and age composition of the population and with the physical environmental conditions. Natality is generally expressed as a rate determined by dividing the number of new individuals produced by a specific unit of time (the *absolute* or *crude natality rate*) or by dividing the number of new individuals per unit time by a unit of population (the *specific natality rate*).

The difference between crude and specific natality or birth rate can be illustrated as follows: Suppose a population of 50 protozoa in a pool increases by division to 150 in one hour. The crude natality is 100 per hour, and the specific natality (average rate of change per unit population) is 2 per hour per individual (of the original 50). Or suppose there were 400 births in one year in a town of 10,000; the crude birth rate is 400 per year, and the specific birth rate is 0.04 per capita (4 per 100, or 4 percent). In human demography, it is customary to express specific birth rates in terms of the number of women of reproductive age rather than in terms of total population. Other considerations that affect natality will be discussed in subsequent sections.

Mortality

Mortality quantifies death of individuals in the population. It is more or less the antithesis of natality. Mortality is equivalent to death rate in human demography. Like natality, mortality may be expressed as the number of individuals dying in a given period (deaths per unit time), or as a specific rate in terms of units of the total population or any part thereof. **Ecological** or **realized mortality**—the loss of individuals under a given environmental condition—is, like ecological natality, not a constant but varies with population and environmental conditions. A theoretical **minimum mortality**—constant for a population—represents the minimum loss under ideal or nonlimiting conditions. Even under the best conditions, individuals would die of old age determined by their *physiological longevity*, which, of course, is often far greater than the average *ecological longevity*. Often, the *survival rate* is of greater interest than the death rate. If the death rate is expressed as a fraction, M, then the **survival rate** is $1 - M$.

Because mortality varies greatly with age, as does natality, especially in the higher organisms, *specific mortalities* at as many different stages of life history as possible are of great interest, because they enable ecologists to determine the forces underlying the crude, overall population mortality. A complete picture of mortality in a population is illustrated systematically by the **life table,** a statistical device developed by students of human populations. Raymond Pearl first introduced the life table into general biology when he applied it to data obtained from laboratory studies of the fruit

Table 6-1 **Life table for the Dall mountain sheep (*Ovis dalli*)**

x*	x'†	d_x‡	l_x§	$1000\ q_x$**	e_x††
0–1	−100	199	1000	199.0	7.1
1–2	−85.9	12	801	15.0	7.7
2–3	−71.8	13	789	16.5	6.8
3–4	−57.7	12	776	15.5	5.9
4–5	−43.5	30	764	39.3	5.0
5–6	−29.5	46	734	62.6	4.2
6–7	−15.4	48	688	69.9	3.4
7–8	−1.1	69	640	108.0	2.6
8–9	+13.0	132	571	231.0	1.9
9–10	+27.0	187	439	426.0	1.3
10–11	+41.0	156	252	619.0	0.9
11–12	+55.0	90	96	937.0	0.6
12–13	+69.0	3	6	500.0	1.2
13–14	+84.0	3	3	1000	0.7

Source: From Deevey 1947; data from Murie 1944, based on known age at death of 608 sheep dying before 1937 (both sexes combined). Mean life span = 7.06 years.

*Age (years)

†Age as percentage deviation from mean life span.

‡Number dying in age interval out of 1000 born.

§Number surviving at beginning of age interval out of 1000 born

**Mortality per 1000 alive at beginning of age interval

††Life expectancy = mean time remaining to attaining age interval (years)

fly, *Drosophila* (Pearl and Parker 1921). Deevey (1947, 1950) assembled data for the construction of life tables for several natural populations, ranging from rotifers to mountain sheep. Since Deevey, numerous life tables have been published for various natural and experimental populations. The life table for an Alaskan population of Dall mountain sheep (*Ovis dalli*), perhaps the most famous life table present in most textbooks, is displayed in Table 6-1. The age of the sheep was determined from the horns (the older the sheep, the more bony rings). When a sheep is killed by a wolf or dies for any other reason, its horns remain preserved for a long period. For several years, Adolph Murie studied the relation between wolves (*Canis lupus*) and mountain sheep living in Mount McKinley National Park, Alaska. He collected many horns, thus providing exhaustive data on the age at which sheep die in an environment subject to all the natural hazards, including wolf predation (but not including predation by humans, as sheep were not hunted in Mount McKinley National Park).

The life table consists of several columns, headed by standard notations, giving l_x, the number of individuals out of a given population (1000 or any other convenient number) that survive after regular time intervals (day, month, year, and so on, given in column x); d_x, the number dying during successive time intervals; q_x, the death rate or mortality during successive intervals (in terms of the initial population at the beginning of the period); and e_x, the life expectancy at the end of each interval. As shown in Table 6-1, the average age of Dall mountain sheep was more than 7 years, and if a sheep survived the first year or so, its chances of survival were good until relatively old age, despite the abundance of wolves and the other vicissitudes of the environment.

Curves plotted from life table data may be very instructive. When the data from column l_x are plotted with the time interval on the horizontal axis and the number of survivors (usually in logarithmic form) on the vertical axis, the resulting curve is called a **survivorship curve.** Figure 6-2 is a survivorship curve for the data tabulated in Table 6-1 for Dall mountain sheep.

Survivorship curves are of three general types, as shown in Figure 6-3. A highly convex curve (Fig. 6-2; Type I in Fig. 6-3) is characteristic of species, such as the Dall mountain sheep, in which the population death rate is low until near the end of the life span. Many species of large animals and, of course, humans exhibit this Type I curve of survivorship. At the other extreme, a highly concave curve (Type III in Fig. 6-3) results when mortality is high during the young stages. Oysters, other shellfish, and oak trees provide examples of the Type III survivorship curve; mortality is extremely high during the free-swimming larval stage or the acorn-seedling stage, but once an individual is well established on a favorable substrate, life expectancy improves considerably. A "stair-step" type of survivorship curve may be expected if survival differs greatly in successive stages of life history, as is often the case in holometabolous insects (insects with complete metamorphosis, such as butterflies). Probably no population in the natural world has a constant age-specific survival rate throughout the whole life span, but a slightly concave curve, approaching a diagonal straight line on a semilog plot (Type II in Fig. 6-3), is characteristic of many birds, mice, rabbits, and deer. In these cases, the mortality is high in the young but lower and more nearly constant in the adult (1 year or older).

The shape of the survivorship curve is often related to the degree of parental care or other protection given to the young. Thus, survivorship curves for honeybees and robins (which protect their young) are much less concave than are those for grasshoppers or sardines (which do not protect their young). The latter species, of course,

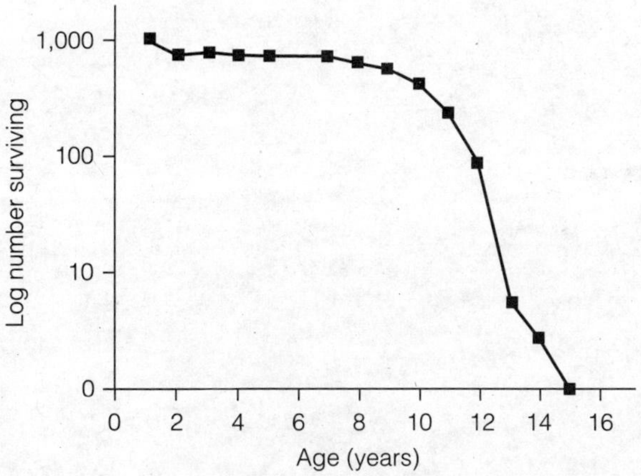

Figure 6-2. Survivorship curve for the data tabulated in Table 6-1 for Dall mountain sheep (data from Deevey 1947).

compensate by laying many more eggs (ratio of maximum to realized natality is high, as noted in the previous section).

The shape of the survivorship curve very often also varies with the density of the population. Survivorship curves for two mule deer (*Odocoileus hemionus*) populations living in the Chaparral of California are shown in Figure 6-4; the survivorship curve of the denser population is quite concave. In other words, deer living in the managed area, where the food supply was increased by controlled burning, have a shorter life expectancy than deer in the unmanaged area, presumably because of increased hunting pressure, intraspecific competition, and so on. From the viewpoint of the hunter, the managed area is the most favorable because of the higher density of game, but from the viewpoint of the individual deer, the less crowded area offers a better chance for a long life. For human populations also, high density tends not to be favorable to the individual. Many ecologists believe that rapid growth and high density in human populations is not so much a threat to survival as it is a threat to the quality of life for the individual. Humans have greatly increased their own ecological longevity because of expanded medical knowledge, increased nutrition, and adequate sanitation. The curve denoting the survival of human beings approaches the sharp-angled, Type I minimum mortality curve.

To prepare the way for mathematical models of population growth to be considered in subsequent sections, it is instructive to add *age-specific natality* (offspring per reproductive female per unit time; m_x) to the life table, so that it is not just a "death" table.

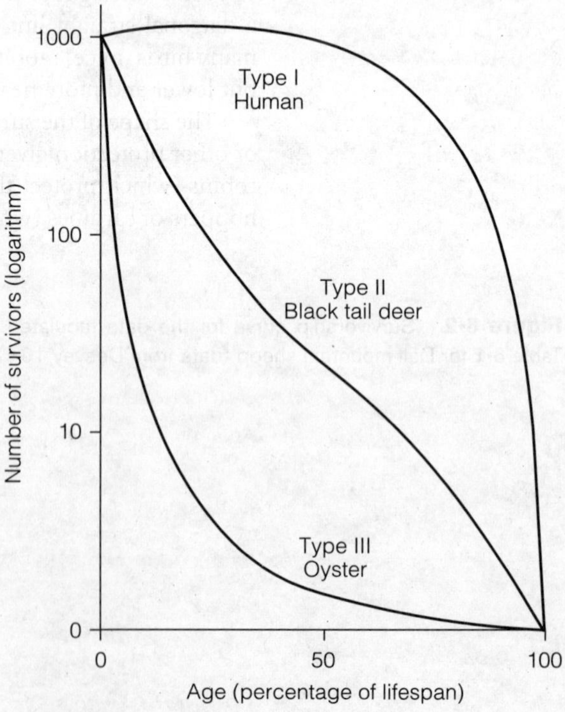

Figure 6-3. Types of survivorship curves. Type I curves represent organisms with a high mortality toward the end of the life span, Type III curves high mortality near the beginning of the life span, and Type II curves uniform mortality throughout life.

Figure 6-4. Survivorship curves for two stable mule deer (*Odocoileus hemionus*) populations living in the Chaparral of California. The high-density population (about 64 deer per 2.6 km^2) is in a managed area where an open shrub and herbaceous cover is maintained by controlled burning, thus providing a greater quantity of browse in the form of new growth. The low-density population (about 27 deer per 2.6 km^2) is in an unmanaged area of old bushes unburned for 10 years (after Taber and Dasmann 1957).

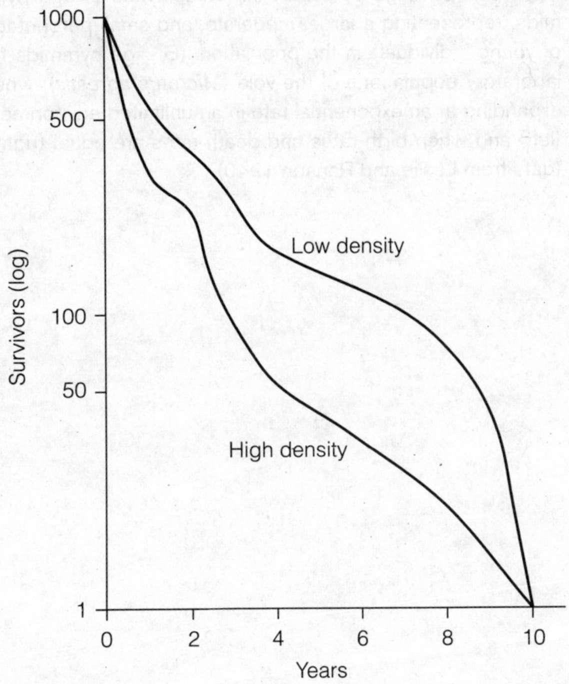

If l_x and m_x are multiplied and the sum of the values is obtained for the different age classes, a net reproductive rate, R_o, is calculated. Thus:

$$R_o = \sum l_x \cdot m_x$$

Under stable conditions in nature, R_o in terms of the total population should be around 1. For example, the calculated R_o value for a sowbug population in a grassland was 1.02, indicating an approximate balance between births and deaths.

The reproductive schedule greatly influences population growth and other population attributes. Natural selection can effect various kinds of change in the life history that will result in adaptive schedules. Thus, selection pressure may change the time when reproduction begins without affecting the total number of offspring produced, or it may affect production or clutch size without changing the timing of the reproduction. These and many other aspects of reproduction can be revealed by life table analyses.

Population Age Distribution

Age distribution, an important attribute of populations, influences both natality and mortality, as shown by the examples discussed in the preceding section. The ratio of the various age groups in a population determines the current reproductive status of the population and indicates what may be expected in the future. Usually, a rapidly expanding population will contain a large proportion of young individuals; a pulsing, stable population will show a more even distribution of age classes; and a

Figure 6-5. Age pyramids. (A) Three types of age pyramids, representing a large, moderate, and small percentage of young individuals in the population. (B) Age pyramids for laboratory populations of the vole (*Microtus agrestis*), when expanding at an exponential rate in an unlimited environment (left) and when birth rates and death rates are equal (right). (data from Leslie and Ranson 1940).

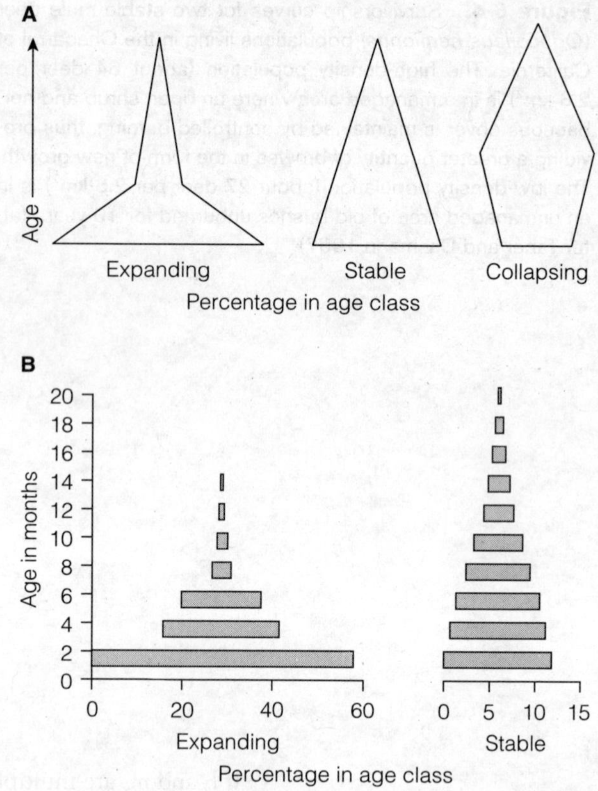

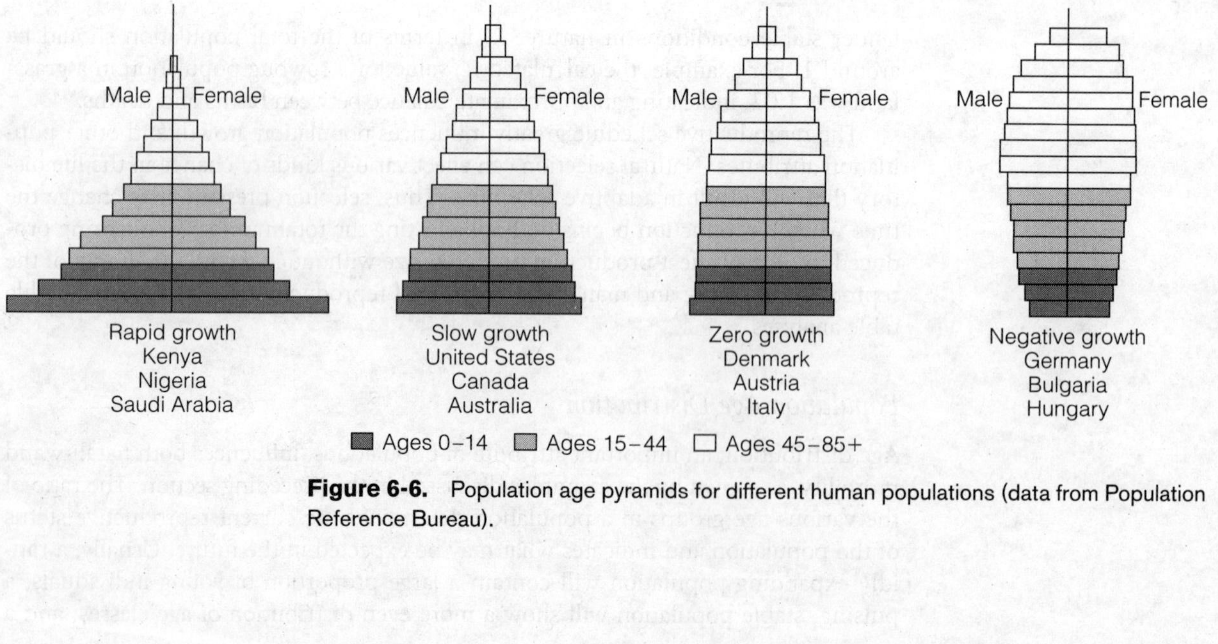

Figure 6-6. Population age pyramids for different human populations (data from Population Reference Bureau).

declining population will have a large proportion of old individuals, as illustrated by the age pyramids in Figures 6-5 and 6-6. A population may pass through changes in age structure without changing in size. There is evidence that populations have a "normal" or stable age distribution toward which actual age distributions are tending, as first proposed by Lotka (1925) on theoretical grounds. Once a stable age distribution is achieved, unusual increases in natality or mortality result in temporary changes, with spontaneous return to the stable situation. As nations go from pioneer conditions of rapidly expanding densities to mature conditions of stable populations, the percentage of individuals in younger age classes decreases, as shown in Figure 6-6. This changing age structure, with an increasing percentage of older individuals, has profound impacts on lifestyles and economic considerations, such as the costs of health care and social security benefits.

The pyramids in Figure 6-6 are based on births and deaths within the population and do not include immigration into countries, which is increasing in Europe and the United States of America. This situation resembles the importance of the ecosystem approach that includes inputs and outputs, as outlined in Chapter 2.

In simplistic form, age structure can be expressed in terms of three ecological ages: *prereproductive, reproductive,* and *postreproductive.* The relative duration of these ages in proportion to the life span varies greatly with different organisms. For humans in recent times, the three ages are relatively equal in length, about a third of a human life falling in each class. Early humans, by comparison, had a much shorter postreproductive period. Many animals, notably insects, have extremely long prereproductive periods, a very short reproductive period, and no postreproductive period. Certain species of mayflies (Ephemeridae) and the 17-year cicada (*Magicicada* spp.) are classic examples. Mayflies require from one to several years to develop in the larval stage in the water, and adults emerge to live for only a few days. Cicada nymphs have an extremely long developmental history (typically 13 and 17 years; Rodenhouse et al. 1997), with adult life lasting less than a single season.

In game birds and fur-bearing mammals, the ratio of first-year animals to older animals, determined during the season of harvest (fall or winter) by examining samples from the population taken by hunters or trappers, provides an index to the population trends. In general, a high ratio of juveniles to adults, as shown in the lower diagram (B) of Figure 6-5, indicates a highly successful breeding season and the likelihood of a larger population the next year, provided that juvenile mortality is not excessive.

A phenomenon known as the **dominant age class** has been repeatedly observed in fish populations that have a very high potential natality rate. When a large year class occurs because of unusually high survival of eggs and larval fish, reproduction is suppressed for the next several years. Hjort's early data on herring in the North Sea provided the classic case, as shown in Figure 6-7 (Hjort 1926). Fish of the 1904 year class dominated the catch from 1909 (when this age class was five years old and large enough to be caught effectively in commercial fish nets) until 1918 (when, at 14 years of age, they still outnumbered fish of younger age groups). The situation produced something of a cycle or pulse in total catch, which was high in 1909 and then declined in subsequent years, as the dominant age class declined before there was replacement from other classes. Fishery biologists continue to conduct research recapitulating what environmental conditions, such as El Niño, result in the unusual survival occurring every now and then.

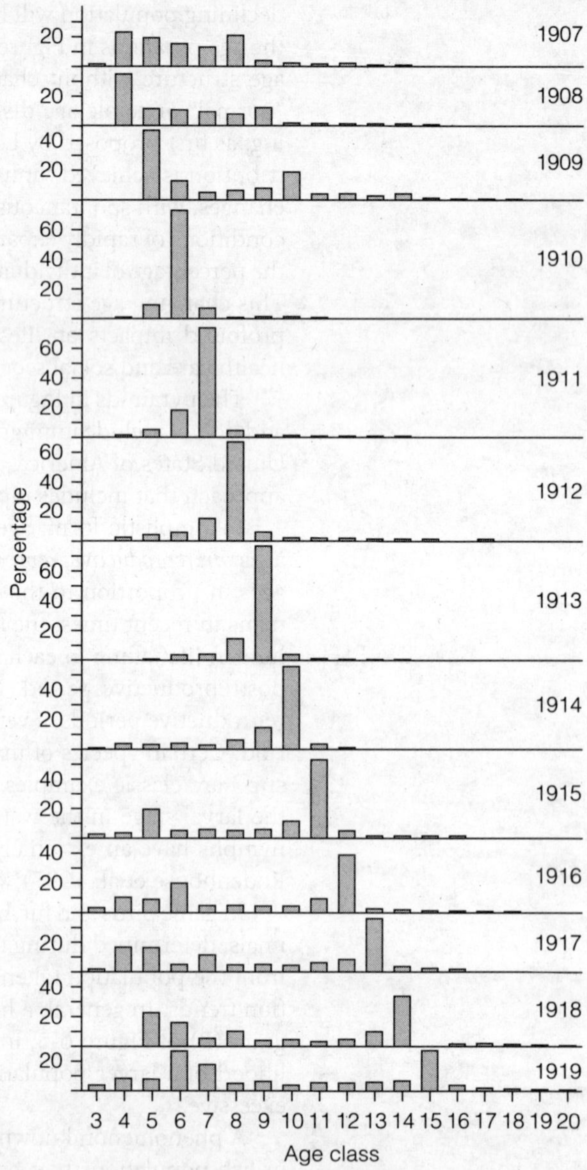

Figure 6-7. Age distribution in the commercial catch of herring in the North Sea between 1907 and 1919, illustrating the dominant age class phenomenon. The 1904 year class was very large and dominated the population for many years. Because fish younger than five years are not caught in the nets, the 1904 class did not show up until 1909. Age of the fish was determined by growth rings on scales, which are laid down annually in the same manner as growth rings on trees (after Hjort 1926).

2 Basic Concepts of Rate

Statement

A population is a changing entity. Even when the community and the ecosystem are seemingly unchanging, density, natality, survivorship, age structure, growth rate, and other attributes of component populations are usually in flux as species constantly adjust to seasons, to physical forces, and to one another. The study of changes in the relative number of organisms in populations and the factors explaining these changes is termed **population dynamics.** Accordingly, ecologists are often more interested in how and at what rate the population is changing than in its absolute size and composition at any one moment. *Calculus,* the branch of mathematics dealing (in part) with the study of rates, thus becomes an important tool in studying population ecology.

Explanation

A **rate** may be obtained by dividing the change in some quantity by the period of time elapsed during the change; such a rate term would indicate the rapidity with which something changes with time. Thus, the number of kilometers traveled by a car per hour is the *speed rate,* and the number of births per year is the *birth rate.* The *per* means "divided by." Remember that, as discussed previously (Chapter 3), productivity is a rate, not a standing state, such as standing crop biomass.

Customarily, the *change* in something is abbreviated by writing the symbol Δ (delta) in front of the letter representing the entity changing. Thus, if N represents the number of organisms and t represents the time, then

$\Delta N =$ the change in the number of organisms.

$\frac{\Delta N}{\Delta t} =$ the average rate of change in the number of organisms per (divided by, with respect to) time.

This is the *growth rate* of the population.

$\frac{\Delta N}{N\Delta t} =$ the average rate of change in the number of organisms per time per organism (the growth rate divided by the number of organisms initially present or, alternatively, by the average number of organisms during the period of time).

This is often termed the *specific growth rate* and is useful when populations of different sizes are to be compared. If multiplied by 100 ($\Delta N/(N\Delta t) \times 100$), it becomes the *percentage growth rate.*

Often, we are interested not only in the *average rate* over a period of time, but also in the theoretical *instantaneous rate* at particular times (that is, the rate of change when Δt approaches zero). In the language of calculus, the letter d (for *derivative*) replaces the Δ when instantaneous rates are being considered. In this case, the preceding notations become:

$\frac{dN}{dt} =$ the rate of change in the number of organisms per time at a particular instant;

$\frac{dN}{Ndt} =$ the rate of change in the number of organisms per time per individual at a particular instant.

Figure 6-8 shows the difference between a growth curve and a growth rate curve. As will be discussed further in Section 4, S-shaped growth curves and humpbacked growth rate curves are often characteristic of populations in pioneer or early growth stages.

On the growth curve, the *slope* (straight line tangent) at any point is the *growth rate.* Thus, in the case of the hypothetical population in Figure 6-8, the growth rate was at

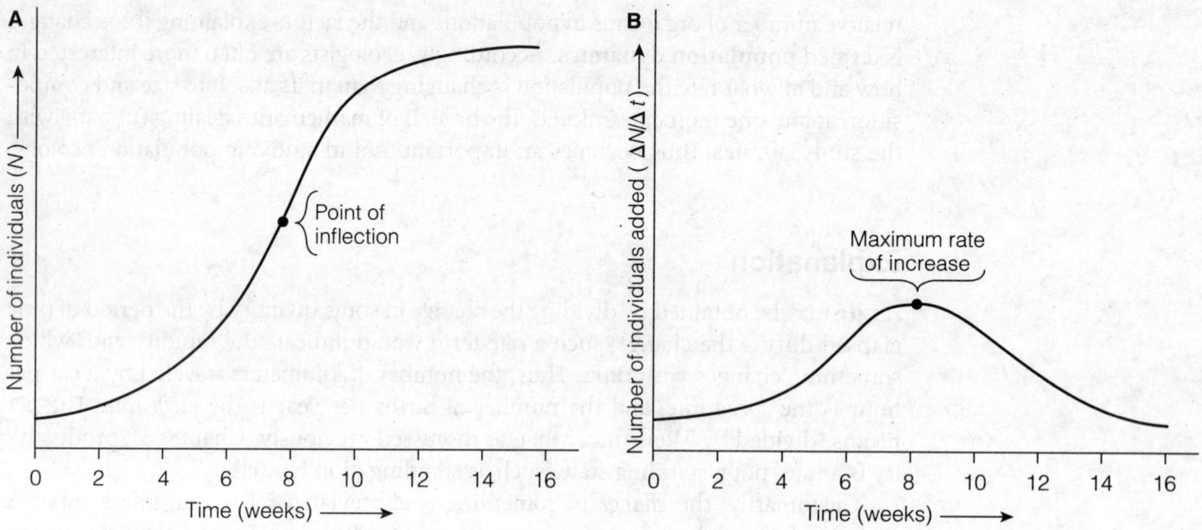

Figure 6-8. (A) population growth curve, and (B) rate of increase growth curve for the same hypothetical population during the same interval of time. Curve A depicts population density (number of individuals per unit area) against time, whereas curve B illustrates rate of change (number of individuals added per unit time) against time for the same population.

a maximum at approximately 8 weeks and fell to zero after 16 weeks. That point where the growth rate is maximum is termed the **point of inflection.** The $\Delta N/\Delta t$ notation serves to illustrate the model for the usual purposes of measurement, but the dN/dt notation must be substituted in most types of actual mathematical models.

The instantaneous rate, dN/dt, cannot be measured directly, nor can $dN/(Ndt)$ be calculated directly from population counts. The rate can be approximated, of course, by taking censuses at very short intervals, connecting these points with lines, and then determining what kind of equation most closely mimics the actual growth curve. The type of population growth curve exhibited by the population would have to be known, and then the instantaneous rate could be calculated from equations, as will be explained in Section 4.

3 Intrinsic Rate of Natural Increase

Statement

When the environment is unlimited (space, food, or other organisms are not exerting a limiting effect), the specific growth rate (the population growth rate per individual) becomes constant and maximum for the existing microclimatic conditions. The value of the growth rate under these favorable population conditions characterizes a particular population age structure and is a single index of the inherent power of a population to grow. It may be designated by the symbol *r*, which is the exponent in the differential equation for population growth in an *unlimited environment* under specified physical conditions:

$$\frac{dN}{dt} = rN \tag{1}$$

This is the same form as used in Section 2. The parameter r can be thought of as an **instantaneous coefficient of population growth.** The exponential integrated form follows automatically by calculus manipulation:

$$N_t = N_0 e^{rt} \tag{2}$$

where N_0 represents the number of individuals at time zero, N_t the number at time t, and e the base of natural logarithms. By taking the natural logarithm of both sides, one converts the equation into the form used in making actual calculations. Thus:

$$\ln N_t = \ln N_0 + rt; \quad \text{or} \quad r = \frac{\ln N_t - \ln N_o}{t} \tag{3}$$

In this manner, the index r can be calculated from two measurements of population size (N_0 and N_t, or at any two times during the unlimited growth phase, in which case N_{t1} and N_{t2} may be substituted for N_0 and N_1 and $(t_2 - t_1)$ substituted for t in the previous equations).

The index r is actually the difference between the instantaneous specific natality rate b (rate per time per individual) and the instantaneous mortality rate d and may thus be expressed as follows:

$$r = b - d \tag{4}$$

The overall growth rate of the population under unlimited environmental conditions, r, depends on the age composition and the specific growth rates due to reproduction of component age groups. Thus, there may be several r values for a species depending on population structure. When a stationary and stable age distribution exists, the specific growth rate is called the **intrinsic rate of natural increase** or r_{max}. This maximum value of r is often called by the less specific but widely used expression *biotic potential* or **reproductive potential.** The difference between the maximum r or biotic potential and the rate of increase that occurs in an actual laboratory or field condition is often taken as a measure of the **environmental resistance,** which is the sum total of environmental factors that prevent the biotic potential from being realized.

Explanation

Natality, mortality, and age distribution all are important, but each tells little by itself about how the population as a whole is growing, about what it would do if conditions were different, and about what its best possible performance is, compared with its everyday performance. Chapman (1928) proposed the term *biotic potential* to designate maximum reproductive power. He defined **biotic potential** as "the inherent property of an organism to reproduce, to survive, . . . to increase in numbers. It is sort of the algebraic sum of the number of young produced at each reproduction, the number of reproductions in a given period of time, the sex ratio and their general ability to survive under given physical conditions." Based on this general definition, biotic potential came to mean different things to different people. To some, it meant a nebulous reproductive power lurking in the population, fortunately never allowed to fully come forth because of the action of the environment (if unchecked, the descendants of a pair of flies would weigh more than Earth in a few years). To others, it meant simply and more concretely the maximum number of eggs, seeds, spores, and so forth that the most fecund individual was known to produce, although this would have little meaning in the population sense, as most populations contain individuals that are incapable of peak production.

Lotka (1925), Dublin and Lotka (1925), Leslie and Ranson (1940), Birch (1948), and others had to translate the rather broad idea of biotic potential into mathematical terms that could be understood in any context. Birch (1948) expressed it well when he said, "If the 'biotic potential' of Chapman is to be given quantitative expression in a single index, the parameter r would seem to be the most effective measure to adopt since it gives the intrinsic capacity of the animal to increase in an unlimited environment." The index r is also frequently used as a quantitative expression of "reproductive fitness" in the genetic sense, as will be noted later.

For the growth curves discussed in Section 2, r is the specific growth rate ($\Delta N/N\Delta t$) when population growth is exponential. Equation 3 in this section's Statement is an equation for a straight line. Therefore, the value of r can be obtained graphically. If growth is plotted as logarithms (or on semilogarithmic paper), the log of population size plotted against time will give a straight line if growth is exponential; r is the slope of this line. Thus, the steeper the slope, the higher the intrinsic rate of increase. The extremely wide differences in biotic potential are especially emphasized when expressed as the number of times the population would multiply itself if the exponential rate continued or as the time required to double the population. *Doubling time* at the maximum intrinsic rate for flour beetles under optimum laboratory conditions is less than a week (Leslie and Park 1949).

The human population reached 6.0 billion during October, 1999, and is expected to reach 8.04 billion by 2025 (Bongaarts 1998). The United Nations projects that the human population will grow from 6.1 billion in 2000 to 9.3 billion in 2050 (L. R. Brown 2001). Most likely, the world's human population will overshoot the life-support capacity of Earth and stop growing sometime during the twenty-first century, hopefully followed by a period of negative growth, to reach an optimum rather than a maximum carrying capacity (Barrett and Odum 2000; Lutz et al. 2001). We will return to this prediction later in this chapter.

Populations in nature often grow exponentially for short periods when there is ample food and no crowding effects, enemies, and so forth, creating "boom-and-bust" patterns. Under such conditions, the population as a whole is expanding at a terrific rate, even though each organism is reproducing at the same rate as before—that is, the specific growth rate is constant. Plankton blooms (mentioned in previous chapters), pest eruptions, or growth of bacteria in new culture media are examples of situations in which growth may be logarithmic. It is obvious that this exponential increase cannot continue for very long; often, it is never realized. Interactions within the population and external environmental resistances soon slow down the rate of growth and play a part in shaping population growth forms in various ways.

4 Concept of Carrying Capacity

Statement

Populations show characteristic patterns of increase, termed *population growth forms.* For comparison, two basic patterns, based on the shapes of arithmetic plots of growth curves, can be designated: the *J-shaped growth form* and the *S-shaped* or *sigmoid growth form.* In the **J-shaped growth form,** density increases rapidly in *exponential* fashion (as shown in Fig. 6-9A) and then stops abruptly as environmental resistance or another limiting factor becomes effective more or less suddenly. This form may be represented by the simple model based on the exponential equation considered in the preceding section:

$$\frac{dN}{dt} = rN$$

In the **sigmoid** or **S-shaped growth form** (Fig. 6-9B), the population increases slowly at first (*establishment* or *positive acceleration phase*), then more rapidly (perhaps approaching a logarithmic phase), but it soon slows down gradually as the environment resistance increases in percentage (the *negative acceleration phase*) until equilibrium is reached and maintained. This form may be represented by the simple logistic model:

$$\frac{dN}{dt} = rN \times \frac{K - N}{K}$$

The upper level, beyond which no major increase can occur, as represented by the constant *K*, is the *upper asymptote* of the sigmoid curve and has been aptly termed the **maximum carrying capacity** (see Barrett and Odum 2000 for details).

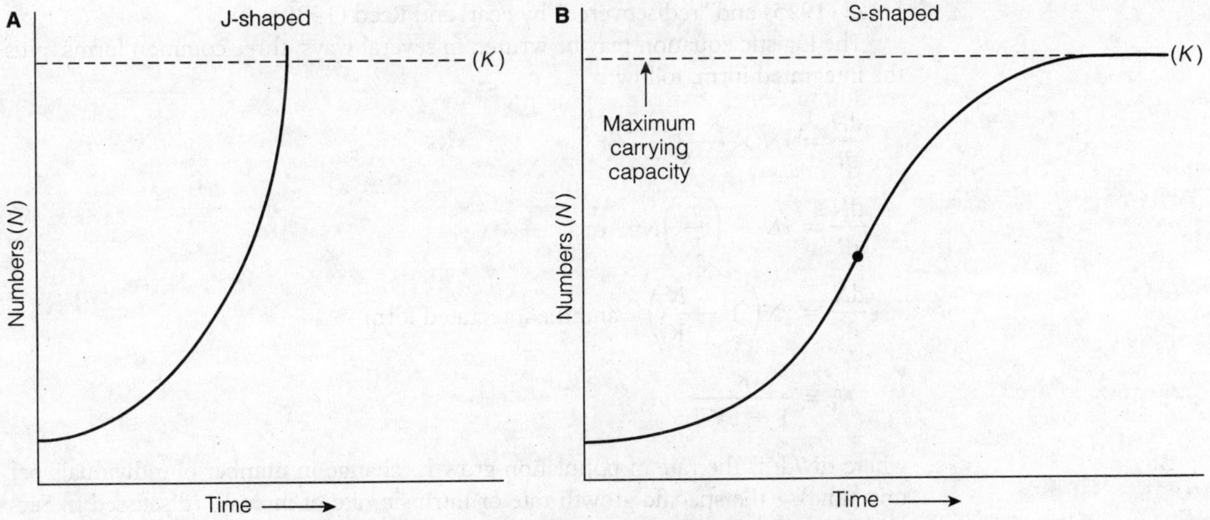

Figure 6-9. Hypothetical examples of (A) J-shaped (*exponential*) and (B) S-shaped (*sigmoid*) growth curves.

Explanation

When a few individuals are introduced into or enter an unoccupied area (for example, at the beginning of a growing season), characteristic patterns of population increase have often been observed. When plotted on an arithmetic scale, the part of the growth curve representing an increase in population often takes the form of a J or an S (Figs. 6-9A and B). It is interesting to note that these two basic growth forms are similar to the two metabolic or growth types that have been described in individual organisms. These patterns of growth and development illustrate processes transcending levels of organization (Barrett et al. 1997). But, as emphasized in Chapter 2, there are no set-point controls for growth at the population level and above; thus, overshooting K is likely.

The equation given previously as a simple model for the J-shaped growth form is the same as the exponential equation discussed in Section 3, except that a limit is imposed on N. The relatively unrestricted growth is suddenly halted when the population runs out of some resource (such as food or space), when frost or any other seasonal factor intervenes, or when the reproductive season suddenly terminates. When the upper limit of N is reached, the density may remain at this level for a time, but usually an immediate decline occurs, producing a relaxation-oscillation (boom-and-bust) pattern in density. Such a pattern in the short term is characteristic of many populations in nature, such as algal blooms, annual plants, some insects, and perhaps lemmings on the Tundra.

The second type of growth form that is also frequently observed follows an S-shaped or *sigmoid* pattern when density and time are plotted on arithmetic scales. The sigmoid curve is the result of the gradually increasing action of detrimental factors (environmental resistance or negative feedback) as the density of the population increases—unlike the J-shaped model, in which negative feedback is delayed until near the end of the increase. A simple case is one in which detrimental factors are linearly proportional to the density. Such a growth form is said to be *logistic* and conforms to the logistic equation used as a basis for the model of the sigmoid pattern. The logistic equation was first proposed by P. F. Verhulst in 1838; it was extensively used by Lotka (1925) and "rediscovered" by Pearl and Reed (1930).

The logistic equation may be written in several ways; three common forms, plus the integrated form, follow:

$$\frac{dN}{dt} = rN \times \frac{K-N}{K} \quad \text{or}$$

$$\frac{dN}{dt} = rN - \left(\frac{r}{K}\right)N^2 \quad \text{or}$$

$$\frac{dN}{dt} = rN\left(1 - \frac{N}{K}\right) \quad \text{and the integrated form}$$

$$N_t = \frac{K}{1 + e^{a-rt}}$$

where dN/dt is the rate of population growth (change in number of individuals per unit time), r the specific growth rate or intrinsic rate of increase (discussed in Section 3), N the population size (number of individuals), a a constant of integration

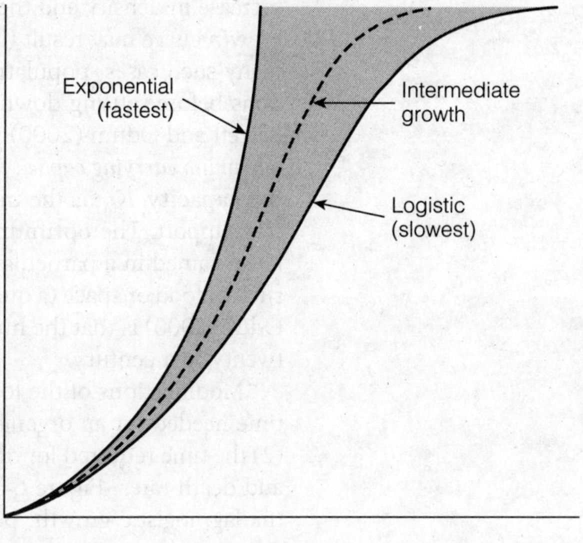

Figure 6-10. Curves showing theoretical upper (*exponential*) and lower (*logistic*) growth forms for any population, with identical maximum growth rates and minimum maintenance densities. The hatched area between the curves represents the area within which the growth forms of most populations would lie (after Wiegert 1974).

defining the position of the curve relative to its origin, and K the maximum population size possible (upper asymptote) or carrying capacity.

This equation is the same as the exponential one in the previous section, with the addition of one of the expressions $(K - N)/K$, $(r/K)N^2$, or $(1 - N/K)$. These expressions are three ways of indicating the **environmental resistance** created by the growing population itself, which brings about an increasing reduction in the potential reproduction rate as the population size approaches the carrying capacity. In word form, these equations simply mean the following:

Rate of population increase	equals	Maximum possible rate of increase (unlimited specific growth rate) times population size	times *or* minus	Degree of realization of maximum rate Unrealized increase

This simple model is a product of three components: a rate constant, r, a measure of population size, N, and a measure of the portion of available limiting factors not used by the population, $(1 - N/K)$. Although the growth of a great variety of populations—representing microorganisms, plants, and animals, and including both laboratory and natural populations—has been shown to follow a sigmoid pattern, it does not necessarily follow that such populations increase according to the logistic equation. As Wiegert (1974) pointed out, the logistic equation represents a sort of minimum sigmoid growth form, as the limiting effects of both space and resources begin at the very beginning of growth. In most cases, we would expect less limited growth at first, followed by a slowing down as density increases. Figure 6-10 illustrates this concept of the logistic as the lowest and the exponential as the highest growth form. Most populations would be expected to follow an intermediate pattern.

In populations of higher plants and animals, which have complicated life histories and long periods of individual development, there are likely to be delays in the increase in density and the impact of limiting factors. In such cases, a more concave growth curve may result (longer period required for natality to become effective). In many such cases, populations overshoot the upper asymptote and undergo oscillations before settling down at the carrying capacity level, as shown in Figure 6-11. Barrett and Odum (2000) presented two types of sigmoid growth, leading either to a *maximum carrying capacity* or to an *optimum carrying capacity.* The **maximum carrying capacity,** K_m, is the maximum density that the resources in a particular habitat can support. The **optimum carrying capacity,** K_o, is a lower-level density that can be sustained in a particular habitat without "living on the edge" regarding resources such as food or space (a quality over quantity parameter). Our prediction (Barrett and Odum 2000) is that the human population will follow the second pattern during the twenty-first century.

Modifications of the logistic growth form include two kinds of time lags: (1) the time needed for an organism to start increasing when conditions are favorable; and (2) the time required for organisms to react to unfavorable crowding by altering birth and death rates. Figure 6-12 illustrates a generalized logistic growth form, showing the lag, logistic growth, point of inflection, environmental resistance, and carrying capacity phases. The **lag phase** illustrates the time lag necessary for a population to become acclimated to its environment. For example, small mammals in a new habitat need to make runways or burrows before becoming reproductively successful; fish in a new pond or tank must adapt to the water chemistry before maximizing their rate of reproduction. Once populations acclimate to habitats where resources, such as food, cover, and space, are abundant, these populations reproduce at an exponential (logarithmic) rate of increase. The maximum rate of increase is termed the *point of inflection.* Demographers and population ecologists seek to determine the point of inflection because just beyond this point in the sigmoid growth curve, the rate of increase begins to decelerate (as opposed to accelerate just before the point of inflec-

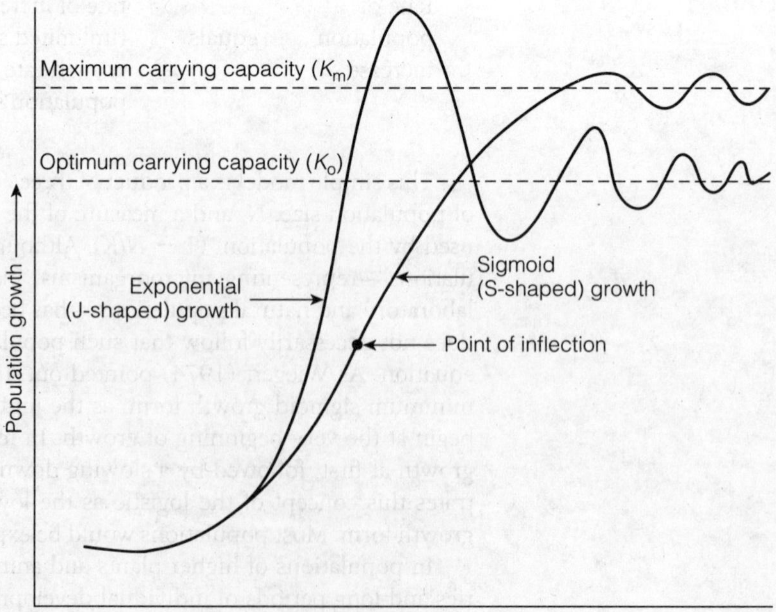

Figure 6-11. Contrasting sigmoid (S-shaped) and exponential (J-shaped) growth form models in relation to the maximum (K_m) and optimum (K_o) carrying capacity concepts. (From Barrett, G. W., and E. P. Odum. 2000. The twenty-first century: The world at carrying capacity. *BioScience* 50:363–368.)

Figure 6-12. Graph depicting the various phases of the modified sigmoid growth curve.

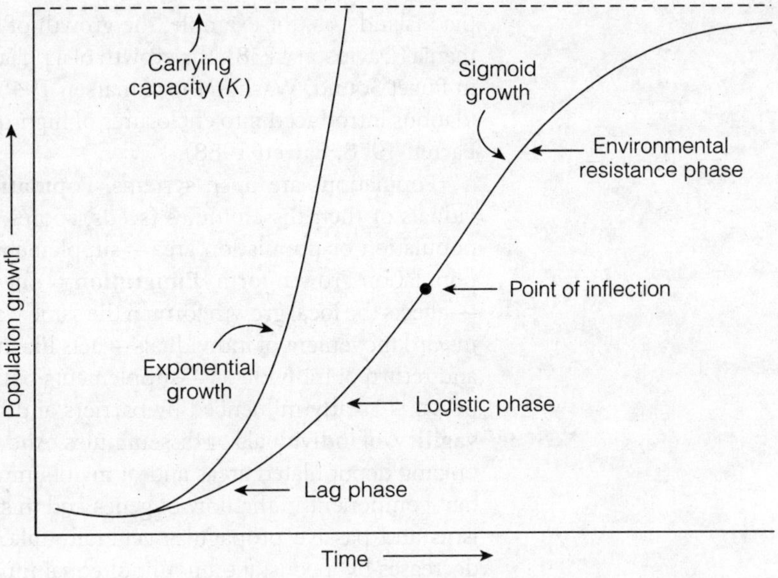

tion). The reason for this deceleration is that a resource, or a set of resources, becomes limiting in the environment. This slowing of population growth due to limiting resources is termed the **environmental resistance phase** of sigmoid growth. Finally, the population reaches carrying capacity conditions, in which the rate of population increase is zero and the population density is maximum, as depicted in Figures 6-11 and 6-12. Humans have yet to reach carrying capacity conditions globally, although there is evidence that this will occur during the twenty-first century (Lutz et al. 2001).

Fast-growing cities, with their dependence on huge external sources of energy, food, water, and life support (natural capital) are especially likely to boom and bust in varying degrees, depending on input factors and the degree to which citizens and governments can anticipate future conditions and plan ahead. Thus, in the early stages of urban growth, when economic conditions are favorable (space and resources available and inexpensive) and when the need for services (water, sewage treatment, streets, schools, and so on) is small, the population grows rapidly (with immigration often providing the major increase), as in a J-shaped growth pattern. Not until some time later (the time lag) do housing and schools become overcrowded, demands for services increase, taxes rise to cover increased maintenance costs, and general diseconomies of scale begin to be felt. In the absence of early negative feedback, such as rational land-use planning, cities will grow too rapidly for their survival and then suffer decline. Thus, there exists a need to couple an economic carrying capacity with an ecological carrying capacity concept in order to guide sustained development (L. R. Brown 2001).

Examples

Even though the simple logistic growth form is probably restricted to small organisms or to organisms with simple life histories, a general sigmoid growth pattern may be observed in larger organisms when they are introduced onto previously unoccupied islands, as, for example, the growth of sheep populations on the island of Tasmania (Davidson 1938), the growth of a pheasant population introduced on an island in Puget Sound, Washington (Einarsen 1945), or the growth of small mammal populations introduced into enclosures of high-quality habitat (Barrett 1968; Stueck and Barrett 1978; Barrett 1988).

Populations are open systems. **Population dispersal**—the movement of individuals or their disseminules (seeds, spores, larvae, and so forth) into or out of the population or population area—supplements natality and mortality in shaping the population growth form. **Emigration**—one-way outward movement of individuals—affects the local growth form in the same way as mortality; **immigration**—one-way inward movement of individuals—acts like natality. **Migration**—periodic departure and return of individuals—supplements both natality and mortality seasonally. Dispersal is greatly influenced by barriers and by the inherent power of movement, or **vagility**, of individuals or disseminules. And, of course, dispersal is the means of colonizing depopulated areas and of maintaining metapopulations. It is also an important component in the flow of genes and in speciation. The dispersal of small organisms and passive propagules generally takes an exponential form, in that density decreases by a constant amount of equal multiples of distance from the source. The dispersal of large, active animals deviates from this pattern and may take the form of "set distance" dispersal, normally distributed dispersal, or other forms. A study conducted by Mills et al. (1975) on the dispersal of big brown bats (*Eptesicus fuscus*) is an example of the interplay of random dispersal and a tendency for southward migration. Within 8 kilometers (5 miles), dispersal was nondirectional (about equal chance that a banded bat would be recovered at any point in any direction), but beyond this distance, dispersal was definitely directional and southward. For general reviews of dispersal patterns, see MacArthur and Wilson (1967), Stenseth and Lidicker (1992), and Barrett and Peles (1999).

5 Population Fluctuations and Cyclic Oscillations

Statement

When populations complete their growth, and $\Delta N/\Delta t$ averages zero over a long period of time, population density tends to pulse or fluctuate above and below the carrying capacity level, because populations are subject to various forms of feedback control rather than set-point controls. Some populations—especially insects, exotic plant species, and pests in general—are *irruptive;* that is, they explode in numbers unexpectedly in a boom-and-bust pattern. Often, such fluctuations result from seasonal or annual changes in the availability of resources, but they may be stochastic (random). Also, some populations oscillate so regularly that they can be classed as *cyclic.*

Explanation

In nature, it is important to distinguish between (1) *seasonal changes* in population size, largely controlled by life-cycle adaptations coupled with seasonal changes in environmental factors; and (2) *annual fluctuations.* For purposes of this analysis, annual fluctuations may be considered under two headings: (1) fluctuations controlled primarily by annual differences in **extrinsic factors** (such as temperature and rainfall) that are outside the sphere of population interactions; and (2) oscillations subject to **intrinsic factors** (biotic factors such as food or energy availability, disease, or predation), controlled primarily by population dynamics. In many cases, year-to-year changes in abundance seem clearly correlated with variation in one or more major extrinsic limiting factors, but some species show such regularity in relative abundance—seemingly independent of obvious environmental cues—that the term *cycles* seems appropriate (species with such regular variation in population size are often known as *cyclic species*). Examples of theories that have been advanced to explain these cycles will be presented later in this chapter.

As has been stressed in earlier chapters, populations modify and compensate for perturbations of physical factors. However, *because of the lack of set-point controls, such balances in mature systems are not steady states but are pulsing-state balances with varying amplitudes of pulsing in theory.* The more highly organized and mature the community, or the more stable the physical environment, or both, the lower will be the amplitude of fluctuations in population density over time.

Examples

Humans are familiar with seasonal variations in population size. We expect that at certain times of the year, mosquitoes or gnats will be abundant, the woods will be full of birds, or the fields will be full of ragweed. At other seasons, the populations of these organisms may dwindle to the vanishing point. Although it would be difficult to find, in nature, populations of animals, microorganisms, and herbaceous plants that do not exhibit some seasonal change in size, the most pronounced seasonal fluctuations occur with organisms that have limited breeding seasons, especially those with short life cycles and those with pronounced seasonal dispersal patterns (such as migratory birds).

An example of an irruptive pattern of population density fluctuations occurred in 1959–1960, when a feral house mice (*Mus musculus*) population in California exploded twice, as depicted in Figure 6-13 (Pearson 1963). These occasional, and usu-

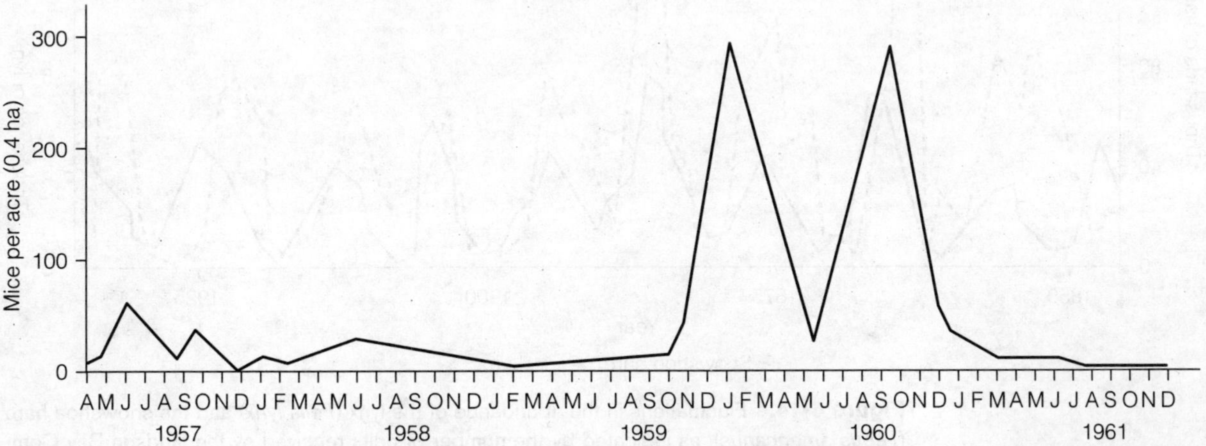

Figure 6-13. Irruptions of a feral house mouse (*Mus musculus*) population in California during 1959 and 1960 (modified after Pearson 1963).

ally unpredictable, population irruptions remain poorly understood, but likely result when a number of favorable conditions (such as weather, abundant food resources, vegetative cover to reduce predation) come together, resulting in a population explosion. Occasionally, the irruptions cover a wide geographical or landscape area, leading ecologists to address general theories regarding population regulation (which will be discussed later in this chapter).

Among the best-known examples of regular "mega-cycles" are the oscillations of certain species of northern mammals and birds that exhibit either a 9- to 10-year or a 3- to 4-year periodicity. A classic example of a 9- to 10-year oscillation is that of the snowshoe hare (*Lepus americanus*) and the lynx (*Felis lynx*), as shown in Figure 6-14. Figure 6-15 depicts the snowshoe hare with summer and winter pelage. Since about 1800, the Hudson Bay Company of Canada has kept records of pelts of fur-bearers trapped each year. When plotted, these records show that the lynx, for example, reached a population peak every 9 to 10 years ($\overline{X} = 9.6$ years) throughout a long period of time. Peaks of abundance were often followed by "crashes," or rapid declines, the lynx becoming scarce for several years. The snowshoe hare (Fig. 6-15) follows the same cycle; its peak abundance generally precedes that of the lynx by a year or more (Keith and Windberg 1978; Keith et al. 1984; Keith 1990). Because the lynx largely depends on the hare for food, it is obvious that the cycle of the predator is related to that of its prey. However, the two cycles are not strictly a cause-and-effect predator-prey interaction, because the cycle of the hare occurs in areas where there are no lynxes. The answer, apparently, is that the hare cycles are a product of the interaction between predation and food supply (Krebs et al. 1995; Krebs, Boonstra, et al. 2001; Krebs, Boutin, et al. 2001).

The shorter, three- to four-year cycle is characteristic of many northern murids (lemmings, mice, voles) and their predators (especially the snowy owl and foxes). The cycle of the lemming of the Tundra and the Arctic fox (*Alopex lagopus*) and the snowy owl (*Nyctea scandiaca*) was first well documented by Elton (1942). Every three

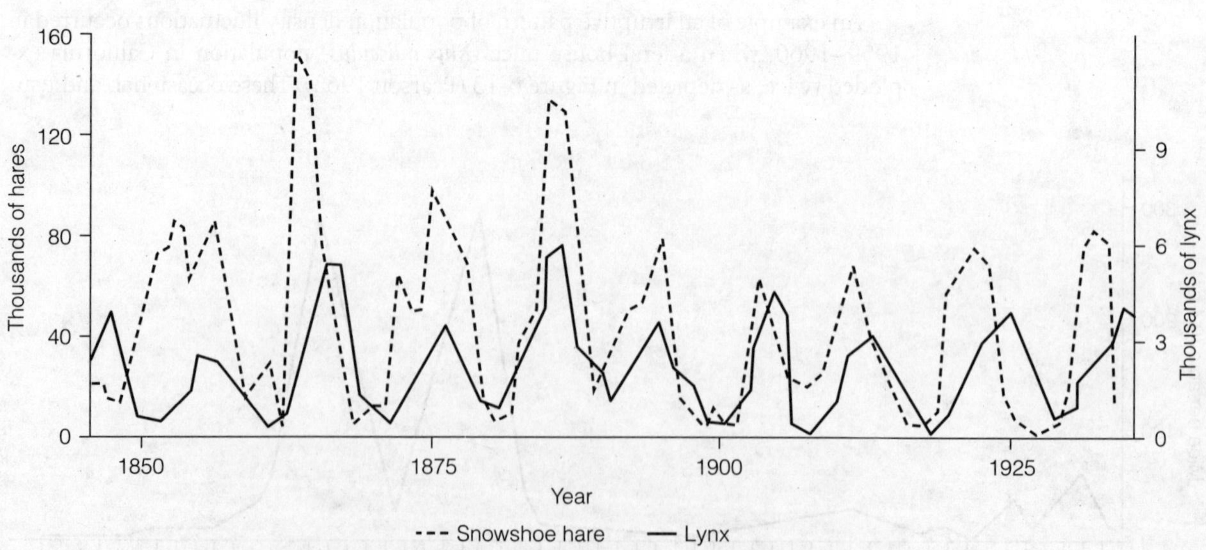

Figure 6-14. Fluctuations in the abundance of the lynx (*Felis lynx*) and the snowshoe hare (*Lepus americanus*), as indicated by the number of pelts received by the Hudson Bay Company (redrawn from MacLulich 1937 and Keith 1963).

Figure 6-15. The snowshoe hare (*Lepus americanus*), with (A) summer and (B) winter pelage. The change from the brown summer pelage to the white winter pelage has been shown to be controlled by photoperiodicity, not by temperature.

or four years, over enormous areas of northern tundra on two continents, the lemmings (two species in Eurasia and one species in North America of the genus *Lemmus,* and one species of *Dicrostonyx* in North America) become abundant only to "crash," often within a single season. Foxes and owls, which increase in numbers as their prey increases, decrease very soon afterward. During crash years, owls may migrate south into the United States in search of food. This irruptive emigration of surplus birds is apparently a one-way movement; few if any owls ever return. The owl population on the Tundra thus crashes as a result of the dispersal movement. So regular is this oscillation that ornithologists in the United States can count on an invasion of snowy owls every three or four years. Because the birds are conspicuous and appear everywhere about cities, they attract much attention and get their pictures in the newspaper. In years between invasions, few or no snowy owls are seen in the United States or southern Canada in winter. Shelford (1943) and Gross (1947) analyzed the records of the owl invasions and showed that they are correlated with the periodic decrease in abundance of the lemming, the chief food of the owl.

In Europe, the lemmings become so abundant at the crest of the cycle that they sometimes emigrate from their overcrowded haunts. Elton (1942) described the famous lemming migrations in Norway. As with the snowshoe hare, the lemming and vole cycles are either predator or resource (food) driven. Interesting enough, the major cause can be determined by the shape of the pulses, as shown in Figure 6-16. The food-driven lemming cycles have sharp peaks, whereas the predator-driven northern vole cycles have rounded peaks (Turchin et al. 2000). Lemmings apparently feed on the Tundra moss (especially in winter under the snow), which regrows slowly. The lemmings deplete their food supply and crash in one year (hence sharp peaks), before predators reach a high enough density to decimate the population. In contrast, the voles feed on foliage that regenerates quickly, allowing two or more years for predators to reduce the population (hence rounded peaks).

In summary, we suggest that most intrinsic fluctuations are either predator or resource driven or both. The exaggerated northern pulses occur where the physical environment is extreme and the diversity of predators, prey, and resources is low. Thus, similar to the explosive increase in snowy owls, the lemming movement is one-way.

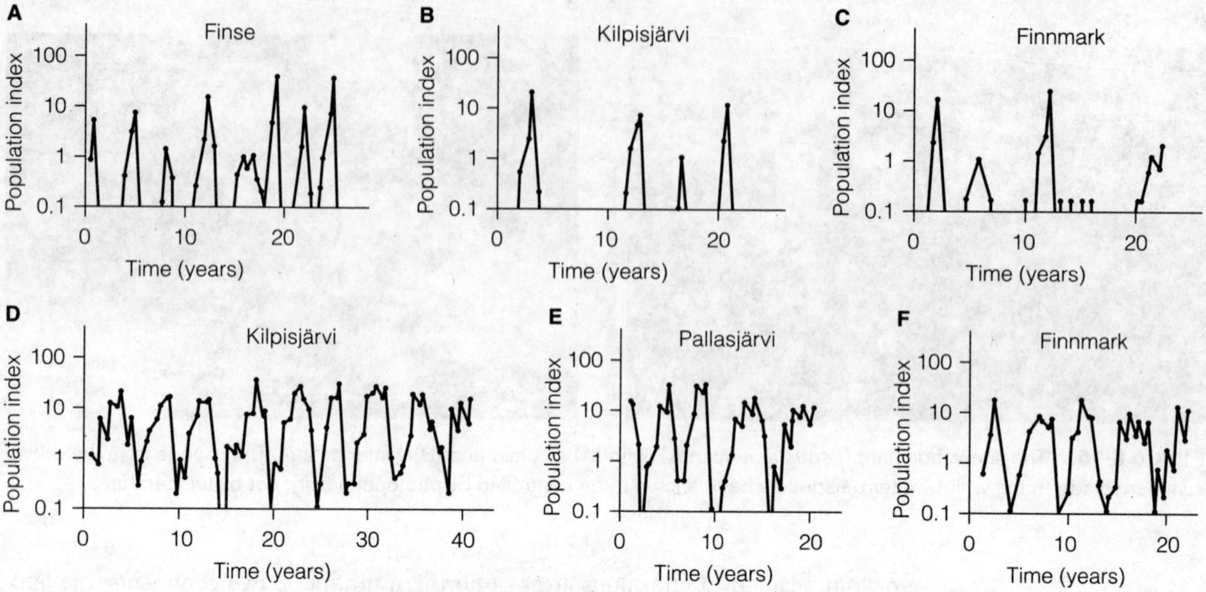

Figure 6-16. Population cycles in (A, B, C) lemmings and (D, E, F) voles in four locations in Scandinavia. Resource-driven lemming cycles show sharper peaks than predator-driven vole cycles. (From Turchin, P., L. Oksanen, P. Ekerholm, T. Oksanen, and H. Henttonen. 2000. Are lemmings prey or predators? *Nature* 405:562–565.)

Spectacular lemming emigrations do not occur at every four-year peak in density, but only during exceptionally high peaks. Often, the population subsides without the animals leaving the tundra or mountains.

Long-term records of violent oscillations in foliage-feeding insects in European forests illustrate a six- to seven-year cycle for the pine looper (*Bupalus piniaria*) in the pine forests of Great Britain (Barbour 1985). Interesting enough, *B. piniaria* exhibits shorter fluctuations at some sites and longer cycles in other areas—the result of a host-parasite interaction with *Dusoma oxyacanthae*. Pronounced cycles have been reported mostly from northern forests, especially pure stands of conifers. Density may vary over five orders of magnitude (log cycles), from less than one to more than 10,000 per 1000 square meters (Fig. 6-17). One can well imagine that with 10,000 potential moths emerging for every 1000 square meters, there are enough caterpillars to defoliate and even kill the trees, as frequently happens. The cycles of defoliating caterpillars are not so regular as the oscillations of the snowshoe hare, and the cycles of different species are not synchronous.

Periodic outbreaks of the spruce budworm (*Choristoneura fumiferana*) and tent caterpillars (*Malacosoma*) are well-known examples of similar patterns in the northern part of North America. Tent caterpillars have been studied by Wellington (1957, 1960) and budworms by Ludwig et al. (1978) and Holling (1980).

The budworm cycles are clearly a phenomenon at the ecosystem level, because the defoliating insect, its parasites and predators, and the conifers (spruce and balsam fir), which often grow in pure stands, are strongly coupled or coevolved. As the forest biomass increases, the large, older trees are vulnerable to a buildup of budworm caterpillars, and many trees are killed by successive defoliation. The death and decomposition of trees and the insect frass and feces return nutrients to the forest floor. Young trees, which are less susceptible to attack, are released from shade suppression and grow rapidly, filling in the canopy in a few years. During this time, parasites and bird predators of the insect combine to reduce the ecological density of the budworm. In the long term, the budworm is an integrated component that periodically rejuvenates the conifer ecosystem, not the catastrophe it seems if one only observes the dead and dying trees at the peak of the cycle.

In fact, after a study of the role of a mountain pine beetle (*Dendroctonus ponderosae*) in lodgepole pine (*Pinus contorta*) mountain forests, Peterman (1978) concluded that the beetle creates forests that are dominated primarily by Douglas fir (*Pseudotsuga menziesii*) at the lower elevations. These forests are more useful to humans than overcrowded stands of lodgepole pine, which are of little value for lumber, wildlife, or recreation. Peterman would have us view the beetle as a management tool rather than a pest, and he suggested that in isolated forests, it would be better to let the periodic outbreaks run their course, thus hastening ecological succession. Such a view, of course, contradicts the traditional management of forest pests, which is to attempt to control insects only when their density becomes high enough to kill trees outright. An alternate strategy might be simply to harvest old-growth trees before the beetles or budworms take them out. This becomes practical now that models have been developed that can predict when an outbreak of insects is likely. Periodic storms perform a function similar to that of the insects; blowing down old and crowded stands of high mountain forest, storms create a patchwork of young and older trees that are constantly being shifted about on the mountainside.

Probably the most famous of all oscillations of an insect population are those involving locusts and grasshoppers. In Eurasia, records of outbreaks of the migratory locust (*Locusta migratoria*) go back to antiquity (J. R. Carpenter 1940). The locusts live in desert or semiarid country and in most years are nonmigratory, eat no crops, and attract no attention. At intervals, however, their population density increases to an enormous extent. Periodic drought also causes crowding of individuals into limited areas. The tactile response of repeated "bumping" causes both males and females to contribute to the production of migratory offspring. Human activities, such as shifting cultivation and overgrazing by cattle, tend to increase rather than decrease

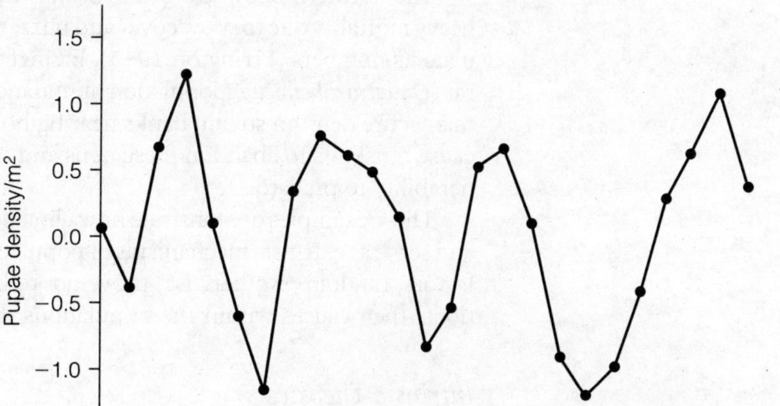

Figure 6-17. Annual estimates in pupal density per m² for the pine looper (*Bupalus piniaria*) at Tentsmuir in Great Britain (after Barbour 1985). Data based on log fluctuations about the equilibrium density of 0.0 (1 pupa per m²).

the chance of an outbreak, because a patchwork or mosaic of vegetation and bare ground (in which the locusts lay eggs) is favorable for an exponential buildup of population. Here, a population explosion is generated by combined instability and simplicity in the environment. As with the lemming, probably not every population maximum is accompanied by an emigration; therefore, the frequency of the plagues does not necessarily represent the true periodicity of the oscillations in density. Even so, outbreaks were recorded at least once every 40 years between 1695 and 1895.

Two striking features of mega-cyclic oscillations are that (1) they are most pronounced in the less complex ecosystems of northern regions and in human-maintained monoculture forests; and (2) although peaks of abundance may occur simultaneously over wide areas, peaks in the same species in different regions do not always coincide. Theories have been advanced to explain the range of mega-cycles throughout the levels-of-organization hierarchy. A very brief summary of some of these theories follows.

Extrinsic Theories

Attempts to relate population density oscillations to climatic factors have so far been unsuccessful (MacLulich 1937). Palmgren (1949) and Cole (1951, 1954) suggested that what appear to be regular oscillations could result from random variations in both biotic and abiotic factors. Lidicker (1988) suggested that population ecologists adopt a multifactorial model in order to comprehend how many extrinsic and intrinsic factors function synergistically to explain changes in population densities.

Some species do, indeed, appear to be regulated by climatic factors For example, there exists a relationship between Gambel's quail (*Callipepla gambelii*) population abundance in southern Arizona and winter rainfall (Sowls 1960). Quail need an abundance of high-quality vegetation and cover in late winter and early spring to supply the nutrients essential to reproduction. In years of low rainfall, the flush of high-quality vegetation fails to appear, and most of the birds fail to reproduce. Thus, the reproductive success of desert quail reflects a density-independent response to rainfall (more will be said about density-dependent and density-independent factors related to population regulation in Section 6).

The northern bobwhite quail (*Colinus virginianus*) has been found to experience heavy mortality due to snow cover and blizzard conditions—extrinsic factors that regulate its numbers (Errington 1945). Errington (1963) also demonstrated that muskrat (*Ondatra zibethicus*) population abundance is affected by drought, because muskrats secure dens on stream banks near high-quality feeding areas. Periods of drought cause muskrats to abandon these dens and seek new dens, thus increasing their vulnerability to predators.

These examples demonstrate how climatic factors affect plant and animal densities and serve as extrinsic mechanisms of population regulation. When climatic (extrinsic) factors, random or otherwise, prove not to be the major cause of the violent oscillations, then causes within the populations themselves (intrinsic factors) are sought.

Intrinsic Theories

Building on a medical theory of stress (the *general adaptation syndrome*) of Hans Selye, John J. Christian and coworkers (see Christian 1950, 1963; Christian and Davis 1964) amassed considerable evidence from both the field and the laboratory to show

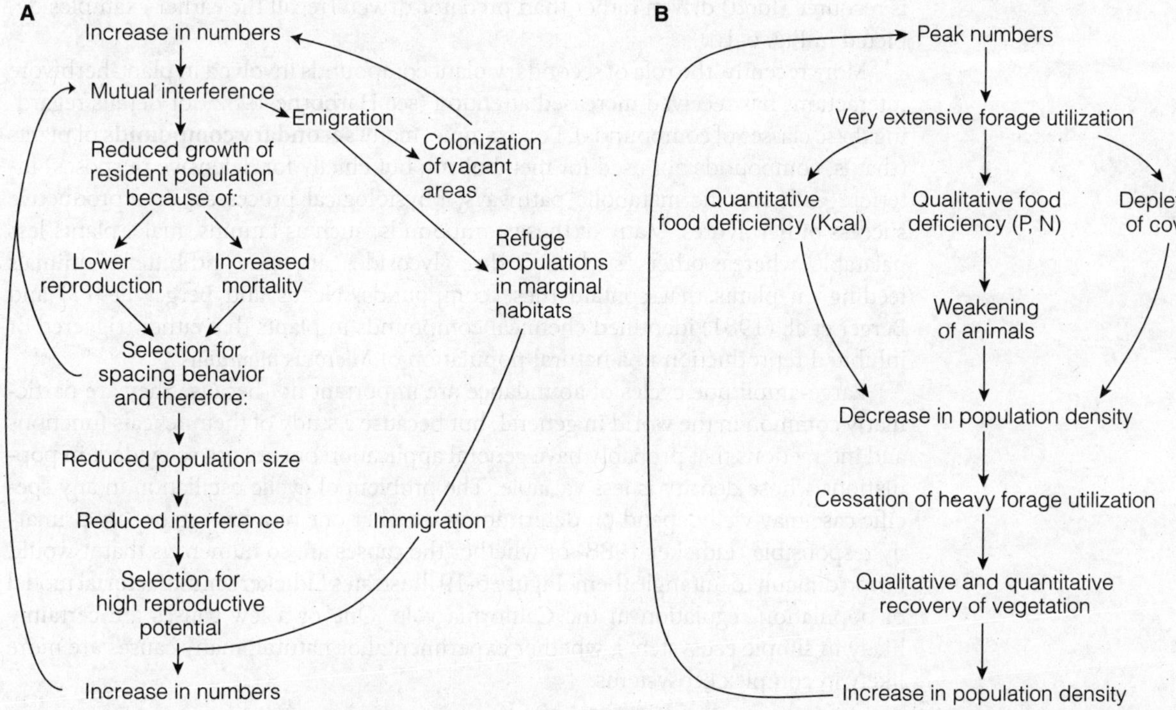

Figure 6-18. Modified versions of the Chitty-Krebs genetic feedback (A) and food quality (B) hypotheses to explain population pulses in microtine rodents (after Krebs et al. 1973 and Pitelka 1973).

that crowding in higher vertebrates results in enlarged adrenal glands. This enlargement is symptomatic of shifts in the neural-endocrine balance that, in turn, cause changes in behavior, in reproductive potential, and in resistance to disease or other stress. Such changes often combine to cause a precipitous decline in population density. This theory has frequently been termed the *adreno-pituitary feedback hypothesis.*

During the 1960s and 1970s, Chitty (1960, 1967), Krebs and Myers (1974), and Krebs (1978) suggested that genetic shifts account for differences in aggressive behavior and survival that are observed at different phases of the cycle for the vole (Fig. 6-18A)—a situation similar to that of the strong and weak races in the tent caterpillar (Wellington 1960).

Another group of theories relies on the idea that cycles of abundance are intrinsic at the ecosystem level rather than at the population level. Certainly, density changes that range over several orders of magnitude must involve not only secondary trophic levels, such as predators and prey (see, for example, Pearson 1963), but also the primary plant-herbivore interactions. An example is the nutrient depletion and recovery hypothesis (Fig. 6-18B) proposed to explain microtine cycles in the Tundra (Schultz 1964, 1969; Pitelka 1964, 1973). This hypothesis is based on documented evidence that heavy grazing during the peak year ties up and reduces the availability of mineral nutrients (especially phosphorus), so that the food is low in nutritional quality. The growth and survival of both adults and young are thus greatly reduced. After two or three years, the recycling of nutrients is restored, plants recover, and the ecosystem can again support a high density of consumers. In other words, the cycle is resource (food) driven rather than predator driven (recall the earlier examples depicted in Fig. 6-16).

More recently, the role of secondary plant compounds involved in plant-herbivore interactions has received increased attention (see Harborne 1982, for details regarding these classes of compounds). For example, many **secondary compounds** of plants (that is, compounds not used for metabolism, but chiefly for defensive purposes) interfere with specific metabolic pathways, physiological processes, or reproductive success of herbivores. Many of these compounds, such as tannins, make plants less palatable, whereas others, such as cardiac glycosides, are toxic and bitter to animals feeding on plants that contain these compounds. Negus and Berger (1977) and Berger et al. (1981) identified chemical compounds in plants that either triggered or inhibited reproduction in a natural population of *Microtus montanus*.

Large-amplitude cycles of abundance are important not because they are particularly common in the world in general, but because a study of them reveals functions and interactions that probably have general application but are not so evident in populations whose density is less variable. The problem of cyclic oscillation in any specific case may well depend on determining whether one or a few factors are primarily responsible (Lidicker 1988) or whether the causes are so numerous that it would be too difficult to untangle them. Figure 6-19 illustrates Lidicker's multifactorial model of population regulation in the California vole. One or a few causes are certainly likely in simple ecosystems, whether experimental or natural; many causes are more likely in complex ecosystems.

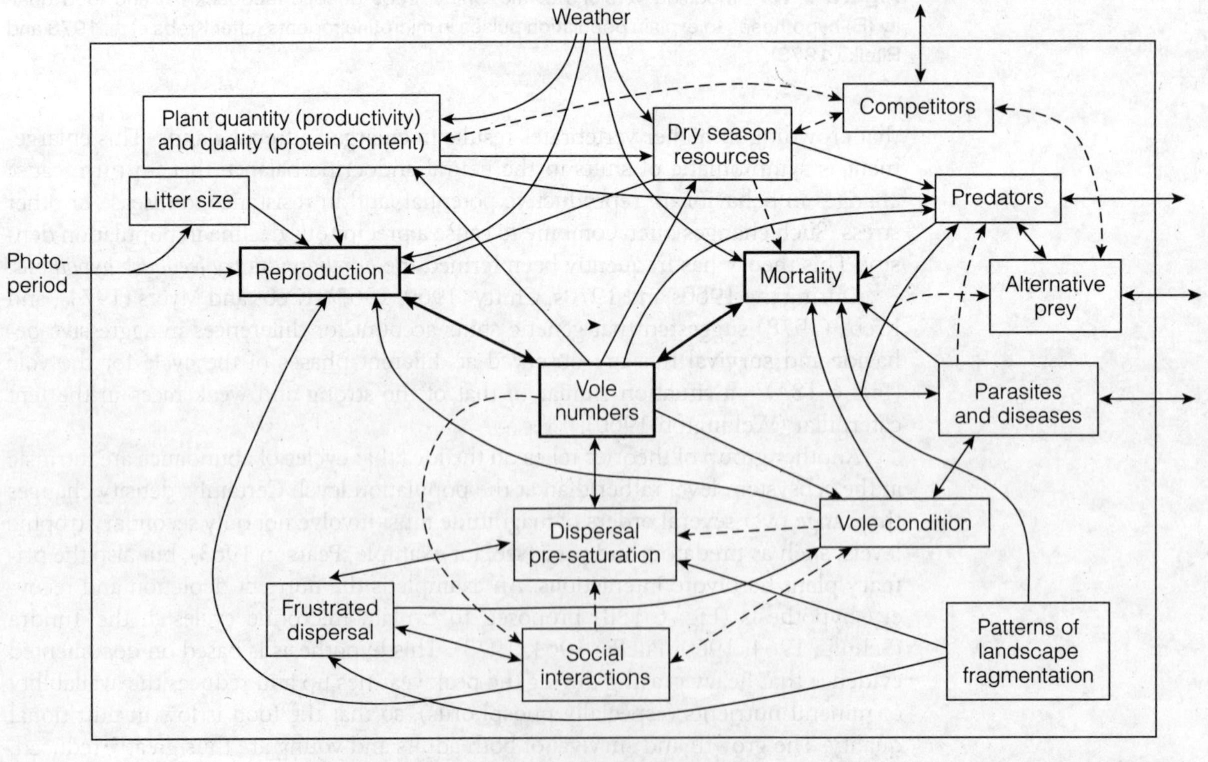

Figure 6-19. Multifactorial model of population regulation in the California vole (*Microtus californicus*) (modified after Lidicker 1988).

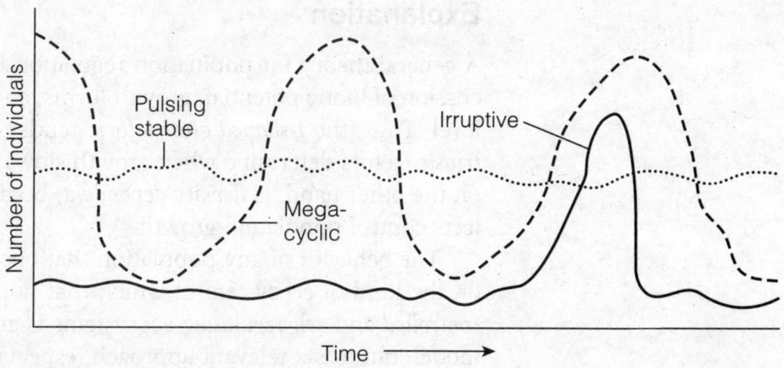

Figure 6-20. Three types of population-level pulsing.

Overview of Cycling

Figure 6-20 models the three basic types of population level fluctuations that can be top down (predator driven), bottom up (resource driven), or both. We suggest that mega-cycles may be exaggerated density pulses that are characteristic of several populations, as proposed by W. E. Odum et al. (1995) in a paper entitled "Nature's Pulsing Paradigm." We suggest that maximum density is achieved when biologically based intrinsic pulses are coordinated with extrinsic, physically based pulses.

Having considered a number of interesting specifics, we can now consider the more general challenge of population regulation.

6 Density-Independent and Density-Dependent Mechanisms of Population Regulation

Statement

In low-diversity, physically stressed ecosystems, or in ecosystems subject to irregular or unpredictable extrinsic perturbations, population size tends to be mainly influenced by physical factors, such as weather, water currents, chemical limiting conditions, and pollution. In high-diversity ecosystems in benign environments (low probability of periodic physical stress such as storms or fire), populations tend to be biologically controlled, and to some extent, at least, their density is self-regulated. Any factor, whether limiting or favorable (negative or positive) to a population, is (1) **density independent** if its effect (change in numbers) is independent of the size of the population; or (2) **density dependent** if its effect on the population is a function of population density. Density-dependent response is usually direct, because it intensifies as the upper limit (carrying capacity) is approached. It may, however, also be inverse (decrease in intensity as density increases). Direct density-dependent factors act like governors on an engine (hence, they can be termed *density governing)* and, for this reason, are considered one of the chief agents in preventing overpopulation. Climatic factors often, but by no means always, act in a density-independent manner, whereas biotic factors (such as competition, parasites, or pathogens) often, but not always, act in a density-dependent manner.

Explanation

A general theory for population regulation logically results from the preceding discussion of biotic potential, growth forms, and variation around the carrying capacity level. Thus, the *J-shaped growth form* tends to occur when *density-independent* or *extrinsic* factors determine when growth slows down or stops. The *sigmoid growth form,* on the other hand, is *density dependent,* because self-crowding and other *intrinsic* effects control population growth.

The behavior of any population that one might wish to select for study depends on the kind of ecosystem of which that population is a part. Contrasting *physically controlled* and *self-regulating* ecosystems is arbitrary and produces an oversimplified model, but it is a relevant approach, especially as human efforts during most of the past century have been directed toward replacing self-maintaining ecosystems with monocultures and stressed systems that require a lot of human care. As the cost (in energy as well as in money) of physical and chemical control has risen, as pest resistance to pesticides has increased, and as toxic chemical by-products in food, water, and air have become more of a threat, *integrative pest management* (IPM) is increasingly being implemented. Evidence for this is the increased interest in a new frontier termed *ecologically based pest management,* which involves efforts to reestablish natural, density-dependent, ecosystem-level controls in agricultural and forest ecosystems (NRC 1996a, 2000a; E. P. Odum and Barrett 2000).

The preceding section showed how physiological and genetic shifts, or the alternation of ecotypes in time, as it were, can dampen oscillations and hasten the return of density to lower levels after it overshoots carrying capacity. However, the question remains how self-regulation at the population level evolves through natural selection at the individual level (see Section 12 regarding population genetics and natural selection for additional details).

Wynne-Edwards (1962, 1965) proposed two mechanisms that can stabilize density at a level lower than saturation: (1) **territoriality,** an exaggerated form of intraspecific competition that limits growth through land-use control (to be discussed more fully in Section 9); and (2) **group behavior,** such as peck orders, sexual dominance, and other behaviors that increase the fitness of offspring but reduce their number. These mechanisms tend to enhance the quality of the environment for the individual and reduce the probability of extinction that might result from overshooting the availability of resources. The importance of such social and behavioral traits is difficult to test experimentally and is much discussed, as indicated in several review books by Cohen et al. (1980), Chepko-Sade and Halpin (1987), Cockburn (1988), and Stenseth and Lidicker (1992).

Density-independent (extrinsic) factors of the environment (such as weather phenomena) tend to cause variations, sometimes drastic, in population density and tend to cause shifting of upper asymptotic or carrying capacity levels. Density-dependent (intrinsic) factors such as competition, however, tend to maintain a population in a stable pulsing state or to hasten the return to such a level. Density-independent environmental factors have a greater role in physically stressed ecosystems; density-dependent natality and mortality become more important in benign environments where extrinsic stress is reduced. As in a smoothly functioning cybernetic system, additional negative feedback control is provided by interactions (both phenotypic and genetic) between populations of different species that are linked together in food chains or by other important ecological relationships.

Examples

Chitty (1960) and Wellington (1960) described how the *quality* of natural populations (voles and tent caterpillars, respectively) changes in relation to population abundance. For example, it appears that reproductive success and population survivorship decrease as vole populations increase in numbers. Likewise, survivorship, foraging behavior, and behavioral construction of "tents" appear to decrease in quality as the population abundance of caterpillars overshoots carrying capacity conditions. These phenomena tend to function in a density-dependent manner, providing a regulatory mechanism for these species. Holling (1965, 1966) emphasized the importance of behavioral characteristics in a series of mathematical models that predict how effectively a given insect parasite will control the insect host at different densities.

Plants exhibit density-dependent population regulation mechanisms just as animals do. Plant populations at high densities undergo a process termed *self-thinning*. When seeds are sown at high densities, the emerging young plants or seedlings compete vigorously. As the seedlings grow, many die, decreasing the density of surviving seedlings. The increasing growth rate of the surviving individual plants results in continuous competition, leading to a decline in the number of surviving plants. When the logarithm of average plant weight is plotted as a function of the logarithm of population density, the data points in the course of the growing season generate a line with a slope of about $-3/2$. Ecologists term this relationship between average weight and plant density a *self-thinning curve.* Because of its regularity among numerous plant species, this relationship is also frequently termed the **−3/2 power law.** Figure 6-21 depicts changes in plant density and mean plant weight during the growing season for redroot pigweed (*Amaranthus retroflexus*) and white goosefoot (*Chenopodium album*) and illustrates the $-3/2$ power law (J. L. Harper 1977). Thus, both plants and animals exhibit density-dependent mechanisms that tend to regulate and control

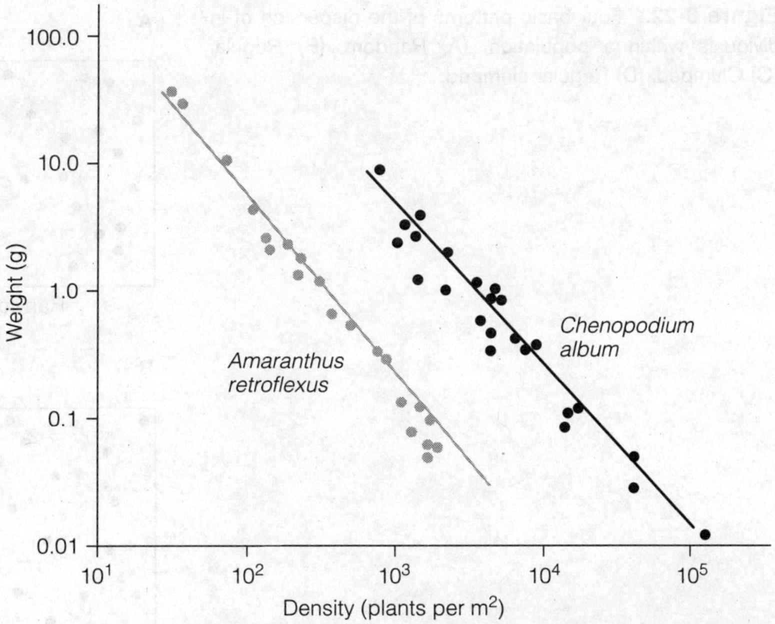

Figure 6-21. Changes in plant density and mean plant weight during the growing season for *Amaranthus retroflexus* and *Chenopodium album*, illustrating the −3/2 self-thinning power law (after J. L. Harper 1977).

population densities to remain at or near the carrying capacity set by the availability of resources and conditions in the environment.

7 Patterns of Dispersion

Statement

Individuals in a population may be dispersed according to four types of board patterns (Fig. 6-22): (1) random; (2) regular; (3) clumped; and (4) regular clumped. All these types are found in nature. **Random** distribution occurs when the environment is very uniform and there is no tendency to aggregate. **Regular** or **uniform** dispersion may occur when competition between individuals is severe or when there is positive antagonism that promotes even spacing; of course, this is also frequently the pattern in monoculture crops and forests. **Clumping** of varying degrees (individuals associate in groups) represents by far the most common pattern. However, if individuals of a population tend to form groups of a certain size—for example, herds of animals or vegetative clones in plants—the distribution of the *groups* may be either random or clumped in a regular pattern. Determination of the type of dispersion is important in the selection of methods of sampling and statistical analysis.

Explanation and Examples

The four patterns of intrapopulation dispersion are depicted in Figure 6-22. Each rectangle contains approximately the same number of individuals. In the case of clumped distribution, the groups could be randomly distributed or uniformly dis-

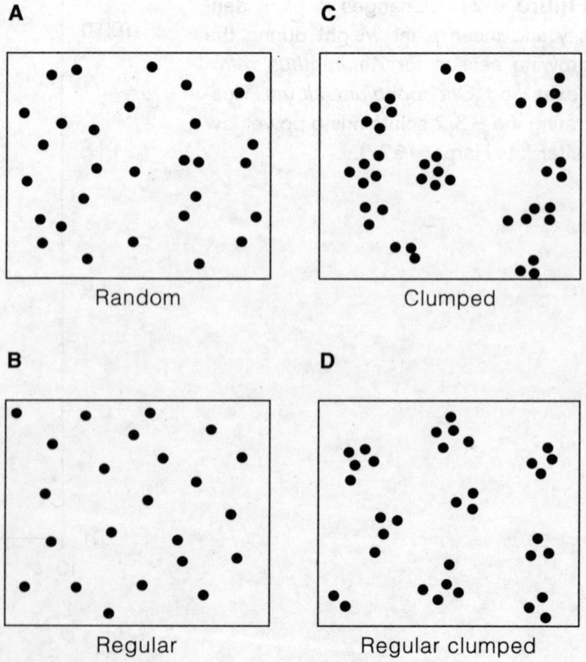

Figure 6-22. Four basic patterns of the dispersion of individuals within a population. (A) Random. (B) Regular. (C) Clumped. (D) Regular clumped.

tributed—that is, themselves clumped in a regular pattern with large unoccupied spaces (Fig. 6-22D). From examining Figure 6-22, one may see that a small sample drawn from the four populations could obviously yield very different results. A small sample from a population with a clumped distribution would tend to give either too high or too low a density when the number in the sample is multiplied to obtain the total population. Thus, clumped populations require larger and more carefully planned sample techniques than nonclumped ones.

Random distribution follows the *normal* or bell-shaped curve on which standard *parametric statistical methods* are based (see Chapter 12 for more details regarding statistical methods). This type of distribution is to be expected in nature when many factors are acting together on the population. When a few major factors are dominating, as is the usual case (recall the principle of limiting factors), and when there is a strong tendency for plants and animals to aggregate for (or because of) reproductive and other functions, there is little reason to expect a completely random distribution. To study such populations, we use *nonparametric statistics,* which are based on nonrandom patterns of distribution; field samples are frequently needed to determine patterns of distribution and, consequently, decide which statistic tests to use when comparing differences among populations. However, nonrandom or "contagious" distributions of organisms are sometimes found to be made up of intermingled random distributions of groups containing various numbers of individuals or, alternatively, of groups that turn out to be uniformly distributed (or at least more regular than random). To take an extreme case, it would be much better to determine the number of ant colonies (using the *colony* as the population unit) by a sample method, and then to determine the number of individuals per colony, than it would be to try to measure the number of individuals directly by random samples.

Several methods have been suggested to determine the type of spacing and the degree of clumping between individuals in a population (where it is not self-evident), but there is much that must still be done in solving this problem. Two methods are mentioned as examples. One method is to compare the actual frequency of occurrence of differently sized groups obtained in a series of samples with a *Poisson series,* which gives the frequency with which groups of 0, 1, 2, 3, 4 . . . *n* individuals will be encountered together if the distribution is random. Thus, if the occurrence of small-sized groups (including blanks) and large-sized groups is more frequent and the occurrence of midsized groups less frequent than expected, the distribution is clumped. The opposite is found in a uniform distribution. Statistical tests can be used to determine whether the observed deviation from the Poisson curve is significant. An example of the use of the Poisson method to test for random distribution in spiders is shown in Table 6-2. In all but 3 of 11 quadrats, spiders were randomly distributed. The nonrandom distributions occurred in quadrats in which the vegetation was least uniform.

Another method to determine dispersion type involves actually measuring the distance between individuals in some standardized way. When the square root of the distance is plotted against frequency, the shape of the resulting *frequency polygon* indicates the distribution pattern. A symmetrical polygon (a normal, bell-shaped curve, in other words) indicates random distribution; a polygon skewed to the right indicates a uniform distribution; and one skewed to the left indicates a clumped distribution (individuals coming closer together than expected). A numerical measure of the degree of skewness may be computed. This method, of course, would be most applicable to plants or stationary animals, but it could be used to determine the spacing between animal colonies or domiciles (such as fox dens, rodent burrows, or bird nests).

Table 6-2 **Number and distribution of wolf spiders (Lycosidae) on 0.1-hectare quadrats in an old-field habitat**

Species	*Quadrat*	*Number per quadrat*	*Chi-square from Poisson distribution*
Lycosa timuqua	1	31	8.90*
	2	19	9.58*
	3	15	5.51
	4	16	0.09
	5	45	0.78
	6	134	1.14
L. carolinensis	2	16	0.09
	5	23	4.04
	6	15	0.05
L. rabida	3	70	17.30*
	4	16	0.09

Source: After Kuenzler 1958.
*Significant at $P \leq 0.01$ level of probability.

Flour beetle larvae are usually distributed randomly throughout their very uniform environment, because their observed distribution corresponds with the Poisson distribution (Park 1934). Lone parasites or predators, such as the spiders in Table 6-2, sometimes show a random distribution (they often engage in random searching behavior for their hosts or prey). Forest trees that have reached sufficient height to form a part of the forest canopy may show a regular uniform distribution, because the competition for sunlight is so great that the trees tend to be spaced at intervals more regular than random. A cornfield, orchard, or pine plantation, of course, would be an even clearer example. Desert shrubs often are very regularly spaced, almost as if planted in rows, because of the intense competition in the low-moisture environment, which may include the production of plant antibiotics that prevent the establishment of near neighbors. A similar more regular than random pattern frequently occurs in territorial animals (see Section 9). The clumped (aggregate) pattern of dispersion is discussed in detail in the following section.

8 The Allee Principle of Aggregation and Refuging

Statement

As noted in the previous section, varying degrees of clumping are characteristic of the internal structure of most populations at one time or another. Such clumping is a result of individuals aggregating (1) in response to local habitat or landscape differences; (2) in response to daily and seasonal weather changes; (3) because of reproductive processes; or (4) because of social attractions (in higher animals). Aggregation may increase competition between individuals for resources such as nutrients, food, or space, but this is often more than counterbalanced by the increased survival of the group because of its ability to defend itself, to find resources, or to modify microclimate or microhabitat conditions. The degree of aggregation—and the overall density—that results in optimum population growth and survival varies with species and conditions; therefore, undercrowding (or lack of aggregation) as well as overcrowding may be limiting. This principle is termed the **Allee principle of aggregation,** named after well-known behavioral ecologist W. C. Allee.

Refuging describes a special type of aggregation in which large, socially organized groups of animals establish themselves in a favorable, central place (*refuge*) from which they disperse and to which they return regularly to satisfy their needs for food or other resources. Some of the most successfully adapted animals on Earth, including starlings and humans, employ this refuging strategy.

Explanation and Examples

In plants, aggregation may occur in response to the first three factors listed in the Statement (habitat, climate, or reproduction). In higher animals, spectacular aggregations may be the result of all four factors, but especially of social behavior—as illustrated, for example, by the herds of reindeer or caribou in the Arctic, the great migratory flocks of birds, or herds of antelope on the East African Savanna that move from one grazing area to another, thus avoiding overgrazing any one part of the range.

In plants in general, it is a well-defined ecological principle that aggregation is inversely related to the mobility of disseminules (seeds or spores), as was brought out in Weaver and Clements' (1929) pioneer textbook *Plant Ecology*. In old fields, cedars, persimmons, and other plants with nonmobile seeds are nearly always clumped near a parent or along fences and other places where birds or other animals have deposited the seeds in groups. On the other hand, ragweeds, grasses, and even pine trees, which have light seeds widely distributed by the wind, are by comparison much more randomly distributed over old fields.

Group survival value is an important characteristic that may result from aggregation (Fig. 6-23). A group of plants may be able to withstand the action of wind better than isolated individuals or may be able to reduce water loss more effectively. With green plants, however, the deleterious effects of competition for light and nutrients generally soon overbalance the advantages of aggregation. The most marked group survival advantages are to be found in animals. Allee (1931, 1951) conducted many experiments in this field and summarized the extensive writings on the subject. He found, for example, that groups of fish could withstand a given dose of poison introduced into the water much better than could isolated individuals. Isolated individuals were more resistant to poison when placed in water formerly occupied by a group of fish than when placed in water not so "biologically conditioned"; in the previously occupied waters, mucus and other secretions aided in counteracting the poisons, thus revealing something of the mechanism of group action in this case. Bees provide another example of group survival value; a hive or cluster of bees can generate and retain enough heat in the mass for survival of all the individuals at temperatures low enough to kill all the bees if each were isolated. Bobwhite quail (*Colinus vir-*

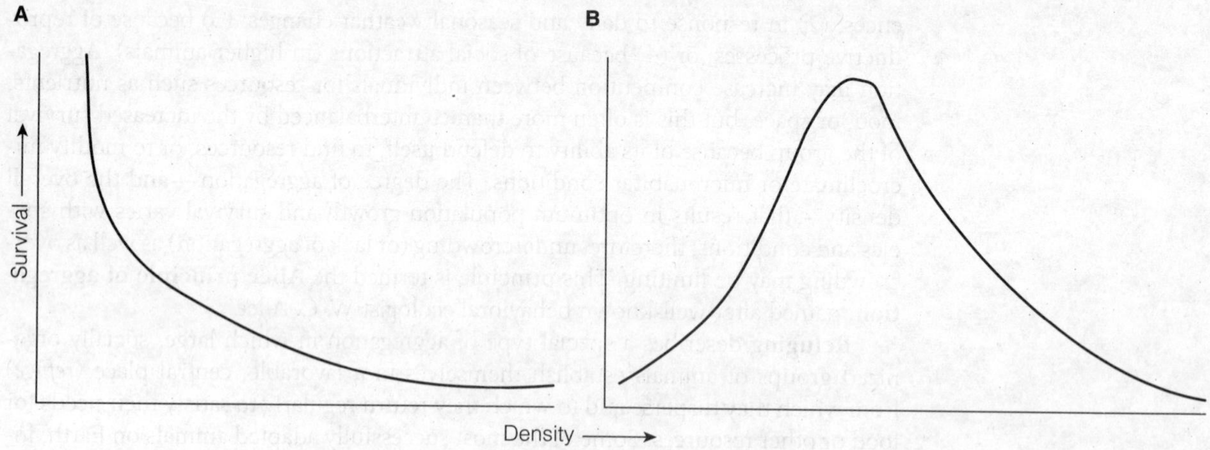

Figure 6-23. Illustration of the Allee principle. In some populations (A), growth and survival is greatest when the population size is small, whereas in others (B), intraspecific cooperation results in an intermediate-sized population being the most favorable. In the latter instance, "undercrowding" is as detrimental as "overcrowding." (after Allee et al. 1949).

ginianus) increase their chances of survivorship by forming a group (*covey*) during the winter months in the midwestern United States; the covey rests in a circle with heads facing outward (Fig. 6-24), thus being able to "flush" in several directions if they are approached by a predator such as a red fox (*Vulpes vulpes*). This social grouping behavior and response to disturbance (such as human hunters) results in at least some of the individuals in the covey escaping harm and, consequently, being able to reproduce in the spring.

Actual social aggregations, as seen in the social insects and vertebrates (as contrasted with passive aggregation in response to some common environmental factor), have a definite organization involving social hierarchies and individual specializations. A *social hierarchy* may be a *peck order* (so termed because the phenomenon was first described in chickens) with clear-cut dominance and subordination between in-

Figure 6-24. A covey of quail (*Colinus virginianus*), illustrating the Allee principle of aggregation as a behavioral strategy.

dividuals, often in linear order (like a military chain of command from general to private), or it may be a more complicated pattern of leadership, dominance, and cooperation, as occurs in well-knit groups of birds and insects that behave almost as a single unit. These kinds of social organizations benefit the population as a whole by preventing density overgrowth.

Among higher animals, a very successful aggregation strategy has been termed *refuging*, as described in detail by W. J. Hamilton and Watt (1970) and Paine (1976). **Refuges** are sites or situations where members of an exploited population have some protection from predators and parasites. Large numbers of individuals resort to a favorable central place or core—for example, a starling roost or a large breeding colony of sea birds. From there, they forage within a large perimeter or life-support area, often daily. Aggregation at a central place is advantageous in ensuring a net energy gain by individuals when good central places are scarce. Disadvantages of refuges are stresses such as excrement pollution and excessive trampling of vegetation or substrate at the central place and increased risk of predation during food-gathering or foraging forays.

The remarkable organizations of social insects are unique in their specialized roles. The most highly developed insect societies are found among termites (Isoptera) and ants and bees (Hymenoptera). In the most specialized species, a division of labor is accomplished by three castes: reproducers (queens and kings), workers, and soldiers. Each caste is morphologically specialized to perform the functions of reproduction, food gathering, and protection, respectively. As will be discussed in the next chapter, this kind of adaptation leads to group selection not only within a species but also in groups of closely linked species.

The Allee principle is relevant to the human condition. Aggregation into cities and urban districts (a refuging strategy) is obviously beneficial, but only up to a point, in connection with the law of diminishing returns. The exploitation of fossil fuels has extended the dispersal of foraging areas to the far reaches of Earth, so cities and other central places have few energy and fuel constraints on the size of the refuging population. But pollution and the cost of maintenance become increasingly limiting as human population density increases. A plot of benefit (y-axis) against city size (x-axis) would theoretically have the same humpbacked shape as curve B in Figure 6-23. Thus, cities, like bee or termite colonies, can get too big for their own good. The optimum size of the aggregation of social insects is determined by the trial and error of natural selection. Because the optimum size of cities cannot as yet be objectively determined, cities tend to overshoot in size and then depopulate when their costs exceed their benefits. According to ecological principles, it is a mistake to maintain or subsidize a city that has grown too large for its life support; however, some wealthy countries do subsidize cities with, for example, federal monies, high taxes, and expensive fossil fuel imports.

9 Home Range and Territoriality

Statement

The forces isolating or spacing individuals, pairs, or small groups in a population are perhaps not as widespread as those favoring aggregation, but these forces are nevertheless important for enhancing fitness and as a mechanism for regulating populations. Isolation usually is the result of (1) competition between individuals for resources in short supply; or (2) direct antagonism, involving behavioral responses in higher animals and chemical isolating mechanisms (antibiotics and allelopathics) in plants, microorganisms, and lower animals. In both cases, either a random or a uniform distribution may result, as outlined in Section 7 (Fig. 6-22), because close neighbors are eliminated or driven away. Individuals, pairs, or family groups of vertebrates and higher invertebrates commonly restrict their activities to a definite area termed the **home range.** If this area is actively defended, so that there is little or no overlap of space used by the antagonistic individuals, pairs, and so on, it is termed a *territory. Territoriality* seems to be most pronounced in vertebrates and in certain arthropods that have complicated reproductive behavior patterns involving nest building, egg laying, and care and protection of young.

Explanation and Examples

Just as aggregation may increase competition but has compensating advantages, so the spacing of individuals in a population may reduce the competition for the necessities of life or provide the privacy necessary for complex reproductive cycles (as in birds and mammals) but comes at the expense of losing the advantages of cooperative group action. Presumably, the pattern that survives through evolution in a particular case depends on which alternative provides the greatest long-term survivorship advantage. In any event, both patterns are frequent in nature; in fact, some species populations alternate from one to the other. Robins (*Turdus migratorius*), for example, isolate into territories during the breeding season but aggregate into flocks in the winter, thus obtaining advantages from both arrangements. Again, different ages and sexes may show opposite patterns at the same time (adults isolated, young aggregated, for instance).

The role of intraspecific competition and "chemical warfare" in bringing about spacing in forest trees and desert shrubs has already been noted in Sections 6 and 7. Isolating mechanisms of this sort are widespread among higher plants. Many animals isolate themselves and restrict their major activities to definite areas or home ranges, which may vary from a few square meters to many square hectares. As home ranges often overlap, only partial spacing is achieved; territoriality achieves the ultimate in spacing. Figure 6-25 compares the home ranges of meadow voles (*Microtus pennsylvanicus*), which overlap (are not defended), with territories of song thrushes (*Turdus philomelos*), which do not overlap (are defended) and are reestablished in successive breeding seasons.

Home range size varies with the size of the animal, as would be expected. Home range size for the grizzly bear (*Ursus horribilis*), for example, was estimated to average 337,000 ha, whereas the home range size for the deer mouse (*Peromyscus maniculatus*) was found to be less than one hectare. See Harris (1984) for a table comparing home range sizes for small and large mammalian species.

The term *territory,* as defined in this section, was first introduced by Elliot Howard in his book *Territory in Bird Life,* published in 1920. Most of the early literature on the subject has dealt with birds. However, the concept of territoriality is now widely recognized for other vertebrates and some arthropods, especially among species in which a parent or parents guard nests and young. **Territory** is defined as that area of the habitat defended by individuals of a particular species—often a breeding pair of individuals—against other members of the same species. **Territoriality**—

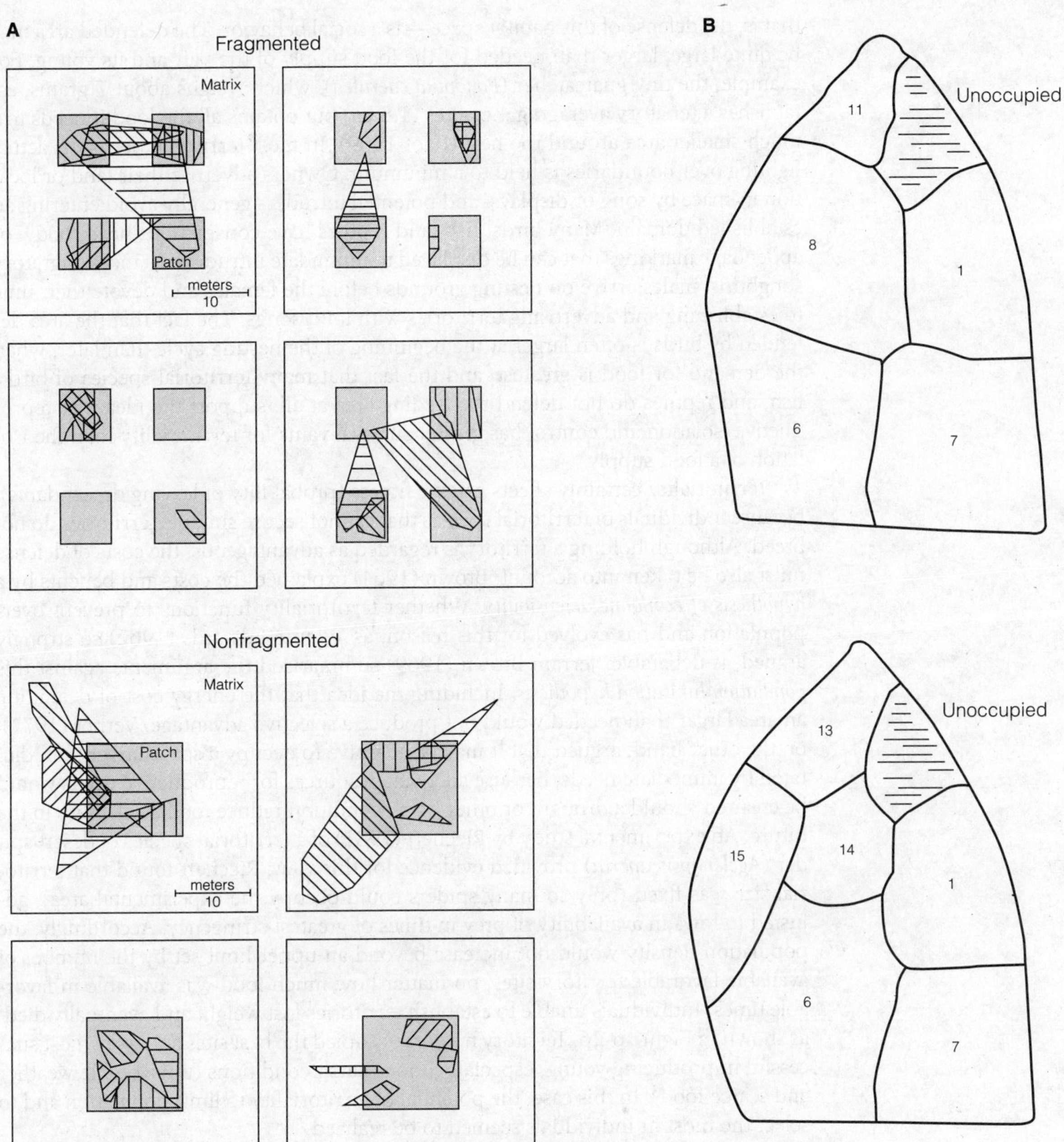

Figure 6-25. (A) Home ranges of meadow voles (*Microtus pennsylvanicus*) in fragmented and nonfragmented habitat patches (after R. J. Collins and Barrett 1997). (B) Territories of song thrushes (*Turdus philomelos*) in two consecutive years (after Lack 1966). Note that individuals 1, 6, and 7 maintained the same territories both years, whereas two individuals holding territories in 1955 failed to return and were replaced by three new individuals.

that is, the defense of this habitat space—is a social behavior. The defended area may be quite large, larger than needed for the food supply of the pair and its young. For example, the tiny gnatcatcher (*Polioptila caerulea*), which weighs about 7 grams, establishes a territory averaging 4.6 acres (1.8 ha) but obtains all the food it needs in a much smaller area around the nest (Root 1969). In most territorial behavior, actual fighting over boundaries is held to a minimum. Owners advertise their land or location in space by song or displays, and potential intruders generally avoid entering an established domain. Many birds, fish, and reptiles have conspicuous head, body, or appendage markings that can be displayed to intimidate intruders. In most migratory songbirds, males arrive on nesting grounds before the females and devote their time to establishing and advertising territories with loud songs. The fact that the area defended by birds is often larger at the beginning of the nesting cycle than later, when the demand for food is greatest, and the fact that many territorial species of birds, fish, and reptiles do not defend the feeding area at all, support the idea that reproductive isolation and control has greater survival value for territoriality than the isolation of a food supply.

Territoriality certainly affects genetic fitness (probability of leaving descendants), because individuals of territorial species that cannot secure suitable territories do not breed. Although holding a territory is regarded as advantageous, the costs of defense must also be taken into account. Brown (1964) explained the costs and benefits by a *hypothesis of economic defensibility*. Whether territoriality functions to prevent overpopulation and has evolved for this reason, as Wynne-Edwards (1962) so strongly argued, is debatable. Jerram Brown (1969) summarized the arguments against this *population limitation hypothesis*, including the idea that the energy cost of defending an area larger than needed would not produce a selective advantage. Verner (1977), on the other hand, argued that it may be adaptive to occupy a space larger than dictated by immediate needs, because adequate resources for reproductive needs would be ensured should a drought or other harsh condition reduce food availability in the future. An experimental study by Riechert (1981) of a territorial species of desert spider (*Agelenopsis aperta*) provided evidence for this view. Riechert found that territorial size was fixed (only so many spiders could occupy the experimental area), adjusted to lows in availability of prey in times of greatest stringency. Accordingly, the population density would not increase beyond an upper limit set by the number of available favorable territory sites, no matter how much food was available in favorable times. Individuals unable to establish territories lost weight and eventually died, as shown in Figure 6-26. Territory holders occupied the best sites and were most successful in producing young, especially under harsh conditions (unfavorable weather and scarce food). In this case, the potential of territoriality to limit population and to select the most fit individuals seemed to be realized.

In migratory songbirds, males usually arrive in the breeding habitat before females in the spring. Loud singing functions to establish a territory to attract mates. Individuals able to establish a territory have no difficulty attracting a mate; individuals unable to establish a territory (*floaters;* see Fig. 6-26) do not breed. Some other functions that have been suggested for territoriality include avoidance of predation or disease through the spacing of individuals and favorable allocation and preservation of resources.

The extent to which humans are territorial by virtue of inherent behavior and the extent to which they can learn land-use control and planning as safeguards against overpopulation are intriguing questions. Certainly, there are some territorial aspects

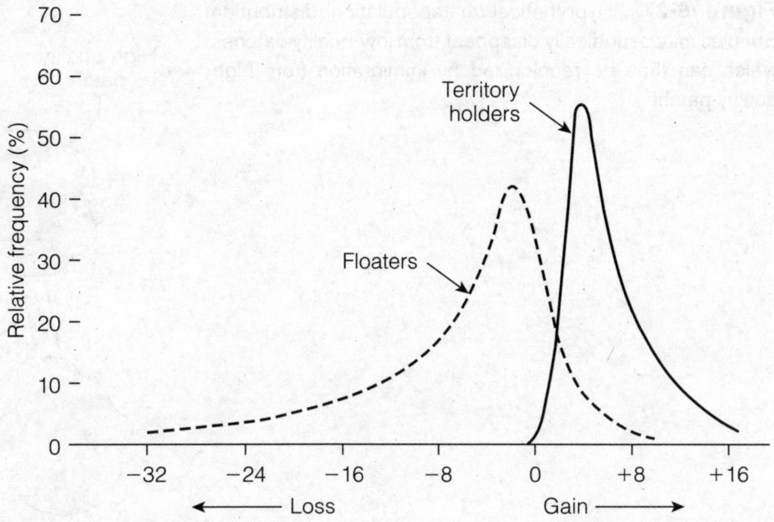

Figure 6-26. Fitness in terms of body weight gained or lost daily of territory-holding spiders compared with individuals unable to establish and hold territories (floaters). In unfavorable seasons, floaters are at an even greater disadvantage; few survive to produce offspring (redrawn from Riechert 1981).

to human behavior, such as the private property imperative and the laws and customs that regard the home as a safe house to be defended against intruders, with weapons if necessary. In a book entitled *The Territorial Imperative* (1967), Robert Ardrey argued optimistically that humans are inherently territorial and will eventually resist crowding and thereby avoid the doomsday of overpopulation. To date, however, there are few data to support this hypothesis.

10 Metapopulation Dynamics

Statement

The *metapopulation* is a level between the organism and the population levels of organization in the ecological hierarchy (see Fig. 1-3). **Metapopulations** may be defined as subpopulations occupying discrete patches or "islands" of suitable habitat that are separated by unsuitable habitat but connected by dispersal corridors. In naturally heterogeneous, fragmented landscapes (especially human-dominated landscapes), groups of individuals in each discrete patch may go extinct at some point in time, but the patch may be recolonized by individuals from a nearby patch, provided there is a navigable corridor connecting the patches. If colonization and extinction balance over a large area of landscape, the total population size may remain about the same. Accordingly, the survival of the species may depend more on dispersal (the ability to migrate from one patch to another) than on births and deaths within the patch.

Explanation and Examples

Figure 6-27 depicts a hypothetical metapopulation in a patchy environment with corridors between patches. If the species fails to reproduce in a patch, especially in a low-quality one, then the metapopulation may survive by receiving immigrants

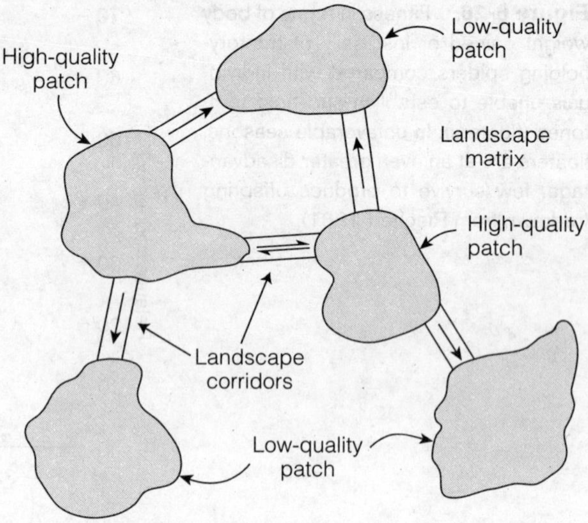

Figure 6-27. Hypothetical metapopulation distribution. Species may periodically disappear from low-quality patches, which can then be recolonized by immigration from high-quality patches.

from a high-quality patch. The metapopulation concept may be considered an aspect of the broader "source-sink" concept (discussed in Chapter 3). The metapopulation concept was first introduced by Levins (1969) and extensively updated by Hanski (1989).

As described by Verboom et al. (1991) the European nuthatch (*Sitta europaea*) persists as a metapopulation because it occurs in isolated subpopulations connected by dispersal corridors. The overall density of the species in its range is much less variable than in any subpopulation.

In another example, Gonzalez et al. (1998) created an experimental metapopulation microcosm on a large, moss-covered rock outcrop by scraping bare the rock to create small patches. Within these isolated patches, a decline in species richness of microarthropods was observed, but when narrow corridors were left connecting the patches, much of the species richness was retained.

The metapopulation concept, along with the theory of island biogeography to be discussed in Chapter 9, provides models not only for the conservation of endangered species but also for wildlife management in general. As with any management strategy, there can be shortcomings and downsides. Too much exchange via conservation corridors can synchronize fluctuations and increase the spread of disease and exotic pests (Simberloff and Cox 1987; Earn et al. 2000). See books by McCullough (1996) and Hanski and Gilpin (1997) for additional readings and examples of metapopulation dynamics as related to wildlife conservation and population genetics, respectively.

11 Energy Partitioning and Optimization: *r*- and *K*-Selection

Statement

Paralleling the partitioning of energy between *P* (production) and *R* (respiration or maintenance) and the concept of net energy for an ecosystem as a whole (discussed in Chapter 3), individual organisms and their populations can grow or reproduce only if they can acquire more energy than is needed for maintenance. **Maintenance energy** consists of the resting or basal rate of metabolism plus a multiple of this to cover minimum activity needed for survival under field conditions. Such *existence energy* must be estimated by time-energy observations in the field, because it varies widely according to whether a species is sedentary or active. The **net energy** required for reproduction and, therefore, for the survival of future generations, entails energy devoted to reproductive structures, mating activities, production of offspring (seeds, eggs, young), and parental care. Through natural selection, organisms achieve as favorable a benefit-cost ratio of energy input minus energy costs of maintenance as possible. For autotrophs, this efficiency involves usable light (convertible to food) minus the energy required to maintain energy-capturing structures (leaves, for example) as a function of the time that light energy is available. For animals, the critical factor is the ratio of usable energy in food minus the energy cost of searching for and feeding on food items. Optimization can be achieved in two basic ways: (1) by minimizing time (by efficient searching or conversion, for example); or (2) by maximizing net energy (by selecting large food items or easily convertible energy sources, for example). Most optimization models indicate that the lower the absolute abundance of food (or other energy sources), the larger the habitat area foraged and the greater the range of food items that should be taken to optimize benefit-cost ratios. However, extrinsic factors such as competition or cooperation with other species can alter this trend.

The ratio of reproductive energy to maintenance energy varies not only with the size of organisms and with life history patterns, but also with population density and carrying capacity. In uncrowded environments, selection pressure favors species with a high reproductive potential (high ratio of reproductive to maintenance effort). In contrast, crowded conditions favor organisms with lower growth potential but better capabilities for using and competing for scarce resources (greater energy investment in the maintenance and survival of the individual). These two modes are known as ***r*-selection** and ***K*-selection**, respectively (and species exhibiting them are designated *r*- and *K*-strategists), based on the *r* and *K* constants in growth equations (described in Section 4).

Explanation

Partitioning or **allocation of energy** among the various activities of an organism reflects balances between the advantages and costs of each activity in producing a change in r_{max}, the intrinsic (genetically determined) rate of increase, to enhance survivorship or fitness. The first consideration, of course, is survival and maintenance of the individual (the *respiratory component*) with additional energy allocated to growth and reproduction (the *production component*). Large organisms, like large cities, must allocate a larger portion of their metabolized energy input to maintenance than small organisms, which do not have so much structure to maintain. Natural selection, that uncompromising master forcing function, requires that all organisms find an optimum balance between the energy spent on future survival and the energy spent on survival in the present.

Figure 6-28 shows four hypothetical allocations of net energy between three major activities: (1) energy expended to cope with competition from other species striving for the same resources; (2) energy expended to avoid being eaten (or grazed) by a predator; and (3) energy expended to produce offspring. When competition and predation have a low impact, a large part of the energy flow may go to reproduction

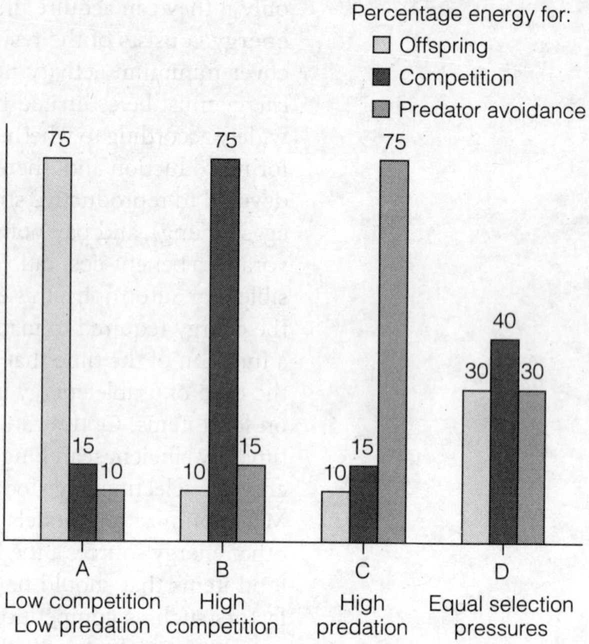

Figure 6-28. Hypothetical allocations of energy to three major activities necessary for survival in four contrasting situations (A–D) where the relative importance of each activity varies (modified from Cody 1966).

and the production of offspring (Fig. 6-28A). Alternatively, competition or antipredator activities may take most of the available energy (Fig. 6-28B and C, respectively). All three demands receive approximately equal allocations in the last example (Fig. 6-28D). Examples A, B, C, and D can represent four different species, or four different communities where selection pressure produces the illustrated pattern in many species. As will be seen in Chapter 8, example A represents a common situation in pioneer or colonizing stages of succession, where *r*-selection predominates, whereas examples B through D are likely patterns in more mature stages, where *K*-selection may predominate.

Schoener (1971), Cody (1974), Pyke et al. (1977), and Stephens and Krebs (1986), in reviewing how energy partitioning and optimization can be analyzed (to determine optimal feeding or foraging strategies), suggest that the problem is analogous to cost-benefit analysis in economics, with the benefit being increased fitness and the costs being energy and time required to ensure future reproductive output. **Optimal foraging** is defined as the maximum possible energy return under a given set of foraging and habitat conditions. A predator, for example, is under selective pressure to increase the ratio between usable energy minus energy cost of obtaining prey and time required to search, pursue, and consume the prey. Increasing the energy available for reproduction can, in theory, be accomplished by (1) selecting larger or more nutritious prey or prey that is easier to catch; or (2) reducing search and pursuit time and effort.

An approach to the optimum partitioning of energy (cost-benefit analysis) involves graphic "strategic analysis," as illustrated in Figure 6-29. A and B are foraging strategy models for a hypothetical species faced with the problem of how many of six potential food items to use (A) or how many isolated feeding areas or "patches" to forage in (B). If only one food item among many available is sought, a greater search effort per item is required, compared with the search effort for feeding indiscriminately on all six items, as shown by the ΔS curve in Figure 6-29A. More pursuit effort is required as difficult-to-catch or smaller prey are sought, as shown by the ΔP curve. In the hypothetical case of Figure 6-29A, the optimum cost-benefit balance comes when the decreasing and increasing trend curves cross at four prey items. Interactions with other species or other environmental factors can shift the optimum in either direction. Competition with other species can force this hypothetical animal to become a *specialist* and feed on only one food item if it has a competitive advantage. Or such selection could be advantageous when food is abundant. Then again, conditions could dictate that becoming a *generalist* is a favorable strategy.

In Figure 6-29B, hunting effort (ΔH) increases as more feeding areas are explored, but this is balanced by a decrease in traveling time per unit food caught (ΔT). Again, the optimum is a compromise between opposing trends—three patches foraged in the situation graphed.

Figure 6-29C is a more general model in which reproductive output per unit energy decreases and nonreproductive output increases (both monotonically) with the time spent in procurement of energy (feeding). The shaded area represents the region where reproductive output is maximized, and optimum feeding time, again, is where the curves cross, providing a favorable balance between the two necessary allocations of energy.

As noted in the Statement, species with a high biotic potential (r) tend to be favored in uncrowded or uncertain environments that are subject to periodic stresses (such as storms or droughts). Species with partition energy in favor of maintenance and enhanced competitive ability do better under K (saturation) densities or stable physical factors (low probability of severe disturbances) and in mature or climax stages of ecological succession (see Chapter 8). To put it another way, species exhibiting the J-shaped population growth form are effective pioneers that can quickly exploit unused or recently accumulated resources, and they are resilient to perturbations. Slower-growing species and populations are better adapted to mature com-

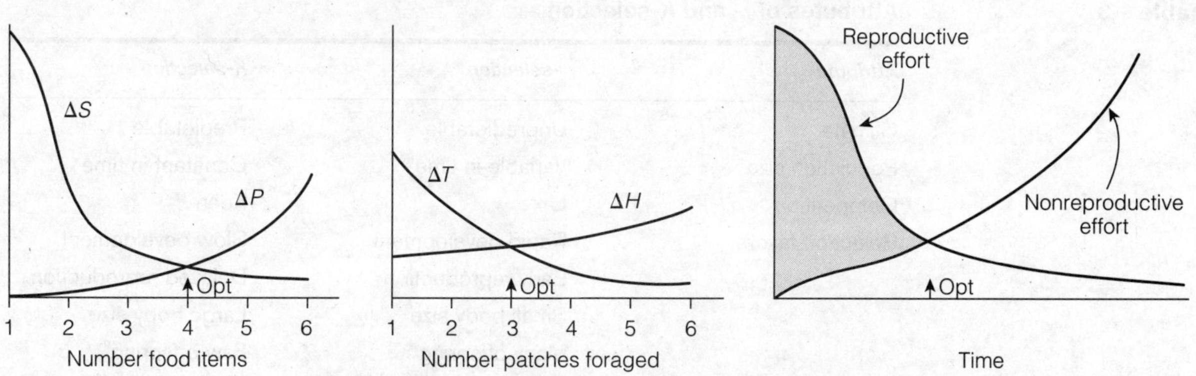

Figure 6-29. Optimization cost-benefit models. (A) Balancing use of food sources. ΔS = energy expended in searching for a preferred food item; ΔP = energy expended in pursuing a particular food item. (B) Balancing use of foraging areas. ΔT = energy expended on traveling between catches; ΔH = energy expended on hunting. (C) Balancing time spent on reproduction and feeding.

munities and are more resistant but less resilient to perturbations (recall the discussion of resistant versus resilient stability in Chapter 2). Table 6-3 summarizes the attributes of *r*- and *K*-selected species.

A general model for *r*- and *K*-selection, as proposed by MacArthur (1972), is shown in Figure 6-30. Although X_1 and X_2 in the diagram were designated as two competing genetic alleles, they can also represent competing species. In region A (to the left of point C), where density is low and food (or sunlight and nutrients, in the case of plants) is abundant, the faster-growing species or allele X_2 wins out; there is *r*-selection. In region B (to the right of point C), species X_1 is growing faster than X_2 and thus wins out; there is *K*-selection. MacArthur noted that *K*-selection prevails in the relatively nonseasonal Tropics, whereas *r*-selection prevails in the seasonal environments of the North Temperate Zone, where population growth is marked by exponential growth followed by catastrophic declines during the winter months.

Clutch size (number of eggs or young per reproductive period) in birds seems not only to reflect mortality and survivorship but also to mirror *r*- and *K*-selection. Opportunistic birds (*r*-strategists) have a larger clutch size than do equilibrium species, as do temperate birds compared with tropical ones.

The *r*- and *K*-strategist designations can be faulted as an oversimplified classification, because many populations have variable or intermediate modes. However, Pianka (1970) found an apparent bimodality in relatively *r*- and relatively *K*-selected organisms in nature related to body sizes and generation times. He argued that "an either/or strategy is usually superior to some kind of compromise."

In his book *Evolution in Changing Environments,* Levins (1968) concluded that environmental uncertainty limits specialization in the evolution of species. Under unstable conditions, for purposes of selection, it is favorable to be a generalist as well as to have a high r_{max}. Also, under such conditions, communities can be only very loosely organized. Specialization and organization can increase to higher levels only if the unpredictability of the environment is low. To what extent can groups of populations and communities, by their concerted action, reduce environmental uncer-

Table 6-3 **Attributes of *r*- and *K*-selection**

Attribute	*r-selection*	*K-selection*
Climate	Unpredictable	Predictable
Population size	Variable in time	Constant in time
Competition	Lax	Keen
Selection favors	Rapid development	Slow development
	Early reproduction	Delayed reproduction
	Small body size	Large body size
	Many offspring	Few offspring
Length of life	Short (<1 year)	Long (>1 year)
Stage in succession	Early	Late (climax)
Leads to	Productivity	Efficiency

Source: Modified after Pianka 1970, 2000.

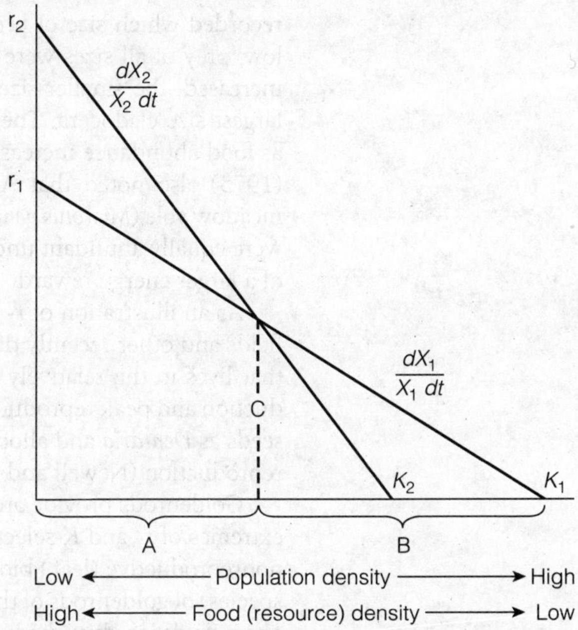

Figure 6-30. MacArthur's (1972) model of *r*- and *K*-selection. Rates of increase of two alleles (or species), X_1 and X_2, are portrayed as functions of population density and resource density. A represents region of *r*-selection, and B region of *K*-selection.

tainty and thereby open the way for organization to proceed to a higher level, as sometimes happens with human societies? That question remains to be answered.

Examples

Table 6-4 compares the allocation of assimilated energy between production, *P* (growth and reproduction), and respiration, *R* (maintenance), in six species representing both predators and herbivores and both vertebrates and invertebrates. In general, predators (marsh wrens, red foxes, and hunting spiders) allocate more of their assimilated energy to maintenance (foraging for food, defending territories, and so on) than do herbivores (cotton rats and pea aphids). Large homeotherms (warm-blooded vertebrates) likewise allocate a greater percentage of assimilated energy to *R* than do small poikilotherms (arthropods).

Comparison of hunting and web-building spiders provides an interesting example of energy partitioning. Because the web has a high protein content, silk formation comes at a high cost in energy, but many spiders recycle the silk by eating it as they rebuild the web, thereby cutting the cost. Peakall and Witt (1976) estimated that silk production in an orbweaver spider, which recycles its web, requires only about one fourth of the total maintenance calories of building the web and keeping it in repair. The total energy cost of the web is about one half of the basal energy consumption, which is less than the energy expended in hunting by some non–web builders. This is a possible lesson for humans: the species that builds expensive, labor-saving devices can reduce energy costs by recycling the materials.

The theory that predators optimize energy cost-benefit by varying the selection of the size of prey according to overall prey abundance has been tested and verified experimentally by Werner and Hall (1974). These investigators presented bluegill sunfish with different combinations of sizes and numbers of cladoceran prey and recorded which size of prey was selected. When the absolute food abundance was low, prey of all sizes were eaten as encountered. When the abundance of prey was increased, the smaller-size classes were ignored, and the fish concentrated on the largest-size cladocera. The fish thus switched from feeding generalists to specialists as food abundance increased (and vice versa when it declined). Barrett and Mackey (1975) also noted that American kestrels (*Falco sparverius*) initially selected the meadow vole (*Microtus pennsylvanicus*) even when deer mice (*Peromyscus maniculatus*) were equally abundant under seminatural aviary conditions, resulting in the capture of a larger energy reward.

As an illustration of *r*- and *K*-selection, ragweed (*Ambrosia*), which grows in old fields and other recently disturbed places, and *Dentaria laciniata*, a herbaceous plant that lives in the relatively stable forest floor, were compared in terms of seed production and peak reproductive effort. The ragweed produced about 50 times as many seeds as *Dentaria* and allocated a much larger percentage of its assimilated energy to reproduction (Newell and Tramer 1978).

Goldenrods provide an example of a range of reproductive strategies between the extremes of *r*- and *K*-selection. In Figure 6-31, reproductive effort is plotted against nonreproductive (leaf) biomass accumulation for six populations (representing four species) of goldenrods of the genus *Solidago*. Population 1, a species that grows in dry, open fields or disturbed sites, maintains a low leaf biomass and allocates about 45 percent of net production to reproductive tissues. In contrast, population 6, which occurs in moist hardwood forests, puts more of its energy into leaves, with only 5 percent allocated to reproduction. The other populations occur in habitats intermediate in moisture and stability and have corresponding intermediate allocations.

Table 6-4 **Allocation of assimilated energy between production (*P*; growth and reproduction) and respiration (*R*; maintenance)**

Trophic level	*Percentage of assimilated energy to production (P)*	*Percentage of assimilated energy to respiration (R)*
Primary Consumer		
Cotton rat (herbivore)	13	87
Secondary Consumers		
Marsh wren (insectivore)	1	99
Red fox (carnivore)	4	96
Raccoon (omnivore)	4	96
Poikilothermic Arthropods		
Pea aphid (herbivore)	58	42
Wolf spider (predator)	25	75

Source: Data after Kale 1965; Vogtsberger and Barrett 1973; Randolph et al. 1975, 1977; Humphreys 1978; and Teubner and Barrett 1983.

Figure 6-31. Reproductive effort (ratio of dry weight of reproductive tissues to total dry weight of aboveground tissue) plotted against nonreproductive biomass (ratio of leaf weight to total weight) in six populations of four species of goldenrods (*Solidago*). Population 1 is a species that occurs in dry, open fields or disturbed soil, whereas population 6 occurs in moist hardwood forests; the other populations occur in habitats intermediate in moisture and stability (after Abrahamson and Gadgil 1973).

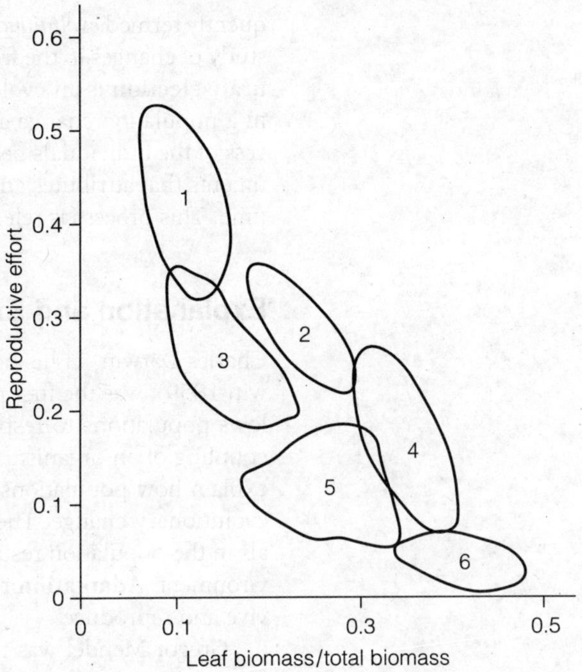

Solbrig (1971) noted that *r*- and *K*-strategists can be found within the same species. The common dandelion (*Taraxacum officinale*), for example, has several strains or variations that differ in the mix of genotypes controlling allocation of energy. One strain grows primarily in disturbed areas and produces more but smaller seeds that ripen earlier in the season, compared with another strain found in a less disturbed area that allocates more energy to leaves and stems and produces fewer seeds that ripen late. The latter strain shades out the more fecund variety when the two are grown together in good soil. Thus, strain 1 is a more effective colonizer of new ground and qualifies as an *r*-strategist; strain 2 is a more effective competitor or *K*-strategist.

Although uncertain or disturbed environments do favor *r*-selection, the *K*-strategist is by no means ruled out. For example, in fire-adapted communities such as the Chaparral of California (see Chapter 5), "resprout" plant species that allocate large energy reserves to underground parts are as well or better adapted to survive periodic fires as are plants that hold their future in seeds. For other examples of energy partitioning from the plant kingdom, vegetation processes, and plant strategies, see Grime (1977, 1979).

12 Population Genetics

Statement

An understanding of population genetics and natural selection is necessary in order to understand how populations evolve and how communities and ecosystems change over time. Population genetics and natural selection underpin the area of study frequently termed *evolutionary biology* or *evolutionary ecology.* **Population genetics** is the study of changes in the frequencies of genes and genotypes within a population. **Natural selection** is an evolutionary process by which the frequencies of genetic traits in a population change as a result of the differential survival and reproductive success of the individuals bearing those traits. The historical record of life on Earth documents that attributes and traits of organisms, populations, and species change over time. This process is referred to as **evolution.**

Explanation and Examples

Charles Darwin, in his book *The Origin of Species by Means of Natural Selection* (Darwin 1859), was the first person to document that the process of natural selection allows populations to respond to changes in the environment, resulting in the close coupling of an organism with its natural environment. Population genetics helps to explain how populations, and consequently communities and ecosystems, undergo evolutionary change. The environment acting on genetic variation among individuals in the population results in the adaptation of the population or species to the environment. **Adaptation** refers to traits of an organism that increase its fitness to survive and reproduce.

Gregor Mendel was the first person to recognize that characteristics are passed from parents to offspring in packets of information that we now term **genes.** Johann Mendel, renamed Gregor when he joined the Augustinian order of monks, was the oldest child of a farmer's family located near Brno, in what is now the Czech Republic. Mendel's early studies included a firm background in the sciences, thanks mainly to the Countess Walpurga Truchsess-Zeil, who ruled the district where Mendel's family lived. As Charles Darwin explored the Galápagos Islands, Mendel, because of his excellent training in mathematics, developed powerful experimental approaches to investigate the natural world. Mendel determined that genes come in alternative forms (**alleles**) that result in variation among genotypes and phenotypes and differences between homozygous and heterozygous genotypes. Some alleles in a population are *dominant,* and the alleles that they suppress are termed *recessive.*

For example, Figure 6-32 shows a picture of an agouti (pigmented phenotype) and an albino (nonpigmented phenotype) meadow vole (*Microtus pennsylvanicus*). The agouti carries the dominant phenotype allele (AA), whereas the albino carries the recessive phenotype allele (aa). Individuals resulting from several generations of breeding confirmed that albinism was inherited as an autosomal recessive trait (Brewer et al. 1993). These investigations hypothesized that albinism would be disadvantageous in old-field and grassland communities, where conspicuousness of coat color would likely result in increased rates of predation. Peles et al. (1995), however, found no significant differences in population densities or rates of recruitment between coat-color (agouti versus albino) treatments conducted in natural old-field experimental plots. The lack of increased rates of predation of albino voles was attributed to high nutritional quality and heavy vegetative cover. In fact, when voles were removed from these experimental enclosures during early winter, more albino than agouti individuals were captured. This example illustrates the close relationship of population genetics to habitat quality (Peles and Barrett 1996).

Building on the preceding example, the proportion of gametes carrying alleles A and a is determined by the individual *genotypes* (the genes received from the parents). Because eggs and sperm generally unite at random, the proportion of offspring of dif-

Figure 6-32. Agouti (left) and albino (right) variants of meadow voles (*Microtus pennsylvanicus*). Meadow voles are herbivorous small mammals that inhabit grassland and old-field communities.

Courtesy of Gary W. Barrett

ferent genotypes can be predicted based on parental genotypes. Offspring (F_1 generation) of a dominant (AA) agouti and a recessive (aa) albino will consist of .25 AA, .50 Aa, and .25 aa. The proportions are termed the **genotype frequencies.** Because the Aa genotype will be expressed phenotypically as agouti (because A is the dominant allele), a 3/1 (agouti/albino) phenotype ratio is predicted.

Will genotype frequencies change through successive generations in a bisexual population? The Hardy-Weinberg equilibrium law helps to identify evolutionary or environmental forces that can change gene frequencies in populations. The **Hardy-Weinberg equilibrium law** is stated as follows: if p is the frequency of allele A (the dominant), and q is the frequency of allele a (the recessive), so that $p + q = 1$, then the genotype frequencies will be $p^2 + 2pq + q^2 = 1$, where p^2 is the frequency of the homozygous individuals (AA), q^2 is the frequency of the homozygous individuals (aa), and $2pq$ is the frequency of the heterozygous individuals (Aa). In the case of the meadow vole population discussed previously, the proportions of the genotypes in the F_1 generation will be $(0.5)^2 + 2(0.5 + 0.5) + (0.5)^2$. The same genotype frequencies will be maintained in the F_2 generation if the conditions of the Hardy-Weinberg law hold. These conditions are (1) mating is random; (2) new mutations do not occur; (3) there is no gene flow from one population to another; (4) no natural selection occurs; and (5) the population size is large. As these conditions are seldom met in reality, what is the value of the Hardy-Weinberg law? Any departure from one or more of these assumptions serves as a null hypothesis against which to test the departure of frequencies away from the Hardy-Weinberg equilibrium model. Changes in allele frequencies due to random variation in allele frequencies in a population over time is termed **genetic drift.** Genetic drift reduces genetic variation in populations by increasing the frequency of some alleles and by reducing the frequency of other alleles.

Genetic drift has more pronounced effects in small populations than in large ones. Population geneticists use the term **effective population size** (N_e) to describe the effects of numbers on drift. **Inbreeding,** defined as mating among close relatives, is frequently the result of small population size. The major genetic effect of inbreeding is an increase in *homozygosity* (a decrease in genetic variation). For example, there is concern that the Florida panther (*Felis concolor coryi*) has too few individuals for an adequate gene pool. Because of generations of inbreeding, 90 percent of the panther sperm is abnormal, which could lead to its extinction (Perry and Perry 1994). Figure 6-33 is a photograph of this magnificent species.

Social behavior is also based on genetic variation, resulting in differential rates of reproduction among lineages of closely related individuals. **Altruistic behaviors** are social behaviors that enhance the fitness of other individuals in the population at the apparent expense of the individual performing them. For example, an alarm call is a behavior by an individual indicating the presence of a predator in the area; the fitness of the neighboring individuals increases, because these individuals have a better chance to escape, whereas the caller's fitness decreases if the alarm call attracts the attention of the predator.

The most widespread examples of altruism are eusociality and cooperative breeding. **Eusociality** is characterized by cooperative caring for the young, division of labor, and an overlap of at least two generations of life stages functioning to contribute to colony or group labor. Examples are colonies of honey bees, ants, and termites. Figure 6-34 shows a large termite mound located near Darwin, Australia. In constructing their huge, complex mounds, termites move considerable amounts of soil and detritus—a task requiring coordinated group labor.

Altruistic behavior in an individual within an extended family that influences the fitness of an individual with which it shares more genes than it does with an individual at random is referred to as **kin selection.** Kin selection is the evolution of a genetic trait expressed by an individual that affects the behavior and genetic fitness of one or more closely related individuals. Kin selection is favored when the increase in fitness of closely related individuals in the population is great enough to compensate for the loss in fitness by the altruistic individual. The Florida jay (*Aphelocoma coerulescens*) provides an example of an altruistic cooperative breeding unit formed through the retention of mature offspring. The Florida jay family (Fig. 6-35) is a monogamous

Courtesy of the Florida Fish and Wildlife Conservation Commission

Figure 6-33. The Florida panther (*Felis concolor coryi*) inhabits the forests and swamps of the southeastern United States.

Figure 6-34. Termite (*Copotermes ascinaciformis*) mound located near Darwin, Australia. Termite colonies may contain over 3 million individuals; termites release over 150 million tons of methane into the atmosphere annually.

Figure 6-35. The Florida scrub jay (*Apelocoma coerulescens*) inhabits the thickets of sand pine and scrub oak along the east and west coasts of Florida.

breeding pair along with some of its offspring, which functions as a unit during the breeding season, providing food for nestlings, guarding the nest, protecting fledglings out of the nest, and defending territory (McGowan and Woolfenden 1989). An individual's own reproductive success *plus* the increased fitness of its relatives, weighted according to the degree of the relationship, is termed **inclusive fitness** (W. D. Hamilton 1964; Smith 1964).

When an individual helps a relative with which it shares many genes, and the reproductive success of the relative increases as a result of the help, that additional reproduction is counted as part of the inclusive fitness of the organism providing the help. Thus, the inclusive fitness of an individual represents its own reproductive success plus a portion of that of close relatives who receive the benefits of its altruistic behavior. This concept explains many aspects of social behavior.

In summary, natural selection, expressed by changes in genotypic and phenotypic frequencies in populations, is a mechanism of adaptation to the environment. The basis of adaptation to the local environment is the genetic variation of individuals in the population. Sources of variation are embedded in genes—specifically, in

DNA molecules. Major sources of genetic variation are the reproductive recombination of genes provided by parents in bisexual populations, and inheritable mutations in the gene or chromosome. Natural selection, acting on this genetic variation, results in increased fitness within the natural environment. **Fitness** is usually measured as the total lifetime reproductive success of an individual. Thus, the direction that change (evolution) takes depends on the genetic structure of those individuals that survive and leave behind reproducing progeny.

Because of habitat fragmentation due to human intervention in the landscape (see Chapter 9), the populations of many species of plants, animals, and microbes are being reduced to small, frequently isolated, populations. These small populations carry only a fraction of the genetic variation of the total population or species—a situation that can increase rates of genetic drift, elicit inbreeding depression, and even lead to extinction. Changes in the genetics of populations are also reflected at the community, ecosystem, and landscape levels of organization. Interactions and evolution among species (coevolution) will be discussed in greater detail in Chapter 7.

13 Life History Traits and Tactics

Statement

Selection pressure resulting from the impact of physical environments and biotic interactions shapes patterns of life history so that each species evolves an adaptive combination of the population traits considered in previous sections of this chapter. Although each species' life history is unique, several basic life history tactics can be recognized, and the combination of traits that is characteristic to organisms living in specified circumstances can, to some extent, be predicted.

Explanation and Examples

Stearns (1976) listed four life history traits that are key to survival tactics: (1) brood size (number of seeds, eggs, young, or other progeny); (2) size of young (at birth, hatching, or germination); (3) age distribution of reproductive effort; and (4) interaction of reproductive effort with adult mortality (especially the ratio of juvenile to adult mortality). The following predictive theories have been summarized by Gadgil and Bossert (1970), Stearns (1976), Pianka (2000), and others:

1. Where adult mortality exceeds juvenile mortality, the species should reproduce only once in a lifetime and, conversely, where juvenile mortality is higher, the organism should reproduce several times.
2. Brood size should maximize the number of young surviving to maturity averaged over the lifetime of the parent. Thus, a ground-nesting bird may require a clutch size of 20 eggs to ensure replacement, whereas a bird nesting in a cavity or other protected place will have a much smaller clutch size.
3. In expanding populations (the growth segment of the population growth curve), selection should minimize age at maturity (r-selected organisms will breed at an early age); in stable populations (at carrying capacity or K level), maturation

should be delayed. This principle seems to hold for human populations; in fast-growing countries, childbirth begins at an early age, whereas in stable countries, on average, people postpone childbearing to a later age.

4. When there is risk of predation, scarcity of resources, or both, size at birth should be large; conversely, size of young should decrease with increasing availability of resources and decreasing predation or competition pressure.
5. For growing or expanding populations in general, not only is the age of maturity minimized and reproduction concentrated early in life, but also brood size should be increased and a large portion of energy flow partitioned to reproduction—a combination of traits recognizable as an *r*-selection tactic. For stable populations, one expects the reverse combination of traits, or *K*-selection.
6. When resources are not strongly limiting, breeding begins at an early age.
7. Complex life histories enable a species to exploit more than one habitat and niche.

Comparison of the floras of extreme deserts and moist tropical forests provides an example of how a particular basic life history trait can predominate throughout an ecosystem type. Annual plants predominate in extreme deserts, where the survival of a perennial plant would be very low because of long periods of drought. Conversely, perennial life histories are favored in the tropical rain forest, where intense competition and seed predation greatly reduce the survival of seedlings. This case can be considered an example of the first predictive theory.

W. P. Carson and Barrett (1988) and Brewer et al. (1994) noted that plant communities that were nutrient enriched for 11 years remained dominated by annuals such as *Ambrosia trifida, A. artemisiifolia,* and *Setaria faberi,* whereas unfertilized plots were dominated by perennials such as *Solidago canadensis, Trifolium pratense,* and *Aster pilosus.* The inverse relationship between nutrient enrichment and plant community biodiversity was documented in Chapter 3. These examples illustrate how plant life histories relate to changes in resource availability in different ecosystem types (see Tilman and Downing 1994, regarding the relationship of biodiversity to stability). Interest in life history strategies among many population ecologists began with a pioneer paper by LaMont Cole (1954) entitled "The Population Consequences of Life History Phenomena," which is recommended reading. Also recommended is Grime's (1979) description of strategies in plant life histories as related to ecological and evolutionary theory.

7

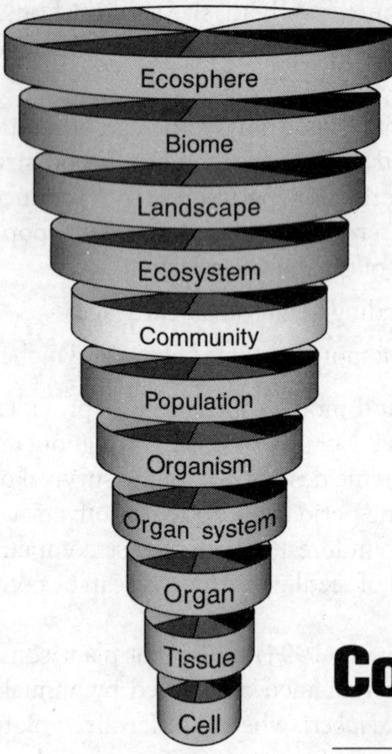

Community Ecology

1 Types of Interaction Between Two Species

2 Coevolution

3 Evolution of Cooperation: Group Selection

4 Interspecific Competition and Coexistence

5 Positive/Negative Interactions: Predation, Herbivory, Parasitism, and Allelopathy

6 Positive Interactions: Commensalism, Cooperation, and Mutualism

7 Concepts of Habitat, Ecological Niche, and Guild

8 Biodiversity

9 Paleoecology: Community Structure in Past Ages

10 From Populations and Communities to Ecosystems and Landscapes

1 Types of Interaction Between Two Species

Statement

Theoretically, populations of two species may interact in basic ways that correspond to combinations of neutral, positive, and negative (0, +, and −) as follows: 0 0, − −, + +, + 0, − 0, and + −. Three of these combinations (+ +, − −, and + −) are commonly subdivided, resulting in nine important interactions and relationships. The terms applied to these relationships in the ecological literature are as follows (see Table 7-1 and Fig. 7-1):

1. **neutralism,** in which neither population is affected by association with the other;
2. **competition, direct interference type,** in which both populations actively inhibit each other;
3. **competition, resource use type,** in which each population adversely affects the other indirectly in the struggle for resources in short supply;
4. **amensalism,** in which one population is inhibited and the other not affected;
5. **commensalism,** in which one population is benefited, but the other is not affected;
6. **parasitism;** and
7. **predation,** in which one population adversely affects the other by direct attack but nevertheless depends on the other;
8. **protocooperation** (also frequently referred to as *facultative cooperation*), in which both populations benefit by the association but their relations are not obligatory; and
9. **mutualism,** in which the growth and survival of both populations is benefited, and neither can survive under natural conditions without the other.

Three trends in the occurrence of these relationships are especially worthy of emphasis:

- Negative interactions tend to predominate in pioneer communities or in disturbed conditions where *r*-selection counteracts high mortality.
- In the evolution and development of ecosystems (succession), negative interactions tend to be minimized in favor of positive interactions that enhance the survival of the interacting species in mature or crowded communities.
- Recent or new associations are more likely to develop severe negative interactions than are older associations.

Explanation

The nine interactions listed in the Statement are analyzed in terms of a two-species relationship at the community level in Table 7-1, and a coordinate model of these interactions is displayed in Figure 7-1. All these population interactions are likely to occur in any large-scale biotic community, such as a large tract of forest, wetland, or grassland. For a given species pair, the type of interaction may change under different conditions or during successive stages in their life histories. Thus, two species might exhibit parasitism at one time, exhibit commensalism at another, and be completely neutral at still another time. Simplified communities (such as mesocosms) and

Table 7-1 **Analysis of two-species population interactions**

Type of interaction	*Species 1*	*Species 2*	*General nature of interaction*
Neutralism	0	0	Neither population affects the other
Competition, direct interference type	−	−	Direct inhibition of each species by the other
Competition, resource use type	−	−	Indirect inhibition when common resource is in short supply
Amensalism	−	0	Population 1 inhibited, 2 not affected
Commensalism	+	0	Population 1, the *commensal,* benefits, while 2, the *host,* is not affected
Parasitism	+	−	Population 1, the *parasite,* generally smaller than 2, the *host*
Predation (including herbivory)	+	−	Population 1, the *predator,* generally larger than 2, the *prey*
Protocooperation	+	+	Interaction favorable to both but not obligatory
Mutualism	+	+	Interaction favorable to both and obligatory

Note: 0 indicates no significant interaction; + indicates growth, survival, or other population attribute benefited (positive term added to growth equation); − indicates population growth or other attribute inhibited (negative term added to growth equation).

laboratory experiments allow ecologists to single out and quantitatively study the various interactions. Also, deductive mathematical models derived from such studies permit ecologists to analyze factors not ordinarily separable from the others.

Growth equation models make definitions more precise, clarify thinking, and allow a determination of how factors operate in complex natural situations. If the growth of one population can be described by an equation, such as the logistic equation, the influence of another population may be expressed by a term that modifies the growth of the first population. Various terms can be substituted according to the type of interaction. For example, in the case of competition, the growth rate of each population is equal to the unlimited rate minus its own self-crowding effects (which increase as its population increases) minus the detrimental effects of the competing species, N_2 (which also increase as the numbers of both species, N and N_2, increase), or

$$\frac{dN}{dt} = rN - \left(\frac{r}{K}\right)N^2 - CN_2N \quad \text{or}$$

Growth rate = Unlimited rate − Self-crowding effects − Detrimental effects of the other species,

C being a constant reflecting the efficiency of the other species.

This equation will be recognized as the logistic equation (see Chapter 6), except for the addition of the last term, "minus detrimental effects of the other species." There are several possible results for this kind of interaction. If the competitive efficiency, *C*, is small for both species, so that the interspecific depressing effects are less than the intraspecific (self-limiting) effects, the growth rate and perhaps the final density of both species will be depressed slightly; but both species will probably be able to live together, because the depressing interspecific effects will be less important than the competition within the species. Also, if the species exhibit exponential growth (with self-limiting factors absent from the equation), interspecific competition might provide the leveling-off function missing from the species' own growth form. However, if *C* is large, the species exerting the largest effect will eliminate its competitor or force it into another habitat. Thus, theoretically, species having similar requirements cannot live together because strong competition will likely develop, causing one of them to be eliminated. These models suggest some of the possibilities; how these possibilities actually work out will be discussed later in this chapter.

When both species of interacting populations have beneficial effects on each other instead of detrimental ones, a *positive* term is added to the growth equations. In such cases, both populations grow and prosper, reaching equilibrium levels that are mutually beneficial. If the beneficial effects of the other population (the positive term in the equation) are necessary for the growth and survival of both populations, the relation is known as *mutualism*. If, on the other hand, the beneficial effects only increase the size or growth rate of the population but are not necessary for growth or survival, the relationship comes under the heading of *protocooperation*. In both protocooperation and mutualism, the outcome is similar; the growth of either population is less or zero without the presence of the other population. When a balance is reached, the two populations pulse together, usually in a definite proportion.

Consideration of population interactions, as shown in Table 7-1 and Figure 7-1, or in terms of the growth equations, avoids the confusion that often results when terms and definitions alone are considered. Thus, the term *symbiosis* is sometimes used in the same sense as *mutualism*. Because **symbiosis** literally means "living together," the word is used in this book in its broad sense, without regard to the exact nature of the relationship. The term *parasitism* and the science of **parasitology** are generally con-

Figure 7-1. Coordinate model of two-species interactions.

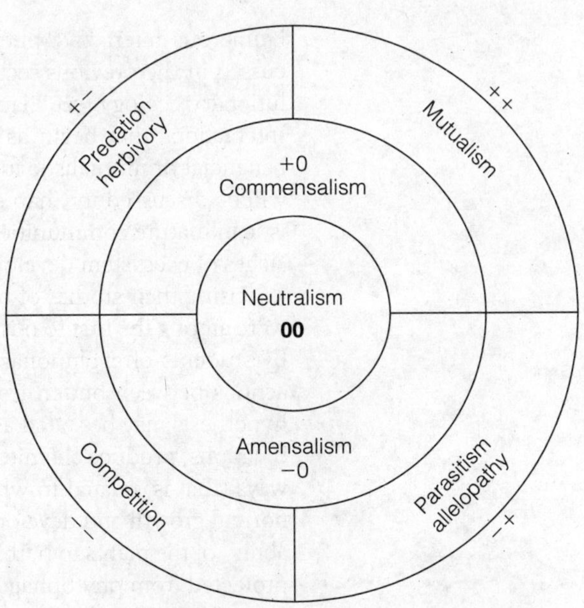

sidered to deal with any small organism that lives on or in another organism, regardless of whether its effect is negative, positive, or neutral. Various nouns have been proposed for the same type of interaction, adding to the confusion. When relations are diagrammed, however, there is little doubt about the type of interaction being considered; the word or label then becomes secondary to the mechanism and its result.

Note that the word "harmful" was not used in describing negative interactions. Competition and predation decrease the growth rate of affected populations, but this does not necessarily mean that the interaction is harmful either to long-term survival or by evolutionary considerations. In fact, negative interactions can increase the rate of natural selection, resulting in new adaptations. Predators and parasites often benefit populations that lack self-regulation, because they prevent overpopulation that otherwise might result in self-destruction.

2 Coevolution

Statement

Coevolution is a type of community evolution (an evolutionary interaction among organisms in which the exchange of genetic information among the kinds is minimal or absent). **Coevolution** is the joint evolution of two or more noninterbreeding species that have a close ecological relationship, such as plants and herbivores, large organisms and their microorganism symbionts, or parasites and their hosts. Through reciprocal selective pressures, the evolution of one species in the relationship depends in part on the evolution of the other.

Explanation

Numerous interactive phenomena occur among sets of interacting species as discussed in the previous section. Indeed, these interactions dominate the field of evolutionary ecology (see Pianka 2000 for a detailed description of this field of study). Interactions that begin as a competitive interaction between species may become beneficial or mutualistic interactions for both species through evolutionary time. As will be discussed in Chapter 8, species interactions appear to become more mutualistic in mature communities and ecosystems compared to young systems in the early stages of ecosystem development.

Using their studies of butterflies and plants as a basis, Ehrlich and Raven (1964) were among the first to outline the theory of coevolution as it is now widely accepted by students of evolutionary biology. This early seminal research focused on interactions between butterflies and the plants on which they feed. Ehrlich and Raven's hypothesis may be stated as follows: plants, through occasional mutations or recombinations, produce chemical compounds not directly related to basic metabolic pathways (that is, related to what is termed *secondary chemistry*) that are not inimical to normal growth and development. Some of these compounds either reduce the palatability of the plants to herbivores or are toxic to herbivores on ingestion. A plant thus protected from phytophagous insects would, in a sense, have entered a new adaptive zone. Evolutionary radiation of these plants might follow, and what began as a chance mutation or recombination might eventually characterize an entire family or group of related families. Phytophagous insects, however, can evolve in response to these physiological obstacles, as shown by the widespread development of immune strains. Indeed, the response to secondary plant substances and the evolution of resistance to insecticides seem to be intimately connected (see Palo and Robbins 1991, *Plant Defenses Against Mammalian Herbivory,* on how these compounds inhibit or reduce mammalian herbivory). If a mutant or recombinant appeared in a population of insects that enabled individuals to feed on the previously protected plant, selection would carry that insect's line into a new adaptive zone, allowing it to diversify in the absence of competition with other herbivores. In other words, the plant and the herbivore evolve together, in the sense that the evolution of each depends on the evolution of the other. The expression *genetic feedback* has been used for this kind of evolution, which leads to population and community homeorhesis within the ecosystem.

Examples

Perhaps coevolution can best be investigated and understood by studying interactions between two sets of species—most frequently species that represent different taxonomic groups. Hummingbird pollinators and the red-flowered plants that they pollinate represent a classic example of coevolution. Bumblebees, widely distributed species of the genus *Bombus,* are very important pollinators of both wild plants and important cultivated crops such as alfalfa, clover, beans, and blueberries. Heinrich (1979, 1980) assessed the interactions of flowers and bumblebees on the basis of energetics. He measured nectar production in terms of sugar available per flower and counted the number of visits by bees and the rate of removal of nectar in relation to time of day and temperature. Bees, unlike butterflies, have a high metabolic rate and must visit flowers frequently to make an energy profit. To attract these necessary pollinators and to ensure their survival, many species of flowers have evolved the mechanism of either blooming synchronously or growing in landscape patches.

Herbivores exert strong selection pressure on the plant species that they graze (that is, the plants evolve to deter grazing). There exists an array of chemical compounds, frequently referred to as *secondary compounds,* that serves to deter herbivores. **Secondary compounds** are organic compounds produced by plants that are used in chemical defense. They are either toxic compounds or compounds, such as tannins, that make the plants less palatable. These compounds appear to represent specific biochemical and physiological adaptations by plants to the selection pressures caused by the herbivores. The herbivores, in turn, frequently adapt to these chemicals by changes in their own genetic or physiological metabolism. Thus, both herbivores and plants coevolve in this "arms race" for increased survivorship.

Grazing has also been shown to stimulate plant growth and increase net primary productivity. Thus, this interaction has evolved to be beneficial to both the herbivore and the select species on which they graze. That reciprocal natural selection is not limited to two-species interactions was shown by Colwell (1973), who described how 10 diverse species—four flowering plants, three hummingbirds, one bird, and two mites—have coevolved to produce a fascinating tropical subcommunity. Coevolution can also involve more than one step in the food chain. Brower and Brower (1964) and Brower (1969), for example, studied the monarch butterfly (*Danaus plexippus*), which is well known to be unpalatable in general to vertebrate predators. They found that monarch caterpillars can sequester the highly toxic cardiac glycosides present in the milkweed plants on which they feed, thereby providing a highly effective defense against bird predators not only for the caterpillar but also for the adult butterfly. Thus, the monarch butterfly has evolved the ability to feed on a plant that is unpalatable to other insects, and it also uses the plant poison for its own protection against predators. It can be assumed that the numerous cases of mutualism described in the next section involve coevolution at various levels.

3 Evolution of Cooperation: Group Selection

Statement

To account for the incredible diversity and complexity of the biosphere, scientists have postulated that natural selection operates beyond the species level and beyond coevolution. **Group selection,** accordingly, is defined as natural selection between groups or assemblages of organisms that are not necessarily closely linked by mutualistic associations. Group selection theoretically leads to the maintenance of traits favorable to populations and communities that may be selectively disadvantageous to genetic carriers within populations. Conversely, group selection may eliminate, or keep at low frequencies, traits unfavorable to the survival of the species but selectively favorable within populations or communities. Group selection involves positive benefits that an organism may exert on the community organization required for that organism's continued survival.

Explanation and Examples

The "struggle for existence" and "survival of the fittest" (T. H. Huxley 1894) are not just a matter of "dog eat dog." In many cases, survival and successful reproduction are based on cooperation rather than competition. How cooperation and elaborate mutualistic relationships get started and become genetically fixed has been difficult to explain in evolutionary theory, because when individuals first interact, it is nearly always advantageous for each individual to act in its own interest rather than to cooperate. Axelrod and Hamilton (1981) analyzed the evolution of cooperation and devised a model based on the *prisoner's dilemma* game and on the theory of reciprocation as an extension of the conventional competition-based, survival-of-the-fittest genetic theory. In the prisoner's dilemma game, two "players" decide whether to cooperate or not on the basis of immediate benefits. On the first encounter, a decision *not* to cooperate (to *defect*) yields the highest reward for each individual, regardless of what the other individual does. However, if *both* choose not to cooperate, they both do worse than if both had cooperated. If individuals continue to interact (the "game" continues), the probability is that cooperation may be selected on a trial basis and its advantages recognized. Deductions from the model show that cooperation based on such reciprocity can get started in an asocial environment and then develop and persist once fully established. Constant close contact between numerous individuals such as microorganisms and plants enhances the possibilities for interaction with mutual benefit, such as has evolved between nitrogen-fixing bacteria and legumes.

It has also been suggested that *altruism*—sacrifice of fitness by one individual for the benefit of another—in related individuals (such as parents and offspring) can be the start of an evolution toward cooperation (even in unrelated species). Once genes favoring reciprocity have become established by kin selection, cooperation can spread into circumstances of less and less relatedness.

David Sloan Wilson (1975, 1977, 1980) stated the case for group selection as follows (1980, p. 97):

> Populations routinely evolve to stimulate or discourage other populations upon which their fitness depends. As such over evolutionary time an organism's fitness is largely a reflection of its own effect on the community and the reaction of the community to that organism's presence. If this reaction is sufficiently strong, only organisms with a positive effect on their community persist.

Wilson argued that selection between "structured demes" (closely knit genetic segments of a population) facilitates group selection. He also drew an analogy between the paradox of individual versus community fitness in biological communities and private benefit compared with public benefit in human communities.

Predator-prey and parasite-host interactions tend to become less negative over time. Gilpin (1975) proposed group selection in the development of a "prudence" trait that leads predators and parasites not to overexploit their prey or hosts, because to do so could lead to the extinction of both species involved in the interaction. The history of the myxomatosis virus introduced to control European rabbits (actually hares) in Australia is an example of selection for reduced virulence. When first introduced, the parasite killed the rabbit within a few days. Subsequently, the virulent strain was replaced by a less virulent one that took two to three times as long to kill the host; hence, the mosquitoes that transmit the virus had a longer time to feed on infected rabbits. Because the avirulent strain did not destroy its food resource (rabbit) as rapidly as the virulent strain, more and more avirulent-type parasites were produced and were available for transmission to new hosts. Thus, interdemic selection favored the avirulent over the virulent strain; otherwise, both parasite and host would eventually have become extinct.

Although few doubt that group selection occurs, its importance in evolutionary history remains controversial. The organized complexity that has developed in the natural world is difficult to explain solely by selection at the individual and species level; hence, higher-level selection and the process of self-organization have to play major roles. For more on group selection, see E. O. Wilson (1973, 1980, 1999), D. S. Wilson (1975, 1977, 1980), and Maynard Smith (1976).

4 Interspecific Competition and Coexistence

Statement

Competition in the broadest sense refers to the interaction of two organisms striving for the same resource. **Interspecific competition** is any interaction that adversely affects the growth and survival of two or more species populations. Interspecific competition can take two forms: (1) *interference competition;* and (2) *exploitation competition.* The tendency for competition to bring about an ecological separation of closely related or otherwise similar species is known as the *competitive exclusion principle.* Simultaneously, competition triggers many selective adaptations that enhance the coexistence of a diversity of organisms in a given area or community.

Explanation and Examples

Ecologists, geneticists, and evolutionists have written much about interspecific competition. Generally, the word *competition* is used in situations in which negative influences are due to a shortage of resources used by both species. Interspecific competition is frequently discussed in terms of direct physical interaction versus exploitation competition. **Interference competition** occurs when two species come into direct contact with each other, such as fighting or defending a territory. **Exploitation competition** occurs when one species exploits a resource, such as food, space, or prey, in common with another species but without direct contact with that species. This indirect exploitation of resources can provide a competitive advantage for one species against another.

The competitive interaction can involve space, food or nutrients, light, waste materials, susceptibility to carnivores, disease, and many other types of mutual interactions. The results of competition are of interest to evolutionary biologists and have been much studied as one of the mechanisms of natural selection. Interspecific competition can result in equilibrium adjustments between two species or, if severe, in one species population replacing another, or forcing the other to occupy another space or to use another food (whatever was the basis of the original competitive action). Closely related organisms having similar habits or morphologies often do not occur in the same places. If they do occur in the same places, they frequently use different resources or are active at different times. The explanation for the ecological separation of closely related (or otherwise similar) species has come to be known as the **Gause principle** (Gause 1932), after the Russian biologist who first observed such separation in experimental cultures (see Fig. 7-2), or the **competitive exclusion principle,** as designated by Hardin (1960).

One of Gause's original experiments with ciliates (Gause 1934, 1935) is a classic example of competitive exclusion (Fig. 7-2). *Paramecium caudatum* and *Paramecium aurelia,* two closely related ciliate protozoans, when grown in separate cultures, exhibited typical sigmoid population growth and maintained a constant population level in culture medium that was maintained with a fixed density of food items (bacteria that did not themselves multiply in the medium and thus could be added at frequent intervals to keep the food density constant). When both protozoans were placed in the same culture, however, *P. aurelia* alone survived after 16 days. Neither organism attacked the other or secreted harmful substances; *P. aurelia* populations simply had a more rapid growth rate (higher intrinsic rate of increase) and thus "outcompeted" *P. caudatum* for the limited amount of food under the existing conditions (a clear case of exploitation competition). By contrast, both *Paramecium caudatum* and *Paramecium bursaria* were able to survive and reach a stable equilibrium in the same culture medium. Although they were competing for the same food, *P. bursaria* occupied a different part of the culture, where it could feed on bacteria without competing with *P. caudatum.* Thus, the habitat proved to be sufficiently different for the two species to coexist, even though their food was identical.

Some of the most widely debated theoretical aspects of competition theory re-

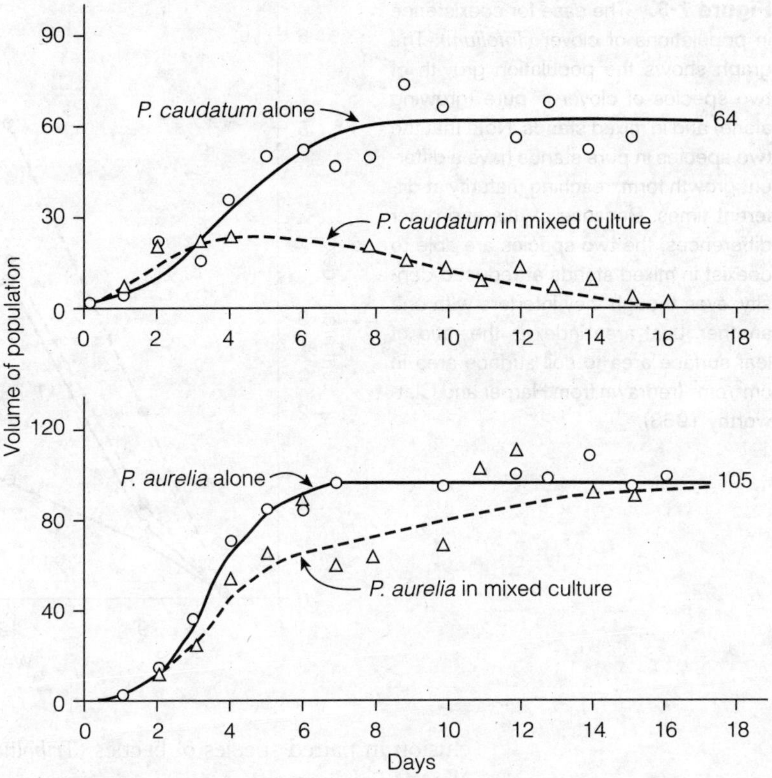

Figure 7-2. Competition between two closely related species of protozoa that have similar niches. When separate, *Paramecium caudatum* and *Paramecium aurelia* exhibit normal sigmoid growth in controlled cultures with constant food supply; when both are cultured together, *P. caudatum* is eliminated (after Gause 1932).

volve around what has become known as the **Lotka-Volterra equations,** so termed because the equations were proposed as models by Lotka (1925) and Volterra (1926) in separate publications. They are a pair of differential equations similar to the one outlined in Section 1. Such equations are useful for modeling predator-prey, parasite-host, competition, or other two-species interactions. In terms of competition within a limited space where each population has a definite K or equilibrium level, the simultaneous growth equations can be written in the following forms, using the logistic equation as a basis:

$$\frac{dN_1}{dt} = r_1 N_1 \left(\frac{K_1 - N_1 - \alpha N_2}{K_1} \right)$$

$$\frac{dN_2}{dt} = r_2 N_2 \left(\frac{K_2 - N_2 - \beta N_1}{K_2} \right)$$

where N_1 and N_2 are the numbers of Species 1 and 2, respectively, α is the *competition coefficient* indicating the inhibitory effect of Species 2 on Species 1, and β is the corresponding competition coefficient signifying the inhibition of Species 1 on Species 2.

To understand competition, one must consider not only the conditions and population attributes that may lead to competitive exclusion but also the situations under which similar species coexist, because large numbers of species do share common resources in the open systems of nature. Figure 7-3 presents what might be termed the *Tribolium-Trifolium model,* which includes an experimental demonstration of ex-

Figure 7-3. The case for coexistence in populations of clover (*Trifolium*). The graph shows the population growth of two species of clover in pure (growing alone) and in mixed stands. Note that the two species in pure stands have a different growth form, reaching maturity at different times. Because of this and other differences, the two species are able to coexist in mixed stands at reduced density, even though they interfere with one another. Leaf area index is the ratio of leaf surface area to soil surface area in cm^2/cm^2 (redrawn from Harper and Clatworthy 1963).

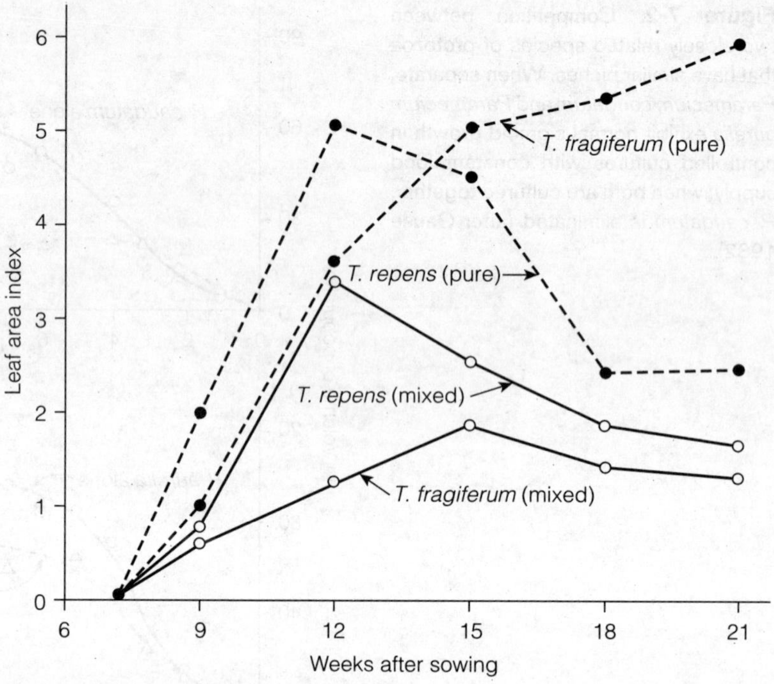

clusion in paired species of beetles (*Tribolium*) and of coexistence in two species of clover (*Trifolium*).

One of the most thorough long-term experimental studies of interspecific competition was carried out at the University of Chicago in the laboratory of Thomas Park (Park 1934, 1954). Park and his students and associates worked with flour beetles, especially those belonging to the genus *Tribolium*. These small beetles can complete their entire life history in a very simple and homogeneous habitat, namely, a jar of flour or wheat bran. The medium in this case is both food and habitat for larvae and adults. If fresh medium is added at regular intervals, a population of beetles can be maintained for a long time. In energy flow terminology, this experimental setup is a stabilized heterotrophic ecosystem in which imports (subsidies) of food energy balance respiratory losses.

The investigators found that when two different species of *Tribolium* were placed in this homogeneous microcosm, one species invariably was eliminated sooner or later, whereas the other continued to thrive. One species always "wins," or to put it another way, two species of *Tribolium* cannot survive in this particular one-habitat microcosm. The relative number of individuals of each species originally placed in the culture (the *stocking rate*) does not affect the eventual outcome, but the climate imposed on the ecosystem does have a great impact on which species of the pair wins out. One species (*T. castaneum*) always wins under conditions of high temperature and humidity, whereas the other (*T. confusum*) always wins under cool and dry conditions, even though either species can live indefinitely in any of the six climates, provided it is alone in the culture. Population attributes measured in one-species cultures help explain some of the outcomes of the competitive action. For example, the species with the highest rate of increase, *r*, under the conditions of existence in question was usually found to win if the species difference in *r* was rather large. If the growth rates differed only moderately, the species with the highest rate did not always win. The presence of a virus in one population could easily tip the balance. Feener (1981) described a case in which a parasitic fly altered the competitive balance between two species of ants. Also, genetic strains within the population may differ greatly in competitive ability.

Some of the most interesting experiments in plant competition were reported by J. L. Harper and associates, researching at the University College of North Wales (see J. L. Harper 1961; J. L. Harper and Clatworthy 1963). The results of one of these studies, shown in Figure 7-3, illustrate how a difference in growth form allows two species of clover to coexist in the same environment (same light, temperature, and soil). Of the two species, *Trifolium repens* grows faster and reaches a peak in leaf density sooner. However, *T. fragiferum* has longer petioles and higher leaves and can overtop the faster-growing species, especially after *T. repens* has passed its peak, and thus avoid being shaded out. In mixed stands, therefore, each species inhibits the other, but both can complete their life cycle and produce seed, even though each coexists at a reduced density. In this case, the two species, although competing strongly for light, could coexist because their morphology and the timing of their growth maxima differed. J. L. Harper (1961) concluded that two species of plants can persist together if the populations are independently controlled by one or more of the following mechanisms: (1) different nutritional requirements (such as legumes and nonlegumes); (2) different causes of mortality (such as differential sensitivities to grazing); (3) sensitivity to different toxins (different responses to secondary chemicals); and (4) sensitivity to the same controlling factor (such as light or water) at different times (like the clover just described).

Brian (1956) was among the first to distinguish between *indirect* or *exploitation competition* and *direct* or *interference competition*. Interference competition appears more frequently as we move up the phylogenetic tree of animal life, from simple filter-feeding protozoans and cladocerans, which usually compete in gathering food, to vertebrates, with their elaborate behavioral patterns of aggression and territoriality. Slobodkin (1964) concluded on the basis of competition experiments with *Hydra* that these two types of competition overlap but that it is useful to distinguish between the two processes on theoretical grounds. A general pattern emerging from the literature on competition is that competition is most severe—and competitive exclusion most likely to occur—in systems where immigration and emigration are absent or reduced, such as in laboratory cultures or mesocosms or on islands or other natural systems with substantial barriers to inputs and outputs. The probability of coexistence is higher in the more typical open systems of nature.

Interspecific competition in plants in the field has been much studied and is generally believed to be an important factor in bringing about a succession of species (as will be described in Chapter 8). The strongest evidence regarding the importance of competition in nature comes from studies of how species respond to experimental additions or removals of potential competitors (Connell 1961, 1972, 1975; Paine 1974, 1984; Hairston 1980). Connell's (1961) classic study of competition among barnacles in natural settings is a splendid example of a well-designed field experiment. Connell's investigation was conducted on the rocky coast of Scotland, where two barnacle species typically occupy different locations in the intertidal zone. The smaller of the two species, *Chthamalus stellatus,* occurs higher in the intertidal zone than the larger species, *Balanus glandula.* Figure 7-4A presents a "barnacle model" based on the experimental studies of J. H. Connell. The intertidal zone on a rocky eastern seacoast

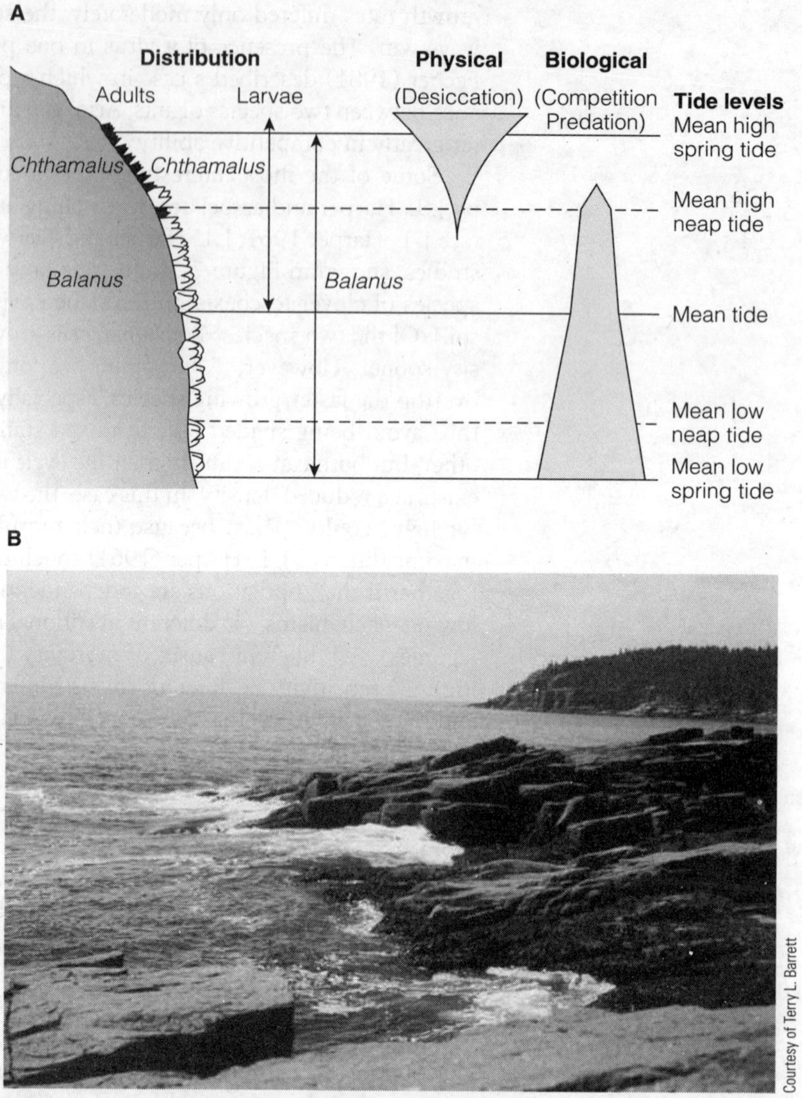

Figure 7-4. (A) Factors that control the distribution of two species of barnacles in an intertidal gradient. The young of each species settle over a wide range but survive to adulthood in a more restricted range. Physical factors such as desiccation control the upward limits of *Balanus*, whereas biological factors such as competition and predation control the downward distribution of *Chthamalus* (after E. P. Odum 1963; from Connell 1961). (B) An example of an intertidal zone located on the East Coast near Bar Harbor, Maine.

provides a miniature gradient from a physically stressed to a more biologically controlled environment (Fig. 7-4B). Connell found that the larvae of two species of barnacles settled over a wide range of the intertidal zone but survived as adults in a much more restricted range. The larger species, *Balanus,* was found to be restricted to the lower part of the zone because it could not tolerate long periods of exposure at low tide. The smaller species, *Chthamalus,* was excluded from the lower zone by competition with the larger species and by predators that are more active below the high tide mark. Accordingly, the physical stress of desiccation was identified as the main controlling factor in the upper part of the gradient, whereas interspecific competition and predation were more important controlling factors in the lower zones. This model can be considered to apply to more extensive gradients such as an Arctic-to-Tropics or a high-to-low-altitude gradient, provided one remembers that all models are, to varying degrees, oversimplifications.

Robert Paine at the University of Washington demonstrated that predation played a major role in shaping the structure of biological communities by influencing the outcome of competitive interactions between prey species. The intertidal zone on an exposed rocky shore on the West Coast of the United States harbors several species of barnacles, mussels, limpets, and chitons. These species are preyed on by the sea star *Pisaster ochraceus.* Following the removal of sea stars from experimental plots, the number of prey species in the removal plots decreased rapidly from 15 to 8 by the end of the investigation. Diversity decreased because populations of barnacles and mussels, superior competitors for space in the absence of predators (sea stars in this case), crowded out many of the prey species. This elegant study demonstrated how predation shaped the biological community and regulated biotic diversity (Paine 1974).

Morphological differences that enhance ecological separation may arise by an evolutionary process termed **character displacement.** For example, in central Europe, six species of titmice (small birds of the genus *Parus*) coexist, segregated partly by habitat and partly by feeding areas and size of prey, which is reflected in small differences in length and width of the bill. In North America, more than two species of titmice are rarely found in the same locality, even though seven species are present on the continent as a whole. Lack (1969) suggested that the American species of titmice are at an earlier stage in their evolution than the European species, and their differences in beak, body size, and feeding behavior are adaptations to their respective habitats and are not yet adaptations for permitting coexistence in the same habitat.

The general theory of the role of competition in habitat selection is summarized in Figure 7-5. The curves represent the range of habitat that can be tolerated by the species, with optimum and marginal conditions indicated. Where there is competition with other closely related or ecologically similar species, the range of habitat conditions that the species occupies generally becomes restricted to the optimum (that is, to the most favorable conditions under which the species has an advantage in some manner over its competitors). Where interspecific competition is less severe, intraspecific competition generally brings about a wider choice of habitats.

Islands are good places to clearly observe the tendency for a wider selection of habitats to occur when potential competitors fail to colonize. For example, meadow voles (*Microtus*) often occupy forest habitats on islands where their forest competitors, the red-backed voles (*Clethrionomys*), are absent.

Just because closely related species are sharply separated in nature does not, of course, mean that competition is actually operating continuously to keep them separated; the two species may have evolved different requirements or preferences that effectively reduce or eliminate competition. For example, in Europe, one species of

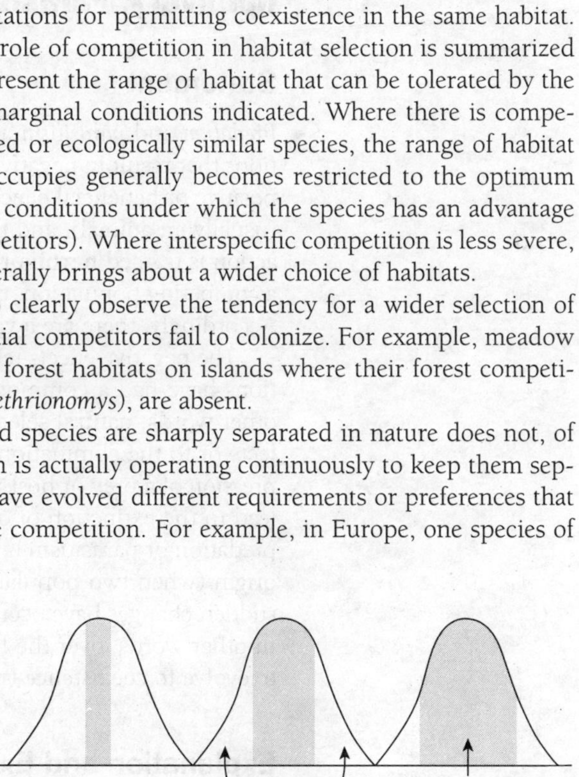

Figure 7-5. The effect of competition on habitat distribution. When intraspecific competition dominates, the species spreads out and occupies less favorable (marginal) areas; where interspecific competition is intense, the species tends to be restricted to a narrower range, representing the optimum conditions.

Rhododendron (*R. hirsutum*), is found on calcareous soils, whereas another species (*R. ferrugineum*) is found on acidic soils. The requirements of the two species are such that neither can live at all in the opposite type of soil, so there is never any actual competition between them. Teal (1958) made an experimental study of habitat selection of species of fiddler crabs (*Uca*), which are usually separated in their occurrence in salt marshes. One species, *U. pugilator,* is found on open, sandy flats; another, *U. pugnax,* is found on muddy substrates covered with marsh grass. Teal found that one species would tend not to invade the habitat of the other even in the absence of the other, because each species would dig burrows only in its preferred substrate. The absence of active competition, of course, does not mean that competition in the past can be ruled out as a factor in originally bringing about the isolating behavior.

Two closely related species of aquatic birds, the shag (*Phalacrocorax aristotelis*) and the cormorant (*P. carbo),* are found together in England during the breeding season, but feed on entirely different kinds of fish. Therefore, they are not in direct competition for food resources (that is, the niche of the two species is different). This is an example of **neutralism** (0 0) as described in Table 7-1 and illustrated in Figure 7-1.

5 Positive/Negative Interactions: Predation, Herbivory, Parasitism, and Allelopathy

Statement

Predation and *parasitism* are familiar examples of interactions between two populations that result in negative effects on the growth and survival of one population and positive or beneficial effects on the other. When the predator is a primary consumer (usually an animal), and the prey or "host" is a primary producer (plant), the interaction is termed **herbivory.** When one population produces a substance harmful to a competing population, the term **allelopathy** is commonly used for the interaction. Accordingly, there are a variety of + − relationships.

The negative effects tend to be quantitatively small when the interacting populations have had a common evolutionary history in a relatively stable ecosystem. In other words, natural selection tends to lead either to a reduction in detrimental effects or to the elimination of the interaction altogether, as the continued severe depression of a prey or host population by the predator or parasite population can only lead to the extinction of one or both populations. Consequently, a severe impact of predation or parasitism is most frequently observed when the interaction is of recent origin (when two populations have just become associated) or when large-scale or sudden changes have occurred in the ecosystem (as might be produced by humans). In other words, over the long term, parasite-host or predator-prey interactions tend to evolve to coexistence (recall the discussion of reward feedback in Chapter 4).

Explanation and Examples

It is frequently difficult for students, or for people in general, to approach the subject of parasitism and predation objectively. People have a natural aversion to parasitic organisms, whether bacteria or tapeworms. Although human effects on Earth as predators and perpetrators of epidemics in nature are extensive, people tend to condemn all other predators without ascertaining whether these predators are actually detrimental to human interests. The idea that "the only good hawk is a dead hawk" is a most uncritical generalization.

A way to be objective is to consider predation, parasitism, herbivory, and allelopathy from the population and community levels of organization, rather than from the individual level. Predators, parasites, and grazers certainly kill or injure the individuals that they feed or secrete toxic chemicals on, and they depress, in some measure at least, the growth rate of their target populations or reduce the total population size. But does this mean that these populations would be healthier without consumers or inhibitors? From the long-term, coevolutionary viewpoint, are predators the sole beneficiaries of the association? As pointed out in the discussion of population regulation (Chapter 6), predators and parasites help keep herbivorous insects at a low density so they will not destroy their own food supply and habitat. In Chapter 3, we discussed how animal herbivores and plants have evolved an almost mutualistic (+ +) relationship.

Deer populations are often cited as examples of populations that tend to irrupt when predator pressure is reduced. The Kaibab deer herd, as originally described by Leopold (1943) on the basis of estimates by Rasmussen (1941), allegedly increased from 4000 (on 700,000 acres on the north side of the Grand Canyon in Arizona) in 1907 to 100,000 in 1924, coincident with a predator removal campaign organized by the U.S. government. Caughley (1970) reexamined this case and concluded that although there is no question that the deer did increase, overgraze, and then decline, there is doubt about the extent of the overpopulation, and there is no evidence that it was due solely to the removal of predators. Cattle and fire may also have played a part. Caughley believed that irruptions of ungulate populations are more likely to result from changes in habitat or food quality, which enable the population to "escape" from the usual mortality control.

One thing is clear: the most violent irruptions occur when a species is introduced into a new area, where there are both unexploited resources and a lack of negative interactions. The population explosion of rabbits introduced into Australia is a well-known example among the thousands of cases of severe oscillations that result when species with high biotic potential are introduced into new areas. An interesting sequel to the attempts to control the irruption of rabbits by introducing a disease organism has provided evidence for group selection in a parasite-host system (as discussed in Section 3).

The most important generalization is that negative interactions become less negative with time if the ecosystem is sufficiently stable and spatially diverse to allow reciprocal adaptations. Parasite-host or predator-prey populations introduced into experimental microcosms or mesocosms usually oscillate violently, with a certain probability of extinction. Pimentel and Stone (1968), for example, have shown experimentally (Fig. 7-6) that violent oscillations occur when a host, such as the house fly (*Musca domestica*) and a parasitic wasp (*Nasonia vitripennis*) are first placed together in a limited culture system. When individuals selected from cultures that had managed to survive the violent oscillations for two years were then reestablished in new cultures, it was evident that through genetic selection, an ecological homeorhesis had evolved in which both populations could now coexist in a much more stable equilibrium.

In the real world of humans and nature, time and circumstances may not favor

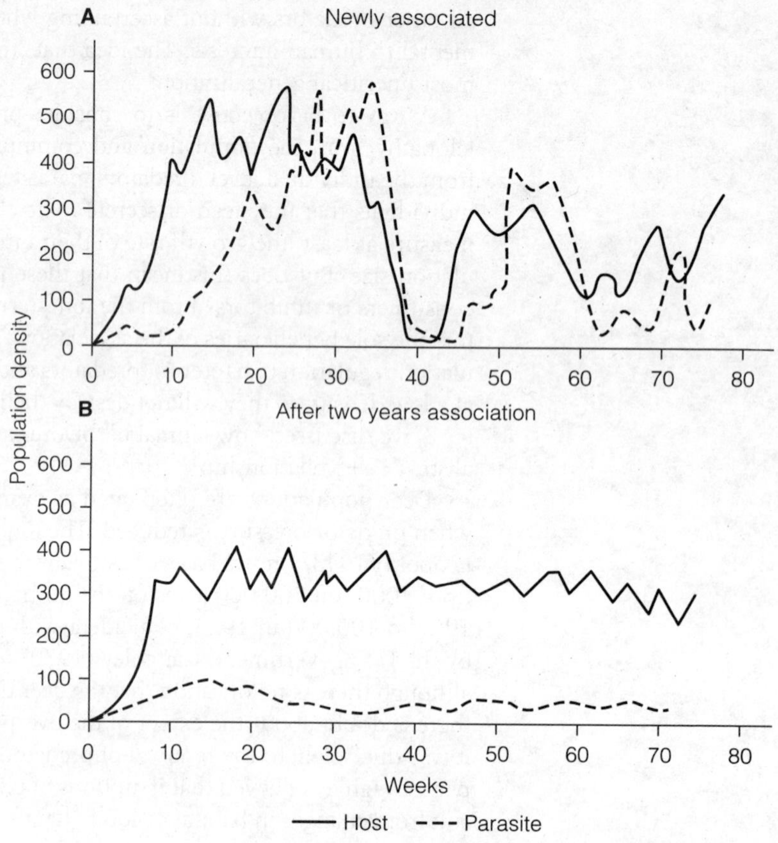

Figure 7-6. Evolution of coexistence in the host-parasite relationship between house fly (*Musca domestica*) and parasitic wasp (*Nasonia vitripennis*) populations in a laboratory investigation. (A) Newly associated populations (wild stocks brought together for the first time) oscillated violently, as first the host (fly) and then the parasite (wasp) density increased and crashed. (B) Populations derived from colonies in which the two species had been associated for two years coexisted in a more stable equilibrium without crashes. The experiment demonstrates how genetic feedback can function as a regulatory and stabilizing mechanism in population systems (redrawn from Pimentel and Stone 1968).

such reciprocal adaptation by new associations. There is always the danger that the negative reaction may be irreversible, in that it leads to the extinction of the host. The chestnut blight in America is a case in which the question of adaptation or extinction hangs in the balance.

Originally, the American chestnut tree (*Castanea dentata*) was an important member of the forests of the Appalachian region of eastern North America, often constituting up to 40 percent of total forest biomass. It had its share of parasites, diseases, and predators. Likewise, the Oriental chestnut trees (*Castanea mollissima*) of China—a different but related species—had their share of parasites, including the fungus *Endothia parasitica,* which attacks the bark of the stems. In 1904, the fungus was accidentally introduced into the United States. The American chestnut tree proved to be nonresistant to this new parasite. By 1952, all the large chestnut trees had been killed, their gaunt, gray trunks becoming a characteristic feature of Appalachian forests (Fig. 7-7). The chestnut tree continues to sprout up from the roots, and such sprouts may produce fruits before they die, but no one can say whether the ultimate outcome will be extinction or adaptation. For all practical purposes, the chestnut tree has been removed as a major influence in the forest.

The preceding examples are not just cases handpicked to prove a point. If the student will do a little research in the library, she or he can find similar examples showing (1) that where parasites or predators have long been associated with their respective hosts or prey, the effect is moderate, neutral, or even beneficial from the long-term viewpoint; and (2) that newly acquired parasites or predators are the most damaging. In fact, a list of the diseases, parasites, and insect pests that cause the greatest loss in agriculture or forestry would include mostly species that have recently been introduced into a new area, such as the chestnut blight, or that have acquired a new host or prey. The European corn ear worm (*Helicoverpa zea*), the gypsy moth (*Lymantria dispar*), the Japanese beetle (*Popillia japonica*), and the Mediterranean fruit fly (*Ceratitis capitata*) are just a few introduced insect pests that belong to this category. The lesson, of course, is to avoid introducing new potential pests and to avoid, wherever possible, the stressing of ecosystems with poisons that destroy useful as well as pest organisms. Much the same principle applies to severe human diseases: the most feared are the newly acquired. For a recent discussion regarding the introduction of new predators and pathogens on resident species at the community, metacommunity, and global levels, see M. A. Davis (2003). Simberloff (2003) described the need for more population biological research in order to control introduced species based on sound ecological theory.

Of special interest are organisms intermediate between predators and parasites—for example, the so-called parasitic or *parasitoid* insects. These organisms often can consume the entire individual prey, as does the predator, yet they have the host specificity, high biotic potential, and small size of the parasite. Entomologists have propagated some of these organisms artificially, using them to control insect pests. In general, attempts to make similar use of large, unspecialized predators have not been successful. For example, the mongoose (*Herpestes edwardsi*) introduced in the Caribbean Islands to control rats in sugarcane fields has more severely reduced ground-nesting birds than rats. If the predator is small, is specialized in its choice of prey, and has a high biotic potential, control can be effective.

Most general theories proposed to explain the trophic structure of plant communities pay little attention to the potentially profound influence of insect herbivores. Indeed, most theories of trophic interactions and community regulation suggest that insect herbivory will have little influence on terrestrial vegetation, particularly on net primary production (see, for example, Hairston et al. 1960; Oksanen 1990). Many ar-

Figure 7-7. Results of the chestnut blight in a southern Appalachian region, illustrating the extreme effect that a parasitic organism (fungus, *Endothia parasitica*) introduced from the Old World had on a newly acquired host (American chestnut tree, *Castanea dentata*).

Figure 7-8. The plot on the left was sprayed with insecticide for eight years and is dominated by a dense stand of the goldenrod *Solidago altissima*. Surrounding plots were left as unsprayed controls. This photograph was taken two years after an outbreak of the chrysomelid beetle *Microrhopala vittata* defoliated numerous stems of *S. altissima*. These outbreaks occur every 5–15 years and typically exert a strong influence on standing crop biomass (after W. P. Carson and Root 2000).

gue that predators and parasites keep insect herbivores from causing major damage to their host plants in terrestrial communities (Strong et al. 1984; Spiller and Schoener 1990; Bock et al. 1992; Marquis and Whelan 1994; Dial and Roughgarden 1995) and that insect herbivores typically consume only a small amount of the available net primary production (Hairston et al. 1960; Strong, Lawton, et al. 1984; Crawley 1989; Root 1996; Price 1997).

A different view holds that insects only damage or consume a small amount of their host plants because most plant species are well defended or have low nutritional value (Hartley and Jones 1997). Lawton and McNeil (1979) suggested that herbivorous insects are caught between the interacting forces of predators and parasites on the one hand and unpalatable or low-quality plants on the other. Regardless of which view one holds, the conclusion is the same: herbivorous insects will have a negligible influence on plant community structure, composition, and productivity (Pacala and Crawley 1992). Pacala and Crawley (1992) concluded that "herbivores often have little effect on communities," although later Crawley (1997) noted that there was an insufficient number of studies of insect herbivory from which to generalize.

More recently, however, it has been found that the removal of arthropods can also cause significant changes in community structure and function. Studies by V. K. Brown (1985) and W. P. Carson and Root (1999) have demonstrated that the exclusion of herbivorous insects with insecticides causes major changes in flowering frequency and plant species composition in old-field communities. W. P. Carson and Root (2000) demonstrated that insect herbivores could have a very strong top-down effect on plant communities, but this occurred primarily during insect outbreaks. Using insecticide-treated and control plots, they examined the long-term (10-year) effects of suppressing insects on the structure and diversity of an old field dominated by the goldenrod *Solidago altissima* (Fig. 7-8). An outbreak of the chrysomelid beetle *Microrhopala vittata,* which specializes on *S. altissima,* occurred during the experiment and persisted for several years. The damage caused by this outbreak dramatically reduced the biomass, density, height, survivorship, and reproduction of *S. altissima.* Herbivore exclusion caused the formation of dense stands of goldenrods with a twofold increase in both standing crop biomass and litter. The understory in these dense stands had significantly lower plant abundance, species richness, flowering shoot production and light levels; these conditions persisted for years following the outbreak. Thus, *M. vittata* functioned as a keystone species. Furthermore, insect herbivory indirectly increased the abundance of invading trees, thereby increasing the rate of succession by speeding up the transition of this old field to a tree-dominated stage.

W. P. Carson and Root (2000) argued that insect outbreaks may be extremely important in community dynamics, but for the most part are ignored in theories of community regulation. The observations (1) that native phytophagous herbivores periodically irrupt (pulse) and reduce the abundance and vigor of dominant plant species; (2) that these outbreaks may occur more readily in dense or lush concentrations of their hosts; and (3) that an outbreak may occur more than once during the life span of a long-lived host, suggest that insect outbreaks may play a very important role in plant community regulation and dynamics.

As manipulators of ecosystems, human beings are slowly learning how to be a prudent predator (when and how much biomass to harvest without damaging the system or relationship). The problem can be approached experimentally by setting up test populations in microecosystems. In one such experimental model, shown in Figure 7-9, guppies (*Lebistes reticulatus*) were used to mimic a commercial fish population being exploited by humans. As shown, the maximum sustained yield was obtained when *one third* of the population was harvested during each reproductive period, which reduced the equilibrium density to slightly less than half the unexploited one. Within the limits of the experiment, these ratios tended to be independent of the carrying capacity of the system, which was varied at three levels by manipulating the food supply.

One-species models often prove to be oversimplifications, because they do not account for competing species that may respond to the reduced density of the harvested species by increasing their own density and using up food or other resources

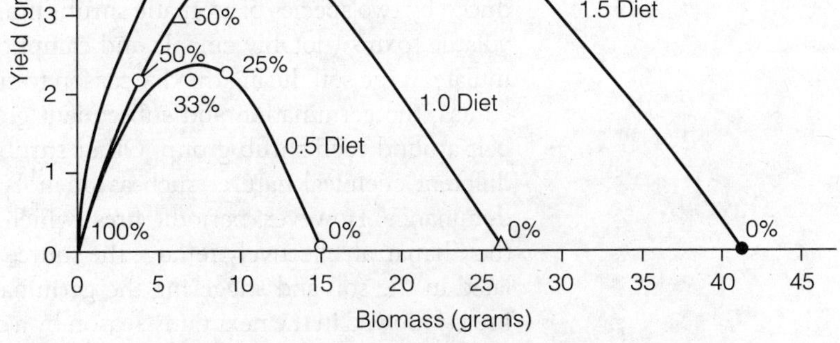

Figure 7-9. Biomass and yield in test populations of the guppy (*Lebistes reticulatus*) exploited at different rates (shown as percentage removal per reproductive period) at three different diet levels. The highest yields were obtained when about one third of the population was harvested per reproductive period and mean biomass was reduced to less than half that of the unexploited population (yield curves skewed to the left) (after Silliman 1969).

needed to sustain the exploited species. Top predators such as humans (or major grazers such as cattle) can easily tip the balance of a competitive equilibrium so that the exploited species is replaced by another species that the predator or grazer may not be prepared to use. In the real world, examples of such shifts are being documented ever more often as human beings strive to become more "efficient" at fishing, hunting, and harvesting plants. This situation poses both a challenge and a danger: *one-species harvest systems and monocultural systems* (such as one-crop agriculture) *are inherently unstable* because, when stressed, they are vulnerable to competition, disease, parasitism, predation, and other negative interactions. Some examples of this general principle are to be found in the fishing industry.

Myers and Worm (2003) evaluated the effects of industrial fishing on fish community biomass and composition for large predatory fishes in four continental-shelf and nine oceanic systems. They estimated that large predatory fish biomass is only about 10 percent of pre-industrial levels. They concluded that the decline of large predators in coastal regions has extended throughout the global ocean, resulting in serious consequences, such as relatively low economic yields. Thus, the reduction of fish biomass to low levels may compromise the sustainability of the fishing industry and will require a global management approach to address its consequences.

The stress of predation or harvest often affects the size of individuals in the exploited populations. Thus, harvesting at the maximum sustained yield level usually reduces the average size of fish, just as maximizing timber yields for volume of wood reduces the size of trees and the quality of the wood. As reiterated many times in this book, a system cannot maximize quality and quantity at the same time. In a classic study, Brooks and Dodson (1965) described how large species of zooplankton are replaced by smaller species when zooplankton-feeding fish are introduced into lakes that formerly lacked such direct predators. In this case, in which the ecosystem is relatively small, both the size and species composition of a whole trophic level may be controlled by one or a few species of predators. The contrast between predator-driven and resource-driven food webs was detailed in Chapter 6.

Amensalism is where one species has a marked negative effect on the other, but there is no detectable reciprocal effect (− 0). Lawton and Hassell (1981) refer to this interaction as *asymmetrical competition.* Amensalism is just one evolutionary step from interactions such as allelopathy (− +).

Classic examples of allelopathy can be cited from the work of C. H. Muller, who studied inhibitors produced by shrubs in the vegetation of the California Chaparral. These investigators have not only examined the chemical nature and physiological action of the inhibitory substances but have also shown that they are important in regulating the composition and dynamics of the community (C. H. Muller et al. 1964, 1968; C. H. Muller 1966, 1969). Figure 7-10 shows how volatile terpenes produced by two species of aromatic shrubs inhibit the growth of herbaceous plants. The volatile toxins (notably cineole and camphor) are produced in the leaves and accumulate in the soil during the dry season to such an extent that when the rainy season comes, the germination and subsequent growth of seedlings is inhibited in a wide belt around each shrub group. Other shrubs produce water-soluble antibiotics of a different chemical nature (such as phenols and alkaloids), which also favor shrub dominance. However, periodic fires, which are an integral part of the ecosystem of the Chaparral, effectively remove the source of the toxins, denaturing those accumulated in the soil and triggering the germination of fire-adapted seeds. Accordingly, fire is followed in the next rainy season by a conspicuous blooming of annuals, which continue to appear each spring until the shrubs grow back and the toxins again be-

Figure 7-10. (A) Aerial view of aromatic shrubs *Salvia leucophylla* and *Artemisia californica* invading an annual grassland in the Santa Inez Valley of California and exhibiting biochemical inhibition. (B) Close-up showing the zonation effect of volatile toxins produced by *Salvia* shrubs seen to the center-left of A. Between A and B is a zone 2 meters wide, bare of all herbs except for a few minute, inhibited seedlings. The root systems of the shrubs that extend under part of this zone are thus free from competition with other species. Between B and C is a zone of inhibited grassland consisting of smaller plants and fewer species than in the uninhibited grassland seen to the right of C.

A

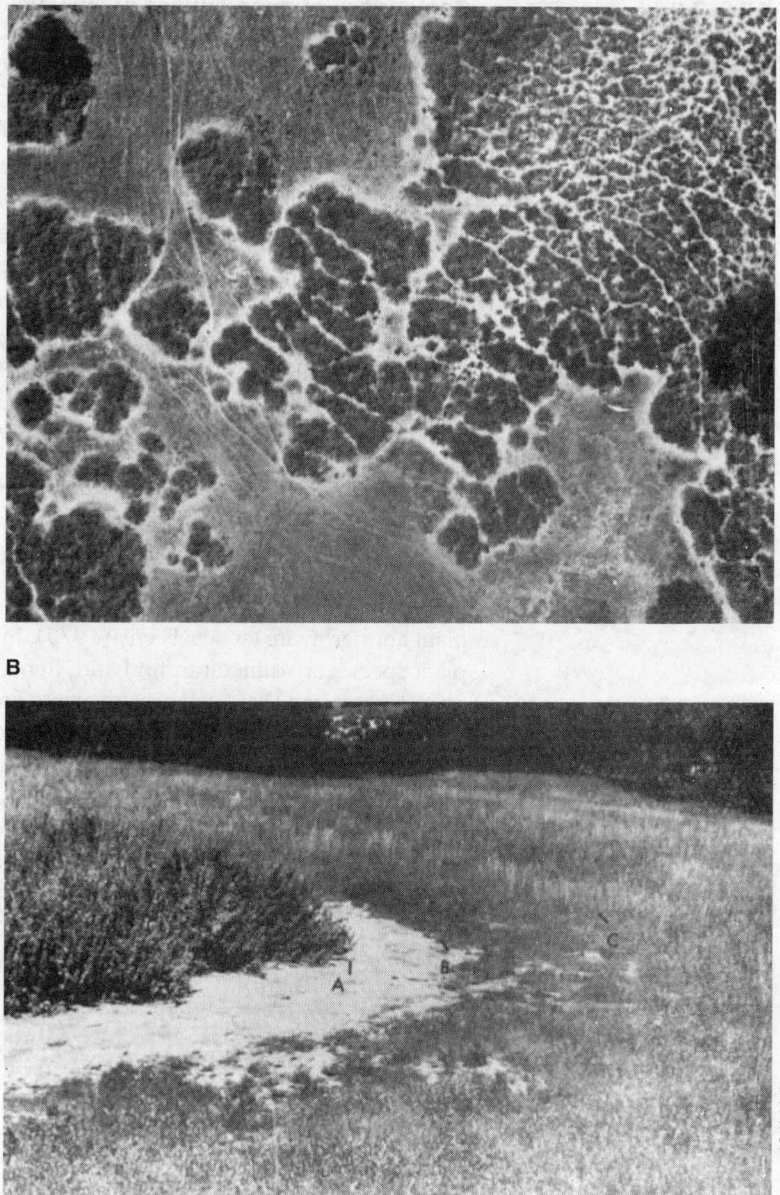

Courtesy of C. H. Muller

B

Courtesy of C. H. Muller

come effective. The interaction of fire and antibiotics thus perpetuates cyclic changes in composition that are the adaptive feature of this type of ecosystem.

Allelopathic effects have a significant influence on the rate and species sequence of plant succession and on the species composition of stable communities. Chemical interactions affect the species diversity of natural communities in both directions; strong dominance and intense allelopathic effects contribute to low species diversity in some communities, whereas a variety of chemical accommodations are part of the basis (as aspects of niche differentiation) of the high species diversity of others.

Table 7-2 **Comparison of apparent and cryptic plants**

Apparent plants	*Cryptic plants*
Common	Rare
Woody perennials	Herbaceous annuals
Slow growing (competitive)	Fast growing (fugitive)
Late seral stages	Early seral stages
Certain to be found by herbivores	Protected from herbivores in time and space
Produce expensive antiherbivore defenses (such as tannins)	Produce inexpensive chemical defenses (such as poisons or toxins)
Quantitative defenses constitute effective ecological barriers to herbivores	Qualitative defenses may be broken down by detoxification mechanisms

Source: After Pianka 2000.

Attempts have been made to generalize about the coevolution of herbivores and plant antiherbivore tactics. Feeny (1975), for example, argued that rare or ephemeral plant species are difficult to find and, hence, are protected in time and space. Moreover, he asserted that such *cryptic* plants have evolved a diversity of **qualitative defenses,** such as chemically inexpensive poisons and toxins, that constitute effective evolutionary barriers to herbivory by the herbivores most likely to find cryptic plants. In contrast, Feeny reasoned, abundant or persistent plant species (*apparent* plants) cannot prevent herbivores from finding them either in ecological or evolutionary time. Such apparent species appear to have evolved more expensive **quantitative defenses,** such as leaves high in tannins and antiherbivore defense chemicals, and adaptations such as tough leaves and thorns.

Table 7-2 summarizes the coevolutionary differences between plants with high and low apparency and quantitative versus qualitative defenses. Readings on this topic include D. F. Rhoades and Cates (1976); Futuyma (1976); Futuyma and Slatkin (1983); Palo and Robbins (1991); Gershenzon (1994); and Hunter (2000).

6 Positive Interactions: Commensalism, Cooperation, and Mutualism

Statement

Associations between two populations of species that result in positive effects are exceedingly widespread and probably as important as competition, parasitism, and other negative interactions in determining the function and structure of populations and communities. Positive interactions may be conveniently considered in an evolutionary series as follows:

- *commensalism*—one population benefits;
- *protocooperation*—both populations benefit; and
- *mutualism*—both populations benefit and become completely dependent on each other.

Explanation

Several decades after Darwin, the Russian Prince Pëtr Alekseevich Kropotkin published a book entitled *Mutual Aid: A Factor of Evolution* (Kropotkin 1902). Kropotkin chided Darwin for his overemphasis on natural selection as a bloody battle (Tennyson's "red in tooth and claw" metaphor). He outlined in considerable detail how survival was often enhanced by—or even dependent on—one individual helping another or one species aiding another for mutual benefit.

Kropotkin's writings were influenced by his personal philosophy of peaceful coexistence. Like Mahatma Gandhi and Martin Luther King, Jr., who made their contributions later, he was a firm believer in nonviolent solutions to human conflict. At the time he wrote *Mutual Aid,* he was a political refugee living in England. A large portion of his book is devoted to documenting the importance of cooperation in primitive human societies, rural villages, and labor union guilds as well as among animals (for more information on Kropotkin, see S. J. Gould 1988; Todes 1989).

Lynn Margulis convinced biologists after a long battle that eucaryotes originated through fusion of an archaebacterium with some eubacteria. It is now understood that mitochondria in all eucaryotes and chloroplasts in plants were once independently living procaryotes. They are examples, much like corals and lichens, of how integrating symbionts evolved into mutualistic entities.

In their book *Acquiring Genomes,* Margulis and Sagan (2002) advanced a theory that speciation is not due to random events and neo-Darwinian processes, such as mutation and natural selection acting through competition; rather, they argued, speciation events are caused by interacting symbionts, cooperation, and reticulation of genomes. Their theory questions some of the central tenets of Darwinism. They speculated that Darwin was wrong to emphasize competition and natural selection as the sole forces shaping speciation and evolution; instead, they postulated that cooperation and mutualism drive evolution. This theory promises to become a fertile area of research during the twenty-first century.

Until recently, positive interactions were not subjected to as much quantitative study as negative interactions. One might reasonably assume that negative and positive relations between populations eventually tend to balance one another, and that both are equally important in the evolution of species and in the stabilization of the ecosystem.

Commensalism is a simple type of positive interaction and perhaps represents the first step toward the development of beneficial relations (see Table 7-1). It is especially common between sessile plants and animals on the one hand and mobile organisms on the other. Practically every worm burrow, shellfish, or sponge contains various "uninvited guest" organisms that require the shelter of the host but do neither harm nor good in return. Oysters, for example, sometimes have a small, delicate crab in the mantle cavity. These crabs are usually commensal, although sometimes they overdo their guest status by partaking of the host's tissues. Dales (1957), in his early review of marine commensalism, listed 13 species that live as guests in the burrows of large sea worms (*Erechis*) and burrowing shrimp (*Callianassa* and *Upogebia*). This array of commensal fish, clams, polychaete worms, and crabs lives by snatching surplus or rejected food or waste materials from the host. Many commensals are not host specific, but some apparently are found associated with only one species of host.

It is but a short step from commensalism to a situation in which both organisms gain by an association or interaction of some kind; this relationship is termed **protocooperation.** W. C. Allee (1951) studied and wrote extensively about this subject. He stressed the importance of cooperation and aggregation among species—a principle frequently termed the *Allee principle of aggregation* (discussed in Chapter 6). He believed that cooperation between species is to be found throughout nature. Returning to the sea for an example, crabs and coelenterates often associate with mutual benefit. The coelenterates grow on the backs of the crabs (and are sometimes "planted" there by the crabs), providing camouflage and protection (as coelenterates have stinging cells). In turn, the coelenterates are transported about and obtain particles of food when the crab captures and eats another animal.

In the preceding example, the crab does not absolutely depend on the coelenterate, nor vice versa. A further step in cooperation results when each population becomes completely dependent on the other. Such cases have been termed **mutualism** or **obligate symbiosis.** Often quite diverse kinds of organisms are associated. In fact, instances of mutualism are most likely to develop between organisms with widely different requirements (organisms with similar requirements are more likely to get involved in competition). The most important examples of mutualism develop between autotrophs and heterotrophs, which is not surprising, as these two components of the ecosystem must ultimately achieve some kind of balanced symbiosis. Examples that would be labeled as mutualistic go beyond general community interdependence to the extent that one particular kind of heterotroph becomes completely dependent on a particular kind of autotroph for food, and the latter becomes dependent on the protection, mineral cycling, or other vital functions provided by the heterotroph. The different kinds of partnerships between nitrogen-fixing microorganisms and higher plants were discussed in Chapter 4. Mutualism is also common between microorganisms that can digest cellulose (and other resistant plant residues) and animals that do not have the necessary enzyme systems for this purpose. As previously suggested, mutualism seems to replace parasitism as ecosystems evolve toward maturity, and it seems to be especially important when some aspect of the environment is limiting (such as water or infertile soil).

Examples

Obligate symbiosis between ungulates (such as cattle) and rumen bacteria is a well-studied example of mutualism. The anaerobic nature of the rumen system is inefficient for bacterial growth (only 10 percent of the energy in grass or hay eaten by the cow is assimilated by the bacteria), but the very nature of this inefficiency constitutes the reason that the ruminant can subsist at all on such a substrate as cellulose. The major portion of the residual energy of microbial action consists of fatty acids that are converted from cellulose but are not further degraded. These end products, however, are directly available for assimilation by the ruminant. Accordingly, the partnership is very efficient for the ruminant, because it gets most of the energy in the cellulose, which it could not obtain without the help of the bacteria. In return, of course, the bacteria get a temperature-controlled culture medium.

Mutualism between cellulose-digesting microorganisms and arthropods is quite common and is often a major factor in detritus food chains. The termite–intestinal flagellate partnership is a classic case, first worked out by Cleveland (1924, 1926). Without the specialized flagellates, many species of termites cannot digest the wood they ingest, as demonstrated by the fact that they starve to death when the flagellates are experimentally removed. The symbionts are so well coordinated with their host that they respond to the termite's molting hormones by encysting, thus ensuring transmission and reinfection when the termite molts its gut lining and then ingests it.

In termites, the symbionts live inside the body of the host. However, even more intimate interdependence may develop with the microorganism partners living outside the body of the animal host, and such associations may actually represent a more advanced stage in the evolution of mutualism (less chance that the relationship might revert to parasitism). One example are the tropical attine ants, which cultivate fungal gardens on the leaves they harvest and store in their nests. The ants fertilize, tend, and harvest the fungal crop in much the same manner as an efficient human farmer would. The ant-fungal system accelerates the natural decomposition of leaves. A succession of microorganisms is normally required to decompose leaf litter, with basidiomycete fungi normally appearing during the late stages of decomposition. However, when leaves are "fertilized" by ant excreta in the fungal gardens, these fungi can thrive on fresh leaves as a rapidly growing monoculture that provides food for the ants. Much ant energy is, of course, required to maintain this monoculture, just as much energy is required in intensive crop culture by humans.

By cultivating a cellulose-degrading organism as a food crop, the ants gain access to the vast cellulose reserves of the rain forest for indirect use as a nutrient. What termites accomplish by their *endosymbiotic* association with cellulose-degrading microorganisms, the attine ants have achieved through their more complex *ectosymbiotic* association with a cellulose-degrading fungus. In biochemical terms, the contribution of the fungus to the ant is the enzymatic apparatus for degrading cellulose. In turn, the fecal material of the ant contains proteolytic enzymes that the fungus lacks, so that the ants contribute their enzymatic apparatus to degrade protein. The symbiosis can be viewed as a *metabolic alliance* in which the carbon and nitrogen metabolisms of the two organisms have been integrated.

Coprophagy, or the reingestion of feces, which appears to be characteristic of detritivores, can probably be viewed as a much less elaborate but much more widespread case of mutualism that couples the carbon and nitrogen metabolism of microorganisms and animals—an "external rumen." Rabbits, for example, reingest their feces, illustrating the role of coprophagy in natural communities.

Ants and *Acacia* trees are involved in another striking tropical mutualistic symbiosis, as first described by Janzen (1966, 1967). The trees house and feed the ants, which nest in special cavities in the branches. In turn, the ants protect the tree from would-be herbivorous insects. When the ants are removed experimentally (as by poisoning with an insecticide), the tree is quickly attacked and often killed by defoliating insects.

Mineral cycling as well as food production are enhanced by mutualistic relationships between microorganisms and plants. Prime examples are **mycorrhizae** ("fungus-root"), mycelia of fungi that live in mutualistic association with the living roots of plants (not to be confused with parasitic fungi, which kill roots). As with nitrogen-fixing bacteria and legumes, the fungi interact with root tissue to form composite "organs" that enhance the ability of the plant to extract minerals from the soil.

In return, of course, the fungi are supplied with some of the plant's photosynthate. So important is the energy flow pathway through mycorrhizae that this route can be listed as a major food chain.

There are two main types of mycorrhizae. In the **ectomycorrhizae,** the fungus forms a *sheath* or network around actively growing roots, from which hyphae grow out into the soil, often for long distances. These mycorrhizae associate mostly with trees, especially pines and other conifers and tropical trees. The **vesicular-arbuscular** or **VA mycorrhizae** (formerly called *endomycorrhizae*) penetrate into root tissue, where they form characteristic vesicle-like structures (hence the name). Hyphae extend out into the soil, as in the ectomycorrhizae. These mycorrhizae colonize all but a few genera of plants, including herbs, shrubs, and trees in all climatic regions.

Mycorrhizae are not generally host-specific, which means that they can often colonize whatever plant root comes into contact with their spores. Some ectomycorrhizae produce large aboveground sporocarps or mushrooms that facilitate dispersal. The VA types produce spores underground, where they may be dispersed by soil-dwelling animals. Mycorrhizae are found in virtually every terrestrial ecosystem, including the rain forests of the Tropics, the prairies of the North Temperate Zone, and the Arctic Tundra. The mycorrhizal relationship between fungi and plants is both ubiquitous and ancient. About 90 percent of plant species, including most crops, form some kind of beneficial association with these fungi (Picone 2002). These benefits include:

- Mycorrhizal fungi increase nutrient uptake, especially of nitrogen and phosphorus. The hyphae extend from the colonized root into the soil. Because of their high rate of surface area to volume, hyphae are highly proficient at absorbing soil nutrients and transporting them to the root. The fungi receive carbohydrates, especially sugars, from the plant in this mutualistic relationship.
- Mycorrhizal fungi help suppress certain weeds. Colonized roots are better able to resist soil pathogens, including nematodes and pathogenic fungi.
- Mycorrhizae play an important role in improving soil texture, and are considered the most important biological agent for aggregating most soil types. This aggregation, termed **tilth,** is what makes for a healthy soil structure. Such soil is loose, permits roots to penetrate, lets water percolate easily, and allows biota, such as earthworms, to burrow unimpeded.

Unfortunately, conventional agricultural practices tend to disturb this beneficial relationship between mycorrhizal fungi and plant colonization—the result being that soil properties and ecosystem processes such as nutrient cycling are impeded (Coleman and Crossley 1996).

Figure 7-11 depicts how conventional agricultural practices affect the soil containing the mycorrhizal community (Picone 2002). Step 1 depicts how tillage crushes soil aggregates, destroys webs of mycorrhizal fungus, and reduces fungus abundance. This disturbance inhibits soil aggregation and tilth formation (Step 2). Thus, the soil becomes compacted and poorly aerated, frequently requiring further tillage (Step 3).

Tillage also promotes dependence on fertilizers due to the disruption of fungal mechanisms of nutrient recycling (Step 4); modern agricultural practices therefore compensate by applying synthetic, commercial fertilizers (Step 5). Synthetic fertilizer, unlike organic fertilizers, reduces the abundance of mycorrhizal fungi and selects for fungi that are ineffective at nutrient uptake (Step 6). Soil with an impoverished mycorrhizal community creates an optimal environment for non-host weeds

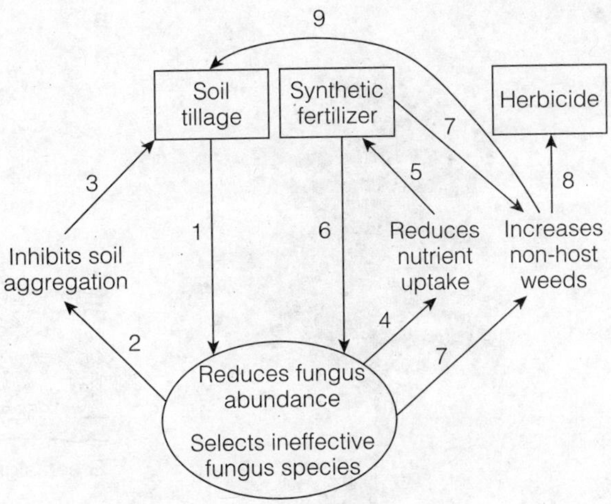

Figure 7-11. A model depicting the effects of conventional agricultural tillage practices on the mycorrhizal soil community (see text for details; with permission of The Land Institute, Salina, Kansas).

(Step 7). Hence, biological mechanisms for weed control (Step 8) and even more tillage (Step 9) are required. Thus, agricultural tillage helps create an industrial cycle rather than the ecological (biotic) cycle that has evolved in natural systems. Farm productivity frequently plummets as a result of this farming practice, until the network of mycorrhizal fungi becomes reestablished and mechanisms of nutrient recycling based on soil health are restored.

Many trees will not grow without mycorrhizae. Forest trees transplanted to prairie soil in a different region often fail to grow unless inoculated with the fungal symbionts. Pine trees with healthy mycorrhizal associates grow vigorously in soil so poor by conventional agricultural standards that corn or wheat could not survive. The fungi can metabolize "unavailable" phosphorus and other minerals by chelation or by other means. When labeled minerals (such as radioactive tracer phosphorus) are added to the soil, as much as 90 percent may be quickly taken up by the mycorrhizal mass, then slowly released to the plant. It is fortunate that the pine tree mycorrhizal system does so well on the millions of acres in the southern United States where topsoil was devastated by the row-crop monoculture and absentee-owner system that persisted for so long; otherwise, many of these eroded acres would be deserts today.

Figure 7-12A depicts mycorrhizal clusters around the roots of a spruce (*Picea pungens*). The role of mycorrhizae in direct mineral recycling, their importance in the Tropics, and the need for crops with such built-in recycling systems were emphasized in Chapter 4. For additional information on mycorrhizal mutualisms, see Mosse et al. (1981) and E. I. Newman (1988).

Ahmadjian (1995) noted that *lichens* are probably the most misunderstood and poorly appreciated organisms in the biological world. Lichens are an important part of the biological web that links all of us together. They are a unique combination of traits, being primarily fungal, but also cyanobacterial. Approximately 8 percent of the terrestrial surface of Earth has lichens as its most dominant life-form. For example, in the boreal forests of North America, Europe, and Russia, vast areas of the ground are covered with reindeer lichen (frequently termed "reindeer moss"), especially species of the genus *Cladina.* It is likely that lichens play a role in regulating the gaseous composition of the atmosphere of Earth, possibly by functioning as a CO_2 sink (Ahmadjian 1995).

Lichens are an association of specific fungi and algae, so intimate in terms of

A

Courtesy of S. A. Wilde, University of Wisconsin

B

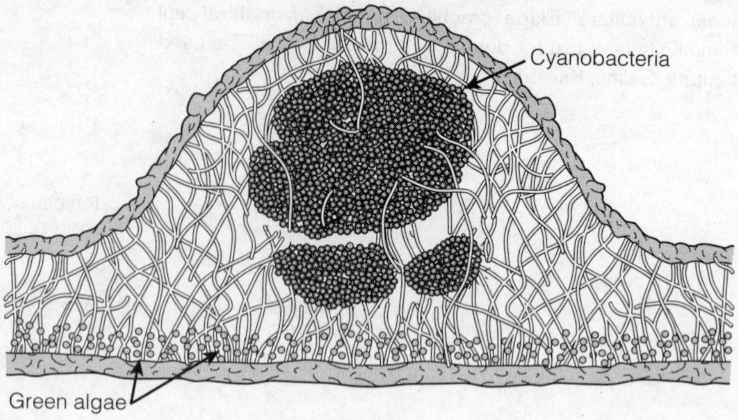

Figure 7-12. (A) Peritrophic mycorrhizae forming clusters or masses around the roots of a spruce (*Picea pungens*) seedling. (B) Principal nitrogen fixer among the epiphytic lichens in the forest canopy community is *Lobaria oregana* (shown in cross-section). *Lobaria* is a mutualistic association of a fungus and a green alga; populations of cyanobacteria are also present in bulges and fix nitrogen at a significant rate. (Adapted from *Scientific American*, June 1973, p. 79. Reprinted with permission of the Estate of Eric Mose.)

functional interdependence and so integrated morphologically that a third kind of organism, resembling neither of its components, is formed. Lichens are usually classified as single species, even though they are composed of two or more unrelated species. In lichens, one sees evidence of an evolution from parasitism to mutualism. In some of the more primitive lichens, for example, the fungi actually penetrate the algal cells and are thus essentially parasites of the algae. In the more advanced species, the fungal mycelia or hyphae do not break into the algal cells, but the two live in close harmony. Figure 7-12B depicts a principal nitrogen fixer among the epiphytic lichens (*Lobaria oregana*) in the forest-canopy community. *Lobaria* is a mutualistic association of a fungus and a green alga, but also contains populations of cyanobacteria that fix nitrogen at a significant rate.

The mutualistic lichen lifestyle has had at least five independent origins in different branches of the fungus family tree; at least 20 percent of all fungal species are lichens (Gargas et al. 1995). Such multiple origins demonstrate that mutualism may be just as important as competition in evolution, as Kropotkin (1902) suggested more than a century ago.

It is evident that mutualism has special survival value when resources become tied up in the biomass, as in a mature forest, or when soil or water is nutrient poor, as in some coral reefs or rain forests. Like corals and other highly organized heterotroph-autotroph mutualistic complexes, lichens are well adapted to natural scarcities and stress, but they are very vulnerable to pollution stress, especially air pollution. With regard to the restoration of landscape in Sudbury, Ontario, which was devastated by air pollution (as mentioned in Chapter 3), the return of lichens is a welcome sign that restoration is working.

For general reviews of symbiotic associations, see Boucher et al. (1982) and Keddy (1990). We also recommend the supplement to *The American Naturalist*, 2003, Volume 162, entitled "Interacting Guilds: Moving Beyond the Pairwise Perspective on Mutualisms." The relationship between humans and cultivated plants and domesticated animals, which might be considered a special form of mutualism, is discussed in Chapter 8. In Chapter 2, coral-algal associations were discussed as an emergent property that enhances the recycling of nutrients and the productivity of the whole ecosystem.

The indirect effects of species on one another may be as important as their direct interactions, and may contribute to *network mutualism.* When food chains function in food web networks, the organisms at each end of a trophic series—for example, plankton and bass in a pond—do not interact directly, but they benefit each other indirectly. Bass benefit by eating planktivorous fish, which are supported by the plankton, whereas plankton benefit when bass reduce the population of the plankton predators. Accordingly, there are both negative (predator-prey) and positive (mutualistic) interactions in a food web network (D. S. Wilson 1986; Patten 1991).

Because of reward feedbacks, as discussed in Chapter 4, and because of the tendency for the severity of negative interactions to decrease with time (see Section 3), it is not too much of a stretch to consider whole food chains as mutualistic (E. P. Odum and Biever 1984). In a study of an alga-herbivore relationship, Sterner (1986) found that the algae grew better when grazed because of the nitrogen regenerated by the grazer.

The ultimate reality is that all positive and negative two-species interactions operate together in food webs at the community and ecosystem levels. The energetics of food chains (as detailed in Chapter 3), combined with what have come to be known as "top-down" and "bottom-up" processes, make the food web a functional system that is more than just a collection of species interactions. *Top-down control,* which includes reward feedback, refers to the role of upstream components—for example, herbivore control over plants and predator control over herbivores. *Bottom-up control* refers to the role of nutrients and other physical factors in determining primary production. Ecologists debate which type of control is most important in a given situation, but most now agree that both are involved to varying degrees in any and all natural situations (Hunter and Price 1992; Polis 1994; de Ruiter et al. 1995; Krebs et al. 1995; Krebs, Boonstra, et al. 2001; Polis and Strong 1996).

Until recently, human beings have generally acted as energy parasites on their autotrophic environment, taking what they want with little concern about the welfare of planet Earth. For example, large cities grow and become parasitic on the countryside, which must somehow supply food and water and degrade huge quantities of wastes for these cities. Human beings must evolve to the stage of mutualism in relationship with nature. If humankind does not achieve mutualism with nature, then, like the "unwise" or "unadapted" parasite, humans may exploit their host to the point of destroying themselves.

7 Concepts of Habitat, Ecological Niche, and Guild

Statement

The **habitat** of an organism is the place where it lives, or the place where one would go to find it. The **ecological niche,** however, includes not only the physical space occupied by an organism but also its functional role in the community (its trophic position, for instance) and its position in environmental gradients of temperature, moisture, pH, soil, and other conditions of existence. These three aspects of the ecological niche can be conveniently designated as the *spatial* or *habitat niche,* the *trophic niche,* and the *multidimensional* or *hypervolume niche.* Consequently, the ecological niche of an organism not only depends on where it lives but also includes the sum total of its environmental requirements. The concept of niche is most useful, and quantitatively most applicable, in terms of *differences* between species (or the same species at two or more locations or times) in one or a few major (operationally significant) features. The dimensions most often quantified are *niche breadth* and *niche overlap* with neighbors. Groups of species with comparable roles and niche dimensions within a community are termed **guilds.** Species that occupy the same niche in different geographical regions (continents and major oceans) are termed *ecological equivalents.*

Explanation and Examples

The term *habitat* is used widely, not only in ecology but elsewhere. Thus, the habitat of the water backswimmer (*Notonecta*) and the water boatman (*Corixa*) is the shallow, vegetation-choked area (*littoral region*) of ponds and lakes; one would go there to collect these particular water bugs. However, the two species occupy very different *trophic niches,* as the backswimmer is an active predator, whereas the water boatman feeds largely on decaying vegetation. The ecological literature is replete with examples of coexisting species that use different energy sources.

If the habitat is the "address" of the organism, *niche* is its "profession," its trophic position in food webs, how it lives and interacts with the physical environment and with other organisms in its community. *Habitat* may also refer to the place occupied by an entire community. For example, the habitat of the sand sage grassland community is the series of ridges of sandy soil occurring along the north sides of rivers in the southern Great Plains of the United States. Habitat in this case consists mostly of physical or abiotic complexes, whereas habitat for the water bugs mentioned previously includes living and nonliving objects. Thus, the habitat of an organism or group of organisms (population) includes other organisms and the abiotic environment.

The concept of ecological niche is not so generally understood outside the field of ecology. Terms such as niche are difficult to define and quantify; the best approach is to consider the component concepts historically. Joseph Grinnell (1917, 1928) used the word *niche* "to stand for the concept of the ultimate distributional unit, within which each species is held by its structural and instinctive limitations . . . no two species in the same general territory can occupy for long identically the same ecological niche." (Incidentally, the latter statement predates Gause's experimental demonstration of the competitive exclusion principle; see Fig. 7-2.) Thus, Grinnell thought of the niche mostly in terms of the microhabitat, or what is now called the **spatial niche.** Charles Elton (1927) was one of the first to begin using the term *niche* in the sense of the "functional status of an organism in its community." Because of Elton's great influence on ecological thinking, it has become generally accepted that niche is by no means a synonym for habitat. Because Elton emphasized the importance of energy relations, his version of the concept is designated the **trophic niche.**

G. E. Hutchinson (1957) suggested that the niche could be visualized as a *multidimensional space* or *hypervolume* within which the environment permits an individ-

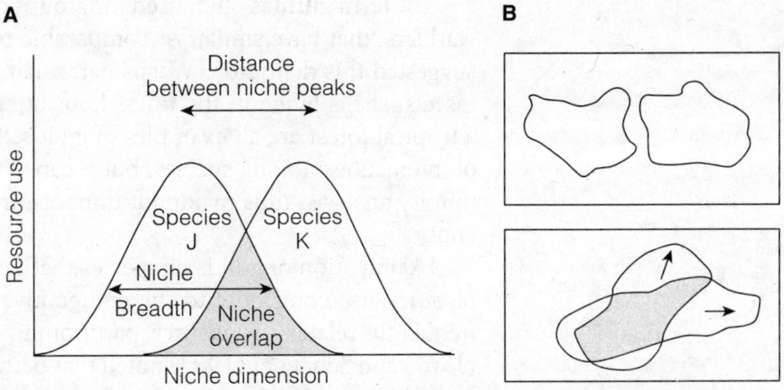

Figure 7-13. Schematic representations of the niche concept. (A) Activity curves for two species along a single resource dimension illustrate the concepts of niche breadth and niche overlap. (B) In the upper diagram, two species occupy nonoverlapping niches, whereas in the lower diagram, niches overlap so much that severe competition results in divergence, as indicated by the arrows.

ual or species to survive indefinitely. Hutchinson's niche, which can be designated the **multidimensional** or **hypervolume niche**, can be measured and mathematically manipulated. For example, two-dimensional climographs, which depict *x*- and *y*-axes of a particular species of bird and a fruit fly, could be expanded as a series of coordinates (*x*-, *y*-, and *z*-axes) to include other environmental dimensions. Hutchinson (1965) also distinguished between the **fundamental niche**—the maximum "abstractly inhabited hypervolume" when the species is not constrained by competition or other limiting biotic interactions—and the **realized niche**—a smaller hypervolume occupied under particular biotic constraints. The concepts of niche breadth and niche overlap are illustrated in two dimensions in Figures 7-13A and B.

Perhaps a simple analogy from everyday human affairs will help to clarify these overlapping and sometimes confusing ecological uses of the term *niche.* To become acquainted with a person in the human community, one would need to know, first of all, his or her address, (where she or he could be found). "Address" would represent *habitat.* To "know" the person, however, one would want to know something about his or her occupation, interests, associates, and role in community life. All this information would be analogous to that person's niche. Thus, in the study of organisms, learning the habitat is just the beginning. To determine the status of the organism within the natural community, one would need to know something of its activities, especially its nutrition; energy sources and resource partitioning; relevant population attributes, such as intrinsic rate of increase and fitness; and finally, the organism's effect on other organisms with which it comes into contact, and the extent to which it modifies or can modify important operations in the ecosystem.

In a classic investigation in the history of ecology, MacArthur (1958) compared the niches of four species of American warblers (Parulidae) that all breed in the same macrohabitat (a spruce forest) and all feed on insects but forage and nest in different parts of the spruce tree. MacArthur constructed a mathematical model, which consisted of a set of competition equations in a matrix from which competition coefficients were calculated for the interaction between each species and any of the other three. Thus, niches of similar species associated together in the same habitat can be precisely compared when only a few operationally significant measurements are involved. Two species proved especially competitive, so that if either were absent, the other might be expected to move into the vacated niche space. The general tendency for niches to narrow with interspecific competition has already been illustrated in Figure 7-5.

The term **guild** is often used for groups or clusters of species, such as MacArthur's warblers, that have similar or comparable roles in the community; Root (1967) first suggested this definition. Wasps parasitizing a herbivore population, nectar-feeding insects, snails living in the forest floor litter, and vines climbing into the canopy of a tropical forest are all examples of guilds. The guild is a convenient unit for studies of interactions among species, but it can also be treated as a functional unit in community analysis, thus making it unnecessary to consider every species as a separate entity.

Examination of guilds or species that fail to coexist can illustrate what aspects of resource use contribute to the competitive exclusion principle. Niche partitioning frequently relates to resource partitioning or resource use. MacArthur and Levins (1967) and Schoener (1983) noted that perhaps the most operational approach to the study of competition and niche overlap is to focus on consumable resources, or factors that serve as surrogates for those resources, such as differences in microhabitats. Winemiller and Pianka (1990) have used this approach to identify nonrandom patterns and clusters regarding the way that species use resources in a guild.

Measurements of morphological features of larger plants and animals can often be used as indices in the comparison of niches. Van Valen (1965), for example, found that variations in the length and breadth of a bird's bill (the bill, of course, reflects the type of food eaten) provide an index of niche width; the coefficient of variation in bill width was found to be greater in island populations of six species of birds than in mainland populations, corresponding with the greater niche width (wider variety of habitat occupied and food eaten) on islands, where competing species are fewer.

Grant (1986) was able to separate feeding niches of Galápagos finches by measuring beak morphology. He found that differences in beak size correlated to differences in diet. Within the same species, competition is often greatly reduced when different stages in the life history of the organism occupy different niches; for example, the tadpole functions as a herbivore and the adult frog as a carnivore in the same pond. Niche segregation may even occur between sexes. In woodpeckers of the genus *Picoides,* males and females differ in bill size and in foraging behavior (Ligon 1968). In hawks, some weasels, and many insects, the sexes differ markedly in size and, therefore, in the dimensions of their food niche.

Both nutrients and toxic chemicals introduced into natural ecosystems can be expected to alter the niche relations of species most severely affected by the perturbation. In a long-term (11-year) experimental study of the effect of applying N-P-K commercial fertilizer and municipal sludge to old-field vegetation, W. P. Carson and Barrett (1988) and Brewer et al. (1994) reported that niche width was significantly enhanced for summer annuals, especially *Ambrosia trifida, A. artemisiifolia,* and *Setaria faberii,* which increased their coverage at the expense of perennials such as *Solidago canadensis.*

Ecologically equivalent species, which occupy similar niches in different geographical regions, tend to be closely related taxonomically in contiguous regions, but are often not related in noncontiguous regions. The species composition of communities differs widely in different floral and faunal regions, but similar ecosystems develop equivalent functional niches wherever physical conditions are similar, regardless of geographical location. The equivalent functional niches are occupied by whatever biological groups happen to make up the flora and fauna of the region. Thus, a grassland ecosystem develops wherever there is a grassland climate, but the species of grass and grazers may be quite different, especially when the regions are widely sep-

Table 7-3

Ecologically equivalent grassland birds in a Kansas field and a Chilean field

Ecologically equivalent pair of species	*Body size (mm)*	*Bill length (mm)*	*Ratio of bill depth to length*
Eastern meadowlark (*Sturnella magna*), Kansas	236	32.1	0.36
Red-breasted meadowlark (*Pezites militaris*), Chile	264	33.3	0.40
Grasshopper sparrow (*Ammodramus savannarum*), Kansas	118	6.5	0.60
Yellow grass finch (*Sicalis luteola*), Chile	125	7.1	0.73
Horned lark (*Eremophila alpestris*), Kansas	157	11.2	0.50
Chilean pipit (*Anthus correnderas*), Chile	153	13.0	0.42

Source: After Cody 1974.

Note: In each field the three species differ in feeding niches as shown by differences in body size and bill dimensions, but each pair of equivalents is very closely matched morphologically indicating very similar niches. The meadowlarks are closely related taxonomically, but the second pair are related only at the family level, and the third pair belong to different families.

Table 7-4

Ecological equivalents in three major niches of four coastal zones of North and Central America

Niche	*Tropical*	*Coast*	*Upper West Gulf Coast*	*Upper East Coast*
Grazer on intertidal rocks (periwinkles)	*Littorina ziczac*	*L. danaxis*, *L. scutulata*	*L. irrorata*	*L. littorea*
Benthic carnivore	Spiny lobster (*Palinurus*)	King crab (*Paralithodes*)	Stone crab (*Menippe*)	Lobster (*Homarus*)
Plankton-feeding fish	Anchovy	Pacific herring, sardine	Menhaden, threadfin	Atlantic herring, alewife

arated by barriers. The large kangaroos of the Australian grassland are the ecological equivalents of the bison and pronghorn of the North American grassland (both now largely replaced by domesticated grazers). Examples of bird ecological equivalents on two continents are listed in Table 7-3. Examples of ecological equivalents in aquatic habitats are shown in Table 7-4.

8 Biodiversity

Statement

In Chapter 2, diversity at the ecosystem level was reviewed, including two major components of diversity—richness and apportionment. In this section, additional aspects of this important subject will be considered, especially as related to the community level of organization. Of the total number of species in a trophic component, or in a community as a whole, often a relatively small percentage are abundant or **dominant** (represented by large numbers of individuals, a large biomass, high rates of productivity, or other indications of importance), and a large percentage are rare (have smaller importance values). Sometimes, though, there are no dominant species but many species of intermediate abundance. As noted previously, the concept of *species diversity* has two components: (1) *richness,* based on the total number of species present; and (2) *apportionment,* based on relative abundance (or other measures of importance) of species and the degree of dominance or lack thereof. Species diversity tends to increase with the size of the area, and from high latitudes to the equator. Diversity tends to be reduced in stressed biotic communities, but it may also be reduced by competition in old communities in stable physical environments. Three other kinds of diversity are also important: (1) **pattern diversity,** which results from zonation, stratification, periodicity, patchiness, food webs, and other arrangements; (2) **genetic diversity,** the maintenance of genotypic heterozygosity, polymorphism, and other genetic variability that is an adaptive necessity for natural populations; and (3) **habitat diversity,** the diversity of habitat or landscape patches, which serves as a basis for metapopulation dynamics (see Chapter 6) and the diversity of species within a particular habitat or community type. Many ecologists are becoming concerned that the reduction in habitat, species, and genetic diversity resulting from human activities is jeopardizing future adaptability in natural ecosystems, agroecosystems, and agrolandscapes.

Explanation

The pattern of a few common or dominant species having large numbers of individuals associated with many rare species having few individuals is characteristic of community structures in high latitudes and in the seasonal Tropics (wet-dry seasons); but in the wet Tropics (with unchanging seasons), one usually finds many species that have low relative abundance. The general trend of an increase in the number of species from high to equatorial latitudes is illustrated in Figure 7-14. Another general trend or natural law is that the number of species increases with the size of an area, and probably also with the evolutionary time that has been available for colonization, niche specialization, and speciation.

In Chapter 2, two broad approaches to measure and analyze species diversity were outlined: (1) *dominance-diversity (relative abundance) curves;* and (2) *diversity indices,* which are ratios of species importance relationships. Comparison of temperate and tropical forest diversity was used to illustrate the graphic approach, and the Shannon-Weaver and Simpson indices were used to illustrate the index methods. An additional graphic method for depicting species diversity is shown in Figure 7-15.

A major component of diversity is *apportionment* (*evenness*), which indicates how

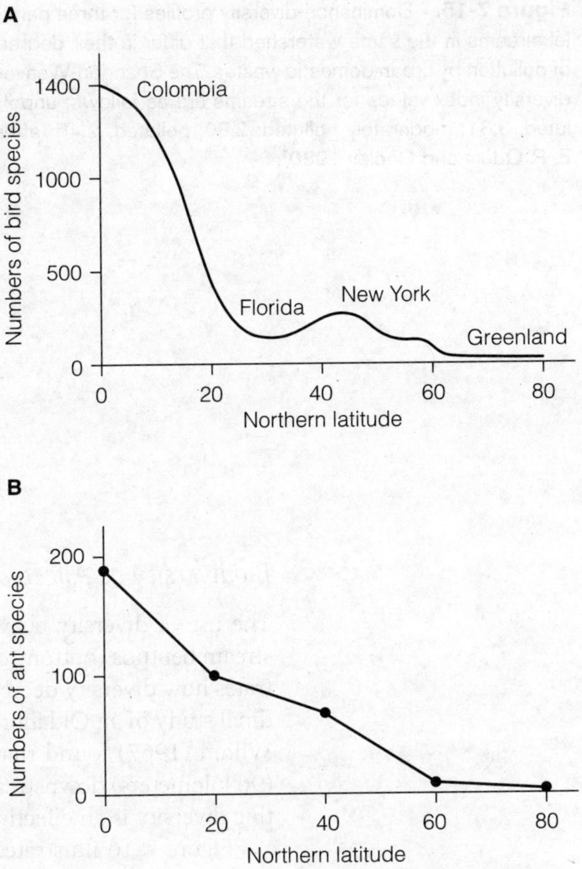

Figure 7-14. Latitudinal gradient in numbers of species of (A) breeding land birds and (B) ants (redrawn from Fischer 1960).

numbers of individuals are distributed (apportioned) among species. For example, two systems, each containing ten species and 100 individuals, have the same S/N richness index (where S equals the number of species and N the number of individuals), but they could differ widely in apportionment, depending on the distribution of the 100 individuals among the ten species—for example, 91, 1, 1, 1, 1, 1, 1, 1, 1, and 1 at one extreme (minimum evenness and maximum dominance) or 10 individuals per species (perfect evenness and no dominance) at the other extreme. Evenness tends to be high and constant in bird populations (probably because of territorial behavior). In contrast, plants and phytoplankton populations tend to average lower in evenness and to exhibit considerable variation in both components.

Because the Shannon-Weaver index is derived from information theory and represents a type of formulation widely used in assessing the complexity and information content of all kinds of systems, it is one of the best to use in making comparisons if one is not interested in separating the two components of diversity. And once $\overline{H}$ is calculated, evenness can be separated out quickly by dividing by the log of the number of species. The Shannon-Weaver index is also reasonably independent of sample size and is normally distributed, provided that all N are integers (Hutcheson 1970); thus, parametric statistical methods can be used to test for the significance of differences between means.

Figure 7-15. Dominance-diversity profiles for three parallel streams in the same watershed that differ in their degree of pollution by urban domestic wastes. The Shannon-Weaver diversity index values for the streams are as follows: unpolluted, 3.31; moderately polluted, 2.80; polluted, 2.45 (after E. P. Odum and Cooley 1980).

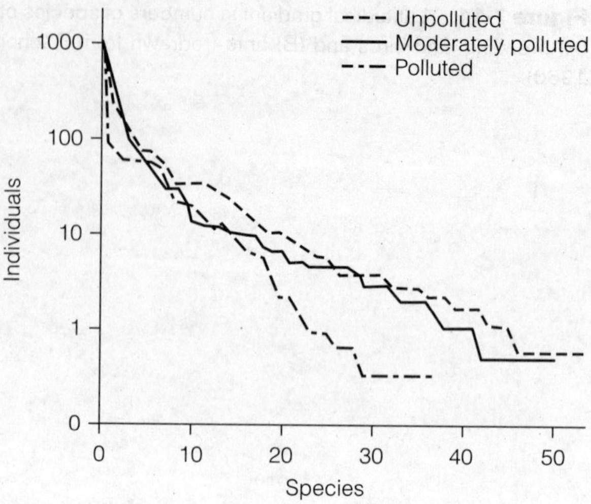

Biodiversity as Affected by Pollution

The use of diversity curves and indices to assess the impact of sewage effluent on stream benthos (bottom-dwelling organisms) is depicted in Figure 7-15, which illustrates how diversity decreases with increasing loads of domestic wastes. In a longitudinal study of an Oklahoma stream receiving inadequately treated municipal wastes, Wilhm (1967) found that benthic diversity was depressed for more than 60 miles (96 kilometers) downstream. From these and many other studies, it is clear that benthic diversity is an effective tool for monitoring water pollution.

Figure 7-16 illustrates how a point source of untreated municipal waste affects species richness. Note that while population densities (especially coliform bacteria and sludge worms) increase, there is a concomitant decrease in species richness (especially of aquatic insects and desirable freshwater fish species). From an ecosystem and landscape perspective (including human societal understanding), it is likely that the diversity of functions (such as aerobic production and respiration) is much more important than species diversity, although species diversity provides the structure for these functional processes. Recall Figure 2-12, which showed how these metabolic processes were affected by the anaerobic conditions following the pollution discharge, illustrating the importance of maintaining aerobic conditions (to prevent pollution) across levels of organization.

Figure 7-17 shows the effect of an acute insecticide stress on the diversity of arthropods in a millet (*Panicum ramosum*) field. Although the richness index (Margalef *d*) was greatly reduced by the treatment, evenness (Pielou *e*) increased and remained elevated for most of the growing season. When the insecticide killed many of the dominant species, a greater evenness (*e*) in the abundance of the surviving populations resulted. The Shannon-Weaver index ($\bar{H}$), which expresses the interaction of richness and apportionment, showed an intermediate response. Although the insecticide used in this experiment remained toxic for only 10 days, and the acute depression lasted only about two weeks, overshoots and oscillations in the diversity ratios were evident for many weeks. This study illustrates several points of interest: (1) it is frequently desirable to separate species richness and relative abundance or apportionment; (2) a generalized or moderate perturbation may increase rather than decrease diversity when there is originally strong dominance; and (3) recovery is rapid when small areas are perturbed, because replacements come in quickly from surrounding areas. Applying insecticides to large areas is quite another matter.

Table 7-5 presents a comparison of the density and diversity of arthropod populations in a grain field and a natural herbaceous community that replaced the field one year later. The values shown are means of ten samples taken over the growing season. After only one year under "nature's management," the following changes had occurred:

- The number of herbivorous (phytophagous) insects was greatly reduced, as was the total arthropod density.
- The richness component of diversity and the index of total diversity were significantly increased for each guild, as well as for the total community of arthropods.
- Evenness increased.
- The number, diversity, and percentage composition of predators and parasites were greatly increased; predators and parasites made up only 17 percent of the population density in the grain field, compared with 47 percent in the natural field (where they actually outnumbered the herbivores).

This comparison gives some clues as to why artificial communities often require chemical or other control of herbivorous insects, whereas natural areas do not re-

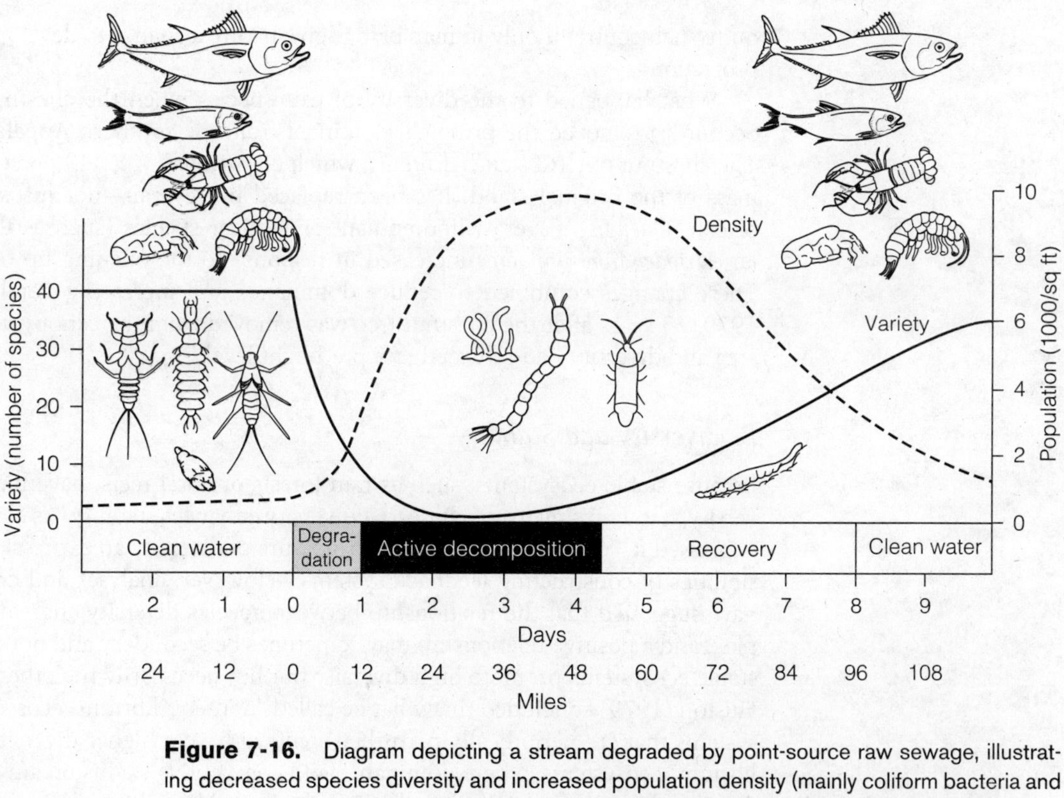

Figure 7-16. Diagram depicting a stream degraded by point-source raw sewage, illustrating decreased species diversity and increased population density (mainly coliform bacteria and sludge worms) in the area of active decomposition, followed by stream recovery to clean water.

Figure 7-17. The effect of a single application of the insecticide Sevin (an organophosphate insecticide that remains toxic for only 10 days) on the arthropod population in a 0.4-hectare plot of a millet (*Panicum ramosum*) field. Two components of diversity (*d, e*) and a general index of total diversity ($\bar{H}$) are based on ten 0.25-m^2 samples taken from the treated area and a control area at weekly or biweekly intervals from early July through September. The semilog plots facilitate a direct comparison of relative deviations resulting from the acute insecticide stress (after Barrett 1968).

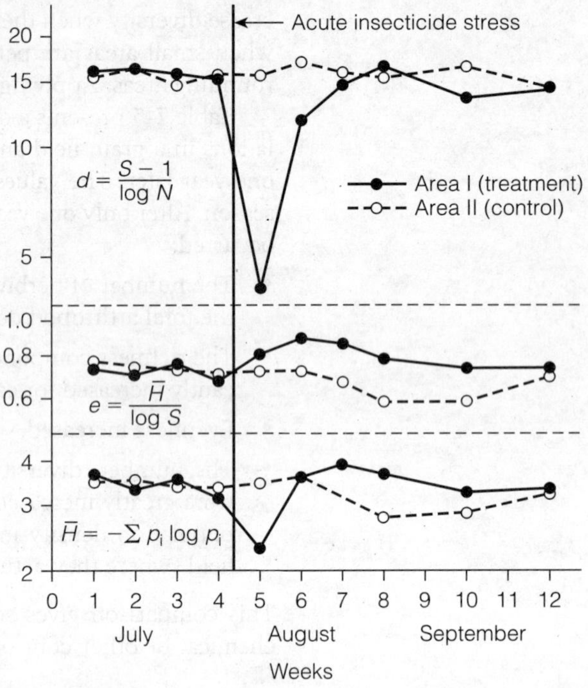

quire such control if only human beings give nature a chance to develop its own self-protection.

What happened to the diversity of tree species when the chestnut blight (see Section 5) removed the principal dominant from the southern Appalachian forest? The chestnut tree (*Castanea dentata*), which constituted 30 to 40 percent of the biomass of the original stand, has been replaced by several—not just one—species of oak (*Quercus*). Several subdominant or pioneer species (such as the tulip poplar, *Liriodendron tulipifera*) increased in response to the opening up of the canopy. These changes combined to reduce dominance and increase general diversity. In 1970, 25 years after the chestnut tree was removed from the canopy, the total basal area and diversity had returned to a pre-blight level.

Biodiversity and Stability

Because stable ecosystems, such as rain forests or coral reefs, have high species diversity, it is tempting to conclude that diversity enhances stability. As Margalef (1968) expressed it, "the ecologist sees in any measure of diversity an expression of the possibilities of constructing feedback systems." However, analyses and critical reviews have suggested that the relationship between species diversity and stability is complex, and a positive relationship may sometimes be secondary and not causal, in that stable ecosystems promote high diversity but not necessarily the other way around. Huston (1979) concluded that what he called "non-equilibrium" ecosystems (that is, systems that are periodically perturbed) tend to have a higher diversity than "equilibrium" ecosystems, where dominance and competitive exclusion are more intense. On the other hand, McNaughton (1978) concluded from his studies of old fields and

East African grasslands that species diversity does mediate functional stability in the community at the primary producer level. W. P. Carson and Barrett (1988) and Brewer et al. (1994) noted that nutrient-enriched old-field communities are significantly less diversified, and apparently less stable, than unenriched communities of equal size and age.

A major problem with studies of species diversity is that, so far, they have dealt only with parts of communities, usually a particular taxonomic segment (such as birds or insects) or a single trophic level. Estimating the diversity of whole communities requires that all the different sizes and niche roles be "weighted" in some manner by some common denominator, such as energy. Stability is likely related more closely to functional than to structural diversity.

Diversity within a habitat or community type is not to be confused with diversity in a landscape or region containing a mixture of habitats and landscape patches. Whittaker (1960) suggested the following terms: (1) *alpha diversity,* for within-habitat

Table 7-5 **Density and diversity of arthropod populations in an unharvested millet crop, compared with the natural successional community that replaced it one year later**

Indices	*Populations*	*Cultivated millet field*	*Natural successional community*
Density (number/m^2)	Herbivores	482	156*
	Predators	82	117
	Parasites	24	51*
	Total arthropods	624	355*
Richness or variety index (Margalef *d*)	Herbivores	7.2	10.6*
	Predators	3.9	11.4*
	Parasites	6.3	12.4*
	Total arthropods	15.6	30.9*
Evenness index (Pielou *e*)	Herbivores	0.65	0.79*
	Predators	0.77	0.80
	Parasites	0.89	0.90
	Total arthropods	0.68	0.84*
Total apportionment index ($\overline{H}$)	Herbivores	2.58	3.28*
	Predators	2.37	3.32*
	Parasites	2.91	3.69*
	Total arthropods	3.26	4.49*

Note: The stand of millet (*Panicum*) was the control plot in the experiment graphed in Figure 7-17. Fertilizer was applied at time of planting in the prescribed agricultural manner but no insecticide or other chemical treatment was applied, and the crop was not harvested. All figures are means of weekly samples taken during growing season (July through September).

*Differences between the two communities significant at the $p < .001$ level of probability.

or within-community diversity; (2) *beta diversity,* for between-habitat diversity; and (3) *gamma diversity,* for diversity of landscape-scale areas.

In well-studied zones (such as benthic aquatic populations) and other *parts* of communities, species diversity is very much influenced by the functional relationships between trophic levels. For example, the amount of grazing or predation greatly affects the diversity of the grazed or prey populations. Moderate predation often reduces the density of dominant species, thus providing less competitive species with a better chance to use space and resources. J. L. Harper (1969) reported that the diversity of herbaceous plants on the English chalk downs declined when grazing rabbits were fenced out. Severe grazing or predation, on the other hand, acts as a stress and reduces the number of species to an unpalatable few. In a classic study, Paine (1966) found that the species diversity of sessile organisms in the rocky intertidal habitat (where space is generally more limiting than food) was higher in both temperate and tropical regions when first- and second-order predators were active. Experimental removal of predators in such situations reduced the species diversity of all sessile organisms, whether they were directly preyed on or not. Paine concluded that "local species diversity is directly related to the efficiency with which predators prevent monopolization of major environmental requisites by one species." This conclusion does not necessarily hold for habitats in which competition for space is less severe. Although human activities tend to reduce diversity and encourage monocultures, they do often increase habitat diversity in the general landscape (openings created in the forest, trees planted in the prairie, or new species introduced). The diversity of small songbirds and plants, for example, is much greater in older, established residential districts than in many natural areas.

Diversity Above and Below the Species Level

The species may not always be the best ecological unit for measurements of diversity, as life history stages within species often occupy different habitats and niches and, thereby, contribute to variety (richness) in the ecosystem. A caterpillar and a butterfly of the same species, or a frog and its tadpole stage, are more diverse in their roles in the community than are two species of butterflies or adult frogs.

Moreover, genetic variability and diversity are hidden when communities are described only in terms of the species present. In the absence of genetic variability, a species will not be able to adapt to new situations and, therefore, will likely become extinct in a changing environment.

Variety of species, life history stages, and genetic types are by no means the only elements involved in community diversity. The structure that results from the distribution of organisms in and their interaction with their environment is termed **pattern diversity.** Many different kinds of arrangements in the standing crop of organisms (that is, nature's architecture) contribute to what might be called pattern diversity. For example:

- *Stratification patterns* (vertical layering, as in vegetation and soil profiles);
- *Zonation patterns* (horizontal segregation, as in mountains or in the intertidal zone);
- *Activity patterns* (periodicity);
- *Food web patterns* (network organization in food chains);
- *Reproductive patterns* (parent-offspring associations or plant clones);

- *Social patterns* (flocks and herds);
- *Coactive patterns* (resulting from competition, antibiosis, or mutualism); and
- *Stochastic patterns* (resulting from random forces).

Diversity is also enhanced by *edge effects*—junctions between patches of contrasting types of vegetation or physical habitats (see Chapter 9 for details).

Biodiversity and Productivity

In low-nutrient natural environments, an increase in biodiversity seems to enhance productivity, as indicated by Tilman (1988) in experimental studies with grasslands, but in high-nutrient or enriched environments, an increase in productivity increases dominance and reduces diversity (W. P. Carson and Barrett 1988). In other words, a biodiversity increase may increase productivity, but a productivity increase almost always decreases diversity. Furthermore, nutrient enrichment (such as nitrogen fertilization and runoff) tends to bring on noxious weeds, exotic pests, and dangerous disease organisms, because these kinds of organisms are adapted to and thrive in high-nutrient environments. For example, when coral reefs are subjected to human-induced nutrient enrichment, we observe an increase in the dominance of smothering filamentous algae and the appearance of previously unknown diseases, either of which can quickly destroy these diverse ecosystems that are adapted to low-nutrient waters.

We venture to suggest that humans, in their efforts to increase productivity to support increasing numbers of people and domestic animals (which in turn excrete huge amounts of nutrients into the environment), are causing a worldwide eutrophication that is the greatest threat to ecosphere diversity, resilience, and stability—essentially a "too much of a good thing" syndrome. Global warming, which results from CO_2 enrichment of the atmosphere, is just one aspect of this overall perturbation. It was appropriate that the first Ecological Society of America Issues in Ecology (ESA 1997) focused on global nitrate enrichment.

There has been much debate regarding the relationship of community productivity to biotic diversity. Tilman et al. (1996), for example, demonstrated in a well-replicated field experiment that ecosystem productivity increased significantly with increased plant biodiversity. This **diversity-productivity hypothesis** (McNaughton 1993; Naeem et al. 1994) is based on the assumption that interspecific differences in the use of resources by plants allow more diverse plant communities to more fully use limiting resources and, thus, attain greater net primary productivity. Furthermore, Tilman (1987) and Tilman and Downing (1994) demonstrated in a long-term study of grasslands that primary productivity in more diverse plant communities is more resistant to and recovers more fully from major perturbations such as drought. These findings support the **diversity-stability hypothesis** (McNaughton 1977; Pimm 1984).

Numerous investigations have demonstrated that increased nutrient enrichment results in increased primary productivity (for example, Bakelaar and Odum 1978; Maly and Barrett 1984; W. P. Carson and Barrett 1988; S. D. Wilson and Tilman 1993; Stevens and Carson 1999). One might deduce that increased nutrient enrichment would lead to increased plant community diversity, which in turn would result in greater ecosystem stability. However, several investigations have shown that long-term nutrient enrichment (such as the addition of fertilizer or municipal sewage sludge to old-field communities) results in decreased plant community diversity (W. P. Carson and Barrett 1988; Brewer et al. 1994; Stevens and Carson 1999).

These findings and hypotheses have resulted in a great deal of confusion regarding the relationship between productivity and diversity. E. P. Odum (1998b) attempted to clarify this relationship by stating that in low-nutrient environments, an increase in community biodiversity seems to enhance or at least to be related to an increase in productivity, but in high-nutrient or enriched environments, an increase in primary productivity results in increased dominance of a few species, resulting in turn in a decrease in species diversity or community richness. This latter relationship appears to be especially relevant where nutrient subsidies, such as fertilizer or sludge, are added to natural communities in significant amounts.

Nutrient enrichment may also affect other aspects of the structure and function of old-field communities, such as changes in species composition (Bakelaar and Odum 1978; Vitousek 1983; Maly and Barrett 1984; Tilman 1987; W. P. Carson and Barrett 1988; W. P. Carson and Pickett 1990; McLendon and Redente 1991; Tilman and Wedin 1991; Cahill 1999). Community responses to nutrient enrichment may also vary with the stage of secondary succession (Tilman 1986), the plant species present at the time of enrichment (Inouye and Tilman 1988), the form of nutrients applied or available (W. P. Carson and Barrett 1988; W. P. Carson and Pickett 1990), and interactions between above- and below-ground competition in old-field species (S. D. Wilson and Tilman 1991; Cahill 1999).

Concern About the Loss of Biodiversity

The word *biodiversity* has become almost synonymous with concern over the loss of species. As is clear from this chapter, this concern should go beyond the species level to include the loss of functions and niches up and down the entire levels-of-organization hierarchical scale. In this context, Wilcox (1984) defined *biodiversity* as "the variety of life forms, the ecological roles they play, and the genetic diversity they contain." We can now include the role of habitat or landscape diversity as well.

It is essential to maintain *redundancy* in an ecosystem—that is, to have more than one species or group of species capable of carrying out major functions or providing links in the food web. In assessing the effects of the removal or addition of species, it is important to know if they might be keystone species. Chapin et al. (2002) defined a **keystone species** as a functional type without redundancy. The loss of such a species or group of species will cause major changes in community structure and ecosystem function. However, as is so often the case, too much of something can also become detrimental. As discussed in Chapter 2, an exotic invader species added to a naturally diverse ecosystem often becomes a keystone species that reduces diversity.

Reduction in species and genetic diversity in historical times has produced short-term benefits in agriculture and forestry, as evidenced by the propagation of specialized, high-yielding varieties over large areas of global cropland and forest. Overdependency on a small number of varieties, however, invites future catastrophe should climates change, should the energy and chemical subsidies needed to maintain these varieties become scarce, or should new diseases and pests attack a vulnerable variety. Agriculturalists are deeply concerned about the loss of crop diversity, as expressed in their efforts to establish nurseries and seed banks as repositories of as many varieties of crop plants as possible. To call attention to the overall threat posed by loss of diversity, some people are organizing gene resources conservation programs. An excerpt from a report of one such program (Gene Resource Conservation Program, Berkeley, California; David Kafton, director) follows:

The biological diversity of animals, plants, and microorganisms is of fundamental importance to human survival. The term "gene resources" may be defined as the genetic diversity that is crucial for meeting societal needs in perpetuity. This diversity is expressed in the differences between species as well as in the variation among individuals that comprise a species. Gene resources include wild and domestic species, including many that have no direct commercial value, but are essential to the survival of those that do. Each year gene resources are utilized to provide billions of dollars worth of new and familiar products (e.g., food, clothing, shelter, pharmaceuticals, energy, and hundreds of industrial products). A broad range of species and their products are required for medical and other research. Agriculture, forestry, and related industries are dependent on appropriate diversity (e.g., resistance to plant diseases). It is this diversity that sets the limits to which both wild and domestic species can successfully adapt to changes involving: (1) weather, insects and disease; (2) technology; (3) demand; and (4) human preferences. Most biological diversity is still found in natural ecosystems whose survival is dependent, in large part, on the diversity within them.

Rhoades and Nazarea (1999) pointed out that pre-industrial agriculture, as still practiced in many parts of the world, forms diverse "gene banks" because of the large number of crop varieties that are kept in cultivation. For example, fifty varieties and several species of potatoes are grown in the Cuzco Valley in the High Andes of Peru.

That the U.S. citizenry is interested, if not yet really concerned, is evident from efforts to save endangered species from extinction. There is strong public opinion, as indicated by legislation enacted by Congress, in favor of preserving rare species, especially if they are as interesting and spectacular as the whooping crane (*Grus americana*)—but not when preservation of the species or its habitat conflicts with development that promises immediate economic benefits. Because a species-by-species approach is bound to fall short in any such conflict, it is time to emphasize the preservation of the diversity of whole ecosystems or landscapes. A stronger case can be made for conservation at this level; a greater diversity of species and gene pools would be saved using such a holistic approach.

To some extent, sound regional planning can compensate for the reduction in local or *alpha diversity* that tends to accompany intensive agricultural, forestry, and urban development. If crop and forest monocultures and tract housing (rows and rows of similar houses on small, lawn-covered lots), are interspersed with more diverse natural or seminatural ecosystems (preserved in perpetuity as parks or nature centers, for instance), and if floodplains and other wetlands together with steep slopes and ravines are left undeveloped, not only will there be a pleasing landscape full of recreational possibilities, but also a high level of beta and gamma diversity will have been safeguarded.

The relationship of natural capital to economic capital needs to be managed by an integrative approach. An example is the National Environmental Research Park (NERP) constituting the 278,000-acre (111,200-ha) Savannah River Site (SRS). Approximately one third of the area once farmed has now been converted into pine plantations, so the alpha diversity of tree species in this part of the area approaches zero (only one species to a plantation, in general). However, because the floodplains and other riparian zones along streams, uplands, hardwood mixed forests, hedgerows, and other original vegetation that compose two thirds of the area have been left more or less intact, the habitat or *gamma diversity* of the whole landscape is quite high

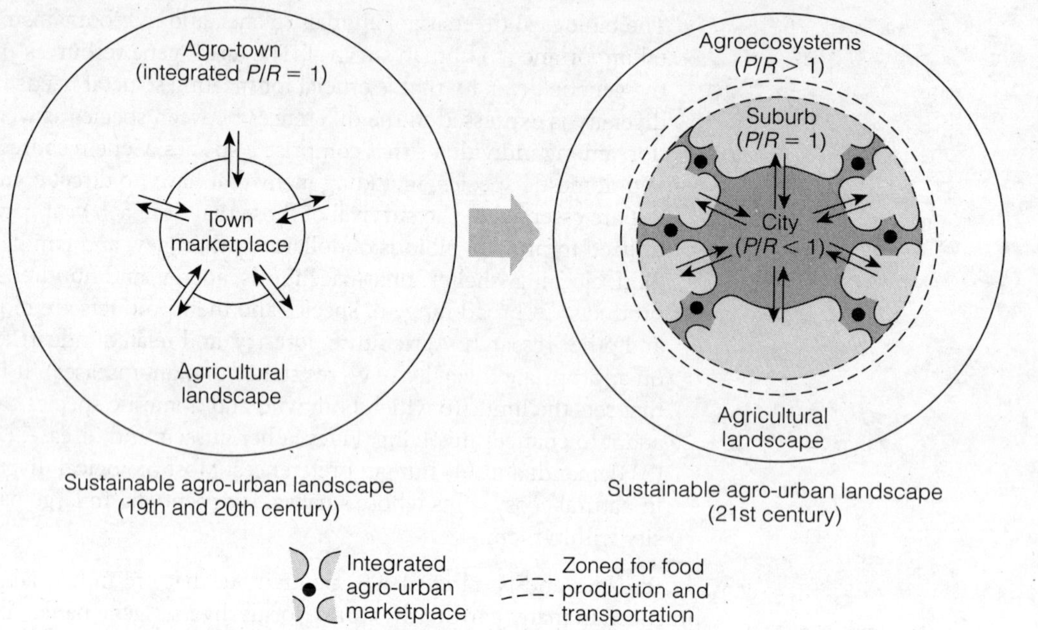

Figure 7-18. The development of a sustainable ($P/R = 1$) agro-urban landscape. The town marketplace historically has been closely linked to the agricultural landscape. Sustainability in the modern agro-urban landscape increasingly must be based on the management of suburban areas (ecotones) as natural linkages between urban and agricultural systems. (Reprinted with permission from Barrett, G. W., T. L. Barrett, and J. D. Peles. 1999. Managing agroecosystems as agrolandscapes: Reconnecting agricultural and urban landscapes. In *Biodiversity in agroecosystems,* ed. W. W. Collins and C. O. Qualset, 197–213. Copyright CRC Press, Boca Raton, Florida. Reprinted with permission.)

(comparable to the diversity of natural soil types). The United States Forest Service, which manages the land on this reservation, initially was tempted to expand the pine plantations at the expense of these other vegetation types to increase the yield of paper pulp and timber. However, such conversion would drastically reduce beta and gamma diversity (thus jeopardizing future maintenance and stability).

Figure 7-18 shows how landscapes can be planned to preserve diversity and yet accommodate urban and agricultural development (Barrett et al. 1999). This model illustrates the need to integrate urban and agricultural landscapes—a pattern common during the nineteenth and twentieth centuries. Columbia, Maryland, is an example of successful planning that was developed within the private, free-market sector, with the blessings of the state and federal governments, but with a minimum of financial assistance from either.

Many ecological concepts relating to diversity are controversial and need more study (Blackmore 1996; Grime 1997; Kaiser 2000), but most ecologists agree that diversity is necessary for the future survival of humans and nature. In Chapter 2, we strongly argued that natural areas must be preserved for their essential role in life support. We can now add a second compelling reason—to preserve and safeguard the diversity required for future adaptation and survival (see *The Future of Life* by E. O. Wilson, 2002, which provides a compelling argument for conserving biodiversity). Biodiversity at the landscape level will be discussed in more detail in Chapter 9.

9 Paleoecology: Community Structure in Past Ages

Statement

We know from the fossil record and from other evidence that organisms were different in past ages and have evolved to their present status. Knowledge of past communities and climates contributes to our understanding of present communities. This is the subject of *paleoecology*, an interface field between ecology and paleontology. **Paleoecology** is the study of the relationships of ancient flora and fauna to their environment. The basic assumptions of paleoecology are (1) that the operation of ecological principles has been essentially the same throughout various geological periods; and (2) that the ecology of fossils may be inferred from what is known about equivalent or related species now living. Findings in DNA fingerprinting have tended to confirm these assumptions. DNA sequencing has become a powerful tool for investigating genetic variation within and among populations—a tool that has helped to merge evolutionary biology with systems ecology.

Explanation

Since Charles Darwin proposed the theory of evolution by natural selection (Darwin 1859), the reconstruction of life in the past through the study of the fossil record has been an absorbing scientific pursuit. The evolutionary history of many species, genera, and higher taxonomic groups has now been pieced together. For example, the story of the skeletal evolution of the horse from a four-toed animal the size of a fox to its present status is pictured in most elementary biology textbooks. But what about the associates of the horse in its developmental stages? What did it eat, and what was its habitat and density? What were its predators and competitors? What was the climate like at the time? How did these ecological factors contribute to the natural selection that must have shaped the horse's structural evolution? Some of these questions, of course, may never be answered completely. However, scientists have been able to determine something of the nature of communities and of their dominant species in the past based on the quantitative study of fossils associated together at the same time and place. Moreover, such evidence, together with evidence of a purely geological nature, has helped to determine climatic and other physical conditions existing at the time. The development of radioactive dating and other geological tools has greatly increased our ability to establish the precise time when a given group of fossils lived.

Until recently, little attention was paid to the questions listed in the preceding paragraph. Paleontologists were busy describing their finds and interpreting them in the light of evolution at the taxonomic level. As information accumulated and became more quantitative, however, it was only natural that interest in the evolution of the group should develop, and thus paleoecology was born.

In summary, the *paleoecologist* attempts to determine from the fossil record how organisms were associated in the past, how they interacted with existing physical conditions, and how communities have changed over time. The basic assumptions of paleontology are that natural laws were the same in the past as they are today, and that organisms with structures similar to those in organisms living today had similar patterns of behavior and similar ecological characteristics. Thus, if the fossil evidence indicates that a spruce forest stood 10,000 years ago where an oak-hickory forest is now, one has every reason to think that the climate was colder 10,000 years ago, because modern species of spruce are adapted to colder climates than are oaks and hickories.

Examples

Fossil pollen provides excellent material for the reconstruction of terrestrial communities that have existed since the Pleistocene. The last great ice sheet, the Laurentide, reached its maximum advance about 20,000 to 18,000 years ago during the *Wisconsin glaciation stage* in North America. Figure 7-19 shows how the nature of postglacial communities and climates can be reconstructed by determining the dominant trees. As the glacier retreats, it often leaves scooped-out depressions, which become lakes. Pollen from plants growing around the lake sinks to the bottom and becomes fossilized in the bottom mud. Such a lake may fill up and become a bog. By taking a core sample from the bog or lake bottom, a chronological record is obtained, from which the percentage of various kinds of pollen can be determined. Thus, in Figure 7-19, the oldest pollen sample comprises chiefly spruce, fir, larch, birch, and pine, indicating a cold climate. A change to oak, hemlock, and beech indicates a warm, moist period several thousand years later, whereas oak and hickory suggest a warm, dry period still later, and a return to slightly cooler and wetter conditions in the most recent part of the profile. Finally, the pollen "calendar" clearly reflects recent human influences. For example, clearing of forests is accompanied by an increase in herbaceous pollen. According to M. B. Davis (1969), pollen profiles in Europe even show the ef-

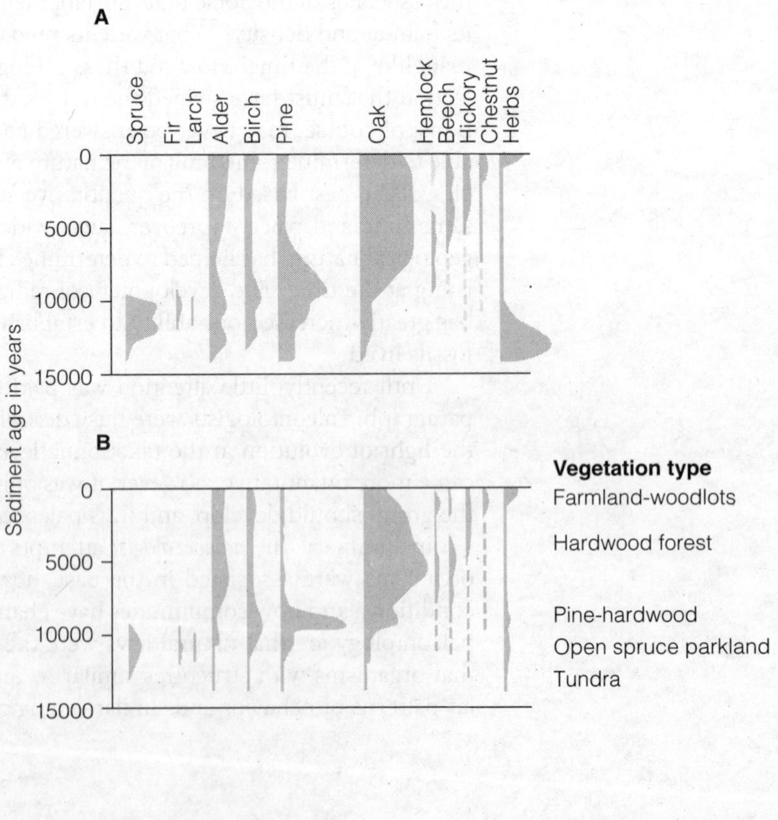

Figure 7-19. Fossil pollen profiles from dated layers in lake sediment cores from southern New England. In (A), the number of pollen grains of each species group is plotted as a percentage of the total number in the sample, whereas in (B), the estimated rate of pollen deposition (10^3 grains · cm^{-2} · $year^{-1}$) for each plant group is plotted. The rate profile gives a better indication of the quantitative nature of the post-Pleistocene vegetation than does the percentage profile (redrawn from M. B. Davis 1969).

fects of the Black Plague, when agriculture declined, resulting in a decrease of herbaceous pollen in sediment layers dated at the same time as the widespread death of human beings. M. B. Davis (1983, 1989) demonstrated how forests in eastern North America have changed with the change in climate, thus helping to reconstruct the history of the deciduous forest biome of eastern North America.

Figure 7-19 also illustrates how improved quantification can change interpretations of the fossil record. When pollen abundance is plotted as a percentage of the total amount in the sample (Fig. 7-19A), one concludes that New England was covered with a dense spruce forest 10,000 to 12,000 years ago. However, when carbon dating made it possible to determine the *rate* of pollen deposition in dated layers, as plotted in Figure 7-19B, it became evident that trees of *all* kinds were scarce 10,000 years ago and that the vegetation existing then was actually an open spruce parkland, probably not unlike the present one along the southern edge of the Tundra. Statisticians often caution against the use of percentage analysis because such analysis may be misleading; this is a clear example.

In the ocean, shells and bones of animals often provide the best record. Shell deposits are especially good for assessing diversity in the past. In past ages, when there was no ice at the poles, there were many more species in northern sea bottoms than there are now. However, in the whole pole-to-equator gradient, there are twice as many species of benthic mollusks now—when the poles are ice covered—presumably because the sharper gradient increases the variety of habitats and, therefore, of niches.

Core samples taken from lake bottoms provide one way to read the recent history of human disturbance of the watershed. For example, a study of fossilized diatoms, midges, and zooplankton, including the chemical composition of dated sections of cores, led Brugam (1978) to identify three stages of human impact in the eastern United States: (1) *early farming* in the late 1700s and 1800s had little effect on the lake; (2) *intensive agriculture* after about 1915 resulted in a flow of agricultural chemicals into the lakes and an increase in eutrophic species of diatoms and midges; and (3) *suburbanization* from the 1960s to the present resulted in still more nutrient enrichment and in soil erosion that brought large amounts of minerals and metals (Fe, Cu) into the sediments. These inputs produced major changes in the composition of the biota, especially the zooplankton.

Hopefully, the study of past communities and events by means of the fossil record will aid humankind in forecasting changes in climate in the future. This need has become especially acute because humans appear to be accelerating the process of climate change.

10 From Populations and Communities to Ecosystems and Landscapes

Statement

The two approaches to studying, understanding, and, where necessary, managing ecosystems are the *holistic approach* (based on the theory that whole entities have a separate existence beyond the mere sum of their parts) and the *reductionist approach*

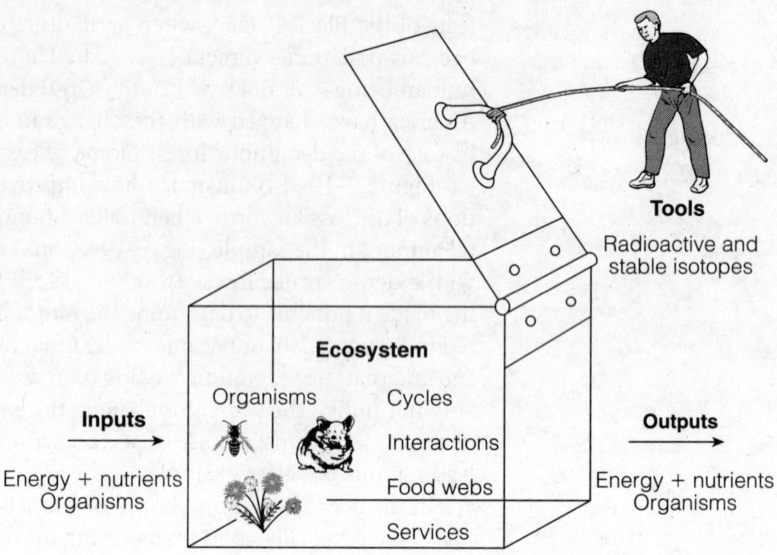

Figure 7-20. In the holistic approach, inputs and outputs to the "black box" of ecosystem science are examined first, and individual components of the system are studied from the perspective of how they fit into the whole. (From Likens, G. E. 2001a. Ecosystems: Energetics and biogeochemistry. In *A new century of biology,* ed. W. J. Kress and G. W. Barrett. Washington, D.C.: Smithsonian Institution Press, 159. Reprinted with permission.)

(based on the theory that every complex system can be explained by analyzing the simplest, most basic parts of that system). In the holistic approach, one first delimits the area or system of interest in some convenient way, as a sort of "black box." Then, the energy and other inputs and outputs are examined (Fig. 7-20), and major functional processes within the system are assessed. Following the parsimony principle (least effort), one then examines operationally significant populations and factors as determined by observing, by modeling, or by perturbing the ecosystem itself. In this general approach, one goes into the details of population components within the box only as far as necessary to understand or manage the system as a whole.

Explanation and Examples

There are so many exciting things to be learned about individual and interacting populations as one advances from the population and community levels of organization to the ecosystem and landscape levels. It is obviously not practical to study every population in detail. Also, as abundantly illustrated in this chapter, populations may behave very differently when functioning in communities than when they are isolated in the laboratory or by enclosures in the field. Once individual components of populations and communities have been studied, how does one reassemble them into ecosystems to consider new holistic properties that may emerge as parts function together in the intact ecosystem or landscape?

An outstanding example of unraveling ecosystem and landscape complexity is illustrated in Figure 7-21. A team of scientists from the Institute of Ecosystem Studies located at Millbrook, New York, investigated the complex interactions among oak trees, deer, white-footed mice, ticks, gypsy moths, and humans with regard to Lyme disease, hunting, timber, and forest aesthetics. Each researcher brought special expertise to bear on this complex ecosystem/landscape problem (Likens 2001).

Abundant acorns from oak trees (*Quercus*) attract white-tailed deer (*Odocoileus virginianus*) to the forest in the northeastern United States. Abundant acorn production occurs every three to four years (mast years). Deer carry adult black-legged ticks (*Ixodes scapularis*), which drop from the deer and lay eggs in the forest floor. The abundance of acorns also attracts white-footed mice (*Peromyscus leucopus*), which rapidly increase in abundance in response to the increased food supply. The following summer, the tick eggs hatch into larvae, which, during the process of obtaining blood meals from mice, pick up the spirochete bacterium (*Borrelia burgdorferi*) that causes Lyme disease in humans.

The tick larvae transform into nymphs a year or so later and pass on the spirochete when they attach to humans or other mammals for blood meals. In general, the risk of Lyme disease in oak forests is correlated with mast years of acorns with a lag of two years (Ostfeld et al. 1996; Ostfeld 1997; Jones et al. 1998; Ostfeld and Jones 1999). Moreover, predation on the pupae of gypsy moths (*Lymantria dispar*) by abundant small mammal populations help to prevent moth outbreaks that, when they occur, can result in defoliation and tree deaths in oaks—thus reducing the production of acorns and, consequently, affecting the risk of Lyme disease (Fig. 7-21). These interactions illustrate how problems and relationships at the population, community, ecosystem, and landscape levels need to be addressed in an interactive manner if large-scale problems are to be solved.

A replicated aquatic mesocosm investigation helps to illustrate how complex interactions occurring at the population and community levels affect changes at the ecosystem and landscape levels. Ecologists at Miami University of Ohio investigated the effects of omnivorous fish on the stability of aquatic food webs in experimental ponds (Vanni et al. 1997; Vanni and Layne 1997; Schaus and Vanni 2000). The study

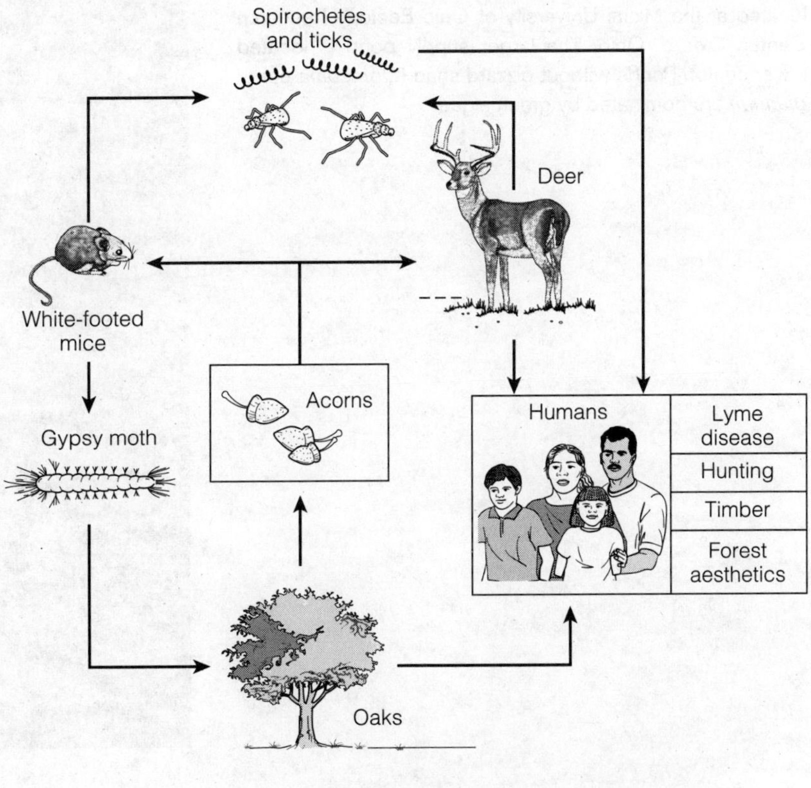

Figure 7-21. Diagram showing the linkages among oak trees, deer, white-footed mice, ticks, gypsy moths, and humans in northeastern U.S. forests (based on Ostfeld et al. 1996; Jones et al. 1998; Ostfeld and Jones 1999; reproduced with permission of R. S. Ostfeld and C. G. Jones).

focused on whether a common omnivorous fish, the gizzard shad (*Dorosoma cepedianum*), could stabilize aquatic ecosystems against perturbations. The gizzard shad is a dominant fish species in many lakes throughout the southern and eastern United States, and feeds on phytoplankton, zooplankton, and sediments (detritus). By consuming sediments and excreting ammonia and phosphate as waste products into the water column, gizzard shad effectively "pump" nutrients to phytoplankton. Moreover, because they feed on many different types of food, these fish increase the number and types of interactions within the food web. Both increased nutrient supply and increased food web complexity have been theorized to stabilize systems, which led these researchers to hypothesize that gizzard shad would stabilize food webs through either one or both of these mechanisms.

To test this hypothesis, six ponds (each 45 × 15 meters) were divided with curtains to yield 12 experimental units (Fig. 7-22). Food webs were then established with contrasting densities of gizzard shad (high, moderate, low, or none). Ponds were filled from a common supply pond (the large, dark pond in Fig. 7-22), and contained similar densities of phytoplankton, zooplankton, and planktivorous fish (bluegill, *Lepomis macrochirus*). Three replications of each of the four density treatments were established by adding the appropriate number of gizzard shad to each pond. Communities were allowed to equilibrate over a six-week pre-perturbation period, and then a large pulse of nitrogen and phosphorus was added. This perturbation simulated a common disturbance experienced by many lakes, which receive large pulses of nutrients following storm events in the form of runoff from surrounding watersheds.

Courtesy of Steven J. Harper

Figure 7-22. Aerial photograph of experimental ponds located at the Miami University of Ohio Ecology Research Center, Oxford, Ohio. The larger supply pond is located lower center. Ponds without gizzard shad (*Dorosoma cepedianum*) are dominated by green algae.

Figure 7-23. Monarch butterfly (*Danaus plexippus*) resting on a milkweed (*Asclepias syriaca*) plant.

The results partially supported the hypothesis that omnivorous gizzard shad stabilize food webs. As predicted, phytoplankton biomass showed increased resistance and resilience to the nutrient pulse in ponds with high gizzard shad density. However, the phytoplankton community composition was less stable in ponds with omnivorous fish. Ponds with gizzard shad became dominated by cyanobacteria after the pulse, whereas those without gizzard shad remained dominated by green algae (bright green ponds), the condition that existed prior to the perturbation. This study demonstrated that it is necessary to study multiple response variables to quantify the stability of food webs. Without conducting controlled, replicated experiments, it may be difficult to accurately predict how systems will respond to disturbance—an important concern of ecologists given the prevalence of perturbations to many natural systems at the landscape scale.

Another example that ranges from the population through the landscape levels of organization is the life history of the monarch butterfly (*Danaus plexippus*). To understand the population dynamics of the monarch, one must also understand its mutualistic relationship with milkweed (*Asclepias syriaca*). Milkweed contains cardiac glycosides that, when absorbed by the monarch butterfly larvae, whose sole source of food is milkweed foliage (Fig. 7-23), make the larvae and the adult butterflies toxic to birds and other predators. The bright orange of the monarch serves as a warning of danger to predators and also as a model for mimic species (Brower 1969).

The monarch has evolved an extraordinary spring and fall migration pattern, permitting it to use the abundant milkweed food supply across the North American continent. Remarkably—and in contrast to vertebrate migrations—the monarch's navigation from the upper midwestern United States to its overwintering sites in the fir forests of the Sierra Transvolcanica mountain range in central Mexico (Fig. 7-24A) is carried out by descendants three or more generations removed from their migrant forebears. That is, the fall migration is completely inherited, with no opportunity for learned behavior (Brower 1994). Survivors from the overwintering colonies begin migrating northward in March to lay their eggs in sprouting milkweed plants (Fig. 7-24B). These adult individuals then die, but their offspring, produced in late

Figure 7-24. (A) Fall migrations of the eastern populations of the monarch butterfly (*Danaus plexippus*) in North America. (B) Spring migrations of the eastern population of the monarch butterfly, including the spring breeding area (after Brower and Malcolm 1991; Brower 1994).

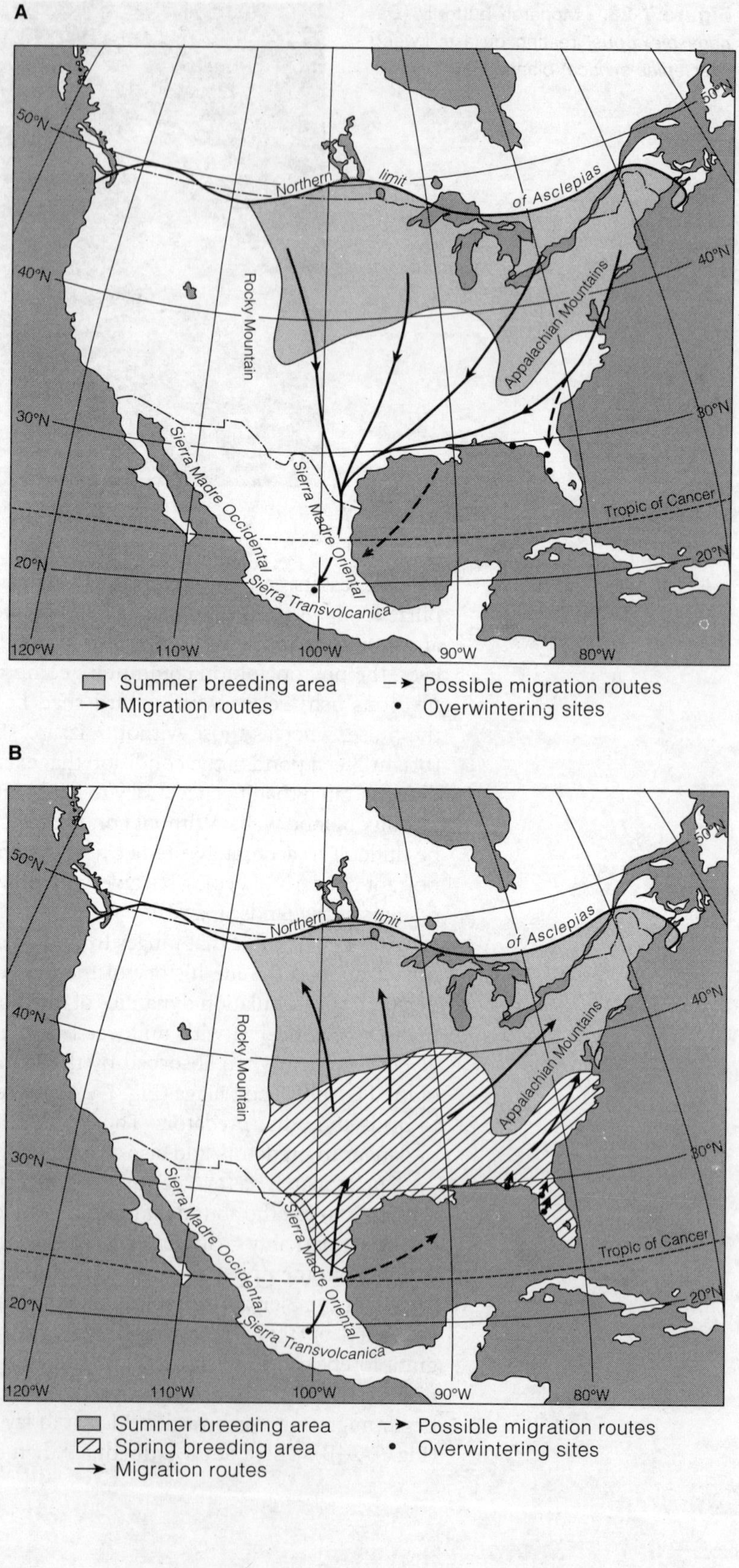

April and early May, continue the migration northward to Canada (the northern limit of *Asclepias*). The temporal and spatial scale of this migration constitutes a unique ecological phenomenon.

The relationship between the parts and the whole may well depend on the level of complexity. At one extreme, ecosystems subjected to severe physical limitations (as on the Arctic Tundra or in a hot spring) have relatively few biotic components. Such "low-numbers" systems can be studied and understood by focusing on the parts, because the whole is probably very close to the sum of the parts, with few, if any, emergent properties. In contrast, "large-numbers" systems (such as the landscape and the ecosphere) have a great many components that act synergistically to produce emergent properties—the whole is definitely not just a sum of the parts. Studying all the parts separately is out of the question, so one must focus on the properties of the whole. Most ecosystems as delimited in practice (a lake or a forest, for example) are "middle-numbers" systems that can best be studied by a multilevel or "black box" approach, as described earlier in this section. For more on levels-of-complexity theory, see T. F. H. Allen and Starr (1982); O'Neill et al. (1986); T. F. H. Allen and Hoekstra (1992); and Ahl and Allen (1996).

How to deal with parts versus wholes has long confounded philosophers and bedeviled societies. Scientists in all disciplines are split on the question of reductionism versus holism. The difficulty in dealing simultaneously with the part and the whole is perhaps best reflected in the conflict between the individual good and the public good. Numerous economic and political approaches designed to deal with this conflict have been suggested or tried, but as yet with little success. In the United States, elected governments over the years have shifted back and forth from strong attention to the individual (the "conservative" stance) to emphasis on public well-being (the "liberal" stance), so the parts (individual) and the whole (public) get attention, but not at the same time. Perhaps the study of how natural ecosystems develop, to be considered in Chapter 8, may help resolve this problem.

8

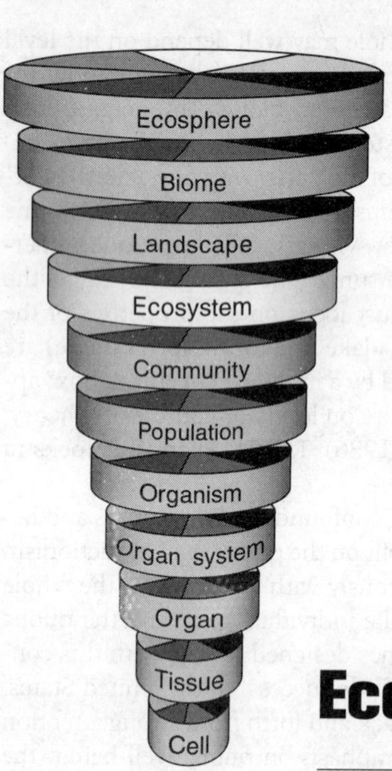

Ecosystem Development

1 **Strategy of Ecosystem Development**

2 **Concept of the Climax**

3 **Evolution of the Biosphere**

4 **Microevolution Compared with Macroevolution, Artificial Selection, and Genetic Engineering**

5 **Relevance of Ecosystem Development to Human Ecology**

1 Strategy of Ecosystem Development

Statement

Ecosystem development, more often known as **ecological succession**, involves changes in energy partitioning, species structure, and community processes over time. When not interrupted by outside forces, succession is reasonably directional and, therefore, predictable. It results from the modification of the physical environment by the community and from competition-coexistence interactions at the population level—that is, succession is community controlled even though the physical environment determines the pattern and the rate of change and often limits the extent of development. If successional changes are largely determined by internal interactions, the process is known as **autogenic** ("self-generated") **succession.** If outside forces in the input environment (such as storms and fire) regularly affect or control change, there is **allogenic** ("externally generated") **succession.**

When new territory is opened or becomes available for colonization (for example, after a volcanic lava flow, in an abandoned crop field, or in a new water impoundment), autogenic succession usually begins with an unbalanced community metabolism, where gross production, P, is either greater than or less than community respiration, R, and proceeds toward a more balanced condition, where $P = R$. The ratio of biomass to production (B/P) increases during succession until a stabilized ecosystem is achieved, in which a maximum of biomass (or high information content) and symbiotic function between organisms are maintained per unit of available energy flow.

The whole sequence of communities that replace one another in a given area is termed the **sere**; the relatively transitory communities during succession are variously termed **seral stages** or **developmental stages.** The initial seral stage is termed the **pioneer stage** and is characterized by early successional *pioneer plant species* (typically annuals), which exhibit high rates of growth, small size, short life span, and production of a great number of easily dispersed seeds. The terminal stage or mature, stabilized system is the **climax,** which persists, in theory, until affected by major disturbances. Succession beginning with $P > R$ is **autotrophic succession**, contrasting with **heterotrophic succession**, which begins with $P < R$. Succession on a previously unoccupied substrate (such as a new lava flow) is termed **primary succession,** whereas succession starting on a site previously occupied by another community (such as a clear-cut forest or abandoned crop field) is known as **secondary succession.**

It is to be emphasized that the mature or *climax stage* is best recognized by the state of the community metabolism, $P = R$, rather than by species composition, which varies widely with topography, microclimate, and disturbance. Even though, as already emphasized, ecosystems are not "superorganisms," their development has many parallels in the developmental biology of individual organisms and in the development of human societies, in that they progress from "youth" toward "maturity."

Explanation and Examples

Descriptive studies of succession on sand dunes, grasslands, forests, marine shores, or other sites—and more recent functional considerations—have led to a partial understanding of the developmental process and generated a number of theories about

Table 8-1 A tabular model for ecological succession of the autogenic type

Ecosystem characteristic	*Trend in ecological development* *Early stage → Climax* *Youth → Maturity* *Growth stage → Pulsing steady state*
Energy flow (community metabolism)	
Gross production (*P*)	Increases during early phase of primary succession; little or no increase during secondary succession
Net community production (yield)	Decreases
Community respiration (*R*)	Increases
P/*R* ratio	$P > R$ to $P = R$
P/*B* ratio	Decreases
B/*P* and *B*/*R* ratios (biomass supported per unit energy)	Increase
Food chains	From linear food chains to complex food webs
Community structure	
Species composition	Changes rapidly at first, then more gradually (relay floristics and faunistics)
Size of individuals	Tends to increase
Species diversity	Increases initially, then stabilizes or declines in older stages as size of individuals increases
Total biomass (*B*)	Increases
Nonliving organic matter	Increases
Biogeochemical cycles	
Mineral cycles	Become more closed
Turnover time and storage of essential elements	Increases
Internal cycling	Increases
Nutrient conservation	Increases
Natural selection and regulation	
Growth form	From *r*-selection (rapid growth) to *K*-selection (feedback control)
Life cycles	Increasing specialization, length, and complexity
Symbiosis (living together)	Increasing mutualism
Entropy	Decreases
Information	Increases
Overall efficiency of energy and nutrient use	Increases
Resilience	Decreases
Resistance	Increases

Source: Tabular model after E. P. Odum 1969, 1997.

A

Courtesy of Gary W. Barrett

B

Courtesy of Gary W. Barrett

Figure 8-1. Photographs of (A) a young old-field community located in Union County, Indiana, and (B) a sugar maple tree. The maple tree is in a mature beech-maple climax forest in Hueston Woods State Park near Oxford, Ohio. The boiled concentrated sap of the sugar maple (*Acer saccharum*) is the commercial source of maple sugar and syrup.

its cause. H. T. Odum and Pinkerton (1955), building on the Lotka law of the maximum energy in biological systems (Lotka 1925), were the first to point out that succession involves a functional shift in energy flows, with increasing energy relegated to maintenance (respiration) as the standing crop of biomass and organic matter accumulates. Margalef (1963b, 1968) documented this bioenergetic basis for succession and extended the concept. The role that population interactions play in shaping the course of *species replacement*—a characteristic feature of ecological succession—was discussed during the 1970s and 1980s (see Connell and Slayter 1977; McIntosh 1980; for reviews). Much of the controversy considering these reviews is reduced if the stages of development are based on energetics rather than on species composition, as noted in the Statement.

Changes that may be expected to occur in major structural and functional characteristics of autogenic development are listed in Table 8-1, in which 24 attributes of ecological systems are grouped for convenience of discussion under four headings. Trends contrast the situation in early and in late development. Figure 8-1A illustrates a young ecosystem (old-field community) in the early stage of development and Figure 8-1B shows a mature ecosystem (beech-maple forest) in the late stage of development. The degree of absolute change, the rate of change, and the time required to reach a mature status may vary not only with different climatic and physiographic conditions but also with different attributes of the ecosystem in the same physical environment. When good data are available, rate-of-change curves are usually convex, with changes occurring most rapidly at the beginning of development, but bimodal or cyclic patterns may also occur.

The trends listed in Table 8-1 represent those that are observed to occur when internal, autogenic processes predominate. The effect of external, allogenic disturbances may reverse or otherwise alter these developmental trends, as will be discussed later.

Bioenergetics of Ecosystem Development

The first seven attributes in Table 8-1 relate to the bioenergetics of the ecosystem. In the early stages of autotrophic succession in an inorganic environment, the rate of primary production or total (gross) photosynthesis, P, exceeds the rate of community respiration, R, so that the P/R ratio is typically greater than 1. The P/R ratio is less than 1 in the special case of an organic environment (such as a sewage pond), so succession in such cases is termed *heterotrophic,* because bacteria and other heterotrophs are the first to colonize the environment. In both cases, however, the theory is that P/R approaches 1 as succession proceeds. In other words, the energy fixed by production tends to be balanced by the energy cost of maintenance (total community respiration) in the mature or climax ecosystem. The P/R ratio, therefore, is a functional index of the *relative maturity* of the system.

As long as P exceeds R, organic matter and biomass, B, will accumulate in the system, with the result that the ratios B/P, B/R, and B/E (where $E = P + R$) will increase (or conversely, the P/B ratio will decrease). Recall that these ratios were discussed in Chapter 3 in terms of the laws of thermodynamics. Theoretically, then, the amount of standing crop biomass supported by the available energy flow, E, increases to a maximum in the mature or climax stage. As a consequence, the net community production, or *yield,* in an annual cycle is large in the early stages and small or zero in mature stages.

A simplified systems (cybernetic) model is shown in Figure 8-2A, in which internal, autogenic processes are considered as inputs that are modified by periodic allogenic inputs. Figure 8-2B, an energy flow model, shows the basic change in energy partitioning between P and R mentioned earlier. As the organic structure builds up, more and more energy is required to maintain this structure and dissipate disorder, and so less energy is available for production. This shift in energy use has parallels in the development of human societies, and it greatly affects attitudes about how the environment is treated, as we shall see later in this chapter. Figure 8-2C is a summary of how the three main factors—production, respiration, and biomass—change over time.

Comparison of Succession in a Laboratory Microcosm and a Forest

One may observe bioenergetic changes by initiating succession in experimental laboratory microecosystems of the type derived from natural systems, as described in Chapter 2. In Figure 8-3, the general pattern of a 100-day autotrophic succession in a typical flask microcosm experiment, based on data from Cooke (1967), is compared with a hypothetical model of a 100-year forest succession presented by Kira and Shidei (1967).

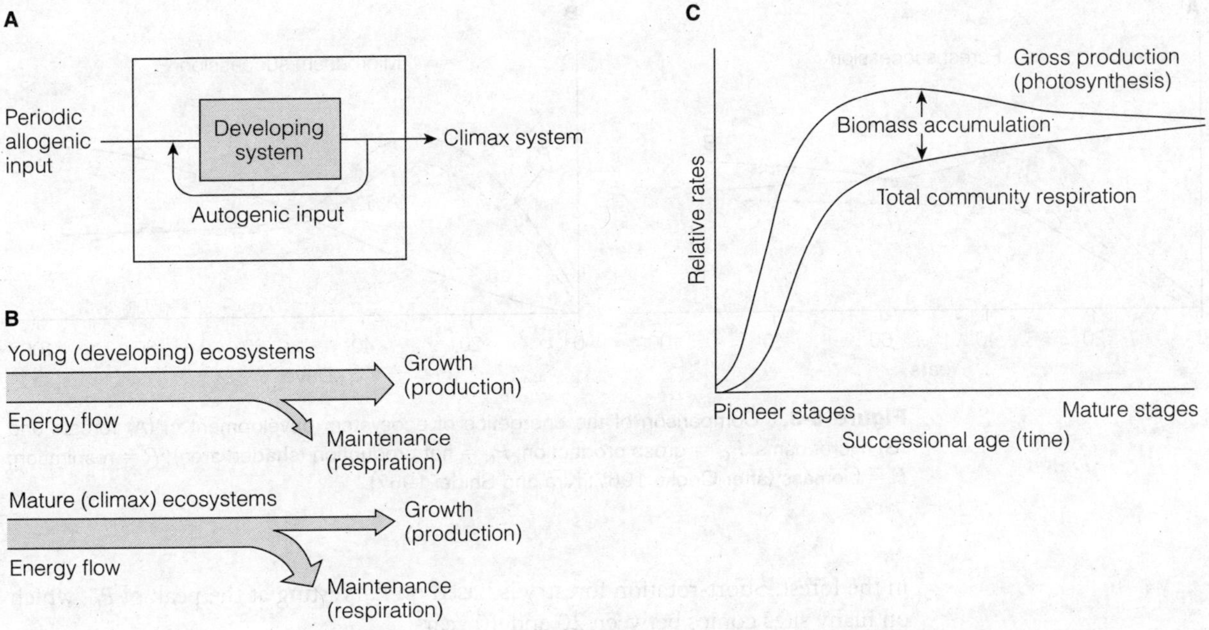

Figure 8-2. Ecosystem development models. (A) Systems (cybernetic) model; (B) Energy flow model; and (C) Production/respiration (*P*/*R*) maintenance model.

During the first 40 to 60 days in the microcosm experiment, daytime net production, *P*, exceeds nighttime respiration, *R*, so that biomass, *B*, accumulates in the system. After an early "bloom" at about 30 days, both rates decline and become approximately equal at 60 to 80 days. The *B*/*P* ratio, in terms of grams of carbon supported per gram of daily carbon production, increases from less than 20 to more than 100 as the steady state is reached. Not only are autotrophic and heterotrophic metabolism balanced in the climax stage, but also a large organic structure is supported by small daily production and respiration rates. The relative abundance of species also changes, so that different kinds of bacteria, algae, protozoa, and small crustaceans dominate at the end than at the beginning of the 100-day succession (Gorden et al. 1969).

Direct projection from small laboratory microcosms to natural systems is not possible, because the former are limited to small organisms with simple life histories and, of necessity, have a reduced species and chemical diversity. Nevertheless, the same basic trends seen in the microcosm are characteristic of succession on land and in large bodies of water. Seasonal succession also often follows the same pattern—an early seasonal bloom, characterized by rapid growth of a few dominant species, followed later in the season by the development of high *B*/*P* ratios, increased diversity, and a relatively steady, although temporary, state in terms of *P* and *R*. Open systems may not experience a decline in total or gross productivity at maturity, as the space-limited microcosms do, but the general pattern of bioenergetic change in microcosms seems to mimic nature quite well.

It is also interesting to note that peak net primary production (P_N), which represents the maximum yield possible, comes at 30 days in the microcosm and at 30 years

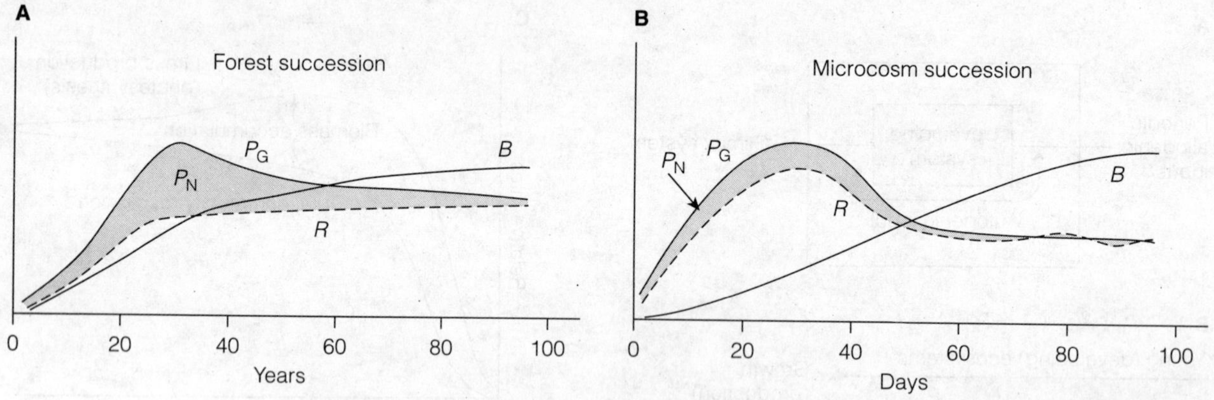

Figure 8-3. Comparison of the energetics of ecosystem development in (A) forests and (B) microcosms. P_G = gross production; P_N = net production (shaded area); R = respiration; B = biomass (after Cooke 1967; Kira and Shidei 1967).

in the forest. Short-rotation forestry is based on harvesting at the peak of P_N, which on many sites comes between 20 and 40 years.

Allogenic Compared with Autogenic Influences

Imported materials or energy, geological forces, storms, and human disturbances can and do alter, arrest, or reverse the trends shown in Table 8-1. For example, eutrophication of a lake, whether natural or cultural, results when nutrients and soil enter the lake from outside—that is, from the watershed. This is equivalent to adding nutrients to a laboratory microecosystem, or fertilizing a field; the system is "set back" in successional terms, to younger, "bloom" states. Brewer et al. (1994) noted, for example, that nutrient enrichment in an old-field community for 11 years resulted in a system that continued to be dominated by annual plant species, rather than the perennials that dominated mature (control) plots. *Allogenic succession* of this type is, in many aspects, the reverse of autogenic succession. When the effect of allogenic processes consistently exceeds that of autogenic ones, as in the case of many ponds and small lakes, the ecosystem not only cannot stabilize but also may become "extinct" by filling up with organic matter and sediments and becoming a bog or a terrestrial community. Such is the ultimate fate of human-made lakes subjected to accelerated erosion within the watershed.

Lakes can and do progress to a more *oligotrophic* (less enriched) condition when nutrient input from the watershed slows or ceases. Thus, there is hope that the troublesome cultural eutrophication, which reduces water quality and shortens the life of the water body, can be reversed if the inflow of nutrients from the watershed can be greatly reduced. An example is the recovery of Lake Washington (Fig. 8-4), located in Seattle, described by W. T. Edmondson (1968, 1970). For 20 years, treated, nutrient-rich sewage was discharged into the lake, which became increasingly turbid and full of nuisance algal blooms. As a result of public outcry, sewage effluent was diverted from the lake, which quickly returned to a more oligotrophic condition (clearer water and no blooms).

The interaction of external and internal forces can be summarized in a general systems model (Fig. 8-2A) of the form first introduced in Figure 1-5. Autogenic forces are depicted as internal input or feedback, which, in theory, tends to drive the system toward some sort of equilibrium state. Allogenic forces are depicted as periodic, external input disturbances, which set back or otherwise alter the developmental trajectory.

Where ecosystem development takes a long time to run its course—as in a forest development starting from bare ground—periodic disturbances will affect the successional process, especially in the variable environments of the temperate zones. Oliver and Stephens (1977) reported on a study of the vegetational history of a small area of the Harvard Forest located in Massachusetts. Fourteen natural and human-caused disturbances of varying magnitudes occurred, at irregular intervals, between 1803 and 1952. There was also evidence of two hurricanes and a fire before 1803. Small disturbances did not bring in new species of trees but often allowed species already in the understory, such as black birch (*Betula lenta*), red maple (*Acer rubrum*), and hemlock (*Tsuga canadensis*), to emerge into the canopy. Large-scale disturbances (such as a hurricane or a large fire) created openings, into which early successional species (such as pin cherry, *Prunus pennsylvanica*) invaded, where a new age class developed from seeds or seedlings already present on or in the forest floor (northern red oak, *Quercus rubra,* was a species that often filled such openings and grew to canopy dominance after several decades). Replacement and succession in forest clearings has been termed **gap phase succession.** Oliver and Stephens concluded from their study that the composition of the forest in the 1970s was more the result of allogenic influences than of autogenic development. In a subsequent review paper, Oliver (1981) concluded that severity and frequency of disturbance are the major factors determining forest structure and species composition in many areas of North America. More recently, Dale et al. (2001) noted that climate change can affect forest structure and function by altering the frequency, intensity, duration, and timing of fire, drought, introduction of exotic species, and insect or pathogen outbreaks.

Figure 8-4. Lake Washington, Seattle, where W. T. Edmondson conducted his classic research in the field of restoration ecology.

Figure 8-5. Wave-generated succession in a balsam fir forest. Bands of different shades represent successive waves of development (Sprugel and Bormann 1981).

If disturbances are rhythmic (come at more or less regular intervals), either because of a cyclic input environment or because of periodicities in the community development itself, the ecosystem undergoes what can logically be termed **cyclic succession.** The historic 1988 fire in Yellowstone National Park, for example, appears to be a cyclic phenomenon, occurring every 280 to 350 years (Romme and Despain 1989; see the November 1989 issue of *BioScience* entitled "Fire Impact on Yellowstone" for details). The fire-chaparral vegetation cycle described previously (Chapter 5) is an example of a self-generated cyclic succession, because the accumulation of undecomposed litter builds up fuel for the periodic fires in the dry season.

Another example of cyclic succession is the *wave-generated succession* in balsam fir (*Abies balsamea*) forests at high altitudes in the northeastern United States (Sprugel and Bormann 1981). As trees reach their maximum height and density in the thin soils, they become vulnerable to strong winds that uproot and kill old trees, thereby starting a secondary succession. As shown in Figure 8-5, a series of bands of young, mature, and dead trees (the latter appearing as light-colored bands in the figure) cover the mountainside. Because of the continuous cyclic succession, the bands move as "waves" across the landscape in the general direction of the prevailing winds. At any one time, all stages of succession are present, providing a variety of habitats for animals and smaller plants. The whole mountainside constitutes a *cyclic climax* in equilibrium with the surrounding environment.

The natural pattern of alternating bands of young and mature stands suggests that strip or patch clear-cutting could prove to be a good commercial harvest procedure for large forested areas, because natural regeneration would be facilitated (thus avoiding expensive tree replanting), and soil and animal populations would be little disturbed compared with the disturbance in a massive clear-cut of the whole forest. Furthermore, mixtures of different successional stages provide an abundance of *edges* (see Chapter 2) that benefit many forms of wildlife.

Still another example of cyclic succession is the cycle of spruce and budworms (described in Chapter 6). In this case, the periodic disturbance is not a physical force but a herbivore that defoliates and kills older growth, thus bringing on a succession of young growth.

The adjective **perturbation dependent** is frequently used to designate ecosystems that are especially adapted to recurrent disturbances by virtue of a makeup of quick recovery processes and species (see Vogl 1980 for a review). In predicting and managing recovery after a disturbance, such as strip mining, one must know in detail the succession pattern and recovery potential of the ecosystem in question, so that reclamation efforts will help and not hinder the natural recovery process (McIntosh 1980). A tentative hypothesis is that older stages of succession, in general, are more resistant to nominal or short-term stress (such as a one-year drought) than are younger stages, but the younger stages are more resilient (recover more quickly) to catastrophic stress, such as a large storm or fire (see Table 8-1).

Nutrient Cycling

Important trends in successional development involve an increase in turnover time, greater storage of materials, and increased biogeochemical cycling of major nutrients, such as nitrogen, phosphorus, and calcium (Table 8-1). The extent to which the conservation of nutrients is a major trend or strategy in ecosystem development is controversial, partly because there are different ways to index it. Figure 8-6 illustrates the problem. Vitousek and Reiners (1975) noted that biogenic nutrients will likely be stored within the system as biomass accumulates during the early stages of succession. According to their theory, the ratio of output, *O*, to input, *I*, drops below 1 as nutrients go into biomass accumulation. The *O/I* ratio then rises again to 1 as output balances input in the mature climax, when there is no further net growth. However, nutrients may continue to accumulate in soil even after plants are no longer adding to their living biomass. Thus, the output/input ratio is not the only way, or perhaps not the best way, to assess the behavior of nutrients. As shown in Figure 8-6, the cycling index (*CI* = ratio of recycled input to output; see Chapter 4) increases steadily as the system matures; accordingly, nutrients are retained for a longer period and reused, thereby reducing the input requirement, even though input and output are balanced. Also, the ratio of amount stored, *S*, to amount lost, *O*, is likely to be low in the early stages and increase in later stages. In summary, there are theoretical reasons and some observational evidence that storage and recycling of nutrients increase during ecosystem development, so that the requirement for input nutrients per unit of bio-

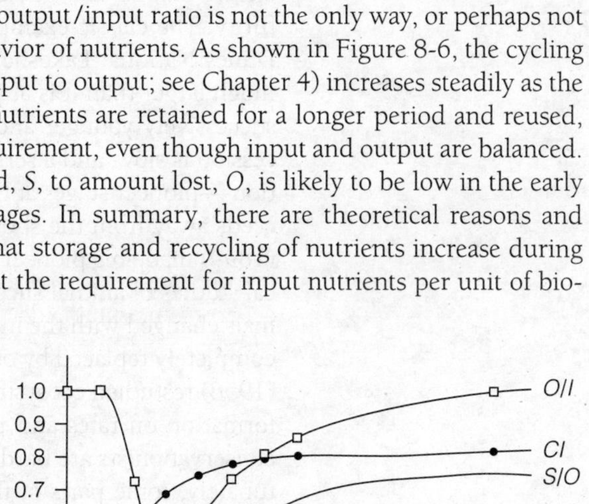

Figure 8-6. Hypothetical trends in output/input (*O/I*) ratio, cycling index (*CI*), and storage/output (*S/O*) ratios of nutrients during succession; BG = bare ground, ES = early stages; MS = middle stages; SS = steady state (from J. T. Finn 1978).

mass supported is reduced. No such conservation would be expected for nonessential or toxic elements.

There may be a shift in nitrogen source from nitrate to ammonia during succession. Theoretically, pioneer plants primarily use nitrate, whereas later stages, particularly forest stages, use ammonia as a nitrogen source. A shift from nitrate to ammonia reduces the amount of energy necessary to recycle nitrogen (see discussion of the nitrogen cycle in Chapter 4) and, thereby, increases the efficiency of energy use. However, Robertson and Vitousek (1981) could not find experimental evidence of the nitrate-to-ammonia shift, so the question remains open.

Also needing more quantitative study is the tendency for nitrogen fixation, mycorrhizal symbiosis, and other mutualisms that enhance the efficiency of nutrient cycling to increase during the course of succession (Table 8-1). It may be that nutrient-conserving mutualisms respond more to demand (nutrient scarcity) than to ecosystem development.

Replacement of Species

A more or less continuous replacement of species over time is characteristic of most successional seres. The changing species composition of vegetation has been termed *relay floristics* by Egler (1954), and, of course, there is also *relay faunistics,* because animal species also replace one another in the sere.

If development begins on an area previously unoccupied by a community (such as a newly exposed rock or sand surface, or a lava flow), the *primary succession* that ensues may be slow to begin and may require a long time to reach pulsing-state maturity. The classic example of primary ecological succession occurs on the Indiana Dunes National Lakeshore at the south end of Lake Michigan. The lake was once much larger than it is at present. In retreating to its present boundaries, the lake left successively younger and younger sand dunes. Because of the sand substrate, succession is slow, and a series of communities of various ages are available for observation—pioneer stages at the lake shore and increasingly older seral stages as one proceeds away from the shore. In this "natural laboratory of succession," H. C. Cowles (1899) made his pioneer studies of plants and V. E. Shelford (1913) made his classical studies of animal succession. Both studies showed that species of plants and animals changed with the increasing age of dunes; species present at the beginning were completely replaced by other, quite different species in the older communities. Olson (1958) restudied ecosystem development on these dunes and provided additional information on rates and processes. Because of the encroachment of heavy industry, conservationists are hard pressed in their efforts to preserve the dune series, but fortunately, some parts of the Indiana Dunes are now protected as the Indiana Dunes National Lakeshore (see Pavlovic and Bowles 1996 for details). Citizens should support such preservation efforts, because these areas not only have a priceless natural beauty that can be readily enjoyed by urban dwellers but also constitute a natural teaching laboratory, in which the visual display of ecological succession is dramatic.

The pioneer colonists on the dunes are beach grasses (*Ammophila, Agropyron, Calammophila*), willow (*Salix*), sand cherry (*Prunus depressa*), and cottonwood trees (*Populus deltoides*); and animals such as long-legged tiger beetles that flit along the sand, burrowing spiders, and grasshoppers. The pioneer community is followed by open, dry forest of jack pine (*Pinus banksiana*), then black oak (*Quercus velutina*), and finally, on the oldest dunes, moist forests of oak and hickory or beech and maple. Although the community began on a very dry and sterile sort of habitat, development eventually results in a closed canopy forest, moist and cool in contrast with the bare dunes. The deep, humus-rich soil, with earthworms and snails, contrasts with the dry sand that it replaced. Thus, the original, relatively inhospitable pile of sand is eventually transformed completely by the action of a succession of communities.

Succession on dunes in the early stages is often arrested when the wind piles up the sand over the plants, and the dune begins to move, entirely covering the vegetation in its path. This is an example of the arresting or reversing effect of allogenic perturbations discussed earlier in this section. Eventually, however, as the dune moves inland, it becomes stabilized, and pioneer grasses and trees again become established. Using carbon dating, Olson (1958) estimated that about 1000 years are required to reach a forest climax on the dunes of Lake Michigan—about five times longer than required for mature forest development starting from a more hospitable site, as seen in the next example.

An example of *secondary succession* is illustrated in Figure 8-7, which shows the sequence of plant communities and bird populations that develop on abandoned upland agricultural fields on the Piedmont of the southeastern United States. Pioneer colonists are *r*-strategist annual plants, such as crabgrass (*Digitaria*), horseweed (*Erigeron*), and ragweed (*Ambrosia*), which spend a large part of their energy on dispersal and reproduction. After two or three years, perennial forbs (asters and goldenrods), grasses (especially broomsedge, *Andropogon*), and shrubs such as blackberry (*Rubus*) move in. If there is a good seed source nearby, pines invade and soon form a closed canopy, shading out the early pioneers. Several species of fast-growing deciduous trees, such as sweetgum and tulip trees, often come in with the pines. Because all these species are long-lived, the pine stage (with scattered broad-leaved trees) persists for a long time, but gradually an understory of shade-tolerant oaks and hickories develops. As pines cannot reproduce under their own shade, the oaks and hickories rise to canopy dominance as the pines die from disease, old age, and storms.

As shown in Figure 8-7, bird populations change with each major seral stage; the most pronounced changes occur as the life-form of the dominant plants changes (herb to shrub to pine to hardwood). Habitat selection by birds is more targeted to vegetative life-form than to species of plant. No species of plant or bird can thrive from one end of the sere to the other. Species have their maxima at different points in the time gradient. Ostfeld et al. (1997) documented a similar relay succession of small mammals and the effect they had on the survival of tree seeds and seedlings in old-field secondary succession. Animals are not just passive agents in community change. Birds and other animals disperse seeds necessary for the establishment of shrub and hardwood stages, and herbivores, parasites, and predators often control the sequence of species.

In shallow-water marine habitats, large animals rather than plants often provide the structural matrix. Glemarec (1979) described a secondary succession of benthic animals off the Brittany coast of France. A period of relative calm followed after storms caused a redistribution of sediments and disruption of bottom fauna. During this period, in the absence of outside interference, a more or less directional and predictable sequence of populations established dominance. First came bivalve suspension feeders, then bivalve deposit feeders, and finally, the benthos became dominated by polychaete worm detritus feeders, thus confirming the theory that uninterrupted succession converts an inorganic environment to a more organic one.

Secondary plant succession is as striking in grassland regions as in forests regions,

Figure 8-7. General pattern of ecological succession on abandoned farmland in the southeastern United States. The graph shows changes in songbird populations that accompany changes in vegetation (after Johnston and Odum 1956; E. P. Odum 1997).

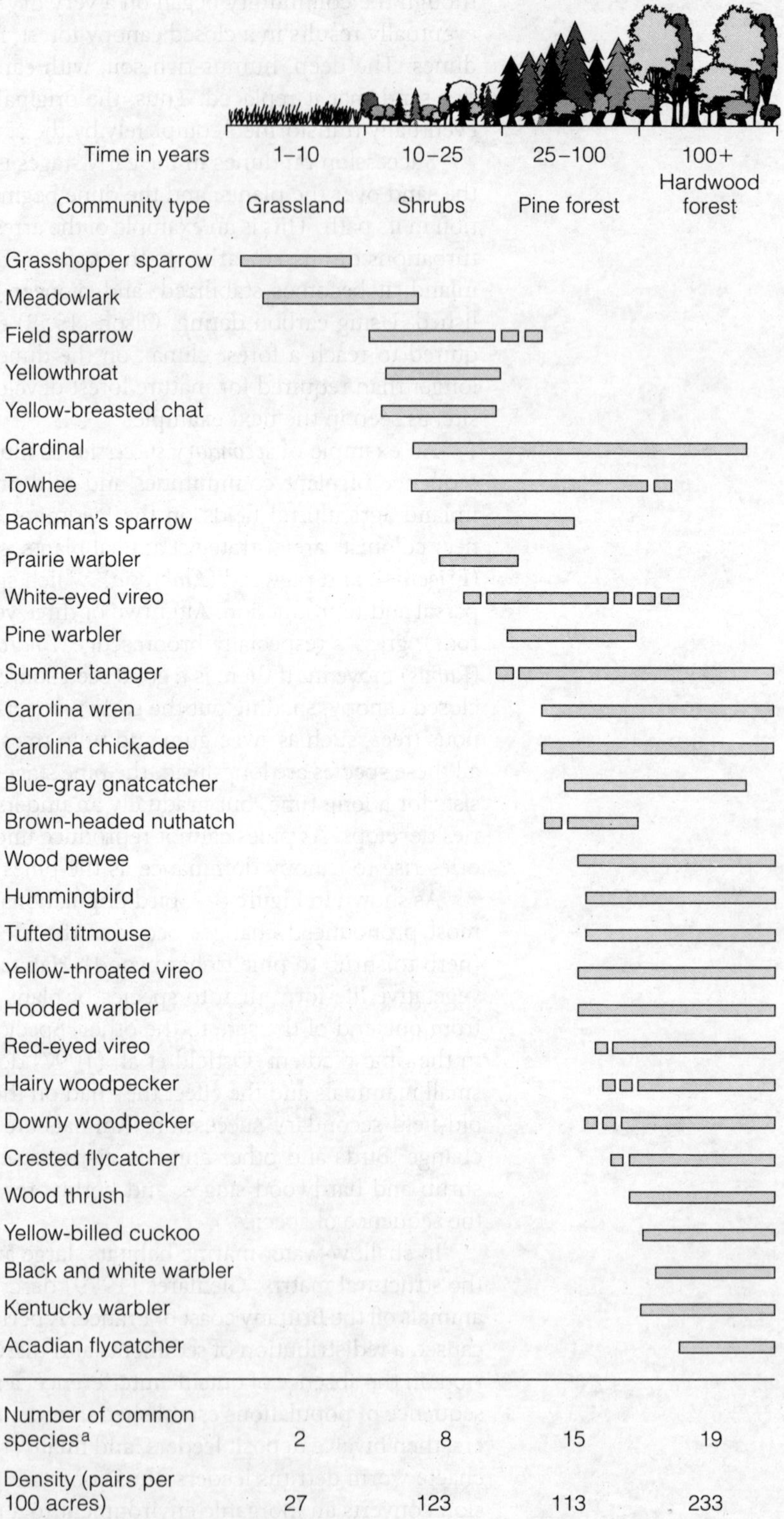

[a] A common species is arbitrarily designated as one with a density of five pairs per 100 acres or greater in one or more of the four community types.

Figure 8-8. The Oregon Trail near Scottsbluff, Nebraska, where a trace etched by the wheels of wagons that carried settlers during the 1840–1860 westward migrations between Missouri and Oregon is still evident.

even though only herbaceous plants are involved. Shantz (1917) described succession on the abandoned wagon roads used by pioneers crossing the grasslands of the central and western United States (Fig. 8-8), and virtually the same sequence has been described many times since. Although the species vary geographically, the same pattern holds everywhere. This pattern involves four successive seral stages: (1) an *annual weed* stage (2–5 years); (2) a *short-lived grass* stage (5–10 years); (3) an *early perennial grass* stage (10–20 years); and (4) a *climax grass* stage (reached in 20–40 years). Thus, starting from bare or plowed ground, 20 to 40 years is required for nature to "build" a climax grassland—the actual time depending on the limiting effect of moisture, grazing, and other factors. A series of dry years or overgrazing causes the succession to revert toward the annual weed stage; how far back depends on the severity of the effect. Increased nutrient enrichment, either with commercial fertilizer or with municipal sludge, will also arrest secondary succession in the annual weed stage of development (W. P. Carson and Barrett 1988; Brewer et al. 1994).

Succession is equally apparent in aquatic as in terrestrial habitats. However, as already emphasized, the community development process in shallow-water ecosystems (ponds and small lakes) is usually complicated by strong inputs of materials and energy from the watershed that may speed up, arrest, or reverse the normal trend of community development that would occur in the absence of such strong allogenic influences. The complex interaction of autogenic and allogenic succession is illustrated by the rapid changes in artificial ponds and impounded lakes. When a reservoir is created by flooding rich soil or an area with a large amount of organic matter (as when a forested area is flooded), the first stage in development is a highly productive bloom stage, characterized by rapid decomposition, high microbial activity, abundant nutrients, low oxygen levels at the bottom, but often rapid and vigorous growth of fish. People who fish are very pleased with this stage. However, when the stored nutrients are dispersed and the accumulated food used up, the reservoir stabilizes at a lower rate of productivity, with higher benthic oxygen and lower fish yields. Those who fish may be displeased with this stage (Fig. 8-9). However, this system and fish yield will likely remain stable on a long-term basis.

If the watershed is well protected by mature vegetation, or if the soils of the water-

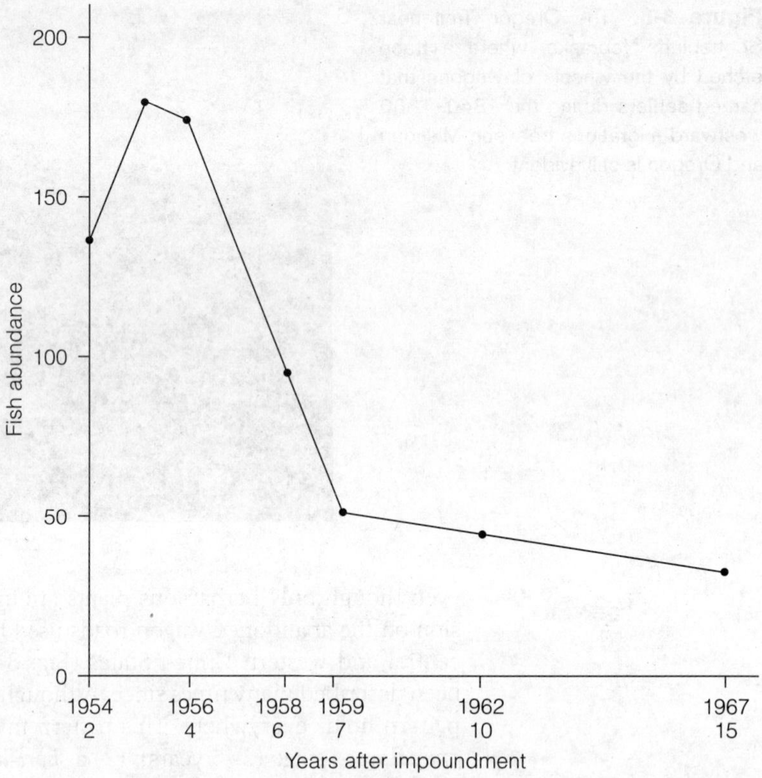

Figure 8-9. Fish abundance in a mainstream reservoir on the upper Missouri River from the second to the fifteenth year after completion of a dam in Lake Francis Case, South Dakota (data from Gasaway 1970).

shed are infertile, the stabilized stage in bodies of water may last for some time—a "climax" of sorts. However, erosion and various human-accelerated nutrient inputs usually produce a continuing series of transient states until the basin fills up. Impoundments in impoverished watersheds or primary, sterile sites will, of course, have a reverse pattern of low productivity at the start. Failure to recognize the basic nature of ecological succession and the relationships between the watershed and the impoundment has resulted in many failures and disappointments in human attempts to maintain such artificial ecosystems.

Because the oceans are, generally speaking, in a mature state and have been chemically and biologically stabilized for centuries, oceanographers have not been concerned with ecological succession. However, with pollution threatening to disturb equilibria in the sea, the interaction of autogenic and allogenic processes is starting to receive greater attention from marine scientists. Successional changes are evident in coastal waters, as already noted in the example of the development of benthic communities after severe storms have disrupted the sea bottom. Changes that occur in such a successional gradient in the coastal water column can be summarized as follows:

- The relative abundance of mobile forms among the phytoplankton increases;
- Productivity slows down;
- The chemical composition of the phytoplankton, as exemplified by the plant pigments, changes;
- The composition of the zooplankton shifts from passive filter feeders to more active and selective hunters, in response to a shift from numerous small suspended

food particles to scarcer food concentrated in bigger units and dispersed in a more organized (stratified) environment; and

- In the later stages of succession, total energy transfer may be lower, but its efficiency seems to improve.

The succession of organisms on artificial substrates in aquatic environments has received a great deal of attention, because of the practical importance of the fouling of ship bottoms and piers by barnacles and other sessile marine organisms. Small replicated substrates, such as glass slides or squares of plastic, wood, or other material, are widely used to assess the effect of pollutants on biota in both fresh and salt water (see Patrick 1954 regarding the early use of this method). Such substrates are a kind of microcosm, on which one would expect ecological succession to occur, but as with any restricted or simplified model, one must be cautious about projecting hypotheses to larger, less space-limited, open systems that possess many kinds of substrates. In general, the first species to colonize these substrates are those that have abundant propagules available in the water when and where the surfaces become available for colonization. Sometimes, these pioneers change the physical or chemical nature of the substrate in ways that may facilitate the invasion of other species, but just as often the pioneers resist encroachment by other species and endure until replaced by a better competitor. As already noted in discussing intertidal communities on rocky coasts (see Chapter 7), negative interaction (competition and predation) plays a greater role than positive interaction (coexistence and mutualism) in determining the replacement of species in confined or space-limited habitats.

Heterotrophic Succession

A laboratory hay-infusion microcosm experiment provides an example of *heterotrophic succession*—and also a laboratory experiment for an ecology class. When a culture medium made by boiling hay is allowed to stand, a thriving culture of bacteria develops. If some pond water (containing seed stock for various protozoa) is then added, a definite succession of protozoan populations with successive dominants occurs, as shown in Figure 8-10. A similar succession of protozoa occurs when unvegetated soil is first exposed to colonization (Bamforth 1997). In the hay infusion experiment, energy and nutrients are maximal at the beginning and then decline. Unless new medium is added, or an autotrophic regime takes over, the system eventually runs down, and all the organisms die or go into resting stages (spores or cysts)—quite different from autotrophic succession, in which energy flow is maintained indefinitely. The hay-infusion microcosm is a model for the kind of succession that occurs in decaying logs, animal carcasses, fecal pellets, and the secondary stages of sewage treatment. It might also be considered a model for the "downhill" succession that must be associated with a society dependent on fossil fuels that is slow to develop alternative energy sources. In all these examples, there is a series of transient stages in a declining energy gradient, with no possibility of achieving a mature climax state.

Heterotrophic and autotrophic successions can be combined in a laboratory microecosystem model if samples from a derived system are added to media enriched with organic matter. The system first becomes "cloudy" as heterotrophic bacteria bloom; then it turns bright green as nutrients and growth substances (especially the vitamin thiamine) required by algae are released by the activities of the bacteria. This

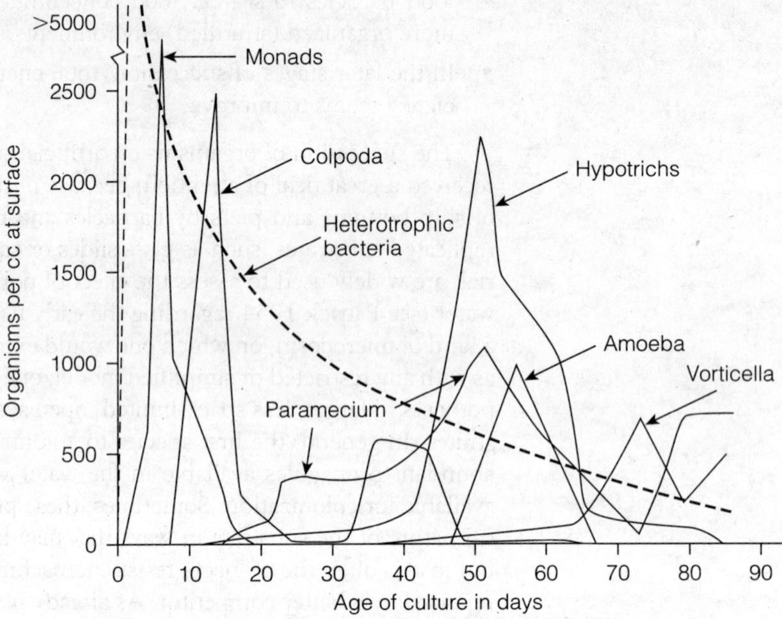

Figure 8-10. Succession in a hay-infusion culture, with dominance by successive species. This is an example of heterotrophic succession (after Woodruff 1912).

succession, of course, is a model of the *cultural eutrophication* resulting from organic pollution, such as inflow of incompletely treated municipal sewage.

Selection Pressure: Quantity Compared with Quality

Stages of the colonization of islands, as first described by MacArthur and Wilson (1967), provide parallels with stages in ecological succession on continents. In the early, uncrowded stages of island colonization, as in the early stages of succession, *r*-selection predominates, so that species with high rates of reproduction and growth are more likely to colonize. In contrast, selection pressure favors *K*-strategist species, with lower growth potential but better capabilities for competitive survival, under the high density of later stages of both island colonization and succession (Table 8-1).

Genetic changes involving the whole biota may be presumed to accompany the successional change from quantity production to quality production, as indicated by the tendency for the size of the individual organism to increase (Table 8-1). For plants, the change in size appears to be an adaptation to the shift of nutrients from inorganic to organic. In a mineral- and nutrient-rich environment, small size is of selective advantage, especially to autotrophs, because of the higher surface-to-volume ratio. As the ecosystem develops, however, inorganic nutrients tend to become more and more tied up in the biomass (that is, to become intrabiotic), so that the selective advantage shifts to larger organisms (larger individuals of the same species, larger species, or both), which have greater storage capacities and more complex life histories and are thus adapted to exploiting seasonal or periodic releases of nutrients or other resources.

Diversity Trends

Although both components of diversity (richness and apportionment) in Table 8-1 almost always increase in the early stages of ecosystem development, the peak of diversity seems to come somewhere in the middle of the sere in some cases and near the end in other situations. Not all trophic or taxonomic groups exhibit the same trend of diversity change with successional time. Nicholson and Monk (1974) determined richness and evenness of plant species for four life-forms—herbs, vines, shrubs, and trees—in major seral stages in the Georgia Piedmont old-field succession already illustrated (Fig. 8-7). Richness increased rapidly in each stratum after its establishment, then decreased throughout the remainder of succession. Evenness, on the other hand, increased to near maximum levels immediately and changed very little thereafter. Dominance-diversity curves for another old-field succession (southern Illinois) are shown in Figure 8-11. The diversity of plant species generally increased with succession, reaching a maximum during the early forest stages. The distribution curves of species are geometric (approaching a straight line in the semilog plot) during the first few years of the secondary succession, and then gradually change to lognormal as more species are added. The process results in a high degree of evenness.

Whether species diversity continues to increase during succession or peaks at some intermediate stage may well depend on whether the increase in potential niches resulting from increased biomass, stratification, and other consequences of biological organization exceeds the countereffects of increasing size of organisms and competitive exclusion by well-adapted, long-lived dominants, which would tend to reduce the richness of species. No one has yet been able to catalogue all the species in any sizable area, much less follow the *total* species diversity in a successional series. Studies on diversity and succession have so far dealt with segments of the community (such as trees, birds, and insects). One would expect that the pattern of change in species composition will vary widely according to the group under consideration and the geographical situation, which determines what species are available for colonization.

As discussed in the section on stability (see Chapter 2), the consensus among ecologists is that changes in species diversity are more an indirect consequence of increasing organic development and complexity than a direct causal factor in succes-

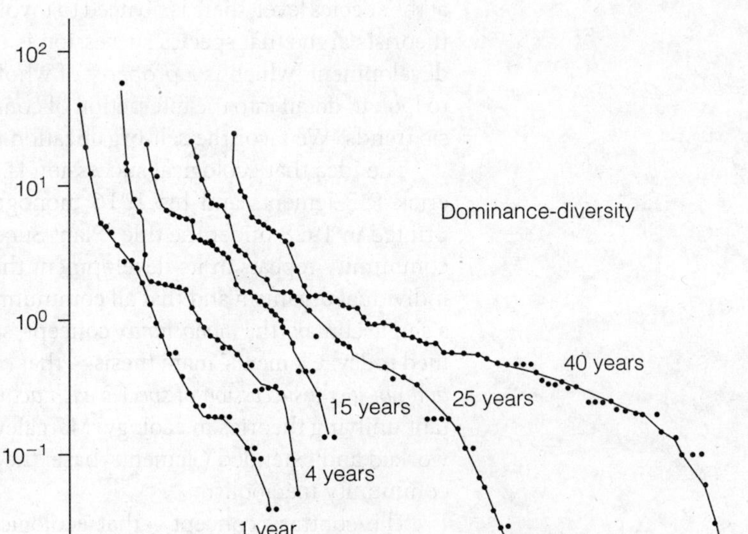

Figure 8-11. Dominance-diversity curves of old fields at five different ages of abandonment in southern Illinois (from Bazzaz 1975).

sion. The level of diversity achieved may well depend on energetics, as the maintenance of high diversity has an energy cost and can be destabilizing (the "too much of a good thing" syndrome again).

Although little studied, aspects of biotic diversity other than species variety and relative abundance would logically be expected to show increasing trends during the course of autogenic ecosystem development. Jeffries (1979), for example, reported that as marine communities mature and become more complex, so do the fatty-acid compositions of the plankton and benthos. During plant succession on land, the variety of antiherbivore chemicals required by long-lived trees to survive increases with succession, thereby countering the tendency for insects to become resistant to pesticides, both natural and human-made. These are examples of an increase in *biochemical diversity* during ecosystem development.

Historical Review of Theoretical Considerations

At the beginning of this section, it was stated that ecosystem development resulted from (1) modification of the physical environment by the community acting as a whole, and (2) the interaction of competition and coexistence between component populations. Although one could logically assume that both ecosystem-level and population-level processes contribute to the many-faceted successional progressions described in this section, some ecologists have chosen to argue either for one level or for the other but not both. Connell and Slayter (1977) compared three theories: (1) the *facilitation model,* in which early seral species change the conditions of existence and thereby prepare the way for later invaders; and (2) the *inhibition model,* in which the pioneer species resist invasion and remain until they are replaced because of competition, predation, or disturbance; and (3) the *tolerance model,* in which a species invades a new habitat and becomes established independently of the presence or absence of other species. Connell and Slayter strongly favored the inhibition model, at least for secondary succession. Proponents of population-level theories of causation essentially argue that if observed successional trends can be explained by interactions at the species level, there is no need to invoke higher-level processes. Conversely, other theorists argue that species succession is only a part of the process of self-organizing development, which is a property of whole ecosystems, and hence there is less need to look in detail into the interaction of component populations in order to explain basic trends. We favor the self-organization theory, as explained in the next section.

The idea that ecological succession is a holistic phenomenon goes back to Frederick E. Clements and his 1916 monograph "Plant Succession" (subsequently reprinted in 1928 under the title "Plant Succession and Indicators"). His notions that a community repeats in its development the sequence of stages of development of an individual organism and that all communities in a given climatic area develop toward a single climax (the *monoclimax* concept; see next section) are deemphasized or modified today. Clements' main thesis—that *ecological succession is a developmental process and not just a succession of species each acting alone*—remains one of the most important unifying theories in ecology. Margalef (1963a, 1968) and E. P. Odum (1969) reworked and extended Clements' basic theory to include functional attributes such as community metabolism.

The contrary concept—that ecological succession does not have an organizational strategy but results from the interactions of individuals and species as they struggle to occupy space—goes back to H. A. Gleason's studies, especially the classic paper "The Individualistic Concept of the Plant Association" (Gleason 1926). Gleason's writings, as reviewed by McIntosh (1975), have provided a point of departure for the development of population-level theories of succession that consider new insights into evolutionary biology and the importance of consumer as well as producer influences. Reviews by Drury and Nisbet (1973) and Horn (1974, 1975, 1981) explored theories of succession that are based on properties of organisms rather than emergent properties of the ecosystem. The basic premise is that evolutionary strategy (Darwinian selection and competitive exclusion) and characteristics of the life cycle determine the position of species in successional gradients that are constantly changing depending on disturbances and physical gradients. Because Clements' holistic theory can also be viewed as an evolutionary theory of population and ecosystem, ecologists may not be so far apart as a reading of their respective position papers might indicate. This position, in general, is the one taken by Whittaker and Woodwell (1972), Whittaker (1975), and Glasser (1982), who noted that although the early colonization phase is often stochastic (chance establishment of opportunistic organisms), later stages are much more organizational and directional.

Sooner or later, theories get tested in the practical world of applied science—for example, in forest management. Foresters, by and large, find that forest succession is directional and predictable. To assess future timber potential, they often develop models that combine natural successional trends with disturbance and management scenarios that modify natural development. For example, on the Georgia Piedmont, the natural forest succession is from pines to hardwoods. Because pines are now more valuable commercially than hardwoods, efforts are made to arrest this succession, so that the pine stages can be retained and regenerated, especially in areas under commercial timber management. It is predicted that hardwood stages will continue to increase in area coverage, although at a slower rate than would be the case if only natural succession were involved. Urbanization and suppression of fire, both of which favor hardwoods over pines, are important factors in future projections. Because the composition of the Piedmont forest is strongly influenced by human management, projected future composition will follow trends of natural succession. The interface between theory and forest management is discussed in detail by Shugart (1984) and Chapin et al. (2002).

Self-Organization, Synergetics, and Ascendancy

A major key to ecosystem development is the concept of *self-organization,* based on Prigogine's theory of non-equilibrium thermodynamics (Prigogine 1962). **Self-organization** can be defined as the process whereby complex systems consisting of many parts tend to organize to achieve some sort of stable, pulsing state in the absence of external interference. The spontaneous formation of well-organized structure, pattern, and behavior from random or unorganized initial conditions—in other words, going from chaos to order—is widespread in nature. Self-organized ecosystems can only be maintained by a constant flow of energy through them; therefore, they are not in thermodynamic equilibrium. The process of many parts working together to achieve order has been termed **synergetics** by Haken (1977). Ulanowicz (1980, 1997) used the term **ascendancy** for the tendency for self-organizing, dissipative systems to develop complexity of biomass and network flows over time, as is seen in the process of ecological succession. Both Holland (1998) and S. Johnson (2001) spoke of the process as **emergence**.

We see that ecosystem development is more than just a succession of species, and more than just evolutionary interactions, such as competition and mutualism; there is an energetic basis. There is a large volume of literature on self-organization, including papers by Eigen (1971), H. T. Odum (1988), and Müller (1997, 1998, 2000); and books by Kauffman (1993), Bak (1996), and Camazine et al. (2001), in addition to those cited earlier. For a less technical discussion of self-organization, see Li and Sprott (2000). Wesson (1991), in his book *Beyond Natural Selection,* argued that what we call *self-ordering* must be added to natural selection to explain the evolution of complex systems. Smolin (1997) extended the self-organization theory to the origin and evolution of the universe that began with the Big Bang and a mass of random-moving molecules but evolved into the current, highly organized system including Earth.

2 Concept of the Climax

Statement

The final $P = R$ community in a developmental series (*sere*) is the **climax community.** In theory, the climax community is self-perpetuating, because it is more or less in equilibrium with itself and with the physical habitat. For a given region, it is convenient, although quite arbitrary, to recognize (1) a **regional** or **climatic climax,** which is determined by the general climate of the region; and (2) a varying number of **local** or **edaphic climaxes,** which are determined by topography and local microclimate (see Fig. 8-12), that would not occur in the absence of disturbance. Succession ends in an edaphic climax when topography, soil, water, and regular disturbances such as fire are such that the development of the ecosystem does not proceed to the theoretical end point.

Explanation and Examples

In terms of species composition, the **polyclimax** concept (choice of climatic and edaphic climaxes) is illustrated by mature forest communities associated with various physical situations, such as in the hilly region located in Ontario, Canada, diagrammed in Figure 8-12A. On level or moderately rolling areas where the soil is well drained but moist, a maple-beech community (*Acer saccharum* and *Fagus grandifolia* being the dominant species) is found to be the terminal stage in succession. Because this type of community is found again and again in the region wherever land configuration and drainage are moderate, the maple-beech community can be designated the *climatic climax* of the region. Where the soil remains wetter or drier than normal (despite the action of communities), different species are dominant in the climax community. Still greater deviations from the climatic climax occur on steep south-facing slopes, where the microclimate is warmer, or on north-facing slopes and in deep ravines, where the microclimate is colder (Fig. 8-12B). These steep-slope climaxes often resemble climatic climaxes found farther south and north, respectively. Accordingly, if you live in eastern North America and wish to see what a climax forest would be like farther north, view an undisturbed, north-facing slope or ravine. Similarly, a

Figure 8-12. Climatic and edaphic climaxes in southern Ontario, Canada. (A) Distribution of climax communities depending on local conditions. (B) Overview of possible climax communities. The maple-beech community is the climatic climax, occurring whenever conditions are moderate. Changes in microclimate conditions lead to various other (edaphic) climaxes. (C) Theoretical development of edaphic climaxes at extremes of moisture (wet or dry) toward a climatic climax at intermediate moisture conditions (simplified from Hills 1952).

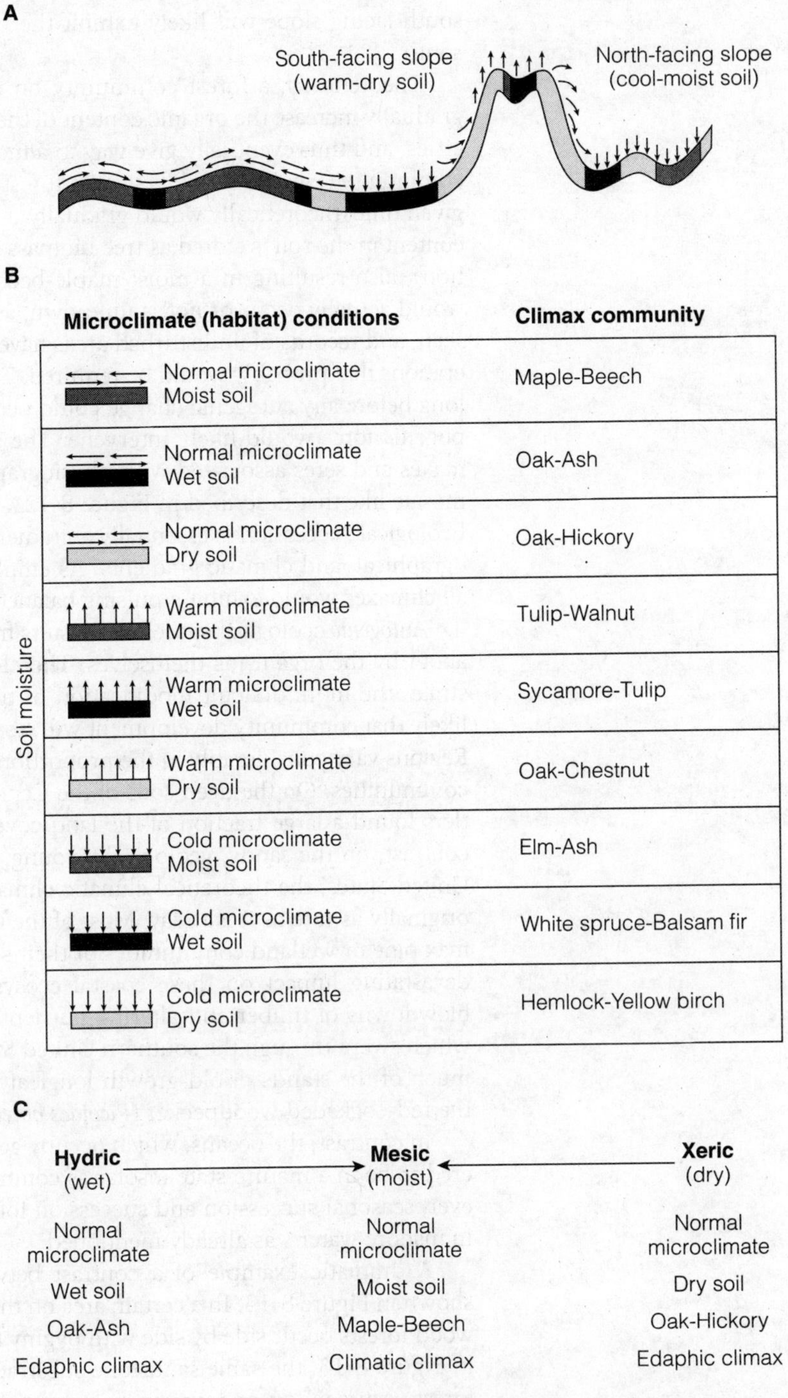

south-facing slope will likely exhibit the type of climax forest to be found farther south.

Theoretically, a forest community on dry soil would, if given indefinite time, gradually increase the organic content of the soil and raise its moisture-holding properties, and thus eventually give way to a more moist forest, such as the maple-beech community (Fig. 8-12C). Likewise, a forest community under wet soil conditions, if given time, theoretically would gradually decrease moisture in the soil as the organic content in the soil is stored as tree biomass (and output as increased plant transpiration), also resulting in a moist maple-beech community. Whether these scenarios would actually occur or not is unknown, as little evidence of such change has been seen, and records of undisturbed areas have not been kept for the many human generations that probably would be required. The question is academic anyway, because long before any autogenic change could occur, some climatic, geological, or anthropogenic force would likely intervene. The alternative to recognizing a series of climaxes and seres associated with physiographic situations in the case of a landscape mosaic like that described in Figure 8-12A would be some form of *gradient analysis*. Ecological succession is essentially a gradient in time that interacts with spatial, topographical, and climatic gradients. As emphasized at the beginning of this chapter, all climaxes would exhibit a pulsing balance between *P* and *R*.

Autogenic ecological succession results from changes in the environment brought about by the organisms themselves. Therefore, the more extreme the physical substrate, the more difficult modification of the environment becomes, and the more likely that community development will stop short of the theoretical regional climax. Regions vary considerably in the proportion of area that can support climatic climax communities. On the deep soils of the Central Plains of the United States, early settlers found a large fraction of the land covered with the same climax grassland. In contrast, on the sandy, geologically young, lower Coastal Plain of the southeastern United States, the theoretical climatic climax (a broad-leaved evergreen forest) was originally as rare as it is today. Most of the Coastal Plain is occupied by edaphic climax pine or wetland communities or their seral stages. Hurricanes frequently have a devastating impact on these coastal ecosystems, causing massive defoliation and blowdowns of timber and altering nutrient cycling. Hurricane Hugo, for example, which swept through the southern United States and Puerto Rico in 1989, destroyed much of the stands of old-growth longleaf pine (*Pinus palustris*)—prime habitat of the red-cockaded woodpecker (*Picoides borealis*).

In contrast, the oceans, which occupy geologically ancient basins, can be considered to be in a mature state insofar as community development is concerned. However, seasonal succession and succession following disturbance do occur, especially in inshore waters, as already mentioned.

A dramatic example of a contrast between regional and edaphic climaxes is shown in Figure 8-13. In a certain area on the coast of northern California, giant redwood forests occur side by side with pygmy forests of tiny, stunted trees. As depicted in Figure 8-13, the same sandstone substrate underlies both forests, but the pygmy forest occurs where an impervious hardpan close to the surface greatly restricts root development and movement of water and nutrients. The vegetation that reaches climax condition in this special situation is almost totally different in species composition and structure from that of adjacent areas that lack the hardpan.

Human beings, of course, greatly affect the progress of succession and the achieve-

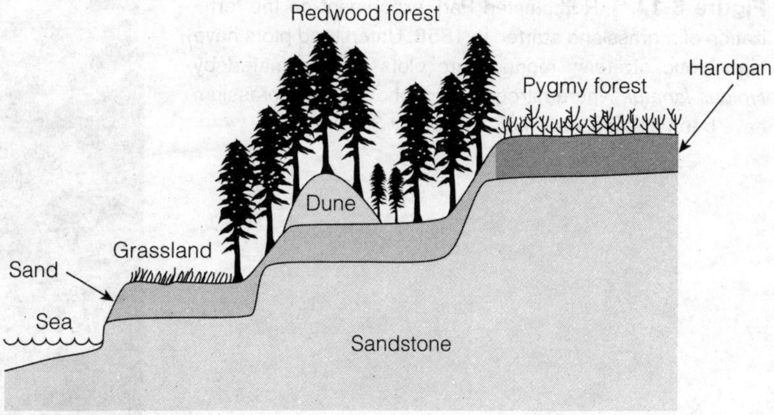

Figure 8-13. Edaphic climaxes on the West Coast of northern California. Forests of tall redwoods and dwarf conifers grow side by side on adjacent marine terraces. The stunted nature of the pygmy forest is due to an iron-cemented, B-horizon hardpan located about 18 inches (0.5 m) below the surface. The soil above the impervious hardpan is extremely acidic (pH 2.8–3.9) and low in Ca, Mg, K, P, and other nutrients (after Jenny et al. 1969).

ment of climaxes. When a biotic community that is not the climatic or edaphic climax for a given site is maintained by people or their domestic animals, it may conveniently be designated a **disclimax** (disturbance climax) or **anthropogenic** (human-generated) **subclimax.** For example, overgrazing by livestock may produce a desert community of creosote bush, mesquite, and cactus where the local climate actually would allow a grassland to maintain itself. The desert community would be the disclimax, and the grassland would be the climatic climax. In this case, the desert community is evidence of poor management by humans, whereas the same desert community in a region with a true desert climate would be a natural climax condition. An interesting combination of edaphic and disturbance climaxes occupies extensive areas of the California grassland region where introduced annual species have almost entirely replaced native prairie grasses.

Agricultural ecosystems (*agroecosystems*) that have been stable for a long time can be regarded as climaxes (or disclimaxes), because on an annual average, imports plus production balance respiration plus exports (harvest), and the agrolandscape remains the same from year to year. Agriculture in the Low Countries (Holland and Belgium) and age-old rice cultures in the Orient are examples of long-term anthropogenic stable climax states. Unfortunately, industrialized crop systems, especially as currently managed in the Tropics and on irrigated deserts, are by no means sustainable, because they are subject to erosion, leaching, accumulation of salt, and irruptions of pests. Maintaining high productivity in such systems requires increasing energy and chemical subsidies, and too much subsidy becomes a stress. For detailed information regarding long-term experiments in agriculture and forestry, see the book edited by R. A. Leigh and A. E. Johnston (1994) entitled *Long-Term Experiments in Agricultural and Ecological Sciences.* Figure 8-14 is a photograph of the long-term (since 1856) fertilization project located at Rothamsted Park. Unfertilized plots are characterized by high biotic diversity, compared to the "monoculture" plots dominated by *Holcus lanatus.*

To summarize, species composition has often been used as a criterion to determine whether a given community is a climax. However, this criterion alone is often not sufficient, because species composition not only varies widely, but also can

Figure 8-14. Rothamsted Park experiment on the fertilization of a grassland started in 1856. Unfertilized plots have high biotic diversity; monoculture plots are dominated by *Holcus lanatus* where nitrogen, phosphorus, and potassium have been applied, causing a soil pH of 3.5.

change appreciably in response to seasons and short-term fluctuations of weather, even though the ecosystem as a whole remains stable. As already indicated, the *P/R* ratio or other functional criteria provide accurate indices of climax communities.

3 Evolution of the Biosphere

Statement

As with short-term ecosystem development, described in Section 1 of this chapter, the long-term evolution of the ecosphere is shaped by (1) *allogenic* (external) *forces,* such as geological and climatic changes; and (2) *autogenic* (internal) *forces,* such as *natural selection* and other processes of self-organization resulting from activities of the organisms in the ecosystem. The first ecosystems, 4.0 billion years ago, were populated by tiny anaerobic heterotrophs that lived on organic matter synthesized by abiotic processes. Then came the origin and population explosion of green bacterial autotrophs, which are believed to have played a dominant role in initiating the conversion of a reducing, CO_2-dominated atmosphere into an oxygenic one. Since then, organisms have evolved through long geological ages into increasingly complex and diverse systems that (1) have achieved control of the atmosphere; and (2) are populated by larger and more highly organized multicellular species. Evolutionary change is believed to occur principally through natural selection at or below the species level, but natural selection above this level is also important, especially (1) **coevolution,** which is the reciprocal selection between interdependent autotrophs and heterotrophs without direct genetic exchanges; and (2) **group** or **community selection,** which leads to the maintenance of traits favorable to the group even when they are disadvantageous to the genetic carriers within the group.

Explanation

We have already briefly outlined the history of life on Earth in connection with the discussion of the Gaia hypothesis in Chapter 2. The broad pattern of the evolution of organisms and the oxygenic atmosphere—two factors that make the biosphere of Earth unique among the planets of our solar system—are depicted in Figure 8-15.

The biogeological clock (Fig. 8-16) shows the entire history of life on Earth, beginning with the origin of Earth about 5 billion years ago and the appearance of the first microbial life about 4 billion years ago. During the long era from 3.5 to 2 billion years ago, the photosynthetic bacteria, especially the cyanobacteria, put oxygen into the atmosphere, paving the way for the origin and evolution of the aerobic, macroscopic eucaryotes. Within what is known as the Phanerozoic eon—the past 570 million years—present-day plants and animals, and finally humans, evolved. The Phanerozoic eon is further divided into the Paleozoic (earliest insects and reptiles), Mesozoic (first birds and mammals), and Cenozoic (first hominids and humans) eras (see Fig. 8-16).

Scientists generally believe that when life began on Earth, the atmosphere contained nitrogen, ammonia, hydrogen, carbon dioxide, methane, and water vapor, but no free oxygen. The atmosphere also contained chlorine, hydrogen sulfide, and other gases that would be poisonous to much of present-day life. The composition of the atmosphere at that time was largely determined by the gases from volcanoes, which were much more active then than they are now. Because of the lack of oxygen, no

Figure 8-15. The evolution of the biosphere and its effect on the atmosphere (E. P. Odum 1997).

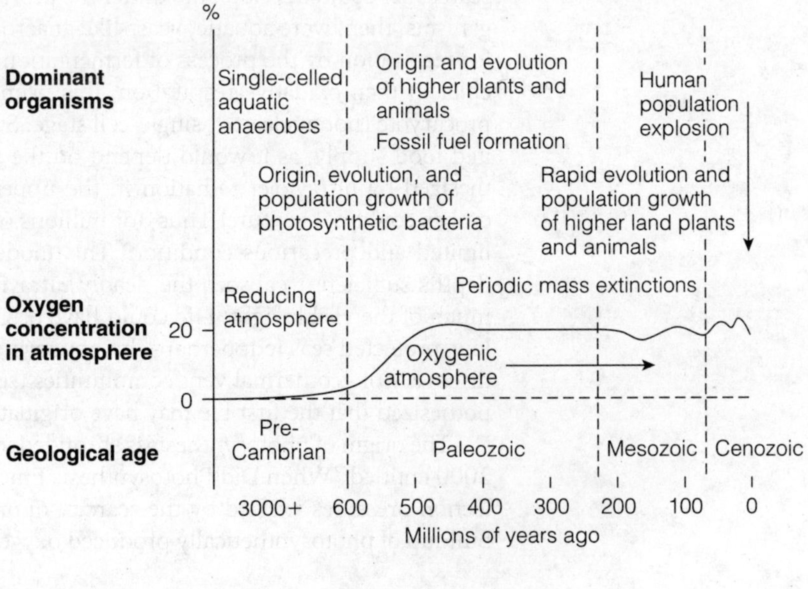

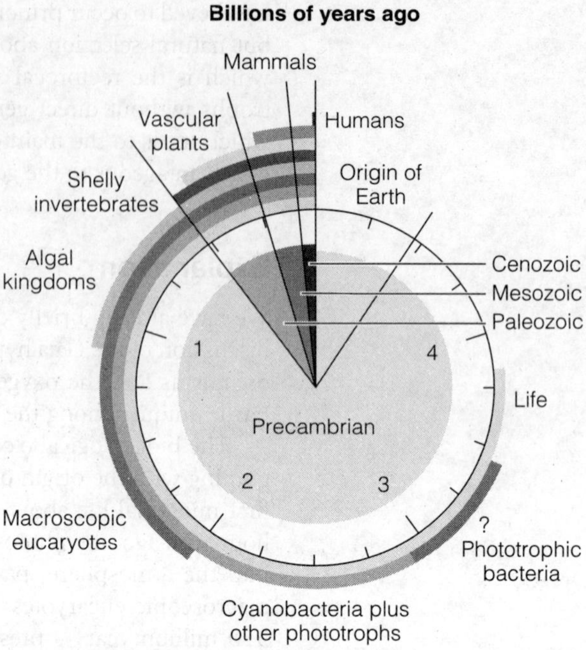

Figure 8-16. Earth's biogeological clock. The great antiquity of our ecosphere contrasts sharply with the relative youth of plants and animals. For almost 2 billion years, microorganisms constituted the only life on Earth, and they continue to dominate basic ecosystem functions, such as material cycling, to this day. (From Des Marais, D. J. 2000. When did photosynthesis emerge on Earth? *Science* [8 September] 289 [5485]: 1703–1705.)

ozone layer (O_2 acted on by short-wave radiation produces O_3 or *ozone,* which absorbs ultraviolet radiation) shielded Earth from the deadly ultraviolet radiation of the Sun. Such radiation would kill any exposed life, but, strange to say, this radiation is thought to have triggered a chemical evolution leading to complex organic molecules such as amino acids, which then became the building blocks for primitive life. The very small amount of nonbiological oxygen produced by ultraviolet dissociation of water vapor may have provided enough ozone to form a slight shield against ultraviolet radiation. However, as long as the atmospheric oxygen and ozone remained scarce, life could develop only under the protective cover of water. The first living organisms, then, were aquatic, yeast-like anaerobes that obtained the energy necessary for respiration by the process of fermentation. Because fermentation is so much less efficient than oxidative respiration, this primitive life could not evolve beyond the procaryote (nonnucleated) single-cell stage. Such primordial life also had a very limited food supply, as it would depend on the slow sinking of organic materials synthesized by ultraviolet radiation in the upper water layers, where the hungry microbes could not venture! Thus, for millions of years, life must have existed in a very limited and precarious condition. This model of primitive ecology calls for water depths sufficient to absorb the deadly ultraviolet, but not so deep as to cut off too much of the visible light. Life could have originated on the bottom of pools or shallow, protected seas fed, perhaps, by hot springs rich in nutrient chemicals. With the discovery of geothermal vent communities (see Chapter 2), some scientists have hypothesized that the first life may have originated there.

The origin of photosynthesis is shrouded in mystery (see the article by Des Marais 2000 entitled "When Did Photosynthesis Emerge on Earth?" for details). Perhaps selection pressures exerted by the scarcity of organic food played a part. The gradual buildup of photosynthetically produced oxygen and its diffusion into the atmosphere about 2 billion years ago (Fig. 8-15) brought about tremendous changes in the geochemistry of Earth and made possible the rapid expansion of life and the development of the eucaryote (nucleated) cell, which led to evolution of larger and more complex living organisms. Many minerals, such as iron, were precipitated from water and formed characteristic geological formations.

As the oxygen in the atmosphere increased, the layer of ozone formed in the upper atmosphere thickened sufficiently to screen out the DNA-disrupting ultraviolet radiation. Life could then move more freely to the surface of the sea. Then followed what Cloud (1978) called the "greening of the lands." Aerobic respiration made possible the development of complex multicellular organisms. It is thought that the first nucleated cells appeared when oxygen reached about 3 to 4 percent of its present level (or about 0.6 percent of the atmosphere, compared with the present 20 percent)—a time now dated to at least 1 billion years ago. Margulis (1981, 1982) has made a strong case for the theory that the eucaryote cell originated as a mutualistic coming together of once independent microbes, analogous to the modern evolution of lichens.

The first multicellular animals (Metazoa) appeared when the atmospheric oxygen content reached about 8 percent, some 700 million years ago (Figs. 8-15 and 8-16). The term *Precambrian* is used to cover that vast period of time when only the small, procaryote single-celled life existed. During the Cambrian period (about 500 million years ago), there was an evolutionary explosion of new life, such as sponges, corals, worms, shellfish, seaweed, and the ancestors of seed plants and vertebrates. Thus, the fact that the tiny green plants of the sea were able to produce an excess of oxygen over the respiration ($P/R > 1$) needs of all organisms allowed the whole of Earth to be populated in a comparatively short time, geologically speaking. In the following periods of the Paleozoic era, the expansion of the biosphere to the whole planet was completed. The developing green mantle of terrestrial vegetation provided more oxygen and food for the subsequent evolution of large creatures, such as dinosaurs, birds, mammals, and eventually humans. At the same time, calcareous and then siliceous forms were added to the organic-walled phytoplankton of the oceans.

When oxygen use finally caught up with oxygen production sometime in the mid-Paleozoic era (about 400 million years ago), the concentration of oxygen in the atmosphere reached its present level of about 20 percent. From the ecological viewpoint, then, the evolution of the biosphere seems to be very much like a heterotrophic succession followed by an autotrophic climax regime, such as one might set up in a laboratory microcosm starting with culture medium enriched with organic matter. During the late Paleozoic era, there appears to have been a decline of O_2 and an increase of CO_2, accompanied by climatic changes. The increase of CO_2 may have triggered the vast autotrophic bloom that created the fossil fuels on which human industrial civilization now depends. After a gradual return to a high O_2–low CO_2 atmosphere, the O_2/CO_2 balance remained in what might be called an *oscillating steady state*. Anthropogenic CO_2, aerosols, and dust pollution may be making this precarious balance still more unsteady (as discussed in Chapters 2 and 4).

The story of the atmosphere, as briefly described here, should be shared with schoolchildren and citizens, because it dramatizes the absolute dependence of human beings on other organisms in the environment. According to the Gaia hypothesis (Chapter 2), homeorhetic control, especially by microorganisms, developed very early in the history of the ecosphere. A contrary hypothesis is that early life induced physicochemical changes and geological processes involved in the cooling down of

Earth. In other words, was the early evolution of life more autogenic than allogenic, or vice versa? Also much debated is whether evolution of life occurs gradually or is strongly pulsed (short periods of rapid change alternating with long periods with little or no change), as is suggested by the fossil record. This question will be considered in the next section.

The constant shifting of the continents through time—a process known as **continental drift** or **plate tectonics**—also has an important bearing on the evolution of life. For a further discussion of this process, see J. T. Wilson (1972) and Van Andel (1994). For an account of the evolution of life, see *Early Life* by Margulis (1982) and Margulis (2001).

4 Microevolution Compared with Macroevolution, Artificial Selection, and Genetic Engineering

Statement

The *species* is a natural biological unit tied together by the sharing of a common gene pool. Evolution involves changing gene frequencies resulting from (1) *selection pressure* from the environment and interacting species; (2) *recurrent mutations;* and (3) *genetic drift* (stochastic or chance changes in gene structure). **Speciation**—the formation of new species and the development of species diversity—occurs when gene flow within the common pool is interrupted by an isolating mechanism. When isolation occurs through geographical separation of populations descended from a common ancestor, **allopatric** (different geographical area) **speciation** may result. When isolation occurs through ecological or genetic means within the same geographical area, **sympatric** (joint geographical area) **speciation** is a possibility. At present, it is uncertain to what extent speciation is a slow, gradual process (**microevolution**) or a matter of periodic, rapid changes (**macroevolution**). It now appears that sympatric speciation and macroevolution are more common than previously thought.

Explanation

Ever since Darwin, biologists have generally adhered to the theory that evolutionary change is a slow, gradual process, involving many small mutations and continuous natural selection of those mutations that provide competitive advantages at the individual level. However, gaps in the fossil record and frequent failure to find transitional forms ("missing links") has led many paleontologists to accept what Gould and Eldredge (1977) have termed the *theory of punctuated equilibria.* According to this theory, species remain unchanged in a sort of evolutionary equilibrium for long periods. Then, once in a while, the equilibrium is "punctuated" when a small population splits off from the parent species and rapidly evolves into a new species without leaving transitional forms in the fossil record. The new species may be sufficiently different to coexist with rather than replace the parent species, or both may become extinct. The punctuated evolutionary theory does not emphasize competition at the individual level as the driving force, but as yet there is no agreed-upon explanation of what might cause a population to suddenly break away to form a new, genetically isolated reproductive unit. For more on macroevolution-microevolution comparisons, see Gould and Eldredge (1977), Rensberger (1982), and Gould (2000, 2002).

Species that occur in different geographical regions or are separated by a spatial barrier are said to be *allopatric;* those occurring in the same area are said to be *sympatric.* Allopatric speciation has been generally assumed to be the primary mechanism by which species arise. According to this conventional view, two segments of a freely interbreeding population become separated spatially, as on an island, or separated by a mountain range. In time, sufficient genetic differences accumulate in isolation so that the segments will no longer be able to interchange genes (interbreed) when they come together again, and thereby coexist as distinct species in different niches. Sometimes, these differences are further accentuated by *character displacement.* When two closely related species have overlapping ranges, because of the selection effects of competition, they tend to diverge in one or more morphological, physiological, or behavioral characteristics in the area of overlap and to converge (to remain or become similar to each other) in parts of their range where each species occurs alone.

Evidence is mounting that strict geographical separation is not necessary for speciation and that sympatric speciation may be more widespread and important than previously believed. Populations can become genetically isolated within the same geographical area as a result of behavioral and reproductive patterns such as colonization, restricted dispersal of propagules, asexual reproduction, selection, and predation. In time, sufficient genetic differences accumulate in the local population segments to prevent interbreeding.

Examples

A classic example of allopatric speciation (resulting from geographical isolation) with subsequent character displacement is the well-documented case of Galápagos finches, which was first described by Darwin, who visited the Galápagos Islands during his famous voyage sailing on the *Beagle.* From a common ancestor, a whole group of species evolved in isolation on the different islands and adaptively radiated so that a variety of potential niches was eventually exploited on reinvasion. These finch species now include slender-billed insect eaters, thick-billed seed-eaters, ground and tree feeders, large- and small-bodied finches, and even a woodpecker-like finch, which, although hardly able to compete with a real woodpecker, survives in the absence of invasion by woodpecker stock. Figure 8-17 shows how beak size varies in one of the Galápagos finches according to whether the species is alone on an island or coexists with two other, closely related species on another, larger island. In the latter situation (Fig. 8-17B), the beak is "displaced" to an increased depth, so that it does not overlap with the beak size of its competitor. As a result, competition for food is reduced, because each species is adapted to feed on different-sized seeds. In addition to the classic book entitled *Darwin's Finches* by David Lack (1947a), this evolutionary process has been reviewed by Grant (1986) and Grant and Grant (1992).

British salt marshes provide an example of sympatric speciation resulting from hybridization and polyploidy. When the American salt marsh grass (*Spartina alterniflora*) was introduced to the British Isles, it crossed with the native species (*S. maritima*), to produce a new, polyploid species (*S. townsendii*), which has now invaded formerly bare tidal mud flats not occupied by native species.

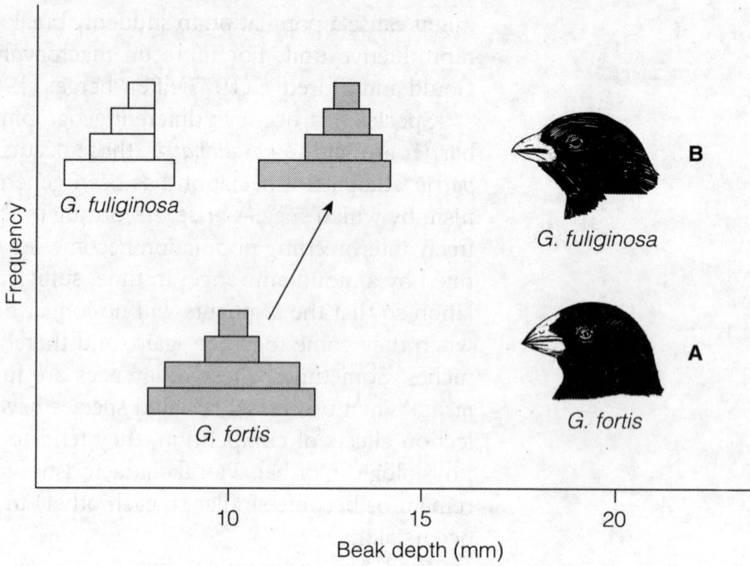

Figure 8-17. Beak size of *Geospiza fortis*, one of Darwin's finches, (A) when alone on Galápagos Island and (B) when competing with *Geospiza fuliginosa* on other islands. Beak size is increased when competitors are present–an example of character displacement (modified after Lack 1947a; Grant 1986).

For decades, what came to be known as "industrial melanism" was thought to be an example of rapid natural selection resulting from industrial pollution. The hypothesis was that dark-pigmented peppered moths (*Biston betularia*) evolved in industrial areas of England where the bark of trees had become greatly darkened by industrial pollution, which kills the lichens that give normal bark a light appearance. Kettlewell (1956) provided evidence that dark moths survived better in dark (polluted) woods, and pale moths survived better in natural rural woods, presumably because of predation by birds for the individuals not protectively colored. Scientists now admit that the real explanation regarding the selective pressures on *B. betularia* is much more complicated than many were first led to believe. We recommend reading Judith Hooper's book *Of Moths and Men* (Hooper 2002) for an excellent overview of the peppered moth controversy. This controversy leads us naturally into the subject of direct or purposeful selection by humans.

Artificial Selection and Domestication

Selection carried out to adapt plants and animals to human needs is known as **artificial selection.** Domestication or cultivation of plants and animals involves more than modifying the genetics of a species, because reciprocal adaptations between the domesticated species and the domesticator are required. The term *cultivation* is preferred by many for the artificial selection of plants. Here we use the term **domestication** in a general sense for both plants and animals. Accordingly, domestication leads to a special form of mutualism. Humans often fall into the trap of thinking that domesticating another organism through artificial selection means merely "bending" nature to suit human purposes. Actually, domestication produces changes (ecological and social, if not genetic) in people as well as in the domesticated organisms. Thus, for example, people are just as dependent on the corn plant as the corn is dependent on them. A society that depends on corn for food develops a very different culture than one that depends on herding cattle. It is a real question as to who becomes more dependent on whom!

Artificial selection in crops—a major basis for the Green Revolution—is an example of interdependence between the domesticated species and the human domesticator. Increased yield is obtained by selecting for an increased **harvest ratio**, which is the ratio of grain (or other edible parts) to supporting tissues (leaves, roots, and stems). As was noted in Chapter 3, increased yield must sacrifice some of the plant's adaptive, self-sustaining capacity. Therefore, highly bred strains require massive subsidies of energy, fertilizers, and pesticides, which bring about profound changes in the social, economic, and political structure of human society (see E. P. Odum and Barrett 2004 for details). Many economically poor countries are finding these socioeconomic and resource requirements to be the greatest obstacle to using high-yielding varieties to increase their food supply. Meanwhile, wealthy countries are finding that the runoff of fertilizers and pesticides produces very serious pollution of waterways (see Vitousek, Aber, et al. 1997 for details).

Throughout history, serious environmental problems have been caused by domesticated plants and animals that escape back into nature (become feral) and become major pests. A *feral organism* differs from its wild ancestor in that it has experienced a period of artificial selection during which some new traits may have been acquired and some of the original "wild" traits have been lost. On returning to the wild, the feral species again comes under natural selection that favors traits necessary for survival on its own. For example, a spotted, pale-color coat with a large body size is selected against when domestic pigs revert to the wild, so the feral pig becomes slender with a deep-color coat. The combination of artificial and natural selection seems to produce plants and animals that thrive in habitats that have been partially altered or disturbed (that is, in areas of extensive habitat fragmentation).

Genetic Engineering in Agroecosystems

Genetic engineering involves the manipulation of DNA and the transfer of genetic material between species. Although there are many potential applications of this new technology, the first large-scale, controversial application is in agriculture, involving generations of genetically modified or *transgenic* crops that are herbicide, disease, or insect resistant. In theory, the introduction of such crops should reduce crop losses to disease, weeds, and pests, and reduce the use of pesticides. However, as is often the case with targeted technology, this theory does not take into consideration the whole picture, including ecological, economic, social, and political implications. Massive planting of transgenic monocultures leads to genetic uniformity, which increases the individual farmer's dependence on multinational corporations that control the innovations. Ecological theory and research in progress show that transgenic monocultures can have serious environmental impacts, ranging from (1) increased gene flow between crops and weed relatives, creating "superweeds"; (2) rapid development of insect resistance; to (3) impacts on soil organisms and other nontarget organisms. For a review of potential impacts, see Altieri (2000).

A report issued by the National Academy of Sciences entitled "Transgenic Plants and World Agriculture" (2000), compiled in collaboration with the science academies of seven other countries, stresses the need for transgenic plants to improve crops in order to feed a hungry world. Perhaps the most ambitious genetic engineering contemplated in agriculture involves trying to increase the efficiency of the photosynthetic enzyme RuBis CO_2 that interacts with CO_2 to initiate the chain of biochemical reactions that converts sunlight to food. One possibility is to transfer the more efficient RuBis CO_2 found in red algae (which are adapted to grow at very low light intensities in the sea) into crop plants such as rice. Another possibility is to transfer C_4 photosynthesis into C_3 crops. As was noted in Chapter 3, C_4 plants produce better in the full sunlight typical of warm, dry climates. See Mann's (1999) review for possibilities and difficulties related to genetic engineering.

5 Relevance of Ecosystem Development to Human Ecology

Statement

The principles of ecosystem development bear profoundly on the relationships between human beings and nature, because developmental trends in both natural systems and human societies involve going from a youth (pioneer) stage to a mature stage in the long term. In the short term, humans strive to prolong the growth stage. The aim of developing increasing structure and complexity per unit of energy flow (a maximum protection strategy) contrasts with the human goal of maximum production (trying to obtain the highest possible yield). Recognizing the ecological basis for this conflict between humans and nature is a first step in establishing rational policies for managing the environment as societies mature.

Explanation

Figures 8-2B and 8-3 depicted a basic conflict between the strategies of humans and of nature. The energy partitioning exhibited in early development, as in the 30-day microcosm or the 30-year forest, illustrates how political and economic leaders think nature should be directed. For example, the goal of industrial agriculture or intensive forestry, as now generally practiced, is to achieve high rates of production for products that may be readily harvested, with little standing crop left to accumulate on the landscape—in other words, a high P/B ratio. Nature's strategy, on the other hand, as seen in the outcome of the successional process, is directed toward the reverse efficiency—a high B/P ratio. Human beings have generally been preoccupied with obtaining as much production from the landscape as possible by developing and maintaining *early successional types* of ecosystems, often monocultures. But, of course, humans do not live by food and fiber alone; humans also need a balanced low CO_2–high O_2 atmosphere, the climatic buffer provided by oceans and masses of vegetation, and clean (that is, oligotrophic) water for cultural and industrial uses. Many essential life-cycle resources, not to mention recreational needs, are best provided by the *less productive* ecosystems and landscapes. In other words, the landscape is not just a supply depot, but it is also the *oikos*—the home—in which humans must live. Until recently, most humans have taken for granted the gas exchange, water purification, nutrient cycling, and other protective functions (natural capital) of self-maintaining ecosystems—that is, until human numbers and human environmental manipulations became great enough to affect regional and global balances. A most beneficial and certainly an enhanced landscape for living is one containing a diversity of crops, forests, lakes, streams, roadsides, marshes, seashores, and "natural areas"—in other

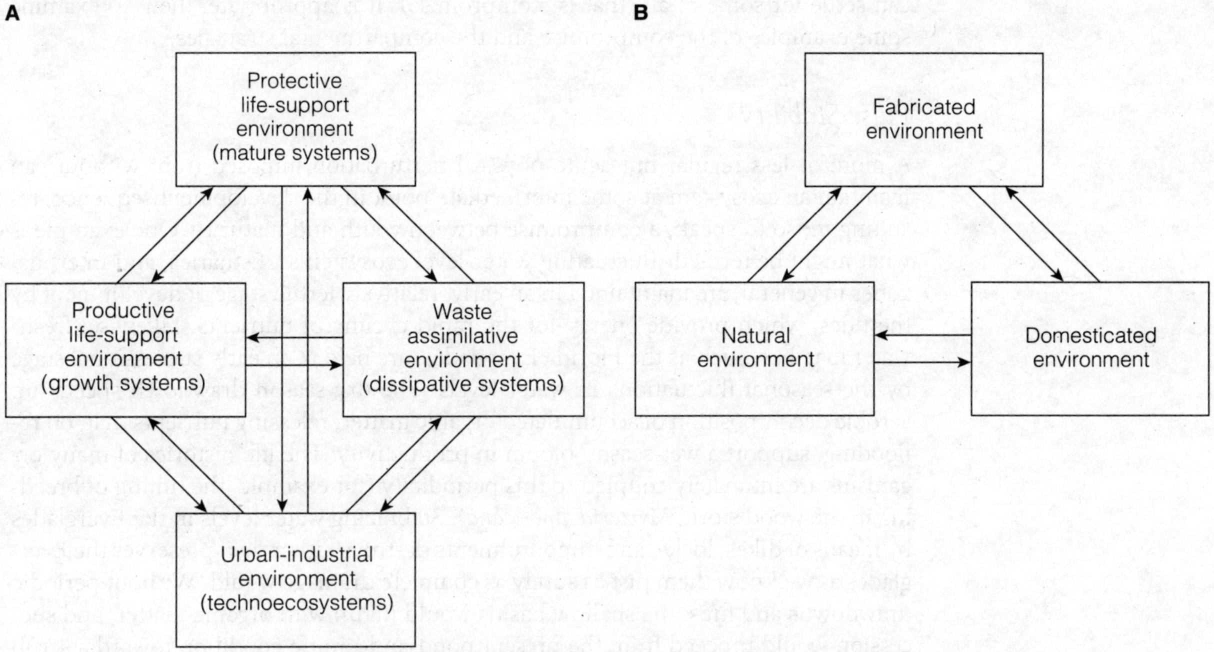

Figure 8-18. Compartment models for landscape-use planning. (A) Partitioned according to ecosystem theory. (B) As viewed by architects and landscape designers.

words, *a mixture of communities at different stages of ecosystem development.* As individuals, humans more or less instinctively select or surround their homes with protective, nonedible cover (trees, shrubs, and grass) and simultaneously strive to coax extra bushels from their fields. For example, a cornfield is a "good thing," of course, but most people would not want to live in the middle of one. It would be suicidal to cover the whole land area of the biosphere with crops, because there would not be the nonedible life-support buffer that is vital for biospheric and aesthetic stability—not to mention that this would invite disaster from epidemic disease.

Because it is impossible to maximize for conflicting uses in the same system, two possible solutions to the dilemma suggest themselves. Either humans can continually *compromise* between quantity of yield and quality of living space, or humans can deliberately *compartmentalize* the landscape to maintain both highly productive and predominantly protective ecosystem types as separate units, subject to different management strategies (ranging from intensive cropping to wilderness management). Figure 8-18 depicts this compartmentalized model. If ecosystem development theory is valid and applicable to planning, then the multiple-use strategy, about which we hear so much, will work only through one or both of these approaches, because in most cases the projected multiple uses conflict with one another. For example, dams on large rivers are often touted as providing a whole range of benefits, such as industrial power generation, flood control, water supply, fish production, and recreation. But these uses actually conflict, because for flood control to be achieved, the water must be drawn down before the flood season—an action that reduces power generation and interferes with recreation. Accordingly, one can maximize for a single use, or perhaps for several closely coupled uses, while reducing other uses, or one can settle for some of all (that is, compromise). It is appropriate, then, to examine some examples of the compromise and the compartmental strategies.

Pulse Stability

A more or less regular but acute physical perturbation imposed from without can maintain an ecosystem at some intermediate point in the development sequence, resulting in, so to speak, a compromise between youth and maturity. One example is what might be termed "fluctuating water-level ecosystems." Estuaries, and intertidal zones in general, are maintained in an early, relatively fertile stage of development by the tides, which provide energy for the rapid cycling of nutrients. Likewise, freshwater marshes, such as the Florida Everglades, are held at an early successional stage by the seasonal fluctuations in water levels. The dry-season drawdown speeds up aerobic decomposition of accumulated organic matter, releasing nutrients that, on reflooding, support a wet-season bloom in productivity. The life histories of many organisms are intimately coupled to this periodicity (for example, the timing of breeding in the wood stork, *Mycteria americana*). Stabilizing water levels in the Everglades by means of dikes, locks, and impoundments destroys rather than preserves the Everglades as we know them just as surely as complete drainage would. Without periodic drawdowns and fires, the shallow basins would fill up with organic matter, and succession would proceed from the present pond-and-prairie condition toward a scrub or swamp forest. At the present time, taxpayer monies are being spent restoring the Everglades by restoring the original pattern of pulsing water flow.

It is unfortunate that humans do not readily recognize the importance of recurrent changes in water level in a natural situation such as the Everglades, even though similar pulses are the basis for some of our most enduring systems of food culture. Alternate filling and draining of ponds has been a standard procedure in fish culture for centuries in Europe and the Orient. The flooding, draining, and soil aeration procedure in rice culture is another example. The rice paddy is thus the cultivated analogue of the natural marsh or intertidal ecosystem.

Fire is another physical factor whose periodicity has been of vital importance over the centuries. As described in Chapter 5, whole biota, such as those of the African grasslands and the California Chaparral, have become adapted to periodic fires, producing what ecologists often term "fire climaxes." For centuries, people have used fire deliberately to maintain such climaxes or to set back succession to some desired point. The fire-controlled forest yields less wood than a *tree farm* (young trees, about the same age, planted in rows and harvested on a short rotation) does, but it provides a greater protective cover for the landscape, wood of higher quality, and a home for game birds (such as quail and wild turkey) that could not survive in a tree farm. The fire climax, then, is an example of a compromise between production and simplicity on the one hand and protection and diversity on the other.

Pulse stability works only if a complete community (including not only plants but also animals and microorganisms) is adapted to the particular intensity and frequency of the perturbation. *Adaptation* (the operation of the selection process) requires times measurable on the evolutionary scale. Most physical stresses introduced by human beings are too sudden, too violent, or too arrhythmic for adaptation to occur, so severe oscillation rather than stability results. In many cases, at least, the modification of naturally adapted ecosystems for cultural purposes would seem preferable to complete redesign.

When Succession Fails

The title of this section is also the title of an essay by Woodwell (1992), who wrote succinctly and with great urgency about the environmental follies of humankind and the need to take action now to deal with global threats such as atmospheric toxification and global warming. Ordinarily, when a landscape is devastated by storms, fires, or other periodic catastrophes, ecological succession is the healing process that restores the ecosystem. However, when landscapes are severely abused over long periods of time (eroded, salinated, stripped of all vegetation, contaminated with toxic wastes, and so on), the land or water becomes so impoverished that succession cannot occur even after the abuse stops. Such sites represent a new class of environment that will remain barren indefinitely unless explicit efforts are made to restore it.

When ecosystem development fails, we have to resort to ecosystem redevelopment. Perhaps this is why there are so many books, journals, and papers dealing with what is termed *restoration ecology.* **Restoration ecology** is the application of ecological theory to the ecological restoration of highly disturbed sites, ecosystems, and landscapes. M. A. Davis and Slobodkin (2004) defined restoration ecology as the process of restoring one or more valued attributes of a landscape. Winterhalder et al. (2004) noted that the goals of restoration ecology require a scientific basis that is ecologically plausible and socially relevant in the long term. Higgs (1997) pointed out that ecological restoration works best as an integrative science, combining expertise from various scientific disciplines and nonscientific fields of study. Ecological restoration requires a transdisciplinary approach in order to maximize the goals of restoration and provides opportunities to learn more about ecosystem structure and function while "rebuilding" the disturbed sites and landscapes (that is, testing ecological ideas and concepts during the process of restoration). The restoration process typically results in restoring ecosystem function rather than in restoring the exact predisturbance structure. The field of restoration ecology ranges from small-scale reclamation to large-scale landscape management challenges and opportunities.

Restoration ecology involves the application of principles, concepts, and mechanisms of ecosystem development to the management and restoration of disturbed systems. This field of applied ecology will assume greater significance as humankind attempts to speed up the recovery of disturbed landscapes (see Cairns et al. 1977, W. R. Jordan et al. 1987; Higgs 1994, 1997; Hobbs and Norton 1996; Meffe and Carroll 1997; W. R. Jordan 2003 for additional discussion).

Network Complexity Theory

The shift in energy use from *growth* to *maintenance* that we have cited as possibly the most important trend in ecological succession has its parallel in growing cities and countries. People and governments consistently fail to anticipate that as population density increases and urban-industrial development intensifies, more and more energy, money, and management effort must be devoted to the services (water, sewage, transportation, and protection) that maintain what is already developed to "pump out the disorder" inherent in any complex, high-energy system. Accordingly, less energy is available for new growth, which eventually can come only at the expense of the development that already exists. Shannon (1950), the "father of information theory," noted that increasing disorder is a property of *all* complex systems. What has come to be known as the **network law** can be stated as follows:

$$C = N\left(\frac{N-1}{2}\right)$$

or approximately $N^2/2$.

In other words, the cost, C, of supporting a network, N, of services is a power function—roughly, a square—of N. That is, when a city or development *doubles* in size, the cost of maintenance may *quadruple*. For more on complex ecology, see Patten and Jørgensen (1995) and Jørgensen (1997).

Compartment Models for Land Use

In thinking about how the principles of ecosystem development relate to the landscape as a whole, consider the compartment models shown in Figure 8-18. Figure 8-18A depicts three types of environments that constitute the life-support systems for the fourth compartment, the urban-industrial technoecosystems, which are in many ways parasitic on the life-support environment (Fig. 8-19). The human productive environment comprises early successional or growth-type ecosystems, such as croplands, pastures, tree plantations, and intensely managed forests that provide food and fiber. Mature ecosystems, such as old-growth forests, climax grasslands, and oceans, are more protective than productive. They stabilize substrates, buffer air and water cycles, and moderate extremes in temperature and other physical factors, while at the same time often providing products. The third category of natural or seminatural ecosystems, which bear the brunt of assimilating the vast wastes produced by the urban-industrial and agricultural systems, consists of waterways (inland and coastal), wetlands, and other intensely stressed environments. Ecosystems in this admittedly arbitrary category are, in the developmental sense, mostly in intermediate, eutrophicated, or arrested stages of succession. All of these components interact continually in terms of input and output (as shown by the arrows in Fig. 8-18).

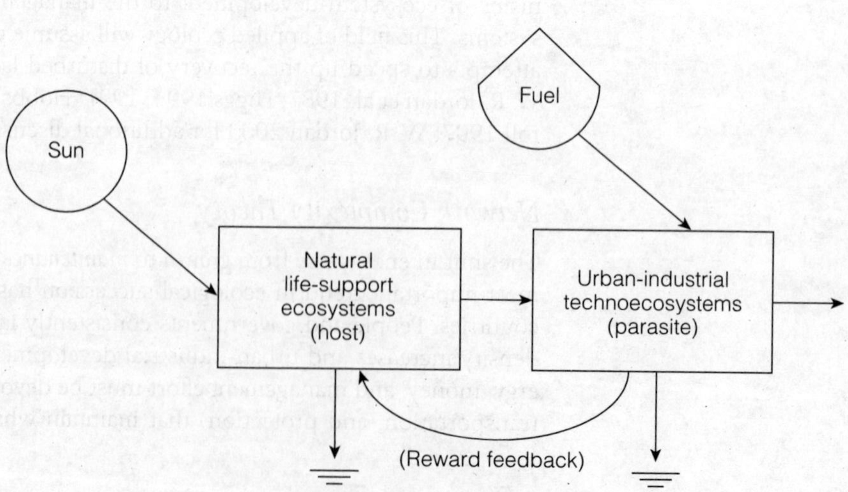

Figure 8-19. Model illustrating the parasitic nature of urban-industrial technoecosystems and the need to link natural life-support ecosystems with these technoecosystems, including a reward feedback loop (modified after E. P. Odum 1997; Barrett and Skelton 2002).

Partitioning the landscape into three environmental components—natural, domesticated, and fabricated—as is traditional with landscape architects (Fig. 8-18B), provides another convenient way to consider the needs of and interrelations between these necessary parts of our household. Although the urbanized or *fabricated* environment is parasitic on the life-support environment (natural and domesticated) for basic biological necessities (breathing, drinking, and eating), it does create and export other, mostly nonbiotic, resources, such as fertilizers, money, processed energy, and goods that both benefit and put stress on the life-support environment. Much more can be done to increase the resource output while reducing the stress and use of subsidies necessary to maintain outputs from high-energy, densely populated "hot spots." However, no known feasible technology can substitute on a global scale for the basic biotic life-support goods and services provided by natural ecosystems.

In Chapters 9, 10, and 11, we will consider the prospects for landscape-level planning and organic development to replace the present, haphazard, mostly short-term–oriented economic and political policies that determine land and water use during the twenty-first century.

9

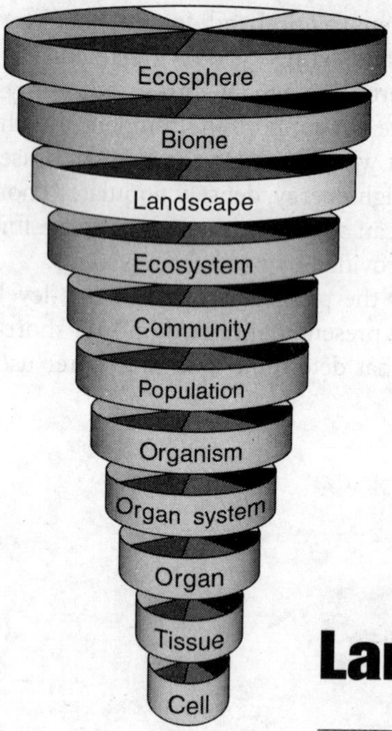

Landscape Ecology

1 Landscape Ecology: Definition and Relation to Levels-of-Organization Concept
2 Landscape Elements
3 Biodiversity at the Community and Landscape Levels
4 Island Biogeography
5 Neutral Theory
6 Temporal and Spatial Scale
7 Landscape Geometry
8 Concept of Landscape Sustainability
9 Domesticated Landscapes

1 Landscape Ecology: Definition and Relation to Levels-of-Organization Concept

Statement

Landscape ecology considers the development and dynamics of spatial heterogeneity, spatial and temporal interactions and exchanges across heterogeneous landscapes, influences of spatial heterogeneity on biotic and abiotic processes, and management of spatial heterogeneity for society's benefit and survival (Risser et al. 1984). Landscape ecology is an integrative field of study that weds ecological theory with practical application; addresses the exchange of biotic and abiotic materials among ecosystems; and investigates human actions as responses to and reciprocal influences on ecological processes.

The relationship between spatial pattern and ecological processes is not restricted to a particular scale. For example, experiments focused at one temporal or spatial scale will likely benefit from experiments at both finer and broader scales, thus providing a greater understanding regarding how plants and animals interact with changes in landscape patterns and processes across scales. Principles and concepts of landscape ecology help to provide theoretical and empirical underpinnings for a variety of applied sciences (such as agroecosystem ecology, ecological engineering, ecosystem health, landscape architecture, landscape design, regional planning, resource management, and restoration ecology).

Explanation

The *levels-of-organization hierarchy* was introduced in Chapter 1 (Fig. 1-3). In this book, we have emphasized the ecosystem as a basic unit (a level of organization) relevant to a holistic understanding of ecology. It has become increasingly recognized, however, that to more fully understand the structure and function of ecosystems, one must focus not only on levels of organization below that of the ecosystem (such as individual organisms, populations, and communities), but also increasingly on levels above that of the ecosystem (such as landscape, ecoregional or biome, and global levels). Chapters 9, 10, and 11 will focus on these higher levels in the hierarchy. This chapter will focus especially on landscape ecology.

Wiens (1992) asked, "What is landscape ecology, really?" The term *landscape,* by its very definition, integrates people and nature (Calow 1999). For example, Merriam-Webster's dictionary defines **landscape** as "the landforms of a region in the aggregate" (*Merriam-Webster's Collegiate Dictionary,* 10th edition, s.v. "landscape"). Landscape ecology appears to have its origins in the late 1930s, when Carl Troll (1939) noted that all the methods of natural science are captured in the area of landscape science (Schreiber 1990). This integrative field of study became widely recognized in central Europe in the 1960s. For example, at the International Association for Vegetation Science meeting in 1963 (Troll 1968), Troll defined *landscape ecology* in accordance with Tansley's concept of the ecosystem (Tansley 1935) as follows:

> Landscape ecology is the study of the entire complex cause-effect network between living communities and their environmental conditions which prevails in a specific section of the landscape (Troll 1968).

Landscape ecology had its beginnings in North America during the 1980s, when Gary W. Barrett, then Ecology Program Director at the National Science Foundation, recommended funding for a workshop held at Allerton Park, Piatt County, Illinois, in April, 1983 (see Risser et al. 1984 for details). This meeting served as a catalyst for annual meetings of the United States International Association for Landscape Ecology (IALE); the first meeting of IALE was held at the University of Georgia in January 1986. This meeting was immediately followed by the first volume of the journal *Landscape Ecology,* published in 1987 with Frank B. Golley serving as editor-in-chief. The now classic book *Landscape Ecology,* by Richard T. T. Forman and Michael Godron, was published in 1986. Thus, the 1980s proved to be the decade when landscape ecology took root in North America.

Numerous benchmark books in the field of landscape ecology are recommended reading, including McHarg (1969), outlining the benefits of designing with nature; Naveh and Lieberman (1984), focusing on theory and application; M. G. Turner (1987), reviewing landscape heterogeneity and disturbance; Hansen and di Castri (1992), discussing the relationship of landscape boundaries to biotic diversity and ecological flows; Forman (1997), discussing the ecology of landscapes and regions; Barrett and Peles (1999), reviewing investigations focused on a model taxonomic group, namely small mammals; Klopatek and Gardner (1999), outlining application of landscape ecology methodologies to management issues; and M. G. Turner et al. (2001), integrating landscape theory to practice and application.

The study of the causes and consequences of spatial patterns in the landscape is the cornerstone of the emerging science of landscape ecology. A landscape perspective in ecology is not new (see Troll 1968); indeed, this is the perspective embodied in *A Sand County Almanac: And Sketches Here and There* by Leopold (1949). It is only in the past couple of decades, however, that principles, concepts, and mechanisms have emerged based on rigorous investigations that have resulted in a theoretically sound basis for understanding landscape-level patterns, processes, and interactions. Thus, this emerging science has also resulted in new emerging properties and landscape-scale understanding, such as the role of gamma diversity (number of species or other taxa occurring on a regional basis); quantifying the spread of disturbance; the importance of source-sink dynamics; species-specific responses of individuals to landscape elements such as corridors; and rates of biotic exchange among ecosystem-types.

To understand landscape *patterns* (such as heterogeneity in an agricultural landscape mosaic) and *processes* (such as the eutrophication of a watershed), theory and application must be integrated into a holistic research and management approach. Integrating concepts include hierarchy theory, sustainability, net energy, patch connectivity, and cybernetic regulatory mechanisms, as discussed in previous chapters (also see Urban et al. 1987 and Barrett and Bohlen 1991 for details). Landscape ecology has now gained general acceptance as a branch of modern ecology dealing with the interrelationship between humans and both natural landscapes and human-built technolandscapes. Landscape ecology provides a scientific basis for such fields as design, planning, management, protection, conservation, and restoration and provides the foundation for natural and human-dominated land management at the regional scale (Hersperger 1994). Landscapes change during the course of history, not only because of ongoing natural processes (that is, because of processes such as ecosystem development, as discussed in Chapter 8), but also because of social, political, and economic processes occurring within these systems. Landscape ecology emphasizes these changing relationships and emphasizes landscape as a system and level of organization. The better one understands landscape-level patterns and processes, the better one will understand processes and phenomena occurring at the organism, population, community, and ecosystem levels as well.

2 Landscape Elements

Statement

The **landscape mosaic** is composed of three major elements: landscape matrices, landscape patches, and landscape corridors. The **landscape matrix** is a large area of similar ecosystem or vegetation types (such as agricultural, prairie, old field, or forest) in which landscape patches and corridors are embedded. A **landscape patch** is a relatively homogeneous area that differs from the surrounding matrix (such as a forest patch or woodlot embedded in an agricultural matrix, or a meadow embedded in a subalpine forest). A landscape patch differs from its surrounding matrix and may be referred to as a *low-quality patch* or a *high-quality patch* depending on its vegetative cover, plant quality (protein content, for example), and species composition.

A **landscape corridor** is a strip of environment that differs from the matrix on either side and frequently connects—either naturally or by design—two or more landscape patches of similar habitat. A stream with its riparian vegetation is an example of a natural landscape corridor. The vegetation of corridors is often similar to the patches they connect but different from the surrounding landscape matrix in which they are embedded. Corridors can be classified into five basic types, based on their origin: *disturbance corridors, planted corridors, regenerated corridors, resource (natural) corridors,* and *remnant corridors.* The function of a corridor depends, among other factors, on its structure (both natural and human-built), size, shape, type, and geographic relation to the surroundings.

Explanation

The **landscape mosaic** can be viewed as a heterogeneous area composed of a variety of different communities or a cluster of ecosystems of different types. The matrix in the landscape mosaic is composed of ecosystems that are quite similar in function and origins. For example, in the grain belt located in the Midwest of the United States, the matrix is most frequently agricultural or cropland. The matrix is frequently pine forests in the Coastal Plain of the southeastern United States; deciduous forests in the northeastern United States; and grasslands in the High Plains. In a landscape mosaic, quilt-like patches that differ from the matrix are frequently embedded. Figure 9-1A illustrates small wooded patches embedded in an agricultural matrix, whereas Figure 9-1B shows larger patches in a similar matrix, but frequently connected by landscape corridors. Patches may be the result of abiotic factors, such as thermal "heat islands" along Interstate 85 in Georgia, South Carolina, and North Carolina (Fig. 9-1C). Naturally, these heat islands influence the physiological and ecological processes of plants and animals inhabiting these landscape patches.

There are numerous natural and human-built (artificial) patches, both terrestrial

A

Courtesy of Gary W. Barrett

B

Courtesy of Gary W. Barrett

C

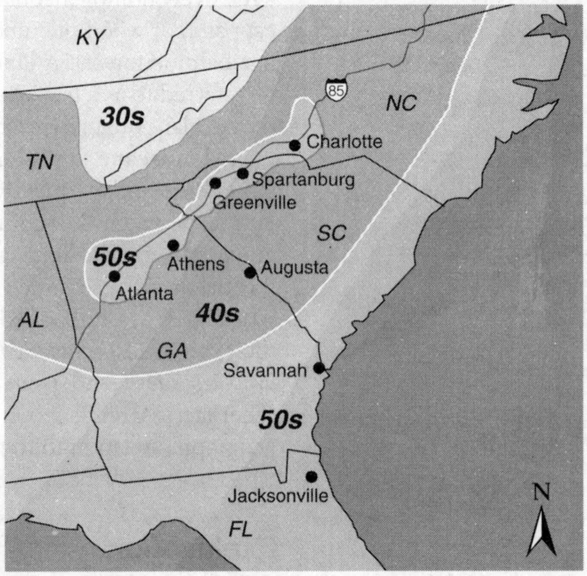

Figure 9-1. (A) Example of landscape patches embedded within an agricultural landscape matrix. (B) Photograph showing how landscape corridors (fence rows) are maintained to connect landscape patches. (C) Diagram depicting a "heat island" along Interstate 85 from Atlanta, Georgia, to Charlotte, North Carolina (diagram from *Athens Banner-Herald*, 23 March 2002).

and aquatic, across the landscape. For example, thousands of ponds, impoundments, and lakes dot the glaciated landscape matrix. Freshwater wetlands are also extremely important patches that pulse. Likewise, there are thousands of woodlots and old-field patches throughout the Midwest in the United States.

Barrett and Barrett (2001) described possible natural, N, and artificial, A, patch-matrix relationships (Fig. 9-2A). For example, it is possible to have a natural wooded patch, N_p, such as a cemetery or urban forest, surrounded by an artificial urban matrix, A_m (Fig. 9-2B); or an artificial, human-made area, A_p, such as a shopping mall or cropland, surrounded by a natural forest matrix, N_m. In Figure 9-2B, for example, N_pA_m represents a natural cemetery patch, dominated by native plant and animal species, located within an artificial urban matrix dominated by human-built structures; whereas A_pN_m may represent an artificial cemetery patch, dominated by horticultural and exotic plant species, located within a natural wooded matrix. Patches such as cemeteries also provide sites for the opportunity to integrate cultural with natural evolutionary processes and human life histories. Recently, conservation biologists, landscape ecologists, resource managers, and restoration ecologists have begun to design investigations and management projects related to these integrative issues (see *Wildlife in Church and Churchyard,* second edition, Cooper 2001 for details).

Patches are frequently used for ecosystem- and landscape-level investigations. For example, Kendeigh (1944) monitored pairs of breeding birds in Trelease Woods, a 55-acre (22-ha) patch of forest located near Urbana, Illinois, to determine whether these individuals belonged primarily to forest-edge or forest-interior species. To evaluate the importance of patch size for breeding bird populations, he compared the data collected from Trelease Woods with data observed from an extensive forest tract at Robert Allerton Park (Kendeigh 1982). He found that 75 percent of the avifauna in Allerton Park favored the closed-forest (interior) habitat, compared to only 20 percent in the smaller, but similar Trelease Woods habitat patch.

A component of spatial distribution related to dispersion is the concept of *grain* (Pielou 1974). Regarding habitat or landscape composition, **grain** relates to the relationship of landscape patch size to an animal's mobility. A habitat or landscape patch is said to be **coarse-grained** for a given animal species if its **vagility**—ability to move freely about—is low relative to the size of habitat patches. A habitat is considered **fine-grained** for a species if that organism has high vagility relative to the size of habitat patches. For example, a wide-ranging predator, such as a red-tailed hawk (*Buteo jamaicensis*), will experience a mosaic of agroecosystem patches (woodlots) as a fine-grained habitat, whereas another species, such as a white-footed mouse (*Peromyscus leucopus*), will likely experience a more coarse-grained agricultural landscape by spending most of its time in a single patch (woodlot).

Landscape patches may be located in a natural landscape matrix or simulated in

A

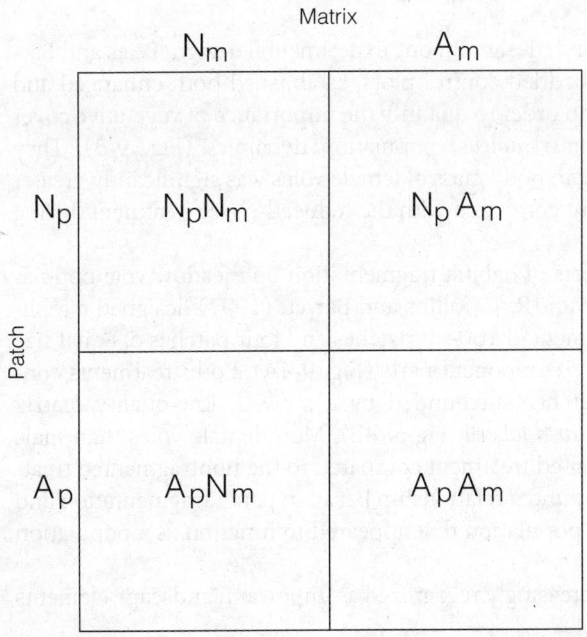

B

Courtesy of the Woodland Cemetery and Arboretum Archives

Figure 9-2. (A) Possible configurations of natural, N, and artificial, A, patch-matrix relationships (after Barrett and Barrett 2001). (B) Aerial photograph of the Woodland Cemetery and Arboretum in Dayton, Ohio, depicting a N_pA_m patch-matrix relationship. Woodland is an older cemetery (>100 years), forming a more natural woodland patch surrounded by an artificial urban matrix (after Barrett and Barrett 2001).

Figure 9-3. Research design depicting twelve 0.04-hectare experimental grassland patches. Four patches contain enhanced cover, four patches contain reduced cover, and four patches are undisturbed controls (after Peles and Barrett 1996).

Courtesy of Gary W. Barrett

a replicated experimental research design. In one experimental design, Peles and Barrett (1996), in addition to embedded control plots, established both enhanced and reduced vegetative cover plots in order to quantify the importance of vegetative cover to meadow vole (*Microtus pennsylvanicus*) population dynamics (Fig. 9-3). They found, for example, that the mean body mass of female voles was significantly greater in the enhanced-cover treatment compared with the reduced-cover treatment during both years of the investigation.

In order to address the effects of habitat fragmentation on meadow vole population dynamics and social behavior, R. J. Collins and Barrett (1997) designed a replicated study using four nonfragmented 160-m^2 patches and four patches of equal size but fragmented into four 40-m^2 component parts (Fig. 9-4A). Both treatments contained high-quality habitat patches surrounded by a mowed, low-quality matrix dominated by giant foxtail (*Setaria faberii;* Fig. 9-4B). More female voles than male voles were found in the fragmented treatment compared to the nonfragmented treatment. Collins and Barrett also found a relationship between patch fragmentation and the social structure of the vole population that appeared to function as a population regulation mechanism.

Landscape corridors are increasingly recognized as important landscape elements that provide a means of enabling animal dispersal, reducing soil and wind erosion, allowing transfer of genetic information between patches, aiding in integrated pest management, and providing habitat for nongame species. Corridors, however, may have negative as well as positive effects (for example, transmission of contagious diseases, spread of disturbances such as fire, and increased exposure of animals to predation; Simberloff and Cox 1987).

Corridors may be classified into several basic types, as noted in the Statement. **Remnant corridors** occur when most of the original vegetation is removed from an area, but a strip of native vegetation is left uncut. Remnant corridors include uncut vegetation along streams, steep land, railroad tracks, or property borders. Remnant patches and corridors also provide "outdoor teaching and learning laboratories" for comparing ecological processes in young and mature systems (Barrett and Bohlen 1991). For example, remnant corridors increase species diversity in the region, improve nutrient cycling, protect natural capital, and provide habitat for *K*-selected edge species, which includes many game species (Figure 9-5A).

A linear disturbance through the landscape matrix produces a **disturbance corridor.** Disturbance corridors disrupt the natural, more homogeneous landscape, but

A

Courtesy of Gary W. Barrett

B

Courtesy of Gary W. Barrett

Figure 9-4. (A) Aerial photograph of study site depicting four high-quality 40-m^2 fragmented patches and four 160-m^2 nonfragmented patches. The experimental matrix consisted of mowed, low-quality habitat dominated by giant foxtail (*Setaria faberii*). (B) Close-up of eight enclosures, each containing fragmented and nonfragmented experimental patches (after R. J. Collins and Barrett 1997).

A

B

Figure 9-5. Examples of landscape corridors. (A) Remnant corridors connecting patches of virgin forest following timbering. Such corridors are valuable regarding movement of forest resources such as wildlife between forest patches. (B) A planted corridor of trees established during the Shelterbelt Project in the 1930s that provide protection against wind, snow drift, and soil erosion. (C) A grassland corridor established to divide soybean (*Glycine max*) patches represents a *planted* corridor. (D) A fence row permitted to develop naturally through time illustrates a *regenerated* corridor.

C

D

Figure 9-5. *(continued)*

provide important habitat for native plant and animal "opportunistic" species adapted to disturbance or for species commonly found during the early stages of secondary succession (see Chapter 8). Power-line corridors that cut through a forest landscape provide an example of a disturbance corridor. Forest-interior species sel-dom use such corridors to nest or reproduce; however, forest-edge wildlife species may flourish in such corridors. Disturbance corridors may act as barriers to the movement of some species, but they provide routes of dispersal for other species, such as the white-footed mouse (*Peromyscus leucopus*) and the eastern chipmunk (*Tamias striatus;* see Henderson et al. 1985). Disturbance corridors can also act as filters for some species but not for others. This **filter effect** can be minimized by providing gaps or nodes of matrix vegetation in the corridor, allowing certain species to cross while restricting others.

Planted corridors are strips of vegetation planted by humans for a variety of economic and ecological reasons. For example, thousands of miles (kilometers) of planted tree corridors were established in the treeless Great Plains as part of the Shelterbelt Project during the 1930s (Fig. 9-5B), to reduce wind erosion and to provide wood and wildlife habitat (Shelterbelt Project 1934). Planted corridors also provide excellent habitat for insectivorous birds and predaceous insects, and they function as dispersal routes for small mammal species.

Planted corridors have also been used in agricultural landscapes for a variety of ecological purposes. For example, Kemp and Barrett (1989) established grassy corridors (Fig. 9-5C) within soybean agroecosystems to retard the movement of adult po-tato leafhoppers (*Empoasca fabae*). Moreover, *Nomuraea rileyi*, a fungal pathogen, in-fected a significantly higher proportion of green cloverworms (*Plathypena scarba*) in plots divided by grassy corridors. This fungus is the primary natural biotic control agent of this major lepidopteran larval pest in the Midwest of the United States.

Planted hedgerows of Osage orange trees (*Maclura pomifera*) and multiflora rose (*Rosa multiflora*) have been planted by farmers and game managers for decades to provide firewood, fenceposts, and wildlife habitat, and to serve as barriers to livestock movement (see Forman and Baudry 1984 for a list of economic and ecological functions of hedgerows at the landscape scale). Unfortunately, in an effort to increase grain yields in the Midwest of the United States, many hedgerows were cut down during the latter part of the twentieth century.

Resource corridors are narrow strips of natural vegetation that extend for long distances across the landscape (such as a gallery forest along a stream). Figure 9-5E illustrates a stream resource corridor including riparian vegetation. Karr and Schlosser (1978) and Lowrance et al. (1984) described how vegetated stream corridors ben-efit the agricultural landscape by intercepting nutrient and sediment runoff from the croplands that otherwise would end up in the streams, contributing to cultural eu-trophication problems. These strips not only improve water quality, but also reduce fluctuations in stream levels and help to conserve the natural biotic diversity within the agricultural landscape mosaic.

A **regenerated corridor** results from the regrowth of a strip of vegetation in a landscape matrix. Hedgerows that develop along fences due to the natu-ral processes of secondary succession provide an excellent example of a regenerated corridor. Birds are common inhabitants of these regenerated corridors; avian species also aid the development and plant species composition of these corridors by providing a mechanism of seed dispersal. Although some animals, especially insect pests, do economic damage by feeding on crops adjacent to corridors, Price (1976) and Forman and Baudry (1984) found that regenerated corridors were often the source of natural enemies that colonize adjacent crops and aid biotic pest control. Forest-edge bird species, and birds that forage in croplands, often nest in wooded regener-ated corridors. These bird species also help to regulate insect species in agricultural crops. Red fox (*Vulpes vulpes*), white-tailed deer (*Odocoileus virginianus*), and wood-chucks (*Marmota monax*) frequently use regenerated corridors. Small mammals have been found to suffer local extinction in relatively isolated patches of habitat (such as forest woodlots) but use regenerated corridors to reestablish the populations and metapopulations (Middleton and Merriam 1981; Henderson et al. 1985; G. Merriam and Lanoue 1990; Fahrig and Merriam 1994; Sanderson and Harris 2000).

Figures 9-6A and B illustrate an array of experimental patch sizes, showing how select patches can be connected by vegetative corridors in order to investigate such parameters as small mammal patterns of movement, behavioral responses to presence or absence of corridors, the relationship of population densities to patch size, and species survivorship. The simulation of landscape elements in replicated, experimental designs provides an elegant and productive research approach to the field of landscape ecology.

Thus, corridors as a landscape element, depending on their origin, provide a multiplicity of functions within the landscape mosaic. G. Merriam (1991) noted that the assessment of connectivity by corridors must come from species-specific empirical studies. More recently, it has been shown that indeed species of small mammals react to corridors in a species-specific manner (Mabry et al. 2003). Numerous studies need to be conducted in diverse landscapes, investigating a diversity of plant and animal species along a gradient of temporal-spatial scales in order to better understand the role of corridors within the landscape mosaic.

Figure 9-6. Experimental vegetation plots designed to examine (A) the effects of patch size and patch quality for studies of movement, behavior, and dynamics of rodent populations, and (B) the effects of patch quality for studies of heavy metal uptake and population dynamics of small mammals.

3 Biodiversity at the Community and Landscape Levels

Statement

Descriptions of how populations and communities are arranged within a given geographical region at the landscape level have featured two contrasting approaches: (1) the *zonal approach,* in which discrete communities are recognized, classified, and listed in a sort of checklist of community types; and (2) the *gradient analysis approach,* which involves the arrangement of populations along a uni- or multidimensional environmental gradient or *axis* with recognition of community based on frequency distributions, similarity coefficients, or other statistical comparisons. The term **ordination** is frequently used to designate the ordering of species populations and communities along gradients, and the term **continuum** or **gradient analysis** is used to designate the gradient containing the ordered populations or communities. In general, the steeper the environmental gradient, the more distinct or discontinuous are communities, not only because abrupt changes are more probable in the physical environment, but also because boundaries are sharpened by competition and coevolutionary processes between interacting and interdependent species.

An area or zone of transition between two or more diverse communities (between forest and grassland or between a soft-bottom and a hard-bottom marine substrate, for example) is known as an **ecotone.** The ecotonal community commonly contains many of the organisms of each of the overlapping communities in addition to organisms characteristic of and often restricted to ecotones (T. B. Smith et al. 1997; Enserink 1997). Often, both the number of species and the population density of some of the species are greater in the ecotone than in the communities flanking it. Frequently, the break between two community types is a sharp demarcation (where a crop field is directly adjacent to a forest, for instance). This narrow zone of habitat transition is frequently termed an **edge.** The tendency for increased variety and density of species at community junctions is known as the **edge effect.** Species that use edges for purposes of reproduction or survival are frequently termed **edge species.**

Explanation and Examples

Deciding where to draw boundaries is not too much of a problem in ecosystem analysis (see Chapter 2) as long as input and output environments are considered part of the system, no matter how the system is delimited, whether by natural features or by arbitrary lines drawn for convenience. Strayer et al. (2003) classified boundaries into four main classes of boundary traits: (1) *origin and maintenance;* (2) *spatial structure* (such as geometric shape); (3) *function* (such as transmissive or permeable); and (4) *temporal dynamics* (such as age and history of the boundary). However, when one delineates biotic communities by their component species populations, there is a problem (see special section entitled "Ecological Boundaries" in the August 2003 issue of *BioScience* regarding the investigation and theory of ecological boundaries).

During the past century, plant ecologists have debated whether land plant communities are to be thought of as discrete units, with definite boundaries, as suggested by Clements (1905, 1916), Braun-Blanquet (1932, 1951), and Daubenmire (1966), or whether populations respond independently to environmental gradients to such an extent that communities overlap in a continuum, so that the recognition of dis-

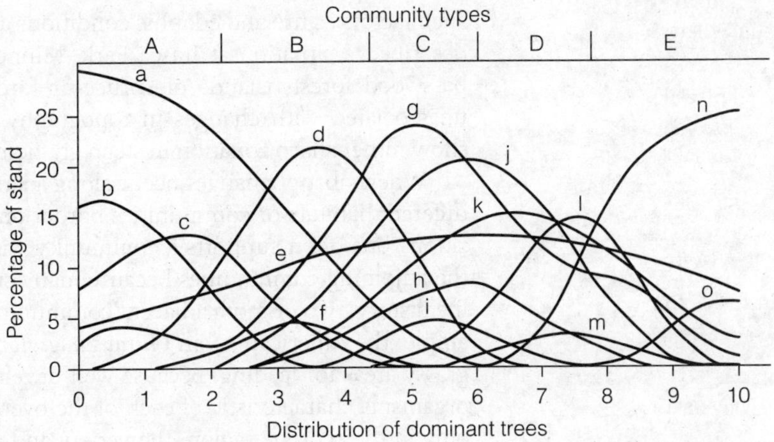

Figure 9-7. Distribution of populations of dominant trees along a hypothetical gradient (0–10), illustrating the arrangement of component populations within a "continuum" community. Each species (a–o) shows a "bell-shaped" distribution with a peak of relative abundance (percentage of stand) at a different point along the gradient. A–E represent different community-types within the large community. Curves have been patterned after Whittaker (1967).

crete units is arbitrary, as viewed by Gleason (1926), Curtis and McIntosh (1951), Whittaker (1951), Goodall (1963), and others. Whittaker (1967) illustrates these contrasting viewpoints with the following example. If, at the peak of autumn coloration in the Great Smoky Mountains National Park, one were to select a vantage point along the highway to obtain a view of the attitudinal gradient from valley floor to ridge top, one would observe five zones of color: (1) a multihued cove forest; (2) a dark green hemlock forest; (3) a dark red oak forest; (4) a reddish-brown oak-heath vegetation; and (5) a light green pine forest on the ridges. These five zones could be viewed as discrete community types, or all five could be considered part of a single continuum to be subjected to a gradient analysis that would emphasize the distribution and response of individual species populations to changing environmental conditions in the gradient. This situation is illustrated in Figure 9-7, which shows the frequency distribution (as hypothetical bell-shaped curves) of 15 species of dominant trees (a through o) that overlap along the gradient, and which presents the somewhat arbitrary designation of five community types, A–E, based on the peaks of one or more dominants. Much can be said for considering the whole slope as one major community, because these community types are linked together by exchanges of nutrients, energy, and animals as a watershed ecosystem. The *watershed* is the smallest ecosystem unit amenable to functional studies and overall human management. On the other hand, recognizing the zones as separate communities is useful to the forester or land manager, for example, as each type of community differs in timber growth rate, timber quality, recreational value, vulnerability to fire and disease, and other aspects.

Ordination techniques often require ecologists to compare the similarity (or dissimilarity) of successive samples taken along an environmental gradient by using an index of the following general form:

$$\text{Similarity index } (S) = \frac{2C}{A + B}$$

where A = number of species in sample A, B = number of species in sample B, and C = number of species common to both samples.

Coactions between populations can contribute to separating one community from another, as in, for example, (1) *competitive exclusion;* (2) *mutualisms* between groups of species that depend on one another; and (3) *coevolution* of groups of species. Also, such factors as fire and edaphic conditions can create sharp boundaries. Buell (1956) described a situation at Itasca Park, Minnesota, where, within the general maple-basswood forest, islands of spruce-fir forest maintained rather sharp boundaries unassociated with changes in topography. Marine benthic communities similarly show rather sharp zonation in steep gradients, as does vegetation on a mountainside.

Where abrupt changes occur along a landscape gradient, or where two distinctly different habitats or communities border one another, the resulting ecotone or transition zone often supports a community with characteristics different from those of the adjoining communities because many species require, as part of their habitat or life history, two or more adjacent communities that differ greatly in structure. For example, the American robin (*Turdus migratorius*) requires trees for nesting and open, grassy areas for feeding. Because well-developed ecotonal communities may contain organisms characteristic of each of the overlapping communities plus species living only in the ecotone region, the variety and density of life are greater in the ecotone. This condition is what is meant by *edge effect*.

In a classic pioneer study, Beecher (1942) found that the population density of birds increased as the number of meters of edge per unit area of community increased. From general observation, most people have observed that the density of songbirds is higher on estates, campuses, residential districts, and similar settings, which have mixed habitats (habitat fragmentation) and, consequently, much edge, than on large, nonfragmented tracts of forest or grassland.

Ecotones may also have characteristic species not found in the communities forming the ecotones. For example, in a study of bird populations along a community developmental gradient, study areas were selected to minimize the influence of junctions (edges) with other communities. Thirty species of birds were found to have a density of at least five pairs per 100 acres in one of these stages. However, about 20 additional species were known to be common breeding birds of the region as a whole; 7 of these were found in small numbers, whereas 13 species were not even recorded in the selected, uniform study areas. Among those not recorded were such common species as robin (*Turdus migratorius*), bluebird (*Sialia sialis*), mockingbird (*Mimus polyglottos*), indigo bunting (*Passerina cyanea*), chipping sparrow (*Spizella passerina*), and orchard oriole (*Icterus spurius*). Many of these species require trees for nest sites or observation posts, yet feed largely on the ground in grass or other open areas; therefore, their habitat requirements are met in ecotones between forest and grass or shrub communities, but not in areas of either alone. Thus, in this case, 40 percent (20 of 50) of the common species known to breed in the region may be considered primarily or entirely ecotonal. T. B. Smith et al. (1997) investigated 12 populations of passerine birds common in rain forest and forest/savanna habitats. Populations in the forest and forest/savanna ecotone were morphologically divergent, despite high gene flow suggesting that ecotone habitats may be a source of evolutionary novelty generating increased biotic diversity.

Hawkins (1940) showed with maps how this change occurred in Wisconsin during the century following the appearance of the first European settlers in 1838. If humans settle on the Plains, they plant and water trees, creating a similar pattern. The preferred habitat of *Homo sapiens* can be said to be a forest edge, because the species likes the shelter of trees and shrubs but mostly gets its food from grassland and cropland. Some of the original organisms of the forests and the plains can survive in the human-made forest edge, whereas those organisms especially adapted to the forest edge, notably many species of weeds, birds, insects, and mammals, often increase in numbers and expand their ranges because humans have created vast new forest-edge habitats.

By and large, game species such as deer, rabbits, grouse, and pheasants can be classified as edge species, so a large part of game management involves *creating edge* by planting food or cover patches, patch clear-cutting, and patch burning. Aldo Leopold, who is generally credited with introducing the concept of edge effect, wrote in his pioneering text on game management (Leopold 1933a) that "wildlife is a phenomenon of edges." Hansson (1979) commented on the importance of landscape heterogeneity to the survival of northern warm-blooded animals that are active all year. Agricultural and other disturbed areas offer more food in winter than do mature, undisturbed forests, which, however, offer more food in spring and summer.

An increase in density in ecotones is by no means a universal phenomenon. Many organisms, in fact, may show the reverse. Thus, the density and diversity of trees is obviously less in a forest-edge ecotone than in the forest interior. Breaking up the vast stretches of tropical rain forest will almost certainly reduce species diversity and cause the extinction of many species adapted to large areas of similar habitat. Ecotones appear to assume their greatest importance where humans have greatly modified natural communities and domesticated the landscape for many centuries, thus allowing evolutionary time for adaptation. In Europe, for example, where most of the forest has been reduced to forest edge, thrushes and other forest birds live in cities and suburbs to a greater extent than do related species in North America. But, of course, many other European species do not adapt and have become rare or extinct.

As with most positive or beneficial phenomena, the subsidy-stress performance curve (see Fig. 3-7) is relevant to edge-diversity relations. Although increasing the amount of edge often increases diversity, *excessive edge* (many small blocks of habitat) causes diminishing returns in diversity. Theoretically, maximum beta species diversity occurs when habitat patches are large, or fairly large, and the total amount of edge in the landscape is also large. These countertrends need to be considered in forest and wildlife management and landscape design in general. Fahrig (1997) discussed the effects of habitat loss and fragmentation on population extinction.

Island biogeographical theory can help in determining just how large the patches should be (the minimum critical size of ecosystems; Lovejoy et al. 1986). See Harris' (1984) book entitled *The Fragmented Forest,* which builds on the theory of island biogeography regarding the management and preservation of biotic diversity.

4 Island Biogeography

Statement

MacArthur and Wilson (1963, 1967) first published the theory of island biogeography. Simply stated, the **island biogeography theory** holds that the number of species on an island is determined by the equilibrium between the immigration of new species and the extinction of those species already present. As rates of immigration and extinction depend on the size of islands and their distance from the mainland, a

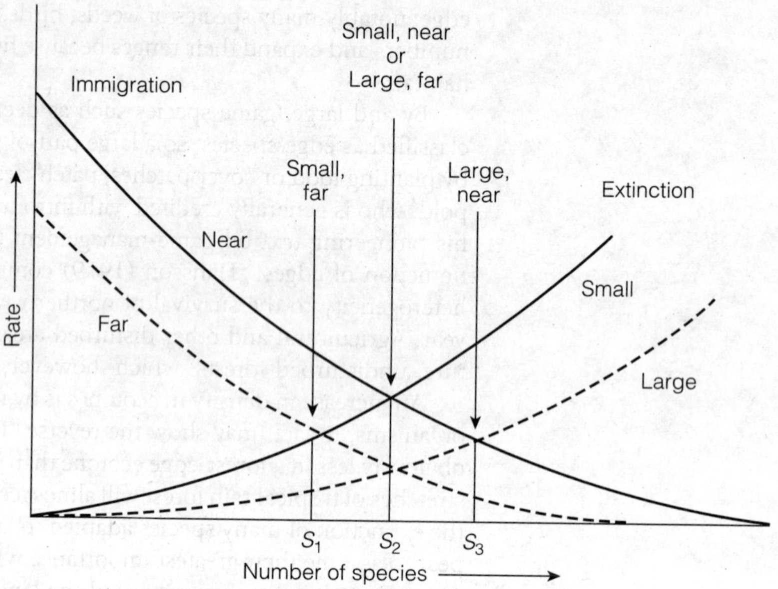

Figure 9-8. Theory of island biogeography. The number of species on an island is determined by the equilibrium between rate of immigration and rate of extinction. Four points of equilibrium are shown, representing different combinations of large and small islands either near or far from continental shores (after MacArthur and Wilson 1963, 1967).

general equilibrium can be diagrammed, as presented in Figure 9-8. Four equilibrium points are shown, representing a small, distant island predicted to have few species, S_1; a small, nearby or a larger, distant island, predicted to be intermediate in terms of species richness, S_2; and a large, nearby island that should support many species, S_3. This model demonstrates the interplay of isolation, natural selection, dispersal, extinction, and speciation that has attracted the attention of population ecologists and evolutionary biologists to island biogeography for more than a century. This model is of fundamental importance in landscape ecology and conservation biology.

Explanation

Islands have fascinated biologists, geographers, and ecologists since Charles Darwin visited the Galápagos Islands. It has also become apparent that landscape patches on the mainland likely function as islands within the landscape mosaic. For example, the Andes Mountains of Ecuador stirred the imagination of Alexander von Humboldt (1769–1859), who laid the foundations of **mountain geoecology** there. Some argue that these Andean landscapes—the cradle of Humboldt's work—should be considered the birthplace of ecology, especially holistic ecology (Sachs 1995; F. O. Sarmiento 1995, 1997). The peaks of such mountains, especially at approximately the same elevations, function as terrestrial islands regarding plant and animal community types. J. H. Brown (1971, 1978) investigated the insular biography of these "islands" in regard to small mammal and bird population diversity and abundances.

These patches, which vary in size—large and small—and distance—near and far—fit the theory of island biogeography as proposed by MacArthur and Wilson (1963). For example, a patch of forest may be located in a "sea" of agricultural cropland (see Fig. 9-1A), isolated from other patches in the landscape. The effect of patch size and isolation appears to have a pronounced influence on the nature and diversity of species within these landscape patches. Preston (1962) formalized the relationship between the area of the island and the number of species present as follows:

$$S = cA^z$$

where S is the number of species, A is the area of the island or patch, c is a constant measuring the number of species per unit area, and z is a constant measuring the slope of the line relating log S and log A (in other words, z is a measure of the change in species richness per unit area).

Thus, the theory of island biogeography states that the number of species of a given taxon (insects, birds, or mammals) present on an island or within a patch represents a dynamic equilibrium between the rate of immigration of new colonizing species of that taxon and the rate of extinction of previously established species (see Fig. 9-8).

Examples

Simberloff and Wilson (1969, 1970) removed all arthropods (by insecticide treatment) from small mangrove islands in the Florida Keys and observed recolonization. The patterns of recolonization of the islands by arthropod populations tended to verify the MacArthur-Wilson dynamic equilibrium model based on the theory of island biogeography. Since that time, similar studies have been conducted (for example, J. H. Brown and Kodric-Brown 1977; Gottfried 1979; Strong and Rey 1982; Williamson 1981), helping to explain the distribution of arthropod, bird, and small mammal species among habitat patches and on islands.

Others have suggested that the theory of island biogeography provides a basis for the design of reserves established to preserve natural diversity, to protect endangered species, or both. Accordingly, a large reserve is preferable to a group of smaller reserves with the same total area. Harris (1984), in his award-winning book *The Fragmented Forest,* also built on the theory of island biogeography by relating it to forest and wildlife management. The idea that corridors should be maintained between reserves or refuges whenever possible was suggested by E. O. Wilson and Willis (1975) based on the equilibrium theory of island biogeography.

Ecological principles based on the theory of island biogeography help planners and resource managers to design nature preserves. When a preserve is to be carved from a homogeneous landscape matrix, the following landscape principles are frequently used regarding the design of the preserve, in order to maximize species richness and to minimize the role of disturbance and edge effects on ecological processes:

- One large patch is better than several smaller patches of the same total size;
- Corridors connecting isolated patches are preferable to a total lack of corridors; and
- Circular or square patches that maximize area-to-perimeter ratios are preferable to elongated, rectangular patches with much edge.

It should be kept in mind that nature preserves must be designed and managed in accordance with plant and animal life histories, special requirements (such as nesting sites, salt licks, and food resources), and the need to minimize the invasion of exotic species.

Simberloff and Cox (1987) provided a model encompassing rates of immigra-

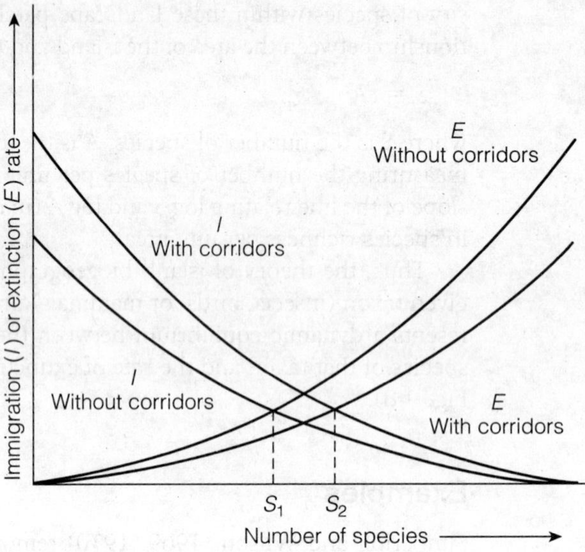

Figure 9-9. Effect of corridors on rate of immigration, *I*, rate of extinction, *E*, and the resulting number of species in equilibrium based on the island biogeographic model. S_1 is the equilibrium number of species without corridors; S_2 is the equilibrium number of species with corridors. (From Figure 1 in Simberloff, D., and J. Cox. 1987. Consequences and costs of conservation corridors. *Conservation Biology* 1:63–71. Copyright 1987 Blackwell Publishing.)

tion and extinction between patches that were either isolated or connected by corridors (Fig. 9-9). Harris (1984), among others, suggested that corridors act by increasing the rate of immigration—thus, the extinction of a dwindling population would be slowed or even halted by an influx of immigrants (the **rescue effect**; J. H. Brown and Kodric-Brown 1977). Furthermore, the individuals of some species, especially large mammals, must range widely and maintain a large home range in order to meet food requirements, and if population sizes are too small, inbreeding depression will ensue and lead to extinction—a concern regarding the small, isolated population of the Florida panther (*Felis concolor coryi*). Figure 9-9 provides a model to test these concerns and hypotheses.

In summary, any patch of habitat isolated from similar habitat by a different, relatively inhospitable terrain or matrix that is navigated with difficulty by organisms of the habitat patch may be considered an *island;* these patches include mountaintops, small lakes, bogs, areas fragmented by human land use, woodlots, or forest patches clear-cut for experimental purposes. How these experimental patches affect the population dynamics of species of small mammals and butterflies has been documented by Bowne et al. (1999), Haddad and Baum (1999), Mabry and Barrett (2002), and Mabry et al. (2003).

5 Neutral Theory

Statement

Neutral theory, in ecology, treats all species as if they had the same per capita rates of birth and death, dispersal, and even speciation. Although this assumption is only a first approximation, neutral theories in ecology are useful in formulating and testing *null hypotheses* about how communities and ecosystems are assembled in landscapes. Recently, neutral theory has been given attention following the publication of *The Unified Neutral Theory of Biodiversity and Biogeography* by Stephen P. Hubbell (2001). Hubbell went beyond the null hypothesis view to suggest that neutral theory might actually give a better explanation for many landscape-level ecological patterns than does current ecological theory.

Explanation

The neutral theory of biodiversity and biogeography (Hubbell 2001) is a generalization and extension of the theory of island biogeography (MacArthur and Wilson 1967). It is termed *neutral theory* because all species are treated as having identical vital rates on a per capita basis (the same birth and death rates, the same rate of dispersal, and the same rate of speciation). Neutral theory applies to communities of organisms that are on the same trophic level and that compete for the same or similar limiting resources. The theory is derived by assuming that the dynamics of communities are a zero-sum game for limiting resources—that is, no species can increase in abundance or biomass without a matching decrease in the collective abundance or biomass of all other competing species. Neutral theory assumes that species are largely substitutable in their use of limiting resources, so that if one species happens to be absent from a community, other species will fully use the resources freed up by its absence. The neutral theory of biodiversity and biogeography extends the MacArthur-Wilson theory by also predicting steady-state patterns for species' commonness and rarity (relative species abundance on islands or in local communities) but by incorporating speciation, a process not included in the original theory of island biogeography.

There is much interest in the neutral theory because it seems to fit many large landscape patterns as well as or better than current ecological theory. The controversy centers on why it performs so well. What does the neutral theory predict? As we have seen, Hubbell built his neutral theory on the foundation of the theory of island biogeography (MacArthur and Wilson 1967), which states that the number of species on islands or habitat patches results from an interplay between the rate at which species immigrate to the island from the mainland—or disperse to a habitat patch from the surrounding landscape—and the rate at which resident species go extinct. The MacArthur-Wilson theory, however, did not explain the equilibrium population sizes of these island species. Hubbell (2001) generalized their theory to predict species commonness and rarity (relative species abundance) on islands or in habitat patches. This generalization was possible because the theory of island biogeography is also a neutral theory. It is neutral because all species are assumed to have the same probability of immigrating to an island or going extinct once there. Hubbell's theory also added a process by which new species could originate (speciation), which was absent from the original theory. Hubbell's theory fits patterns of relative species abundance in a number of ecological communities, but some of the closest fits are related to the abundance of tree species in a number of large, permanent plots of tropical rain forest.

Figure 9-10 shows the relative abundance of tree species in a tropical forest (50-ha plot) located in Southeast Asia. The relative abundance data are displayed as a dominance-diversity curve. A dominance-diversity curve arranges species on the *x-axis* in rank order of abundance, from the most common species at the lowest rank

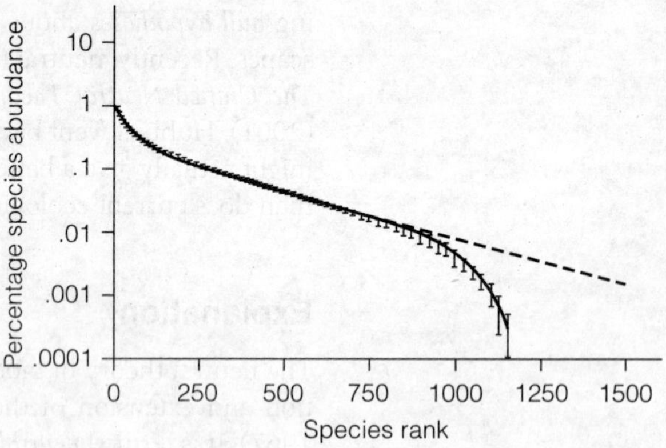

Figure 9-10. Dominance-diversity curve for a sample of 324,592 trees and shrubs of 1175 species in a 50-hectare forest plot in Lambir Hills National Park, Sarawak, Borneo. The dashed line is the curve expected in a much larger area (the metacommunity), with an estimated value of $\Theta = 310$. The solid line is the observed dominance-diversity curve. The line with error bars (± 1 standard deviation of the mean) is fitted for an immigration rate $m = 0.15$ per birth. Rare species are rarer than predicted from the metacommunity dominance-diversity curve because they are more extinction-prone locally than common species, and once locally extinct, they take longer to re-immigrate.

position (left end) to the rarest species at the highest rank position (right end). The logarithm of the relative abundance of the species (usually plotted as the log of the species percentage or some other measure of species importance in the community) is plotted on the *y-axis.* The neutral theory predicts that reducing the immigration rate will make rare species rarer and fewer in number in island communities than in their mainland community counterparts. The diagonal line that does not bend at the rare-species end is the fitted *expected curve* for the source ("mainland") area. The observed curve for the 50-ha plot of rain forest drops away from the source area curve at the rare-species end, as expected based on neutral theory. From fitting the neutral theory to dominance-diversity curves, one can estimate the immigration rate that can explain the observed further rarefaction of the already rare species in a community. The immigration rate (denoted by the parameter *m*) estimated for this rain forest plot is 15 percent, meaning that 15 percent of the trees in this plot are calculated to have originated as immigrants from the forest surrounding the plot. The neutral theory estimates the dominance-diversity curve for the source area, termed the **metacommunity** in neutral theory, based on a number called theta. *Theta* (Θ) in the neutral theory is a fundamental biodiversity measure that characterizes species diversity at the metacommunity equilibrium between the rates of speciation and extinction of species in the metacommunity. Theta is the product of two parameters—one specifying the size of the metacommunity and the other the rate of speciation. Although the speciation rate and the size of the metacommunity are generally not known, remarkably, their product (Θ) can be estimated from relative species abundance data (Hubbell 2001; Volkov et al. 2003).

Neutral theory has not gone unchallenged. A number of papers and books have argued that nonneutral theories can do as well or better. Neutral theory assumes that large landscape patterns are the result of random speciation, dispersal, and random drift in population size of individual species, known as *demographic stochasticity.* Sugihara et al. (2003) and Chase and Leibold (2003) argued in support of **niche assembly theory**—the hypothesis that ecological communities are equilibrium assemblages of niche-differentiated, competing species, coexisting because each species is the most effective competitor in its own niche. Contrary to MacArthur and Wilson (1967) and Hubbell (2001), niche assembly theory asserts that dispersal is not very important in determining which species are present or absent from a particular ecological community.

Sugihara et al. (2003) and McGill (2003) argued for continuing to describe patterns of relative species abundance using the widely fitted *lognormal distribution,* first used by Preston (1948) to describe patterns of bird species abundance. Sugihara et al. (2003) suggested that if niches are nested hierarchically and communities are in equilibrium, a lognormal pattern of relative species abundance can be obtained. McGill (2003) argued that the lognormal curve fits better than the neutral theory.

Volkov et al. (2003) argued in favor of the neutral theory and in disfavor of a return to the lognormal distribution on both biological and mathematical grounds. On biological grounds, they argued that the parameters of the neutral theory all have straightforward biological interpretations, like birth and death rates, immigration rates, speciation rates, and the size of the community. In contrast, the parameters of the lognormal distribution are generic—a mean, a variance, and a modal species frequency—and have no clear biological derivation. On mathematical grounds, Volkov et al. (2003) pointed out that, although the lognormal curve can be fit to static data, it can never be the foundation for successful *dynamic* hypothesis testing for ecological communities, because the variance of the lognormal distribution increases indefinitely through time. They also rebutted McGill's claim by showing that the neutral theory actually fits relative abundance data as well as or better than the lognormal curve, based on analytical solutions to neutral theory.

Chave et al. (2002) argued that factors other than dispersal assembly could explain patterns of relative species abundance in communities. One possibility is density and frequency dependence, or *rare species advantage.* If populations of rare species grow faster than populations of common species, then rare species will tend to grow out of rare prevalence categories, resulting in fewer rare species at equilibrium. However, Banavar et al. (2004) showed that density and frequency dependence are not inconsistent with neutral theory, so long as all species with the same degree of abundance experience the same rare species advantage. For this reason, neutral theories are also said to be *symmetrical.* As long as each species obeys the same ecological rules, neutral theories can incorporate quite complex and biologically interesting ecological processes. Neutral theory has stimulated ecologists to ask to what extent ecological communities can be treated as approximately symmetrical and when, how, and under what circumstances this symmetry is broken. Neutral theory has highlighted the value in ecology of starting with the simplest hypothesis and only adding complexity when the collected data force one to do so. Neutral theory offers a reexamination of ecological hypotheses about how landscapes are put together.

Neutral theory also makes many predictions about species-area relationships and patterns in phylogeny and **phylogeography**—the study of the patterns of speciation as they are embedded in biogeographic landscapes. The neutral theory's basic assumptions of symmetry and species equivalence in per-capita vital rates still await thorough and rigorous testing. Many fundamental questions remain: When are species similar enough in terms of their functional roles in ecosystems and landscapes to be ecologically substitutable? When do ecological niche differences matter to the assembly of natural ecological communities? A reexamination of these questions in basic ecology should help to resolve why the neutral theory works so well despite making only a few, simple assumptions. One possibility is that on large landscape scales, communities and ecosystems may exhibit a kind of emergent, average statistical-mechanical behavior that can be described by far simpler theories than might have been expected from species diversity and complexity indices.

6 Temporal and Spatial Scale

Statement

Ecological processes vary in their effects or importance at different spatial and temporal scales. For example, biogeochemical processes may be relatively unimportant in determining local patterns but may have major effects on regional or landscape patterns. At the population and community levels, processes leading to population decline or reduced biodiversity may produce extinction at the local scale, but at the landscape level, the same processes may appear only as spatial redistributions or alterations. The **concept of scale** encourages analyses at different levels of organization (different levels in the hierarchical system). For example, a landscape might appear to be heterogeneous at one scale but quite homogeneous at another scale. Thus, when ecologists select a scale for investigation, they must understand how a change in temporal or spatial scale can affect patterns, processes, and emergent properties across scales.

Explanation

Kenneth E. F. Watt, in his textbook entitled *Principles of Environmental Science* (Watt 1973), listed five categories of fundamental ecological variables that need to be understood in order to understand large temporal-spatial processes, including the interactions among these five categories of world resources. These resources (variables) are *energy* (covered in Chapter 3), *matter* (in Chapter 4), *diversity* (in Chapters 1, 7, and 9), *time,* and *space* (both covered in this chapter). Since 1973, numerous books and publications have focused on the importance of time and space as fundamental world resources—indeed, temporal-spatial relationships underpin the emerging disciplines of conservation biology, ecosystem health, landscape ecology, and restoration ecology, among others.

Figure 9-11 depicts how processes and patterns change at different temporal and spatial scales (see Urban et al. 1987 for details). Figure 9-11A illustrates short-term disturbances that involve increasing spatial but not temporal scales, ranging from treefalls to large-scale fires and floods. Figure 9-11B illustrates forest processes that involve both increasing temporal and spatial scales, ranging from the role of local seed banks to speciation and extinction. Figure 9-11C depicts environmental constraints, scaled from microhabitat conditions to glacial cycles; and Figure 9-11D depicts changes in vegetation patterns, scaled from ground cover to major biomes (to be discussed in the next chapter). Each of these space-time relationships can best be understood when viewed as changes in temporal and spatial scales.

Both temporal and spatial scales are involved in the ecological/economic hierarchy of farming. Figure 9-12 depicts the hierarchical nature of agricultural systems. The success of the crop depends not only on field conditions (small temporal-spatial scale), but also on the sustainability of the whole farm (ability to profit), on water from the watershed, and on a market for the crop (large temporal-spatial scale). Problems such as pest control, stream eutrophication, and economic constraints change with scale, ranging from small-scale (and frequently short-term) problems at the field or single-crop level to large-scale (and frequently long-term) challenges at the land-

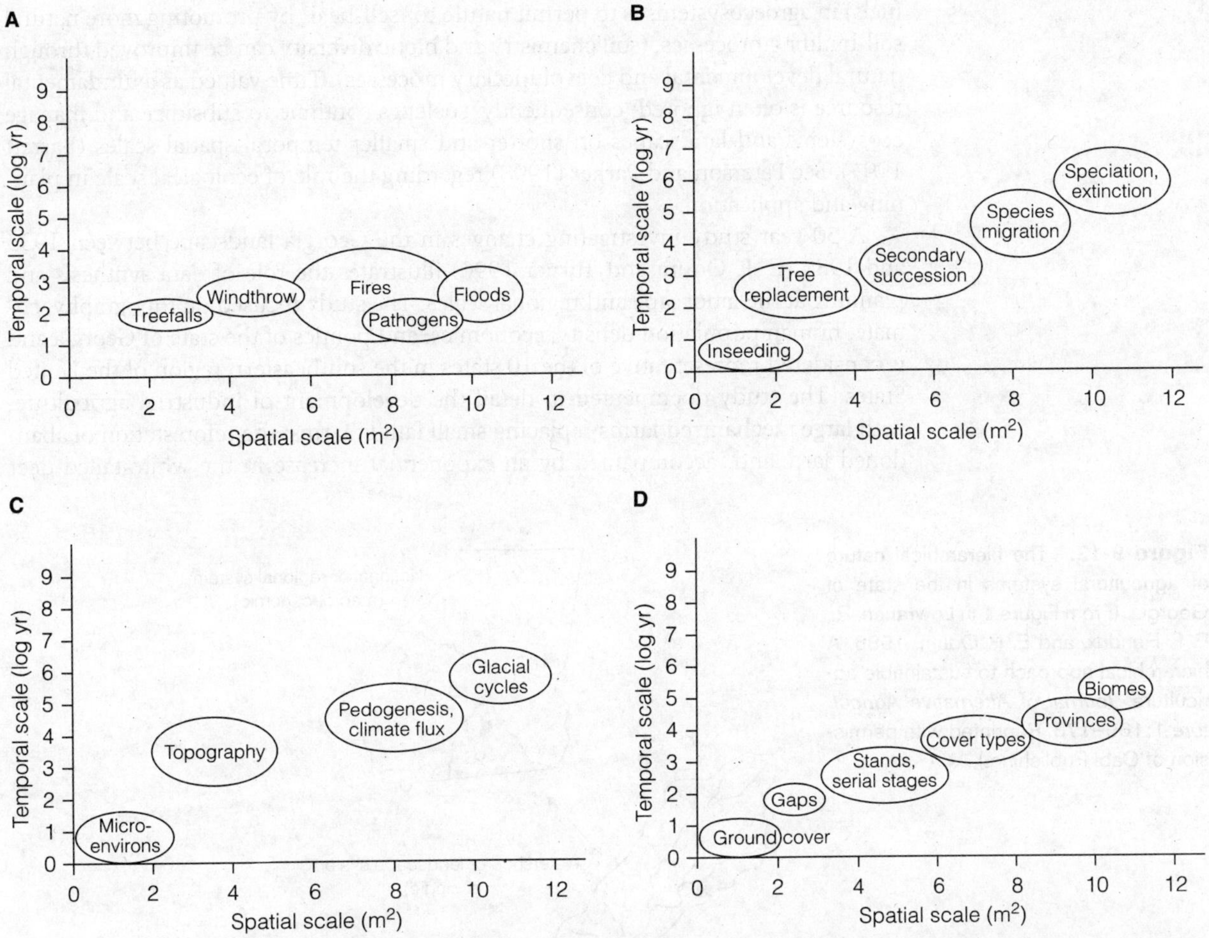

Figure 9-11. Examples of changes at different temporal-spatial scales in (A) disturbance regimes; (B) forest processes; (C) environmental constraints; and (D) vegetation patterns. (From Urban, D. L., R. V. O'Neill, and H. H. Shugart, Jr. 1987. Landscape ecology. *BioScience* 37:119–127. Reprinted with permission.)

scape or national levels. For example, a specific pesticide applied at the field or crop level may control a particular insect species at a critical time during the growing season, whereas only the implementation of large-scale, ecologically based pest management programs at the landscape level improve the problem on a long-term basis (NAS 2000; E. P. Odum and Barrett 2000).

The impact of natural disturbances or human-originated perturbations on population dynamics and community structures is also a matter of scale, involving an assessment of the size of the area affected and the duration and intensity of the perturbation. Connell (1979) suggested that the intensity of the perturbation is maximal at intermediate levels of disturbance in regard to the maximum number of species in a community, ecosystem, or landscape (Fig. 9-13). This relationship of species richness to intensity of disturbance is termed the **intermediate disturbance hypothesis.**

One way to reduce the excessive use of subsidies (energy, pesticides, and fossil fuels) in agroecosystems is to permit nature to "self-heal" by promoting more natural soil-building processes, (soil chemistry and biotic diversity can be improved through natural developmental and coevolutionary processes). Time valued as a fundamental resource is often ignored; consequently, societies continue to subsidize and manage ecosystems and landscapes on shorter and smaller temporal-spatial scales (Barrett 1994). See Peterson and Parker (1998) regarding the role of ecological scale in planning and application.

A 50-year study investigating changes in the Georgia landscape between 1935 and 1985 (E. P. Odum and Turner 1990) illustrates the role of data synthesis and planning at the landscape and regional scales. The study focused on topography, climate, human population density, economics, and politics of the state of Georgia and is considered representative of the 10 states in the southeastern region of the United States. The study documented in detail the development of industrial agriculture, with large mechanized farms replacing small family farms; the reforestation of abandoned farmland, accompanied by an exponential increase in the white-tailed deer

Figure 9-12. The hierarchical nature of agricultural systems in the state of Georgia. (From Figure 1 in Lowrance, R., P. F. Hendrix, and E. P. Odum. 1986. A hierarchical approach to sustainable agriculture. *Journal of Alternative Agriculture* 1:169–173. Reprinted with permission of Cabi Publishing.)

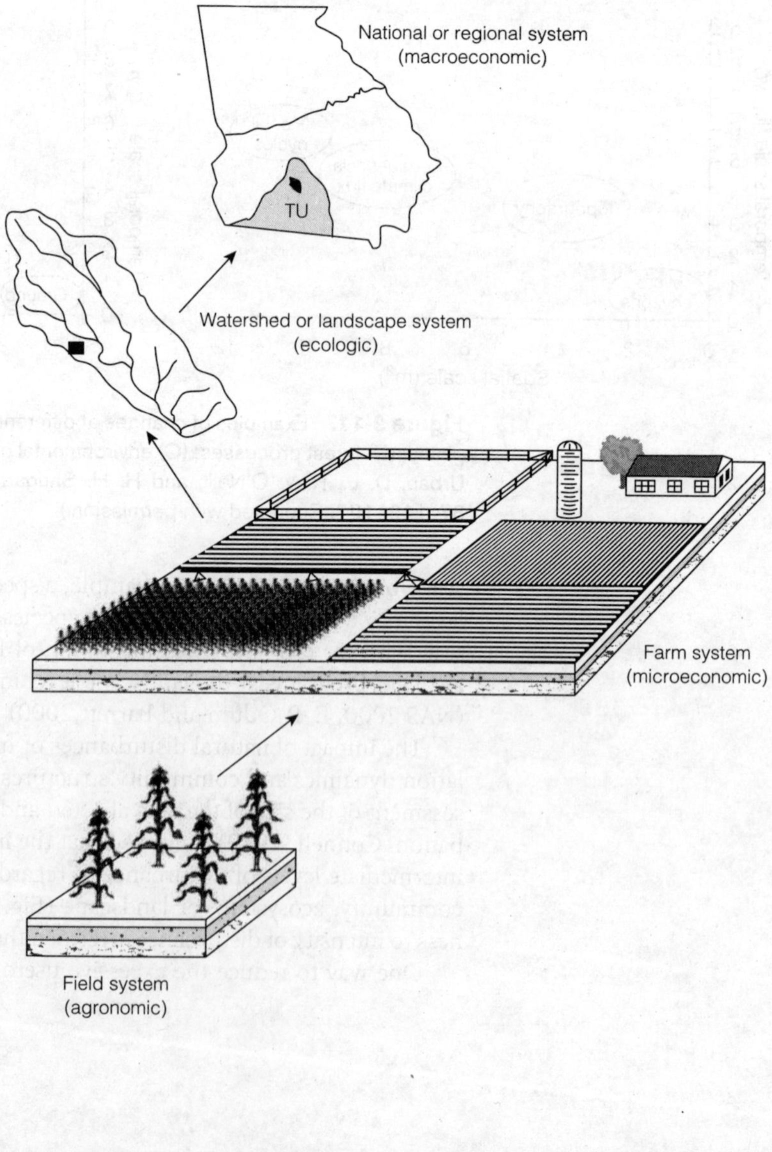

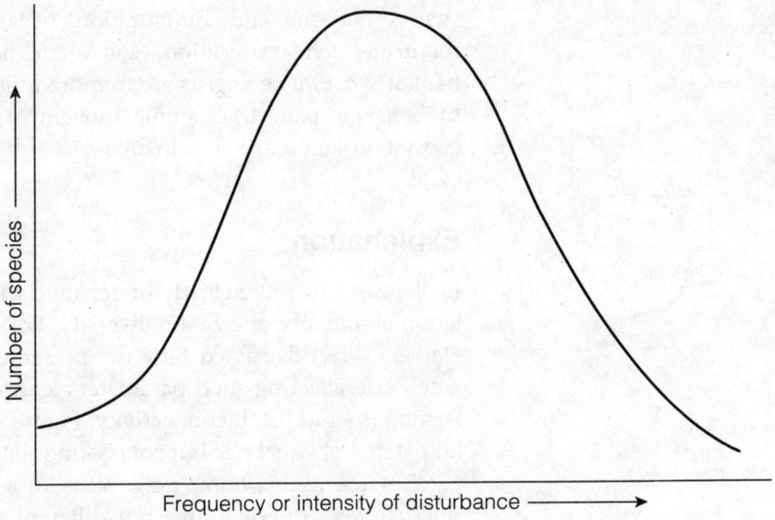

Figure 9-13. Diagram depicting the intermediate disturbance hypothesis (J. H. Connell 1979). The number of species in a community tends toward a maximum at intermediate levels of disturbance.

(*Odocoileus virginianus*) population; and the rapid increase in urbanization, human population growth, transportation needs, and industrialization.

The yields of field crops (cotton, corn, soybeans, and peanuts) increased twofold, but vegetable production decreased fourfold, as more food was imported from other regions and countries. Point-source municipal pollution declined, as indicated by a steady reduction of coliform bacteria in rivers, but nonpoint pollution increased, as indicated by the steady increase of nutrients, pesticides, and industrial wastes in all rivers in the state. The study pointed to the urgent need for serious land-use planning to replace the current haphazard development on a statewide and landscape scale. Recommendations included the need for more greenbelt buffers, cluster development, an urban-zone vegetable agriculture, source reduction of water and air pollution, and increased environmental literacy education at all grade levels. Planning can be more effective when accomplished at the landscape and regional scales and needs to include the interaction of urban and rural areas and cultures (Barrett et al. 1999).

7 Landscape Geometry

Statement

Just as the size and quality of landscape patches and corridors (amount of vegetative cover and food quality, for example) affect ecological processes and plant and animal abundances, the *geometry and configuration of landscape elements* affect ecological processes at the population and community levels. It is now widely recognized that the size and shape of landscape patches influence biotic diversity, home range size and shape, animal dispersal behavior, and species abundance. Recent investigations have addressed the role that landscape geometry plays in plant and animal survivorship, source-sink dynamics, rates of species invasion, and edge-habitat dynamics. Future investigations need to address the role of **landscape geometry** (the study of the shapes, patterns, and configurations of landscape elements) and **landscape architecture** (patch stratification, "soft versus hard" edges, and three-dimensional use of habitat space) if ecologists are to more fully understand such phenomena as dispersal behavior, patterns of animal movement, bioenergetics at the landscape scale, and ecosystem-landscape sustainability.

Explanation

Ecologists now more clearly understand the relationship of ecosystem size to population abundance and biotic diversity. For example, in *The Fragmented Forest,* Larry Harris (1984) discussed how the preservation of biotic diversity is dependent on patch size affecting such parameters as home range size, island biogeography (see Section 4), and patch connectivity. Figure 9-14A depicts eight 1600-m^2 experimental patches of equal size but contrasting shapes. S. J. Harper et al. (1993), using these patches, found that home range sizes for a small mammal species (*Microtus pennsylvanicus*) were of equal area but different shapes due to the contrasting shapes of patches. Thus, the geometry of habitat patches may affect population dynamics owing to differences in edge-to-area ratios for patches of different sizes and shapes. This change in home range shape, however, did not affect survivorship or age structure. S. J. Harper et al. (1993) concluded that plasticity of behavior (adapting to changes in the home range shape) appears to have prevented differences in population den-

A

B

Courtesy of Gary W. Barrett

Courtesy of Terry L. Barrett

Figure 9-14. (A) Aerial photograph of eight 1600-m^2 experimental patches of contrasting shape, $n = 4$ replications per shape (after S. J. Harper et al. 1993). (B) An example of radial-arm irrigation, which creates circular landscape patterns.

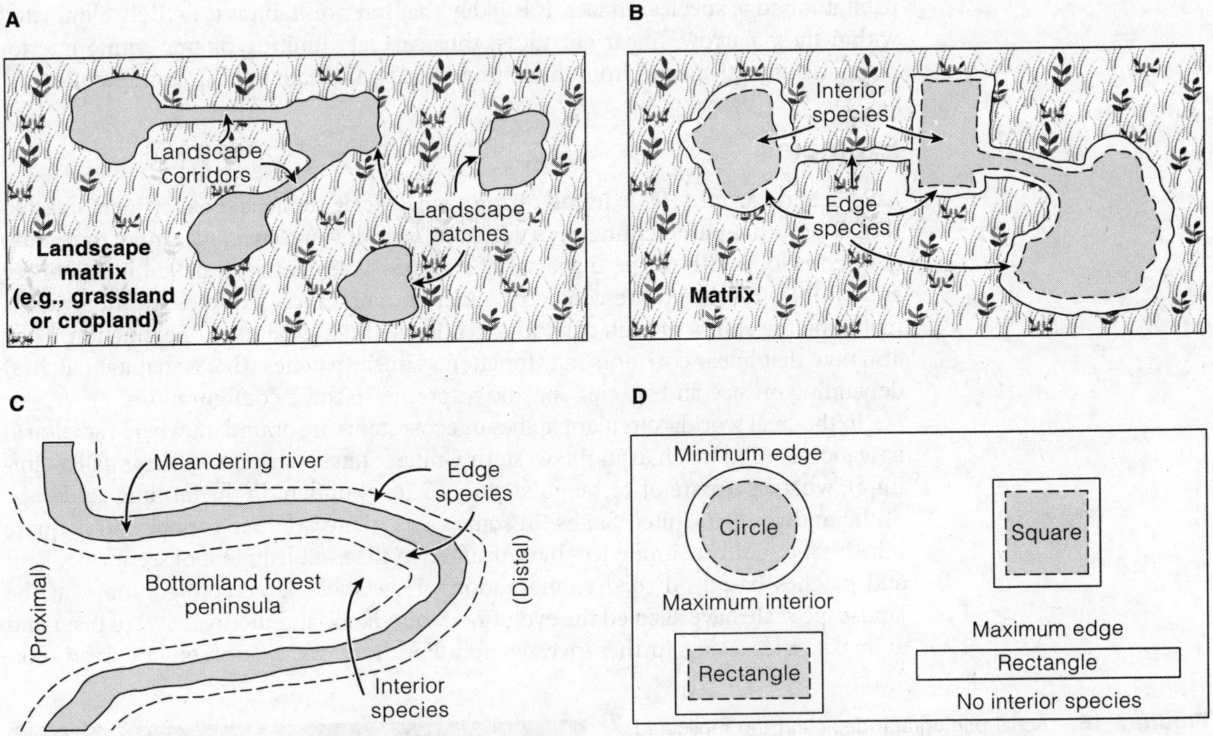

Figure 9-15. Diagram depicting (A) the three major landscape elements (patch, corridor, and matrix); (B) the relative abundances of edge and interior species in patches and corridors; (C) a bottomland forest peninsula and; (D) examples of contrasting patch shapes of equal size but different geometry.

sity, survival, and age structure between the different patch shapes. The shape of habitat patches, however, did affect the number of meadow voles that dispersed when population densities were low, but not when densities were high.

Figure 9-15A summarizes the three major landscape elements (patches, corridors, and matrices) found within the landscape mosaic. Figure 9-15B illustrates how the abundances of interior and edge species are related to patch and corridor shape and size. Bird edge species, for example, include the indigo bunting (*Passerina cyanea*), the eastern bluebird (*Sialia sialis*), and the cardinal (*Cardinalis cardinalis*); examples of interior bird species include the wood thrush (*Hylocichla mustelina*), the red-eyed vireo (*Vireo olivaceus*), and the downy woodpecker (*Picoides pubescens*). Kendeigh (1944) noted that patches of eastern deciduous forest larger than 26 hectares (65 acres) warranted consideration as valid avifaunal census patches to evaluate forest edge versus interior bird species. Although two of the patches on the left of Figure 9-15B are connected by a narrow corridor, the corridor is *too* narrow to permit the movement of interior species between patches. Figure 9-15C illustrates a bottomland forest peninsula, created by a meandering stream or river. Peninsulas may be large in scope (such as the Baja Peninsula of California) or small in scope (such as the neck of a forest extending into farmland). Figure 9-15D illustrates several patches of equal size but of different shapes. A circular patch maximizes habitat for interior species (see Fig. 9-14B as an example), whereas a long linear, narrow patch maximizes habitat for edge species. In fact, it is likely that interior habitat is entirely eliminated within these narrow, linear corridors, thus severely limiting or preventing interior plant and animal species from inhabiting a landscape patch of this configuration.

Examples

LaPolla and Barrett (1993) found, using a small-scale replicated experiment in Ohio (Fig. 9-16), that **patch connectivity** (that is, corridor presence) was more important than corridor width to the dispersal of meadow voles between patches. Rosenberg et al. (1997) discussed the geometry, function, and efficacy of biological corridors, including the ability of such corridors to mitigate high local rates of extinction. They also note that linear corridors may function as linear patches (that is, habitat patches) depending on size and species-specific responses to these configurations.

In the "real world," circular patches or ecosystems are abundant where radial-arm irrigation is used (such as in the western United States, where annual rainfall is limiting), whereas square or rectangular patches are abundant in the farming landscape of the midwestern United States. In some cases, there are roads about every square mile (that is, cultural influences have resulted in the establishment of sections of land and patches based on road configuration). How these large configurations at the landscape scale have affected the evolution (abundance and biodiversity) of plant and animal species awaits further investigation. For more information on how road ecol-

Figure 9-16. Aerial photograph depicting the replicated research design used to evaluate the effects of corridor width and presence on the population dynamics of the meadow vole (*Microtus pennsylvanicus*) (after LaPolla and Barrett 1993).

Courtesy of Gary W. Barrett

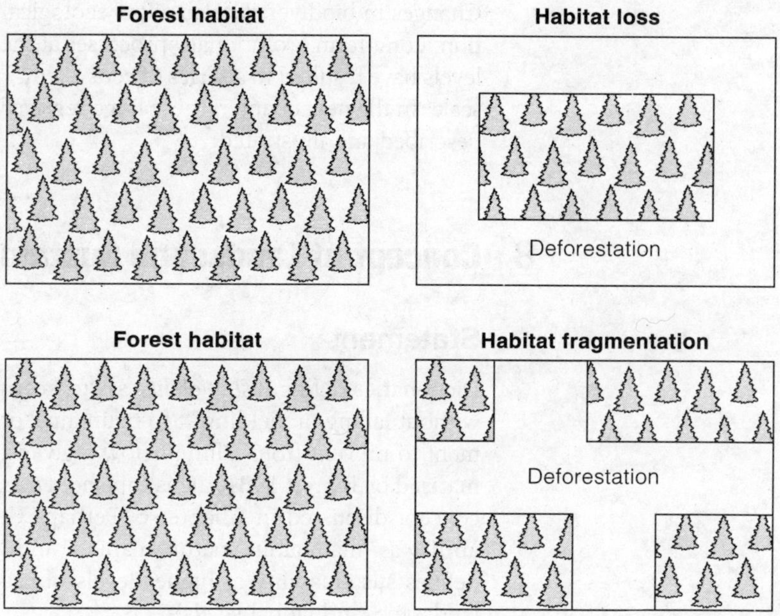

Figure 9-17. Diagram depicting the difference between habitat loss (top) and habitat fragmentation (bottom). The same total area of deforestation can result in the loss of species requiring large home ranges when habitat is fragmented, for example.

ogy and systems are related to ecological processes and wildlife abundance, see Forman et al. (2003).

Because the amount of *edge habitat* changes with patch shape, it is only natural that landscape ecologists have increasingly investigated the role of edge habitat within the landscape mosaic. Because the variety and diversity of life is often greatest in and about edges and ecotones (a phenomenon termed the *edge effect*), it follows that changes in patch shape, resulting in increasing or decreasing amounts of edge habitat, also result in increasing or decreasing species abundance and biotic diversity.

Ostfeld et al. (1999) summarized how interactions between meadow voles (*Microtus pennsylvanicus*) and white-footed mice (*Peromyscus leucopus*) at forest–old-field edges affected tree invasion in old-field ecosystems. Meadow voles prey on seedlings of sugar maple (*Acer saccharum*), white ash (*Fraxinus americana*), and tree of heaven (*Ailanthus altissima*), whereas white-footed mice prefer seeds of red oak (*Quercus rubra*) and white pine (*Pinus strobus*). Voles have their strongest effects on seedling mortality at sites more than 10 meters away from the forest edge, whereas seed predation by white-footed mice is strongest in the zone less than 10 meters from the forest edge. Thus, not only the amount of edge habitat, but the ecosystem types involved and the species living within these edge habitats cause long-term effects on rates of secondary succession and plant community composition.

The previous sections attempted to illustrate how habitat fragmentation has resulted in a landscape mosaic dominated by changes in the size, shape, and frequency of landscape elements (patches, corridors, and matrix). These changes resulting from habitat fragmentation—changes caused by both natural and human-dominated processes—have affected plant and animal diversity and abundance, numbers of edge versus interior species, and changes in micro- and macroevolutionary processes. Figure 9-17 illustrates the difference between habitat loss and habitat fragmentation. Equal amounts of deforestation can result in changes not only in the geometry of the landscape, but also in species diversity, because numerous avian and mammalian species require large home ranges for their reproductive success and survivorship.

Changes in biodiversity also affect ecological processes at several levels of organization. Long-term evolutionary processes at the organism, population, and community levels have resulted in a series of ecosystem, landscape, and biome types at the global scale. In the next chapter, these large regional landscape units, termed *biomes,* will be described and illustrated.

8 Concept of Landscape Sustainability

Statement

Dictionaries define *sustainability* as "to keep in existence," "to support," "to endorse without failing or yielding," "to maintain," or "to supply with necessities or nourishment to prevent from falling below a given threshold of health or vitality" (as summarized by Barrett 1989). This threshold could be viewed as the *carrying capacity* (*K*) concept discussed in Chapter 6. Perhaps Goodland's (1995) definition of **sustainability** as "maintaining natural capital" and "maintenance of resources" more clearly defines sustainability at higher levels of ecological organization (at the ecosystem, landscape, and global levels).

Explanation

There have been several attempts to summarize and discuss the benefits supplied to human societies by natural ecosystems (see Daily et al. 1997; Costanza, d'Arge, et al. 1997; Hawken et al. 1999 for a more detailed discussion of natural capital). **Natural capital**, in contrast to *economic capital,* are those benefits supplied to human societies by natural ecosystems and landscapes (see Table 1-1 for a summary of perceived differences between ecology and economics). In addition to the production of goods (such as timber, game species, fruits, and nuts), ecosystem and landscape services support life through such functions as the purification of air and water, the cycling of nutrients, the pollination of crops, the preservation and renewal of soil fertility, the partial stabilization of climate, the maintenance of biodiversity, the provision of aesthetics, and the control of pests, among others. These services are so fundamental to life that they are, unfortunately, easily taken for granted. Thus, these services—which underpin the concept of landscape sustainability—are greatly undervalued and poorly understood by human societies. As noted in Chapters 1 and 2, energy would be a better currency to value these goods and services than economic currencies. If awareness and understanding are not increased and current trends continue, humankind will dramatically alter Earth's natural ecosystems and landscapes, to the extent that these services of natural capital will greatly diminish.

9 Domesticated Landscapes

Civilization seems to reach its most intense development in what was originally forest and grassland, especially in temperate regions. Consequently, most temperate forests and grasslands have been greatly modified from their primeval condition, but the basic nature of these ecosystems has by no means changed. Humans, in fact, tend to combine features of both grasslands and forests into habitats that might be termed *forest edge.* A **forest edge** may be defined as an ecotone between forest and grass or shrub communities. When humans settle in grassland regions, we plant trees around our homes, villages, and farms, so that small patches of forest become dispersed in what may have been treeless country. Likewise, when humans settle in a forest, we replace most of it with grasslands and croplands (as lesser amounts of human food can be obtained from a forest), but leave patches of the original forest on farms and around residential areas. Many of the smaller plants and animals originally found in both forest and grassland are able to adapt and thrive in close association with humans, as are domestic or cultivated species. The American robin (*Turdus migratorius*), for example, once a bird of the forest, has become so well adapted to the human-made forest edge that it has not only increased in numbers, but has extended its geographic range. Most forest birds (the thrushes in Europe, for example) have switched from the forest to gardens, cities, and hedgerows—or else they have become extinct, because there are no longer many tracts of unbroken forest. Most native species that persist in regions heavily settled by humans become useful members of the forest-edge landscape, but a few become pests.

If we consider croplands and pastures as modified grasslands of early successional types, then we can say that we depend on grasslands for food but like to live and play in the shelter of the forest, from which we also garner useful wood products. At the risk of oversimplifying the situation, we might say that humans, in common with other heterotrophs, seek production and protection from the landscape. In many cases, the monetary value of wood, if harvested all at once, is less than the value of the intact forest that provides recreation, watershed protection, and other life-support services such as home sites and a sustainable harvest of wood (see Bergstrom and Cordell 1991).

Agroecosystems and Agrolandscapes

Agroecosystems are domesticated ecosystems that are in many ways intermediate between natural ecosystems, such as grasslands and forests, and fabricated ecosystems, such as cities (E. P. Odum 1997; Barrett et al. 1999). Like natural ecosystems, agroecosystems are solar-powered, but they differ from natural systems in several ways; processed fossil fuels, along with human and animal labor, provide them with auxiliary energy sources that enhance productivity but also increase pollution; diversity is greatly reduced by human management in order to maximize the yield of specific foods or other products; the dominant plants and animals are under artificial rather than natural selection; and control is external and goal-oriented, rather than internal via subsystem feedback, as in natural ecosystems.

The simplified graphic models in Figure 9-18 depict the stages in the development of agriculture, namely, *pre-industrial* (similar to a natural ecosystem); *industrial;* and *reduced-input conservation tillage* (E. P. Odum and Barrett 2004). The food web in pre-industrial agriculture—as still practiced in less developed countries—is very similar to that in natural ecosystems, with domestic animals replacing wild animals in the grazing food chain. Pre-industrial agriculture is energy efficient and diverse, with a variety of crops grown together, including fish fed on plant residues, and it feeds local villages but does not produce enough surplus for export, or to feed cities. In industrial agricultural monocultures, frequently termed *conventional agriculture* (Fig. 9-18), the microbial recycling loop is almost eliminated and replaced by large subsidies of fertilizer, pesticides, and water, which increases yields but produces widespread pollution, erosion, and loss of habitat and soil quality. Crop and meat

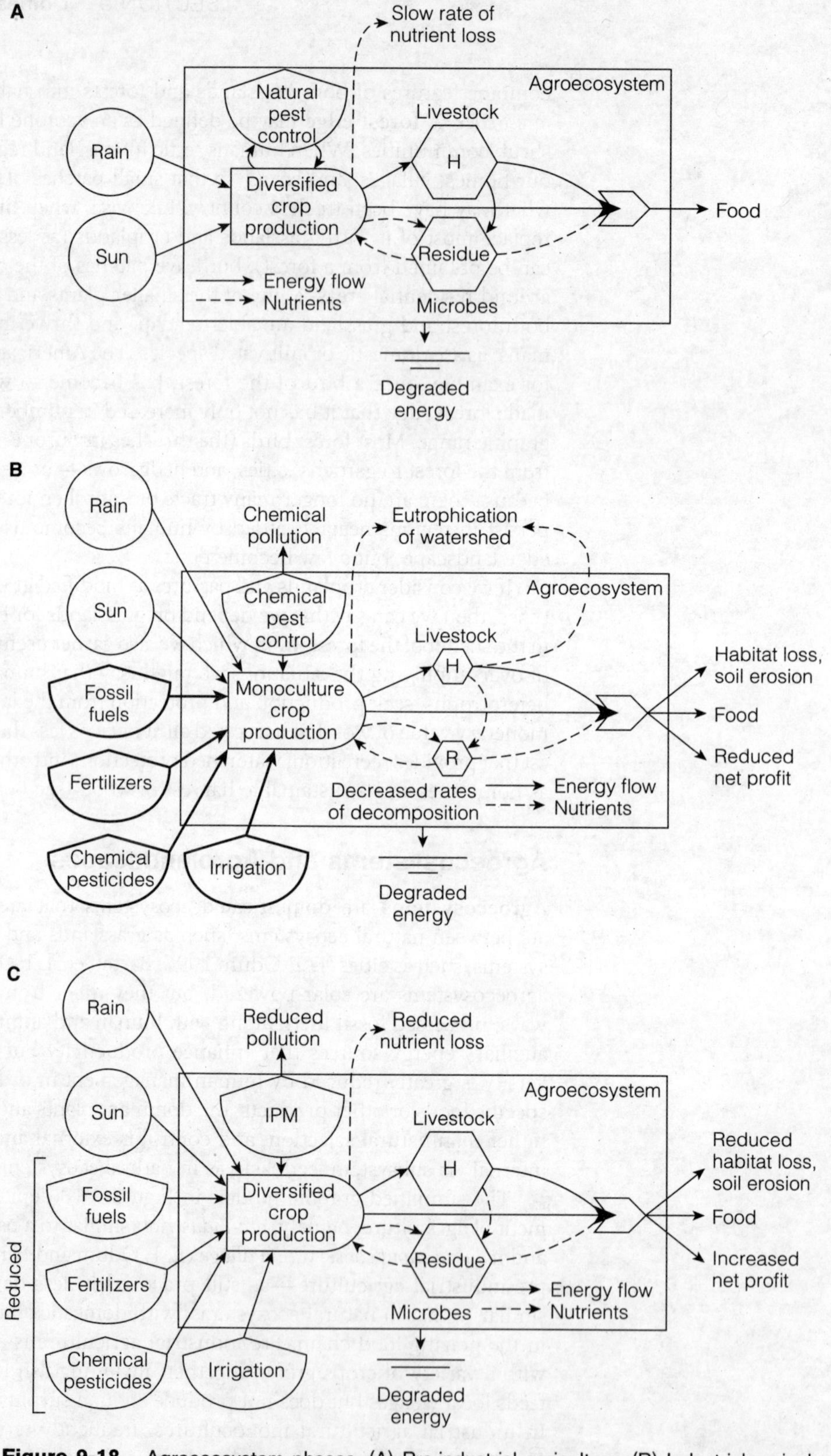

Figure 9-18. Agroecosystem phases. (A) Pre-industrial agriculture. (B) Industrial agriculture. (C) Low-input sustainable agriculture. IPM = integrated pest management.

production are decoupled with the development of feed lots, so manure becomes a pollutant rather than a fertilizer resource (Brummer 1998). Because industrial agriculture is not sustainable, it is being redesigned along the lines of conservation tillage, frequently termed *low-input sustainable agriculture.*

It is important that we review the history of industrial agriculture in order to gain perspective on current problems and research needs. The development of conventional agriculture in the midwestern United States can be described in four stages (see Table 2-2 for details).

From 1833 to 1934, some 90 percent of prairie land, 75 percent of wetlands, and most forest land on good soils were converted to croplands, pastures, and woodlots. Natural vegetation was restricted to steep land and shallow, infertile soils. However, farms were generally small, crops diversified, and human and animal labor extensive, so the impact of farming on water, soil, and air quality was not deleterious overall. Farming in the early 1800s had little effect on watershed and lake dynamics, but the intensification of agriculture after about 1915 caused eutrophication of lakes, resulting from an inflow of agricultural chemicals.

From 1935 to circa 1960, an intensification of farming associated with inexpensive fuel and chemical subsidies, mechanization, and an increase in crop specialization and monoculture occurred. Total cropland acreage decreased and forest cover increased 10 percent, as more food was harvested from less area by fewer farmers.

From 1961 to 1980, energy subsidy, the size of farms, and farming intensity on the best soils all increased, with emphasis on continuous monoculture of grain and soybean cash crops, much of which are grown for export trade. These changes essentially put the small family farm out of business and created pollution that in many cases equaled the worst industrial pollution on a global scale. From 1980 to the present, rapid urbanization and increased farming intensity resulted in increased *cultural eutrophication* as agroindustrial wastes and extensive erosion brought large amounts of soil, heavy metals, and other toxic substances into the watershed. During the 1980s and 1990s, however, farming practices begin to change, with increased emphasis on **alternative agriculture** (NRC 1989) and on increased biodiversity in agroecosystems (W. W. Collins and Qualset 1999). New major crops, such as sunflower (*Helianthus annuus*), have helped to diversify the agricultural landscape (Fig. 9-19). The

Figure 9-19. Field of sunflower (*Helianthus annuus*) in the agricultural Midwest of the United States.

Courtesy of Terry L. Barrett

perspective of agriculture also changed as farmers and other stakeholders viewed challenges and opportunities at greater temporal and spatial scales; Barrett et al. (1999) and Barrett and Skelton (2002) termed this an **agrolandscape perspective** rather than solely an agroecosystem perspective.

In summary, current unsustainable industrial agriculture can be redesigned to sustain soil quality and high yields by combining the new technologies of residue management, conservation tillage, and polyculture with the soil-building processes of natural ecosystems and pre-industrial agroecosystems. There is increased emphasis on managing agroecosystems as agrolandscapes, with greater focus on reconnecting agricultural and urban landscapes (Barrett et al. 1999; Barrett and Skelton 2002). There is also increased interest in **agroforestry**—a practice that involves the cultivation of small, fast-growing trees and food crops in alternate rows (MacDicken and Vergara 1990)—in ecologically based pest management (NRC 2000a), and in opportunities to integrate soil, crop, and weed management in low-input farming systems (Liebman and Davis 2000).

Urban-Industrial Technoecosystems

The concepts of the *technoecosystem* and the *ecological footprint* were introduced in Chapter 2. Cities, suburbs (that is, metropolitan districts), and industrial development zones are the major technoecosystems. They are small but very energetic islands with large ecological footprints in the matrix of natural and agricultural landscapes. Realistically, these urban-industrial environments are parasites on the biosphere in terms of life-support resources (Fig. 9-20).

The planned cities of ancient Greece, the city-states of the Middle Ages, and the towering skyscrapers of contemporary cities are achievements of humankind. Technological utopias, such as the ones designed by Walt Disney, inspire vision in some. Unfortunately, many cities today are in disorder and decline, as globally more people immigrate into cities seeking richer economic lives. The growth of cities is especially rapid in less developed countries. As of the year 2000, Mexico City and São Paulo, Brazil, already had a population of more than 25 million people each—far more than any city in the industrialized world, except possibly Tokyo, and approaching twice as many as New York City. It is predicted that by the year 2010, between 50 and 80 percent of people worldwide will be living in urban environments (see the United Nations report *Mega-City Growth and the Future,* edited by Fuchs et al. 1994).

As documented in Chapter 6, anything that grows rapidly and haphazardly (without plan or control) and without regard to life support will outstrip the infra-

Figure 9-20. Model illustrating the need to link natural, life-support ecosystems with urban-industrial ecosystems, including a reward-feedback loop necessary to provide for a sustainable landscape (modified after Odum 2001; Barrett and Skelton 2002).

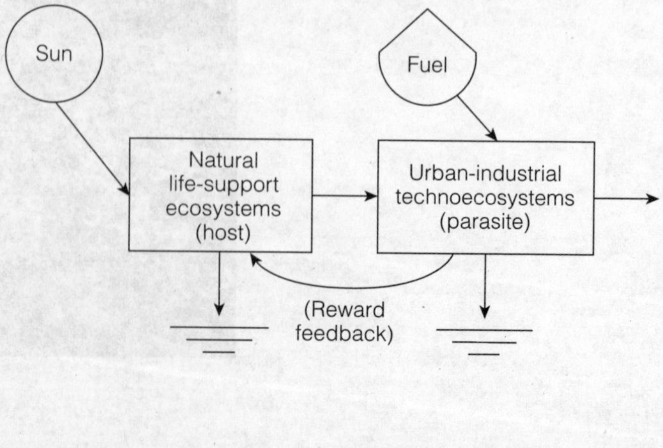

A

B

Figure 9-21. (A) Model diagram depicting a home site on a stream peninsula, illustrating an increased diversity of ecosystem, landscape, and biome habitats. (B) Photograph of residential development located in Henry County near Atlanta, Georgia.

structure necessary to maintain its growth, thereby bringing on boom-and-bust cycles. Citizens must engage in serious urban planning to correct this course. Most of the literature dealing with the plight of the cities focuses on internal problems, such as deteriorating infrastructure and crime, but as Lyle (1993) pointed out, cities of the future will have to "embrace the ecology of the landscape, rather than set themselves apart." Figure 9-21A depicts a diversified home site situated on a stream peninsula

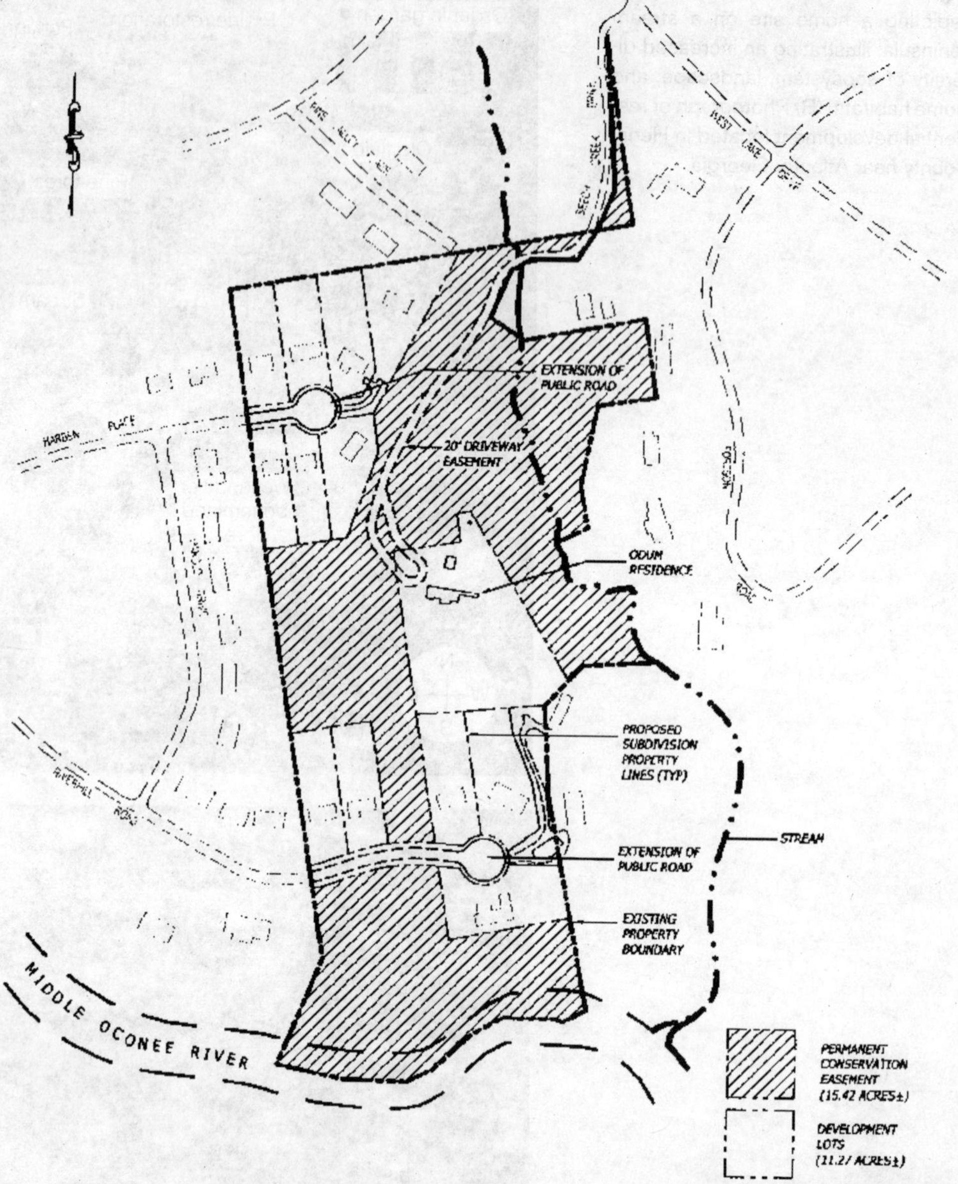

Figure 9-22. Map outlining the Odum conservation easement plan for Beech Creek Reserve, located in Athens, Georgia. Note that more acres (hectares) are in permanent conservation protection than are set aside for the development of residential lots, thus providing habitat for wildlife and preventing stream erosion (Robinson Fisher Associates 2004).

surrounded by a diversity of habitat types and, consequently, recreational, resource, and cultural opportunities. Figure 9-21B is a photograph of a highly clustered residential development near Atlanta, Georgia. This architectural development could be viewed as similar to monoculture agricultural systems in the Midwest of the United States.

Figure 9-22 depicts the master conservation easement plan for the Beech Creek Reserve, a 26.7-acre (10.8-ha) area of land including the long-time Martha H. and Eugene P. Odum residence, located in Athens, Georgia. Note that more than 50 percent of this urban area is in permanent conservation easement (N_pA_m) providing habitat for wildlife, watershed protection, natural privacy, and aesthetic beauty. This model plan illustrates how natural and economic capital can be integrated within urban areas.

Urban regeneration will depend more and more on reconnecting the city to the life-supporting land and water bodies because, as we pointed out earlier in this book, a parasite prospers only if its host remains in good condition (see also Haughton and Hunter 1994). What we can learn from ecology about dealing with this situation will be discussed in Chapter 11.

10

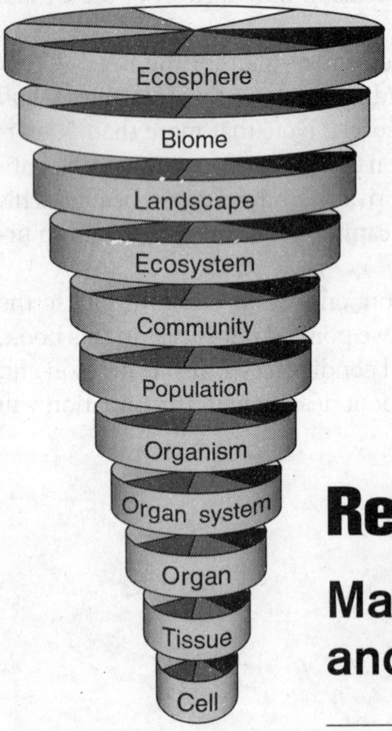

Regional Ecology

Major Ecosystem Types and Biomes

1 **Marine Ecosystems**
2 **Freshwater Ecosystems**
3 **Terrestrial Biomes**
4 **Human-Designed and Managed Systems**

For the most part, in this book, we have based our approach to ecology on the analysis of units of the landscape as ecological systems. Principles and common denominators that apply to any and all situations—whether aquatic or terrestrial, natural or human-made—have been emphasized. The importance of the natural environment as the life-support module for planet Earth and of the driving force of energy have been stressed. In Chapters 5, 6, and 7, another useful approach was introduced—that of concentrating study on organisms, populations, and communities—which describes mechanisms for evolutionary change. Still another useful approach is *geographic,* involving the study of the Earth forms, climates, and biotic communities that make up the ecosphere. In this chapter, we list and briefly characterize the major ecological formations (*biomes*) or easily recognized ecosystem types (Table 10-1), with an emphasis on the geographic and biological differences that underlie the remarkable diversity of life on Earth. In this manner, we hope to establish a global frame of reference for Chapter 11, which deals with the human challenge to solve problems on a large scale.

Table 10-1 **Major biomes, ecosystem types, and habitat types of the biosphere**

Marine ecosystems	Open ocean (pelagic)
	Continental shelf waters (inshore waters)
	Upwelling regions (fertile areas with productive fisheries)
	Deep sea hydrothermal vents (geothermally powered ecosystems)
	Estuaries (coastal bays, sounds, river mouths, salt marshes)
Freshwater ecosystems	Lentic (standing water): lakes and ponds
	Lotic (running water): rivers and streams
	Wetlands: marshes and swamp forests
Terrestrial biomes	Tundra (arctic and alpine)
	Polar and mountaintop ice caps
	Boreal coniferous forests
	Temperate deciduous forests
	Temperate grassland
	Tropical grassland and savanna
	Chaparral (winter rain–summer drought regions)
	Desert (herbaceous and shrub)
	Semi-evergreen tropical forest (pronounced wet and dry seasons)
	Evergreen tropical rain forest
Habitat types	Mountains
	Caves
	Cliffs
	Forest-edge habitats
	Riparian habitats

On land, *plants* are a conspicuous element composing the matrix of the landscape, so ecosystem types can be identified and classified as biotic regions or *biomes* based on the *dominant mature vegetation.* In contrast, plants are very small and inconspicuous in most aquatic environments—especially large rivers, lakes, and oceans—so *ecosystem types* in these environments are more easily identified by *physical attributes.*

We would do well to start our world tour with the ocean, the largest and most stable macroecosystem. The ocean, presumably, was the first ecosystem, for life is now thought to have originated in the saltwater milieu.

1 Marine Ecosystems

Statement

The major oceans (Antarctic, Arctic, Atlantic, Indian, and Pacific) and their connectors and extensions cover approximately 70 percent of the surface of Earth. Physical factors dominate life in the ocean (Fig. 10-1A). Waves, tides, currents, salinity, temperature, pressure, and light intensity largely determine the makeup of the biological communities that, in turn, have considerable influence on the composition of bottom sediments and gases in solution and in the atmosphere. Most important, oceans play a major role in shaping the weather and climate over the entire Earth.

Explanation and Examples

The study of the biology, chemistry, geology, and physics of the ocean are combined into a sort of "superscience" called **oceanography**, which is becoming a necessity as a basis for international cooperation. Although the exploration of the ocean is not quite as expensive as the exploration of outer space, considerable funds for ships, shore laboratories, equipment, and specialists are required. Most research beyond the seashores is, of necessity, carried out by a relatively few large institutions backed by government subsidies, mostly from the affluent nations.

The Ocean

The food chains of the ocean begin with the smallest known autotrophs and end with the largest of animals (giant fish, squid, and whales). Tiny green flagellates, algae, and bacteria—"*picoplankton*"—that are too small to be captured by a plankton net are more important as a base for the ocean food web than the larger "*net plankton*" previously believed to fill that niche (Pomeroy 1974b, 1984). Because a large portion of primary production is in the dissolved or particulate organic matter (DOM and POM), organic matter food chains are important in the open ocean. A diversity of filter feeders, ranging from protozoa to pelagic mollusks that spin mucus nets to entrap microbes and detritus particles, provide the links between the small autotrophs and the large consumers.

To fully appreciate both the promise and the problems involved in human interaction with the ocean, we need to look at the contour of the ocean bottom, which is also used in standard oceanographic nomenclature for the zones of the ocean. Because there are likely to be phytoplankton under every square meter of water, and because life in some form extends to the greatest depths, the oceans are the largest three-dimensional ecosystems. They are also biologically very diverse, because many of the major taxonomic groups (*phyla*) are found only in the ocean.

The surprisingly high diversity and evolutionary adaptations of the deep sea fauna have been noted by numerous investigators. Deep sea fish are a curious lot: some produce their own light (lantern fish); others have a luminous-tipped movable spine that is used as bait to attract prey (angler fish); and many have enormous mouths and can swallow prey larger than themselves (viperfish, gulpers). Meals are few in the dark depths, but fish are adapted to make the best of their opportunities. Because of the lack of light (no net primary productivity) at this depth, deep sea eco-systems are dependent on detritus raining in from above.

Continental Shelf

Marine life is concentrated near the shore, where nutrient conditions are favorable. No other area has such a variety of life as the continental shelf—not even the tropical rain forests. The inshore zooplankton is enriched with much **meroplankton** (tem-porary or seasonal organisms that "grow out" of the planktonic phase), consisting of pelagic larvae of *benthic organisms* (such as crabs, marine worms, and mollusks), in sharp contrast to freshwater and the open ocean, where most of the floating life is **holoplankton** (organisms whose entire life cycle is planktonic). Pelagic larvae have been shown to have a remarkable ability to locate the kind of bottom suitable for sur-vival as sedentary adults. When ready to metamorphose, the larvae do not settle at random, but only in response to particular chemical conditions of the substrate. The benthos has two vertical components: (1) the **epifauna,** organisms living on the sur-face, either attached or moving freely about; and (2) the **infauna,** which dig into the substrate or construct tubes and burrows (Fig. 10-1B). Benthic aggregations world-wide occur in what Thorson (1955) called "parallel level bottom communities" dom-inated by ecologically equivalent species, often of the same genus.

The great commercial fisheries of the world are almost entirely located on or near the continental shelf, especially in regions of cold water upwellings (see next section and Fig. 10-2). Figure 10-2 also depicts the major ocean currents in both the Northern and Southern Hemispheres. Relatively few species make up the bulk of the commercial fishes, including anchovies, herring, cod, mackerel, pollock, pilchard, flatfishes (flounders, halibut), salmon, and tuna. It is now certain that the world catch has peaked, and many areas are now overfished. Fishing, especially when involving long-distance trawling or seining, is energy expensive. Increasing food production from the sea may well depend on mariculture (food or fish farming in enclosures in bays and estuaries).

A recent review of marine fisheries concluded that a startling 90 percent of the world's large predatory fish, including tuna, swordfish, cod, halibut, and flounder, have disappeared in the past 50 years. This 10-year study by Myers and Worm (2003) at Canada's Dalhousie University attributed the decline to a growing demand for seafood, coupled with an expanded global fleet of technologically efficient boats. Once thought to be inexhaustible, the fisheries of the world are now showing their vulnerability. The United Nations Food and Agriculture Organization (FAO) estimates that three quarters of the world's oceanic fisheries are being fished at or beyond their sustainable yields. The innovations that have allowed us to pull more fish out of the oceans—larger and more powerful boats (some with on-deck processing facilities), improved fishing gear, and enhanced navigational and fish-finding technologies—may undermine the presumed resilience of the oceans.

Data show that once large boats target a fishery, they can deplete populations in a matter of years. Within 15 years, some 80 percent of the large fish are lost. Smaller species may initially flourish, but often their populations soon crash too, either because of a limited food supply, overcrowding, and disease, or because they become targets for those who are "fishing down the food web." The average size of top preda-tory fish is now only one fifth to one half that in the past, in part because the fish that are left to breed are the ones small enough to escape from nets. Another problem is that slow-maturing fish are often caught before they are old enough to reproduce.

The deterioration of oceanic fisheries can be reversed. Granting fishers an ownership stake in fish stocks is a way to help them appreciate that the more productive their fishery is, the more valuable their share. For example, fishers located in Iceland and New Zealand have used marketable quotas, allowing them to sell catch rights, since the late 1980s. The upshot is smaller but more profitable catches, and rebounding fish populations. The classic "tragedy of the commons" problem (Kennedy 2003) is thus averted.

Because of the complexity of marine ecosystems, some scientists are advocating the management of whole ecosystems rather than single species. Moreover, studies have shown that well-positioned and fully protected marine reserves, known as *fish parks,* can help replenish an overfished area. By giving fish a refuge to breed and mature in, reserves can increase the size and total number of fish both in the reserve and in the surrounding waters. For example, a network of reserves established near Saint Lucia in 1995 has raised the catch of adjacent small-scale fisheries by up to 90 percent. Preservation of nursery habitats such as coral reefs, kelp forests, and coastal wetlands is integral to keeping fish in the sea for generations to come.

Because of their accessibility and richness of life, seashores are the most studied part of the continental shelf region. No biologist, not to mention any amateur natu-

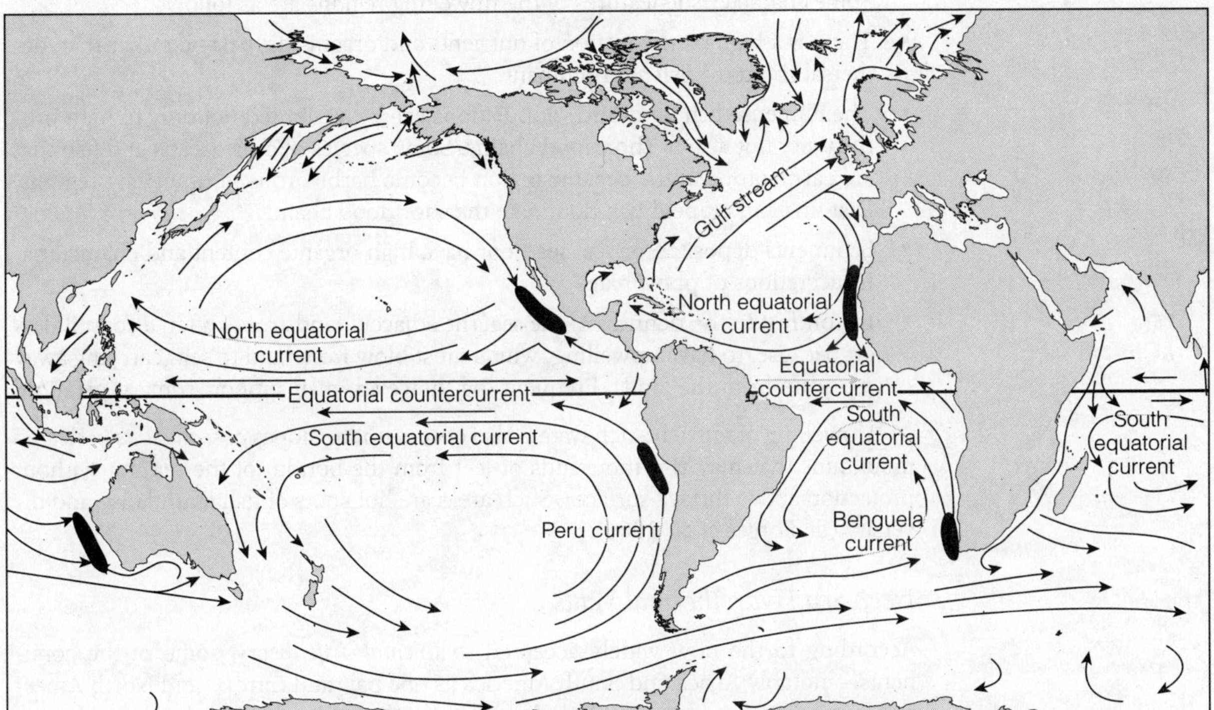

Figure 10-1. The major upwellings (dark areas) and ocean currents driven by the winds and by Earth's rotation (after Duxbury 1971).

ralist, considers his or her education complete without study at the seashore. As on mountains, communities in the intertidal zone are arranged in distinct levels. Some aspects of this zonation on two contrasting seashores, a sandy beach and a rocky shore, are shown in Figure 10-2. Ecologists who study these physically demanding habitats are especially impressed with the role of competition and predation (see Chapter 7).

The physical energy level of breaking waves, surf, and tides is a major input factor to which organisms must adapt. A low-energy coast with gentle water flow will be populated by more and different species than will a high-energy coast subject to strong waves.

Upwelling Regions

An important process termed **upwelling** occurs where winds consistently move surface water away from precipitous coastal slopes, bringing to the surface cold water rich in nutrients that have been accumulating in the depths. Upwelling creates the most productive of all marine ecosystems that support large fisheries. Upwellings are located largely on western coasts, as shown in Figure 10-2. Besides fish, upwelling supports large populations of seabirds that deposit countless tons of nitrate- and phosphate-rich guano on their nesting grounds on the coastal shores and islands. Before the industrial production of nitrogen was developed, these guano deposits were mined and shipped all over the world to be used as fertilizer.

Some characteristic features of the upwelling regions are as follows:

- There is a high concentration of nutrients and organisms; pelagic rather than demersal (bottom) fish are dominant.
- The immense fish (and bird) populations can be attributed not only to high productivity but also to short food chains. Some species of crustaceans and fish that are carnivorous in the oceanic region become herbivorous in upwelling regions. Diatoms and clupeid fish dominate the short food chain.
- Sediments deposited on the sea floor have high organic content and characteristic accretions of phosphate.
- In contrast to the richness of the sea, the adjacent land area is often a coastal desert because, to have upwelling, winds must blow from land to sea (carrying away moisture from the land). Frequent fog, however, may support some vegetation.

Upwelling of nutrient-rich waters also occurs up the slopes of *seamounts* (volcanic mountains that may rise thousands of feet from the bottom of the sea but without projection above the sea surface). Such areas are hot spots of fish abundance and diversity (de Forges et al. 2000).

Deep Sea Hydrothermal Vents

According to the now widely accepted *continental drift theory,* some of the continents—notably Africa and South America as one pair and Europe and North America as another—were once single land masses, but drifted apart through the ages. The *mid-oceanic ridges* (Fig. 10-3) are, according to this theory, lines of former contact be-tween continents now thousands of miles apart. Along these ridges and elsewhere, spreading tectonic plates create vents, hot sulfurous springs, and seeps. These **hydro-thermal vents** support unique geothermally powered communities unlike anything else so far discovered in the ocean, as described and illustrated in Chapter 2 (see also

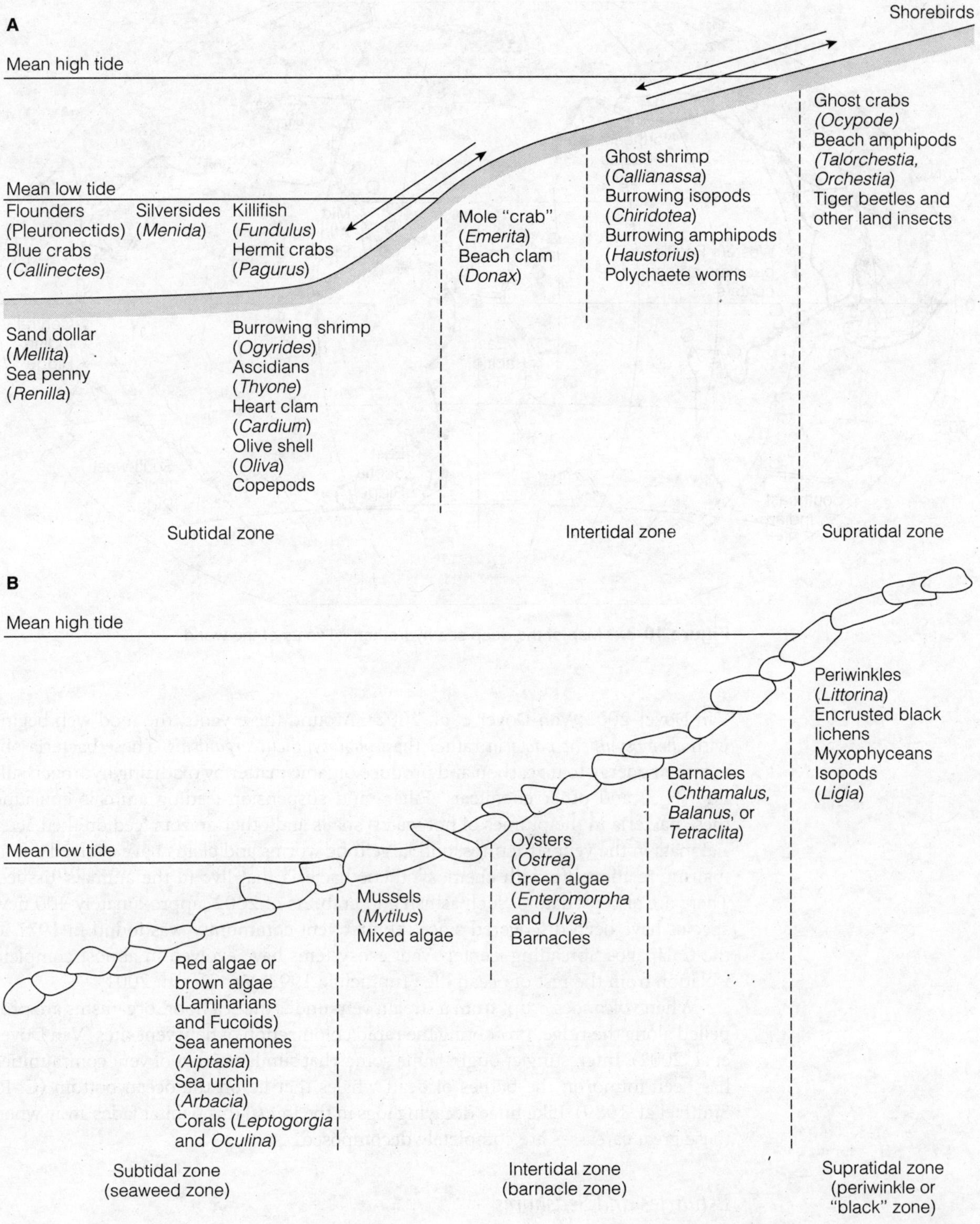

Figure 10-2. Transects of (A) a sandy beach (based on data from Pearse et al. 1942) and (B) a rocky shore (based on data from Stephenson and Stephenson 1952) at Beaufort, North Carolina, showing zones and characteristic dominant species.

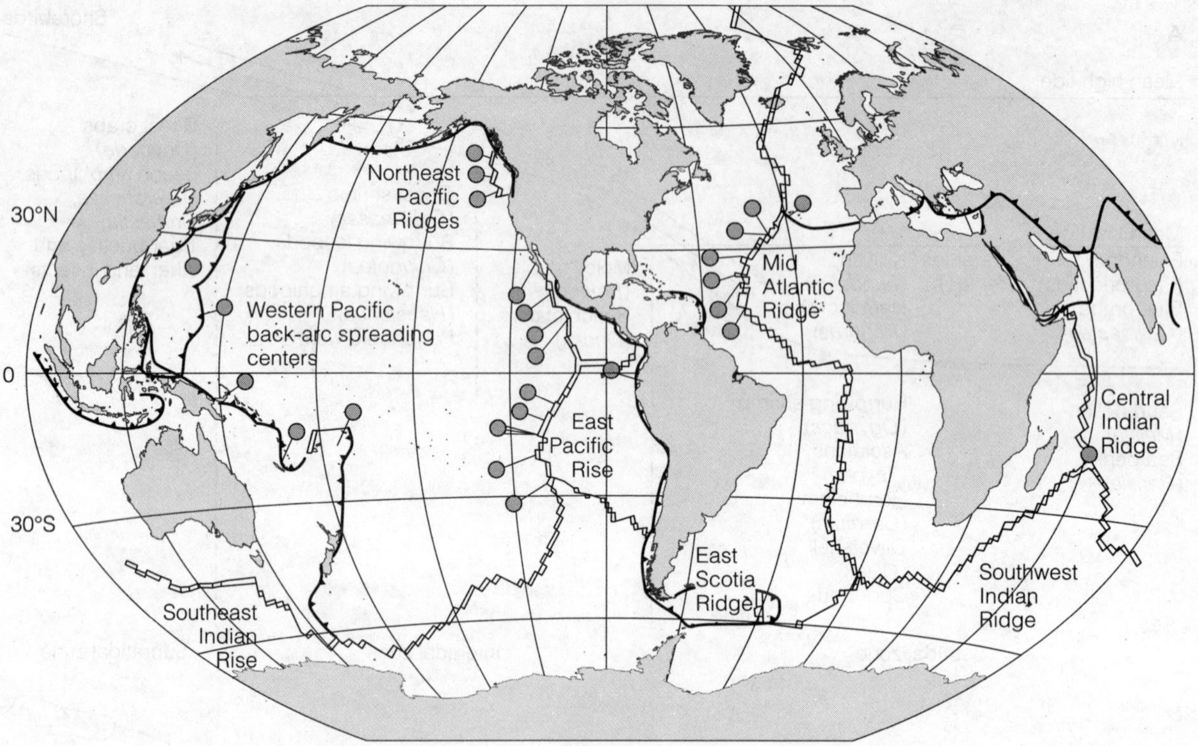

Figure 10-3. Map of the deep sea hydrothermal vents of the world.

Van Dover 2002; Van Dover et al. 2002). Around these vents, the food web begins with *chemosynthetic bacteria* rather than *photosynthetic organisms.* These bacteria obtain their energy to fix carbon and produce organic matter by oxidizing hydrogen sulfide (H_2S) and other chemicals. Filter- and suspension-feeding animals consume these bacteria in the plumes of hot water; snails and other grazers feed on the bacterial mats in the vent structures; and large tube worms and clams have evolved a mutualistic relationship with chemosynthetic bacteria that live in the animals' tissues. There are also predators, such as fish and crab. As of 2002, approximately 400 new species have been discovered since the first vent community was found in 1977 at the Galápagos Spreading Center. Vent ecosystems have evolved in almost complete isolation from the rest of ocean life (Tunnicliffe 1992; Von Damm 2001).

When volcanoes erupt from a stream vent under the sea floor, organisms are propelled along the ridge, promoting the rapid colonization of new vent sites (Van Dover et al. 2002). Interesting enough, fauna somewhat similar to that of vent communities has been found on the bodies of dead whales that lie on the ocean bottom (C. R. Smith et al. 1989). Like huge decaying logs in the forest, this habitat fades away when these great carcasses are completely decomposed.

Estuaries and Seashores

Between the oceans and the continents lies a band of diverse ecosystems. These are not just transition zones but have ecological characteristics of their own (they are true ecotones). Although physical factors such as salinity and temperature are much more variable near the shore than in the ocean itself, food is so plentiful that these regions are packed with life. Along the shore live thousands of adapted species that are not to be found in the open sea, on land, or in freshwater. Two kinds of marine inshore ecosystems—a rocky shore and a tidal estuary dominated by salt marshes—are shown in Figure 10-4.

The word **estuary** (from Latin *aestus,* "tide") refers to a semi-enclosed body of water, such as a river mouth or a coastal bay, where the salinity is intermediate between salt- and freshwater, and where tidal action is an important physical regulator and energy subsidy. Estuaries and inshore marine waters are among the most naturally fertile in the world. Major life-forms of autotrophs are often intermixed in an estu-

Figure 10-4. Two types of coastal ecosystems. (A) A rocky shore on the California coast, characterized by underwater seaweed beds, tide pools containing colorful invertebrates, sea lions, and sea birds. (B) Aerial view of a Georgia salt marsh, showing how it is dissected by small feeder creeks through which the water flows in and out with the tide. Oysters live in reefs in these creeks, and ribbed mussels live at the creek heads. Blue crabs, fishes, and even bottlenosed dolphins swim up the larger feeder creeks to forage at high tide.

A

B

ary and fill varying niches, maintaining a high gross production rate: *phytoplankton; benthic microflora* (algae living in and on mud, sand, rocks, and bodies or shells of animals); and *macroflora* (large attached plants, including seaweeds, submerged eel grasses, emergent marsh grasses, and, in the Tropics, mangrove trees). Estuaries provide the "nursery grounds" (places for young stages to grow rapidly) for most coastal shellfish and fish that are harvested not only in the estuary but offshore as well. Although estuaries and salt marshes do not support a great diversity of species, their net primary productivity is very high. Indeed, some of the most productive fisheries occur in these ecosystem types, and both aquatic and terrestrial species select them as nursery grounds for their young. Organisms have evolved many adaptations to cope with tidal cycles, thereby enabling them to exploit the many advantages of living in an estuary. Some animals, such as fiddler crabs, have internal biological clocks that help to time their feeding activities to the most favorable part of the tidal cycle. If such animals are experimentally removed to a constant environment, they continue to exhibit rhythmic activity synchronous with the tides.

An estuary is often an efficient nutrient trap that is partly physical (differences in salinity retard the vertical but not the horizontal mixing of water masses) and partly biological. This property enhances the capacity of the estuary to absorb nutrients in wastes, provided that organic matter has been reduced by secondary treatment. Estuaries have traditionally been used as free sewage treatment areas for some coastal cities. Since 1970, both awareness of and research on the values of estuaries have greatly increased. Most states bordering estuaries in the United States have enacted legislation designed to protect these natural capital values.

Mangroves and Coral Reefs

Two very interesting and distinctive communities found in tropical and subtropical land-sea ecotones are *mangroves* and *coral reefs*. Both are potential "land builders" that help to form islands and to extend seashores.

Mangroves are among the few emergent woody plants that tolerate the salinity of the open sea. A succession of species often forms a zone from open water to the upper intertidal region (Fig. 10-5A). Extensive prop roots penetrate deeply into the anaerobic mud, bringing oxygen to its depths and providing surfaces for the attachment of clams, oysters, barnacles, and other marine animals. In Central America and Southeast Asia (Vietnam, for example), mangrove forests can have a biomass equal to that of a terrestrial forest. The wood is very hard and commercially valuable. In much of the Tropics, mangrove forests replace salt marshes as intertidal wetlands, and they have many of the same values—for example, serving as nursery grounds for fish and shrimp (W. E. Odum and McIvor 1990).

Coral reefs, depicted in Figure 10-5B, are widely distributed in warm, shallow waters. They form barriers along continents (such as the Great Barrier Reef of Australia), fringing islands, and atolls (horseshoe-shaped ridges that develop on top of extinct underwater volcanoes). Coral reefs are among the most productive and diverse of biotic communities. We suggest that at least once, everyone should don a face mask and snorkel and explore one of these colorful and prosperous "natural cities."

Coral reefs can prosper in nutrient-poor waters because of water flow and a large investment in mutualism. Coral is a plant-animal *superorganism*, as algae, called zooxanthellae, grow inside the tissues of the animal polyp. The animal component gets

A

© Frank B. Golley

B

© Cousteau Society/Getty Images

Figure 10-5. (A) Red mangrove (*Rhizophora mangle*) is typically the outermost species found in the mangrove forest ecotone. Its prop roots provide a substrate for oysters and numerous other marine organisms. (B) Photograph of a coral reef, showing the branching coral structures that develop where currents are gentle.

its "vegetables" from the algae growing in its body and obtains its "meat" by extending its tentacles at night to fish for zooplankton in the water flowing past its limestone house, which the colony builds by depositing calcium carbonate from raw material that is plentiful in the ocean. The plant component of this partnership gets protection and nitrogen and other nutrients from the animal component.

Because humans must learn to prosper in a world of declining resources, the coral ecosystem serves as an example of how to efficiently retain, use, and recycle resources (Muscatine and Porter 1977). However, like a complex and energetic city, a finely tuned coral reef is neither resistant nor resilient to perturbations such as pollution or a rise in water temperature. In recent years, coral reefs around the world have shown signs of stress that may be early warnings of global warming and oceanic pollution. An early sign of stress is the *bleaching* that occurs when the green symbiotic algae leave the coral animal. If the mutualism is not restored, the coral slowly dies of starvation. One theory, yet to be verified, is that the coral ejects the algae, which can survive as free-living plankton, in order to colonize a different strain that is better adapted to the changed environment. In other words, the bleaching is a survival strategy (Salih et al. 2000; Baker 2001).

2 Freshwater Ecosystems

Statement

Freshwater habitats may be conveniently considered in three groups as follows:

- Standing-water or **lentic** (from *lenis*, "calm") ecosystems: lakes and ponds;
- Running-water or **lotic** (from *lotus*, "washed") ecosystems: springs, streams, and rivers; and
- **Wetlands,** where water levels fluctuate up and down, often seasonally as well as annually: marshes and swamps.

Examples of lentic and lotic habitats are pictured in Figure 10-6.

Groundwater, although a large freshwater reservoir and an essential resource for humans, is not generally thought of as an ecosystem because it contains little or no life (except sometimes bacteria). Groundwater does interconnect with all three major aboveground ecosystems and is thereby an important part of the input and output environments of lentic, lotic, and wetland ecosystems.

Explanation and Examples

Freshwater habitats occupy a relatively small portion of the surface of Earth compared with marine and terrestrial habitats, but their importance to humans is greater than their relative area for the following reasons:

- They are the most convenient and cheapest source of water for domestic and industrial needs (we can and probably will get more water from the sea, but at a great cost in terms of energy required to desalinate and the salt pollution created in this process).
- The freshwater components are the "bottleneck" in the hydrological cycle.
- Freshwater ecosystems, along with estuaries, provide more convenient and economical tertiary waste disposal systems. Almost without exception, the largest cities in the world are located near rivers, lakes, or estuaries that serve as free sewage treatment (in other words, located to tap natural capital, rather than economic capital, to provide beneficial services to urban populations). Because this natural resource is so abused, a major effort to reduce this stress should come quickly—otherwise, water could become *the* limiting factor for our species.

Water has several unique thermal properties that combine to minimize temperature changes; thus, the range of temperature variation is smaller and temperature changes occur more slowly in water than in air. The most important of these thermal properties are as follows:

- Water has its maximum density at 4° C; it expands, and hence becomes lighter, both above and below this temperature. This unique property prevents lakes from freezing solid.
- Water has a high specific heat—that is, a relatively large amount of heat is involved in changing the temperature of water. For example, one gram-calorie (gcal) of heat is required to raise the temperature of 1 milliliter (or 1 gram) of water by 1 degree Centigrade (between 15° and 16° C, for instance).

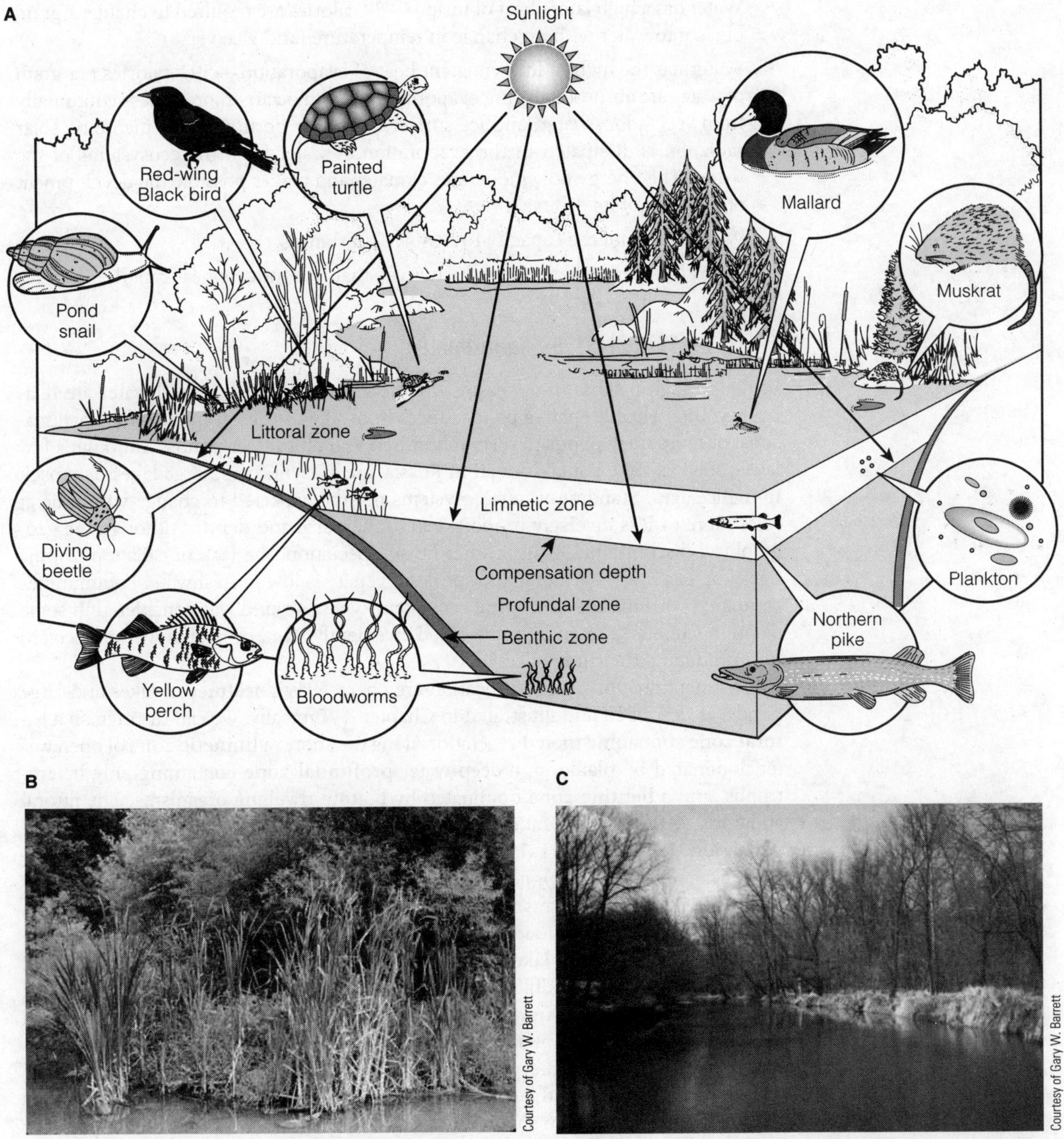

Figure 10-6. (A) Diagram depicting a lentic (pond) ecosystem. The system is composed of five zones (littoral, limnetic, profundal, compensation depth, and benthic). (B) A pond depicting the littoral zone dominated by cattails (*Typha*). (C) A stream (lotic ecosystem) located in Butler County, Ohio.

- Water has a high latent heat of fusion—80 calories are required to change 1 gram of ice into water with no change in temperature (and vice versa).
- Water has the highest known latent heat of evaporation—597 calories per gram of water are absorbed during evaporation, which occurs more or less continually from vegetation, water, and ice surfaces. A major portion of the incoming solar radiation is dissipated in the evaporation of water from the ecosystems of the world. This energy flow moderates climates and makes possible the development of life in all of its diversity.
- Water has immense capacity to dissolve substances.
- Water has a high thermal conductivity—that is, conducts heat rapidly.

Lentic Ecosystems (Lakes and Ponds)

In the geological sense, most basins that now contain standing freshwater are relatively young. The life span of ponds ranges from a few weeks or months for small seasonal or temporary ponds to several hundred years for larger ponds. Although a few lakes, such as Lake Baikal located in Russia, are ancient, most large lakes date back to the ice ages. Standing-water ecosystems may be expected to change with time at rates more or less inversely proportional to their size and depth. Although the geographical discontinuity of freshwater favors speciation, the lack of isolation in time does not. Generally speaking, the species diversity is low in freshwater communities compared with marine or tropical ecosystem-types. A pond was considered in some detail in Chapter 2 as an example of a conveniently sized ecosystem (or mesocosm) for introducing the study of ecology.

Distinct zonation and stratification are characteristic features of lakes and large ponds, as described and illustrated in Chapter 2. Typically, we can distinguish a **littoral zone** containing rooted vegetation along the shore; a **limnetic zone** of open water dominated by plankton, a deep-water **profundal zone** containing only heterotrophs, and a **benthic zone** dominated by bottom-dwelling organisms. The littoral and limnetic zones have a ratio of $P/R > 1$; the profundal zone $P/R < 1$; and the **compensation depth** $P/R = 1$. Life-forms in a pond include *plankton* (free-floating organisms, such as diatoms), *nekton* (free-swimming organisms, such as fish), *benthos* (bottom-dwelling organisms, such as clams), *neuston* (organisms at the surface film of water, such as water striders), and *periphyton* (attached organisms, such as hydras).

In temperate regions, lakes often become *thermally stratified* during summer and again in winter, owing to differential heating and cooling. The warmer, upper part of the lake, or **epilimnion** (from Greek *limnion*, "lake"), becomes temporarily isolated from the cooler, deeper water, or **hypolimnion**, by a **thermocline** that acts as a barrier to the exchange of materials (Figs. 2-3C and D illustrate this zonation and stratification). Consequently, the supply of oxygen in the hypolimnion and of nutrients in the epilimnion may run short. During spring and fall, as the entire body of water approaches the same temperature, mixing occurs. Blooms of phytoplankton often follow these seasonal turnovers, as nutrients from the bottom become available in the photic zone. The **photic zone** is the lighted portion of a lake or ocean inhabited by phytoplankton. In warm climates, mixing may occur only once a year (in winter); in temperate biomes, mixing typically occurs twice a year (*dimictic*).

Primary production in standing-water ecosystems depends on the chemical nature of the basin and on the nature of imports from streams or land (that is, inputs from the watershed) and is generally inversely related to depth. Accordingly, the yield of fish per unit of water surface area is greater in shallow than in deep lakes, but the deep lakes may have larger individual fish. Lakes are often classified as either **oligotrophic** (low in nutrients) or **eutrophic** (high in nutrients) on the basis of productivity. Because a biologically poor lake is preferable to a fertile one from the standpoint of water quality for domestic use and recreation, there is a paradox. In some parts of the biosphere, humans are increasing its fertility to feed themselves, whereas in other places, they are preventing its fertility (by removing nutrients or poisoning plants) to maintain what some societies traditionally have considered an aesthetic environment. For example, a fertile green pond that can produce many fish is not always considered a desirable swimming pool.

By constructing artificial ponds and lakes, termed **impoundments,** humans have changed the landscape in regions that lack natural bodies of water. In the United States, almost every farm now includes at least one farm pond, and large impoundments have been constructed on practically every river. Much of this activity is beneficial, but the impoundment strategy of covering fertile land with a body of water that cannot yield much food may not always be the best long-term land use. Standing waters are generally less efficient at oxidizing waste than are running waters. Unless the watershed is well vegetated, erosion may fill up an impoundment in a human generation.

The heat budget of impoundments may differ greatly from that of natural lakes, depending on the design of the dam. If water is released from the bottom—as is the case with dams designed for hydroelectric power generation—cold, nutrient-rich but oxygen-poor water is exported downstream, whereas warm water is retained in the lake. The impoundment then becomes a *heat trap* and *nutrient exporter,* in contrast to natural lakes, which discharge from the surface and, therefore, function as *nutrient traps* and *heat exporters.* Accordingly, the type of discharge greatly affects downstream conditions.

Lotic Ecosystems (Streams and Rivers)

Differences between running and standing water generally revolve around a triad of conditions: (1) current is much more of a major controlling and limiting factor in streams; (2) land-water interchange is relatively more extensive in streams, resulting in a more "open" ecosystem and a heterotrophic type of community metabolism when the size of the stream is small; and (3) oxygen tension is generally high and more uniform in streams, and there is little or no thermal or chemical stratification, except in large, slow-moving rivers.

The *river continuum concept* (Cummins 1977; Vannote et al. 1980), involving longitudinal changes in community metabolism, biotic diversity, and particle size from headwater to river mouth (see Chapter 4; Fig. 4-9), describes how biotic communities adjust to changing conditions. In a given stretch of stream, two zones are generally apparent:

1. A **rapids zone** has a current great enough to keep the bottom clear of silt or other loose material, thus providing a firm substrate. This zone is occupied by specialized organisms that become firmly attached or cling to the substrate (such as black fly and caddis fly larvae) or, in the case of fish, that can swim against the current or cling to the bottom (such as trout or darters).
2. A **pool zone** has deeper water, where the velocity of the current is reduced, so that sand and silt settle, providing a soft bottom favorable for burrowing and

swimming animals, rooted plants, and, in large pools, plankton. In fact, the communities of pools in large rivers resemble those of ponds.

Rivers in their upper reaches are generally eroding; they cut into the substrate, so a hard bottom predominates. As rivers reach base level in the lower reaches, sediments are deposited and floodplains and deltas that are often extremely fertile are built up. In terms of the chemical composition of the water, lotic systems can be divided into two types: (1) *hard-water* or *carbonate* rivers, with 100 or more ppm dissolved inorganic solids; and (2) *soft-water* or *chloride* rivers, with less than 25 ppm dissolved solids. The water chemistry of carbonate rivers is controlled largely by rock weathering, whereas atmospheric precipitation is the dominant factor in chloride rivers. *Humic* or *black-water streams,* with high concentrations of dissolved organic material, represent still another class of streams that are found in warm lowlands. Several studies and reviews of the food-chain energetics of streams, with emphasis on fish, have been compiled by, for example, Cummins (1974), Cummins and Klug (1979), and Leibold et al. (1997).

Springs hold a position of importance as study areas that is far out of proportion to their size and number. Some of the classic whole-system studies on springs, for example, are on the large limestone springs located in Florida (H. T. Odum 1957), the small cold-water springs located in New England (Teal 1957), and the hot springs located in Yellowstone (Brock and Brock 1966; Brock 1967).

Freshwater Wetlands

A **freshwater wetland** is defined as any area covered by shallow freshwater for at least part of the annual cycle; accordingly, wetland soils are saturated with water continually or for part of the year. The key factor that determines the productivity and species composition of the wetland community is the **hydroperiod**—that is, the periodicity of water-level fluctuations. Freshwater wetlands can thus be classified as "pulse-stabilized, fluctuating water level ecosystems," as are intertidal marine and estuarine ecosystems.

Wetlands tend to be very open systems and can be conveniently classified according to their interconnections with deep water or upland ecosystems, or both, as follows:

- **Riverine wetlands** are located in low-lying depressions (oxbows) and floodplains associated with rivers. The bottomland hardwood forests on the floodplains of large rivers are among the most productive of natural ecosystems, as are the freshwater tidal marshes along the lower reaches of large rivers in the Coastal Plains of the United States.
- **Lacustrine** (from *lacus,* "lake") **wetlands** are associated with lakes, ponds, or dammed river channels. They are periodically flooded when these deeper bodies of water overflow.
- **Palustrine** (from *palus,* "marsh") **wetlands** include what are variously called marshes, bogs, fens, wet prairies, and temporary ponds that occur in depressions not directly connected with lakes or rivers (although they may be in old riverbeds or filled ponds or lake basins). *Fen wetlands* are only slightly acidic and are typically dominated by sedges, whereas *bog wetlands* are very acidic, characterized by accumulation of peat, and dominated by sphagnum moss. Such wet-

Figure 10-7. Microbial decomposi-tion and recycling in wetland sediments. The four major anaerobic decomposers gasify–and thereby recycle into the atmosphere–carbon, nitrogen, and sulfur. Phosphate is also converted from insoluble sulfide forms to soluble forms that are more easily available to organisms (after E. P. Odum 1979).

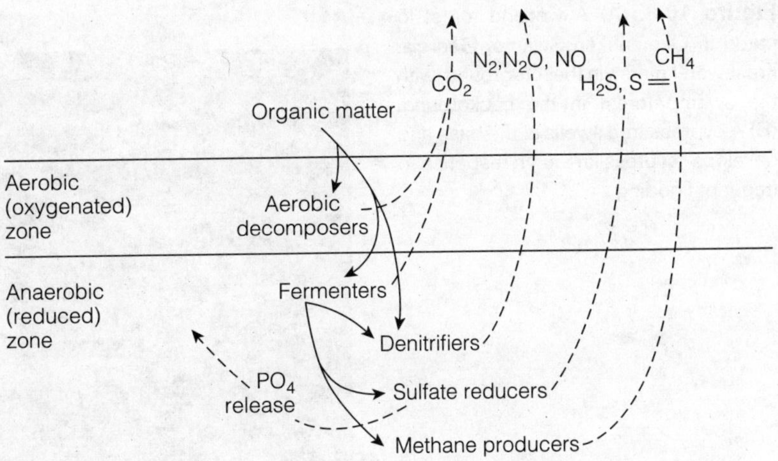

lands are widely scattered across the landscape, especially in formerly glaciated regions. They are generally vegetated with various submerged aquatic macrophytes, emergent marsh plants, and shrubs. *Palustrine marshes,* dominated by emergent herbaceous vegetation, are often a prime breeding habitat for waterfowl and other aquatic or semiaquatic vertebrates. Wetlands dominated by woody vegetation or forested wetlands are commonly called *swamps* in the United States. A deep water swamp dominated by bald cypress (*Taxodium distichum*), tupelo (*Nyssa sylvatica*), and swamp oak (*Quercus bicolor*) is an example.

Although wetlands occupy only about 2 percent of the surface area of Earth, they are estimated to contain 10 to 14 percent of the carbon (Armentano 1980). Wetland soils, such as the histosols, may contain up to 20 percent carbon by weight, and of course the peats are even more carboniferous. Draining of wetlands for conversion to agriculture releases large quantities of CO_2 to the atmosphere, thus contributing to the "CO_2 problem" (see Chapter 4). The aerobic-anaerobic stratification of wetland sediments (including saltwater marshes) is also important out of proportion to their area for the part they play in the global cycling of sulfur, nitrogen, and phosphorus as well as carbon. Figure 10-7 summarizes key aspects of microbial decomposition and recycling in wetland and shallow marine sediments.

During the latter part of the twentieth century, public attitudes toward wetlands changed dramatically as ecological and economic studies revealed previously unrecognized values. No longer are wetlands always viewed as wastelands to be destroyed or modified. Although some progress has been made in preservation, especially of the coastal wetlands, much remains to be done in the legal and political arenas.

It is significant that rice culture, one of the most productive and dependable of agricultural systems yet devised by humans, is actually a type of freshwater marsh ecosystem. The flooding, draining, and careful rebuilding of the rice paddy each year has much to do with the maintenance of the continuous fertility and high production of the rice plant, which itself is a kind of cultivated marsh grass. This flooding process is similar to the hydroperiod (duration, frequency, and depth) of natural wetlands. The hydroperiod influences seed germination, plant composition, and productivity of wetland ecosystems.

Figure 10-8. (A) A wetland forest located in Okefenokee Swamp, Georgia. Freshwater marsh in the foreground, with the swamp forest in the background. (B) A cypress tree swells at the base and develops "cypress knees" in response to frequent flooding.

A

Courtesy of Eugene P. Odum

B

© Eastcott Momatiuk/National Geographic/Getty Images

Forested Wetlands

Swamp and floodplain forests occur in river bottoms, often intermixed with marshes, especially where large rivers cross coastal plains. They are also found in large depressions (see Okefenokee Swamp, Fig. 10-8A), lime sinks, and other low-lying areas that are flooded at least some of the time. As with marshes, hydrology plays a major role in determining species composition and productivity. Bald cypress (*Taxodium distichum*) and water tupelo (*Nyssa aquatica*) are the trees best adapted to flooding, whereas bottomland hardwoods (lowland species of oak, ash, elm, and gum) do best where flooding is pulsed, as on floodplains. The knees of bald cypress trees conduct air from the atmosphere to roots when a swamp is flooded and the waterlogged sediments contain little or no free oxygen (Fig. 10-8B). The greatest productivity occurs where the soil surface is flooded in winter or spring and is relatively dry during most of the growing season.

Tidal Freshwater Marshes

In low-lying coastal plains, tides extend inland on large rivers. For example, meter-amplitude tides occur on the Potomac River beyond Washington, D.C., and on the James River as far inland as Richmond, Virginia, creating a unique habitat of freshwater wetland. The biota benefit from the tidal pulses but do not have to contend with salt stress. Fleshy, low-fiber vegetation is produced in freshwater tidal marshes in summer, which then decomposes back to mud in winter, in contrast to salt marshes

A

Courtesy of Lawrence Pomeroy

B

Courtesy of Lawrence Pomeroy

C

Courtesy of Lawrence Pomeroy

Figure 10-9. Freshwater and salt-water tidal marshes. (A) A northern salt marsh, located in Tuckerton, New Jersey, dominated by *Spartina alterniflora.* Open areas of standing water on the high marsh are typical. (B) Low tide on a tidal creek in a Georgia salt marsh. The bare mud banks are covered by a "secret garden" of microscopic algae—so small that they are visible only as a brown film over the gray mud—which are a major base of the estuarine food chain consisting of benthic microinvertebrates, shrimp, and fishes. The tall grass, *Spartina alterniflora,* on the higher marsh is grazed only by a few insects, so it mostly dies in place and is the basis of a detritus food chain. (C) A freshwater tidal marsh with cypress trees bordering on the Altamaha River, Georgia. Marsh soils denitrify much of the nitrogen that enters the river upstream from cities and agricultural landscapes.

where the more fibrous marsh grass remains standing all year round (Figs. 10-9A and B). The anaerobic microbial populations are dominated by *sulfide reducers* in salt marshes and by *methane producers* in freshwater marshes (Fig. 10-9C). In general, primary production in salt marshes culminates in aquatic animals, such as fish, shrimp, and shellfish. In contrast, animals in freshwater tidal marshes tend to be semi-aquatic amphibians, reptiles (alligators), birds (ducks, herons, and other wad-ers), and furbearers. For further comparisons of these two marsh types, see W. E. Odum (1988).

3 Terrestrial Biomes

Statement

F. E. Clements and V. E. Shelford (1939) introduced the *biome* concept as a classification of world vegetation patterns. They considered the biome concept to include both the major plant formations and their associated animal life as a biotic unit or level of ecological organization. *Biome* is typically defined as a major regional ecological community of plants and animals. We define **biome** as the level of organization between the landscape and global (ecosphere) levels of organization (see Fig. 1-3).

Before the biome concept became widely accepted by ecologists, C. Hart Merriam developed a *life zone* classification. His **life zone** concept (C. H. Merriam 1894) was based on the relationship between climate and vegetation, and was best applied to mountainous regions, where temperature changes accompany both changes in altitude and vegetation.

Other methods of classification include Whittaker's (1975) *patterns of world plant formations,* based on the relationship of mean annual temperature to mean annual precipitation (Fig. 10-10). Whittaker plotted the boundaries of major vegetation types with respect to average temperature and precipitation in climates between those of forest and desert community types. He suggested that factors such as fire, soil, and climatic seasonality determine whether grassland, shrubland, or woodland develops as a major community type.

Holdridge (1947, 1967) presented a more detailed and sophisticated approach to the relationship of vegetation to climate (Fig. 10-11). The **Holdridge life zone system,** used for classifying plant formations, is determined by a gradient of mean annual biotemperatures with dimensions of latitude and altitude, the ratio of potential evapotranspiration to annual precipitation, and total annual precipitation. There are three levels of classification within the Holdridge system: climatically defined *life zones;* subdivisions of life zones termed *associations,* based on local environmental conditions; and local subdivisions based on actual cover or land use. Holdridge defined an **association** as a unique ecosystem type or natural unit of vegetation, often dominated by particular species, thus providing a relatively uniform species composition. The Holdridge life zone classification differs from other classifications in that it specifically defines a relationship between climate and vegetation (ecosystem) distribution.

Another classification for mapping the biotic world is the concept of the *ecoregion* developed by Bailey (1976, 1995, 1998). **Ecoregions** are defined as ecosystems based

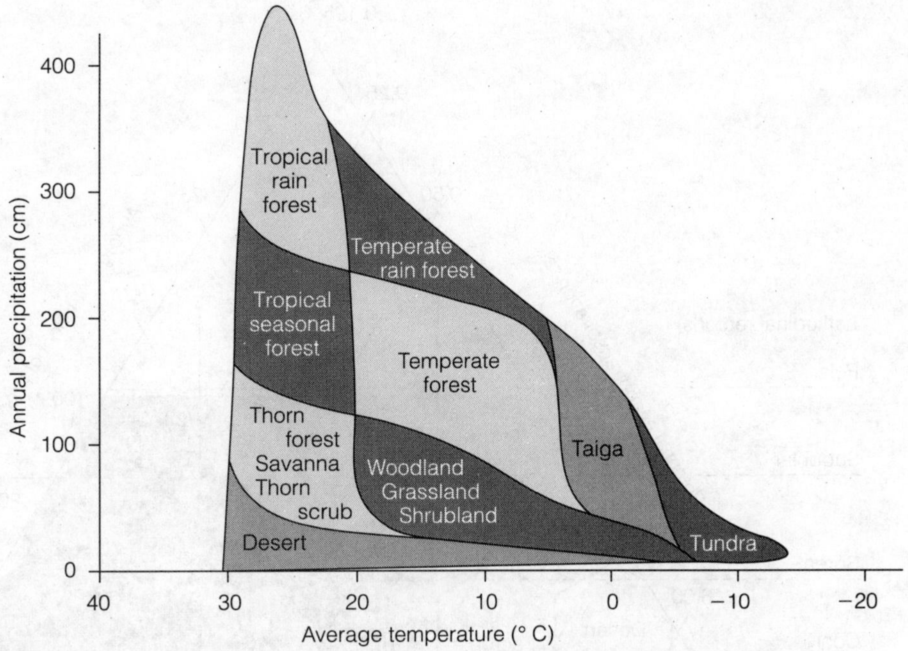

Figure 10-10. Whittaker's (1975) patterns of world plant formations, based on the relationship of mean annual precipitation (cm) to mean annual temperature (° C).

on a continuous geographic or landscape area across which the interactions of climate, soil, and topography are sufficiently uniform to permit the development of similar types of vegetation. Classification units include *domains, divisions,* and *provinces.* An advantage of the ecoregion concept is that it embraces both the terrestrial and the oceanic (see Bailey 1995 for a description and hierarchy of the ecoregions of North America). Because all systems operate within the hierarchy of larger systems, a knowledge of larger systems allows ecologists to better understand the smaller systems. An understanding of this hierarchy gives ecologists a higher degree of predictability regarding land-use management and natural resource development.

Students of ecology need to be familiar with an array of large-scale classification systems depending on their descriptive and sampling needs. We have elected to focus on the biome concept to illustrate climatic climax vegetation. The biome concept is especially important because it encompasses both major plant and animal relationships at larger scales.

Explanation and Examples

The **life-form** (grass, shrub, deciduous tree, coniferous tree, and so on) of the climatic climax vegetation is the key to delimiting and recognizing terrestrial biomes. Thus, the climax vegetation of the grassland biome is grass, although the species vary topographically in different parts of the biomes and on different continents. The climatic climax vegetation is the key to classification, but edaphic climaxes and developmental stages that may be dominated by other life-forms are an integral part of a

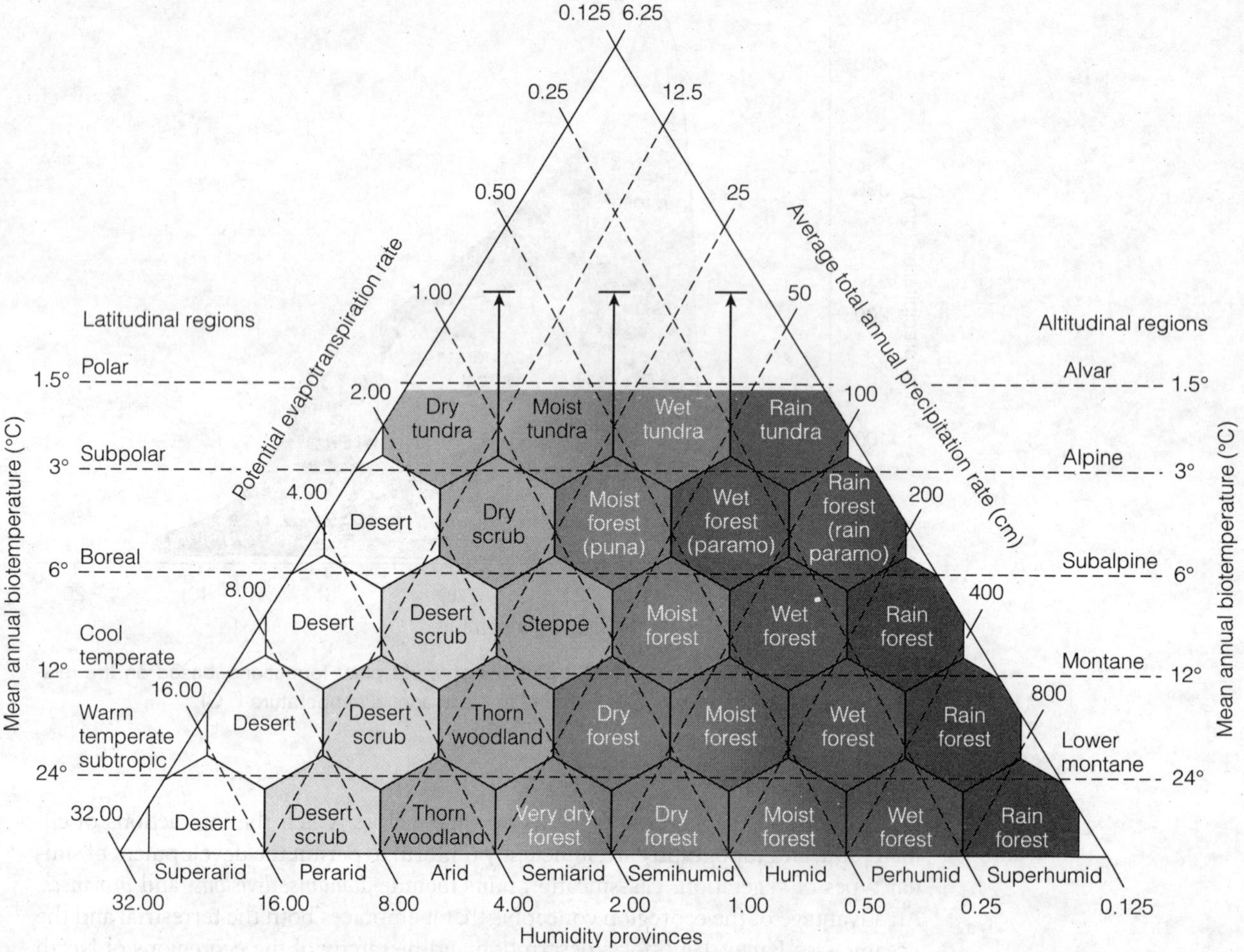

Figure 10-11. The Holdridge life zone classification system for classifying plant formations and associations. (Reprinted from diagram based on Holdridge, L. R. 1967. Determination of world plant formations from simple climatic data. *Science* 130:572. Copyright 1967 AAAS.)

biome. For example, grassland communities may be developmental stages in a forest biome, and riparian forests may be part of a grassland biome.

Mobile animals help to integrate different vegetation strata and stages. Birds, mammals, reptiles, and many insects move freely between subsystems and between developmental and mature stages of vegetation, and migratory birds move seasonally between biomes on different continents. In many cases, life histories and seasonal behavior are organized so that a given animal species will occupy several—often quite different—vegetative types. Large mammalian herbivores—deer, moose, caribou, antelope, bison, domestic bovine—are a characteristic feature of terrestrial biomes. Many of these herbivores are *ruminants,* which possess a remarkable nutrient-regenerating microecosystem (microcosm): the *rumen,* in which anaerobic microorganisms can break down and enrich the lignocellulose that constitutes a large part of terrestrial plant biomass. Likewise, the detritus food chain, featuring fungi and soil decomposer animals, is a major energy flow pathway, as is the mutualism between plant roots and mycorrhizae, nitrogen fixers, and other microorganisms.

The biomes of the world are shown in Figure 10-12A, and the major biomes of North America are depicted in Figure 10-13B. Climographs for six major biomes are compared in Figure 10-13.

During the 1970s, and continuing into the 1980s, several of the major biomes (grassland, temperate deciduous forest, northern coniferous forest, tundra, and desert) were subjected to interdisciplinary team research as part of the U.S. contribution to the International Biological Program (IBP; for a general review of this program, see Blair 1977; Loucks 1986). Ecosystem- and landscape-level research sponsored by the National Science Foundation's Long-Term Ecological Research (LTER) program has continued on many of the IBP sites (see Callahan 1984; Hobbie 2003; for details regarding the LTER research sites).

In terms of flora and fauna, biogeographers divide the world into five or six

A

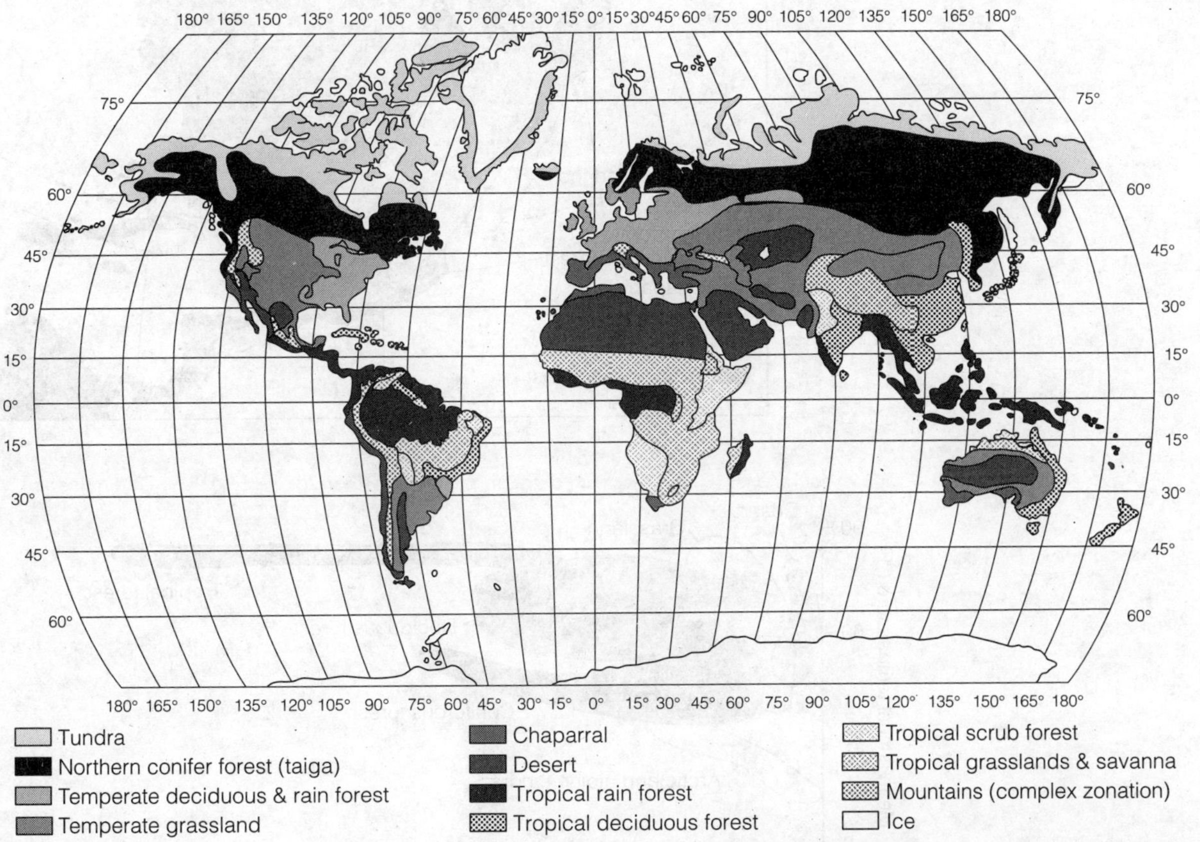

Figure 10-12. (A) Major biomes of the world. Several biomes in different parts of the world may be isolated in different biogeographical regions and, therefore, may be expected to have ecologically equivalent but often taxonomically unrelated species. (B) Schematic map of the major biomes of North America. *(continued)*

Figure 10-13. (*continued*)

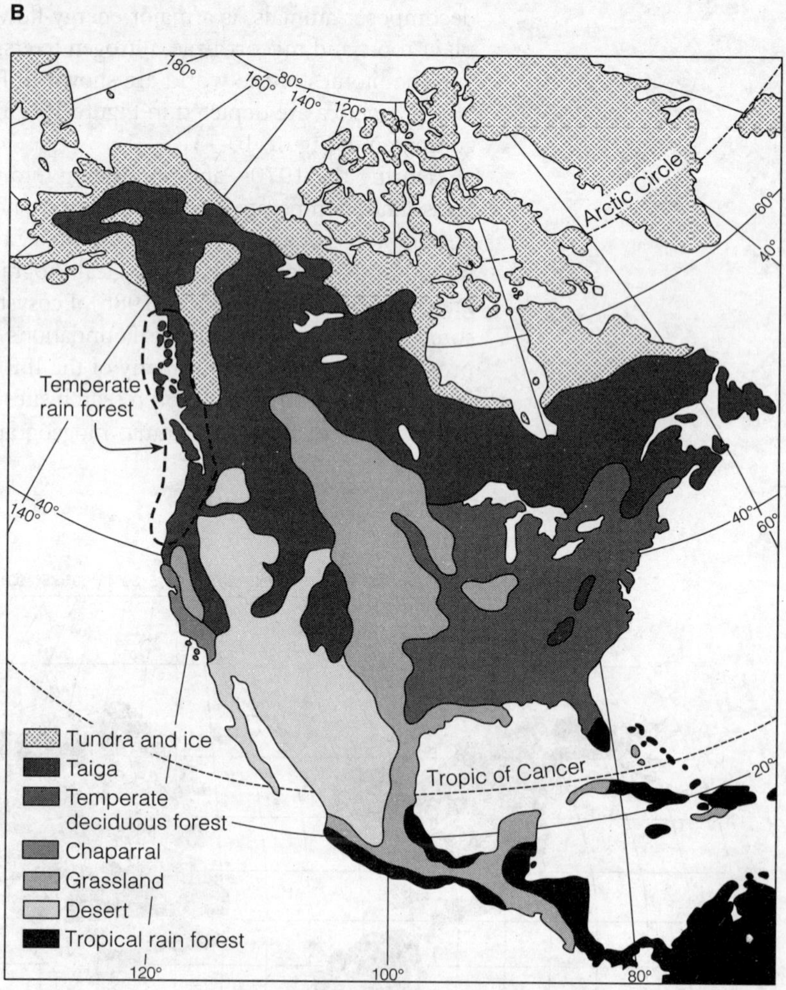

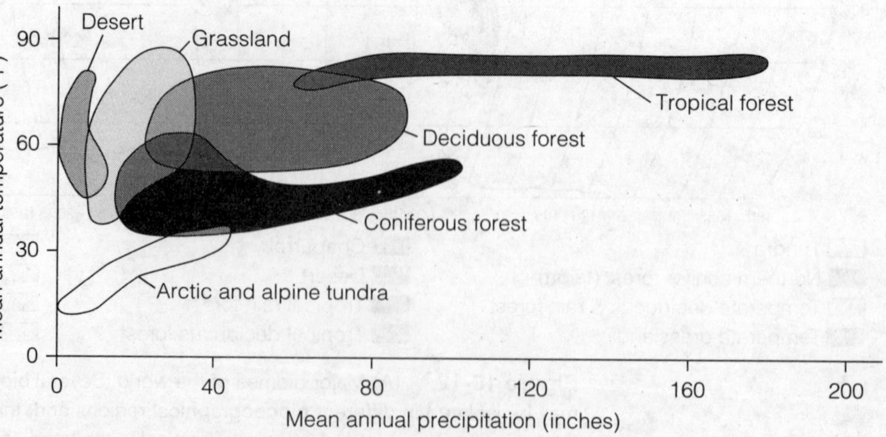

Figure 10-13. Distribution of six major biomes in terms of mean annual temperature and mean annual rainfall. (From the National Science Foundation.)

major regions that correspond roughly to the major continents. Australia and South America are the most isolated regions. Accordingly, ecologically equivalent species in biomes on these continents can be expected to be taxonomically quite different (see Chapter 7).

Tundra—Arctic and Alpine

Between the forests to the south and the polar ice caps to the north lies a circumpolar band of about 5 million acres (>2 million ha) of treeless country (Fig. 10-14). Smaller but ecologically similar regions found above tree limitation altitudes on high

A

Courtesy of R. E. Shanks, E. E. Clebsch, and J. Koranda

B

Courtesy of R. E. Shanks, E. E. Clebsch, and J. Koranda

Figure 10-14. Two views of the Tun-dra in July on the Coastal Plain near the Arctic Research Laboratory, Point Barrow, Alaska. (A) A broad swale at the head of a stream about 3 kilometers from the coast, dominated by Arctic grass (*Dupontia fischeri*) and sedge (*Carex aquatilis*) rooted in a peaty layer of half-bog soil. This area is frequently regarded as climax on low tundra sites near the coast. Note the sample quadrats and an exclosure (fenced area) to keep out lemmings. (B) A site about 16 kilometers inland, showing characteristic polygonal ground; ice wedges underlying the troughs contribute to the raised polygons. The white fruits seen in the foreground are cottongrass (*Eriophorum scheuchzeri*).

mountains, even in the Tropics, are termed **alpine tundra.** In both North America and Eurasia, the boundary between tundra and forest lies farther north in the west, where the climate is moderated by warm westerly winds.

Low temperatures and short growing seasons are the major limiting factors to life in the Tundra; precipitation may also be low but is not limiting because of the low evaporation rate. All but the upper few centimeters of ground remain frozen during the summer. The permanently frozen deeper soil is called **permafrost.** The Tundra is essentially a wet arctic grassland, with vegetation consisting of grasses, sedges, dwarf woody plants, and lichens ("reindeer moss") on drier locations. **Low tundra** (as on the Alaskan Coastal Plain) is characterized by a thick, spongy mat of living and very slowly decaying vegetation, often saturated with water and dotted with ponds on which numerous species of migratory shorebirds and other waterbirds breed during the short summer. **High tundra,** especially where there is considerable relief, is covered by a much scantier growth of lichens and grasses. Although the growing season is short, the long sunny photoperiods allow a respectable amount of primary production (up to 5 grams dry matter/day) at favorable sites such as Point Barrow, Alaska.

Combined aquatic (including fertile arctic waters) and terrestrial productivity supports not only large numbers of migratory birds and insects during the open season but also permanent residents that remain active throughout the year. Large animals, such as musk ox, caribou, reindeer, polar bears, wolves, foxes, and predatory birds, such as the snowy owl, along with lemmings and other small animals that tunnel about in the vegetation mantle, are some of the permanent residents. The large animals are highly migratory, whereas many of the smaller animals "cycle" in abundance, as described in Chapter 6, because there is not enough net production in any one area to support them all the year round. Where humans "fence in" the animals or select nonmigratory strains for domestication, as with domestic reindeer, overgrazing is inevitable, unless rotation of pastures is employed to substitute for migratory behavior. The special fragility of the Tundra needs to be recognized as mineral exploitation and other human impacts increase; the thin living mat is easily broken and is slow to recover. The building of the Alaskan pipeline provided many object lessons. Pressures continue from the oil industry to drill for oil in the 19.8 million acres (7.9 million ha) of the Arctic National Wildlife Refuge as we move into the twenty-first century.

Polar and High-Mountain Ice Caps

Ice caps are extreme environments, but they are not quite lifeless. Green *ice algae* and a variety of heterotrophic microorganisms live in and under the ice. Survival under these conditions requires a complex suite of physiological and metabolic adaptations, which suggests that there may be similar life-forms on ice-covered extraterrestrial bodies (Thomas and Dieckmann 2002). Under the Antarctic ice cap, there are unfrozen lakes that are known to contain microorganisms. One such lake, known as Lake Vostok, lies 4000 meters below the ice cap surface and as of 2002 has yet to be explored (Gavaghan 2002).

All the highest mountains of Earth, even in the Tropics (such as Mount Kilimanjaro located in Africa), have permanent ice caps that contain similar microbial life. The mountaintop ice caps have an additional source of energy in the form of detritus blown up by the wind from the vegetation below the cap. Swan (1992) suggested that these ice caps be termed the "Aeolian biome" (from *aeolus,* "wind").

A record of past climates can be obtained from cores drilled in the polar regions. Humanity, of course, has a vital interest in polar ice caps, because if they melted, the sea would rise, to the detriment of at least half the world's human population.

Northern Coniferous Forest Biomes

Stretching as broad belts all the way across both North America and Eurasia are the vast northern evergreen forest regions. This biome is also frequently termed the **Taiga** or the **Northern Circumpolar Boreal Forest.** Extensions occur in the mountains, even in the Tropics. The identifying life-forms are the coniferous, needle-leaved tree species, especially evergreen spruce (*Picea*), fir (*Abies*), and pine (*Pinus*), as well as the deciduous, coniferous larch (*Larix*). A dense shade thus exists the year around, often resulting in poor development of shrub and herb layers. However, the continuous blanket of chlorophyll results in a fairly high annual production rate despite low temperatures during half of the year. The coniferous forests are among the great lumber-producing regions of the world. Coniferous needles decay very slowly, and the soil develops a highly characteristic podzolic profile. The soil may contain a fair population of small organisms, but few larger ones compared with deciduous forest or grassland soils. Many of the larger, herbivorous vertebrates, such as the moose, snowshoe hare, and grouse, depend, at least in part, on broad-leaved developmental communities for their food. The seeds of the conifers provide essential food for many animals, such as squirrels, siskins, and crossbills.

As in the Tundra, seasonal periodicity in the Taiga is pronounced, and populations tend to pulse or cycle. Coniferous forests are also subject to outbreaks of bark beetles and defoliating insects (such as sawflies and budworms), especially where stands have only one or two dominant species. However, as noted in Chapter 6, such outbreaks are part of the continuous cycle of development to which coniferous forest ecosystems are adapted.

Coniferous forests of a distinctive type occur along the West Coast of North America from central California to Alaska, where temperatures are higher, seasonal range is relatively small, and the humidity is very high (Fig. 10-16). Although dominated by conifers, these forests are quite different florally from the northern coniferous forest. Accordingly, these "temperate rain forests," as they are often known, could well be considered a separate biome (see Fig. 10-13B).

Western hemlock (*Tsuga heterophylla*), western arborvitae (*Thuja plicata*), grand fir (*Abies grandis*), and Douglas fir (*Pseudotsuga menziesii*)—the latter on drier sites or subclimax on wet sites—are the four most dominant trees in the Puget Sound area, where the temperate rain forest reaches its greatest development. Southward, the magnificent coastal redwoods (*Sequoia sempervirens*) are found, and northward, the Sitka spruce (*Picea sitchensis*) is prominent. Unlike the drier and more northern coniferous forests, the understory vegetation is well developed wherever any light filters through; mosses and other moisture-loving lesser plants are abundant. Epiphytic mosses are the ecological equivalent of the epiphytic bromeliads of moist tropical forests. The standing crop of producers is indeed impressive, and as can be imagined, the production of lumber per unit area is potentially very high if the harvest regeneration and nutrient cycles can be maintained. As with all ecosystems in which such a large percentage of nutrients may be tied up in the biomass, overexploitation may reduce future productivity.

Other subdivisions of the northern coniferous forest biome that could be considered separate biomes are the piñon-juniper woodland, or "pygmy conifers," and the

Figure 10-15. Three kinds of coniferous forests. (A) A high-altitude forest of Engelmann spruce and subalpine fir in Colorado. Note the large amount of litter that accumulates because of low temperatures and long seasonal periods of snow cover. (B) A spruce forest in Idaho with one of its chief developmental stages, aspen, a broad-leaved species whose leaves turn golden yellow in the fall (light-colored stand, left and center of the photograph). (C) An example of a moist coniferous forest, often termed *temperate rain forest,* in Humboldt County, California. Note the large size of the trees and luxuriant ground cover of ferns and other herbaceous plants.

ponderosa pine (*Pinus ponderosa*) forests that occur at intermediate altitudes between the grassland and spruce zones located in Colorado, Utah, New Mexico, and Arizona. These stands are generally open and have a considerable amount of herbaceous growth, especially where fire is part of the environment.

Temperate Deciduous Forests

Deciduous forest communities (Fig. 10-15) occupy areas with abundant, evenly distributed rainfall—30 to 60 inches (75 to 150 cm)—and moderate temperatures that have a distinct seasonal pattern. Temperate deciduous forests originally covered eastern North America, most of Europe, and part of Japan, Australia, and the tip of South America. Deciduous forest biomes are thus more isolated from one another than the

Tundra and northern coniferous forests, and their species composition will, of course, reflect this degree of isolation. The contrast between winter and summer is great, because the trees and shrubs are without leaves for part of the year. Herb and shrub layers of stratification tend to be well developed, as are the soil biota. Many plants produce pulpy fruits and nuts, such as acorns and beechnuts. Animals of the original forest of North America included the white-tailed deer, bear, gray and fox squirrels, gray and red foxes, bobcat, and wild turkey. The red-eyed vireo, wood thrush, tufted titmouse, ovenbird, and several species of woodpeckers are characteristic small birds of the mature stages. The development of temperate deciduous forests involves lengthy ecological successions, as reviewed in Chapter 8. Transient grassland or old-field vegetation is characteristic of early seral stages, and pines are often prominent in intermediate and edaphic stages, especially in the southeastern United States.

The deciduous forest biome is a crucial biotic region of the world, because West-

A

Courtesy of Gary W. Barrett

B

Courtesy of Eugene P. Odum

C

Courtesy of Tall Timbers Research Station

Figure 10-16. (A) A virgin stand of deciduous temperate forest in Hueston Woods State Park, near Oxford, Ohio. This American beech–sugar maple (*Fagus grandifolia–Acer saccharum*) climax forest appears rather open due to the heavy top canopy. (B) An excellent stand of mixed hardwoods (oak-hickory-poplar) located in Anne Arundel County, Maryland. Note the underbrush cover. (C) A remnant, virgin longleaf pine (*Pinus palustris*) forest located near Thomasville, Georgia. Frequent controlled burning has maintained the open, park-like condition and prevents the invasion of fire-tender hardwood trees.

ern civilization has achieved extensive development in these areas. This biome is therefore greatly modified, and much of it has been replaced by cultivated and forest-edge communities.

The deciduous forest biome of North America has many important subdivisions, which have different mature forest types. Some of these are,

- The *beech-maple forest* of the north central region.
- The *maple-basswood forest* of Wisconsin and Minnesota.
- The *oak-hickory forest* of the western and southern regions.
- The *oak-chestnut forest* of the Appalachian Mountains (currently an oak forest because the chestnut was wiped out by fungus diseases).
- The *diverse mixed mesophytic forest* of the Appalachian plateau.
- The *pine edaphic forest* of the southeastern coastal plain, maintained by fire and nutrient-poor sandy soils where evergreen forests have a competitive advantage over deciduous forests.

Each of these has distinctive features, but many organisms, especially the larger animals, range through two or more of these subdivisions. Deciduous forests of western Europe have relatively few species (diversity premixed mesophytic type, not glaciated), and comparable forests in eastern Asia are the richest in species of the temperate forests of the world.

Harvesting the Forests

Timber production and the practice of forestry pass through two phases. The first phase involves the harvest of net production that has been stored as wood over a period of many years. When this accumulated old growth of the past has been used up, we must adjust forestry practice to harvest no more than the annual growth if we expect to have wood products in the future. In much of the western United States, the first phase is still under way; that is, the annual timber harvest exceeds the annual growth. In contrast, the second phase has been reached in much of the eastern United States. Most of the old timber has been cut; hence, forestry practices are increasingly concerned with young forests, where the harvest now balances or is less than the annual growth. Globally, outside of the tropical forests, forest cover and growth are on the increase (see review by Moffat 1998b).

Although the annual net production in a young forest is often greater than that of an old forest, the quality of the wood for lumber use is not as high, because the wood of fast-growing young trees is not as dense as that of slow-growing older trees. As in so many situations, the dichotomy between quantity and quality has to be recognized; rarely can we have both. How timber is cut may depend on how timber is taxed. If standing timber is taxed on its market value each year, the owner (an individual or a timber company) is encouraged to harvest early in order to reduce taxes. If, on the other hand, the timber is not taxed until it is cut (and revenue obtained), there is an incentive to let the stand grow higher-quality wood.

In recent years, a controversy has been raging over massive clear-cuts of old growth forests in the Pacific Northwest, especially in the national forests. Jerry Franklin of the University of Washington is a spokesperson for a new generation of foresters who advocate managing a forest not just as a wood production system, but as a complex ecosystem with multiple uses and values, including watershed protection and life support. Reducing soil disturbance, leaving logs and debris in the forest, leaving uncut buffers along streams, leaving seed trees in stands being harvested, and arranging clear-cuts so as to provide corridors for animal use are some of the procedures being suggested. All of these ideas provide alternatives to the "stark choice between tree farms and total preservation" (Franklin 1989).

Temperate Grasslands

Temperate grasslands occur where rainfall is intermediate between that of desert lands and that of forest lands—that is, where annual precipitation is between 10 and 30 inches (25 to 75 cm), depending on temperature, seasonal distribution of the rainfall, and the water-holding capacity of the soil. Soil moisture is a key factor, especially because it limits microbial decomposition and recycling of nutrients. Large grassland areas occupy the interior of the North American and Eurasian continents, southern South America (the Argentine pampas), and Australia.

In North America, the grassland biome is divided into east-west zones of tallgrass, mixed-grass, short-grass, and bunch-grass prairies. These zones are determined by the rainfall gradient, which is also a gradient of decreasing primary productivity. Some of the important perennial species, classified according to the height of the aboveground parts, are as follows:

- *Tallgrasses* (2–3 meters)—big bluestem (*Andropogon gerardii*), switchgrass (*Panicum virgatum*), Indian grass (*Sorghastrum nutans*), and, in the bottomlands, sloughgrass (*Spartina pectinata*).
- *Midsize grasses* (1–2 meters)—little bluestem (*Andropogon scoparius*), needlegrass (*Stipa spartea*), dropseed (*Sporobolus heterolepis*), western wheatgrass (*Agropyron smithii*), June grass (*Koeleria cristata*), and Indian ricegrass (*Oryzopsis hymenoides*).
- *Short grasses* (0.1–0.5 meters)—buffalo grass (*Buchloe dactyloides*), blue grama (*Bouteloua gracilis*), other gramas (*Bouteloua* spp.), the introduced bluegrass (*Poa* spp.), and cheat grass (*Bromus* spp.).

The roots of most species penetrate deeply (up to 2 meters), and the mass of roots of healthy climax perennials will be several times that of the aboveground parts. The growth form of the roots is important. Some of the aforementioned species—for example, big bluestem, buffalo grass, and wheatgrass—have underground rhizomes and are thus *sod formers*. Other species—such as little bluestem, June grass, and needlegrass—are *bunch grasses,* which grow in clumps. These two life-forms may be found in all zones, but bunch grasses predominate in the drier regions, where grassland grades into desert.

Forbs (nongrass herbs, such as composites and legumes) generally constitute only a small part of the producer biomass in climax grasslands, but they are consistently present. Certain species are of special interest as indicators of stress. Increased grazing or drought, or both, tends to increase the percentage of forbs, which are also prominent in early seral stages. Secondary succession in the grassland biome and the rhythmic changes in vegetation during wet and dry cycles were described in Chapter 8. The automobile traveler through the middle of the United States should bear in mind that the conspicuous annual forbs at the roadsides, such as the Russian thistle "tumbleweed" (*Salsola kali*) and sunflowers (*Helianthus* spp.), owe their luxuriance to the continual disturbance of the soil by highway maintenance machinery.

Very extensive areas of grassland, especially the tallgrass prairies, have now been replaced by grain agriculture or cultivated pastures or have been invaded by woody vegetation. Original or virgin tallgrass prairie is difficult to find, and where preserved for study (as in the University of Wisconsin Arboretum and the Konza Prairie LTER site; Fig. 10-16), it must be burned to preserve its prairie character.

A well-developed grassland community contains species with different temperature adaptations—one group growing in the cool part of the season (spring and fall) and another in the hot part (summer). The grassland as a whole "compensates" for temperature fluctuations, thus extending the period of primary production. The role of the C_3 and C_4 types of photosynthesis in this adaptation was discussed in Chapter 2.

The grassland community builds an entirely different type of soil than that of a forest, even when both start with the same mineral parent material. Because grass plants are short-lived compared with trees, a large amount of organic matter is added to the soil. The first phase of decay is rapid, resulting in little litter but much humus—in other words, humification is rapid, but mineralization is slow. Consequently, grassland soils may contain 5 to 10 times as much humus as forest soils. The dark, humus-rich grassland soils of Iowa, for example, are among those best suited for the growing of corn, wheat, and other grains, which, of course, are species of cultivated grasses.

Fire helps maintain grassland vegetation in competition with woody vegetation in warm or moist regions, such as the longleaf pine (*Pinus palustris*) and wiregrass (*Aristida stricta*) savanna located in the Coastal Plains of the southeastern United States (Fig. 10-17). Large herbivores are a characteristic feature of grasslands. The ecologi-cal equivalence of bison, pronghorn, and kangaroos in grasslands exists in different biogeographical regions. The mammalian grazers come in two "life-forms"—running forms, such as those just mentioned, and burrowing types, such as ground squirrels and prairie dogs.

When natural grasslands become pasture, the native grazers are replaced by domestic species (such as cattle, sheep, and goats). Because grasslands are adapted to heavy energy flow along the grazing food chain, such a switch is ecologically sound. However, humans persistently suffer from the "tragedy of the commons" by allowing

Courtesy of Donald W. Kaufman

Figure 10-17. Konza Prairie, a 3487-hectare native tallgrass prairie, located in the Flint Hills region of northeastern Kansas, is dominated by perennial, warm-season grasses such as big bluestem, little bluestem, Indian grass, and switchgrass, frequently reaching 2.5 meters in height.

Figure 10-18. A prescribed fire in longleaf pine (*Pinus palustris*) and wire-grass (*Aristida stricta*) savanna located in southwestern Georgia. The dominant species are among the most fire tolerant in North America. Burning eliminates invasive deciduous tree seedlings, while favoring a species-rich groundcover. Only 3 percent of these relic longleaf pine woodlands remain in a fragmented modern landscape, and they require human-ignited fires because lightning ignitions are too infrequent and unpredictable to functionally conserve this ecosystem.

overgrazing and overplowing. Many former grasslands, therefore, are now human-made deserts that are difficult to restore to the grassland state. For example, Morello (1970) reported that intensive cattle grazing on the Argentine pampas reduced the combustible matter to such an extent that fires, which are necessary to maintain grass cover, can no longer burn. As a result, thorny shrubs, formerly kept in check by periodic fires, take over. The only way to restore grazing productivity is to expend fuel energy in the mechanical removal and burning of woody vegetation. This is an example of an anthropogenic vegetation change reversible only at great cost.

Tropical Grasslands and Savannas

Tropical savannas (grasslands with scattered trees or clumps of trees) are found in warm regions with 40 to 60 inches (100 to 150 cm) of rainfall but with one or two pro-longed dry seasons, when fires are an important part of the environment (Fig. 10-19). The largest area of this type is in central and eastern Africa, but sizable tropical savan-nas or grasslands also occur in South America and Australia. Because both trees and grass must be resistant to drought and fire, the number of species in the vegetation is not large, in sharp contrast to adjacent forests. Grasses belonging to such genera as *Panicum, Pennisetum, Andropogon,* and *Imperata* provide the dominant cover, whereas the scattered trees are of entirely different species from those of the tropical forest. In Africa, the thorny and picturesque *Acacias* and other leguminous trees and shrubs, the large-trunk baobab trees (*Adansonia digitata*), arborescent euphorbias (the eco-logical equivalent of cacti), and palms dot the landscape. Often, single species of both grass and trees may be dominant over large areas.

In number and variety, the population of hoofed mammals of the African Savanna is not exceeded anywhere in the world. Antelope (numerous species, including the wildebeest), zebra, and giraffe graze or browse and are sought by lions and other predators in areas in which the "big game" has not been replaced by humans and cattle. Figure **9-19** shows a wildebeest migration in the African Savanna. Mutualism

© Donald I. Ker, Ker & Downey Safaris Limited, Nairobi, East Af-

Figure 10-19 A view of the tropical Savanna of East Africa. Grass, scattered trees with picturesque shapes, dry-season fires, and numerous species of large mammalian herbivores (Thomson's gazelle, *Gazella thomsonii,* is the species shown here) are unique features of this biome.

Courtesy of Eugene P. Odum

Figure 10-20. Migration of wildebeest (*Connochaetes taurinus*) across the Savanna in Africa.

between grazers and grass and the concept of reward feedback was discussed in some detail in Chapter 3. Insects are most abundant during the wet season, when most birds nest, whereas reptiles may be more active during the dry season. Thus, seasons are regulated by rainfall, rather than by temperature as in the temperate grasslands.

The earliest traces of human existence have been found in East Africa; however, it is not certain whether this region was wetter or drier at the "dawn of humans" than it is at present.

Chaparral and Sclerophyllous Woodland

In mild temperate regions with abundant winter rainfall but dry summers, the vegetation consists of trees, shrubs, or both with hard, thick, evergreen leaves (Fig. 9-20). Under this heading is included a range of vegetation from the Coastal Chaparral, in which shrubs predominate, to broad sclerophyllous woodland dominated by small to medium-sized piñon-juniper evergreen trees, such as scrub oak (*Quercus dumosa*) and chamise (*Adenostoma fasciculatum*). Chaparral communities are extensive in California and Mexico, along the shores of the Mediterranean Sea, in Chile, and along the southern coast of Australia. Many plant species may serve as dominants, depending on the region and local conditions. All species have mycorrhizae, and some have nitrogen-fixing actinomycete nodules. Both mutualisms enhance survival under harsh conditions. Fire is an important factor that tends to perpetuate shrub dominance at the expense of trees. Chaparral is a fire-type ecosystem and was noted in Chapters 5, 7, and 8 as an example of a cyclic climax.

In California, about 5 to 6 million acres (2–2.5 million ha) of slopes and canyons are covered with chaparral. Chamise (*Adenostoma*) and manzanita (*Arctostaphylos*) are common shrubs that often form dense thickets, and several evergreen oaks are characteristic as either shrubs or trees. The rainy growing season generally extends from November to May. Mule deer and many birds inhabit the Chaparral during this period, then move north or to higher elevations during the hot, dry summer. Resident vertebrates are generally small and dull-colored to match the dwarf forest; the small brush rabbits (*Sylvilagus bachmani*), wood rats, chipmunks, lizards, wren-tits, and brown towhees are characteristic. The population density of breeding birds and insects is high as the growing season comes to a close, then decreases as the vegetation dries out in late summer. At this time, fires frequently sweep the slopes with incredible swiftness (and into the southern California suburban areas). After a fire, the chaparral shrubs sprout vigorously with the first rains and may gain maximum size in 15 to 20 years.

Very similar to the West Coast Chaparral is the sclerophyllous woodland of the winter rain areas of the Mediterranean region, locally called *maquis;* similar vegetation in Australia, in which trees and shrubs of the genus *Eucalyptus* are dominant, is called *mallee scrub*. It is not a surprising adaptation that Australian "eucalypts" grow well in California, where they have been widely introduced and have largely replaced the native woody vegetation in urban areas.

© Rich Reid/National Geographic/Getty Images

Figure 10-21. A view of the chapar-ral-covered hills of southern California. The Chaparral is a Mediterranean-type shrubland biome that is highly prone to fire.

Deserts

Regions having less than 10 inches (25 cm) of rainfall— or sometimes regions with greater rainfall that is very unevenly distributed—are generally classified as **deserts** (Fig. 10-22). The scarcity of rainfall may be due to (1) high subtropical pressure, as in the Sahara and Australian deserts; (2) geographical position in rain shadows, as in the western North American deserts; or (3) high elevation, as in the Tibetan, Bolivian, and Gobi deserts. Most deserts receive some rain during the year and have at least a sparse cover of vegetation, unless the edaphic conditions of the substrate happen to be especially unfavorable (as with moving sand dunes). Apparently the only places where little or no rain falls are located in the central Sahara Desert and northern Chile. A picture of the desert climate of Arizona is shown in Figure 10-21.

When deserts are irrigated, and water is no longer a limiting factor, the type of soil becomes a prime consideration. When the texture and nutrient content of the soil are favorable, irrigated deserts can be extremely productive, because of the large amount of sunlight they receive. However, the cost per kilogram of food produced may be high because of the high cost of developing and maintaining irrigation systems. Very large volumes of water must flow through the system; otherwise, salts may accumulate in the soil (as a result of the rapid evaporation rate) and become limiting. As the irrigated ecosystem ages, increased water demands may bring on an "inflationary spiral," requiring the building of more aqueducts, higher costs of production, and greater exploitation of the underground or mountain water sources. Old World deserts are full of the ruins of irrigation systems. At least, these ruins warn us that the irrigated desert will not continue to bloom indefinitely without due attention to the basic laws of ecology.

Three life-forms of plants are adapted to deserts: (1) the *annuals,* which avoid drought by growing only when there is adequate moisture; (2) the *stem succulents,* such as the cacti, which have the moisture-conserving CAM photosynthesis (see Chapter 2) and also store water; and (3) the *desert shrubs* (Fig. 10-22A), which have numerous branches ramifying from a short basal trunk bearing small, thick leaves that are shed during the prolonged dry periods. The desert shrub presents very much the same appearance throughout the world, even though species may belong to diverse taxa (another striking example of ecological equivalence due to ecological convergence).

Desert shrubs have a highly characteristic regularly spaced distribution in which individual plants are scattered thinly, with large bare areas in between. In some cases, secondary chemicals act as allelopathic substances to keep plants spaced apart (see Chapter 7 for details). In any event, the spacing reduces competition for a scarce resource; otherwise, intense competition for water might result in the death or stunting of all the vegetation.

Although arbitrary, it is convenient to recognize two types of desert on the basis of temperature: *hot deserts* and *cool deserts* (Fig. 10-22). In North America, the creosote bush (*Larrea tridentata*) is a widespread dominant of the southwestern hot desert, and sagebrush (*Artemisia*) is the chief plant over large areas of the more northern cool deserts of the Great Basin. Bur sage (*Franseria*) is also widespread in southern ar-eas, whereas at higher altitudes, where there is a little more moisture, the giant sa-guaro cactus (*Carnegiea gigantea*) and paloverde (*Parkinsonia*) are conspicuous. East-ward, a considerable amount of grass is mixed with desert shrubs to form a desert grassland type. In the cool deserts—especially on the alkaline soils of the internal

Figure 10-22. Two types of deserts in western North America. (A) A low-altitude "hot" desert near South Mercury, Nevada, dominated by creosote bush (*Larrea*). Note the characteristic growth form of the desert shrub (numerous branches ramifying from ground level) and the rather regular spacing. (B) An Arizona desert at a somewhat higher elevation, with several kinds of cacti and a greater variety of desert shrubs and small trees.

drainage regions—saltbushes of the family Chenopodiaceae, such as shadscale (*Atriplex confertifolia*), hop sage (*Grayia spinosa*), winterfat (*Eurotia lanata*), and greasewood (*Sarcobatus vermiculatus*), occupy extensive zones. The succulent life-forms, including the cacti and the arborescent yuccas and agaves, reach their greatest development in the Mexican Desert. Some species of this type extend into the shrub deserts of Arizona and California, but this life-form is unimportant in the cool deserts. In all deserts, annual forbs and grasses may make quite a show during brief wet periods. The extensive "bare ground" in deserts is not necessarily free of plants. Mosses, algae, and lichens may be present, and on sands and other finely divided soils, they may form a stabilizing crust. The cyanobacteria (often associated with lichens) are also important as nitrogen fixers (see Evenari 1985 for a review of world deserts).

Desert animals and plants are adapted in various ways to the lack of water. Reptiles and some insects are "preadapted" because of their relatively impervious integuments and dry excretions (uric acid and guanine). Desert insects are "waterproofed" with substances that remain impermeable at high temperatures. Although evaporation from respiratory surfaces cannot be eliminated, it is reduced to a minimum in insects by the internally invaginated spiracle system. It should be pointed out that the production of metabolic water (from the breakdown of carbohydrates)—often the only water available—is not in itself an adaptation; it is the *conservation* of this water that is adaptive, as is the ability to produce more metabolic water at low humidities in the case of tenebrionid beetles (a characteristic desert group). Mammals, by contrast, are not very well adapted as a group (because they excrete urea, which involves the loss of much water), yet certain species have developed remarkable secondary adaptations. Among these desert mammals are rodents of the families Heteromyidae and Dipodidae, especially the kangaroo rat (*Dipodomys*) and the pocket mouse (*Perognathus*) of the New World deserts and the jerboa (*Dipus*) of Old World deserts. These animals can live indefinitely on dry seeds and do not require drinking water. They remain in burrows during the day and conserve water by excreting very concentrated urine and by not using water to regulate their body temperature. Thus, the adaptation to deserts by these rodents is as much behavioral as physiological. Other desert rodents—wood rats (*Neotoma*), for example—cannot live solely on dry food but survive in parts of the desert by consuming succulent cacti or other plants that store water. Even the camel (*Camelus*) must drink water, but camels can endure long periods without water because their body tissues can tolerate an elevation of body temperature and a degree of dehydration that would be fatal to most animals. Despite popular belief, camels do not store water in their humps.

Semi-Evergreen Seasonal Tropical Forests

Tropical seasonal forests, including the monsoon forests of tropical Asia, occur in humid tropical climates with a pronounced dry season, during which some or all of the trees lose their leaves (depending on the length and severity of the dry season). The key factor is the strong seasonal pulse of a fairly large annual rainfall. Where wet and dry seasons are of approximately equal length, the seasonal appearance is the same as that of a temperate deciduous forest, with "winter" corresponding to the dry season. In the Panamanian seasonal forest shown in Figure 10-24, the tall emergent trees lose their leaves during the dry season, but palms and other understory trees retain theirs (hence the term *semi-evergreen*). Seasonal tropical forests have a species richness second only to that of the tropical rain forests.

Tropical Rain Forests

The diversity of life perhaps reaches its culmination in the broad-leaved evergreen tropical rain forests that occupy low-altitude zones near the equator. Rainfall exceeds 80 or 90 inches (200 to 225 cm) a year and is distributed over the year, usually with one or more relatively "dry" seasons (5 inches per month or less). Rain forests occur

Figure 10-23. View of a lowland seasonal tropical forest in Brazil. The tall emergent trees (with white trunks), which lose their leaves in the dry season, project above the general canopy of broad-leaved evergreen hardwoods and palms.

in three main areas: (1) the Amazon and Orinoco Basins in South America (the largest continuous area) and the Central American Isthmus; (2) the Congo, Niger, and Zambezi Basins of central and western Africa and Madagascar; and (3) the Indo-Malay, Borneo, and New Guinea regions. These areas differ from each other in the species present (because they occupy different biogeographical regions), but the forest structure and ecology are similar in all three areas. The variation in temperature between winter and summer is less than that between night and day. *Seasonal periodicity* in breeding and other activities of plants and animals is largely related to variations in rainfall or is regulated by inherent rhythms. For example, some trees of the family Winteraceae apparently show continuous growth, whereas other species in the same family show discontinuous growth, with the formation of tree rings. Rain-forest birds may also require periods of "rest," because their reproduction often exhibits periodicity unrelated to the seasons.

The rain forest is highly stratified. Trees generally form five layers: (1) scattered, very tall *emergent* trees that project 50 to 60 meters above the general level of the (2) *canopy layer,* which forms a continuous evergreen carpet 25 to 35 meters tall; (3) a lower-tree *understory* stratum, 15 to 24 meters high, that becomes dense only where there is a break in the canopy; (4) *poorly developed shrubs and young trees* in deep shade; and (5) a *ground layer* composed of tall herbs and ferns. The tall trees are shallow rooted and often have swollen bases or flying buttresses. A profusion of *climbing plants,* especially woody lianas and epiphytes, often hides the outline of the trees. The "strangler figs" and other arborescent vines are especially noteworthy. The number of species of plants is very large; often, there are more species of trees in a few hectares than in the entire flora of Europe or North America. For example, Peter Ashton of the Arnold Arboretum found 700 species of trees in 10 selected 1-hectare plots located in Borneo—the same number of species as in all of North America (Wilson 1988).

A much larger proportion of animals lives in the upper layers of the vegetation in rain forests than in temperate forests, where most life is near ground level. For example, more than 50 percent of the mammals in British Guiana are arboreal. Besides the arboreal mammals, there is an abundance of chameleons, iguanas, geckos, arboreal snakes, frogs, and birds. Ants, Orthoptera, and Lepidoptera are ecologically important. Symbiosis between animals and epiphytes is widespread. Like the flora, the fauna of the rain forest is incredibly rich in species. For example, in a 6-square-mile area on Barro Colorado, a well-studied bit of rain forest located in the Panama Canal Zone, there are 20,000 species of insects, compared with only a few hundred in all of France. E. O. Wilson recovered 43 species of ants belonging to 26 genera from a single leguminous tree in Tambopata Reserve located in Peru—about equal to the entire ant fauna that could be found in the British Isles (E. O. Wilson 1987). Numerous archaic types of animals and plants survive and fill the multitude of niches in this unchanging environment. Many scientists believe that the rate of evolutionary change and speciation is especially high in the tropical forest regions, which, therefore, have been a source of a number of species that have invaded more northern communities. The need to preserve large areas of tropical forests as a *gene resource* is a matter of increasing concern in the scientific community.

Fruit and termites are staple foods for animals in the tropical rain forest. One reason why birds are often abundant is that so many of them, such as fruit-eating parakeets, toucans, hornbills, contingas, trogons, and birds of paradise, are herbivorous. Because the "attics" of the jungle are crowded, many bird nests and insect cocoons are of a hanging type, enabling their inhabitants to escape from army ants and other predators. Although some spectacularly bright-colored birds and insects occupy the more open areas, the majority of rain-forest animals are inconspicuous, and many are nocturnal.

In the mountainous areas of the Tropics, there is a variant of the lowland rain forest, the **montane rain forest**, which has some distinctive features. The forest becomes progressively less tall with increasing elevation, and epiphytes make up an increasingly larger proportion of the autotrophic biomass, culminating in the dwarf **cloud forest.** A functional classification of rain forests can be based on *saturation deficit*, because this determines transpiration, which in turn determines root biomass and the height of trees. Still another variant of the rain forest occurs along banks and floodplains of rivers and is called the **gallery forest**, or sometimes the **riverine forest.**

Efficient *direct nutrient cycling* by mutualistic microorganisms is a remarkable property of rain forests that enables them to be as luxuriant on poor soils as on more fertile sites. In general, with poor soils, the fertility is in the biomass, not in the soil (as it is in temperate forests and grasslands), so when a forest is removed, the land makes poor pasture or cropland (Fig. 10-25).

When the rain forest is removed, a secondary forest often develops that includes softwood trees, such as *Musanga* (Africa), *Cecropia* (America), and *Macaranga* (Malaysia). The secondary forest looks lush, but it is quite different from the virgin forest in both ecology and flora. The climax rain forest is usually very slow to return, especially on sandy or other nutrient-poor sites, because most of the nutrients in the original forest were lost with the removal of the biomass and the disruption of the microbial recycling networks.

How to manage rain forests for human use continues to be a controversial and frustrating question to those who look on these great forests as one of the last frontiers for colonization and as a potential source of wealth. The large size of the trees fooled the early European explorers of the Amazon into believing that the soil was rich. There followed many unsuccessful attempts to convert the rain forest to agricultural and commercial forestry. Despite these failures, developers and governments

A

B

Figure 10-24. (A) A virgin tropical raw rain forest. (B) An area of tropical rain forest following slash-and-burn (swidden) agriculture. When cleared, the land often makes poor pasture or cropland, because the fertility was not in the soil but in the biomass that has been removed.

continue to try to transfer temperate-zone agricultural and forest technology to the region. These ventures were—and are—definitely not appropriate technology. In the late 1960s, for example, billionaire D. K. Ludwig acquired an area of the Brazilian Amazon (Para) about the size of Connecticut. He floated in a pulp mill and converted mature forests to plantations of exotic species. C. F. Jordan (1985; C. F. Jordan and Russell 1989) commented on this expensive venture, noting that the failure to establish productive plantations was blamed on mismanagement rather than on a lack of understanding of the ecological limitations imposed by the infertile Amazonian soils. A more effective way to use the forest resources, Jordan suggested, would be to develop a system of strip cutting in such a way that the retention of nutrients by the root mat in cut areas is not greatly disturbed and that seedlings from adjacent uncut areas can become quickly established in cleared areas. Horticulture might be similarly organized. Clearly, humans should design with, not against, natural adaptations in regional ecosystems.

Tropical Scrub or Thornwoods

Where moisture conditions in the Tropics are intermediate between desert and savanna on the one hand and seasonal or rain forest on the other, tropical scrub or thorn forests may be found. These cover large areas in central South America, south-western Africa, and parts of Southwest Asia. The key climatic factor is the imperfect, irregular distribution of a moderate total rainfall. Thorn forests, which are often re-ferred to as *the bush* in Africa or Australia and the *caatinga* in Brazil, contain small hardwood trees, often grotesquely twisted and thorny; the leaves are small and are shed during dry seasons. Thorn trees may occur in dense stands or in a scattered or clumped pattern. In some locations, it is not certain whether thorn woodlands are natural or the product of generations of use by pastoralists.

Mountains

Mountains occupy about 20 percent of the land area of Earth, and 10 percent of humans live in the mountains. Where flat terraces have been constructed on steep slopes in the Old World, farming has been sustained for many generations (Fig. 10-25). The zonation of eleven natural vegetation types in five biomes in western North Amer-ica has been depicted in Figure 2-3A. There is closer contact (narrow ecotones) and more exchanges between biomes than occur in nonmountainous regions. On the other hand, similar communities are more isolated. In general, many species charac-teristic of a biome in its nonmountainous location are also found in the belt-like ex-tensions in the mountains. Because of isolation and topographical differences, many other species and varieties are unique to the mountain communities.

Each of the major continents has one or more high mountain ranges. No matter where one lives, mountains are important in terms of available surface water supply,

Figure 10-25. Example of Old World terrace farming in the mountains in Central Madagascar.

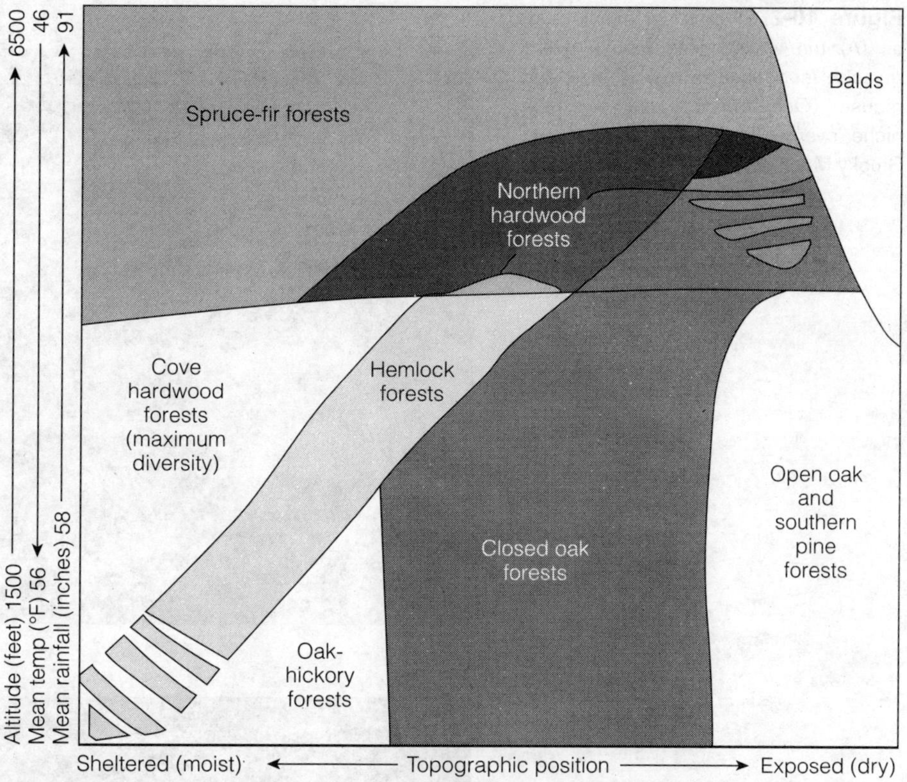

Figure 10-26. The pattern of forest vegetation in the Great Smoky Mountains National Park as related to temperature and moisture gradients (after Whittaker 1952).

because most major rivers originate in mountains, where rainfall and snowfall are usually much greater than in the plains below.

A place in which to observe the clear pattern of forests in relation to climate and substrata is the Great Smoky Mountains National Park, located along the Tennessee–North Carolina border. At sea level, one would have to travel hundreds of miles to observe the variety of climates present in a small geographic area in the Smokies. Fig-ure 10-26 shows a diagram of the complex zonation of communities in the Smokies to help view the landscape through the eyes of the ecologist. The altitude change pro-duces a north-south temperature gradient, whereas the valley and ridge topography provides a gradient of soil moisture conditions at any given altitude. Contrasts in the pattern of vegetation along the gradients are pronounced in May and early June (when floral displays are also spectacular), but the remarkable way in which forests adapt to topography and climate is evident at any time of year.

As Figure 10-26 shows, the forests of the Smokies range from open oak and south-ern pine stands, found on the drier, warmer slopes at low altitudes, to northern conif-erous forests of spruce and fir on the cold, moist summits. The southern pine stands extend upward along the exposed ridges, and the northern hemlock forest extends downward in the protected ravines, where moisture and local temperature condi-tions are like those of higher altitudes. The maximum diversity of tree species oc-curs in sheltered (that is, moist) locations about midway in the temperature gradient. The Smokies have a high diversity of wildflowers, birds, and mammals. Representa-

Figure 10-27. Small mammals such as (A) the white-footed mouse (*Peromyscus leucopus*) and (B) the golden mouse (*Ochrotomys nuttalli*) exhibit niche overlap in forests in the Great Smoky Mountains National Park.

tive small mammals include the white-footed mouse (*Peromyscus leucopus*) and the golden mouse (*Ochrotomys nuttalli*; Fig. 10-27). For more information on mountains, see Messerli and Ives (1997).

Caves

There exist numerous habitat types that, although they are not biomes or major ecosystem types, contain unique flora and fauna. These habitat types include caves, cliffs, and forest-edge (discussed earlier) habitats. These habitat types provide opportunities for exploratory investigations and learning experiences. For example, very interesting heterotrophic communities are found in caves, which are numerous in many parts of the world. Organisms, many of them endemic (that is, found nowhere else in the world) live for the most part on organic matter washed into the cave. Sarbu et al. (1996) explored a cave in which a geothermal, chemoautotrophic ecosystem had evolved. This system is similar to the undersea vent systems in that hydrogen sulfide rather than sunlight provides the energy source.

4 Human-Designed and Managed Systems

Agroecosystems

Agroecosystems are domesticated ecosystems that are in many ways intermediate between natural ecosystems, such as grasslands and forests, and fabricated ecosystems, such as cities. They are solar powered, as are natural ecosystems, but they differ from natural ecosystems in several ways: (1) the auxiliary energy sources that enhance productivity, but also increase pollution, are processed fossil fuels, along with human and animal labor; (2) diversity is greatly reduced by human management in order to maximize the yield of specific crops or other products; (3) the domestic plants and animals are under artificial rather than natural selection; and (4) control is external and goal oriented, rather than internal via subsystem feedback as in natural ecosystems.

Many long-standing traditional agricultural practices in less developed countries are receiving increasing attention because they are energy efficient, ecologically sustainable, and provide adequate food for local people. These practices, however, do not produce *excess* grain or other products that can feed large populations of people or be exported as commodities in exchange for currency. Therefore, many small countries are pushing to replace their traditional agroecosystems with industrial ones, with little consideration of their downsides—namely, pollution and the displacing of the independent farm family.

Thailand is an example of such a country. In rice culture agroecosystems, flood energy input is being replaced by industrial inputs of fossil fuel, fertilizers, and pesticides, resulting in increased yields per unit area but a reduction in energy efficiency (ratio of energy input to grain output; Gajaseni 1995). Another example is Bali, where 1000-year-old gravity irrigation systems are being replaced by fuel-powered pumping systems, along with increased use of synthetic chemicals, for the sake of a short-term gain in market economics (Steven 1994). Engineers designing these systems need to recognize and understand that natural systems have a long evolutionary history reflecting energy efficiency, high biotic diversity, and a mutualistic relationship with human culture.

Plantation forests, like croplands, are agroecosystems—"tree farms" designed to increase wood and fiber production per unit area. Sooner or later, the same management problems that affect other crops—loss of soil quality, pest control, or artificial fertilizers to replace nutrients removed with the harvest—have to be dealt with. There is increasing interest in *agroforestry,* a practice that involves the cultivation of small, fast-growing trees and food crops in alternate rows (MacDicken and Vergara 1990).

Urban-Industrial Technoecosystems

The concept of the *technoecosystem* was introduced and discussed in Chapters 2 and 9. Cities, suburbs, and industrial developments are major technoecosystems. They are energetic islands with large ecological footprints in the matrix of natural and agricultural landscapes. These urban-industrial environments are parasitic in nature, depending on the biosphere in terms of life-support resources (see Fig. 9-20).

We are concerned that these technoecosystems, which frequently grow rapidly and haphazardly and without regard to life support, will outstrip the infrastructure necessary to maintain their growth. Societies need to engage in some serious urban and landscape planning to maintain the quality of these systems. Most of the voluminous literature dealing with the plight of the cities focuses on internal problems, such as deteriorating infrastructure and crime. Urban planners now need to embrace the ecology of the landscape (for example, the regional watershed) rather than set themselves apart. Urban regeneration will depend more and more on reconnecting the city to the life-supporting land and water bodies.

The long-term management and restoration of these human-designed systems will require an understanding of *conservation* and *restoration ecology*—both fields of study that are based on the principles, concepts, and mechanisms of basic ecology. These fields of applied ecology provide exciting challenges in the management and development of sustainable systems.

Conservation Ecology

Any discussion of the higher levels of organization (ecosystems in Chapter 8; landscapes in Chapter 9; biomes in this chapter) illustrates the biotic richness, natural capital, and ecological aesthetics manifested by these large temporal-spatial scale systems. The field of **conservation biology** or **conservation ecology** provides an integrative approach and field of study, focusing on the protection and management of biodiversity based on the principles of both applied and basic ecology. In recent decades, the social sciences—especially sociology, economics, ethics, and philosophy—have become prominent components of these protection and management processes.

For an introduction to the field of conservation biology, we refer you to papers by Soulé (1985, 1991) and Soulé and Simberloff (1986). Books by C. F. Jordan (1995), Meffe and Carroll (1997), and Primack (2004) provide an overview of the principles, threats to diversity, case studies, and management strategies in the field of conservation biology.

11

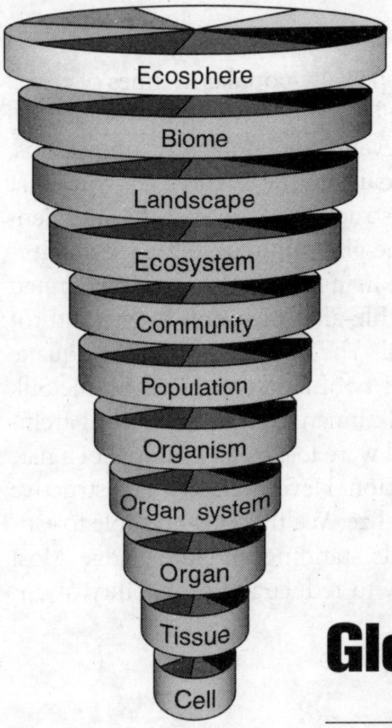

Global Ecology

1 **The Transition from Youth to Maturity: Toward Sustainable Civilizations**

2 **Ecological-Societal Gaps**

3 **Global Sustainability**

4 **Scenarios**

5 **Long-Term Transitions**

1 The Transition from Youth to Maturity: Toward Sustainable Civilizations

Statement

Predicting the future is a fascinating game that is especially popular in times of crisis. Actually, one cannot predict the future in any detail, or with any degree of precision —there are too many unknowns, kaleidoscopic events, technological innovations, and other variables that cannot be foreseen. Events such as the terrorist activities that were responsible for the loss of life and material destruction of the World Trade Center in New York City on 11 September 2001, or the grid failure that caused an electrical power outage that affected 50 million people living in the northeastern United States on 14 and 15 August 2003, are examples of life-altering events due to anthropogenic failures that were not effectively predicted. The 6.6 magnitude earthquake on 26 December 2003 that destroyed the ancient city of Bam located along the Silk Road in Iran, in which approximately 25 thousand human lives and historical architectural treasures such as the 2000-year-old citadel were lost, is an example of a natural phenomenon that still eludes adequate prediction. Nevertheless, it is instructive to consider a range of possibilities that could actualize. We then may be able to estimate their probability given current conditions, understanding, and knowledge. Most important, we might be able to do something now to reduce the probability of undesirable futures and losses.

Explanation

During the twenty-first century, about the only certainties are that humans will continue to increase in numbers, at least until well into the century; that something will have to be done about the fouling of our life-support systems (especially the atmosphere and water); that humanity will have to make a major transition in energy use from predominantly fossil fuels to other, less certain, and probably less lucrative, sources; and finally, because there are no set-point controls (see Fig. 1-4), that humanity will likely overshoot its optimal carrying capacity, as we seem to be already doing with many resources, bringing on boom-and-bust cycles. The challenge of the future, then, will be not how to *avoid* the overshoot but how to *survive* it by downsizing growth, resource consumption, and pollution (Barrett and Odum 2000).

We must begin to reduce our current prodigious waste and become more efficient in order to do more with less high-quality energy and reduce the pollution that results from the waste of energy and industrial resources. Most people also agree that reducing per capita energy consumption in the industrialized countries would not only improve the quality of life locally (H. T. Odum and E. C. Odum 2001) but also help to improve the quality of life globally.

Most students would agree that rapid growth should be avoided, if for no other reason than that it tends to create social and environmental problems faster than they can be dealt with. Rapid population growth and urban-industrial development combine to create a momentum that is very difficult to control (for a review of early warnings, see National Academy of Sciences 1971; Catton 1980). It is significant that in 1992, the prestigious United States National Academy of Sciences and the Royal Society of London issued the following joint proclamation:

World population is growing at the unprecedented rate of almost 100 million people every year, and human activities are producing major changes in the global environment. If current predictions of population growth prove accurate and patterns of human activity remain unchanged, science and technology may not be able to prevent either irreversible degradation of the environment or continued poverty for much of the world.

There is no shortage of studies, "think tank" reports, and popular books assessing the current predicament of humankind. Many of these paint a rather grim picture of present global problems, but others are optimistic about the future. The ways in which scholars, and people in general, view the future range from complete confidence in business as usual and in new technology (a "more of the same" philosophy) to a belief that society must completely reorganize, "power down," and develop new international political and economic procedures in order to deal with a world of finite resources. The late Herman Kahn (*The Next 200 Years,* 1976) and economist Julian Simon (*The Ultimate Resource,* 1981) are well-known spokesmen for the business-as-usual view, whereas Paul Ehrlich (*The Population Bomb,* 1968), E. F. Schumacher (*Small Is Beautiful,* 1973), Fritjof Capra (*The Turning Point,* 1982), economists Herman Daly and John Cobb (*For the Common Good,* 1989), Herman Daly and Kenneth Townsend (*Valuing the Earth,* 1993), Paul Hawken (*The Ecology of Commerce,* 1993), Edward Goldsmith (*The Way,* 1996), and E. O. Wilson (*Consilience,* 1998; *The Future of Life,* 2002) are among those arguing for the need for fundamental changes—a position that is becoming more and more of a consensus among world leaders. Then there are the cornucopian ("horn of plenty") technologists, who are optimistic that an efficient and clean hydrogen economy (to replace carbon-based, "dirty" fossil fuels), reduced-input agriculture, waste-free industry, and other future technologies will enable nine billion or more people to coexist with enough natural environment to provide life support, preservation of endangered species, and enjoyment of nature (see Ausubel 1996).

Historical Perspectives

Anthropologist Brock Bernstein (1981) noted that in many isolated cultures that must survive on local resources alone, actions that would be detrimental to the future are perceived and avoided. Such *local feedback* in decision making is lost when isolated cultures are incorporated into large and complex industrial societies. As Bernstein said, "Economics must develop a coherent theory of decision-making behavior that is applicable at all levels of group organization. This will necessitate defining self-interest in terms of survival rather than consumption. Such a shift would bring economic behavior under something akin to natural selection, which has worked so well to insure the perpetuation of life on Earth over the eons."

One of the obstacles to avoiding overshoots in resource use is what Garrett Hardin (1968) termed "the tragedy of the commons." By *commons,* he meant that part of the environment that is open to use by anyone and everyone, with no one person responsible for its welfare. A pasture or open range shared by many herders is an example. Because it is to the advantage of each herder to graze as many animals as possible, the capacity of the range to sustain grazing will be exceeded unless restrictions are agreed on and enforced by the community as a whole. Prior to the Industrial Revolution, many commons were protected by community-enforced restrictions and customs. Herding societies solved the problem by moving their animals from one place to another on a regular basis before overgrazing occurred at any one place. Many European cities have a long tradition of maintaining commons in the form of large parks and greenbelts. The "tragedy" in these modern times is that local restrictions, as might be embodied in zoning ordinances, are so easily overturned by the influence of "big money"—that is, the *economic capital* that is available for the kind of development that yields large short-term profits, often at the expense of *natural capital,* thus affecting the local quality of life.

In another book, Hardin (1985) raised a most intriguing question: would the Industrial Revolution have gotten off the ground without the exploitation of people and environment in the beginning? Recall Dickens' novels describing labor abuse and the complete inattention to air and water pollution in the nineteenth century. Certainly, the exploitation of people (as in industrial sweatshops) and the unrestricted pollution of the environment greatly accelerated the capital accumulation on which the present affluence of the industrial world is based. Most people now realize, however, that we are at a turning point in history. We cannot continue "commonizing the costs and privatizing the profits" (Hardin 1985) and postponing the environmental and human costs of rapid growth and development without incurring widespread damage to our global life-support systems. Donald Kennedy, in an editorial in *Science* (12 December 2003), noted that the big question in the end is not whether science can help to solve problems at large scales (such as global warming); rather, it is whether scientific evidence can successfully overcome social, political, and economic resistance. That was Hardin's big question 35 years ago, and it is ours now.

In the closing decades of the twentieth century, numerous papers and books were published suggesting ways to deal with the "commons," including another book by Hardin entitled *Living Within Limits* (1993); and a book by Hawken et al. (1999) entitled *Natural Capitalism,* proposing a common ground among economics, environment, and society for the twenty-first century.

Global Models

Some of the most comprehensive futuristic reports include those prepared by the Club of Rome and the global models produced by the United States and other governments and the United Nations. The Club of Rome was a group of scientists, economists, educators, humanists, industrialists, and civil servants brought together by Italian industrialist Arillio Peccei, who felt the urgent need to prepare a series of books on the future predicament of humankind. Its first and best-known book, *The Limits to Growth* (Meadows et al. 1972), predicted on the basis of models that if our political and economic methods continue unchanged, severe boom-and-bust cycles will occur. Essentially, this first Club of Rome study employed a modern systems approach to the older "warnings to humankind" classics, such as George Perkins Marsh's *Man and Nature* (1864, reissued in 1965), Paul Sears' *Deserts on the March* (1935), William Vogt's *Road to Survival* (1948), Fairfield Osborn's *Our Plundered Planet* (1948), and Rachel Carson's *Silent Spring* (1962). The report denounced society's obsession with growth, in which at each level (individual, familial, corporate, and national) the goal is to get richer and bigger and more powerful, with little consideration for basic human values and the ultimate cost of unrestricted, unplanned consumption of resources and stress on environmental life-support goods and services.

The Limits to Growth was followed by a series of additional reports that attempted not only to describe possible future trends in more detail, but also to suggest actions that should be taken to avoid a boom-and-bust doomsday. These studies were published in book form with titles such as *Mankind at the Turning Point; Reshaping the International Order; Goals for Mankind; Wealth and Welfare;* and *No Limits to Learning: Bridging the Human Gap* (all published by Pergamon Press, New York). A variety of distinguished scholars contributed to these efforts, including economists, educators, engineers, historians, and philosophers. Laszlo (1977) assessed the overall impact of these reports as follows:

> Thanks largely to the efforts of the Club of Rome, international awareness of the world problematique has rapidly grown. The Club pioneered the way (to use a medical analogy) from diagnosis to prescription but very little was accomplished in the way of therapy. To use another metaphor, the Club helped point the way but did little to generate the will to take it.

Between 1971 and 1981, a number of other global models were developed. These models were computerized mathematical simulations of the world's physical and socioeconomic systems and made projections into the future that were logical consequences of the data and the assumptions that went into the model. It should be emphasized that each model remains unique with respect to the assumptions that motivated it. These models were reviewed and compared as a group in a report issued by the Congressional Office of Technology Assessment (OTA 1982), and in a book by Meadows et al. (1982). Despite differing assumptions and biases, the modelers agreed on some points, namely:

- Technological progress is expected, and is vital, but social, economic, and political changes will also be necessary.
- Populations and resources cannot grow forever on a finite planet that is not growing larger.
- A sharp reduction in the growth rates of population and urban-industrial development will greatly reduce the seriousness of overshoots or major breakdowns in life-support systems.
- Continuing "business as usual" will not lead to a desirable future, but rather will result in further widening of undesirable gaps (such as between rich and poor).
- Cooperative long-range approaches will be more beneficial for all parties than competitive short-term policies.
- Because the interdependence among peoples, nations, and the environment is much greater than commonly acknowledged, decisions should be made in a holistic (systems) context. Actions to alter current undesirable trends (such as atmospheric toxification), taken soon (within the next couple of decades), will be more effective and less costly than actions taken later. This calls for strong civic awareness of needs that will force strong political action and changes in education, because by the time a problem is obvious to everyone, it may be too late.

In the 1990s, Donella Meadows and colleagues came out with a sequel to *The Limits to Growth*, entitled *Beyond the Limits* (1992). They concluded that global conditions are worse than predicted in 1972; however, they still envisioned a sustainable future if the six points just outlined are taken seriously and acted on. They commented that what the world needs is good old-fashioned "love" that will enable people to work together for common causes. In Chapter 8, we discussed a parallel trend in ecosystem development, in which mutualism increases when resources are scarce.

Models are good for integrating data and trends, but they are not able to factor in human resolve and ingenuity (or lack of it). Next, we will explore several approaches to understanding our predicament and what we can learn from the study of ecology that will help us to address current problems, including those that will likely arise in the future.

Ecological Assessment

The wisdom of the many contributors to the aforementioned reports, models, and global assessments conforms rather well to basic ecosystem theory, especially to three of its paradigms: (1) a *holistic* approach is necessary when dealing with complex systems; (2) *cooperation* has greater survival value than competition when limits (resources or otherwise) are approached; and (3) orderly, high-quality development of human communities, like that of biotic communities, requires *negative* as well as *positive feedback* mechanisms. It should be noted that these scholars' conclusions also conform to the age-old human wisdom found in common-sense proverbs such as "look before you leap," "do not put all your eggs in one basket," "haste makes waste," "an ounce of prevention is worth a pound of cure," and "power corrupts."

A civilization is a system, not an organism, contrary to what Arnold Toynbee (1961) said in *A Study of History.* Civilizations do not necessarily have to grow, mature, become senescent, and die, as organisms do—even though this process has happened in the past (as with the rise and fall of the Roman Empire). According to geographer Karl Butzer (1980), civilizations become unstable and break down when the high cost of maintenance results in a bureaucracy that makes excessive demands on the productive sector. Such a view coincides with ecological theories of *P*/*R* ratios (see Chapter 3), resource recycling (see Chapter 4), carrying capacity (see Chapter 6), complexity (see Chapter 8), and habitat fragmentation (see Chapter 9). As we continue to point out, the study of ecology can help us deal with human predicaments.

Coming Full Circle

Human societies go from pioneer to mature status in a manner parallel to the way that natural communities undergo ecosystem development. As we have already noted, there are many strategies and behaviors that are appropriate and necessary for survival during the youth or pioneer stage but that become inappropriate and detrimental at maturity. Continuing to act on a short-term, one-problem–one-solution basis as society grows larger and more complex leads to what economist A. E. Kahn (1966) called "the tyranny of small decisions." Increasing the height of smokestacks—a quick fix for local smoke pollution—is one example in which many such "small decisions" lead to the larger problem of increased regional air pollution. W. E. Odum (1982) gave another example: no one purposefully planned to destroy 50 percent of the wetlands along the northeastern coast of the United States between 1950 and 1970, but it happened as a result of hundreds of small decisions to develop small tracts of marshland. Finally, the state legislatures woke up to the fact that valuable life-support environment was being destroyed, and one by one, each legislature passed wetlands protection acts in an effort to save the remaining wetlands. It is human nature to avoid long-term or large-scale actions until there is a threat that is perceived by a majority of the population.

What all this means for the future is that the transition time for human communities has come—or will be coming soon—that necessitates "coming full circle" or "doing an about-face" on many previously acceptable concepts and procedures.

In the next several sections, we will present the growth-maturity theme from different but interrelated viewpoints, with suggestions for solutions to the dilemmas raised. We will conclude with a table summarizing the prerequisites for achieving maturity.

2 Ecological-Societal Gaps

Statement

A good way to assess the predicament of humankind is to consider the *gaps* that must be narrowed if humans and the environment (as well as nations) are to be brought into more harmonious relationships. Among these gaps are the following:

- The *income gap* between the rich and the poor, both within nations and between the industrialized nations (30 percent of the world population) and the nonindustrialized nations (70 percent of the world population).
- The *food gap* between the well-fed and the underfed.
- The *value gap* between market and nonmarket goods and services.
- The *education gap* between the literate and the illiterate, the skilled and the underskilled.
- The *resource management gap* between development and stewardship.

None of these gaps has been narrowed very much during the past several decades; in fact, the income and value gaps have gotten much worse. Seligson (1984) noted that the gap in per capita income between rich and poor nations had grown from $3,617 to $9,648 between 1950 and 1980; this gap has continued to grow. By 2000, the per capita gross domestic product (GDP) had reached $35,000 in western countries, compared with $1,000 in many other countries—a gap of $34,000 (United Nations statistic; see Zewail 2001). In terms of resource consumption per capita, paper consumption in 1999, for example, was 135 kilograms/year in the United States and 4 kilograms/year in India (Abramovitz and Mattoon 1999).

Explanation and Examples

Well-meaning efforts by rich nations to help poorer ones have too often failed, because the deleterious cultural and environmental impacts of the aid were not anticipated. For example, the construction of a reservoir in a fertile valley may provide benefits initially, but it may also force farmers to move upstream to less suitable land, resulting in severe erosion and deforestation of the watershed and subsequent silting of the reservoir. Morehouse and Sigurdson (1977) pointed out more than two decades ago that the transfer of industrial technology from industrialized to developing nations too often benefits the small modern sector but not the masses of rural poor in the traditional society. Wealth does not "trickle down" when there are profound cultural, educational, and resource differences within the population that have not changed in the past quarter century. As noted earlier (see Chapter 3), one cannot transfer a high-energy industrial or agricultural technology to a less developed country without also providing the high-quality energy needed to sustain it. Likely it is better to enhance current "low-tech" operations (that is, use "appropriate" technologies) until the nation can go "high-tech" on its own. As in nature, one gets to a mature state only after the way is prepared by younger developmental stages.

Social Traps

A situation in which a short-term gain is followed in the long term by a costly or deleterious situation not in the best interest of either the individual or society has been termed a **social trap** (Cross and Guyer 1980). An analogy is a trap that entices an animal into it with an attractive bait; the animal enters the trap in the hope of an easy meal, but then finds it difficult or impossible to escape. Substance abuse is an example of a *behavioral* social trap, whereas hazardous waste dumping, destruction of wetlands (or other life-support environments), and proliferation of weapons of mass destruction are examples of *environmental* social traps. Edney and Harper (1978) suggested using a simple game played with poker chips to illustrate the relationship of social traps to the tragedy of the commons. A pool of poker chips is established, and each player has a choice of removing from one to three chips. The pool is renewed after each round in proportion to the number of chips remaining. If players think only of their immediate, short-term gains and remove the maximum of three chips, the renewable resource of the common pool becomes smaller, and ultimately the resource pool is gone. Removing only one chip each round sustains the renewable resource.

Dominion versus Stewardship

There is an interesting parallel to the youth-maturity theme in the Scriptures. Early in the Biblical text, followers are told to "be fruitful and multiply" and take dominion over Earth (Genesis 1:28). One interpretation of this message is that it was meant to apply to all living things, not just humans (Bratton 1992). Elsewhere in the Scriptures (for instance, Luke 12:42) readers are admonished to become "stewards" (keepers of the house) and take care of Earth. A reasoned interpretation is that these messages are not contradictory, nor a matter of right or wrong, but constitute a *sequence* in time (in other words, there are times and places for each).

In the early development of civilizations, taking dominion over the environment and exploiting resources (such as clearing the land for crops and mining the earth for materials and energy) and high birth rates are necessary for human survival. However, as societies become more crowded, resource-demanding, and technologically complex, various limitations are reached that should encourage humans to turn to stewardship in order not to destroy our life-supporting "house."

Stephen R. Covey, in his national best-selling book *The 7 Habits of Highly Effective People* (Covey 1989), noted that both ecological systems and social systems go through three stages of growth and development. He referred to the youth stage as *dependent,* the next stage as *independent,* and finally, the mature stage as *interdependent.* Thus, just as ecosystem growth and development leads to increased mutualistic relationships among species, increased biotic diversity, and system regulation, social system growth and development leads to interdependent relationships among members of a society, increased educational and cultural opportunities, and inter- and intragenerational stability.

Environmental Ethics and Aesthetics

Maintaining and improving environmental quality requires an ethical underpinning. Not only must it be against the law to abuse nature's life-support systems, it should be understood to be *unethical* as well. One of the most widely read and cited essays on the subject of environmental ethics is Aldo Leopold's "The Land Ethic," first published in 1933 and included in his classic book, *A Sand County Almanac: And Sketches Here and There* (1949). Leopold spent his early years as a forester in a part of the western United States that could be reached only on horseback and where the howl of the wolf could still be heard. Later, he pioneered the field of game management and became a professor at the University of Wisconsin. He and his family spent as much time as possible at a cabin ("the shack," which has since become a historical site frequently visited by conservationists) located on an old worn-out farm they bought and restored as a place of natural beauty in Sand County, Wisconsin. Aldo Leopold will be best remembered for the writing he did there—writings about the wild and the beautiful, reminiscent of the work of Henry David Thoreau (for more on the legacy of Aldo Leopold and the land ethic paradigm, see A. Carl Leopold 2004).

Human rights have received increasing ethical as well as legal and political attention. But what about the rights of other organisms and the environment? Leopold defined an **ethic** ecologically as "a limitation on freedom of action in the struggle for existence," and philosophically as "a differentiation of social from antisocial conduct." He suggested that the extension of ethics over time is a sequence as follows: First, there is the development of religion as a human-to-human ethic. Then comes democracy, as a human-to-society ethic. Finally, there is a *yet to be developed* ethical relationship between humans and their environment; in Leopold's words, "the land-human relation is still strictly economic, entailing privileges but not obligations."

As we have attempted to document in this book, there are strong scientific and technical reasons for the proposition that an expansion of ethics should include the life-support environment necessary for human survival. There are many legal mechanisms, such as **conservation easements,** available that can encourage landowners to trade their options to develop property for tax relief or other favorable economic considerations. It is also encouraging that in the past two decades, there has been a great increase in the number of articles, books, college courses, and journals that deal with environmental ethics (Callicott 1987; Potter 1988; Hargrove 1989; Ferré and Hartel 1994; Dybas 2003; Ehrlich 2003; A. C. Leopold 2004).

3 Global Sustainability

Statement

In 1987, a World Commission on Environment and Development issued a report entitled *Our Common Future,* which has come to be known as the "Brundtland Report," after the former Prime Minister Brundtland of Norway, who was chair of the commission. The report concluded that the current trends of economic development and the accompanying environmental degradation are unsustainable. Irrevocable damage to planetary ecosystems is depressing the economic status of much of the world's population. Survival depends on *changes now.* The first step in bringing about these changes is to seek ways to enhance cooperation between nations, so that they can work together toward global sustainability. The report is important not so much for what it says as for the fact that a group of 23 political leaders and scientists from both developed and less developed countries could agree that the health of the global environment is essential for the future of everyone.

Explanation and Examples

In 1991, UNESCO issued a report entitled "Environmentally Sustainable Economic Development: Building on Brundtland" (Goodland et al. 1991). This report made a distinction between economic *growth,* which involves getting larger (quantitative growth), and economic *development,* which involves getting better (qualitative growth) without increasing the total consumption of energy and materials beyond a level that is reasonably sustainable. The report concluded that "a five-to-tenfold expansion of anything remotely resembling the present economy (which some economists say is necessary to reduce poverty worldwide) would simply speed us from today's long-run unsustainability to imminent collapse." Therefore, the economic growth required for poverty reduction (especially in the less developed countries) "must be balanced by negative throughput growth for the rich."

In 1992, world leaders convened an *Earth Summit* in Rio de Janeiro, Brazil, in search of international agreements that could help save the world from pollution, poverty, and the waste of resources. Confrontation between the wealthy "North" and the poor "South" dominated the proceedings, and few meaningful agreements were reached. However, the concept of sustainable development did emerge as a means of combining economic and ecological needs. Many who attended the summit came away with the feeling that a pathway had been opened for future cooperation among nations.

Another Earth Summit was convened from 26 August to 4 September 2002 in Johannesburg, South Africa. Negotiators for 191 countries agreed on an action plan to alleviate poverty and conserve the natural resources of Earth. For an overview of decades of depletion of resources, and for suggestions on how things might change in the future, we suggest four consecutive issues of *Science,* beginning with the "State of the Planet" (14 November 2003). However, summit delegates agreed to drop targets and timetables for the installation of renewable energy—a loss for the European Union, which had been pushing for a target of 15 percent of global energy coming from renewable sources by the year 2015. Positive agreements included Brazil and the World Bank agreeing to protect 50 million hectares of tropical forests; the United States announcing a partnership package of 10 billion dollars in pledges to alleviate poverty and promote health care and education; and Costa Rica announcing a moratorium on offshore oil exploration. Summit ministers also agreed to reaffirm the "Rio Principles," including the **precautionary principle**, which asks people to take action before effects are seen, as in the case of ozone depletion in the upper atmosphere.

The human disposition being what it is—that is, most people wait until a problem gets *really* bad before taking action—it often takes a crisis or a disaster to bring about environmental planning and start the transitions we have been discussing. The following example describes how a local disaster was followed by economic development without an increase in the size of the city (Flanagan 1988):

> In 1972, Rapid City, South Dakota, suffered a devastating flash flood of Rapid Creek that damaged 160 million dollars in property, destroyed 1200 buildings,

and killed 238 people. Through the leadership of the mayor, Don Barnett, the community instituted a national prototype floodplain acquisition program, removed the damaged homes from the floodplain, and created a six-mile long, quarter-mile wide urban greenway through the center of the city. The greenway now contains parks, recreation trails, and golf courses. Rapid Creek was stocked with game fish and is now the most popular recreational fishing stream in the entire state. Rapid City stands as a creative example of enlightened leadership, turning a disaster into a multi-use community asset that benefits all aspects of the city, including commerce and a tourist industry.

Dual Capitalism

A recurrent theme in this book has been the contention that the overly narrow economic theories and policies that dominate world markets and world politics are major obstacles to achieving a reasonable, commonsense balance between our need for nonmarket and market goods and services. Around the turn of the twentieth century, a group of scholars calling themselves *holistic economists* formed a school critical of economic models of the day (Gruchy 1967). Efforts to establish a holistic economy at that time, however, were drowned, as it were, in the flood of oil that spawned rapid growth in monetary and material wealth. Classic growth theory served well enough as long as the supply of oil exceeded demand. As the oil supply is now peaking and global pollution and overshoots are rampant, it seems that the time has come to redevelop some kind of *holoeconomics* that includes cultural and environmental values along with monetary ones—in other words, an economics that gives equal consideration to market capital and natural capital.

It should be possible, with appropriate regulatory and incentive measures, to develop a *dual capitalism* gradually over a period of time (E. P. Odum 1997). Under such a **dual capitalism**, a business or industry would not only consider the market possibilities for a new product or service, but also plan how to produce the product or service with an efficient use of resources and with as much recycling and as little pollution as possible. It would also consider how to internalize source reduction and waste management costs, so that the consumer—rather than the taxpayer—pays for waste management.

It is encouraging to see that, beginning with the first international conference on the integration of economy and ecology in 1982 (Jansson 1984), the economics-ecology interface has been receiving increasing attention. For reviews, see Costanza (1991), Daly and Townsend (1993), Costanza, Cumberland, et al. (1997), Hawken et al. (1999), and H. T. Odum and E. C. Odum (2001); see also the special issue of *BioScience* (April 2000), entitled "Integrating Ecology and Economics," organized by Barrett and Farina (2000).

Paradox in Technological Development

Just about every technological advance that is intended to improve well-being and prosperity has its dark side as well as its bright side. As Paul Gray, engineer and former president of the Massachusetts Institute of Technology, stated in 1992, "A paradox of our time is the mixed blessing of almost every technological development." We have described a number of examples of this paradox in previous chapters, including the mixed blessings of plant pest and disease control technology and the

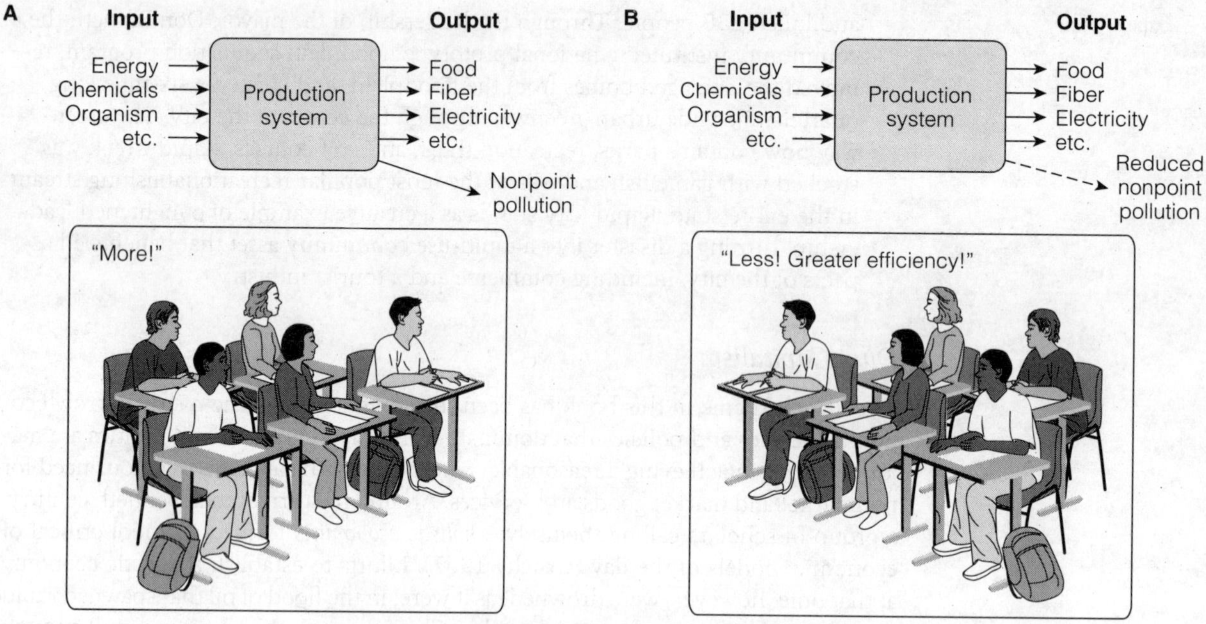

Figure 11-1. The about-face needed in the management of production systems. (A) Focus on output (yield) with consequences such as increased nonpoint-source pollution. (B) The shift (about-face) to input management, with focus on efficiency and reduction of costly and environmentally damaging inputs, so as to reduce nonpoint-source pollution (after Odum 1989).

Green Revolution technology. The bright side of these technologies has been the increased yield of food produced with less labor. Their dark side is the heavy use of fertilizers, pesticides, and machinery, which has resulted in widespread air, water, and soil pollution; the development of resistant strains of pests; and serious rural unemployment. Another example is the coal-fired power plants that provide electricity for most of the United States, but also make a major contribution to acid rain.

The point to be made here is that as we seek new technologies, we must be aware that they *will* have dark sides that must not only be anticipated, but also dealt with based on sound ecological theory and research. Often, what is needed is a *countertechnology* that will at least ameliorate the detrimental effects. In agriculture, for example, conservation tillage or alternative agriculture is a countertechnology that is being widely adopted. In the case of power plants, "clean coal" technology that would eliminate acid emissions is available.

Input Management

"The solution to pollution has been dilution" (find some place to dump it). "Now the solution must be source reduction" (E. P. Odum 1989, 1998a, 1998c). The strategy of managing inputs rather than outputs was first mentioned in Chapter 1 as an "about-face" that is necessary in order to reduce pollution. *Input management* of production systems (such as agriculture, power plants, and manufacturing) is a practical and economically feasible approach to improving and sustaining the quality of life-support systems. This concept is illustrated in Figure 11-1. As Figure 11-1A shows, attention in the past has focused on increasing *outputs*—that is, yields—by pouring in resources (such as fertilizers and fossil fuels) without much regard to efficiency or to the amount of unwanted outputs (such as nonpoint pollution) created. Input management involves an about-face, as shown in Figure 11-1B, with the goal of reducing *inputs* to only those that can be efficiently converted to the desired product. Input management can also be called *top-down management* because it involves assessing inputs to the whole system (such as the external forcing functions) *first,* then internal dynamics and outputs *second.* Applying the concept to wastes means that *waste reduction takes precedence over waste disposal.*

Professor Luo Shi Ming of South China Agricultural University suggested that the proper path for less developed countries to take in developing their agriculture is to bypass the wasteful, high-input stage and go directly from traditional agriculture to new reduced-input practices, as shown in the graphic model depicted in Figure 9-18C. Why not employ the same strategy for industrial development? To reduce and eliminate toxic wastes at their source will require a combination of regulations and incentives in order to establish a sustainable society (Barrett 1989).

Restoration Ecology Revisited

Because so much of the environment has been damaged beyond nature's ability to repair it, restoring damaged ecosystems is becoming big business. **Restoration ecology** focuses on restoring landscape heterogeneity and patterns that have been altered by human disturbance. Figure 11-2 illustrates the Dust Bowl of the 1930s caused by human disturbance. A better understanding of soil ecology, appropriate agricultural practices, and a renewed conservation ethic have largely restored these damaged landscapes. This focus also increases the conservation value of fragmented landscapes and develops a close relationship with the field of landscape ecology. Ways and means of restoring wetlands that were drained or destroyed before their value

Figure 11-2. An extreme example of soil erosion by wind occurred during the Dust Bowl in the 1930s.

as life-support buffers was recognized are an especially active area of research. A pioneer in developing the new field of restoration ecology is John Cairns, Jr. Since 1971, he has written and edited several large books on the subject. He presented a number of case studies in a report that he organized for the National Academy of Sciences (NAS; 1992). In reviewing such environmental engineering projects, it is evident that they are most successful when four key groups work together in a coordinated manner—namely, (1) civic interests; (2) government agencies (provincial, regional, and federal); (3) science and technology; and (4) business interests. When any one of these groups is not strongly involved, restoration and environmental engineering projects usually fail to achieve their long-term goals. There are three journals in the field of environmental engineering: *Environmental Engineering, Ecological Engineering,* and *Restoration Ecology.*

Pulsing Paradigm Revisited

In emphasizing youth–maturity development, we need to reemphasize that everything in the environment and in society *pulses.* The weather and the climate fluctuate, often in a rhythmic manner. The density of all animal populations fluctuates over time. Anything that approaches a long-term steady-state seems to be rare. Therefore, we ask, does the mature state pulse within itself, or does it age, as do individuals, and is it eventually replaced? C. S. Holling's *panarchy* or *cyclic hierarchy concept* (Holling and Gunderson 2002) suggests that after r develops to K (and ecological succession is complete), the K or mature organization becomes less resilient and adaptive and collapses, to be followed by a new r to K development. **Panarchy** is a theory intended to transcend the boundaries of scale and discipline, and be capable of organizing understanding of economic, ecological, and institutional systems (in other words, it must explain situations where all three systems interact). Its focus is to rationalize the interplay between change and persistence. This cyclic or numerical model seems to occur in forests, and in some civilizations or cities in the past, but seems less applicable when it comes to oceans or large countries. Scale does seem to matter in this regard.

4 Scenarios

Statement

A **scenario** is an outline of a sequence of scenes or events, which, as used here, refers to possible future sequences of events that will determine the quality of survival of humanity. As examples, we present several scenarios that contrast optimistic and pessimistic futures. We do have choices.

Examples

We selected the following books and articles as examples of scenarios regarding possible future relationships between humankind and nature.

Global Bioethics: Building on the Leopold Legacy, *by V. R. Potter (1988)*

In Figure 11-3, the left-hand sequence starts with the assumption that we will continue to take the short-term view and restrict ethics and law to protecting and promoting the welfare of the individual (with little regard for the public welfare, assuming that what is good for the individual is always good for society and the world). The logical consequences of placing value only on the individual are continued rapid expansion of world population and stressed and degraded life-support ecosystems. Together, these will lead to a less than satisfactory life for all but perhaps a few very rich people, as air, food, and water will be increasingly poor in quality and food and water will be in short supply.

The alternative scenario (the right-hand sequence in Fig. 11-3) is based on the assumption that humans will turn more and more to the long-term view, with value placed on species survival and on maintaining healthy ecosystems worldwide. The logical consequences of extending ethics and law to the species, ecosystem, and landscape levels are reduced population growth (with stabilization and downsizing in this century) and healthy life-support systems, leading to favorable survival for all people and for all life.

"Managing Planet Earth," by W. C. Clark (1989)

Because human predicaments differ dramatically in different parts of the world, priorities for seeking solutions must differ accordingly. In terms of economic conditions and population densities, W. C. Clark (1989) divided the world into the following four regions:

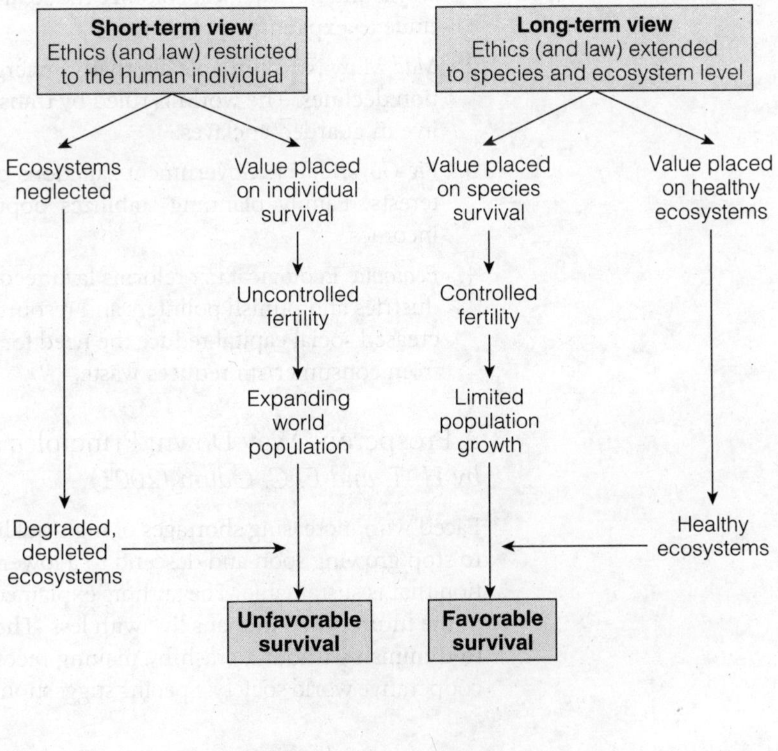

Figure 11-3. A survival model of two contrasting scenarios, a short-term view focused on the individual and a long-term view including species and ecosystem levels (after Potter 1988).

- Low-income, high-density regions (such as India and Mexico);
- Low-income, low-density regions (such as Amazonia and Malaya/Borneo);
- High-income, low-density regions (such as the United States, Canada, and the oil-rich desert kingdoms);
- High-income, high-density regions (such as Japan and northwestern Europe).

In low-income countries, the first priority is reducing poverty, which means promoting sustainable economic growth. In high-density countries, reducing birth rates by family planning or other means should be paramount. Reducing waste and overconsumption of resources should be the first priority of high-income countries, which means shifting from quantitative to qualitative growth (true economic development). Most high-income countries are already well into the demographic transition (reduction of their population growth rates).

The essay "Sustainability Science" by Kates et al. (2001) reduced these four regions to two—the rich "North" and the poor "South." Closing this divide is going to be difficult and will take time.

"Visions, Values, Valuation, and the Need for an Ecological Economics," by R. Costanza (2001)

In an article entitled "Visions, Values, Valuation, and the Need for an Ecological Economics," Costanza (2001) outlined four alternative visions of the future. These four visions of the future are based on whether resources are unlimited or limited, and on two worldviews—technological optimism and technological skepticism. These four visions are as follows:

1. *Star Trek.* Fusion energy becomes a practical solution to most economic and environmental problems. Leisure time increases because robots will do most of the work, and humans will colonize the solar system, so human populations can continue to expand.
2. *Mad Max.* No affordable alternative energy sources emerge as fossil fuel production declines. The world is ruled by transnational corporations whose employees live in guarded enclaves.
3. *Big Government.* Government sanctions companies that fail to pursue public interests. Family planning stabilizes population and progressive taxes equalize incomes.
4. *Ecotopia.* Ecological tax reforms favor ecologically beneficial technologies and industries and punish polluters and resource depleters. Habitation patterns and increased social capital reduce the need for transportation and energy. A shift away from consumerism reduces waste.

A Prosperous Way Down: Principles and Policies, *by H. T. and E. C. Odum (2001)*

Faced with increasing shortages of high-quality energy, the world economy will have to stop growing soon and descend to a lower level of resource and energy consumption that is sustainable. The authors explained how our world can thrive and prosper in the future when humans live with less. They charted a way for modern civilization to diminish without a crash by making recommendations for a more equitable and cooperative world society. Specific suggestions are based on their evaluations of trends in global populations, wealth distribution, energy sources, urban development, capitalism and international trade, information technology, and education.

"Closing the Ecological Cycle: The Emergence of Integrative Science," by G. W. Barrett (2001)

As Figure 11-4 shows, ecology had its roots mainly in the biological sciences, accompanied by an understanding of physics, chemistry, and mathematics. During the past two decades, a proliferation of emerging fields of study has evolved (an increase in "academic fragmentation"), which can be attributed to advances in the medical and ecological sciences, increased needs to implement policies focused on resource management, challenges regarding environmental literacy, and the continuing development of innovative approaches to regional and global planning. Unfortunately, most institutions of higher learning are poorly equipped to administer and to provide resources to ensure the success of these challenges and opportunities. During the twenty-first century, institutions of higher learning, with an infrastructure based primarily on disciplinary education, will increasingly need to foster and establish integrative and transdisciplinary programs—termed *integrative science*—that are necessary to ensure a sustainable future. How will this academic fragmentation be addressed within colleges and universities to increase academic benefits, including a fair reward system for its participants? Will resources be made available from a plethora of governmental and private organizations, especially during a time of increased environmental illiteracy, to address these issues and to meet these challenges?

Colleges and universities that evolved based on disciplinary cornerstones (Fig. 11-4) now face the challenge to administer emerging, integrative programs in a cost-effective and intellectually challenging manner. Interesting enough, ecology—the study of the "home" or "total environment"—appears best equipped to guide higher learning (teaching, research, and service) into a world requiring a holistic approach to problem solving, global planning, ecological literacy, and resource management. Will ecology provide an underpinning for a transdisciplinary, integrative science in the future? A sustainable future likely depends on meeting this challenge.

The Future of Life, *by E. O. Wilson (2002)*

Edward O. Wilson, in his book *The Future of Life* (Wilson 2002), noted two bottlenecks that must be addressed during the coming decades—namely, the increase of human populations and the rate of consumption of natural resources. He noted that humanity managed collaterally to decimate the natural environment and to draw down the nonrenewable resources of planet Earth. Wilson suggested that the challenge of the twenty-first century is how best to shift to a *culture of permanence,* both for ourselves and for the ecosphere that sustains us. Wilson, among others, noted that we have entered the "Century of the Environment."

As the annual rate of human population increase is about 1.4 percent, this adds 200,000 humans each day to the human population, or the equivalent of the population of a large city each week. During the twentieth century, more people were added to the planet than in all of previous human history. When *Homo sapiens* passed the six billion mark on 12 October 1999, we as a species had already exceeded approximately 100 times the biomass of any other large animal species that ever existed on land. All life cannot afford another century like that.

The constraints of the biosphere are fixed; thus, the bottleneck of consumption,

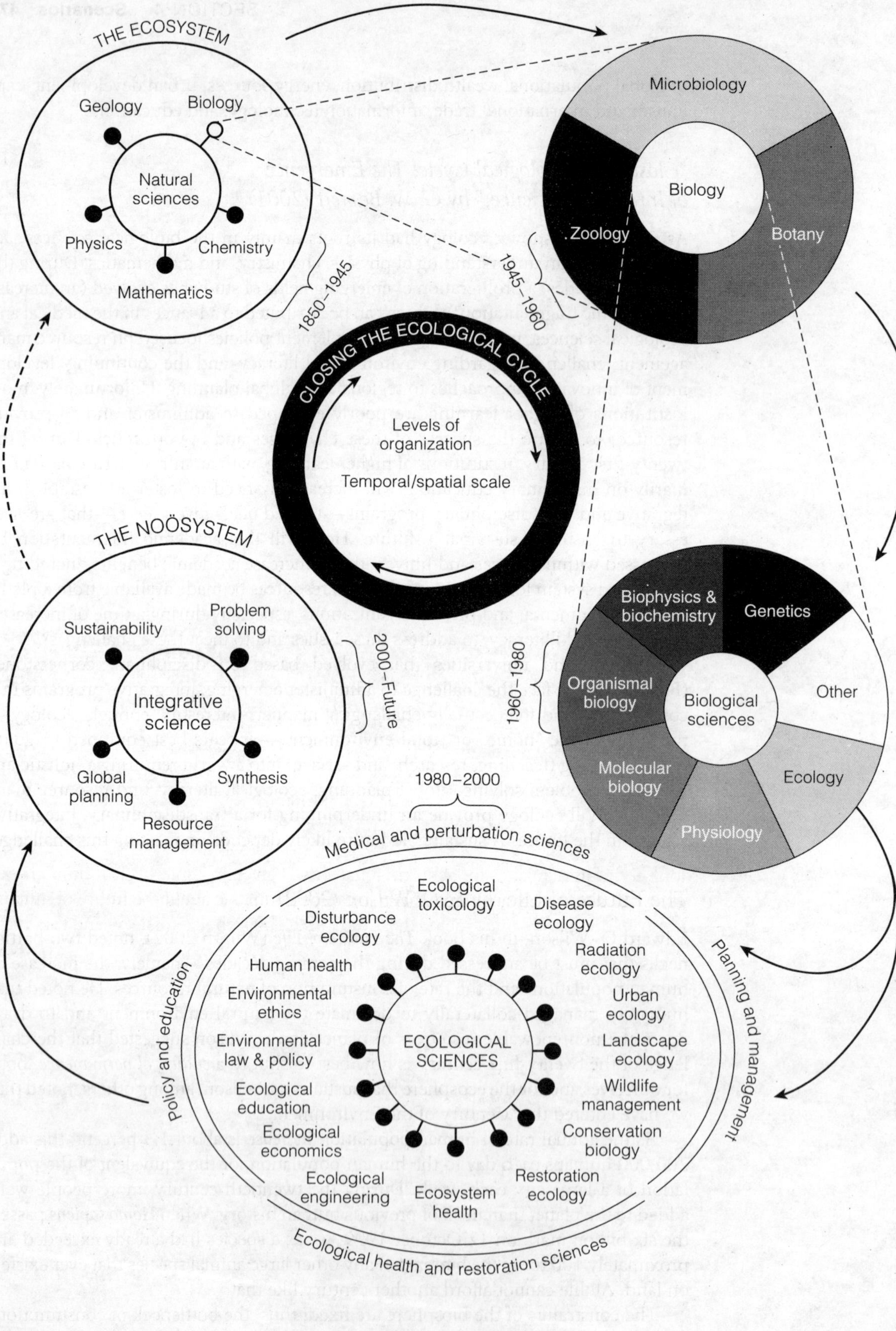

THE ECOSYSTEM
Geology
Biology
Natural sciences
Physics
Chemistry
Mathematics
Microbiology
Biology
Zoology
Botany
1850–1945
1945–1960
CLOSING THE ECOLOGICAL CYCLE
Levels of organization
Temporal/spatial scale
THE NOÖSYSTEM
Sustainability
Problem solving
Integrative science
Global planning
Synthesis
Resource management
2000–Future
1960–1980
Biophysics & biochemistry
Genetics
Organismal biology
Biological sciences
Other
Molecular biology
Ecology
Physiology
1980–2000
Medical and perturbation sciences
Ecological toxicology
Disease ecology
Disturbance ecology
Radiation ecology
Human health
Environmental ethics
Urban ecology
Environmental law & policy
ECOLOGICAL SCIENCES
Landscape ecology
Ecological education
Wildlife management
Ecological economics
Conservation biology
Ecological engineering
Ecosystem health
Restoration ecology
Ecological health and restoration sciences
Policy and education
Planning and management

or the capacity of Earth to support our species, is real. For example, humankind already appropriates 40 percent of the planet's organic matter produced by green plants. Wilson considered the People's Republic of China as the epicenter of environmental change. Its human population was 1.2 billion in 2000 (one fifth of the world's total population). The people of China have exhibited highly intelligent and innovative solutions for survival. Today, for example, China and the United States are the two leading grain producers of the world. However, the huge population of China is on the verge of consuming more than it can produce. The U.S. National Intelligence Council (NIC) predicts that China will need to import 175 million tons of grain annually by 2025. Constraints such as water for irrigation are being addressed by the Chinese government by building the Xiaolangdi Dam, which will be exceeded in size only by the Three Gorges Dam on the Yangtze River. The problem (bottleneck), resource experts agree, cannot be solved entirely by hydrological engineering. Changes must include shifts in agricultural practices, strict water conservation measures, and addressing problems related to water pollution.

Wilson stressed that *environmentalism*—his definition of the guiding principle of those devoted to the health of the planet—needs to become a worldview. He suggested that indifference to the environment springs from deep within human nature, and noted that the human brain evidently evolved to commit itself emotionally only to a small piece of geography, a limited band of kinship, and only two or three generations into the future. The great dilemma of environmental reasoning stems from the conflict between short-term and long-term values. To combine both visions in order to create a *universal environmental ethic* is very difficult. A universal environmental ethic may be a guide by which all life can be safely conducted through the bottleneck into which humans have blundered.

5 Long-Term Transitions

A summary of youth–maturity parallels between human and ecological systems and long-term prerequisites for achieving maturity is given in Table 11-1. If human society can make these transitions, then we can be optimistic about the future of humankind. To do this, we must merge the "study of the household" (*ecology*) and the "management of the household" (*economics*), and *ethics* must include environmental values with human values. Accordingly, bringing together the three "E's" of ecology, economics, and ethics creates a holism commensurate with the great challenge for the future. The late Donella Meadows summarized the future of humanity under the heading "Just So Much and No More":

> The first commandment of economics is: Grow. Grow forever. Companies must get bigger. National economies need to swell by a certain percent each year. People should want more, make more, earn more, spend more—ever more.

Figure 11-4. (*opposite*) A historical perspective (1850–2000) depicting the evolution of ecology and related ecological sciences. This diagram illustrates the disciplinary fragmentation of biology and ecology in a clockwise fashion, leading to a more transdisciplinary and integrated science. (From Barrett, G. W. 2001. Closing the ecological cycle: The emergence of integrative science. *Ecosystem health* 7:79–84. Copyright 2001 Blackwell Publishing.)

Table 11-1 Youth–maturity parallels between individual, biotic community, and society; prerequisites for achieving maturity

Youth–maturity parallels	*Prerequisites for maturity (sustainability)*	*The ultimate measure*
Individual: Transition is called *adolescence.* Biotic community: Transition is called *ecological succession.*	Market capitalism → dual capitalism	*Ecosystem:* $P = R$ (When R exceeds P, the system cannot be sustained.)
Energy flow shift: In early development, the major flow of energy must be directed to *growth.* In later development, an increasing proportion of available energy must be directed to *maintenance* and control of disorder.	Quantitative growth → qualitative growth	*Population:* Natality = mortality
Competition vs. Cooperation: In nature and society, cooperation pays when systems get complex and resources are limiting.	Haphazard development → regional or landscape-level land-use planning	*Biotic community:* Production = respiration
Network "Law": The cost of maintenance (C) increases as a power function–roughly, as a square of the number of network services (N). For example, as a city doubles in size, the cost of maintenance more than doubles	Competition → mutualism (confrontation → cooperation)	*Society:* Production = maintenance

> The first commandment of the Earth is: Enough. Just so much and no more. Just so much soil. Just so much water. Just so much sunshine. Everything born of the Earth grows to its appropriate size and then stops. The planet does not get bigger, it gets better. Its creatures learn, mature, diversify, evolve, create amazing beauty and novelty and complexity, but live within absolute limits. (Meadows 1996)

To bring about the needed change and merger of the three E's, we need to add the two "C's"—*consensus* and *coalition.* Finally, if we can "dualize" current capitalism by combining human-produced goods and services with life-support goods and services (in other words, combining human and natural capital), we can really be optimistic about the future. For an overview of frontiers in ecology and suggestions to strengthen the science of ecology, we refer students to Thompson et al. (2001) and Belovsky et al. (2004), respectively.

12

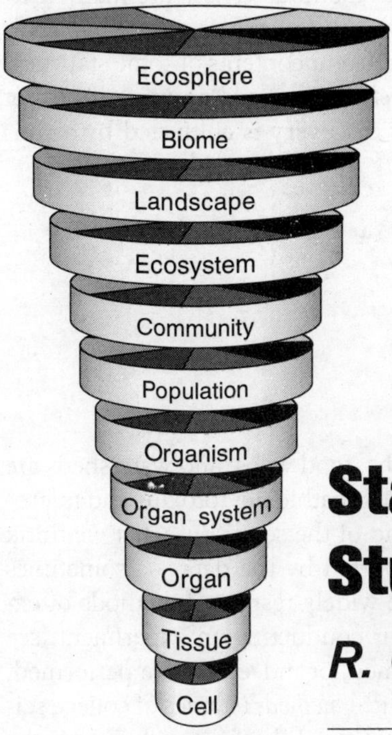

Statistical Thinking for Students of Ecology

R. Cary Tuckfield

1 Ecosystems and Scale
2 Theory, Knowledge, and Research Design
3 The Ecological Study Unit
4 Inference Methods and Reliability
5 Experimental versus Observational Method in Ecology
6 Statistical Thinking in Ecology
7 The Nature of Evidence
8 Evidence and Hypothesis Testing
9 Formulating the Right Problem
10 Publish or "Parish"?
11 The Evidence-Oriented Alternative
12 The Two Ways of Discovery
13 The Weight of Evidence Paradigm

This chapter is designed to challenge students with statistical thinking. It will demonstrate the value of data and statistics as tools for assembling evidence in ecosystem science and may well apply to ecology in a broader context. It is not intended to run the gamut of fundamental statistical methods. There are many textbooks that do that, of which Steel and Torrie (1980) is the exemplary reference. Ecology students should also be familiar with past and current contents of some statistical journals, such as the *American Statistician* or *Biometrics.* But read no more if further inquiry is not an exciting prospect. A passion for discovery is cultivated by a willingness to be temporarily ignorant.

1 Ecosystems and Scale

Statement

Ecology is a matter of scale. For instance, autotrophs, food webs, and watersheds are ecosystem features at the systemic scale, whereas allelopathic phytotoxins and microparasites are features of ecosystems at the other end of the scale. But what we think we know about nature at any scale must be supported by the data—a sometimes daunting statistical challenge even when using the widely respected methods of **experimental design** (that is, statistical planning for conducting an experiment; see Scheiner 1993). Such methods are seldom performed, or indeed can be performed, at the whole ecosystem scale. Thus, standard statistical *methods* (staples of college science) may not always apply, but the *principles* of statistical thinking will. That is the basic premise of this chapter.

Explanation

Understanding our environment at large scales requires thinking about nature within the context of a system—the overarching message of the first edition of *Fundamentals of Ecology* (E. P. Odum 1953). In the preface of the second edition, E. P. Odum (1959) defined *ecosystem science* with enviable brevity as "nature's population problems and . . . use of Sun energy." The capacity to measure the behavior of populations, communities, ecosystems, or landscapes and the flow of solar energy through these levels of organization provides the basis for the hopeful mitigation of human impacts on the regulation of these systems (see Chapters 2 and 3). Measurement data are the intellectual currency for sustaining that compelling cause, and they must be obtainable, interpretable, and sufficiently free of equivocation before advances in ecological understanding will find value in human economics.

Consider, for instance, the current controversies surrounding the global warming of Earth. Walther et al. (2002) suggested that there is "ample evidence" for ecosystem impacts from recent and global climate change. However, not every scientist agrees with the evidence in support of this purportedly large-scale trend. Michaels and Balling (2000) contended that the published recommendations on what to do about global warming do not verify the long-term data trends. Variation happens. After all, what goes up, must come down, and Michaels and Balling believe that planetary "greenhouse gases" have been doing that for millennia. To them, the **evidence** (that is, the interpreted summary of the data) for a justifiable concern about global warming due to anthropogenic activity is equivocal at best and, if anything, supports a view that global warming has resulted in *positive* environmental outcomes. Michaels and Balling (2000) cited, for example, increases in agricultural production and decreases in human mortality in recent years. Their principal concern was that government funding policies for scientific inquiry are misguided by the rush to be on the "side" of the environment. Taking action without knowledge is a dangerous thing. Large sums of money could be spent doing something before we know the appropriate thing to do, and if science is to be a guide, data and politics must have equal footing, at least, in the decision. The ever-present imperative is to summarize the preponderance of the evidence. Suffice it to say that there are reputable scientists on both sides of nearly every large-scale issue.

Methods for getting **unequivocal results** (that is, reliable and consistently interpreted data) are sometimes controversial and often limited by the scale at which observations are made. Studies at microbial scales will undoubtedly yield much information about microbes, but much less about ecosystems, unless we study ecosystem processes (such as rates of decomposition) and patterns among the microbes in a geographic frame of reference. Again, it is a matter of scale, whether in time or in space. An entire hectare of rain forest in Puerto Rico (H. T. Odum 1970), for example, was enclosed to measure ecosystem properties, rates, and throughputs. Certainly teamwork among colleagues (see, for example, Likens et al. 1977) is essential for research projects with such a large scope. But consider the logistics or even the feasibility of a goal to compare by experimental design the evapotranspiration rates between single cubic hectares of rain forest on tropical islands and the Interior Amazon rain forest. This is the problem of using statistical methods developed at one scale that do not translate to another. A brief history of modern statistical methods will be discussed later.

Fundamental concepts of experimental design include **treatment** (that is, action to produce an effect), **control** (no action to confirm no effect), **experimental unit** (the item acted on), and **randomization** (a chance-mediated assignment of items to treatments, to reduce researcher bias). Consider if controlled experimental designs should apply to every scale of investigation, whether microcosm or macrocosm, organismal or landscape, subatomic or cosmological. If the concept of treatment versus control is essential to planning an experiment, can we randomly assign entire ecosystems to a treatment and others to the control? If assigning a whole ecosystem (the *experimental unit*), not to mention an entire galaxy, to the control is out of the question, can efforts to understand these systems still be called science? *Yes*—on condition that we obtain reliable data. The methods for studying galaxies and old fields may be different and are certainly scale dependent, but we are justified in calling these investigations science if we carefully collect the data to which our hypotheses direct us.

Horgan (1996) gave an intriguing account of his interview with Roger Penrose, the well-known cosmologist, and of his subsequent attendance at a philosophy symposium at Gustavus Adolfus College in Minnesota. The fundamental tenet of both discussion forums was the disquieting prospect that we are fast approaching the "end of science" (as we know it, of course)! Astrophysicists, cosmologists, and other universally big thinkers in the scientific community—the ones whose quest it is to search out the deep things of physical reality—are all perilously close to drowning in their speculative hypotheses—hypotheses that generate few if any testable predictions. Horgan called this predicament "ironic science." Whether in physical science or natural science, the bedrock of believability is the same—data. If guesses about the structure of nature recommend no measurable quantity for inquiry, then no matter how exciting the prospect evoked by these guesses, no matter how weird or wonderful the views of nature they promote; they are more likely just good science fiction. In other words, if the real explanation cannot be made evident, we cannot know it. Ecosystem-level explanations on how Earth works require data to be obtainable and interpretable for these explanations to be believable or useful.

2 Theory, Knowledge, and Research Design

Statement

Measurement and observation are crucial activities for obtaining knowledge, but it is not enough to collect and assemble data for the purpose of describing what is there. Mere data do not lead to knowledge—a viewpoint held and fostered by the world-renowned statistician W. Edwards Deming (1993). Data in "raw" or even summarized condition are just information, like yesterday's NBA game scores. Knowledge comes from the theory that provides the incentive to collect the data. Without prior predictability, without an examined expectation for what should be found in the ecosphere, we cannot claim to know much about it.

Explanation

It is one thing to report the NBA game scores, and quite another to have predicted them. The latter demonstrates knowledge convincingly, and the key to that knowledge is theory, expectation, and hypothesis—the creative, tentative explanations for why nature is the way it is and tends to stay that way. Data, therefore, become the exciting prospect for demonstrating that we know what we think we know. This leads to a fundamental principle of data collection, which I refer to as the **Tuckfield maxim**—namely, that *data collected for some general or unspecified purpose can answer very few specific questions.* We are cheerfully free to generate as many explanations as we want, but only the specific data can indicate which, if any, explanation may be right. Here, then, is no small challenge for students of ecosystem ecology—namely, determining the data that *must* be obtained and then interpreting the data that *are* obtained to see how well they match our predictions.

On one occasion, Henry Eyring, a distinguished chemist from the University of Utah, visited Battelle Memorial Institution (Ralph Thomas, personal communication). The chemists of the institution were assembled and asked to give brief presentations of their work. Eyring asked one young man just what he expected to find by conducting his experiment. He replied that he did not know; that was why he was doing the experiment. Eyring then asked the aspiring chemist how he could learn anything from the outcome of his experiment if he had no expectation regarding what he *should* find—the point being that it is what one thinks one is going to learn that forms the basis for what one will learn, regardless of the outcome. Eyring, Deming, and others hold that there is no knowledge without theory.

Theory leads to the important work of **study** or **research design,** which is also scale dependent. One cannot interpret the data one obtains without a prior research design and a corresponding expectation of the results. The complication is that large-scale research often limits design features such as **sample size** (the number of experimental units measured). Sample size is a standard feature of research design, as is the **power** (the probability of detecting a significant treatment effect when in fact there is one) of the subsequent statistical tests that derive from the design. Given a specific *sample size,* **effect size** (magnitude of treatment effect), and **error acceptance criterion** (the risk one is willing to take of making a wrong conclusion), one can calculate the statistical power of the study design. Most textbooks in applied statistical methods (see Steel and Torrie 1980) illustrate this principle. The careful researcher, however, will not focus exclusively on sample size—an all too common shortcoming of many field ecologists.

Hurlbert (1984) has discussed a frequent research design mistake termed **pseudoreplication.** The most common form of pseudoreplication occurs when researchers attempt to increase sample size by measuring the same experimental unit at short successive time intervals and treating all such observations as **independent** (uncorrelated), **identically distributed** (according to the same mathematical probability function), **random variables** (quantities that vary in measurement value due only to chance or treatment). Such measurements are not **eureplicates** (true replicates), meaning different experimental units with a completely separate treatment response. The measurement of an experimental unit at time $t + 1$ will be more or less correlated with a measurement of that same experimental unit at time t, depending on the length of the intervening time interval. The consequence of pseudoreplication is an underestimate of **random variation** (an estimate of the measurement dispersion around the most likely value of a random variable) among experimental units treated alike and results in declaring a statistically significant treatment effect more often than we should.

Recent authors (Heffner et al. 1996; Cilek and Mulrennan 1997; J. Riley and Edwards 1998; GarciaBerthou and Hurlbert 1999; Morrison and Morris 2000; Ramirez et al. 2000; Kroodsma et al. 2001) have amplified this design principle of *pseudoreplicate avoidance* in diverse scientific disciplines. Others (Oksanen 2001; Schank 2001; Van Mantgem et al. 2001), however, have questioned its relevance as an ordering concept in research design, especially regarding large-scale ecological research. Researchers frequently give too much attention to verifying the parametric statistical assumption of **normality** (data conforming to a bell-shaped curve of the normal probability distribution) when the more vital concern is ensuring statistical *independence* among sample measurements, which avoids pseudoreplication. A careful sampling plan promotes a reliable estimate of the variability among different, completely independent experimental units. Without such a reliable estimate, subsequent **statistical inferences** (conclusions based on mathematical summaries of the data) are not meaningful. Departures from normality can often be amended after the sampling is completed; departures from statistical independence cannot. But we will explain later why pseudoreplication at the ecosystem scale is a comparatively lesser issue.

3 The Ecological Study Unit

Statement

The physical size of the ESU (**ecological study unit**) demands careful consideration *before* a study begins. Forging ahead with traditional statistical design principles when the ESU is large may not yield reliable or expected results. This is a common problem in research design that often results from consulting a statistician only after the study is over. This represents a latent feedback approach to scientific investigation—latent because we are likely to discover features that could have been included in the experiment.

Explanation

Because experiments are costly enterprises, it is understandable why students and researchers alike desire a definitive outcome for their effort. But collecting some data to see what they tell one is a practice that should repeatedly evoke skepticism. This is simply trial and error, not design. Research design is a "feed-forward" process. It is the initial phase of every experiment and should drive the subsequent data collection effort—always. How one will later analyze the data is also part of the design and avoids the typical *post hoc* decision quandary about choosing a method that will yield statistical significance.

For constructing hypothesis tests about why some ants are red and others are black, for example (G. E. P. Box et al. 1978), a controlled experimental design is a useful method. Finding enough ESUs should not be a problem. But should we expect effective designs at one scale to apply to all scales? For example, should we expect a design used to discover why the upland sandpipers (*Bartramia longicauda*) do not breed on their wintering grounds in Argentina (where it is the breeding season for native birds) to be equally effective in testing hypotheses about greater carbon sequestration in tropical compared to temperate forest ecosystems? Even if the same experimental design could be applied, getting enough ESUs to allow reliable statistical inferences becomes increasingly unlikely with increasing scale. It is not that ecosystem ecologists lack sufficient determination, nor can they be accused of being small thinkers. It is just a property of nature that large-scale emergent patterns and effects are difficult to identify from small-scale investigations. Collecting enough data to make a strong inference at the intended research scale is a problem with modern statistical methods. Simply put, ants are more numerous than sandpipers, sandpipers are more numerous than rain forests, and rain forests are more numerous than . . . well, planets—at least for treatment purposes. With each increment in scale, the statistical requirements of standard experimental designs are less and less achievable. Design implementation is always fraught with practical impediments that can only be addressed in the hour of need and that multiply with the scale of the ESU. Unless we can devise data analysis methods that are scale independent—that generate equally reliable results regardless of sample size—research designs are likely to be as unique to the scale of research as they are common to the scientific discipline. The ant and sandpiper scales permit the use of experimental controls, but it is unlikely that the same could be said for rain forests and planets.

Ecologists are keenly aware of this problem. E. P. Odum (1984) advocated the use of the field-scale *mesocosm* as an extension of the bench-scale *microcosm* to understand more fully how *macrocosms* behave (see Chapter 2). The intent was to bridge the gap between the closed, laboratory-sized experimental systems and the large, complex, open systems of living things powered by the energy of the Sun and modified by regional geology. This concept is certainly an advance in research scale and gives the opportunity to observe, quantify, and elucidate patterns of functionality that are otherwise unobservable. The suggestion is obvious: the next task is to create **metacosms** (worlds between the meso and macro scale) for ecosystem research. It is another step up in scale—one that would identify patterns and principles of ecosystem function and that would require even smaller leaps of extrapolation about what is going on in the ecosphere. The problem, of course, is that the larger the scale, the greater the sampling complexity, and the fewer the observations that we can feasibly acquire. The latter means that the widely extolled and reliable statistical methods developed for small-scale experimental research will be increasingly difficult to apply to large-scale investigations.

Moreover, the larger the ESU, the more similar individual ESUs will be to one another (J. Vaun McArthur, personal communication). This results in treating unlike things alike, whether biotic or abiotic—the very antithesis of the principle of *experimental design.* The more alike ESUs are in the experiment, the more measurement precision we acquire in detecting *treatment effects.* This can be illustrated clearly in the simplest of experimental designs, which have only one **main effect** (treatment) with multiple **treatment levels** (graduated quantities or types of treatment), a so-called "one-way design." Typically, the *less* similarity among ESUs, the *more* response variation there will be to a treatment. The variation between treatments is less likely to dominate, and the treatment effect, therefore—if there is one—will not be detectable. The only recourse is to increase the *sample size* (n_j) for all *treatment levels* ($j = 1, \ldots, k$). This is because the unbiased estimate ($\hat{\sigma}^2$) of the true response variance (σ^2) for all ESUs in the population is calculated as the near-average sum of squared deviations from the response **mean,** or

$$\hat{\sigma}^2 = \sum_{j=1}^{k} \hat{\sigma}_j^2 \qquad (1)$$

and

$$\hat{\sigma}_j^2 = \frac{\sum_{i=1}^{n} (X_{ij} - \overline{X}_j)^2}{n_j - 1} \qquad (2)$$

where

($\hat{\sigma}_j$) is the sample response variation for the *j*th treatment and an estimate of the corresponding population response variation (σ_j),

X_{ij} is the ith response measurement for the *j*th treatment,

$\overline{X}_j$ is the sample average (mean) response measurement and is an estimate of μ_j, the population average for the *j*th treatment,

and

n_j is the number of experimental units (that is, replicates) measured for the *j*th treatment.

Note that as n_j increases, equation (2) decreases. The worst case is when the experimental units or ESUs are so dissimilar as to make it impossible to distinguish the response differences due to treatments from response differences due to variance between ESUs. This undesirable outcome is called **confounding** and is precisely what designed experiments ideally attempt to avoid.

To illustrate, let us hypothetically compare a treatment method for reversing the effects of freshwater lake eutrophication with a control. Freshwater lakes are well-studied and distinct ecosystems. Our ESU is a lake. Suppose that we want to restrict our **scope of inference** (target population and breadth of conclusion generality) to eutrophic lakes in the southeastern United States, in order to reduce travel costs and to reduce extraneous sources of variation due to differing climates and watershed chemistry. We will be careful to randomly assign each eutrophic lake selected for study to either the treatment or the control condition. The study will begin in the fall, and we will measure the success of this treatment method by measuring the phosphorus concentrations in 12 epilimnion water samples collected during spring turnover (see Chapter 2 on lake dynamics). Twelve sample measurements per treatment would seem to be enough. The problem here is that we want to say something about lake-to-lake responses to the treatments, not sample-to-sample responses. We will need more lakes (more *eureplicates*) in each treatment category and fewer repeated measurements within lakes. Aside from logistical and technical feasibility, the fact is that lakes with very similar surface areas and volumetric measurements can be very different biologically, as reflected in species richness and biodiversity measures at a variety of levels of organization (see Chapter 2). Even if we could show a treatment effect, the reliability of the predicted response for lakes other than the ones treated in the study is questionable. This is an undesirable consequence for the ecologist, who wants to say something more general about eutrophic lakes and term it knowledge.

4 Inference Methods and Reliability

Statement

The axiom of nature revealed in the previous section may be termed the **inference-reliability tradeoff**, and is illustrated in Figure 12-1. The *scope of inference* is inversely proportional to the scale of the ESU. An experimental treatment may be applied to large-scale ($>10^6$ m^2 or $>10^5$ m^3) ESUs, should the research funding be sufficient, but any inference regarding a typical measurement response from another ESU on the same scale is questionable. Broad conclusions can only come from the narrowest of differences among the ESUs themselves.

Explanation

The aforementioned axiom simply means that we will demonstrate a statistically significant treatment effect at traditionally selected **significance levels** (maximum acceptable percentage chance of an erroneous conclusion) when the p value is at a $p < .05$ level of probability among large-scale ESUs treated alike. If we let p represent the probability of getting the data we did when in fact the treatment has no ef-

Figure 12-1. An illustration of the ecological axiom called the inference-reliability tradeoff, where generalization about the treatment response of other ecological study units (ESUs) based on the results of a controlled experimental design are increasingly diminished in scope with increasing size of the ESU.

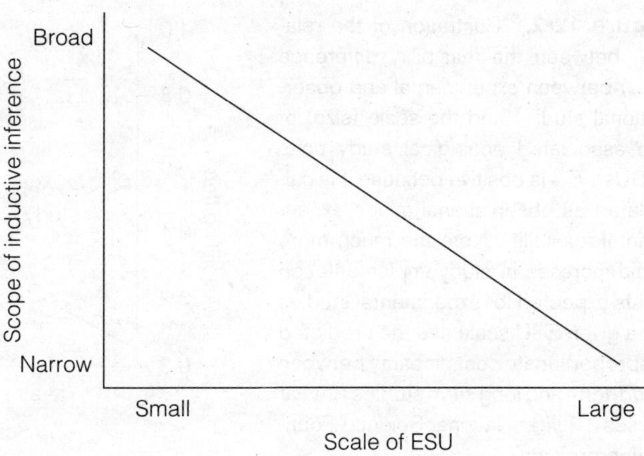

fect, we use a *p*-value to represent the probability of being wrong—of concluding that treatment effects are real when they are not. For instance, $p < .05$ means that we would expect to obtain data like those we collected only 5 percent of the time if our null hypothesis is true. This is fundamental thinking in the discipline of statistics.

Modern statistical inference and data analysis methods were largely developed in the twentieth century (Salsburg 2001) to test hypotheses about relatively small-scale, observable phenomena. Many investigations were undertaken in Great Britain and in the United States to discover more effective crop varieties, rates of fertilizer application, or growing conditions where the *experimental unit* was the plot of soil or even the individual plant. **Experimental units,** like ESUs, are the entities measured in response to the experimental treatment. In agricultural science, however, experimental units are relatively small and very much alike. In fact, the more alike they are, the less response variation there will be to the same treatment, which means that treatment-to-treatment variation in the response should dominate—if the treatment is effective. Treating like things alike, as discussed earlier, is the key to a successful experiment, at least at scales well below entities like ecosystems or landscapes, and maybe even landscape patches (see Chapter 9).

Salsburg (2001) described the impact of one statistician, R. A. Fisher, on the development of statistics as a modern scientific discipline, portraying the state of affairs among experimenters in the late nineteenth and early twentieth centuries as a "mess of confusion and vast troves of unpublished and useless data." This was particularly true of agricultural research. Salsburg (2001) stated that,

> The Rothamsted Agricultural Experimental Station, where Fisher worked . . . had been experimenting with different fertilizer components (called "artificial manures") . . . for almost ninety years before he arrived. In a typical experiment, the workers would spread a mixture of phosphate and nitrogen salts over an entire field, plant grain, and measure the size of the harvest, along with the amount of rainfall that summer. . . Fisher then examined the data on rainfall and crop production . . . and showed that the effects of different weather from year to year were far greater than any effect of different fertilizers. To use a word Fisher developed later in his theory of experimental design, the year-to-year differences in weather and year-to-year differences in artificial manures were "confounded." This means that there was no way to pull them apart using data from

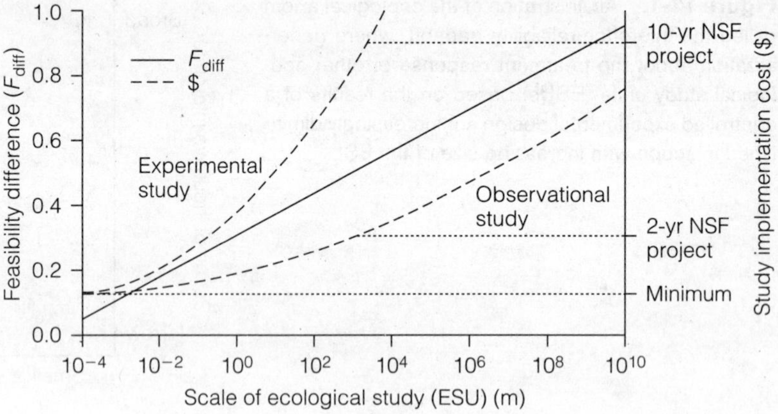

Figure 12-2. Illustration of the relation between the feasibility difference (F_{diff}) between experimental and observational studies and the scale (size) of the associated ecological study units (ESUs). F_{diff} is positive because it is calculated as observational minus experimental feasibility. Note the much more rapid increase in study implementation costs expected for experimental studies for a given ESU scale and the predicted disproportionate cost disparity between short-term and long-term studies (2-year versus 10-year National Science Foundation projects).

these experiments. Ninety years of experimentation and twenty years of dispute had been an almost useless waste of effort! Agricultural scientists recognized the great value of Fisher's work on experimental design. And [his] methods were soon dominating schools of agriculture in most of the English-speaking world.

Experimental design, as a statistical method for obtaining knowledge, has been adopted by many if not most branches of natural science. The boon of this methodology to science is the structure it provides for making reliable **inductive** (reasoning from specific to general conclusions) **inferences.** However, a comparable statistical method for partitioning the "noise" from the "signal" at the system, cycle, or global scale does not exist, principally because the experimental units are dissimilar, difficult to define, and too few in number to yield a sufficiently powerful statistical inference.

Ecosystem ecology, however, is like another holistic science, namely, cosmology. It is no more feasible to apply the designed experimental methods of R. A. Fisher to study solar systems than it is to apply them to the study of large-scale ecosystems or landscapes. *Holistic science,* incidentally, is considered nearly an oxymoron by most statisticians, given the deep historical roots of Baconian tradition and *reductionist* culture in Western civilization. Ecosystem ecology attempts to grasp the whole of what is going on biotically and abiotically in nature, but whether one upscales or downscales the research to find out what is going on, there is always a tradeoff.

Figure 12-2 illustrates the relationship between inference scale and study feasibility. Let us define **feasibility** as the degree of assurance that a particular study design is logistically accomplishable, ranging in value from 0 to 1. We will distinguish between the concepts of *experimental* and *observational feasibility*—the former based on the principles of *controlled experimental design* and the latter based on the principles of the **hypothetico-deductive** (reasoning from general to specific conclusions) **inference method** (Popper 1959), a method for verifying hypotheses by directed observation. The human capacity to observe nature is greater than the human capacity to control it. In fact, without seemingly incredible technological advances (which most continue searching for), tractable experimental control can only be exerted on relatively small-scale ESUs. *The larger the scale, the less the experimental control.* Experimental feasibility is much more dependent on scale than observational feasibility. The latter, because of our capacity to observe, is always nearly 1—relatively speaking. Therefore, the simple difference of observational minus experimental feasibility (F_{diff}) is nearly 0 at small scales and nearly 1 at large scales. It is also likely that study implementation costs rise even faster with increasing study scale, as demonstrated by the willingness of the United States federal government to allocate billions of dollars to the National Aeronautics and Space Administration for the management of a space program (7.02 billion dollars for NASA in fiscal year 2003) compared with the million-dollar funding sources to the National Science Foundation for the study of ecosystems (0.09 billion dollars for NSF in fiscal year 2003).

Figure 12-2 also illustrates the ever-widening gap between the cost of experimental and observational studies as a function of scale. It may be argued that experimental studies are preferred over observational studies regardless of cost, because experimental studies can eliminate equivocal results. However, uncertainty prevails and variation rules in every human endeavor. Frequently, the results of one experimental study do not agree with another, as is the case in medical studies concerning alcohol consumption and heart disease.

5 Experimental versus Observational Method in Ecology

Statement

An alternative to controlled ecological experiments, which are nearly infeasible at large scales, is the diligent search for predicted results by verifiable observations. There is power in observation—power to grasp by rumination over what we can discern. Besides, there is not much choice. The alternative is no data—an unacceptable outcome if inquiries are to remain in the realm of science.

Explanation

Observation can identify undiscovered patterns and can generate ideas about what we should expect to find. Robert H. MacArthur, a gifted population ecologist and able mathematician, held this viewpoint. In his book *Geographical Ecology,* MacArthur (1972) said that "to do science is to search for patterns, not simply to accumulate facts." He attempted to counter a public misconception about the seemingly cold and calculated character of science that supposedly robbed nature of aesthetic quality:

> Doing science is not such a barrier to feeling or such a dehumanizing influence as is often made out. It does not take the beauty from nature. The only rules of scientific method are honest observations and accurate logic. To be great science it must also be guided by judgement, almost an instinct, for what is worth studying.

Feeling, beauty, judgment? MacArthur's point was that the human endeavor called science is, at the root, observation driven by the passion for insight and understanding.

Consider the power of a single observation made in 1919 that confirmed an intriguing prediction—a measurement that forever removed Newtonian physics from its pedestal. Arthur S. Eddington, a noted British astronomer, was one of a few scientists who grasped the fundamentals of Einstein's general theory of relativity. He proposed to test a prediction of the theory—that light could be bent by the gravitational force of massive celestial objects like the Sun. He organized an expedition to Principe Island located near the coast of West Africa in May 1919 to observe a solar eclipse. Knowing the positions of certain stars relative to Earth as it revolves around the Sun, a Newtonian astronomer can calculate the exact time of a specific appearance of a star as its starlight emerges tangentially from behind the obscuring solar disk. If this star were to appear in a photograph before its calculated time of appearance (because its light was bent by the Sun's gravity), it would be consistent with the prediction of Einstein. Eddington's photographic plates did indeed capture such starlight (McCrea 1991) and provided evidence in favor of Einstein's theory. It was not a controlled experiment, but it was an *observational experiment*—one that is eminently repeatable by anyone in possession of the same measurement technology. In short, the power of this observation was in turning the seemingly intestable into the verifiable.

6 Statistical Thinking in Ecology

Statement

Gaining knowledge about ecosystems and landscapes and about large-scale interrelationships between biotic and abiotic phenomena has more to do with statistical thinking than with statistical tests and more with sampling uncertainty than with sample size. **Statistical thinking** is a concept borrowed from Snee (1990). It is offered as a practical paradigm (example, pattern, or archetype) for the pursuit of ecological knowledge, founded on just three principles:

- All natural processes occur within the context of a system.
- Process measurement is fraught with variation and uncertainty.
- The *weight of evidence* from multiple observations and different measurement collections, effectively and visually portrayed, is crucial to judging the merit of any explanation or hypothesis about the natural processes for which the data were collected.

Without such an approach, neither students, colleagues, nor the public will be convinced that anything is known about ecosystems, global warming, or continental carrying capacity (see Chapters 2, 4, and 6), and the influence of ecology on the social stewardship of the environment will diminish.

Explanation

Let us briefly review what a knowledge of statistics provides as part of science education. The intent here is simply to assess what is pertinent in statistical methods to increase our understanding about ecosystems.

First, statistics has become as essential to a balanced education in ecology as evolutionary biology or genetics. This is largely due to the influence of Karl Pearson, who founded the science of biometrics (Salsburg 2001), and to the impact of R. A. Fisher on multiple scientific disciplines. Fisher's (1935) work on the design of experiments, his methods for analysis of variance (ANOVA), the notion of statistical significance, and the principle of randomizing experimental units to treatments have become the staples of careful research in the natural sciences, whether in the classroom, laboratory, or field. Of equal importance to the future of ecology was Fisher's work *The Genetical Theory of Natural Selection* (Fisher 1930). More than any other of its day, this book was responsible for the synthesis of Mendelian genetics and Darwinian evolutionary theory by cogent argument based on statistics and mathematics. Provine (1971) stated that,

> Darwin's assumption of blending inheritance was that "the heritable variance is approximately halved in every generation." Thus, Darwin's theory required the appearance of an enormous amount of new variation each generation. Fisher showed that Mendelian inheritance offered a solution to this problem in Darwin's theory because it conserved the [statistical] variance in the population.

Fisher had put statistical methods on the map for inquiring students of ecology. It was good timing as well for other emerging scientific disciplines at the beginning of the twentieth century (Barrett 2001).

The essential concept that powerfully underlies modern statistical methods is that of the *population*. The **population** refers to the collection of all ESUs and is the usually uncountable "ideal" of our scientific inquiry (see Chapter 6). As in Plato's well known assertion from the *Republic* (Book VII), the population is the immeasurable entity—the ponderous truth that we cannot know with certainty by our prescient capacity and that can only be glimpsed from the shadows we observe of it. In short, it is the unknowable ideal of reality that we call nature. The "shadows," incidentally, are called *samples* (small subsets of the population), from which we gain hints and intimations about the population that is really there. Fisher's (1930) foundational statistical precept was that large populations have more genetic variation than small populations in response to the environment; in consequence, genetic variation tends to preserve large populations and to threaten small populations. The quandary is that we can seldom measure every member of the population. We may redefine the population to preserve countability, but we trivialize our intended inference in the process. Herein lies the brilliance and the limitation of statistical methods. Because it is too costly and often impractical to count or measure every member of the population, we must be content with taking a **sample**—a representative, small number of individuals (or ESUs) from which we may predict, envision, and infer the character and dynamics of the population. We trade the foolish quest for certainty for a wiser outcome with **uncertainty** (a measurable quantity of variation).

As stated earlier, statistical methods may not always apply, but statistical thinking will. The underlying, incontrovertible construct of this truism is that *variation happens.* Variation is as much a property of nature as is the boiling point of water at sea level (100° C). We simply cannot measure anything without error or *uncertainty.* On the one hand, we *posit* what we think must be true about nature—that certain but unknowable ideal of reality. These are often bold assertions that derive from innovative thinking and are buttressed by mathematical rigor. On the other hand, we seek to *test* these assertions about nature by measuring what we can of nature—by collecting and analyzing data. Einstein's viewpoint leaned toward the theoretical ideal. Uncertainty was not a property of interest in the physical world he studied, because he sought the Platonic ideal—what must be true and ultimately portrayable—by the spare language of mathematics. Einstein's conviction was that all things must derive from fixed properties and cosmic constants, regardless of how well we can measure them. However, when it comes to corroborating knowledge about the natural world, it is requisite to match explanation with metric and expectation with measurement.

Variation happens, whether among experimental units treated alike or in the observational measurement of greenhouse gases in the atmosphere. There are a variety of causes of variation, most of which are unknown and, perhaps, unknowable. These we lump into the category of *random variation.* They are causes that cannot be managed or manipulated—forces of chance that are a function of the sampling process, much like reaching into an urn containing black and white marbles and counting the number of black ones in successive handfuls. That number will vary with each handful, but some values are likely to be more frequent than others. *Systematic variation* occurs, for example, when one researcher uses a NIST standard 1-liter bucket to add water to a mesocosm experiment. When this field technician goes on vacation, his or her untrained replacement uses a different bucket to continue the watering regimen. In the end, the recorded number of liters of water added differs from the actual number by a fixed amount for each bucket added during vacation and systematically perpetuated after the technician's return. *Distinguishing the difference between random and systematic variation is indispensable to the application of statistical thinking.*

7 The Nature of Evidence

Statement

Scientific theories are often depicted as **deterministic** (that is, without a context in probability or measurement uncertainty), as, for example, the theory of the proportionality of mass and energy or the theory of ecological efficiency. However, the corroboration or refutation of the theory rests with empirical results—the data—and one role of statistics in ecological science is to ensure that the research design will provide pertinent data with measurable uncertainty and, therefore, sufficient evidence to test the theory (Tuckfield 2004).

Explanation

Note the word **corroboration** used in the Statement. It means "to support by additional evidence." This is a feature of what the philosopher Thomas Kuhn called "normal" science (Kuhn 1970). Empirical evidence is amassed during those periods of human history when everyone is shoring up and corroborating the current or normal view of the world. Karl Popper, another well-known philosopher of science, saw the concept of corroboration (Popper 1959) like driving piles into the muck. You do not have to drive them so deep that they strike bedrock, but deep enough to support the superstructure—the theory, the model, the current explanation for why things are the way they are. Certainty, however, is out of the question. The lack of success from repeated attempts to falsify a hypothesis only drives the piles deeper and deeper and portrays the understanding of nature as uncertain. Under the weakness of insufficient evidence, the superstructure could topple.

This *falsification* approach to doing science (Popper 1979) is like peeling leaves from an artichoke to discard the falsehoods and expose, by successive refutations, the heart and veritable center of what *must* be true. Others hold that verifiable evidence accumulates until each of us has sufficient evidence to warrant belief in the theory. Most scientists find greater satisfaction in verifying hypotheses than in falsifying them. In fact, very few of the articles published in the journal *Ecology* that report the results of a statistical test are offerings toward hypothesis falsification rather than verification. What is summarized as verifiable insight is not depicted without judgment. What we are willing to accept as knowledge, if not by appeal to authority, is based on the extent of corroborated results—the **weight of evidence.**

Examples

Let us examine the widely acknowledged, fundamental tenet of the *theory of ecological efficiency,* which states that only 10 percent of the energy at one trophic level is transferred to the next trophic level (Phillipson 1966; Kormondy 1969; Pianka 1978; Slobodkin 1980). Slobodkin (1980) defined the ecological efficiency of "trophic level x, which is fed upon by trophic level x + 1, [as] the food consumption of trophic level x + 1 divided by the food consumption of trophic level x." He cited an earlier study by Patten (1959) in which 10 different field estimates of ecological efficiency ranged from 5.5 percent to 13.3 percent. Kormondy (1969) quoted Slobodkin (evidently from personal communication) as believing that gross ecological efficiency "is of the order of 10 percent." Pianka (1978) reported that "after standardization per unit area and unit time," most ecologists would estimate ecological efficiency to be between 10 and 20 percent. Slobodkin's belief and Pianka's contention are simply assertions about a property of nature with loosely defined features and unquantified variability. But the data are what provides the *evidence* for judging the merit of the theory. The first table in Patten's (1959) paper summarizes the energy (gcal · cm^{-2} · $year^{-1}$) intake at each of four trophic levels (producer, P; herbivore, H; carnivore, C; and top carnivore; TC) for each of four aquatic ecosystems—Lake Mendota, Wisconsin (Juday 1940); Cedar Bog Lake, Minnesota (Lindeman 1942); Root Spring, Maine (Teal 1957); and Silver Springs, Florida (H. T. Odum 1957). Although the sample size is small, the estimated average ecological efficiency among these four ESUs was 90.5 percent for the H-P comparison; 11.7 percent for the C-H comparison; and 4.6 percent for the TC-C comparison. Obviously, ecological efficiency is *not* 10 to 20 percent regardless of the trophic levels involved. If this theory is important to ecosystem management, then more reliable estimates are needed, which will likely depend on latitude as well as trophic level.

Let us examine some evidence from real lake experiments. Schindler (1990) provided an excellent summary of findings for the experimental manipulation of lacustrine ecosystems in the Experimental Lake Area (ELA) of the Canadian Shield. Although the project began in 1968 to test hypotheses about lake eutrophication management practice, threats from oil and hydroelectric power production and industry-mediated acidification kept the project going well into the 1980s. This research also provided the opportunity to address the predictions of a more general theory of ecosystem response to stress developed by E. P. Odum (1985). Of the 18 predictions tabulated in Odum's (1985) article, data from the ELA studies over a 20-year period were used to evaluate each prediction, save 3. Schindler (1990) provided a table of these same predictions in one column, matching the ELA research outcomes relevant to each corresponding prediction in two other columns—one for lake acidification experiments and one for lake eutrophication experiments. Let us assign a rating value of 1 to these results when they confirm the prediction, -1 if they refute the prediction, and 0 if the results were equivocal. The overall sum of these values was 1 for the acidification experiments and 6 for the eutrophication experiments. The maximum and minimum possible values for both are 15 and -15, respectively. A prediction test ratio, R_{pt}, indicating the *weight of evidence* in favor of Odum's (1985) general theory of ecosystem stress response can be calculated as the rating value sum divided by the number of predictions, which will always vary between -1 and 1. The R_{pt} was 0.07 for the acidified ELA lake data and 0.40 for the eutrophic ELA lake investigation data.

Conclusions? Evidence supporting this general theory of stressed ecosystems is nearly equivocal (0.07) in substantiating the predicted behavior of acidified lake ecosystems, and weak (0.40) in substantiating the predicted behavior of eutrophic lake ecosystems. This leads to some interesting questions: Why did these aquatic ecosystems respond differently to each type of stress? If anthropogenic stresses need to be categorized, is the theory really a "general" theory? Should a general theory of ecosystems focus solely on perturbations? Would a general theory be more general if it were to focus on the mechanism of ecosystem resilience or recovery (homeorhesis; see Chapter 1)? The preceding queries beg the ultimate question: exactly what is the theory of ecosystems? This question is intriguing and certainly fires the imagination, but it is beyond the scope of this chapter. Nevertheless, it is worth noting that this ultimate question arose as a consequence of weighing the evidence in the data without the aid of statistical significance testing.

8 Evidence and Hypothesis Testing

Statement

The evidence obtained during any ecological investigation—whether experimental or observational—pertains to the *scientific hypothesis*. Almost without exception, the standard statistical methods require converting the *scientific* hypothesis into an equivalent *statistical* one, which may not resemble the hypothesis that one thought one was going to test. This is necessary for clear thinking and logical outcomes to prevail. The data speak for themselves, however, and the task for the student of ecology is to present the data in a manner that directly addresses the scientific hypothesis.

Explanation

The same standard college education in statistics will present the concept of the null hypothesis (H_0). The **null hypothesis** is an initial surmise that a sample of size N individuals drawn from a population in nature will have come from a population with a known characteristic or parameter, such as the mean (μ) or average value. When only one sample of size N is drawn, the null hypothesis is H_0: $\mu = \mu_0$. That is, the population mean is equal to some specific hypothesized value μ_0. An **alternative hy-**

pothesis can be stated as $H_1: \mu > \mu_0$. These are statistical hypotheses in the *Neyman-Pearson style* (Royall 1997). We might wish to know, for example, if black-throated sparrows (*Amphispiza bilineata*) in southern New Mexico differ in body weight (g) from those in central Nevada, where they are sympatric with sage sparrows (*A. belli*), their larger congeners (Tuckfield 1985). If we state that we believe black-throated sparrows to be larger in Nevada than in New Mexico, that is a scientific hypothesis. In order to use statistical methods to test this scientific hypothesis, we must translate it into something like $H_0: \mu_1 - \mu_2 = 0$ (with μ_1 being the mean body weight in Nevada and μ_2 the mean body weight in New Mexico, for example), which is referred to as a *two-sample null hypothesis*. It simply means than unless there is sufficient evidence to the contrary, we will accept the tentative conclusion that black-throated sparrows are *no* different in body weight between these two locations.

We could be wrong in our conclusions even with the most careful of interpretations from the data. We may erroneously conclude that the two populations are different when in fact they are not—that is, a **Type I error** (false positive conclusions) —or that they are not different when in fact they are—that is, a **Type II error** (false negative conclusions). It is the nature of conclusions that you cannot be certain. Ideally, we would like to reduce the chances of making such erroneous conclusions, but depending on the situation this can be a costly enterprise. We may, for example, tolerate a risk of making a false positive error 5 percent of the time or less, but we simply may not have sufficient funds or human resources to collect all the samples needed to minimize both types of errors. The merit of the Neyman-Pearson style of hypothesis testing is the apparent objectivity. The outcomes are black or white. Either $H_0: \mu_1 - \mu_2 = 0$, or $H_1: \mu_1 - \mu_2 \neq 0$. The latter is a *two-sample alternative hypothesis*.

Experience suggests that simple displays of ecological data often provide more interpretable evidence than these sorts of either-or test outcomes. Even so, formal statistical inference methods have led to practical and vital treatments in medical and pharmaceutical research. In these disciplines, even small improvements relative to standard treatment regimens, if statistically significant, could benefit a large number of people. The scale of measurement and the experimental unit are both small relative to an ecosystem and are well suited to the Neyman-Pearson style of statistical hypothesis testing.

The heavy-handed influence of this style, however, is the portrayal of science as proceeding inexorably and *objectively* toward the truth under the skillful hands of the erudite practitioner of statistical inference, regardless of the scientific discipline. This is illogical. *If there is no place for judgment, one has no knowledge,* and data displays offered as evidence permit both flexible interpretation and judgment regarding the scientific hypothesis. Scientific knowledge is, for the most part, a widely regarded and effective model of the truth—but it is a model nonetheless. It persists by peer-reviewed consensus and possesses reasonably quantifiable uncertainty (excluding deterministic physics, of course). What is knowledge to a paleontologist, after all, if not solid-rock judgment in the face of *absolute* uncertainty? It is this *subjective process* of weighing the evidence that determines how much one knows about what is really going on in nature—and weighing means judging.

In a tribute to the late Robert H. MacArthur, Stephen Fretwell (1975) stated, in effect, that *truth is what is. Knowledge is what we think we know about the truth, and wisdom is knowing the difference.* Wisdom is likely more than this, but it is certainly wise to know the difference.

9 Formulating the Right Problem

Statement

Just as a respectable curriculum in ecology is incomplete without a knowledge of statistics, aspiring ecology students are not without knowledge of, say, how to perform an ANOVA (see just about any statistics text). Students often focus on acquiring a broad range of skills for analyzing data—as well they should. But more important than the development of a big statistical "tool chest," it is the formulating of the right question or problem that will accelerate understanding.

Explanation

This perspective has to do with asking *answerable questions* and discovering the problem to be solved. Problem identification and formulation are the precursors to method selection, not vice versa. To do otherwise is to risk committing **Type III errors**—that is, producing an elegant solution to the wrong problem (Kimball 1957). The latter is probably more of a risk to students in statistics who are developing consulting skills than to students in ecology, because ecologists are a quantitative lot and take a great deal of pride and initiative in analyzing the data *themselves*. However, if the focus is on the method, what happens if a suitable method does not exist? Well, one should rely on the local statistician at this point, but that does not mean that the student of ecology has reached an impasse. The task is to summarize the data in such a way as to provide evidence regarding the scientific hypothesis. Herein lies the strength of simple data displays for visualizing results. Thus, we return to the fundamental issue of this chapter—the issue of using data and statistical thinking, not just statistical methods, to gain more knowledge about ecosystems and ecology.

10 Publish or "Parish"?

Statement

"Publish or parish" is a variant of the common aphorism *publish or perish,* well known among students who aspire to careers in ecological science. It refers to a *bona fide* decision for the young Charles Darwin, who had to choose between pursuing his interests in science or in theology (Irvine 1955). This career decision signifies a real challenge facing most students—toward securing a position within their profession. Practicality is advisable and often unavoidable, but the passionate pursuit of understanding is at the root of most ecological aspirations.

Explanation

Discovery is a very gratifying outcome and is worth pursuing. As with all scientific vocations, the profession of ecology is a means to perpetuate the excitement of learning and, more specifically, learning about what nature retains within itself for discovery.

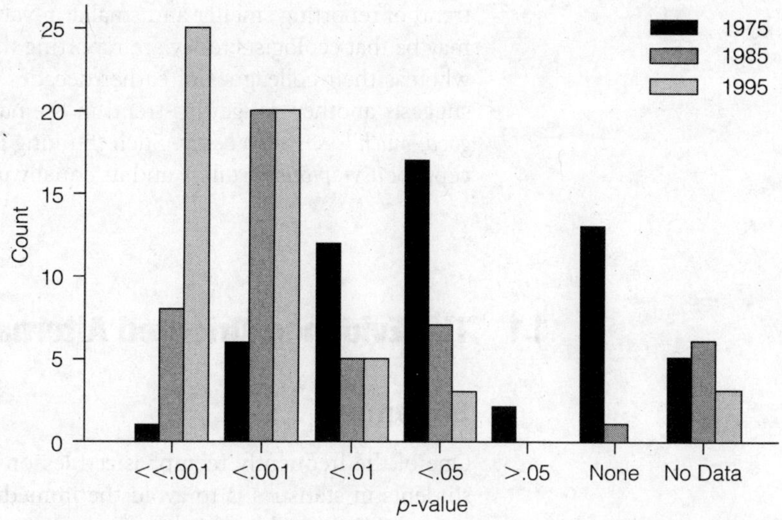

Figure 12-3. Frequency distribution among categories of *p*-values reported in articles published in three volumes of the journal *Ecology* during 1975, 1985, and 1995. Note the gradual skewing toward smaller *p*-values over time.

Publishing one's discoveries is essential to a scholarly profession. Darwin's youthful penchant for hunting and collecting beetles, birds, and barnacles was a typical naturalist activity, but from his father's perspective was a risky livelihood (Himmelfarb 1962). However, it was probably regarded as an insightful avocation for a parish priest—a respectable living in that context. Whatever Darwin's aptitude as a country squire, he turned his penchants and talents to pen and paper, and his writings have influenced biological and ecological science ever since. In the current context, achieving a *p*-value of .05 seemingly has become imperative to publication acceptance. It is generally regarded as reasonable evidence in favor of the *alternative hypothesis* if the *p*-value is .05 or less—that is, when the frequency of random occurrence of the data obtained is about 1 in 20 or smaller. If $p < .05$ is considered reasonable evidence, $p < .01$ would be considered stronger evidence, and $p < .001$ even stronger.

Figure 12-3 shows the results of an investigation concerning published *p*-values in the journal *Ecology* at 10-year intervals from 1975 through 1995. As there were 47 articles published in the 1975 volume, the first 47 articles were also examined in the 1985 and 1995 volumes for comparison. The number of occurrences of a specific reported *p*-value was recorded for each volume. Note that in 1975, the frequency of reported *p*-values reached a maximum in the <.05 category. Three articles reported *p*-values > .05, an unusual allowance by today's standards. Interesting enough, 13 articles were published containing data without accompanying statistical analysis; the *weight of evidence* must have been beyond repudiation in the eyes of the reviewers of those articles. By 1985, the frequency distribution had shifted to smaller *p*-values, reaching a maximum in the <.001 category. Finally, by 1995, the distribution was thoroughly **skewed** (nonsymmetrical) to the right and obtained a maximum in the <<.001 category. The implication is that by 2005, uncertainty may be a thing of the past!

Ecologists are in peril of spinning into the vortex of *p-value reliance.* In the current age of desktop computing capacity, we are more and more capable of performing statistical inference tests in short order, but the incorporation of successively improved versions of the same statistical methods is not a likely explanation for this trend of reporting smaller and smaller *p*-values. How, then, can we account for it? It may be that ecologists today are reporting their actual and often very small *p*-values, whereas their colleagues of earlier decades would simply report $<.05$. If true, this suggests another dangerous trend in the nature of human belief—namely, *if less is good, much less is even better.* Such thinking fosters the erroneous, nonstatistical concepts of *asymptotic certainty* and its equally invalid counterpart, *vanishing uncertainty.*

11 The Evidence-Oriented Alternative

Statement

One of the frequently reemphasized lessons taught to undergraduate and graduate students in statistics is to avoid the immediate use of powerful statistical inference procedures. *Data description* (such as summary tables, plots, and graphs) is always the first step in data analysis. Most students react with a sort of dismissive acknowledgment, knowing full well that the qualifying or final exams will require next to nothing of this simplistic admonition. In their subsequent role of statistical consultants, however, beginning with the simple methods of data display is absolutely essential.

Explanation

Data display provides a pictorial representation of the evidence, summarized for convincing visual examination. There is no substitute. As testimony to this fact, the pioneering work of John Tukey on exploratory data analysis has secured a solid foothold in current statistical computing software. Tukey's (1977) most successful inventions for displaying data were the *box-and-whisker plot* and the *stem-and-leaf display.* They are widely used and are apt summary illustrations of the information in the data. It is the comparison of such plots that produces evidence. Subjective interpretation now becomes a criterion for inference—a confirmation, at least, of the seemingly objective mathematical conclusion.

Examples

Consider the data analysis portrayed in Figure 12-4. Measurements of cesium-137 (pCi/g) were collected from the muscle tissue of white-tailed deer (*Odocoileus virginianus*) from a variety of locations on and off the U. S. Department of Energy Savannah River Site (SRS) located in South Carolina and in Georgia (Wein et al. 2001). Cesium-137 is a common radionuclide in worldwide fallout due to past nuclear weapons testing. The question of interest was to find other locations that represented "background" conditions for comparison to the SRS data. This required SRS-like locations, but study sites without the past radionuclide production activities of SRS. Data were used from six other locations where 20 or more individual deer could be measured. Figure 12-4A is a **box-and-whisker plot.** The top (or right) end of the box represents the *upper quartile* (75th percentile) and the bottom (or left) end represents the *lower quartile* (25th percentile) of the sample measurements. The top

Figure 12-4. Box-and-whisker plots of ^{137}Cs measurements in muscle tissue sampled from deer in seven different locations: the Savannah River Site (SRS), two military bases (one in South Carolina and one in Georgia), and four other localities near SRS. The data presented on the original scale (A) show very little that is useful for a population-to-population comparison, but the data presented on the common logarithmic scale (B) reveal the similarity of the two military installations to SRS. Data points on both plots are "jittered" to show the density of measurements in a particular region of the measurement scale (for example, the data cluster at the lower detection limit in the northeast quadrant offsite population).

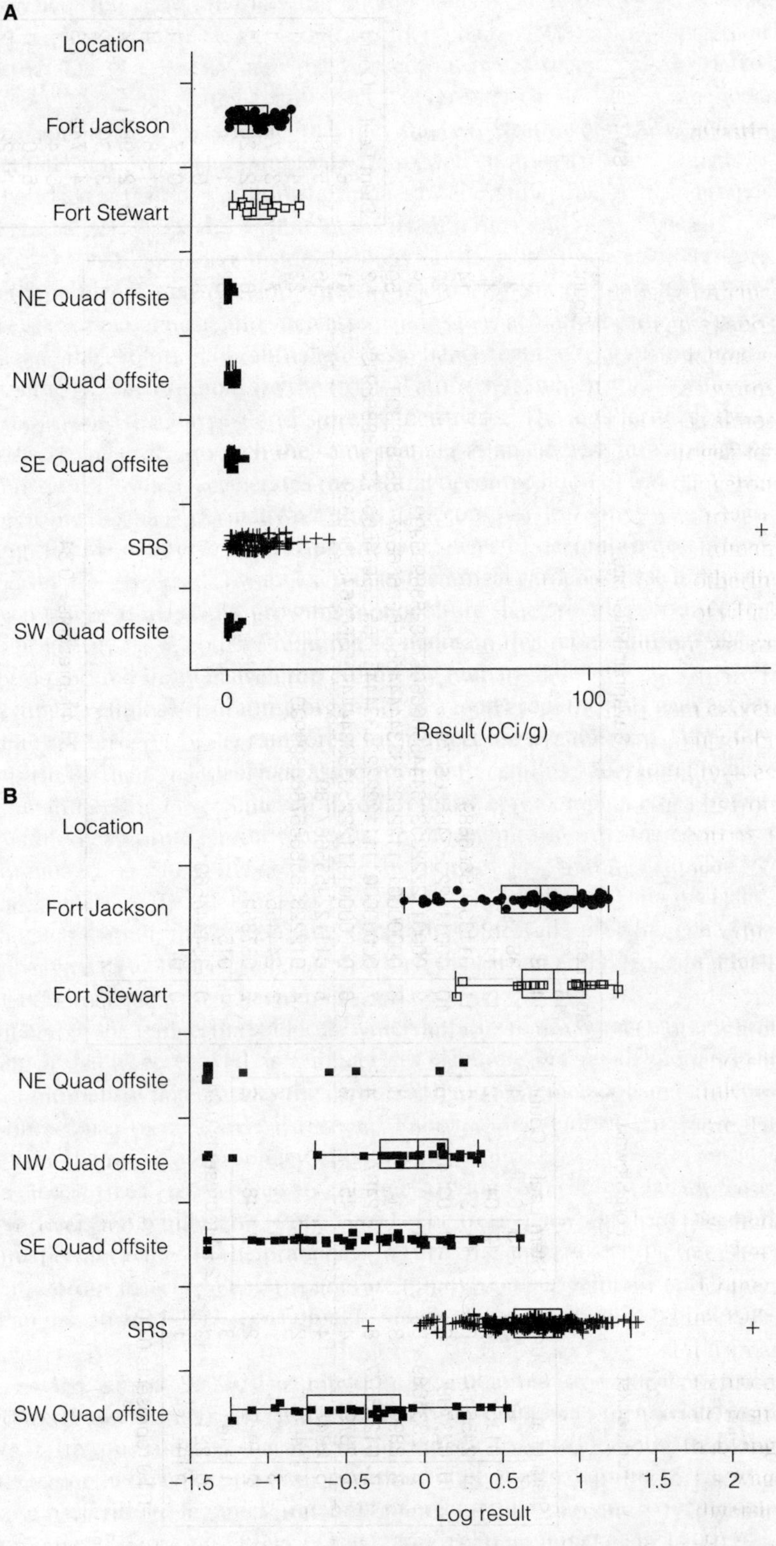

Fort Jackson

Stem	Leaf	Count
13		
12	0115	4
11	245599	6
10	023788	6
9	12335689	8
8	389	3
7	01225677	8
6	2379	4
5	2347	4
4	14	2
3		
2	49	2
1	114	3
0	224	3
−0	0	1

−0 | 0 represents 0.00

Fort Stewart

Stem	Leaf	Count
13	0	1
12	01	2
11	3	1
10	033479	6
9		
8	1	1
7	2467	4
6	457	3
5		
4		
3		
2	48	2

2 | 4 represents 0.24

Savannah River site

Stem	Leaf	Count
15		
14	3	1
13	38	2
12	1125	4
11	01133445666889	14
10	00011122333455566679	20
9	000011222233333444556666667788888999999	39
8	000000011112222233334444455555666666777888888999	50
7	00000000111112222223334445555555566666677777888889999	53
6	00000001111222333333444445556666678888899	41
5	0111224555667789	16
4	01222335566667888899999	23
3	2444557	7
2	037889	6
1	015	3
0	334444669	8
−0	0	1
−1		
−2		
−3	4	1

−3 | 4 represents −0.34

SW Quad

Stem	Leaf	Count
6		
5	5	1
4	5	1
3	7	1
2	26	2
1		
0		
−0	7	1
−1	8732	4
−2	9841	4
−3	30	2
−4	740	3
−5	4	1
−6	0	1
−7	22	2
−8	20	2
−9	62	2
−10		
−11	00	2
−12	2	1

−12 | 2 represents −1.22

Figure 12-5. A series of stem-and-leaf plots that illustrate the similarity of ^{137}Cs measurements among deer taken from two military bases (Fort Jackson, South Carolina, and Fort Stewart, Georgia) compared to the Savannah River Site (SRS) and the southwest quadrant offsite to SRS. All measurements came from muscle tissue samples and are plotted as the common logarithms of the original measurements (pCi/g).

and the bottom whiskers beyond the box end at the upper and lower quartiles plus 1.5 times the *interquartile range* (upper minus lower quartile), respectively. Note that on this measurement scale (Fig. 12-4A), little insight is obtained by comparing locations. However, when the common logarithm of each measurement is plotted instead (Fig. 12-4B), it becomes clear that both the Fort Jackson, South Carolina, and Fort Stewart, Georgia, locations are more similar to SRS than are the other locations.

Transforming the data from one scale to another is a common statistical practice. It reduces the influence of **outliers** (suspicious extreme values, beyond the whiskers), stabilizes the variance among treatment levels, and often preserves the assumption of normality. Other features of these box-and-whisker plots, which are evidence oriented, are color and jitter. *Color* can be used to aid our visual assessment of the display and to promote a convincing comparison. *Jitter* is the off-center depiction of the data points along the central vertical axis of each box. This method allows for a more obvious illustration of the sample size corresponding to each box. Figure 12-4B also shows two outliers for SRS—one large and one small—that do not appear consistent with the remainder of the data and that will inordinately influence *summary statistics* like the mean. There are more **robust** (less influenced) measures of the central tendency of a population, such as the **median** (50th percentile), that enable more useful comparisons among treatments. As an illustration, compare the horizontal centerline in each box of Figure 12-4B. These centerlines are the corresponding *median* ^{137}Cs measurements. The two military bases have protected and secured environmental settings, like those at SRS, whereas the quadrants surrounding SRS are largely rural and agricultural environments. Thus, the question arising from the data display becomes: why are ^{137}Cs measurements in deer tissue substantially lower in the rural and agricultural locations surrounding SRS than in the non-agricultural areas? But this is a new question—one that derives from the insightful display of the data.

Estimating the population parameters for the distribution of ^{137}Cs concentrations in SRS deer may be informative, but is a statistical test of our original hypothesis really necessary (see Fig. 12-4B)? Is a formal inference procedure required to convince us that the ^{137}Cs concentrations among Fort Jackson, Fort Stewart, and SRS deer are not substantially different? In fact, this exercise was partly exploratory in nature. In such instances, there is seldom a formal null hypothesis. When having some *a priori* suspicion about what one will find, the primary aim is pattern recognition, if any pattern exists. Our suspicion, in this case, was that the two military bases ought to be more like SRS than the other, rural study locations surrounding SRS.

Figure 12-5 shows corroborating evidence supporting a similar conclusion. It is a **stem-and-leaf display** of the data values, where the stem separates the tenths position of the decimal number (logarithm of ^{137}Cs measurement) from the hundredths position. The tenths position is to the left of the stem, and the hundredths position is added to the right of the stem for every measured value, arranged in ascending order. The dashed line is a reference line, showing the comparability of the SRS, Fort Stewart, and Fort Jackson data relative to the dissimilar southwest quadrant data set. The utility of this display is that it contains the actual measurement values, as opposed to plotting characters, but the conclusion is the same and a formal statistical test is not essential.

Some principles of graphics science (Tufte 1997) are also pertinent. One of these is to incorporate both space and time into data displays for more convincing por-

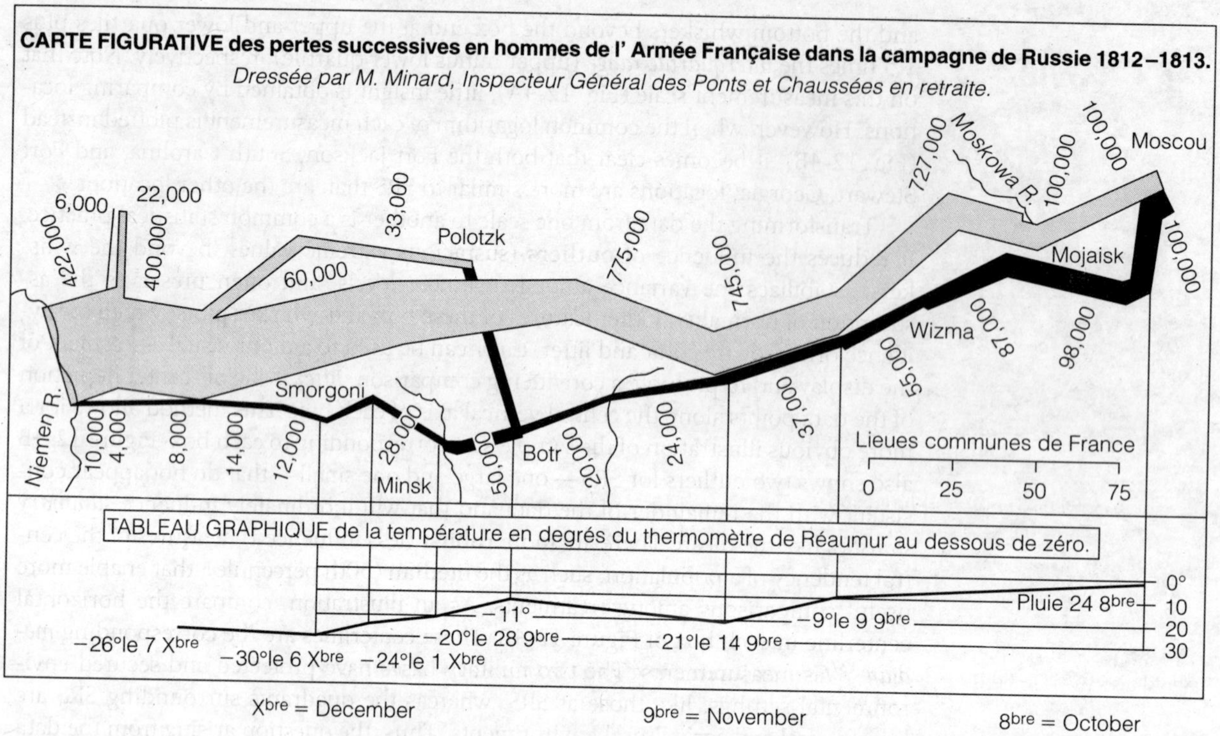

Figure 12-6. Graphic depiction in both space and time by Charles Joseph Minard (1781–1879) of the devastating losses sustained by Napoleon's army during its march to Moscow in the winter of 1812–1813. This nineteenth-century French graphic portrays data on six variables–namely, troop strength, geographic location, spatial dimension, temporal dimension, directional troop movement, and temperature (1 degree Réaumur = 1.25° C) on selected dates during the army's return march (courtesy of Graphics Press).

trayals of the evidence derived from the data. Consider two figures (Figs. 12-6 and 12-7), which Tufte used to illustrate this point. Figure 12-6 tells the story of Napoleon's fateful march to Moscow during the campaign of 1812–1813 (from Minard 1869). It is easy to see the progressive loss of life among the troops as they marched to Moscow. The width of the line marking the geographic route of the army indicates the number of troops still alive at that point during the march. One is able to see at a glance the dramatic decline over both space and time from an initial troop strength of 422,000 to the tragic final troop strength of nearly 10,000.

Figure 12-7 illustrates the life cycle of the Japanese beetle *Popillia japonica* (from Newman 1965). Note the change in soil depth as the larval stage of this organism matures by pupation to the plant-feeding adult stage, with the adults emergent from early June through mid-September. Here, both space and time progress within the data display, the former from bottom to top and the latter from left to right.

Other examples of a *weight of evidence* approach to scientific study can be found in issues of the journal *Landscape Ecology.* Twenty issues of this journal, which contained 143 articles published from January 1999 through August 2001, were reviewed. The titles of each article were searched for keywords indicating whether the article was an attempt to advance knowledge by an appeal to hypothesis testing or not.

Some keywords were simply informative, like *pattern, effects, influence,* and *impact.* Others were seemingly knowledge laden, like *predict, evidence, test,* and *hypothesis.* Among these articles, 18 had the word *pattern* in the title, 10 had the word *effects,* 8 had the word *predict,* 2 had the word *test,* 2 had the word *evidence,* and 1 had the word *hypothesis.* Two articles were based on evidence that suggested additional study.

Palmer et al. (2000) presented evidence that stream invertebrates respond to the type and spatial arrangement of patches in a streambed landscape. Their first hypothesis was that stream fauna should respond to differences among patches that vary structurally in microbial abundance (potential food). They considered the null hypothesis to be that a particular species should occur in a given patch type (such as sand or leaf debris) in proportion to the relative abundance of that patch type. They collected data for chironomids and copepods from among several sections of a northern Virginia stream. Figure 12-8 shows the data display presented in their paper and offered as a test of their first hypothesis. The solid line in this figure represents the "null expectation" and has a slope of 1.0. As the proportion of streambed covered by leaves increases, the proportion of animals in the leaf patches (as opposed to in the sand patches) should increase with one-to-one correspondence (if the null hypothesis is true). What one sees instead is a disproportionately greater number of animals collected in leaf patches than in sand patches. In short, chironomids and copepods prefer leaf patches. Palmer et al. (2000) report that "sign tests . . . showed that proportion of both taxa in the leaves was greater than that based on the null expectation." Were sign tests really necessary? The evidence portrayed in the data display is substantially convincing by itself. It is a commendable portrayal of the data to distinguish

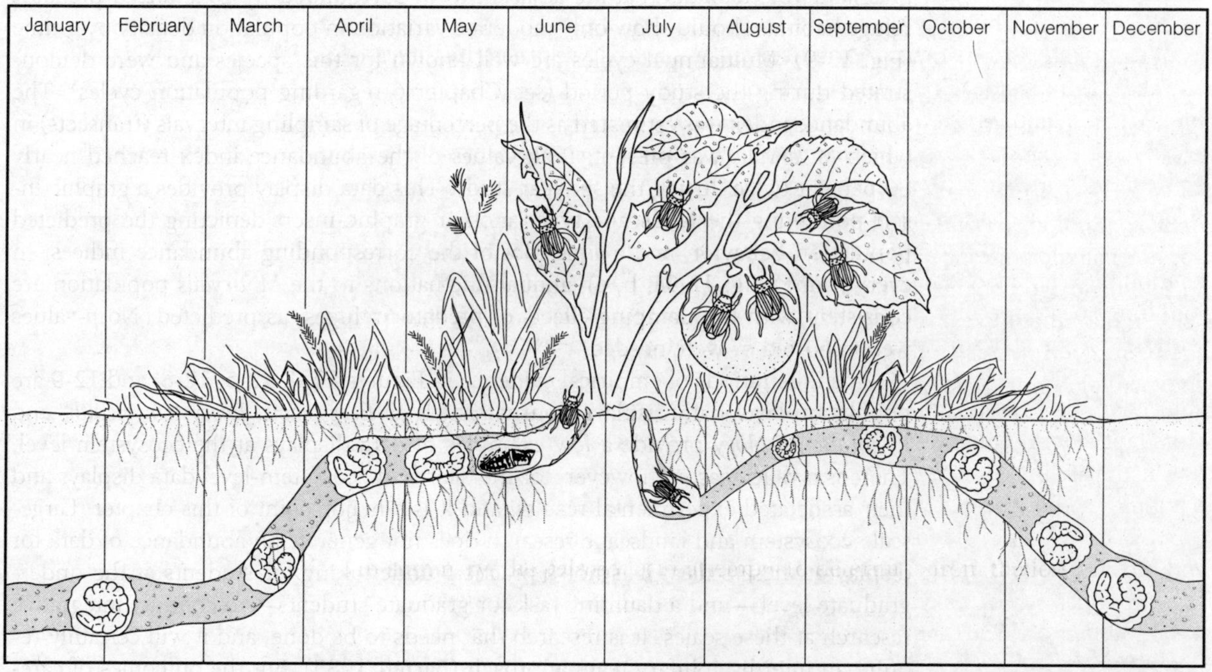

Figure 12-7. Life cycle of the Japanese beetle (*Popillia japonica*), cleverly depicted in both space and time (courtesy of Graphics Press).

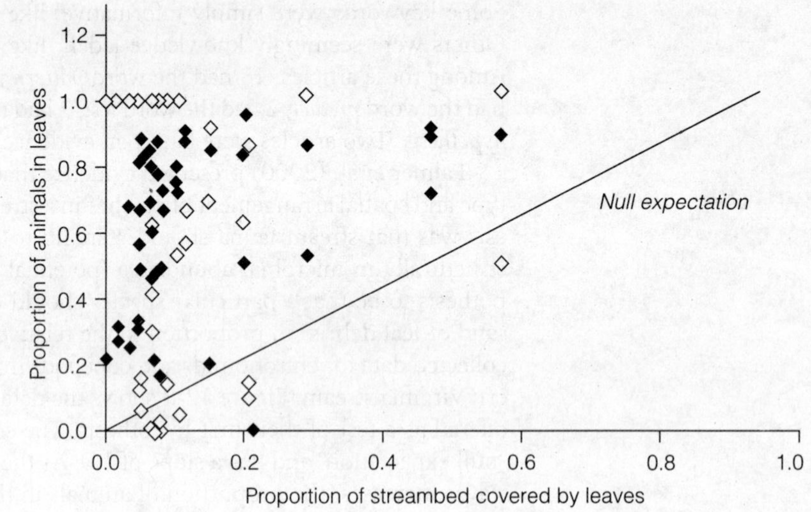

Figure 12-8. Relationship between the proportion of animals found in one of two patch types (leaf debris) and the proportion of the streambed covered by this patch type for each of two invertebrate species: chironomids (solid diamonds) and copepods (open diamonds). (From Kluwer Academic Publishers, *Landscape ecology,* vol. 14, pp. 401–412, figure 4. With kind permission of Kluwer Academic Publishers.)

between two possible outcomes that allow one to judge for oneself. Subsequent statistical tests in this instance may not be necessary, but they are certainly confirmatory.

Delattre et al. (1999) presented evidence from a six-year study of vole (*Microtus arvalis*) outbreaks among landscape formations in the Jura Mountains of eastern France. The intent of the study was to test several landscape-based hypotheses regarding vole outbreaks developed by Lidicker (1995). One hypothesis was that the grassland vole habitats that are adjacent to or surrounded by generalized predator refuge habitats should show only moderate variation in population density over time (Fig. 12-9). Multiannual cycles are well known for this species and were demonstrated during the study period (see Chapter 6 regarding population cycles). The abundance index was estimated as the percentage of sampling intervals (transects) in which *M. arvalis* was present. Peak values of the abundance index reached nearly 80 percent twice during the six-year study. This data display provides a graphic insert portraying the landscape type, another graphic insert depicting the predicted population behavior, and a time plot of the corresponding abundance indices. In every frame (Fig. 12-9a, b, c), annual fluctuations in the *M. arvalis* population are consistent with the damping effects of predator refuges, as predicted. No *p*-values were reported or were needed.

The illustrations of empirical evidence in Figures 12-4, 12-5, 12-8, and 12-9 are population-level data displays. Journals such as *Ecosystems, Ecosystem Health,* and *Restoration Ecology* provide a few examples of data displays at the ecosystem level. There is a valid reason, however, for the dearth of ecosystem-level data displays and their associated experimental results, and it is a major point of this chapter. Large-scale ecosystem and landscape research does not generate an abundance of data for display at these scales. It is not a practical undertaking for students at the undergraduate level—and a daunting task for graduate students—to conduct integrative research at these scales. It is research that needs to be done, and it will certainly require an interdisciplinary team approach (Barrett 1985). But the outcomes of carefully designed studies will require convincing data displays, because the small-scale methods of twentieth-century statistics are not likely to be effective or applicable.

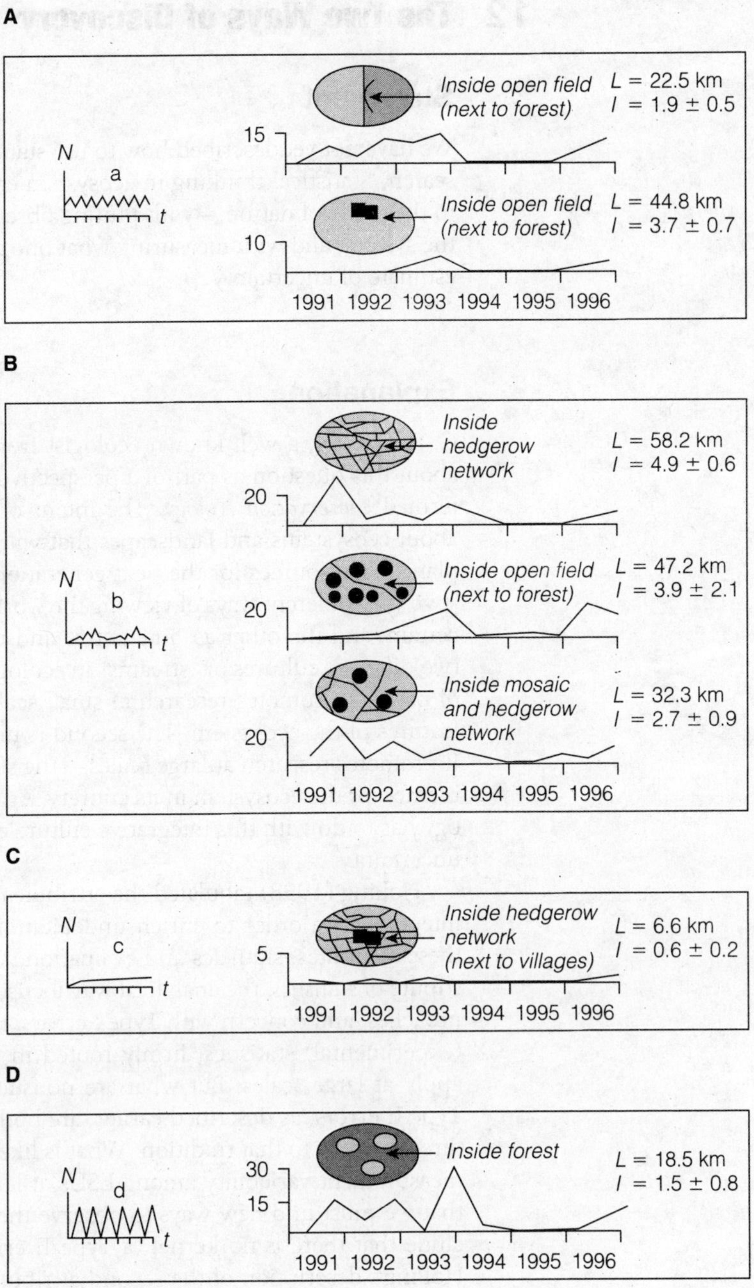

Figure 12-9. *Microtus arvalis* population dynamics as measured in or next to generalized predator refuge habitats (A, B, C) or in a patch surrounded by a barrier (D). Left-hand diagrams (a, b, c, d) show predicted population abundances (*N*); right-hand diagrams show measured population abundances over time. *L* = total length of sampled habitat over the six-year study period. *l* = average length of sampled habitat per sampling period (after Delattre et al. 1999; from Kluwer Academic Publishers, *Landscape Ecology,* vol. 14, pp. 401–412, figure 4. With kind permission of Kluwer Academic Publishers).

12 The Two Ways of Discovery

Statement

We have not yet described how to use statistical thinking in large-scale ecological research. Statistical thinking in ecosystem ecology has to do with holistic approaches to the study of nature—with putting observation and study results in the context of the system, and with measuring what one can and reporting it with the most reliable estimate of uncertainty.

Explanation

C. S. Holling, a well-known ecologist from the University of Florida, has thought about this question as part of a perspective in ecology—a transition discipline—he termed *conservation ecology.* The intent of this discipline is to acquire knowledge about ecosystems and landscapes that will feed policy making for the protection of planetary resources for the next generation and beyond. Holling (1998) pointed to "two very different ways of viewing the world." He described one as "reductionist and certain" and the other as "integrative and uncertain." He likened these two ways to two different cultures or "streams" in ecology. The first is analytical and is the science of parts. It promotes research at small scales—that is, the study of relatively small features of the ecosystem. The second is integrative and is the science of the whole. It promotes research at large scales—the study of large features, such as landscapes, biomes, or the ecosystem in its entirety. Again, statistical thinking in large-scale ecology has to do with this integrative culture and with the *a priori* acknowledgement of uncertainty.

Holling (1998) tabulated the attributes of these two ways, the analytical and the integrative, in order to enrich understanding by comparison (Table 12-1). Two of these attributes, statistics and evaluation goal, provide pertinent insight. For the attribute of *statistics,* the analytical way focuses on standard statistics, experimental approaches, and concern with Type I errors. Holling acknowledged that the "standard" (experimental) statistics, firmly rooted in the classical, Fisherian tradition, do not apply at large scales. But what are nonstandard statistics? Technically, Type I and Type II errors, as described earlier, are both vital concepts of standard statistics and directly apply to that tradition. What is likely meant by nonstandard statistics is that measurement variability among ESUs at large scales is so potentially overwhelming that we must find new ways to observe the kernel of truth at the core, lest we conclude that there is no kernel (a Type II error). This interpretation is confirmed by Holling's description of the second attribute, namely the *evaluation goal.* In his estimation, the analytical goal is often viewed as "peer assessment to reach ultimate *unanimous agreement,*" and the integration goal as "peer assessment [by] judgement to reach a *partial consensus.*" If peer assessment were unanimous—a likely unattainable outcome—the tenet would be conventionally regarded as certain. If only partial consensus was achieved instead, it would be regarded as . . . well, uncertain, and in need of more study. Holling stated that,

> In principle, therefore, there is an inherent unknowability, as well as unpredictability, concerning ecosystems and the societies with which they are linked.

Table 12-1 **A comparison of the two cultures of biological ecology**

Attribute	*Analytical*	*Integrative*
Philosophy	Narrow and targeted	Broad and exploratory
	Disproof by experiment	Multiple lines of converging evidence
	Parsimony is the rule	Requisite simplicity is the goal
Perceived organization	Biotic interactions	Biophysical interactions
	Fixed environment	Self-organization
	Single scale	Multiple scales with cross-scale interactions
Causation	Single and separable	Multiple and only partially separable
Hypothesis	Single hypothesis; null rejection of false hypothesis	Multiple, competing hypotheses
Uncertainty	Eliminate uncertainty	Incorporate uncertainty
Statistics	Standard statistics	Nonstandard statistics
	Experimental	Observational
	Concern with Type I error	Concern with Type II error
Evaluation goal	Peer assessment to reach ultimate unanimous agreement	Peer assessment, judgment to reach a partial consensus
The danger	Exactly the right answer to the wrong question	Exactly the right question, but a useless answer

Source: Adapted from Holling 1998.

> There is, therefore, an inherent unknowability and unpredictability to sustaining the foundations for functioning systems of people and nature.

And, as a consequence of this kind of uncertainty,

> information and decisions are vulnerable to being manipulated by powerful interests. While scientists do not thereby need to become politicians, they do have to be sensitive to political and human realities, and to recognize how theories, different modes of inquiry, and different rules of evidence can facilitate, hinder, or destroy the development of constructive policy and action. Recommendations have to be based on responsible judgement and interpretation of the *burden of evidence.*

This last concept is similar if not tantamount to the *weight of evidence* discussed in the next section.

Holling's (1998) ideas suggest that inference from small-scale ESUs is inductive, whereas inference from large-scale ESUs is deductive. *Inductive* inference attempts to generalize from a specific outcome, and *deductive* inference attempts to predict specifically from a general hypothesis. The former is largely experimental; the latter observational and correlative. The concepts of Type I and Type II error still apply at large scales of deductive statistical thinking, but they are directed at predictable phenomena that derive from the theoretical "big picture." It is important to the integrative scientist that data from the predicted outcome be used directly to test the hypothesis—which leads us back to the power of data displays and to the largely unused methods of Tukey (1977) to convince colleagues or professors regarding findings. Astronomers are deductively oriented, integrative scientists. So are ecosystem scientists. Students of ecosystem science will benefit by recognizing this distinction.

13 The Weight of Evidence Paradigm

Statement

What this chapter advocates is not a new discipline, but a different perspective—one that places more value on that well-known but underused aspect of descriptive statistics called *data display.* Data displays are often perceived as part of the preliminary efforts, prefacing the real data analysis and hypothesis testing that comes later. Data displays are, in fact, crucial to hypothesis testing; the acknowledgment of this point can be called the **weight of evidence paradigm.**

Explanation

Figure 12-10 illustrates the features of this ordering concept. Note that formal statistical inference tests are a part of this paradigm, because of the prowess of parametric and nonparametric statistical methods when they apply to the research scale selected. Even then, however, we are trading in the hypothesis we want to test for one that we can test statistically—one with greater objectivity, perhaps, but one that requires a bridge back from mathematical abstraction. The other empirical skills besides inference tests (Fig. 12-10) are (1) determining which data are needed and how to reliably collect them; (2) freely exploring relationships among the data; and (3) portraying the data insightfully to convince one's colleagues and critics. These three skills constitute the applied methods of statistical thinking. All are weighting skills for making sense of the data—that is, for directly testing the scientific hypothesis of in-

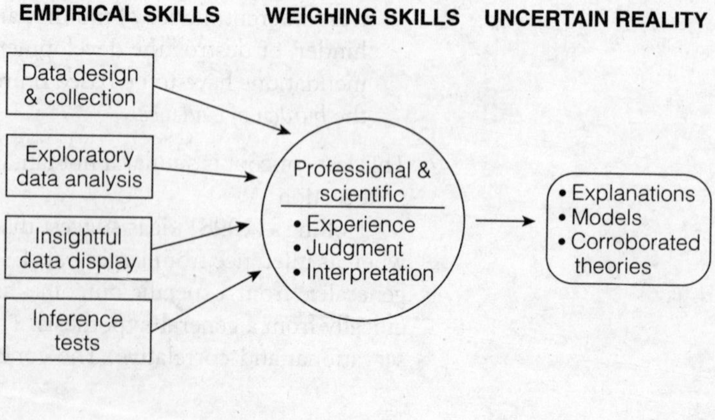

Figure 12-10. Illustration of the weight of evidence paradigm for pursuing an understanding of nature. Note that in addition to formal statistical inference tests (when the scale of research permits), the three other empirical skills that collectively constitute statistical thinking are equally important to this process.

terest, not a statistical one that is tantamount to it. The weight of evidence paradigm is a proposal to loosen the structure of scientific pursuit in ecology and admit more freedom to the process of getting knowledge that is quintessential to human nature.

Here are a few additional recommendations for students of ecology and ecosystem scientists.

For undergraduate students:

- Minimize the use of tables.
- Discover the ideas of graphic science (Tufte 1997, 2001). Do not rely strictly on the current crop of statistical computing software; new ideas will come from thinking about new ways to display data.
- Rediscover the seminal ideas of John Tukey (1977). Learn more about the thinking behind the methods of exploratory data analysis.
- Relinquish the fixation on reporting sample means; the median is a more robust statistic and could be considered a more useful measure of central tendency.

For ecosystem scientists:

- Recognize the value of replicated studies for corroborating and adding to the weight of evidence.
- Increase the emphasis on data display and minimize p-value reliance. Frequently, $p < .10$ can provide as much insight into ecosystem processes and dynamics as relying on $p < .05$.
- Incorporate statistical thinking; it always leads to research design.
- Acknowledge that large-scale research means hypothetico-deductive inference via predicted observational outcomes.
- Advocate a *weight of evidence* orientation to students and in ecological issues vital to the stewardship of nature.

In summary, a practical approach to statistics should be more "data-philic" than "parameter loving," more evidence oriented than confidence oriented, more insightful than informative, and more scale translational than scale dependent. As students of ecology, we must not lose sight of the aim for which statistical tools were developed, to avoid finding ourselves too focused on the tools themselves. The evidence is more than the devices used to obtain it.

If understanding nature is the aim of ecological science (Scheiner 1993), then the process of understanding is accelerated by *statistical thinking* more than by traditional statistical methods, especially when the scale of ecological research is beyond the scope of common statistical practice.

The message to the anxious student and to the experienced ecologist is the same. There is no certainty in statistics, and it is wise to regard one's ideas as either more or less corroborated based foremost on a presentation of the data, and on inference test results when necessary. Truth is what we seek; tested explanations are what we get, and the only certainty is in knowing that the two will never be one and the same. It is the *weight of evidence* that counts.

Glossary

A

abiotic nonliving components of an ecosystem, such as water, air, light, and nutrients

abyssal relating to bottom waters of oceans

acid rain anthropogenic emissions of hydrogen sulfide and nitrogen oxides from fossil fuel combustion that interact with water vapor to produce dilute sulfuric and nitric acids, causing widespread acidification of cloud and rainwater

aerobic refers to life in the presence of free oxygen, either as a gas in the atmosphere or dissolved in water

aestivation dormancy in animals during periods of drought

age distribution ratio of each age group (prereproductive, reproductive, and postreproductive) in a population

aggregate dispersion distribution of individuals in a clumped or aggregate pattern of dispersion (such as herds, coveys, or schools)

A horizon surface stratum of soil, characterized by maximum accumulation of organic matter and biological activity

Allee principle of aggregation a special type of density dependence, first identified by W. C. Allee in 1931, in which a degree of aggregation results in optimum population growth and survival

allele frequency commonness of an allele in a population

allelopathy direct inhibition of one plant species by another using noxious or toxic compounds

allochthonous (from Greek *chthonos,* "of Earth," and *allos,* "other") refers to organic materials not generated within the community or ecosystem

allogenic succession successional changes that are largely the result of external forces or perturbations, such as fire or flooding

allopatric speciation speciation (from a common ancestor into distinct species) resulting from the geographical separation of populations

alpine tundra tundra-like conditions found above tree lines on high mountains

altruism sacrifice of fitness by one individual for the benefit of another

altruistic behavior social behavior that apparently enhances the fitness of other individuals in the population at the expense of the individual performing the behavior

amensalism a relationship between two species in which one population is inhibited and the other not affected

anaerobic refers to life or processes that occur in the absence of free oxygen

anthrosol human-created urban soil type, containing an abundance of pulverized concrete, dust, debris, and "fill" materials

applied ecology application of ecological theory, principles, and concepts to resource management

aquifer porous underground strata (limestone, sand, or gravel) bounded by impervious rock or clay, containing significant quantities of water

ascendancy the tendency for self-organizing, dissipative systems to develop network flows over time

association natural unit of vegetation, often dominated by a particular species, thus providing a relatively uniform vegetative composition

atoll circular or semicircular group of islands encircling a lagoon, formed by coral reefs growing on the submerged slopes of a seamount

aufwuchs plants and animals attached to or moving about on submerged surfaces; also frequently termed *periphyton*

autochthonous refers to photosynthesis or organic materials generated within the community or ecosystem

autogenic succession successional changes largely determined by internal (self-generated) interactions

autotrophic producing its own food (as photosynthetic plants); production (P) is greater than respiration (R)

autotrophic succession succession beginning when $P/R > 1$

B

Batesian mimicry benign species that resemble a noxious or dangerous species

behavioral ecology the branch of ecology that focuses on the behavior of organisms in their natural habitat

benthic zone lowermost region or bottom of a freshwater lake or aquatic ecosystem

benthos bottom-dwelling organisms that inhabit the bottom of rivers, lakes, and the sea

B horizon a stratum of soil characterized by minerals, in which organic matter in the A horizon has been converted by decomposers into inorganic compounds such as silica and clay

biennial plant that requires two years to complete its vegetative and reproductive growth cycle

biocoenosis term used in European and Russian literature to denote biotic communities inhabiting a defined space at the same time; the biotic component of an ecosystem

biodiversity diversity of life forms, the ecological roles they perform, and the genetic diversity they contain; term used to describe all aspects of biological diversity (genetic, species, habitat, and landscape)

biogeochemistry the branch of science that focuses on the movement of elements or nutrients through organisms and their environment; the study of natural cycles of elements and their movement through biological and geological compartments

biogeocoenosis term used in European and Russian literature equivalent to the term *ecosystem,* or biocoenosis together with its abiotic components

biological clock endogenous, physiological mechanism that keeps time independent of external events, enabling organisms to respond to daily, lunar, seasonal, or other periodicities

biological magnification increase in the concentration of a chemically stable substance or element (such as pesticides, radioactive materials, or heavy metals) as it moves up a food chain

biological oxygen demand indicator of pollution caused by an effluent, related to the uptake of dissolved oxygen by microorganisms that decompose organic matter present in the effluent

biomass weight of living material, typically expressed as dry weight per unit area or volume

biome large regional or subcontinental system characterized by a particular major vegetation type (such as a temperate deciduous forest); biomes are distinguished by the predominant plants associated with a particular climate (especially temperature and precipitation)

biosphere that part of the environment of Earth in which living organisms are found

biotic refers to the living components of an ecosystem

biotic potential maximum reproductive potential of an organism

bog wetland ecosystem characterized by acidic conditions and accumulation of peat, dominated by sphagnum moss

bottom-up regulation regulation of a community or ecosystem trophic structure related to increased productivity of the producer trophic level; influence of producers on the trophic levels above them in the food web

breeding dispersal movement of individuals out of a population prior to initiation of the breeding season

C

C_3 plant plant that produces a 3-carbon compound (phosphoglyceric acid) as the first step in photosynthesis; pathway of carbon fixation common in plants adapted to low temperatures, average light conditions, and adequate water supply

C_4 plant plant that produces a 4-carbon compound (malic or aspartic acid) as the first step in photosynthesis; pathway of carbon fixation common in plants adapted to high temperatures, strong light, and low water supply

calorie amount of heat needed to raise the temperature of 1 gram of water by 1° C, usually from 15° C to 16° C

calorific value energy content of biological materials, expressed in calories or kilocalories per gram dry weight

cannibalism intraspecific predation

carbon cycle movement of carbon, C, between the atmosphere, hydrosphere, and biosphere, and the transformations (such as photosynthesis and respiration) between its different chemical forms

carrying capacity maximum population of a given species that a particular environment or ecosystem can sustain; the *K* value for an S-shaped sigmoid growth curve

Chaparral biome type dominated by broadleaf scrubs and sclerophyllous woodland located in regions of Mediterranean climate; a fire-dependent ecosystem that tends to perpetuate scrub dominance at the expense of trees

character displacement changes in the physical characteristics of similar species occupying overlapping ranges, resulting in reduced interspecific competition; divergence of the characteristics of two species occupying overlapping ranges

chelation (from Greek *chele,* "claw") a complex formation of organic matter with metal ions (for example, chlorophyll is a chelate compound in which the metal ion is magnesium)

C horizon stratum of soil beneath the A and B horizons that is relatively unmodified by biological activity or soil-forming processes (the stratum of parent material)

circadian rhythm ability of an organism to time and repeat functions at about 24-hour intervals even in the absence of conspicuous environmental cues such as daylight

climatic climax stable seral stage in equilibrium determined by the general climate of the region

climax term introduced by F. E. Clements in 1916, representing the final stage of ecological succession; a stage of vegetation where $P = R$ that is self-perpetuating in the absence of major disturbance

climograph chart in which one major climatic factor is plotted against another

coarse-grained refers to a habitat or landscape patch in which the vagility of a given animal species is low relative to the size of the patch

coevolution a type of community evolution in which evolutionary interactions occur among organisms in which exchange of genetic information among different populations is minimal or absent; the joint evolution of one species in a non-interbreeding relationship partially depending on the evolution of the other through reciprocal selective pressures

coexistence two or more species living together in the same habitat

cohort group of individuals of the same age class

collective properties summation of the properties of the parts (for example, birth rate, which is the sum of individual births within a designated time period)

commensalism relationship between two species in which one population is benefited but the other is not affected

community includes all the populations inhabiting a specific area at the same time

compensation depth depth in a lake where light penetration is so reduced that oxygen production by photosynthesis balances oxygen consumption by respiration (that is, the depth in a lake where $P/R = 1$)

competition relationship between two species that is mutually detrimental to both populations

competitive exclusion principle principle stating that no two species can permanently occupy the same ecological niche

conservation biology field of science concerned with the protection and management of biodiversity based on the principles of basic and applied ecology

conservation easement legal mechanism to designate personal property for ecological, economic, and conservation benefits

continuum gradient of environmental conditions reflecting changes in community composition

control experimental condition in which no action is taken; designed to confirm no effect

coprophagy feeding on feces

coral reefs colonial groups of cnidarians that secrete an external skeleton of calcium carbonate, usually in a mutualistic relationship with algae; three types of coral reefs are atolls, barrier reefs, and fringing reefs

corridor connection between two patches of landscape habitat

corroboration support by additional evidence

crassulacean acid metabolism (CAM) method of CO_2 fixation that conserves water in certain succulent, drought-resistant desert plants

crude density number of individuals per unit area

cultural eutrophication overfertilization of freshwater ecosystems by nutrients, mainly nitrogen and phosphorus, from anthropogenic sources

curie the amount of material in which 3.7×10^{10} atoms disintegrate each second

cyanobacteria group of bacteria that possess chlorophyll *a* and that carry out photosynthesis; regarded as the first oxygen-producing organisms responsible for generating oxygen in the atmosphere, thus profoundly influencing the course of biosphere evolution

cybernetics (from Greek *kybernetes,* "pilot" or "governor") the science that deals with communication systems and system controls; in ecology and the life sciences, the study of feedback controls in homeostasis

cyclic succession succession caused by periodic, rhythmic disturbances in which the sequence of seral stages is repeated, preventing the development of a climax or stable plant community

D

deciduous forest forest composed of trees that drop leaves during unfavorable, winter conditions; temperate deciduous forests are a major biome-type in eastern North America

decomposers organisms, typically bacteria and fungi, that obtain energy from the breakdown of dead organic matter

decomposition breakdown of complex organic materials into simpler products

deductive reasoning from general to specific conclusions

denitrification reduction of nitrates to atmospheric nitrogen by microorganisms

density dependence regulation of population size or growth by mechanisms whose effectiveness increases as population size increases; effect on population size as a function of density

density independence regulation of population growth not tied or related to population density; change in population numbers is independent of population size

desert biome with less than 10 inches (25 centimeters) of rainfall per year, dominated by stem succulents, such as cacti, and desert shrubs that are frequently regularly spaced in their distribution

deterministic without a concept of probability or measurement uncertainty

detritivores organisms that feed on dead or decaying organic matter (such as earthworms)

detritus dead or partially decomposed plant and animal matter; nonliving organic matter

detritus food chain food chain in which the primary producers are not consumed by grazing herbivores, but where dead and decaying plant parts form litter (detritus) on which decomposers (bacteria and fungi) and detritivores feed, with subsequent transfer of energy through the detritus food chain

dimictic refers to mixing or inversion of a lake twice a year

disclimax seral stage of succession maintained by anthropogenic (human-generated) disturbance

dispersal leaving an area of birth for another area; movement of individuals or their disseminules (seeds, larvae, spores) into or out of a population or area

dispersion pattern of distribution of individuals within a population over an area

disturbance corridor a linear disturbance through the landscape matrix

diversity-productivity hypothesis hypothesis based on the assumption that interspecific differences in the use of resources by plants allow more diverse plant communities to more fully use limiting resources and, thereby, attain greater net primary productivity

diversity-stability hypothesis hypothesis based on the assumption that primary productivity in more diverse plant communities is more resistant to and recovers more fully from major perturbations such as drought

domestication (from Latin *domus,* "home") evolutionary changes in plants and animals brought about under human artificial selection

dominance-diversity curve number or percentage of each species plotted in sequence from the most abundant to the least abundant in a defined habitat; an expression or graph of species diversity based on species importance

dystrophic refers to a body of water, such as a shallow freshwater lake, with a high content of humic matter, with deeper water depleted of oxygen

E

ecological density number of individuals per habitat space (that is, the area of habitat that can actually be colonized by that population)

ecological economics field of study that attempts to integrate economic capital (goods and services provided by humankind, or the human workforce) with natural capital (goods and services provided by nature)

ecological equivalent species that occupy the same ecological niche in different geographical regions

ecological footprint the area of productive ecosystems outside a city that is required to support life in the city

ecological study unit (ESU) physical size of the experimental study area, mesocosm, or patch required in order to achieve proper replication

ecological succession process of change and development whereby previous seral stages are replaced by subsequent seral stages until a mature (climax) community is established

ecology (from Greek *oikos,* "household," and *logos,* "study of") branch of science dealing with interactions and relationships between organisms and the environment; the study of goods and services provided by natural ecosystems, including the integration of these nonmarket services with the economic market

economics (from Greek *oikos,* "household," and *nomics,* "management"; translates as the "management of the household") branch of science dealing with the goods and services provided by humankind, encompassing the integration of market services with nonmarket services provided by natural ecosystems

economic capital goods and services provided by humankind, or the human workforce, typically expressed as gross national product (GNP)

ecophysiology that branch of ecology concerned with the responses of individual organisms to abiotic factors such as temperature, moisture, atmospheric gases, and other factors of the environment

ecoregion classification of major vegetation types or ecosystems developed by R. W. Bailey in 1976, based on a continuous land area in which the interaction of climate, soil, and topography permit the development of similar types of vegetation

ecosphere all the living organisms of Earth interacting with the physical environment as a whole

ecosystem a biotic community and its abiotic environment functioning as a system (first used by A. G. Tansley in 1935); a discrete unit that consists of living and nonliving parts interacting to form an ecological system

ecotone zone of transition from one type of community or ecosystem to another (a transition from a forest to a grassland, for example)

ecotypes subspecies or local populations adapted to a particular set of environmental conditions

ectomycorrhizae relationship between a fungus and plant roots in which the fungus forms a network structure around root cells

edaphic climax stable plant community in equilibrium dependent on soil, topography, and local microclimatic conditions (as opposed to the general climate)

edge a site where two or more structurally different communities or ecosystems meet (such as the edge of a pond or lake)

edge effect response of plants and animals to the site where two or more communities or ecosystems meet (typically creating an increase in biotic diversity along the edge site)

edge species species that inhabit edge or boundary habitats; species that use edges for reproductive and survivorship purposes

effective population size effective size of a population as a measure of how many genetically distinct individuals actually participate in the next generation; the minimum population size, below which a given species might lose its evolutionary potential

emergence theory theory that the whole possesses properties not possessed by the individual components

emergent property refers to properties at various levels of organization that cannot be derived from lower-level systems studied in isolation

eMergy amount of one type of energy required to develop the same amount of another type; amount of energy already used directly or indirectly to create a service or product

emigration one-way movement of individuals out of a population

endemic refers to species restricted to certain specialized habitats and found nowhere else

energy budget rate at which an organism or population consumes energy relative to the rate at which an organism or population expends energy

energy flow exchange and dissipation of energy through the food chain trophic levels of an ecosystem

energy subsidies subsidies from outside the system (such as fertilizer, pesticides, fossil fuels, or irrigation) that enhance growth or rates of reproduction within the system

entropy index of disorder associated with energy degradation; transformation of energy to a more random and disorganized state

environmental resistance sum total of environmental limiting factors, both biotic and abiotic, that prevent the biotic potential (r_{max}) of a population from being realized

environmentalism guiding principle of those devoted to the health and sustainability of planet Earth

ephemerals annuals that persist as seeds during periods of drought, but sprout and produce seeds quickly when moisture is favorable; term meaning short-lived or of short duration

epilimnion warmer, oxygen-rich upper part of a lake when thermally stratified during summer

epiphyte plant that grows on another plant but is not parasitic (such as an orchid living on a tree)

error acceptance criterion risk one is willing to take of reaching an incorrect conclusion

estuary (from Latin *aestus,* "tide") partially enclosed embayment, such as a river mouth or coastal bay, where freshwater and salt (sea) water meet and where tidal action is an important physical regulator and energy subsidy

eury- prefix meaning "wide," derived from Greek *eurus*

eusociality characterized by cooperative caring for young, division of labor, and an overlap of at least two generations of life stages functioning to contribute to group success

eutrophic refers to a body of water high in nutrients and productivity

eutrophication process of nutrient enrichment (typically of phosphates and nitrates) in aquatic ecosystems, resulting in increased primary productivity

evapotranspiration total loss of water by evaporation from an ecosystem, including water loss from the surface of plants, mainly through the stomata

evenness index index expressing equitability in the distribution of individuals among a group of species; measure of the extent to which species are equally represented in a community

evidence interpreted data summaries

evolutionary ecology that branch of ecology dealing with the natural selection of and changes in gene frequencies in a population through time

experimental design statistical plan for conducting an experiment, ensuring that causes associated with effects can be evaluated by carefully controlling all appropriate variables

exploitative competition relationship between two species in which one population exploits a resource, such as food, space, or common prey, to the extent that it adversely affects the other population (as opposed to *interference competition*)

extrinsic factors factors such as temperature and rainfall that are outside the sphere of population interactions

F

facilitation model model for succession in which previous seral stages prepare or facilitate the way for the next seral stage of community development

factor compensation the ability of organisms to adapt and modify the physical environment to reduce limiting factors, stress, or other physical conditions of existence

fecundity number of disseminules (eggs or seeds) produced by an organism

fen wetland ecosystem that receives part of its nutrient input through a flow of groundwater; wetland that is only slightly acidic, dominated by sedges

filter effect influence of gaps in landscape corridors that allow certain organisms to cross but restricts the movement of others

fine-grained refers to a habitat or landscape patch in which the vagility of a given animal species is high relative to the size of the patch

first law of thermodynamics although energy may be transformed from one form to another, it can never be created or destroyed (in other words, no gain or loss of energy occurs)

fitness genetic contribution by an individual's descendants to future generations; a measure of expected reproductive success

floating reserve individuals in a population that do not hold territories and remain unmated, but are available to refill a territory vacated by the death of its holder

flood pulse concept description of changes in a river, both laterally and longitudinally (both of the river and its associated riparian floodplain), especially during heavy rain or flooding (pulse) conditions

flow pathways movement of matter or energy from one compartment to another

food web summary or model of the feeding relationships within an ecological community; a representation of energy flow through populations in a community

foraging strategy manner in which individuals seek food and allocate their time and energy in obtaining it

forcing function independent or extrinsic variables that cause a system to respond but are not themselves affected by the system

founder effect population founded by a small number of colonists, often resulting in genetic variation markedly different from the parent population

frequency of occurrence percentage of sample plots occupied by a particular species

frugivore organism that feeds on fruit

functional response increase in the rate of feeding by a predator that occurs in response to an increase in prey availability

fundamental niche niche determined in the absence of competitors or other biotic interactions such as predation; total range of environmental conditions under which a species could survive

G

Gaia hypothesis (from Greek *Gaia,* the Earth goddess) hypothesis formulated by James Lovelock in 1968, which holds that organisms, especially microorganisms, have evolved with the physical environment to provide control (self-regulation) and to maintain conditions favorable to life on Earth

gallery forest rain forest that occurs along banks and floodplains of rivers; riverine forest

gap phase succession successional development at a disturbed site within a stable plant community; replacement and succession in a gap in a forest caused by a disturbance such as wind or disease

Gause principle principle (first demonstrated in 1932 by G. F. Gause, a Russian biologist) that states that no two species with the same ecological requirements can coexist (see *competitive exclusion principle*)

genetic diversity diversity or maintenance of genotypic heterozygosity, polymorphism, and other genetic variability in a natural population

genetic drift changes in allele frequencies due to random variation or chance fluctuation in allele frequencies in a population over time

genetic engineering manipulation of DNA, including the transfer of genetic material between species

genotype frequency frequency of different genotypes in a population

global climate change modification of the global climate resulting from the increased proportion of greenhouse gases, especially CO_2, emitted as by-products of human activities

global positioning system (GPS) system that determines locations on the surface of Earth, including longitude, latitude, and altitude, using radio signals from satellites

gradient analysis graph depicting vegetative response to a gradient (moisture, temperature, or elevation)

grain the relationship of the size of landscape patches to an animal's vagility

granivorous food chain food chain originating with feeding on seeds

grazing food chain food chain in which green plants (primary producers) are eaten by grazing herbivores (primary consumers), with subsequent energy transfer up the food chain to carnivores (secondary and tertiary consumers)

greenhouse effect absorption of infrared radiation by greenhouse gases, especially CO_2, in the atmosphere that was reradiated from the surface of Earth

gross primary productivity (GPP) rate at which radiant energy is fixed by photosynthesis of producer organisms; the sum of net primary productivity plus respiration by autotrophs ($GPP = NPP + R$)

group selection natural selection between groups or assemblages of organisms that are not necessarily closely linked by mutualistic associations; elimination of a group of individuals by another group of organisms possessing superior genetic traits

guild group of species that make their living by exploiting the same class of resources in a similar way

H

Haber process industrial catalytic process for synthesizing ammonia from nitrogen and hydrogen, discovered by Fritz Haber, a German chemist

habitat place where an organism lives

habitat diversity diversity of habitat patches in a landscape, which serves as a basis for metapopulation dynamics

habitat fragmentation analysis determining how the landscape has changed by humans affecting the size, shape, and frequency of landscape elements (patches, corridors, and matrices)

Hardy-Weinberg equilibrium law law, discovered independently in 1908 by G. H. Hardy and W. Weinberg, stating that in a population mating at random in the absence of evolutionary forces, allele frequencies will remain constant

harvest method technique for measuring net primary productivity of herbaceous, terrestrial vegetation (such as old fields or grasslands); harvests are made periodically by clipping the vegetation at ground level from randomized sample sites, sorting to species, and then drying to a constant dry weight

harvest ratio the ratio of grain (or other edible parts) to supporting plant tissue

herbivore organism that feeds on plant material

heterogeneity mixed genetic or environmental composition

heterotroph individual unable to manufacture its own food from inorganic matter, which therefore consumes other organisms for its source of energy

heterotrophic depending on other organisms for food or nourishment; a system where respiration (*R*) exceeds production (*P*)

heterotrophic succession succession beginning with $P/R < 1$; successional process on dead organic matter

hierarchy arrangement into a graded series, such as levels of biological organization

holism (from Greek *holos*, "whole") theory that whole systems cannot fully be understood by investigating their individual parts or properties; theory that whole entities have an existence, rather than being a mere sum of their parts

holological refers to studies that investigate the ecosystem as a whole, rather than examining each component part

homeorhesis (from Greek "maintaining the flow") tendency of a system to maintain itself in a pulsing state of equilibrium

homeostasis tendency of a system to resist change and maintain itself in a state of stable equilibrium

homeotherm organism that uses metabolic energy to maintain a relatively constant body temperature

home range area over which an individual ranges throughout the year; empirical determination of home range typically involves monitoring the movements of an individual by plotting them on a map, then joining the outermost points to form a minimum convex polygon

homing ability of an individual to navigate a long distance in order to find its way back to its home area

homozygous containing two identical alleles at the same loci of a pair of chromosomes

human ecology study of the impact of humankind on and integration with natural systems

humus organic matter derived from partial decay of plant and animal matter

hydrological cycle flow and cycling of water in its various states and reservoirs through the terrestrial, aquatic, and atmospheric environments

hydrology study of water, both quantitatively and qualitatively, as it moves through the hydrological cycle

hydroperiod periodicity of water-level fluctuations

hydrothermal vents sites in the ocean bed, usually near a mid-oceanic ridge, releasing geothermally heated water that is rich in dissolved sulfides, which are oxidized by chemosynthetic bacteria, providing organic compounds that support animal communities

hypervolume niche multidimensional space concept proposed by G. E. Hutchinson in 1957, in which the niche of a species is represented as a point or volume in a hyperspace whose axes correspond to attributes of that species

hypolimnion cold, oxygen-poor, bottom part of a lake when thermally stratified during summer; zone of a lake below the thermocline

hypothesis idea or concept that can be tested by experimentation

I

ideal free distribution distribution of individuals across resource patches of different intrinsic quality that equalizes the net rate of gain when competition is taken into account

immigration movement of new individuals into a population or habitat

importance value sum of the relative density, relative dominance, and relative frequency of a species in a community on a scale from 0 to 300; the higher the importance value, the more dominant is that species in that particular community

inbreeding mating between close relatives

inbreeding depression detrimental effects of inbreeding, resulting in an inadequate gene pool

inclusive fitness an individual's own reproductive success, plus the increased fitness of its relatives, weighted according to the degree of the relationship

individualistic concept concept of community development, first proposed by H. A. Gleason in 1926, stating that species of plants are distributed individually with respect to biotic and abiotic factors; thus, associations result only from similar requirements

inductive reasoning from specific to general conclusions

inhibition model model of succession proposing that dominant plant species occupying a site prevent colonization of that site by other plant species

input management strategy of managing system inputs rather than system outputs; source reduction approach to reducing pollution

insectivore heterotrophic organism that feeds mainly on insects

integrative pest management (IPM) program or management strategy designed to address or control a particular pest problem with only a minimum application of pesticides; use of different methods and farming strategies (chemical, biological, physical, or cultural) to suppress pest populations below their economic threshold

interdisciplinary approaches resulting in cooperation among members of different scientific disciplines when addressing a higher-level concept, problem, or question

interference competition competition between two species in which both populations actively inhibit each other; competition in which access to a resource is limited directly by the presence of the other species (as opposed to *exploitative competition*)

interior species individuals, chiefly birds and mammals, that inhabit the interior of a forest or grassland ecosystem rather than its edges

interspecific competition competition between individuals of different species

intraspecific competition competition between individuals of the same species

intrinsic factors population fluctuations controlled primarily by regulatory mechanisms (genetic, endocrine, behavioral, disease, and so on) within the population

intrinsic rate of natural increase the maximum per capita rate of population increase (r_{max}) based on a stable age distribution and freedom from competition and limiting resources

island biogeography theory that the number of species on an island is determined by the equilibrium between the immigration of new species and the extinction of species already present

J

J-shaped growth curve J-shaped pattern of population growth that occurs when the population density increases in exponential fashion

K

keystone species functional group or population without redundancy; a species (such as a predator) having a dominating influence on the structure and functioning of a community or ecosystem

kinetic energy energy associated with motion

kin selection selection in which individuals increase their inclusive fitness by helping increase the survival and reproductive success of relatives (kin) that are not their offspring

***K*-selection** selection under carrying capacity, *K*, conditions with a high level of competition; species characterized by *K*-selection tend to dominate mature stages of ecological succession

L

landscape heterogeneous area composed of a cluster of interacting ecosystems that are repeated in a similar manner throughout the area; landforms of a region in the aggregate; a regional level of organization between the ecosystem and the biome

landscape architecture study of the structure and three-dimensional use of habitat space in the field of landscape ecology

landscape corridor strip of vegetation that differs from the matrix and frequently connects two or more patches of similar habitat

landscape ecology the branch of ecology that focuses on the development and dynamics of spatial heterogeneity, the influences of spatial heterogeneity on biotic and abiotic processes among ecosystems, and the management of spatial heterogeneity at the landscape scale

landscape geometry study of shapes, patterns, and configurations at the landscape scale

landscape matrix large area of similar ecosystem or vegetation types (agricultural or forest, for example) in which landscape patches are embedded

landscape mosaic quilt-like patches of different types of vegetative cover across a landscape; cluster of ecosystems of different types

landscape patch relatively homogeneous area that differs from the surrounding matrix (such as a woodlot embedded in an agricultural matrix)

law of diminishing returns as an ecosystem becomes larger and more complex, the proportion of gross productivity that must be respired to sustain growth increases (in other words, the proportion of productivity that can go into further growth declines)

lentic (from Latin *lenis*, "calm") refers to standing-water ecosystems such as lakes and ponds

levels of organization a hierarchical arrangement of order ranging from the ecosphere (or beyond) to cells (or beyond) illustrating how each level manifests emergent properties that are best explained at that particular level of organization

lichen organism that consists of a fungus (the *mycobiont*) and an alga or cyanobacterium (the *phycobiont*) living in a mutualistic association; a lichen may be crustose, foliose, or fruticose according to species

Liebig law of the minimum concept, first stated by Baron J. von Liebig in 1840, that the essential material or resource most closely approaching the minimum need tends to be the limiting one

life table tabulation of mortality and survivorship schedules of a population based on an initial cohort

life zone concept early classification of major vegetation types put forth by C. H. Merriam in 1894, based on the relationship between climate and vegetation

light saturation level that value of photosynthetically active radiation (PAR) at which any further increase results in no further increase in photosynthesis

limiting factor resource that limits the abundance, growth, and distribution of an organism or species

limits of tolerance upper and lower limits to the range of particular environmental factors (such as light or temperature) within which an organism or species can survive

limnetic zone the open water of a lake beyond the littoral zone, where $P/R > 1$

limnology study of freshwater ecosystems such as lakes

Lincoln index mark-recapture method use to estimate the total population density; an index devised in 1930 by the American ornithologist Frederick C. Lincoln to estimate population density

littoral zone zone containing rooted floating and emergent vegetation along the shore of a lake or pond, where $P/R > 1$

logistic growth pattern of population growth producing a sigmoidal (S-shaped) curve that levels off at the carrying capacity

lotic (from Latin *lotus,* "washed") running-water ecosystems such as streams and rivers

Lotka-Volterra equations model developed in the 1920s by Alfred Lotka, an American mathematician, and Vito Volterra, an Italian mathematician, based on the logistic equation, expressing interspecific competition (such as predator-prey relationships)

M

macrocosm (from Greek *makros,* "large") a large-sized experimental or natural ecosystem

macroevolution evolution of rapid, major phenotypic changes, resulting in changed lineage of descendants into distinct new taxons

macronutrients elements required by living organisms in substantial amounts (such as nitrogen and phosphorus)

macrophytes rooted or large floating plants (such as water lilies)

maintenance energy resting or basal rate of metabolism plus the energy necessary to cover minimum activities under field conditions

mangroves emergent woody plants that tolerate the salinity of the open sea; trees that dominate tropical intertidal forests

mariculture fish or other food farming in enclosures (mesocosms) in bays and estuaries

marsh wetland ecosystem with periodically waterlogged mineral soils, dominated by cattails and sedges

maximum carrying capacity (K_m) maximum density that the resources in a particular habitat can support

maximum sustained yield the point in a population growth curve where harvested biomass would be most rapidly replaced; the maximum rate at which individuals in a population can be harvested without reducing its size

median fiftieth percentile or middle value in a distribution

merological (from Greek *meros,* "part") refers to studies that investigate component parts first in an attempt to understand the system as a whole

mesic describes moist habitat conditions

mesocosm (from Greek *mesos,* "middle") a midsized experimental ecosystem

metacommunity the dominance-diversity curve in neutral theory that characterizes species diversity at the metacommunity equilibrium between rates of speciation and extinction of species

metapopulation group of subpopulations of a population living in separate locations, but with active exchange of individuals among subpopulations

microcosm (from Greek *mikros,* "small") a small, simplified, experimental ecosystem

microevolution evolution of small changes within a population that occur over time by natural selection

micronutrients elements required by living organisms only in small or trace amounts

mid-oceanic ridges undersea areas in which spreading tectonic plates create vents, hot sulfurous springs, and seeps

migration periodic departure and return of the same individuals in a population

mimicry resemblance of one species to another as a mechanism evolved to deceive predators

minimum known alive (MKA) mark-recapture method used to estimate the percentage of the total population known to be alive on a particular date; frequently termed the *calendar-of-catches method,* based on the capture history for a particular species

mode the value or item occurring most frequently in a series of observations or statistical data

model formulation that mimics a real-world phenomenon; a simplified representation of the real world that aids understanding

montane related to mountains

mortality death of individuals in a population

Mullerian mimicry physical resemblance of various members of a group of noxious or dangerous species (compare *Batesian mimicry*)

mutualism relationship between two species in which the growth and survival of both populations benefit

mycorrhizae mutualistic associations between fungi and the roots of plants

N

natality ability of a population to increase by reproduction; production of new individuals in a population

natural capital benefits and services supplied to human societies by natural ecosystems

natural selection evolutionary process involving differences between individuals in their rates of survival or reproduction, resulting in some types of individuals being represented more frequently than others in the next generation

nectar food chain food chain originating from the nectar of flowering plants, frequently dependent on insects and other animals for pollination

negative feedback information that inhibits the growth or output process of a system

nekton free-swimming organisms, such as fish

neretic refers to regions of marine environments where land masses extend outward as a continental shelf

net community productivity rate of storage of organic matter in an ecosystem that is not used by heterotrophs during the period of measurement (usually a growing season or a year)

net energy energy remaining after metabolic losses that is available for growth and reproduction; yield beyond the energy cost of sustaining the conversion system

net primary productivity (NPP) rate of storage of organic matter in plant tissues exceeding the respiratory use by the plants during the period of measurement ($NPP = GPP - R$)

neutralism relationship between two species in which neither population is affected by association with the other

neutral theory extension of the theory of island biogeography, proposed by Stephen P. Hubbell in 2001, in which all species are treated as if they possess the same per capita rates of birth, death, dispersal, and speciation

niche functional role of a species in a biotic community or ecosystem

niche assembly theory hypothesis that ecological communities are equilibrium assemblages of competing species, coexisting because each species is the most effective competitor in its own niche

niche overlap overlap or sharing of niche space by two or more species

niche width range of a niche dimension occupied by a population or species

nitrogen cycle description or model depicting the movement of nitrogen-containing compounds as they cycle among the atmosphere, soil, and living matter; movement of nitrogen, N, among the atmosphere, biosphere, and hydrosphere, including transformations between different chemical forms

nonparametric statistics statistical tests that do not require a normal or random pattern of distribution but can be carried out on qualitative or ranked information

nonreducible properties properties of the whole that are not reducible to the sum of the properties of the parts

noösphere (from Greek *noös,* "mind") a system dominated or managed by the human mind, as proposed by the Russian scientist V. I. Vernadskij in 1945

normality data conforming to a bell-shaped curve of the normal probability distribution

null hypothesis hypothesis that a sample of individuals drawn from a population in nature will come from a population with a known characteristic or parameter

numerical response change in the size of a population of predators in response to a change in the density of prey

nutrient substance required by an organism for normal growth, health, and vitality

nutrient cycling biogeochemical pathway in which an element or nutrient moves through ecosystems from assimilation by organisms to release by decomposers (bacteria and fungi), to be taken up by producers and recycled through the trophic levels once again

nutrient spiraling model of nutrient dynamics in a stream or river that, because of the downstream displacement of organisms or materials, are best represented by a spiral

O

oceanography the branch of ecology dealing with the biology, chemistry, geology, and physics of the ocean

oikos Greek term meaning "household" or "place to live"

oligotrophic refers to a body of water that is low in nutrients and productivity

omnivore organism that consumes both plant and animal matter

optimal foraging maximum possible energy return under a given set of foraging and habitat conditions; tendency of individuals to select food sizes or food patches that result in maximum food intake per energy expended

optimum carrying capacity (K_o) a lower level of population density at carrying capacity conditions than the maximum carrying capacity (K_m), which can be sustained in a particular habitat without "living on the edge" relative to resources such as food and space

optimum sustained yield level or amount of harvest that can be removed from a population that will result in the greatest yield that can be sustained indefinitely

ordination ordering of species populations and communities along a gradient; process in which communities are positioned graphically in order to have distances between species reflect community composition

P

paleoecology study of the relationships of ancient flora and fauna to their environment by means of the fossil record

parametric statistics statistical tests that require quantitative data or observations based on a normal or random pattern of distribution

parasitism relationship between two species in which one population (the *parasite*) benefits, but the other (the *host*) is harmed (although not usually killed)

parasitoid insect larva that kills its host, usually another insect, by consuming the soft tissue of the host before metamorphosis into an adult

parasitology the branch of science dealing with small organisms (*parasites*) living on or in other organisms (*hosts*), regardless of whether the effect of the parasite on the host is negative, positive, or neutral

pattern diversity biotic diversity based on zonation, stratification, periodicity, patchiness, or other criteria

pedology study of soil in all its aspects

pelagic refers to the open sea

perennial plant that lives for a number of years

periphyton organisms attached to natural substrates, such as plant stems, in the littoral or benthic zones of a lake (see also *aufwuchs*)

permafrost permanently frozen soil, characteristic of the Tundra biome

perturbation-dependent succession succession depending on allogenic, recurrent perturbations, such as fire or storms

phagotrophs heterotrophic organisms that ingest other organisms or particulate organic matter; macroconsumers

phenology study of the seasonal changes in plant and animal life and the relationship of these changes to weather and climate

pheromone chemical substance or compound secreted by individual organisms for communicating with other members of their species

phosphorus cycle movement of phosphorus, P, among the lithosphere (the dominant reservoir), hydrosphere, and biosphere, including the transformations between its different chemical forms

photochemical smog reaction of hydrocarbons with molecules of nitrogen oxide in the presence of ultraviolet radiation in sunlight, producing complex organic molecules of peroxyacetyl nitrates (PAN), resulting in atmospheric smog

photoperiod day-length period or cue by which organisms time their seasonal activities

phycosphere halo of bacteria that surrounds living algal cells in the marine environment

phylogeography the study of the patterns of speciation as embedded in biogeographic landscapes

physiological ecology the branch of ecology dealing with the physiological functioning of organisms in relation to their environment

physiological longevity maximum life span of an individual in a population under optimum environmental conditions

physiological natality maximum number of young that a female is capable of producing physiologically during her lifetime

phytoplankton small, floating plant life in aquatic systems

pin frame device used for obtaining a quantitative estimate of vegetative cover

pioneer stage initial seral stage of a sere characterized by early successional plant species (typically annuals)

planted corridor a strip of vegetation planted by humans for economic or ecological purposes (such as a strip of trees planted as a windbreak)

poikilotherm organism whose body temperature varies directly with the temperature of its environment

point of inflection the point on a sigmoid (S-shaped) growth curve where the growth rate is maximal

pollution frequently defined as *resources misplaced*

polyclimax concept theory that the final seral stage of succession is controlled by one of several local environmental forces or conditions such as soil, fire, or climate

population group of individuals of the same species living in a given area or habitat at a given time

population density number of individuals in a population within a defined unit of space

population dynamics study of those factors and mechanisms that cause changes in the number and density of populations in time and space

population ecology study of the relationship of a particular population or species with its environment

population equivalent amount of energy consumed by domestic animals that is equivalent to the amount of energy consumed by an average human being

population genetics study of changes in the frequencies of genes and genotypes within a population

population regulation mechanisms or factors within a population that cause its density to decrease when high (above its carrying capacity) and to increase when the density is low (below its carrying capacity)

positive feedback information that generates growth processes within a system

potential energy energy available to perform work due to position or chemical bonding

power probability of detecting a significant treatment effect when in fact there is one

precautionary principle encourages humankind to take action before effects are seen

predation relationship between two species in which one population serves as a food source for the other; a relationship whereby a predator kills its prey (depends on the prey as its food resource)

prescribed burning fires managed by humankind that favor particular organisms and ecosystems, such as grasslands and longleaf pines

primary consumers first-order consumers (herbivores) feeding directly on living plants or plant parts

primary producers organisms that produce food from simple inorganic substances (that is, photosynthetic plants)

primary production production of biological materials (biomass) by photosynthetic or chemosynthetic autotrophs

primary productivity rate at which primary producers produce biomass

primary succession ecological succession beginning on a barren, previously unoccupied substrate (such as a lava flow)

producers autotrophic organisms (such as green plants that can manufacture food via photosynthesis)

profundal zone the deep-water area of a lake that lies beyond the depth of effective light penetration, where $P/R < 1$

protocooperation relationship between two species in which both populations benefit from the relationship, but the association is not obligatory; frequently termed *facultative cooperation*

pseudoreplication occurs when researchers attempt to increase sample size by increased sampling or measuring of the same experimental treatment unit or mesocosm, rather than replicating the number of treatment units or mesocosms; three common types of pseudoreplication are spatial, temporal, and sacrificial

pulse stability property of systems adapted to a particular intensity and frequency of perturbations; populations that oscillate near the carrying capacity of a particular set of environmental conditions

punctuated speciation theory of evolution in which a species evolves in short bursts of rapid change early on but quite slowly thereafter

pyramid of biomass model or diagram that depicts the amounts of standing crop biomass at different trophic levels of an ecosystem

pyramid of energy model or diagram that depicts the rates of energy flow through the different trophic levels of an ecosystem

pyramid of numbers model or diagram that depicts the number of organisms present at each trophic level of an ecosystem; frequently termed the *Eltonian pyramid,* attributed to Charles Elton, a British ecologist

Q

quadrat a basic sampling unit, typically 1 m^2, used to sample grassland and old-field plant communities

qualitative chemical defenses chemically inexpensive poisons and toxins in plants that constitute effective barriers to herbivory

quantitative chemical defenses chemically expensive compounds in plants, such as tannins, that constitute barriers to herbivory by reducing plant quality or palatability

R

rad unit of radiation defined as an absorbed dose of 100 ergs of energy per gram of tissue

radiation ecology the branch of ecology concerned with the effects of radioactive materials on living systems and with the pathways by which these materials are dispersed within ecosystems

radionuclides isotopes of elements that emit ionizing radiations

rain shadow dry area on the lee side of mountains

random distribution distribution of individuals in a random pattern independent of all other individuals

randomization chance-mediated assignment of items or units to treatments to reduce researcher bias

random variation estimate of measurement dispersion around the most likely random variable

realized niche niche determined in the presence of competitors and other biotic interactions, such as predation

recessive describes an allele or phenotype that is expressed only in its homozygous state

reductionism theory that complex systems can be explained by analyzing the simple, basic parts of those systems

refuge site where individuals of an exploited population find protection from predators and parasites; isolated area where plants and animals find refuge from unfavorable environmental conditions; location where flora and fauna that were once widespread, but now considerably diminished in area, remain present

regenerated corridor regrowth of a strip of natural vegetation (such as a hedgerow that develops along fences due to the natural processes of secondary succession)

regular distribution distribution of individuals in a pattern more evenly spaced than would be expected by chance; uniform spacing or dispersion

relative dominance basal area for a particular species divided by the total basal area for all species; a value typically used to describe the dominance of tree species in a forest community

relative humidity percentage of water vapor present compared with saturation under existing temperature-pressure conditions

remnant corridor a strip of native vegetation left uncut after the surrounding vegetation is removed

rescue effect the concept that extinction is prevented by an influx of immigrants

resilience ability of a system to return to its original state or condition following a perturbation

resilience stability ability of a system to recover from a perturbation when the system is disrupted

resistance capacity of a system to maintain its structure and function during a perturbation

resistance stability ability of a system to resist perturbations and to maintain its structure and function intact

resource corridor a strip of natural vegetation that extends across the landscape (such as a galley forest along a stream)

resource partitioning differential use of resource types by a species or category of organisms

resource-use competition competition between two species in which each population adversely affects the other indirectly in the struggle for resources in short supply

resprout species fire-dependent species of plants that put more energy into underground storage organs and less into reproductive structure

restoration ecology the branch of ecology that deals with the restoration of a disturbed site of a plant and animal community or ecosystem to conditions that existed prior to the disturbance; the branch of ecology focusing on the application of ecological theory to restoration of highly disturbed sites, ecosystems, and landscapes

reward feedback positive feedback (increased growth or survivorship) by an organism or trophic level that sustains the survival of its food resource

rheotrophic describes wetlands (such as fens) that obtain much of their nutrient input from groundwater

rhizosphere region of soil activity immediately surrounding roots

R horizon parent rock below the C horizon of soil

riparian along banks of rivers and streams

river continuum concept model depicting a continuum of changes in physical structure, dominant organisms, and ecosystem processes along the length of a river system

Rosenzweig rule states that the logarithm of net primary productivity is linearly related to the logarithm of evapotranspiration

***r*-selection** selection at or near intrinsic growth rate, *r*, characterized by high rates of reproduction under conditions of low competition; *r*-selection tends to dominate in early stages of ecological succession

S

sample subset of all observations or individuals, *N*, in a population or sampling universe

saprotroph organism that feeds on dead organic matter; organism that absorbs organic nutrients from dead plant or animal matter

saturation dispersal movement of individuals out of a population that has reached or exceeded carrying capacity

Savanna tropical grassland on which trees and woody shrubs are widely spaced

sclerophyllous refers to woody plants with leathery, evergreen leaves that prevent moisture loss

scope of inference target population and breadth of conclusion

secondary compounds chemical compounds not used for metabolism, but chiefly for defensive purposes; compounds that interfere with specific metabolic pathways, physiological processes, palate response, or reproductive success of herbivores

secondary productivity rate of storage of organic matter by heterotrophs; rate at which heterotrophs (primary or secondary consumers) accumulate biomass by the production of new somatic or reproductive tissues

secondary succession succession taking place on a site that had previously supported life (such as an abandoned crop field)

second law of thermodynamics when energy is transformed from one form to another, there occurs a degradation of energy from a concentrated form to a dispersed form (in other words, no transformation of energy into potential energy is 100 percent efficient)

sedimentary cycle any global cycle where geological processes, such as the weathering of existing rock, erosion, and sedimentation, dominate or originate the cycle; calcium and potassium exemplify this type of cycle

self-thinning form of density-dependent plant mortality in an even-aged monoculture plant community

self-thinning rule (also called *−3/2 rule*) rule stating that plotting the average weight of individual plants in a stand against density often produces a line with an average slope of approximately −3/2

seral stage one of the successional stages in a sere

sere series of successional stages at a particular site leading to a mature, climax community

Shannon-Weaver index measure of the apportionment of species in a community based on information theory, published in 1949 by Claude E. Shannon and Warren Weaver

Shelford law of tolerance law, proposed by V. E. Shelford in 1911, stating that the presence and success of an organism or species depends on both the maximum and minimum resource or set of conditions

shredders stream invertebrates, typically aquatic insects, that feed on coarse particulate organic matter (CPOM)

sigmoid growth curve S-shaped pattern of population growth, with population size leveling off at the carrying capacity of that particular habitat

sink habitat where local mortality exceeds local reproductive success

soil conservation ethic practice used by farmers and stakeholders to prevent the loss of soil or decrease in soil quality

soil erosion removal of soil particles by water and wind, frequently accelerated by human disturbance

soil profile distinct strata or layering of horizons in the soil

soil texture relative proportions of sand, silt, and clay particles in the soil

solar constant rate at which sunlight reaches the atmosphere of Earth, equal to 1.94 gcal $\cdot$ $cm^{-2} \cdot min^{-1}$

solar powered refers to systems run and maintained by solar energy

source population a habitat where the reproductive success of a population provides individuals for sink habitats

spatial niche functional status of a species in its habitat expressed as a spatial dimension

species apportionment apportionment of one species in a defined area compared to the apportionment (numbers, biomass) of other species in the same area

species richness number of different species in a defined area

standing crop biomass weight of living biological materials in a particular area at a specific time

state variables sets of numbers used to represent the state or condition of a system at a particular time

statistical inference conclusion based on a mathematical summary of the data

steno- prefix meaning "narrow," derived from Greek *stenos*

stochastic refers to patterns arising from random factors or effects

stream order numerical classification of stream drainage based on stream structure and function from the headwaters to the mouth of the stream

stream spiraling movement and cycling of essential elements (such as carbon, nitrogen, and phosphorus) between organisms and available pools as they move downstream

subsidy-stress gradient gradient of response of a system to a perturbation, either positive (increased productivity) or negative (growth or reproductive retardation), through time

succession replacement of one community or seral stage by another

sulfur cycle movement of sulfur, S, between the lithosphere (the dominant reservoir) and the atmosphere, hydrosphere, and biosphere, and the transformations between different chemical forms

survivorship curve graphic description of the pattern of survival of individuals in a population from birth to the maximum age attained by each individual

sustainability ability to meet the needs of the present generation without compromising the ability to meet the needs of future generations; maintaining natural capital and resources to supply with necessities or nourishment to prevent falling below a given threshold of health or vitality

symbiosis two dissimilar species living together in close association

sympatric speciation speciation that occurs in the absence of geographic isolation

synergism result of combined factors, each of which influences a process in the same direction, but that, when combined, provide a greater effect than they would acting separately

system collection of interdependent components functioning within a defined boundary

systems ecology the branch of ecology focusing on general systems theory and application

T

Taiga circumpolar northern boreal forest biome

technoecosystem human-built system such as urban, suburban, and industrial development

temperate grasslands biomes dominated by grasses, such as *Andropogon, Panicum,* and *Bouteloua,* where annual precipitation is between 10 and 30 inches (25–75 centimeters) per year

territoriality defense of habitat space by an individual or a social group

territory area within a habitat defended by an individual

thermocline layer of water in a thermally stratified lake where the temperature profile changes rapidly relative to the body of water as a whole; zone of water in a thermally stratified lake between the epilimnion and hypolimnion

3/4 power law the metabolic rate of an individual animal tends to increase as the 3/4 power of its weight

tilth aggregation of mycorrhizae and soil particles that constitutes a healthy soil structure

tolerance model model of succession that proposes that succession leads to a community composed of those plant species that are most efficient in exploiting resources

top-down regulation regulation of a community or ecosystem trophic structure by increased predation; influence of secondary consumers on the sizes of the trophic levels below them in the food web

transdisciplinary refers to approaches involving multilevel large-scale cooperation focusing on entire educational or innovative systems

transpiration efficiency ratio of net primary production to water transpired

treatment experimental action designed to produce an effect

trophic dynamics transfer of energy from one trophic level or part of an ecosystem to another

trophic level position in a food chain as determined by the number of energy transfer steps to that level (primary producer to secondary consumer, for example); functional classification of organisms in an ecosystem according to feeding relationships

trophic niche functional status of a species based on trophic level or energy relationships

Tuckfield maxim data collected for a general or unspecified purpose can answer very few specific questions

Tundra biome characterized by mosses, lichens, sedges, and forbs, but absence of trees; area of permanently frozen soil dominated by treeless vegetation

turnover time time required to replace a quantity of a substance or resource equal to the amount of that component present in the system

U

ultraviolet radiation (UV) electromagnetic radiation at wavelengths between 100 and 400 nanometers, lying just beyond the high-energy (violet) end of the visible light band of the solar spectrum

understory layer of vegetation below the canopy of a forest

upwelling movement of deep ocean water to the surface or into the euphotic zone; occurs most commonly along the west coasts of continents (for example, the Peru Current along the coast of South America)

V

vagility inherent ability to move about freely

vector organism that transmits a pathogen from one organism to another

vernal pool temporary pond or shallow pool filled in the spring

vesicular-arbuscular mycorrhizae (VAM) relationship between fungal mycelia and plant roots, in which the fungus enters and grows within the root cells of the host and extends into the surrounding soil

virulence the aspect of parasites that measures the harm done to the host

W

water potential capacity of water to perform work, determined by its free energy content

watershed catchment or drainage basin of a river; the total area above a given point on a stream or river that contributes water to the flow at that point

wave-generated succession secondary succession initiated in plant communities vulnerable to strong winds when waves of trees or vegetation are uprooted

weathering physical and chemical breakdown of rock and its component minerals at the soil interface

wetlands habitats that are perpetually or periodically flooded

wildfires intense fires that destroy most of the vegetation and some soil organic matter

wildlife management the branch of ecology dealing with the management and conservation of indigenous wildlife

woodland tree-covered land including associated plant and animal habitats

X

xeric characterized by dry conditions

xerophyte plant with special adaptions (such as sunken stomata) for surviving prolonged periods of drought

xerosere term used to describe succession on dry land or rock surface

Y

yield output or return, usually expressed in energy or mass, of natural resources (such as fish, timber, or game) from an aquatic or terrestrial ecosystem

Z

zonation distribution of vegetation along an environmental gradient, such as latitudinal, altitudinal, or horizontal zones within a landscape

zoogeography study of the abundance and distribution of animals

zooplankton floating or weakly swimming animals in marine or freshwater ecosystems

References

A

Abrahamson, W. G., and M. Gadgil. 1973. Growth form and reproductive effort in goldenrods (*Solidago*, Compositae). *American Naturalist* 107:651–661.

Abramovitz, J. N., and A. T. Mattoon. 1999. *Paper cuts: Recovering the paper landscape.* Worldwatch Paper 149. Washington, D.C.: Worldwatch Institute.

Adkisson, P. L., G. A. Niles, J. K. Walker, L. S. Bird, and H. B. Scott. 1982. Controlling cotton's insect pests: A new system. *Science* 216:19–22.

Ahl, V., and T. F. H. Allen. 1996. *Hierarchy theory: A vision, vocabulary, and epistemology.* New York: Columbia University Press.

Ahmadjian, V. 1995. Lichens are more important than you think. *BioScience* 45:124.

Allee, W. C. 1931. *Animal aggregations: A study in general sociology.* Chicago: University of Chicago Press.

———. 1951. *Cooperation among animals: Social life of animals.* New York: Schuman.

Allee, W. C., A. E. Emerson, O. Park, T. Park, and K. P. Schmidt. 1949. *Principles of animal ecology.* Philadelphia: W. B. Saunders.

Allen, J. 1991. *Biosphere 2: The human experiment.* New York: Penguin Books.

Allen, T. F. H., and T. W. Hoekstra. 1992. *Toward a unified ecology. Complexity in ecological systems.* New York: Columbia University Press.

Allen, T. F. H., and T. B. Starr. 1982. *Hierarchy: Perspectives for ecological complexity.* Chicago: University of Chicago Press.

Altieri, M. A. 2000. The ecological impacts of transgenic crops on agroecosystem health. *Ecosystem Health* 6:13–23.

American Institute of Biological Sciences (AIBS). 1989. *Fire impact on Yellowstone.* Special Issue. *BioScience* 39:667–722.

Andreae, M. O. 1996. Raising dust in the greenhouse. *Nature* 380:389–390.

Andrewartha, H. G., and L. C. Birch. 1954. *The distribution and abundance of animals.* Chicago: University of Chicago Press.

Ardrey, R. 1967. *The territorial imperative.* New York: Atheneum.

Armentano, T. V. 1980. Drainage of organic soils as a factor in the world carbon cycle. *BioScience* 30:825–830.

Ausubel, J. H. 1996. Can technology spare the earth? *American Scientist* 84:166–178.

Avery, W. H., and C. Wu. 1994. *Renewable energy from the ocean: A guide to OTEC.* Oxford: Oxford University Press.

Axelrod, R., and W. D. Hamilton. 1981. The evolution of cooperation. *Science* 211:1390–1396.

Ayala, F. J. 1969. Experimental invalidation of the principle of competitive exclusion. *Nature* 224:1076.

———. 1972. Competition between species. *American Scientist* 60:348–357.

Azam, F. 1998. Microbial control of oceanic carbon flux: The plot thickens. *Science* 280:694–696.

Azam, F., T. Fenchel, J. G. Field, J. S. Gray, L. A. Meyer-Reil, and F. Thingstad. 1983. The ecological role of water-column microbes in the sea. *Marine Ecology Progress Series* 10:257–263.

B

Bailey, R. G. 1976. *Ecoregions of the United States.* Washington, D.C.: Forest Service, Fish and Wildlife Service, United States Department of Agriculture. Reprint, 1978, *Description of the ecoregions of the United States.* Ogden, Utah: Forest Service Intermountain Region, United States Department of Agriculture.

———. 1995. *Descriptions of ecoregions of the United States,* 2nd ed. Washington, D.C.: Forest Service, United States Department of Agriculture.

———. 1998. *Ecoregions: The ecosystem geography of the oceans and continents.* New York: Springer Verlag.

Bak, P. 1996. *How nature works: The science of self-organized criticality.* New York: Springer Verlag.

Bakelaar, R. G., and E. P. Odum. 1978. Community and population level responses to fertilization in an old-field ecosystem. *Ecology* 59:660–665.

Baker, A. C. 2001. Reef corals bleach to survive change. *Nature* 411:765–766.

Bamforth, S. S. 1997. Evolutionary implications of soil protozoan succession. *Revista de la Sociedad Mexicana de Historia Natural* 47:93–97.

Banavar, J. R., S. P. Hubbell, and A. Meritan. 2004. Symmetric neutral theory with Janzen-Connell density dependence explains patterns of beta diversity in tropical tree communities. *Science,* in review.

Barbour, D. A. 1985. Patterns of population fluctuation in the pine looper moth *Bupalus piniaria L.* in Britain. In *Site characteristics and population dynamics of Lepidopteran and Hymenopteran forest pests,* D. Bevan and J. T. Stoakley, Eds. Edinburgh: Forestry Commission Research and Development Paper 135:8–20.

Barrett, G. W. 1968. The effects of an acute insecticide stress on a semi-enclosed grassland ecosystem. *Ecology* 49:1019–1035.

———. 1985. A problem-solving approach to resource management. *BioScience* 35:423–427.

———. 1988. Effects of Sevin on small mammal populations in agricultural and old-field ecosystems. *Journal of Mammalogy* 69:731–739.

———. 1989. Viewpoint: A sustainable society. *BioScience* 39:754.

———. 1990. Nature's model. *Earth•Watch* 9:24–25.

———. 1992. Landscape ecology: Designing sustainable agricultural landscapes. *Journal of Sustainable Agriculture* 2:83–103.

———. 1994. Restoration ecology: Lessons yet to be learned. In *Beyond preservation: Restoring and inventing landscapes,* D. Baldwin, J. de Luce, and C. Pletsch, Eds. Minneapolis: University of Minnesota Press, pp. 113–126.

———. 2001. Closing the ecological cycle: The emergence of integrative science. *Ecosystem Health* 7:79–84.

Barrett, G. W., and T. L. Barrett. 2001. Cemeteries as repositories of natural and cultural diversity. *Conservation Biology* 15:1820–1824.

Barrett, G. W., and P. J. Bohlen. 1991. Landscape ecology. In *Landscape linkages and biodiversity,* W. E. Hudson, Ed. Washington, D.C.: Island Press, pp. 149–161.

Barrett, G. W., and A. Farina. 2000. Integrating ecology and economics. *BioScience* 50:311–312.

Barrett, G. W., and G. E. Likens. 2002. Eugene P. Odum: Pioneer of ecosystem ecology. *BioScience* 52:1047–1048.

Barrett, G. W., and K. E. Mabry. 2002. Twentieth-century classic books and benchmark publications in biology. *BioScience* 52:282–285.

Barrett, G. W., and C. V. Mackey. 1975. Prey selection and caloric ingestion rate of captive American kestrels. *The Wilson Bulletin* 87:514–519.

Barrett, G. W., and E. P. Odum. 2000. The twenty-first century: The world at carrying capacity. *BioScience* 50:363–368.

Barrett, G. W., and J. D. Peles, Eds. 1999. *Landscape ecology of small mammals.* New York: Springer Verlag.

Barrett, G. W., and R. Rosenberg, Eds. 1981. *Stress effects on natural ecosystems.* New York: John Wiley.

Barrett, G. W., and L. E. Skelton. 2002. Agrolandscape ecology in the twenty-first century. In *Landscape ecology in agroecosystems management,* L. Ryszkowski, Ed. Boca Raton, Fla.: CRC Press, pp. 331–339.

Barrett, G. W., T. L. Barrett, and J. D. Peles. 1999. Managing agroecosystems as agrolandscapes: Reconnecting agricultural and urban landscapes. In *Biodiversity in agroecosystems,* W. W. Collins and C. O. Qualset, Eds. Boca Raton, Fla.: CRC Press, pp. 197–213.

Barrett, G. W., J. D. Peles, and S. J. Harper. 1995. Reflections on the use of experimental landscapes in mammalian ecology. In *Landscape approaches in mammalian ecology and conservation,* W. Z. Lidicker, Jr., Ed. Minneapolis: University of Minnesota Press, pp. 157–174.

Barrett, G. W., J. D. Peles, and E. P. Odum. 1997. Transcending processes and the levels of organization concept. *BioScience* 47:531–535.

Barrett, G. W., N. Rodenhouse, and P. J. Bohlen. 1990. Role of sustainable agriculture in rural landscapes. In *Sustainable agricultural systems,* C. A. Edwards, R. Lai, P. Madden, R. H. Miller, and G. House, Eds. Ankeny, Iowa: Soil and Water Conservation Society, pp. 624–636.

Baskin, Y. 1999. Yellowstone fires: A decade later—ecological lessons in the wake of the conflagration. *BioScience* 49:93–97.

Bazzaz, F. A. 1975. Plant species diversity in old-field successional ecosystems in southern Illinois. *Ecology* 56:485–488.

Beecher, W. J. 1942. *Nesting birds and the vegetation substrate.* Chicago: Chicago Ornithological Society.

Bell, W., and R. Mitchell. 1972. Chemotactic and growth responses of marine bacteria to algal extracellular products. *Biological Bulletin* 143:265–277.

Belovsky, G. E., D. B. Botkin, T. A. Crowl, K. W. Cummins, J. F. Franklin, M. L. Hunter Jr., A. Joern, D. B. Lindenmayer, J. A. MacMahon, C. R. Margules, and J. M. Scott. 2004. Ten suggestions to strengthen the science of ecology. *BioScience* 54:345–351.

Belt, C. B., Jr. 1975. The 1973 flood and man's constriction of the Mississippi River. *Science* 189:681–684.

Benner, R., A. E. Maccubbin, and R. E. Hodson. 1984. Anaerobic biodegradation of the lignin and polysaccaride components of lignocellulose and synthetic lignin by sediment microflora. *Applied and Environmental Microbiology* 47:998–1004.

Berger, P. J., N. C. Negus, E. H. Sanders, and P. D. Gardner. 1981. Chemical triggering of reproduction in *Microtus montanus*. *Science* 214:69–70.

Bergstrom, J. C., and H. K. Cordell. 1991. An analysis of the demand for and value of outdoor recreation in the United States. *Journal of Leisure Research* 23:67–86.

Bernstein, B. B. 1981. Ecology and economics: Complex systems in changing environments. *Annual Review of Ecology and Systematics* 12:309–330.

Bertalanffy, L. 1950. An outline of general systems theory. *British Journal of Philosophy of Science* 1:139–164.

———. 1968. *General systems theory: Foundations, development, application*. Rev. ed., 1975, New York: G. Braziller.

Beyers, R. J. 1964. The microcosm approach to ecosystem biology. *American Biology Teacher* 26:491–498.

Beyers, R. J., and H. T. Odum. 1995. *Ecological microcosms*. New York: Springer Verlag.

Birch, L. C. 1948. The intrinsic rate of natural increase of an insect population. *Journal of Animal Ecology* 17:15–26.

Blackmore, S. 1996. Knowing the Earth's biodiversity: Challenges for the infrastructure of systematic biology. *Science* 274:63–64.

Blair, W. F. 1977. *Big biology: The US/IBP*. Stroudsburg, Pa.: Dowden, Hutchinson, and Ross.

Bock, C. E., J. H. Bock, and M. C. Grant. 1992. Effect of bird predation on grasshopper densities in an Arizona grassland. *Ecology* 73:1706–1717.

Bolen, E. G. 1998. *Ecology of North America*. New York: John Wiley.

Bolliger, J., J. C. Sprott, and D. J. Mladenoff. 2003. Self-organization and complexity in historical landscape patterns. *Oikos* 100:541–553.

Bongaarts, J. 1998. Demographic consequences of declining fertility. *Science* 282:419–420.

Bormann, F. H., and G. E. Likens. 1967. Nutrient cycling. *Science* 155:424–429.

———. 1979. *Pattern and process in a forested ecosystem*. New York: Springer Verlag.

Bormann, F. H., D. Balmori, and G. T. Geballe. 2001. *Redesigning the American lawn: A search for environmental harmony*. New Haven, Conn.: Yale University Press.

Bormann, F. H., G. E. Likens, and J. M. Melillo. 1977. Nitrogen budget for an aggrading northern hardwood ecosystem. *Science* 196:981–983.

Boucher, D. H., S. James, and K. H. Keeler. 1982. The ecology of mutualism. *Annual Review of Ecology and Systematics* 13:315–347.

Boulding, K. 1962. *A reconstruction of economics*. New York: Science Editions.

———. 1966. The economics of the coming spaceship Earth. In *Environmental quality in a growing economy*, H. Jarrett, Ed. Baltimore, Md.: Johns Hopkins Press for Resources for the Future, pp. 3–14.

———. 1978. *Ecodynamics: A new theory of societal evolution*. Beverly Hills, Calif.: Sage.

Bowne, D. R., J. D. Peles, and G. W. Barrett. 1999. Effects of landscape spatial structure on movement patterns of the hispod cotton rat, *Sigmodon hispidus*. *Landscape Ecology* 14:53–65.

Box, E. 1978. Geographical dimensions of terrestrial net and gross productivity. *Radiation and Environmental Biophysics* 15:305–322.

Box, G. E. P., W. G. Hunter, and J. S. Hunter, Jr. 1978. *Statistics for experimenters*. New York: John Wiley.

Boyle, T. P., and J. F. Fairchild. 1997. The role of mesocosm studies in ecological risk analysis. *Ecological Applications* 7:1099–1102.

Bratton, S. 1992. *Six billion and more: Human population regulation and Christian ethics.* Louisville, Ky.: Westminster/John Knox Press.

Braun-Blanquet, J. 1932. *Plant sociology: The study of plant communities,* G. D. Fuller and H. S. Conard, Trans., Rev., and Eds. New York: McGraw-Hill.

———. 1951. *Pflanzensoziologie: Grundzüge der vegetationskunde.* Vienna: Springer Verlag.

Bray, J. R., and E. Gorham. 1964. Litter production in forests of the world. *Advances in Ecological Research* 2:101–157.

Brewer, S. R., and G. W. Barrett. 1995. Heavy metal concentrations in earthworms following long-term nutrient enrichment. *Bulletin of Environmental Contamination and Toxicology* 54:120–127.

Brewer, S. R., M. Benninger-Truax, and G. W. Barrett. 1994. Mechanisms of ecosystem recovery following eleven years of nutrient enrichment in an old-field community. In *Toxic metals in soil-plant systems,* S. M. Ross, Ed. New York: John Wiley, pp. 275–301.

Brewer, S. R., M. F. Lucas, J. A. Mugnano, J. D. Peles, and G. W. Barrett. 1993. Inheritance of albinism in the meadow vole, *Microtus pennsylvanicus. American Midland Naturalist* 130: 393–396.

Brian, M. W. 1956. Exploitation and interference in interspecies competition. *Journal of Animal Ecology* 25:335–347.

Brillouin, L. 1949. Life, thermodynamics, and cybernetics. *American Scientist* 37:354–368.

Brock, T. D. 1967. Relationship between primary productivity and standing crop along a hot spring thermal gradient. *Ecology* 48:566–571.

Brock, T. D., and M. L. Brock. 1966. Temperature options for algal development in Yellowstone and Iceland hot springs. *Nature* 209:733–734.

Brooks, J. L., and S. I. Dodson. 1965. Predation, body size, and composition of plankton. *Science* 150:28–35.

Brower, L. P. 1969. Ecological chemistry. *Scientific American* 220:22–29.

———. 1994. A new paradigm in conservation of biodiversity: Endangered biological phenomena. In *Principles of conservation biology,* G. K. Meffe and C. R. Carroll, Eds. Sunderland, Mass.: Sinauer, pp. 104–109.

Brower, L. P., and J. Brower. 1964. Birds, butterflies, and plant poisons: A study in ecological chemistry. *Zoologica* 49:137–159.

Brower, L. P., and S. B. Malcolm. 1991. Animal migrations: Endangered phenomena. *American Zoologist* 31:265–276.

Brown, H. S. 1978. *The human future revisited: The world predicament and possible solutions.* New York: W. W. Norton.

Brown, J. H. 1971. Mammals on mountaintops: Nonequilibrium insular biogeography. *American Naturalist* 105:467–478.

———. 1978. The theory of insular biogeography and the distribution of boreal birds and mammals. *Great Basin Naturalist Memoirs* 2:209–227.

Brown, J. H., and A. Kodric-Brown. 1977. Turnover rates in insular biogeography: Effect of immigration on extinction. *Ecology* 58:445–449.

Brown, J. L. 1964. The evolution of diversity in avian territorial systems. *The Wilson Bulletin* 76:160–169.

———. 1969. Territorial behavior and population regulation in birds, a review and reevaluation. *The Wilson Bulletin* 81:293–329.

Brown, L. R. 1980. *Food or fuel: New competition for the world croplands.* Worldwatch Paper 35. Washington, D.C.: Worldwatch Institute.

———. 2001. *Eco-economy: Building an economy for the Earth.* New York: W. W. Norton.

Brown, V. K. 1985. Insect herbivores and plant succession. *Oikos* 44:17–22.

Brugam, R. B. 1978. Human disturbance and the historical development of Linsley Pond. *Ecology* 59:19–36.

Brummer, E. C. 1998. Diversity, stability, and sustainable American agriculture. *Agronomy Journal* 90:1–2.

Brundtland, G. H. 1987. *Our common future.* World Commission on Environment and Development. New York: Oxford University Press.

Buell, M. F. 1956. Spruce-fir and maple-basswood competition in Itasca Park, Minnesota. *Ecology* 37:606.

Bullock, T. H. 1955. Compensation for temperature in the metabolism and activity of poikilotherms. *Biological Reviews* 30:311–342.

Butcher, S. S., R. J. Charlson, G. H. Orians, and G. V. Wolfe, Eds. 1992. *Global biogeochemical cycles.* London: Academic Press.

Butzer, K. W. 1980. Civilizations: Organisms or systems? *American Scientist* 68:517–523.

C

Cahill, J. F., Jr. 1999. Fertilization effects on interactions between above- and belowground competition in an old field. *Ecology* 80:466–480.

Cairns, J. 1992. *Restoration of aquatic ecosystems: Science, technology, and public policy.* Washington, D.C.: National Academy Press.

———. 1997. Global coevolution of natural systems and human society. *Revista de la Sociedad Mexicana de Historia Natural* 47:217–228.

Cairns, J., K. L. Dickson, and E. E. Herrick, Eds. 1977. *Recovery and restoration of damaged ecosystems.* Charlottesville: University of Virginia Press.

Callahan, J. T. 1984. Long-term ecological research. *BioScience* 34:363–367.

Callicott, J. B., Ed. 1987. *Companion to a Sand County Almanac.* Madison: University of Wisconsin Press.

Callicott, J. B., and Freyfogle, E. T., Eds. 1999. *For the health of the land: Previously unpublished essays and other writings/Aldo Leopold.* Washington, D.C.: Island Press.

Calow, P., Ed. 1999. *Blackwell's concise encyclopedia of ecology.* Oxford: Blackwell.

Camazine, S., J. L. Deneubourg, N. R. Franks, J. Sneyd, G. Theraulaz, and E. Bonabeau. 2001. *Self-organization in biological systems.* Princeton, N.J.: Princeton University Press.

Cannon, W. B. 1932. *The wisdom of the body.* 2nd ed., 1939, New York: W. W. Norton.

Capra, F. 1982. *The turning point: Science, society, and the rising culture.* New York: Simon and Schuster.

Carpenter, J. R. 1940. Insect outbreaks in Europe. *Journal of Animal Ecology* 9:108–147.

Carpenter, S. R., and J. F. Kitchell. 1988. Consumer control of lake productivity. *BioScience* 38:764–769.

———. 1993. *The trophic cascade in lakes.* Cambridge, U.K.: Cambridge University Press.

Carpenter, S. R., J. F. Kitchell, and J. R. Hodgson. 1985. Cascading trophic interactions and lake productivity. *BioScience* 35:634–639.

Carroll, S. B. 2001. Macroevolution: The big picture. *Nature* 409:669.

Carson, R. 1962. *Silent spring.* New York: Houghton Mifflin.

Carson, W. P., and G. W. Barrett. 1988. Succession in old-field plant communities: Effects of contrasting types of nutrient enrichment. *Ecology* 69:984–994.

Carson, W. P., and S. T. A. Pickett. 1990. The role of resources and disturbance in the organization of an old-field plant community. *Ecology* 71:226–238.

Carson, W. P., and R. B. Root. 1999. Top-down effects of insect herbivores during early succession: Influence on biomass and plant dominance. *Oecologia* 121:260–272.

———. 2000. Herbivory and plant species coexistence: Community regulation by an outbreaking phytophagous insect. *Ecological Monographs* 70:73–99.

Caswell, H., F. Reed, S. N. Stephenson, and P. A. Werner. 1973. Photosynthetic pathways and selected herbivory: A hypothesis. *American Naturalist* 107:465–480.

Catton, W. R., Jr. 1980. *Overshoot: Ecological basis of revolutionary change.* Urbana: Illinois University Press.

———. 1987. The world's most polymorphic species. *BioScience* 37:413–419.

Caughley, G. 1970. Eruption of ungulate populations, with emphasis on Himalayan thar in New Zealand. *Ecology* 51:53–72.

Chapin, F. S., III, P. A. Matson, and H. A. Mooney. 2002. *Principles of terrestrial ecosystem ecology.* New York: Springer Verlag.

Chapin, F. S., E. D. Schulze, and H. A. Mooney. 1992. Biodiversity and ecosystem processes. *Trends in Ecology and Evolution* 7:107–108.

Chapman, R. N. 1928. The quantitative analysis of environmental factors. *Ecology* 9:111–122.

Chase, J. M., and M. A. Leibold. 2003. *Ecological niches.* Chicago: University of Chicago Press.

Chave, J., H. Muller-Landau, and S. A. Levin. 2002. Comparing classical community models: Theoretical consequences for patterns of species diversity. *American Naturalist* 159:1–23.

Chepko-Sade, B. D., and Z. T. Halpin, 1987. *Mammalian dispersal patterns: The effects of social structure on population genetics.* Chicago: University of Chicago Press.

Chitty, D. 1960. Population processes in the vole and their relevance to general theory. *Canadian Journal of Zoology* 38:99–113.

———. 1967. The natural selection of self-regulatory behavior in animal populations. *Proceedings of the Ecological Society of Australia* 2:51–78.

Christian, J. J. 1950. The adreno-pituitary system and population cycles in mammals. *Journal of Mammalogy* 31:247–259.

———. 1963. Endocrine adaptive mechanisms and the physiologic regulation of population growth. In *Physiological mammalogy,* W. V. Mayer and R. G. van Gelder, Eds. New York: Academic Press, pp. 189–353.

Christian, J. J., and D. E. Davis. 1964. Endocrines, behavior, and populations. *Science* 146:1550–1560.

Cilek, J. E., and J. A. Mulrennan. 1997. Pseudo-replication: What does it mean, and how does it relate to biological experiments? *Journal of the American Mosquito Control Association* 13:102–103.

Clark, E. H., J. A. Haverkamp, and W. Chapman. 1985. *Eroding soils: The off-farm impacts.* Washington, D.C.: The Conservation Foundation.

Clark, W. C. 1989. Managing planet Earth. *Scientific American* 261:47–54.

Clausen, J. C., D. D. Keck, and W. M. Hiesey. 1948. Experimental studies on the nature of species. Volume 3. Environmental responses to climatic races of *Achillea. Carnegie Institution of Washington Publication* 581:1–129.

Clean Air Act 1970. 42 *United States Code Annotated* Sections 7401–7671, 1990 Amendment.

Clements, F. E. 1905. *Research methods in ecology.* Lincoln, Nebr.: University Publishing Company.

———. 1916. Plant succession: Analysis of the development of vegetation. Washington, D.C.: *Publications of the Carnegie Institute* 242:1512. Reprint, 1928, *Plant succession and indicators.* New York: Wilson.

Clements, F. E., and V. E. Shelford. 1939. *Bio-ecology.* New York: John Wiley.

Cleveland, L. R. 1924. The physiological and symbiotic relationships between the intestinal protozoa of termites and their host with special reference to *Reticulitermes fluipes* Kollar. *Biology Bulletin* 46:177–225.

———. 1926. Symbiosis among animals with special reference to termites and their intestinal flagellates. *Quarterly Review of Biology* 1:51–60.

Cloud, P. E., Jr., Ed. 1978. *Cosmos, Earth, and man: A short history of the universe.* New Haven, Conn.: Yale University Press.

Cockburn, A. 1988. *Social behaviour in fluctuating populations.* New York: Croom Helm.

Cody, M. L. 1966. A general theory of clutch size. *Evolution* 20:174–184.

———. 1974. Optimization in ecology. *Science* 183:1156–1164.

Cohen, M. N., R. S. Malpass, and H. G. Klein. 1980. *Biosocial mechanisms of population regulation.* New Haven, Conn.: Yale University Press.

Cole, L. C. 1951. Population cycles and random oscillations. *Journal of Wildlife Management* 15:233–251.

———. 1954. The population consequences of life history phenomena. *Quarterly Review of Biology* 29:103–107.

———. 1966. Protect the friendly microbes. In *The fragile breath of life,* Tenth Anniversary Issue, Science and Humanity Supplement. *Saturday Review,* 7 May 1966, pp. 46–47.

Coleman, D. C. 1995. Energetics of detritivory and microbiology in soil in theory and practice. In *Food webs: Integration of patterns and dynamics,* G. A. Polis and K. O. Winemiller, Eds. New York: Chapman and Hall, pp. 39–50.

Coleman, D. C., and D. A. Crossley, Jr. 1996. *Fundamentals of soil ecology.* San Diego, Calif.: Academic Press.

Coleman, D. C., P. F. Hendrix, and E. P. Odum. 1998. Ecosystem health: An overview. In *Soil chemistry and ecosystem health,* P. H. Wang, Ed. Madison, Wisc.: *Soil Science Society of America Special Publication* 52:1–20.

Collins, R. J., and G. W. Barrett. 1997. Effects of habitat fragmentation on meadow vole, *Microtus pennsylvanicus* population dynamics in experimental landscape patches. *Landscape Ecology* 12:63–76.

Collins, S. L., A. Knapp, J. M. Briggs, J. M. Blair, and E. M. Steinauer. 1998. Modulation of diversity by grazing in native tallgrass prairie. *Science* 280:745–747.

Collins, W. W., and C. O. Qualset, Eds. 1999. *Biodiversity in agroecosystems.* Boca Raton, Fla.: CRC Press.

Colwell, R. K. 1973. Competition and coexistence in a simple tropical community. *American Naturalist* 107:737–760.

Connell, J. H. 1961. The influence of interspecific competition and other factors on the distribution of the barnacle, *Chthamalus stellatus. Ecology* 42:133–146.

———. 1972. Community interactions on marine rocky intertidal shores. *Annual Review of Ecology and Systematics* 3:169–192.

———. 1975. Some mechanisms producing structure in natural communities: A model and evidence from field experiments. In *Ecology and evolution of communities,* M. L. Cody and J. Diamond, Eds. Cambridge, Mass.: Harvard University Press, pp. 460–490.

———. 1978. Diversity in tropical rain forests and coral reefs. *Science* 199:1302–1310.

———. 1979. Tropical rainforests and coral reefs as open nonequilibrium systems. In *Popu-*

lation dynamics, R. M. Anderson, B. D. Turner, and L. R. Taylor, Eds. Oxford: Blackwell, pp. 141–163.

Connell, J. H., and R. O. Slayter. 1977. Mechanism of succession in natural communities and their role in community stability and organization. *American Naturalist* 111:1119–1144.

Cooke, G. D. 1967. The pattern of autotrophic succession in laboratory microecosystems. *BioScience* 17:717–721.

Cooper, N. 2001. *Wildlife in church and churchyard,* 2nd ed. London: Church House.

Costanza, R., Ed. 1991. *Ecological economics: The science and management of sustainability.* New York: Columbia University Press.

———. 2001. Visions, values, valuation, and the need for an ecological economics. *BioScience* 51:459–468.

Costanza, R., J. Cumberland, H. Daly, R. Goodland, and R. Norgaard. 1997. *An introduction to ecological economics.* Boca Raton, Fla.: St. Lucie Press.

Costanza, R., R. d'Arge, R. de Groot, S. Farber, M. Grasso, B. Hannon, K. Limburg, S. Naeem, R. V. O'Neill, J. Paruelo, R. G. Raskin, P. Sutton, and M. van den Belt. 1997. The value of the world's ecosystem services and natural capital. *Nature* 387:253–260.

Council on Environmental Quality (CEQ). 1981a. *Environmental quality.* 12th Annual Report. Washington, D.C.: United States Government Printing Office.

———. 1981b. *Global future—time to act,* G. Speth, Ed. Washington, D.C.: United States Government Printing Office.

Covey, S. R. 1989. *The seven habits of highly effective people: Restoring the character ethic.* New York: Simon and Schuster.

Cowles, H. C. 1899. The ecological relations of the vegetation of the sand dunes of Lake Michigan. *Botanical Gazette* 27:95–117; 167–202; 281–308; 361–391.

Crawley, M. J. 1989. Insect herbivores and plant population dynamics. *Annual Review of Entomology* 34:531–564.

———. 1997. Plant-herbivore dynamics. In *Plant ecology,* M. Crawley, Ed. Cambridge, Mass.: Blackwell, pp. 401–474.

Cross, J. G., and M. J. Guyer. 1980. *Social traps.* Ann Arbor: University of Michigan Press.

Crowner, A. W., and G. W. Barrett. 1979. Effects of fire on the small mammal component of an experimental grassland community. *Journal of Mammalogy* 60:803–813.

Cummins, K. W. 1967. Biogeography. *Canadian Geographic* 11:312–326.

———. 1974. Structure and function of stream ecosystems. *BioScience* 24:631–641.

———. 1977. From headwater streams to rivers. *The American Biology Teacher* 39:305–312.

Cummins, K. W., and M. J. Klug. 1979. Feeding ecology of stream invertebrates. *Annual Review of Ecology and Systematics* 10:147–172.

Currie, R. I. 1958. Some observations on organic production in the northeast Atlantic. *Rapports et Procès-Verbaux des Réunions. Conseil International pour l'Exploration de la Mer* 144: 96–102.

Curtis, J. T., and R. P. McIntosh. 1951. An upland forest continuum in the prairie forest border region of Wisconsin. *Ecology* 32:476–496.

D

Daily, G. C. 1997. *Nature's services: Societal dependence on natural ecosystems.* Washington, D.C.: Island Press.

Daily, G. C., S. Alexander, P. R. Ehrlich, L. Goulder, J. Lubchenco, P. A. Matson, H. A. Mooney, S. Postel, S. H. Schneider, D. Tilman, and G. M. Woodwell. 1997. Ecosystem services: Benefits supplied to human societies by natural ecosystems. *Ecological Society of America Issues in Ecology* 2:2–15.

Dale, V. H., L. A. Joyce, S. McNulty, R. P. Neilson, M. P. Ayres, M. D. Flannigan, P. J. Hanson, L. C. Irland, A. E. Lugo, C. J. Peterson, D. Simberloff, F. J. Swanson, B. J. Stocks, and B. M. Wotton. 2001. Climate change and forest disturbances. *BioScience* 51:723–734.

Dales, R. P. 1957. Commensalism. In *Treatise on marine ecology and paleoecology*, J. W. Hedgpeth, Ed. Volume 1. Boulder, Colo.: Geological Society of America, pp. 391–412.

Daly, H. E., and J. B. Cobb. 1989. *For the common good: Redirecting the economy towards community, the environment, and a sustainable future.* Boston: Beacon Press.

Daly, H. E., and K. N. Townsend, Eds. 1993. *Valuing the Earth: Economics, ecology, ethics.* Cambridge: Massachusetts Institute of Technology Press.

Darwin, C. 1859. *The origin of species by means of natural selection.* Reprint, 1998, New York: Modern Library.

Daubenmire, R. 1966. Vegetation: Identification of typal communities. *Science* 151:291–298.

———. 1974. *Plants and environment: A textbook of plant autecology,* 3rd ed. New York: John Wiley.

Davidson, J. 1938. On growth of the sheep population in Tasmania. *Transactions of the Royal Society of South Australia* 62:342–346.

Davis, M. A. 2003. Biotic globalization: Does competition from introduced species threaten biodiversity? *BioScience* 53:481–489.

Davis, M. A., and L. B. Slobodkin. 2004. The science and values of restoration ecology. *Restoration Ecology* 12:1–3.

Davis, M. B. 1969. Palynology and environmental history during the Quaternary period. *American Scientist* 57:317–332.

———. 1983. Quaternary history of deciduous forests of eastern North America and Europe. *Annals of the Missouri Botanical Garden* 70:550–563.

———. 1989. Retrospective studies. In *Long-term studies in ecology: Approaches and alternatives.* G. E. Likens, Ed. New York: Springer Verlag, pp. 71–89.

Day, F. P., Jr., and D. T. McGinty. 1975. Mineral cycling strategies of two deciduous and two evergreen tree species on a southern Appalachian watershed. In *Mineral cycling in southeastern ecosystems*, F. C. Howell, J. B. Gentry, and M. H. Smith, Eds. United States Department of Commerce. Washington, D.C.: National Technical Information Service, pp. 736–743.

Day, J. W., G. P. Shaffer, L. D. Britsch, D. J. Reed, S. R. Hawes, and D. Cahoon. 2000. Pattern and process of land loss in the Mississippi Delta: A spatial and temporal analysis of wetland habitat change. *Estuaries* 23:425–438.

Dayton, R. K. 1971. Competition, disturbance, and community organization: The provision and subsequent utilization of space in a rocky intertidal community. *Ecological Monographs* 41:351–389.

———. 1975. Experimental evaluation of ecological dominance in a rocky intertidal algal community. *Ecological Monographs* 45:137–389.

Deevey, E. S., Jr. 1947. Life tables for natural populations of animals. *Quarterly Review of Biology* 22:283–314.

———. 1950. The probability of death. *Scientific American* 182:58–60.

de Forges, B. R., J. A. Koslow, and G. C. B. Poore. 2000. Diversity and endemism of the benthic seamount fauna in the southwest Pacific. *Nature* 405:944–947.

Delattre, P., B. De Sousa, E. Fichet-Calvet, J. P. Quere, and P. Giraudoux. 1999. Vole outbreaks in a landscape context: Evidence from a six year study of *Microtus arvalis. Landscape Ecology* 14:401–412.

Deming, W. E. 1993. *The new economics.* Cambridge: Massachusetts Institute of Technology Press.

de Ruiter, P. C., A. M. Neutel, and J. C. Moore. 1995. Energetics, patterns of interactive strengths and stability in real ecosystems. *Science* 269:1257–1260.

Des Marais, D. J. 2000. When did photosynthesis emerge on Earth? *Science* 289:1703–1705.

Dial, R., and J. Roughgarden. 1995. Experimental removal of insectivores from rainforest canopy: Direct and indirect effects. *Ecology* 76:1821–1834.

Drury, W. H., and I. C. T. Nisbet. 1973. Succession. *Journal of the Arnold Arboretum* 54: 331–368.

Dublin, L. I., and A. J. Lotka. 1925. On the true rate of natural increase as exemplified by the population of the United States, 1920. *Journal of the American Statistical Association* 20: 305–339.

Dunlap, J. C. 1998. Common threads in eukaryotic circadian systems. *Current Opinion in Genetics and Development* 8:400–406.

Duxbury, A. C. 1971. *The Earth and its oceans.* Reading, Mass.: Addison-Wesley.

Dwyer, E., J. M. Gregoire, and J. P. Malingrean. 1998. A global analysis of vegetation fires using satellite images: Spatial and temporal dynamics. *Ambio* 27:175–181.

Dybas, C. L. 2003. Bioethics in a changing world: Report from AIBS's 54th annual meeting. *BioScience* 53:798–802.

Dyer, M. I., A. M. Moon, M. R. Brown, and D. A. Crossley, Jr. 1995. Grasshopper crop and midgut extract effect on plants: An example of reward feedback. *Proceedings of the National Academy of Sciences* 92:5475–5478.

Dyer, M. I., C. L. Turner, and T. R. Seastedt. 1993. Herbivory and its consequences. *Ecological Applications* 3:10–16.

E

Earn, D. J. D., S. A. Levin, and P. Rohani. 2000. Coherence and conservation. *Science* 290: 1360–1364.

Edmondson, W. T. 1968. Water-quality management and lake eutrophication: The Lake Washington case. In *Water resource management and public policy,* T. H. Campbell and R. O. Sylvester, Eds. Seattle: University of Washington Press, pp. 139–178.

———. 1970. Phosphorus, nitrogen, and algae in Lake Washington after diversion of sewage. *Science* 169:690–691.

Edney, J. J., and C. Harper. 1978. The effect of information in resource management: A social trap. *Human Ecology* 6:387–395.

Edwards, C. A., R. Lal, P. Madden, R. H. Miller, and G. House, Eds. 1979. Lake Washington and the predictability of limnological events. *Archiv für Hydrobiologie. Beiheft: Ergebnisse der Limnologie* 13:234–241.

———. 1990. *Sustainable agricultural systems.* Ankeny, Iowa: Soil and Water Conservation Society.

Effland, W. R., and R. V. Pouyat. 1997. The genesis, classification, and mapping of soils in urban areas. *Urban Ecosystems* 1:217–228.

Egler, F. E. 1954. Vegetation science concepts. Volume 1. Initial floristic composition—a factor in old-field vegetation development. *Vegetatio* 4:412–417.

Ehrlich, P. R. 1968. *The population bomb.* New York: Ballantine Books.

———. 2003. Bioethics: Are our priorities right? *BioScience* 53:1207–1216.

Ehrlich, P. R., and P. H. Raven. 1964. Butterflies and plants: A study of coevolution. *Evolution* 18:586–608.

Eigen, M. 1971. Self-organization of matter and the evolution of biological macromolecules. *Naturwissenschaften* 58:465.

Einarsen, A. S. 1945. Some factors affecting ring-necked pheasant population density. *Murrelet* 26:39–44.

Ekbom, B., M. E. Irwin, and Y. Robert, Eds. 2000. *Interchanges of insects between agricultural and surrounding landscapes.* Dordrecht: Kluwer.

Elton, C. S. 1927. *Animal ecology.* London: Sidgwick and Jackson.

———. 1942. *Voles, mice, and lemmings: Problems in population dynamics.* Oxford: Clarendon Press.

———. 1966. *The pattern of animal communities.* London: Methuen.

Engelberg, J., and L. L. Boyarsky. 1979. The noncybernetic nature of ecosystems. *American Naturalist* 114:317–324.

Enserink, M. 1997. Life on the edge: Rainforest margins may spawn species. *Science* 276: 1791–1792.

Errington, P. L. 1945. Some contributions of a fifteen-year local study of the northern bobwhite to a knowledge of population phenomena. *Ecological Monographs* 15:1–34.

———. 1963. *Muskrat populations.* Ames: Iowa State University Press.

Esch, G. W., and McFarlane, R. W., Eds. 1975. *Thermal ecology II.* Energy Research and Development Administration. Springfield, Va.: National Technical Information Center.

Evans, E. C. 1956. Ecosystem as the basic unit in ecology. *Science* 123:1127–1128.

Evans, F. C., and S. A. Cain. 1952. Preliminary studies on the vegetation of an old-field community in southeastern Michigan. *Contributions from the Laboratory of Vertebrate Biology of the University of Michigan* 51:1–17.

Evenari, M. 1985. The desert environment. In *Hot deserts and arid shrublands.*, M. Evenari, I. Noy-Meir, and D. W. Goodall, Eds. Ecosystems of the World, Volume 12A. Amsterdam: Elsevier, pp. 1–22.

F

Fahrig, L. 1997. Relative effects of habitat loss and fragmentation on population extinction. *Journal of Wildlife Management* 61:603–610.

Fahrig, L., and G. Merriam. 1994. Conservation of fragmented populations. *Conservation Biology* 8:50–59.

Farner, D. S. 1964a. The photoperiodic control of reproductive cycles in birds. *American Scientist* 52:137–156.

———. 1964b. Time measurement in vertebrate photoperiodism. *American Naturalist* 98: 375–386.

Feener, D. H. 1981. Competition between ant species: Outcome controlled by parasitic flies. *Science* 214:815–817.

Feeny, P. P. 1975. Biochemical coevolution between plants and their insect herbivores. In *Coevolution of animals and plants,* L. E. Gilbert and P. H. Raven, Eds. Austin: University of Texas Press, pp. 3–19.

Ferré, F., and P. Hartel, Eds. 1994. *Ethics and environmental policy: Theory meets practice.* Athens: University of Georgia Press.

Field, C. B., J. G. Osborn, L. L. Hoffmann, J. F. Polsenberg, D. D. Ackerly, J. A. Berry, O. Bjorkman, Z. Held, P. A. Matson, and H. A. Mooney. 1998. Mangrove biodiversity and ecosystem function. *Global Ecology and Biogeography Letters* 7:3–14.

Finn, J. T. 1978. Cycling index: A general definition for cycling in compartment models. In *Environmental chemistry and cycling processes,* D. C. Adriano and I. L. Brisbin, Eds. United States Department of Energy Symposium Number 45. Springfield, Va.: National Technical Information Center, pp. 138–164.

Fischer, A. G. 1960. Latitudinal variations in organic diversity. *Evolution* 14:64–81.

Fisher, R. A. 1930. *The genetical theory of natural selection.* Oxford: Oxford University Press.

———. 1935. *The design of experiments.* Edinburgh: Oliver and Boyd.

Flader, S. L., and Callicott, J. B., Eds. 1991. *The river of the Mother of God and other essays by Aldo Leopold.* Madison: University of Wisconsin Press.

Flanagan, R. D. 1988. Planning for multi-purpose use of greenway corridors. *National Wetlands Newsletter* 10:7–8.

Folke, C., A. Jansson, J. Larsson, and R. Costanza. 1997. Ecosystem appropriation by cities. *Ambio* 26:167–172.

Food and Agricultural Organization (FAO). 1997. *Production Yearbook.* Rome: FAO.

Forbes, S. A. 1887. The lake as a microcosm. *Bulletin Scientifique.* Reprint, 1925, Peoria, Ill.: *Natural History Survey Bulletin* 15:537–550.

Force, J. E., and G. E. Machlis. 1997. The human ecosystem, Part 2. *Society and Natural Resources* 10:369–382.

Forman, R. T. T. 1997. *Land mosaics: The ecology of landscapes and regions.* Cambridge, U.K.: Cambridge University Press.

Forman, R. T. T., and L. E. Alexander. 1998. Roads and their major ecological effects. *Annual Review of Ecology and Systematics* 29:207–231.

Forman, R. T. T., and J. Baudry. 1984. Hedgerows and hedgerow networks in landscape ecology. *Environmental Management* 8:495–510.

Forman, R. T. T., and M. Godron. 1986. *Landscape ecology.* New York: John Wiley.

Forman, R. T. T., D. Sperling, J. A. Bissonette, A. P. Clevenger, C. D. Cutshall, V. H. Dale, L. Fahrig, R. France, C. R. Goldman, K. Heanue, J. A. Jones, F. J. Swanson, T. Turrentine, and T. C. Winter. 2003. *Road ecology: Science and solutions.* Washington, D.C.: Island Press.

Fortescue, J. A. C. 1980. *Environmental geochemistry: A holistic approach.* Ecology Series 35. New York: Springer Verlag.

Franklin, J. 1989. Towards a new forestry. *American Forests* November/December, pp. 37–44.

French, N. R. 1965. Radiation and animal population: Problems, progress, and projections. *Health Physics* 11:1557–1568.

Fretwell, S. D. 1975. The impact of Robert MacArthur on ecology. *Annual Review of Ecology and Systematics* 6:1–13.

Fuchs, R. L., E. Prennon, J. Chamie, F. C. Lo, and J. L. Vito. 1994. *Mega-city growth and the future.* New York: United Nations University Press.

Futuyma, D. J. 1976. Food plant specialization and environmental predictability in Lepidoptera. *American Naturalist* 110:285–292.

Futuyma, D. J., and M. Slatkin, Eds. 1983. *Coevolution.* Sunderland, Mass.: Sinauer.

G

Gadgil, M., and W. H. Bossert. 1970. Life historical consequences of natural selection. *American Naturalist* 104:1–24.

Gajaseni, J. 1995. Energy analysis of wetland rice systems in Thailand. *Agriculture, Ecosystems, and Environment* 52:173–178.

Garcia-Berthou, E., and S. H. Hurlbert. 1999. Pseudo-replication in hermit crab shell selection experiments: Comment. *Bulletin of Marine Science* 65:893–895.

Gargas, A., P. T. DePriest, M. Grube, and A. Tehler. 1995. Multiple origins of lichen symbiosis in fungi suggested by SSUrDNA phylogeny. *Science* 268: 1492–1495.

Gasaway, C. R. 1970. Changes in the fish population of Lake Francis Case in South Dakota in the first sixteen years of impoundment. Technical Paper 56. Washington, D.C.: Bureau of Sport Fisheries and Wildlife.

Gates, D. M. 1965. Radiant energy, its receipt and disposal. *Meteorological Monographs* 6:1–26.

Gause, G. F. 1932. Ecology of populations. *Quarterly Review of Biology* 7:27–46.

———. 1934. *The struggle for existence.* New York: Hafner. Reprint, 1964, Baltimore, Md.: Williams and Wilkins.

———. 1935. Experimental demonstration of Volterra's periodic oscillations in the numbers of animals. *Journal of Experimental Biology* 12:44-48.

Gavaghan, H. 2002. Life in the deep freeze. *Nature* 415:828–830.

Gershenzon, J. 1994. Metabolic costs of terpenoid accumulation in higher plants. *Journal of Chemical Ecology* 20:1281–1328.

Gessner, F. 1949. Der chlorophyllgehalt in see und seine photosynthetische valenz als geophysikalisches problem. *Schweizerische Zeitschrift für Hydrologie* 11:378–410.

Gibbons, J. W., and R. R. Sharitz. 1974. Thermal alteration of aquatic ecosystems. *American Scientist* 62:660–670.

———, Eds. 1981. *Thermal ecology.* United States Atomic Energy Commission. Springfield, Va.: National Technical Information Center.

Giesy, J. P., Ed. 1980. *Microcosms in ecological research.* United States Department of Energy Symposium Number 52. Springfield, Va.: National Technical Information Center.

Gilpin, M. E. 1975. *Group selection in predator-prey communities.* Princeton, N.J.: Princeton University Press.

Glasser, J. W. 1982. On the causes of temporal change in communities: Modification of the biotic environment. *American Naturalist* 119:375–390.

Gleason, H. A. 1926. The individualistic concept of the plant association. *Bulletin of the Torrey Botanical Club* 53:7–26.

Gleick, P. H. 2000. *The world's water 2000–2001.* The Biennial Report on Freshwater Resources. Washington, D.C.: Island Press.

Glemarec, M. 1979. Problèmes d'écologie dynamique et de succession en Baie de Concerneau. *Vie Milou,* Volume 28–29, Fasc. 1, Ser. AB, pp. 1–20.

Gliessman, S. R. 2001. *Agroecosystem sustainability: Developing practical strategies.* Boca Raton, Fla.: CRC Press.

Goldman, C. R. 1960. Molybdenum as a factor limiting primary productivity in Castle Lake, California. *Science* 132:1016–1017.

Goldsmith, E. 1996. *The way: An ecological world-view.* Totnes, U.K.: Themis Books.

Golley, F. B. 1993. *A history of the ecosystem concept in ecology: More than the sum of the parts.* New Haven, Conn.: Yale University Press.

Gonzalez, A., J. H. Lawton, F. S. Gilbert, T. M. Blackburn, and I. Evans-Freke. 1998. Metapopulation dynamics, abundance, and distribution in a microecosystem. *Science* 281: 2045–2047.

Goodall, D. W. 1963. The continuum and the individualistic association. *Vegetatio* 11: 297–316.

Goodland, R. 1995. The concept of environmental sustainability. *Annual Review of Ecology and Systematics* 26: 1–24.

Goodland, R., H. Daly, S. E. Serafy, and B. von Droste, Eds. 1991. *Environmentally sustainable economic development: Building on Brundtland.* Paris: United Nations Education, Scientific, and Cultural Organization (UNESCO).

Gopal, B., and U. Goel. 1993. Competition and allelopathy in aquatic plant communities. *Botanical Review* 59: 155–210.

Gorden, R. W., R. J. Beyers, E. P. Odum, and E. G. Eagon. 1969. Studies of a simple laboratory microecosystem: Bacterial activities in a heterotrophic succession. *Ecology* 50: 86–100.

Gore, J. A., and E. D. Shields, Jr. 1995. Can large rivers be restored? *BioScience* 45: 142–152.

Gornitz, V., S. Lebedeff, and J. Hansen. 1982. Global sea level trend in the past century. *Science* 125: 1611–1614.

Gosselink, J. G., E. P. Odum, and R. M. Pope. 1974. *The value of the tidal marsh.* LSU-SG-74-03, Louisiana State University. Baton Rouge: Center for Wetland Resources.

Gotelli, N. J., and D. Simberloff. 1987. The distribution and abundance of tallgrass prairie plants: A test of the core-satellite hypothesis. *American Naturalist* 130: 18–35.

Gottfried, B. M. 1979. Small mammal populations in woodlot islands. *American Midland Naturalist* 102: 105–112.

Gould, J. L., and C. G. Gould. 1989. *Life at the edge.* New York: W. H. Freeman.

Gould, S. J. 1988. Kropotkin was no crackpot. *Natural History* 97: 12–21.

———. 2000. Beyond competition. *Paleobiology* 26: 1–6.

———. 2002. *The structure of evolutionary theory.* Cambridge, Mass.: Harvard University Press.

Gould, S. J., and N. Eldredge. 1977. Punctuated equilibria: The tempo and mode of evolution reconsidered. *Paleobiology* 3: 115–151.

Graedel, T. E., and P. J. Crutzen. 1995. *Atmosphere, climate, and change.* New York: Scientific American Library.

Grant, P. R. 1986. *Ecology and evolution of Darwin's finches.* Princeton, N.J.: Princeton University Press.

Grant, P. R., and B. R. Grant. 1992. Demography and the genetically effective sizes of two populations of Darwin's finches. *Ecology* 73: 766–784.

Graves, W., Ed. 1993. *Water: The power, promise, and turmoil of North America's fresh water.* National Geographic Special Edition, Volume 184. Washington, D.C.: National Geographic Society.

Gray, P. E. 1992. The paradox of technological development. In *Technology and environment,* Washington, D.C.: National Academy Press, pp. 192–205.

Grime, J. P. 1977. Evidence for the existence of three primary strategies in plants and its relevance to ecological and evolutionary theory. *American Naturalist* 111: 1169–1194.

———. 1979. *Plant strategies and vegetation processes.* New York: John Wiley.

———. 1997. Biodiversity and ecosystem function: The debate deepens. *Science* 277: 1260–1261.

Grinnell, J. 1917. Field test of theories concerning distributional control. *American Naturalist* 51: 115–128.

———. 1928. Presence and absence of animals. *University of California Chronicles* 30:429–450.

Gross, A. O. 1947. Cyclic invasion of the snowy owl and the migration of 1945–1946. *Auk* 64:584–601.

Gruchy, A. G. 1967. *Modern economic thought: The American contribution.* New York: A. M. Kelley.

Gunderson, L. H. 2000. Ecological resilience—in theory and application. *Annual Review of Ecology and Systematics* 31:425–439.

Gunderson, L. H., and C. L. Holling. 2002. *Panarchy: Understanding transformations in human and natural systems.* Washington, D.C.: Island Press.

Gunn, J. M., Ed. 1995. *Restoration and recovery of an industrial region: Progress in restoring the smelter-damaged landscape near Sudbury, Canada.* New York: Springer Verlag.

H

Haberl, H. 1997. Human appropriation of net primary production as an environmental indicator: Implications for sustainable development. *Ambio* 26:143–146.

Haddad, N. M., and K. A. Baum. 1999. An experimental test of corridor effects on butterfly densities. *Ecological Applications* 9:623–633.

Haeckel, E. 1869. Über entwickelungsgang und aufgabe der zoologie. *Jenaische Zeitschrift für Medizin und Naturwissenschaft* 5:353–370.

Hagen, J. B. 1992. *An entangled bank: The origins of ecosystem ecology.* New Brunswick, N.J.: Rutgers University Press.

Haines, E. B., and R. B. Hanson. 1979. Experimental degradation of detritus made from salt marsh plants *Spartina alterniflora, Salicornus virginia,* and *Juncus roemerianus. Journal of Experimental Marine Biology and Ecology* 40:27–40.

Haines, E. B., and C. L. Montague. 1979. Food sources of estuarine invertebrates analyzed using $^{13}C/^{12}C$ ratios. *Ecology* 60:48–56.

Hairston, N. G. 1980. The experimental test of an analysis of field distributions: Competition in terrestrial salamanders. *Ecology* 61:817–826.

Hairston, N. G., F. K. Smith, and L. B. Slobodkin. 1960. Community structure, population control, and competition. *American Naturalist* 94:421–425.

Haken, H., Ed. 1977. *Synergetics: A workshop.* Proceedings of the International Workshop on Synergetics at Schloss Elmau, Bavaria, 2–7 May 1977. Berlin: Springer Verlag.

Hall, A. T., P. E. Woods, and G. W. Barrett. 1991. Population dynamics of the meadow vole, *Microtus pennsylvanicus,* in nutrient-enriched old-field communities. *Journal of Mammalogy* 72:332–342.

Hamilton, W. D. 1964. The genetical evolution of social behavior, I and II. *Journal of Theoretical Biology* 7:1–52.

Hamilton, W. J., III, and K. E. F. Watt. 1970. Refuging. *Annual Review of Ecology and Systematics* 1:263–286.

Hansen, A. J., and F. di Castri. 1992. *Landscape boundaries: Consequences for biotic diversity and ecological flows.* New York: Springer Verlag.

Hanski, I. A. 1982. Dynamics of regional distribution: The core and satellite species hypothesis. *Oikos* 38:210–221.

———. 1989. Metapopulation dynamics: Does it help to have more of the same? *Trends in Ecology and Evolution* 4:113–114.

Hanski, I. A., and M. E. Gilpin, Eds. 1997. *Metapopulation biology: Ecology, genetics, and evolution.* San Diego, Calif.: Academic Press.

Hansson, L. 1979. On the importance of landscape heterogeneity in northern regions for breeding population density of homeotherms: A general hypothesis. *Oikos* 33:182–189.

Harborne, J. B. 1982. *Introduction to ecological biochemistry,* 2nd ed. London: Academic Press.

Hardin, G. J. 1960. The competitive exclusion principle. *Science* 131:1292–1297.

———. 1968. The tragedy of the commons. *Science* 162:1243–1248.

———. 1985. *Filters against folly: How to survive despite economists, ecologists, and the merely eloquent.* New York: Viking Press.

———. 1993. *Living within limits: Ecology, economics, and population taboos.* New York: Oxford University Press.

Hargrove, E. C. 1989. *Foundations of environmental ethics.* Englewood Cliffs, N.J.: Prentice Hall.

Harper, J. L. 1961. Approaches to the study of plant competition. In *Mechanisms in biological competition.* Symposium of the Society for Experimental Biology, Number 15, pp. 1–268.

———. 1969. The role of predation in vegetational diversity. In *Diversity and stability in ecological systems,* G. M. Woodwell and H. H. Smith, Eds. Brookhaven Symposium on Biology, Number 22. Upton, N.Y.: Brookhaven National Laboratory, pp. 48–62.

———. 1977. *Population biology of plants.* New York: Academic Press.

Harper, J. L., and J. N. Clatworthy. 1963. The comparative biology of closely related species. Volume VI. Analysis of the growth of *Trifolium repens* and *T. fragiferum* in pure and mixed populations. *Journal of Experimental Botany* 14:172–190.

Harper, S. J., E. K. Bollinger, and G. W. Barrett. 1993. Effects of habitat patch shape on meadow vole, *Microtus pennsylvanicus* population dynamics. *Journal of Mammalogy* 74: 1045–1055.

Harris, L. D. 1984. *The fragmented forest: Island biogeography theory and the preservation of biotic diversity.* Chicago: University of Chicago Press.

Harrison, S., and N. Cappuccino. 1995. Using density-manipulation experiments to study population regulation. In *Population dynamics: New approaches and synthesis,* N. Cappuccino and P. W. Price, Eds. San Diego, Calif.: Academic Press, pp. 131–147.

Hartley, S. E., and C. G. Jones. 1997. Plant chemistry and herbivory, or why the world is green. In *Plant ecology,* M. Crawley, Ed. Cambridge, Mass.: Blackwell, pp. 284–324.

Harvey, H. W. 1950. On the production of living matter in the sea off Plymouth. *Journal of the Marine Biology Association of the United Kingdom* 29:97–137.

Haughton, G., and C. Hunter. 1994. *Sustainable cities.* London: Jessica Kingsley.

Hawken, P. 1993. *The ecology of commerce: A declaration of sustainability.* New York: HarperBusiness.

Hawken, P., Lovins, A., and L. H. Lovins. 1999. *Natural capitalism: Creating the next industrial revolution.* Boston: Little, Brown.

Hawkins, A. S. 1940. A wildlife history of Faville Grove, Wisconsin. *Transactions of the Wisconsin Academy of Sciences, Arts, and Letters* 32:39–65.

Hazard, T. P., and R. E. Eddy. 1950. Modification of the sexual cycle in the brook trout, *Salvelinus fontinalis* by control of light. *Transactions of the American Fisheries Society* 80:158–162.

Heezen, B. C., C. M. Tharp, and M. Ewing. 1959. *The floors of the ocean. Volume 1. North Atlantic.* New York: Geological Society of America Special Paper 65.

Heffner, R. A., M. J. Butler, IV, and C. K. Reilly. 1996. Pseudo-replication revisited. *Ecology* 77: 2558–2562.

Heinrich, B. 1979. *Bumblebee economics.* Cambridge, Mass.: Harvard University Press.

———. 1980. The role of energetics in bumblebee-flower interrelationships. In *Coevolution of animals and plants,* L. E. Gilbert and P. H. Raven, Eds. Austin: University of Texas Press, pp. 141–158.

Henderson, M. T., G. Merriam, and J. F. Wegner. 1985. Patchy environments and species survival: Chipmunks in an agricultural mosaic. *Biological Conservation* 31:95–105.

Hendrix, P. F., R. W. Parmelee, D. A. Crossley, Jr., D. C. Coleman, E. P. Odum, and P. Groffman. 1986. Detritus food webs in conventional and no-tillage agroecosystems. *BioScience* 36:374–380.

Hersperger, A. M. 1994. Landscape ecology and its potential application to planning. *Journal of Planning Literature* 9:14–29.

Hickey, J. J., and D. W. Anderson. 1968. Chlorinated hydrocarbons and egg shell changes in raptorial and fish-eating birds. *Science* 162:271–272.

Higgs, E. S. 1994. Expanding the scope of restoration ecology. *Restoration Ecology* 2:137–146.

———. 1997. What is good ecological restoration? *Conservation Biology* 11:338–348.

Hills, G. A. 1952. The classification and evaluation of site for forestry. *Ontario Department of Lands and Forests Research Report* 24.

Himmelfarb, G. 1962. *Darwin and the Darwinian revolution.* New York: W. W. Norton.

Hjort, J. 1926. Fluctuations in the year classes of important food fishes. *Journal du Conseil Permanent International pour l'Exploration de la Mer* 1:1–38.

Hobbie, J. E. 2003. Scientific accomplishments of the Long-Term Ecological Research program: An introduction. *BioScience* 53:17–20.

Hobbs, R. J., and D. A. Norton. 1996. Toward a conceptual framework for restoration ecology. *Restoration Ecology* 4:93–110.

Holdridge, L. R. 1947. Determination of wild plant formations from simple climatic data. *Science* 105:367–368.

———. 1967. Determination of world plant formations from simple climatic data. *Science* 130:572.

Holland, J. H. 1998. *Emergence: From chaos to order.* Reading, Mass.: Addison-Wesley.

Holling, C. S. 1965. The functional response of predators to prey density and its role in mimicry and population regulation. *Memoirs of the Entomological Society of Canada* 45.

———. 1966. The functional response of invertebrate predators to prey density. *Memoirs of the Entomological Society of Canada* 48.

———. 1973. Resilience and stability of ecological systems. *Annual Review of Ecology and Systematics* 4:1–23.

———. 1980. Forest insects, forest fires, and resilience. In *Fire regimes and ecosystem properties,* H. Mooney, J. M. Bonnicksen, N. L. Christensen, J. E. Lotan, and W. E. Reiners, Eds. United States Department of Agriculture, Forest Service General Technical Report WO-26.

———. 1998. Two cultures of ecology. *Conservation Ecology* [online] 2:1–4. URL: http://www.consecol.org/vol2/iss2/art4

Holling, C. S., and L. H. Gunderson. 2002. Resilience and adaptive cycles. In *Panarchy: Understanding transformations in human and natural systems,* L. H. Gunderson and C. S. Holling, Eds. Washington, D.C.: Island Press, pp. 25–62.

Hooper, J. 2002. *An evolutionary tale of moths and men: The untold story of science and the peppered moth.* New York: W. W. Norton.

Hopkinson, C. S., and J. W. Day. 1980. Net energy analysis of alcohol production from sugarcane. *Science* 207:302–304.

Horgan, J. 1996. *The end of science.* Reading, Mass.: Addison-Wesley.

Horn, H. S. 1974. The ecology of secondary succession. *Annual Review of Ecology and Systematics* 5:25–37.

———. 1975. Forest succession. *Scientific American* 232:90–98.

———. 1981. Succession. In *Theoretical ecology,* 2nd ed., R. M. May, Ed. Sunderland, Mass.: Sinauer, pp. 253–271.

Howard, H. E. 1948. *Territory in bird life.* London: London Collins.

Hubbell, S. P. 1979. Tree dispersion abundance and diversity in a tropical dry forest. *Science* 203:1299–1309.

———. 2001. *The unified neutral theory of biodiversity and biogeography.* Princeton, N.J.: Princeton University Press.

Hulbert, M. K. 1971. The energy resources of the earth. *Scientific American* 224:60–70.

Humphreys, W. F. 1978. Ecological energetics of *Geolycosa godeffroyi* (Araneae: Lycosidae) with an appraisal of production efficiency of ectothermic animals. *Journal of Animal Ecology* 47:627–652.

Hunter, M. D. 2000. Mixed signals and cross-talk: Interactions between plants, insect herbivores and plant pathogens. *Agricultural and Forest Entomology* 2:155–160.

Hunter, M. D., and P. W. Price. 1992. Playing chutes and ladders: Heterogeneity and the relative roles of bottom-up and top-down forces in natural communities. *Ecology* 73: 724–732.

Hurd, L. E., and L. L. Wolf. 1974. Stability in relation to nutrient enrichment in arthropod consumers of old-field successional ecosystems. *Ecological Monographs* 44:465–482.

Hurlbert, S. H. 1984. Pseudo-replication and the design of ecological field experiments. *Ecological Monographs* 54:187–211.

Huston, M. 1979. A general hypothesis of species diversity. *American Naturalist* 113:81–101.

Hutcheson, K. 1970. A test for comparing diversities based on the Shannon formula. *Journal of Theoretical Biology* 29:151–154.

Hutchinson, G. E. 1944. Nitrogen and biogeochemistry of the atmosphere. *American Scientist* 32:178–195.

———. 1948. On living in the biosphere. *Scientific Monthly* 67:393–398.

———. 1950. Survey of contemporary knowledge of biogeochemistry. Volume 3. The biogeochemistry of vertebrate excretion. *Bulletin of the American Museum of Natural History* 95:554.

———. 1957. *A treatise on limnology. Volume 1. Geography, physics, and chemistry.* New York: John Wiley.

———. 1964. The lacustrine microcosm reconsidered. *American Scientist* 52:331–341.

———. 1965. The niche: An abstractly inhabited hyper-volume. In *The ecological theatre and the evolutionary play.* New Haven, Conn.: Yale University Press, pp. 26–78.

Huxley, J. 1935. Chemical regulation and the hormone concept. *Biological Reviews* 10:427.

Huxley, T. H. 1894. *Evidence of man's place in nature, man's place in nature, and other anthropological essays.* New York: D. Appleton. Reprint, 1959, Ann Arbor: University of Michigan Press.

I

Inouye, R. S., and G. D. Tilman. 1988. Convergence and divergence in vegetation along experimentally created gradients of resource availability. *Ecology* 69:12–26.

Irvine, W. 1955. *Apes, angels, and Victorians.* New York: McGraw-Hill.

J

Jackson, D. L., and L. L. Jackson, Eds. 2002. *The farm as natural habitat: Reconnecting food systems with ecosystems.* Washington, D.C.: Island Press.

Jahnke, R. A. 1992. The phosphorus cycle. In *Global biogeochemical cycles,* S. S. Butcher, R. J. Charlson, G. H. Orians, and G. V. Wolfe, Eds. London: Academic Press, pp. 301–315.

Jansson, A. M., Ed. 1984. *Integration of economy and ecology: An outlook for the eighties.* Proceedings of the Wallenberg Symposia, Stockholm.

Jansson, A., C. Folke, J. Rockström, and L. Gordon. 1999. Linking freshwater flows and ecosystem services appropriated by people: The case of the Baltic Sea drainage basin. *Ecosystems* 2:351–366.

Jantsch, E. 1972. *Technological planning and social futures.* New York: John Wiley.

———. 1980. *The self-organizing universe.* Oxford: Pergamon Press.

Janzen, D. H. 1966. Coevolution of mutualism between ants and acacias in Central America. *Evolution* 20:249–275.

———. 1967. Interaction of the bull's horn acacia, *Acacia cornigera* L. with an ant inhabitant, *Pseudomyrmex ferruginea* F. (Smith) in eastern Mexico. *University of Kansas Science Bulletin* 57:315–558.

———. 1987. Habitat sharpening. *Oikos* 48:3–4.

Jeffries, H. P. 1979. Biochemical correlates of a seasonal change in marine communities. *American Naturalist* 113:643–658.

Jenkins, J. H., and T. T. Fendley. 1968. The extent of contamination, detention, and health significance of high accumulation of radioactivity in southeastern game populations. *Proceedings of the Annual Conference of Southeastern Association Game and Fish Commissioners* 22:89–95.

Jenny, H., R. J. Arkley, and A. M. Schultz. 1969. The pygmy forest-podsol ecosystem and its dune associates of the Mendocino coast. *Madrono* 20:60–75.

Johnson, B. L., W. B. Richardson, and T. J. Naimo. 1995. Past, present, and future concepts in large river ecology. *BioScience* 45:134–152.

Johnson, S. 2001. *Emergence: The connected lives of ants, brains, cities, and software.* New York: Scribner.

Johnston, D. W., and E. P. Odum. 1956. Breeding bird populations in relation to plant succession on the Piedmont in Georgia. *Ecology* 37:50–62.

Jones, C. G., R. S. Ostfeld, M. P. Richard, E. M. Schauber, and J. O. Wolff. 1998. Chain reactions linking acorns, gypsy moth outbreaks, and Lyme-disease risk. *Science* 279:1023–1026.

Jordan, C. F. 1985. Jari: A development project for pulp in the Brazilian Amazon. *The Environmental Professional* 7:135–142.

———. 1995. *Conservation.* New York: John Wiley.

Jordan, C. F., and R. Herrera. 1981. Tropical rain forests: Are nutrients really critical? *American Naturalist* 117:167–180.

Jordan, C. F., and C. E. Russell. 1989. Jari: A pulp plantation in the Brazilian Amazon. *GeoJournal* 19:429–435.

Jordan, W. R., III. 2003. *The sunflower forest: Ecological restoration and the new communion with nature.* Berkeley: University of California Press.

Jordan, W. R., III, M. E. Gilpin, and J. D. Aber, Eds. 1987. *Restoration ecology: A synthetic approach to ecological research.* Cambridge, U.K.: Cambridge University Press.

Jørgensen, S. E. 1997. *Integration of ecosystem theories: A pattern.* Dordrecht: Kluwer.

Juday, C. 1940. The annual energy budget of an inland lake. *Ecology* 21:438–450.

———. 1942. The summer standing crop of plants and animals in four Wisconsin lakes. *Transactions of the Wisconsin Academy of Sciences, Arts, and Letters* 34:103–135.

Junk, W. J., P. B. Bayley, and R. E. Sparks. 1989. The flood pulse concept in river-floodplain systems. *Canadian Special Publication of Fisheries and Aquatic Sciences* 106:110–127.

K

Kahn, A. E. 1966. The tyranny of small decisions: Market failures, imperfections, and the limits of economies. *Kyklos* 19:23–47.

Kahn, H., W. Brown, and L. Martel. 1976. *The next 200 years: A scenario for America and the world.* New York: William Morrow.

Kaiser, J. 2000. Rift over biodiversity divides ecologists. *Science* 289:1282–1283.

———. 2001. How rain pulses drive biome growth. *Science* 291:413–414.

Kale, H. W. 1965. Ecology and bioenergetics of the long-billed marsh wren, *Telmatodytes palustris griseus* (Brewster) in Georgia salt marshes. Publication Number 5. Cambridge, Mass.: Nuttall Ornithological Club.

Karlen, D. L., M. J. Mausbach, J. W. Doran, R. G. Cline, R. F. Harris, and G. E. Schuman. 1997. Soil quality: A concept, definition, and framework for evaluation. *Soil Science Society of America Journal* 61:4–10.

Karr, J. R., and I. J. Schlosser. 1978. Water resources and the land-water interface. *Science* 201:229–234.

Kates, R. W., W. C. Clark, R. Corell, J. M. Hall, C. C. Jaeger, I. Lowe, J. J. McCarthy, H. J. Schellnhuber, B. Bolin, N. M. Dickson, S. Faucheux, G. C. Gallopin, A. Grübler, B. Huntley, J. Jäger, N. S. Jodha, R. E. Kasperson, A. Mabogunje, P. Matson, H. Mooney, B. Moore III, T. O'Riordan, and U. Svedin. 2001. Sustainability science. *Science* 292:641–642.

Kauffman, S. A. 1993. *The origins of order: Self organizing and selection in evolution.* New York: Oxford University Press.

Keddy, P. 1990. Is mutualism really irrelevant to ecology? *Bulletin of the Ecological Society of America* 71:101–102.

Keith, L. B. 1963. Wildlife's ten-year cycle. Madison: University of Wisconsin Press.

———. 1990. Dynamics of snowshoe hare populations. *Current Mammalogy* 4:119–195.

Keith, L. B., and L. A. Windberg. 1978. A demographic analysis of the snowshoe hare cycle. *Wildlife Monographs* 58.

Keith, L. B., J. R. Cary, O. J. Rongstad, and M. C. Brittingham. 1984. Demography and ecology of declining snowshoe hare population. *Journal of Wildlife Management* 90:1–43.

Kelner, K., and L. Helmuth, Eds. 2003. Obesity. Special Section. *Science* 299:845–860.

Kemp, J. C., and G. W. Barrett. 1989. Spatial patterning: Impact of uncultivated corridors on arthropod populations within soybean agroecosystems. *Ecology* 70:114–128.

Kendeigh, S. C. 1944. Measurement of bird populations. *Ecological Monographs* 14:67–106.

———. 1982. *Bird populations in east central Illinois: Fluctuations, variations and development over half a century.* Champaign: University of Illinois Press.

Kennedy, D. 2003. Sustainability and the commons. *Science* 302:1861.

Kettlewell, H. B. D. 1956. Further selection experiments on industrial melanism in the Lepidoptera. *Heredity* 10:287–301.

Killham, K. 1994. *Soil ecology.* New York: Cambridge University Press.

Kimball, A. 1957. Errors of the third kind in statistical consulting. *Journal of the American Statisticians Association* 52:133–142.

Kira, T., and T. Shidei. 1967. Primary production and turnover of organic matter in different forest ecosystems of the western Pacific. *Japanese Journal of Ecology* 17:70–87.

Klopatek, J. M., and R. H. Gardner, Eds. 1999. *Landscape ecological analysis: Issues and applications.* New York: Springer Verlag.

Knapp, A. K., J. M. Briggs, D. C. Hartnett, and S. L. Collins. 1998. *Grassland dynamics: Long-term ecological research in tallgrass prairie.* New York: Oxford University Press.

Koestler, A. 1969. Beyond atomism and holism: The concept of holon. In *Beyond reductionism.* The Alpbach Symposium, 1968. London: Hutchinson, pp. 192–232.

Kormondy, E. J. 1969. *Concepts of ecology.* Englewood Cliffs, N.J.: Prentice Hall.

Kozlowski, T. T., and C. E. Ahlgren, Eds. 1974. *Fire and ecosystems.* New York: Academic Press.

Krebs, C. J. 1978. A review of the Chitty hypothesis of population regulation. *Canadian Journal of Zoology* 56:2463–2480.

Krebs, C. J., R. Boonstra, S. Boutin, and A. R. E. Sinclair. 2001. What drives the 10-year cycle of snowshoe hares? *BioScience* 51:25–35.

Krebs, C. J., S. Boutin, and R. Boonstra, Eds. 2001. *Ecosystem dynamics of the boreal forest: The Kluane project.* New York: Oxford University Press.

Krebs, C. J., S. Boutin, R. Boonstra, A. R. E. Sinclair, J. N. M. Smith, M. R. T. Dale, K. Martin, and R. Turkington. 1995. Impact of food and predation on the snowshoe hare cycle. *Science* 269:1112–1115.

Krebs, C. J., and K. T. DeLong. 1965. A *Microtus* population with supplemental food. *Journal of Mammalogy* 46:566–573.

Krebs, C. J., M. S. Gaines, B. L. Keller, J. H. Myers, and R. H. Tamarin. 1973. Population cycles in small rodents. *Science* 179:35–41.

Krebs, C. J., and J. H. Meyers. 1974. Population cycles in small mammals. *Advances in Ecological Research* 8:267–349.

Kroodsma, D. E., B. E. Byers, E. Goodale, S. Johnson, and W. C. Liu. 2001. Pseudoreplication in playback experiments, revisited a decade later. *Animal Behavior* 61:1029–1033.

Kropotkin, P. A. 1902. *Mutual aid: A factor of evolution.* New York: McClure and Phillips. Reprint, 1937, Harmondsworth, U.K.: Penguin Books.

Kuenzler, E. J. 1958. Niche relations of three species of Lycosid spiders. *Ecology* 39:494–500.

———. 1961a. Phosphorus budget of a mussel population. *Limnology and Oceanography* 6:400–415.

———. 1961b. Structure and energy flow of a mussel population. *Limnology and Oceanography* 6:191–204.

Kuhn, T. S. 1970. *The structure of scientific revolutions.* Chicago: University of Chicago Press.

L

Lack, D. L. 1947a. *Darwin's finches.* New York: Cambridge University Press.

———. 1947b. The significance of clutch size. *Ibus* 89:302–352.

———. 1966. *Population studies of birds.* Oxford: Clarendon Press.

———. 1969. Tit niches in two worlds or homage to Evelyn Hutchinson. *American Naturalist* 103:43–49.

Lal, R. 1991. Current research on crop water balance and implication for the future. In *Proceedings of the International Workshop on Soil Water Balance in the Sudano-Sahelian Zone, Niamey.* Wallingford, U.K.: IAHS Press, pp. 34–44.

Langdale, G. W., A. P. Barnett, L. Leonard, and W. G. Fleming. 1979. Reduction of soil erosion by the no-till system in the southern Piedmont. *Journal of Soil and Water Conservation* 34: 226–228.

LaPolla, V. N., and G. W. Barrett. 1993. Effects of corridor width and presence on the population dynamics of the meadow vole, *Microtus pennsylvanicus. Landscape Ecology* 8:25–37.

Laszlo, E., Ed. 1977. *Goals for mankind: A report to the Club of Rome on the new horizons of global community.* New York: Dutton.

Laszlo, E., and H. Margenau. 1972. The emergence of integrating concepts in contemporary science. *Philosophy of Science* 39:252–259.

Lawton, J. H. 1981. Moose, wolves, daphnia, and hydra: On the ecological efficiency of endotherms and ectotherms. *American Naturalist* 117:782–783.

Lawton, J. H., and M. P. Hassell. 1981. Asymmetrical competition in insects. *Nature* 289: 793–795.

Lawton, J. H., and S. McNeil. 1979. Between the devil and the deep blue sea: On the problem of being a herbivore. In *Population dynamics,* B. D. Turner and L. R. Taylor, Eds. London: Blackwell, pp. 223–244.

Leffler, J. W. 1978. Ecosystem responses to stress in aquatic microcosms. In *Energy and environmental stress in aquatic systems,* J. H. Thorp and J. W. Gibbons, Eds. United States Department of Energy. Springfield, Va.: National Technical Information Center, pp. 102–119.

Leibold, M. A., J. M. Chase, J. B. Shurin, and A. L. Downing. 1997. Species turnover and the regulation of trophic structure. *Annual Review of Ecology and Systematics* 28:467–494.

Leigh, R. A., and A. E. Johnston, Eds. 1994. *Long-term experiments in agricultural and ecological sciences.* Proceedings of the 150th Anniversary of Rothamsted Experimental Station, Rothamsted 14–17 July 1993. Oxford: CAB International.

Leopold, A. 1933a. *Game management.* New York: Scribner.

———. 1933b. The conservation ethic. *Journal of Forestry* 31:634–643.

———. 1943. Deer irruptions. *Wisconsin Conservation Bulletin.* Reprint, August 1943, *Wisconsin Conservation Department Publication Technical Bulletin* 321:3–11.

———. 1949. The land ethic. In *A Sand County almanac: And sketches here and there,* A. Leopold. New York: Oxford University Press, pp. 201–226.

Leopold, A. C. 2004. Living with the land ethic. *BioScience* 54:149–154.

Leslie, P. H., and T. Park. 1949. The intrinsic rate of natural increase of *Tribolium castaneum* Herbst. *Ecology* 30:469–477.

Leslie, P. H., and R. M. Ranson. 1940. The mortality, fertility, and rate of natural increase of the vole (*Microtus agrestis*) as observed in the laboratory. *Journal of Animal Ecology* 9:27–52.

Levine, M. B., A. T. Hall, G. W. Barrett, and D. H. Taylor. 1989. Heavy metal concentrations during ten years of sludge treatment to an old-field community. *Journal of Environmental Quality* 18:411–418.

Levins, R. 1966. The strategy of model building in population biology. *American Scientist* 54: 421–431.

———. 1968. *Evolution in changing environments.* Princeton, N.J.: Princeton University Press.

———. 1969. Some demographic and genetic consequences of environmental heterogeneity for biological control. *Bulletin of the Entomology Society of America* 15:237–240.

Li, B. L., and P. Sprott. 2000. Landscape ecology: Much more than the sum of parts. *The LTER Network News* 13:12–15.

Lidicker, W. Z., Jr. 1988. Solving the enigma of microtine "cycles." *Journal of Mammalogy* 69:225–235.

———. 1995. The landscape concept: Something old, something new. In *Landscape approaches in mammalian ecology and conservation,* W. Z. Lidicker, Jr., Ed. Minneapolis: University of Minnesota Press, pp. 3–19.

Liebig, J., Baron von. 1840. *Organic chemistry in its application to agriculture and physiology.* Reprint, 1847, *Chemistry in its application to agriculture and physiology,* L. Playfair, Ed. Philadelphia: T. B. Peterson.

Liebman, M., and A. S. Davis. 2000. Integration of soil, crop and weed management in low-external-input farming systems. *Weed Research* 40:27–47.

Lieth, H., and R. H. Whittaker, Eds. 1975. *Primary productivity of the biosphere.* New York: Springer Verlag.

Ligon, J. D. 1968. Sexual differences in foraging behavior in two species of *Dendrocopus* woodpeckers. *Auk* 85:203–215.

Likens, G. E. 1998. Limitations to intellectual progress in ecosystem science. In *Successes, limitations, and frontiers in ecosystem science,* M. L. Pace and P. M. Groffman, Eds. New York: Springer Verlag, pp. 247–271.

———. 2001a. Ecosystems: Energetics and biogeochemistry. In *A new century of biology,* W. J. Kress and G. W. Barrett, Eds. Washington, D.C.: Smithsonian Institution Press, pp. 53–88.

———. 2001b. Eugene P. Odum, the ecosystem approach, and the future. In *Holistic science: The evolution of the Georgia Institute of Ecology 1940–2000,* G. W. Barrett and T. L. Barrett, Eds. New York: Taylor and Francis, pp. 309–328.

Likens, G. E., and F. H. Bormann. 1974a. Acid rain: A serious regional environmental problem. *Science* 184:1176–1179.

———. 1974b. Linkages between terrestrial and aquatic ecosystems. *BioScience* 24:447–456.

———. 1995. *Biogeochemistry of a forested ecosystem,* 2nd ed. New York: Springer Verlag.

Likens, G. E., F. H. Bormann, R. S. Pierce, J. S. Eaton, and N. M. Johnson. 1977. *Biogeochemistry of a forested ecosystem.* New York: Springer Verlag.

Likens, G. E., C. T. Driscoll, and D. C. Busco. 1996. Long-term effects of acid rain: Response and recovery of a forest ecosystem. *Science* 272:244–246.

Lindeman, R. L. 1942. The trophic-dynamic aspect of ecology. *Ecology* 23:399–418.

Lodge, T. E. 1994. *The Everglades handbook: Understanding the ecosystem.* Delray Beach, Fla.: St. Lucie Press.

Loomis, L. R., Ed. 1942. *Five great dialogues.* Roslyn, N.Y.: Walter J. Black.

Lotka, A. J. 1925. *Elements of physical biology.* Reprint, 1956, *Elements of mathematical biology.* New York: Dover.

Loucks, O. L. 1986. The United States IBP: An ecosystems perspective after 15 years. In *Ecosystem theory and application,* N. Polunin, Ed. New York: John Wiley, pp. 390–405.

Lovejoy, T. E., R. O. Bierregaard, Jr., A. B. Rylands, J. R. Malcolm, C. E. Quintela, L. H. Harper, K. S. Brown, Jr., A. H. Powell, G. V. N. Powell, H. O. R. Shubart, and M. B. Hays. 1986. Edge and other effects of isolation on Amazon forest fragments. In *Conservation biology: The science of scarcity and diversity,* M. E. Soule, Ed. Sunderland, Mass.: Sinauer, pp. 257–285.

Lovelock, J. E. 1979. *Gaia: A new look at life on Earth.* New York: Oxford University Press.

———. 1988. *The ages of Gaia: A biography of our living Earth.* New York: W. W. Norton.

Lovelock, J. E., and S. R. Epton. 1975. The quest for Gaia. *New Scientist* 65:304–306.

Lovelock, J. E., and L. Margulis. 1973. Atmospheric homeostasis by and for the biosphere: The Gaia hypothesis. *Tellus* 26:1–10.

Lowrance, R., P. F. Hendrix, and E. P. Odum. 1986. A hierarchical approach to sustainable agriculture. *American Journal of Alternative Agriculture* 1: 169–173.

Lowrance, R., R. Todd, J. Fail, Jr., O. Hendrickson, Jr., R. Leonard, and L. Asmussen. 1984. Riparian forests as nutrient filters in agricultural watersheds. *BioScience* 34:374–377.

Luck, M. A., G. D. Jenerette, J. Wu, and N. B. Grimm. 2001. The urban funnel model and the spatially heterogeneous ecological footprint. *Ecosystems* 4:782–796.

Ludwig, D., D. D. Jones, and C. S. Holling. 1978. Qualitative analysis of insect outbreak systems: The spruce budworm and forest. *Journal of Animal Ecology* 47:315–332.

Lugo, A. E., E. G. Farnworth, D. Pool, P. Jerez, and G. Kaufman. 1973. The impact of the leaf cutter ant *Atta columbica* on the energy flow of a tropical wet forest. *Ecology* 54: 1292–1301.

Luo, Y., S. Wan, D. Hui, and L. L. Wallace. 2001. Acclimatization of soil respiration to warming in a tall grass prairie. *Nature* 413:622–625.

Lutz, W., W. Sanderson, and S. Scherbov. 2001. The end of world population growth. *Nature* 412:543–545.

Lyle, J. T. 1993. Urban ecosystems. *In Context* 35:43–45.

M

Mabry, K. E., and G. W. Barrett. 2002. Effects of corridors on home range sizes and interpatch movements of three small-mammal species. *Landscape Ecology* 17:629–636.

Mabry, K. E., E. A. Dreelin, and G. W. Barrett. 2003. Influence of landscape elements on population densities and habitat use of three small-mammal species. *Journal of Mammalogy* 84:20–25.

MacArthur, R. H. 1958. Population ecology of some warblers of northeastern coniferous forest. *Ecology* 39:599–619.

———. 1972. *Geographical ecology: Patterns in the distribution of species.* New York: Harper and Row.

MacArthur, R. H., and R. Levins. 1967. The limiting similarity, convergence, and divergence of coexisting species. *American Naturalist* 101:377–385.

MacArthur, R. H., and E. O. Wilson. 1963. An equilibrium theory of insular zoogeography. *Evolution* 17:373–387.

———. 1967. *The theory of island biogeography.* Princeton, N.J.: Princeton University Press.

MacDicken, K. G., and N. T. Vergara. 1990. *Agroforestry: Classification and management.* New York: John Wiley.

MacElroy, R. D., and M. M. Averner. 1978. *Space ecosynthesis: An approach to the design of closed ecosystems for use in space.* NASA Tech. Memo. 78491. National Aeronautics and Space Administration. Moffet Field, Calif.: Ames Research Center.

Machlis, G. E., J. E. Force, and W. R. Burch. 1997. The human ecosystem, Part 1. The human ecosystem as an organizing concept in ecosystem management. *Society and Natural Resources* 10:347–367.

MacLulich, D. A. 1937. Fluctuations in the numbers of the varying hare, *Lepus americanus. University of Toronto Studies, Biology Series* 43.

Maly, M. S., and G. W. Barrett. 1984. Effects of two types of nutrient enrichment on the structure and function of contrasting old-field communities. *American Midland Naturalist* 111: 342–357.

Mann, C. C. 1999. Genetic engineers aim to soup up crop photosynthesis. *Science* 283: 314–316.

Margalef, R. 1958. Temporal succession and spatial heterogeneity in phytoplankton. In *Perspectives in marine biology,* Buzzati-Traverso, Ed. Berkeley: University of California Press, pp. 323–347.

———. 1963a. On certain unifying principles in ecology. *American Naturalist* 97:357–374.

———. 1963b. Successions of populations. New Delhi: Institute of Advanced Science and Culture. *Advance Frontiers of Plant Science* 2:137–188.

———. 1968. *Perspectives in ecological theory.* Chicago: University of Chicago Press.

Margulis, L. 1981. *Symbiosis in cell evolution: Life and its environment on the early Earth.* San Francisco: W. H. Freeman.

———. 1982. *Early life.* Boston: Science Books International.

———. 2001. Bacteria in the origins of species: Demise of the Neo-Darwinian paradigm. In *A new century of biology,* W. J. Kress and G. W. Barrett, Eds. Washington, D.C.: Smithsonian Institution Press, pp. 9–27.

Margulis, L., and J. E. Lovelock. 1974. Biological modulation of the earth's atmosphere. *Icarus* 21:471–489.

Margulis, L., and L. Olendzenski. 1992. *Environmental evolution: Effects of the origin and evolution of life on planet Earth.* Cambridge: Massachusetts Institute of Technology Press.

Margulis, L., and C. Sagan. 1997. *Slanted truths: Essays on Gaia, symbiosis, and evolution.* New York: Copernicus.

———. 2002. *Acquiring genomes: A theory of the origins of species.* New York: Basic Books.

Marks, P. L. 1974. The role of the pin cherry, *Prunus pennsylvanica* in the maintenance of stability in northern hardwood ecosystems. *Ecological Monographs* 44:73–88.

Marquet, P. A. 2000. Invariants, scaling laws, and ecological complexity. *Science* 289:1487–1488.

Marquis, R. J., and C. J. Whelan. 1994. Insectivorous birds increase growth of white oak through consumption of leaf-chewing insects. *Ecology* 75:2007–2014.

Marsh, G. P. 1864. *Man and nature: or physical geography as modified by human action.* New York: Scribner.

Martin, J. H., R. M. Gordon, and S. E. Fitzwater. 1991. The case for iron. *Limnology and Oceanography* 36:1793–1802.

Max-Neef, M. 1995. Economic growth and quality of life: A threshold hypothesis. *Ecological Economics* 15:115–118.

Maynard Smith, J. 1976. A comment on the Red Queen. *American Naturalist* 110:325–330.

McCormick, F. J. 1969. Effects of ionizing radiation on a pine forest. In *Proceedings of the second national symposium on radioecology,* D. Nelson and F. Evans, Eds. United States Department of Commerce. Springfield, Va.: Clearinghouse of the Federal Science Technical Information Center, pp. 78–87.

McCormick, F. J., and F. B. Golley. 1966. Irradiation of natural vegetation: An experimental facility, procedures, and dosimetry. *Health Physics* 12:1467–1474.

McCrea, W. 1991. Arthur Stanley Eddington. *Scientific American* 264:66–71.

McCullough, D. R. 1979. *The George Reserve deer herd: Population ecology of a K-selected species.* Ann Arbor: University of Michigan Press.

———. 1996. *Metapopulations and wildlife conservation.* Washington, D.C.: Island Press.

McGill, B. 2003. A test of the unified neutral theory of biodiversity. *Nature* 422:881–885.

McGowan, K. J., and G. E. Woolfenden. 1989. A sentinel system in the Florida scrub jay. *Animal Behaviour* 37:1000–1006.

McHarg, I. L. 1969. *Design with nature.* Garden City, N.Y.: Natural History Press.

McIntosh, R. P. 1975. H. A. Gleason—individualistic ecologist, 1882–1975. *Bulletin of the Torrey Botanical Club* 102:253–273.

———. 1980. The relationship between succession and the recovery process in ecosystems. In *The recovery process in damaged ecosystems,* J. Cairns, Ed. Ann Arbor, Mich.: Ann Arbor Sciences, pp. 11–62.

McLendon, T., and E. F. Redente. 1991. Nitrogen and phosphorus effects on secondary succession dynamics on a semi-arid sagebrush site. *Ecology* 72:2016–2024.

McMillan, C. 1956. Nature of the plant community. Volume 1. Uniform garden and light period studies of five grass taxa in Nebraska. *Ecology* 37:330–340.

McNaughton, S. J. 1976. Serengeti migratory wildebeest: Facilitation of energy flow by grazing. *Science* 191:92–94.

———. 1977. Diversity and stability of ecological communities: A comment on the role of empiricism in ecology. *American Naturalist* 111:515–525.

———. 1978. Stability and diversity in grassland communities. *Nature* 279:351–352.

———. 1985. Ecology of a grazing ecosystem: The Serengeti. *Ecological Monographs* 55: 259–294.

———. 1993. Grasses and grazers, science and management. *Ecological Applications* 3:17–20.

McNaughton, S. J., F. F. Banyikwa, and M. M. McNaughton. 1997. Promotion of the cycling of diet-enhancing nutrients by African grazers. *Science* 278:1798–1800.

McPhee, J. 1999. The control of nature: Farewell to the nineteenth century—the breaching of the Edwards Dam. *The New Yorker* 75:44–53.

McShea, W. J., and W. M. Healy, Eds. 2002. *Oak forest ecosystems: Ecology and management for wildlife.* Baltimore, Md.: Johns Hopkins University Press.

Meadows, D. H. 1982. Whole earth models and systems. *CoEvolution Quarterly,* Summer 1982, pp. 98–108.

———. 1996. The laws of the earth and the laws of economics. White River Junction, Vt.: *The Valley News,* "The Global Citizen," syndicated biweekly column, 14 December 1996.

Meadows, D. H., D. L. Meadows, and J. Randers. 1992. *Beyond the limits: Confronting global collapse, envisioning a sustainable future.* Post Mills, Vt.: Chelsea Green.

Meadows, D. H., D. L. Meadows, J. Randers, and W. W. Behrens. 1972. *The limits to growth: A report for the Club of Rome's project on the predicament of mankind.* New York: Universe Books.

Meadows, D. H., J. Richardson, and G. Bruckmann. 1982. *Groping in the dark: The first decade of global modelling.* New York: John Wiley.

Meentemeyer, V. 1978. Macroclimate and lignin control of litter decomposition rates. *Ecology* 59:465–472.

Meffe, G. K., and C. R. Carroll, Eds. 1994. *Principles of conservation biology.* Sunderland, Mass.: Sinauer.

———, Eds. 1997. *Principles of conservation biology.* Sunderland, Mass.: Sinauer.

Menzel, D. W., and J. H. Ryther. 1961. Nutrients limiting the production of phytoplankton in the Sargasso Sea with special reference to iron. *Deep Sea Research* 7:276–281.

Merriam, C. H. 1894. Laws of temperature control of the geographic distribution of terrestrial animals and plants. *National Geographic Magazine* 6:229–238.

Merriam, G. 1991. Corridors and connectivity: Animal populations in heterogeneous environments. In *Nature conservation: The role of corridors,* D. Saunders and R. Hobbs, Eds., Chipping Norton, Australia: Surrey Beatty and Sons, pp. 133–142.

Merriam, G., and A. Lanoue. 1990. Corridor use by small mammals: Field measurements for three experimental types of *Peromyscus leucopus. Landscape Ecology* 4:123–131.

Merriam-Webster's collegiate dictionary, 10th ed. 1996. Springfield, Mass.: Merriam-Webster.

Mertz, W. 1981. The essential trace elements. *Science* 213: 1332–1338.

Mervis, J. 2003. Bye, bye, Biosphere 2. *Science* 302: 2053.

Messerli, B., and J. D. Ives, Eds. 1997. *Mountains of the world: A global priority.* New York: Parthenon.

Michaels, P. J., and R. C. Balling. 2000. *The satanic gases.* Washington, D.C.: Cato Institute.

Middleton, J., and G. Merriam. 1981. Woodland mice in a farmland mosaic. *Journal of Applied Ecology* 18: 703–710.

Mills, R. S., G. W. Barrett, and M. P. Farrell. 1975. Population dynamics of the big brown bat, *Eptesicus fucus* in southwestern Ohio. *Journal of Mammalogy* 56: 591–604.

Minard, C. J. 1869. Carte figurative. Reprinted in *The visual display of quantitative information,* E. Tufte (1983). Cheshire, Conn.: Graphics Press, p. 41.

Mitchell, R. 1979. *The analysis of Indian agroecosystems.* New Delhi: Interprint.

Moffat, A. S. 1998a. Global nitrogen overload problem grows critical. *Science* 279: 988–989.

———. 1998b. Ecology—Temperate forests gain ground. *Science* 282: 1253.

Montague, C. L. 1980. A natural history of temperate western Atlantic fiddler crabs, genus *Uca* with reference to their impact on the salt marsh. *Contributions in Marine Science* 23: 25–55.

Mooney, H. A., and P. R. Ehrlich. 1997. Ecosystem services: A fragmentary history. In *Nature's services: Societal dependence on natural ecosystems,* G. C. Daily, Washington, D.C.: Island Press, pp. 11–19.

Morehouse, W., and J. Sigurdson. 1977. Science, technology, and poverty. *Bulletin of Atomic Science* 33: 21–28.

Morello, J. 1970. Modelo de relaciones entra pastizales y lenosas colonzodores en el Chaco Argentino. *Idia* 276: 31–51.

Morrison, D. A., and E. C. Morris. 2000. Pseudo-replication in experimental designs for manipulation of seed germination treatments. *Australian Ecology* 25: 292–296.

Mosse, B., D. P. Stribley, and F. Letacon. 1981. Ecology of mycorrhizae and mycorrhizal fungi. *Advances in Microbial Ecology* 5: 137–210.

Mulholland, P. J., J. D. Newbold, J. W. Elwod, L. A. Ferren, and J. R. Webster. 1985. Phosphorus spiraling in a woodland stream. *Ecology* 66: 1012–1023.

Muller, C. H. 1966. The role of chemical inhibition (allelopathy) in vegetational composition. *Bulletin of the Torrey Botanical Club* 93: 332–351.

———. 1969. Allelopathy as a factor in ecological process. *Vegetatio* 18: 348–357.

Muller, C. H., R. B. Hanawalt, and J. K. McPherson. 1968. Allelopathic control of herb growth in the fire cycle of California chaparral. *Bulletin of the Torrey Botanical Club* 95: 225–231.

Muller, C. H., W. H. Muller, and B. L. Haines. 1964. Volatile growth inhibitors produced by aromatic shrubs. *Science* 143: 471–473.

Müller, F. 1997. State-of-the-art in ecosystem theory. *Ecological Modelling* 100: 135–161.

———. 1998. Gradients in ecological systems. *Ecological Modelling* 108: 3–21.

———. 2000. Indicating ecosystem integrity—theoretical concepts and environmental requirements. *Ecological Modelling* 130: 13–23.

Mullineaux, C. W. 1999. The plankton and the planet. *Science* 283: 801–802.

Mumford, L. 1967. Quality in the control of quantity. In *Natural resources, quality and quantity,* Ciriacy-Wantrup and Parsons, Eds. Berkeley: University of California Press, pp. 7–18.

Munn, N. L., and J. L. Meyer. 1990. Habitat-specific solute retention in streams. *Ecology* 71: 2069–2032.

Murie, A. 1944. Dall sheep. In *Wolves of Mount McKinley.* Fauna Series 5. Washington, D.C.: National Park Service.

Muscatine, L. C., and J. Porter. 1977. Reef corals: Mutualistic symbioses adapted to nutrient-poor environments. *BioScience* 27:454–456.

Myers, R. A., and B. Worm. 2003. Rapid worldwide depletion of predatory fish communities. *Nature* 423:280–283.

N

Naeem, S., K. Hakansson, J. H. Lawton, M. J. Crawley, and L. J. Thompson. 1996. Biodiversity and plant productivity in a model assemblage of plant species. *Oikos* 76:259–264.

Naeem, S., L. J. Thompson, S. P. Lawler, J. H. Lawton, and R. M. Woodfin. 1994. Declining biodiversity can alter the performance of ecosystems. *Nature* 368:734–737.

Naiman, R. J., and H. Décamps, Eds. 1990. *The ecology and management of aquatic-terrestrial ecotones.* Park Ridge, N.J.: Parthenon.

National Academy of Sciences (NAS). 1971. *Rapid population growth: Consequences and policy implications,* R. Revelle, Ed. Baltimore, Md.: Johns Hopkins University Press.

———. 2000. *Transgenic plants and world agriculture.* Washington, D.C.: National Academy Press.

National Academy of Sciences/The Royal Society of London. 1992. *Population growth, resource consumption, and a sustainable world.* Joint statement.

National Research Council (NRC). 1989. *Alternative agriculture.* Washington, D.C.: National Academy Press.

———. 1993. *Soil and water quality: An agenda for agriculture.* Washington, D.C.: National Academy Press.

———. 1996a. *Ecologically based pest management: New solutions for a new century.* Washington, D.C.: National Academy Press.

———. 1996b. *Use of reclaimed water and sludge in food crop production.* Washington, D.C.: National Academy Press.

———. 2000a. *Professional societies and ecologically based pest management.* Washington, D.C.: National Academy Press.

———. 2000b. *The future role of pesticides in U. S. agriculture.* Washington, D.C.: National Academy Press.

Naveh, Z. 1982. Landscape ecology as an emerging branch of human ecosystem science. In *Advances in ecological research.* Volume 12. New York: Academic Press, pp. 189–209.

———. 2000. The total human ecosystem: Integrating ecology and economics. *BioScience* 50: 357–361.

Naveh, Z., and A. S. Lieberman. 1984. *Landscape ecology: Theory and application.* New York: Springer Verlag.

Nee, S., A. F. Read, J. J. D. Greenwood, and P. H. Harvey. 1991. The relationship between abundance and body size in British birds. *Nature* 351:312–313.

Negus, N. C., and P. J. Berger. 1977. Experimental triggering of reproduction in a natural population of *Microtus montanus. Science* 196:1230–1231.

Newell, S. J., and E. J. Tramer. 1978. Reproductive strategies in herbaceous plant communities in succession. *Ecology* 59:228–234.

Newman, E. I. 1988. Mycorrhizal links between plants: Their functioning and significance. *Advances in Ecological Research* 18:243–270.

Newman, L. H. 1965. Man and insects. Reprinted in *The visual display of quantitative information,* E. Tufte (1983). Cheshire, Conn.: Graphics Press, pp. 104–105.

Nicholson, S. A., and C. D. Monk. 1974. Plant species diversity in old-field succession on the Georgia piedmont. *Ecology* 55:1075–1085.

Nicolis, G., and I. Prigogine. 1977. *Self-organization in non-equilibrium systems: From dissipative structures to order through fluctuations.* New York: John Wiley.

Nixon, S. W. 1969. A synthetic microcosm. *Limnology and Oceanography* 14:142–145.

Novikoff, A. B. 1945. The concept of integrative levels of biology. *Science* 101:209–215.

O

Odum, E. P. 1953. *Fundamentals of ecology.* Philadelphia: W. B. Saunders.

———. 1957. The ecosystem approach in the teaching of ecology illustrated with sample class data. *Ecology* 38:531–535.

———. 1959. *Fundamentals of ecology,* 2nd ed. Philadelphia: W. B. Saunders.

———. 1963. Primary and secondary energy flow in relation to ecosystem structure. Washington, D.C.: *Proceedings of the Sixteenth International Congress of Zoology,* pp. 336–338.

———. 1968. Energy flow in ecosystems: A historical review. *American Zoologist* 8:11–18.

———. 1969. The strategy of ecosystem development. *Science* 164:262–270.

———. 1977. The emergence of ecology as a new integrative discipline. *Science* 195:1289–1293.

———. 1979. The value of wetlands: A hierarchical approach. In *Wetland functions and values: The state of our understanding,* P. E. Greeson, J. R. Clark, and J. E. Clark, Eds. Minneapolis, Minn.: American Water Resources Association, pp. 16–25.

———. 1983. *Basic ecology,* 3rd ed. Philadelphia: W. B. Saunders.

———. 1984. The mesocosm. *BioScience* 34:558–562.

———. 1985. Trends expected in stressed ecosystems. *BioScience* 35:419–422.

———. 1989. Input management of production systems. *Science* 243:177–182.

———. 1992. Great ideas in ecology for the 1990s. *BioScience* 42:542–545.

———. 1997. *Ecology: A bridge between science and society.* Sunderland, Mass.: Sinauer.

———. 1998a. *Ecological vignettes: Ecological approaches to dealing with human predicaments.* Amsterdam: Harwood.

———. 1998b. Productivity and biodiversity: A two-way relationship. *Bulletin of the Ecological Society of America* 79:125.

———. 1998c. Source reduction, input management, and dual capitalism. In *Ecological vignettes: Ecological approaches to dealing with human predicaments,* E. P. Odum. Amsterdam: Harwood, pp. 235–236.

———. 2001. The technoecosystem. *Bulletin of the Ecological Society of America* 82:137–138.

Odum, E. P., and G. W. Barrett. 2000. Pest management: An overview. In *Professional societies and ecologically based pest management.* National Research Council Report. Washington, D.C.: National Academy Press, pp. 1–5.

Odum, E. P., and G. W. Barrett. 2004. Redesigning industrial agroecosystems: Incorporating more ecological processes and reducing pollution. *Journal of Crop Improvements,* in press.

Odum, E. P., and L. J. Biever. 1984. Resource quality, mutualism, and energy partitioning in food chains. *American Naturalist* 124:360–376.

Odum, E. P., and J. L. Cooley. 1980. Ecosystem profile analysis and performance curves as tools for assessing environmental impacts. In *Biological evaluation of environmental impacts.*

Washington, D.C.: Council on Environmental Quality and Fish and Wildlife Service, pp. 94–102.

Odum, E. P., and M. G. Turner. 1990. The Georgia landscape: A changing resource. In *Changing landscapes: An ecological perspective,* I. S. Zonneveld and R. T. T. Forman, Eds. New York: Springer Verlag, pp. 137–164.

Odum, H. T. 1957. Trophic structure and productivity of Silver Springs, Florida. *Ecological Monographs* 27:55–112.

———. 1970. Summary: An emerging view of the ecological system at El Verde. In *A tropical rainforest,* H. T. Odum and R. F. Pigeon, Eds. Oak Ridge, Tenn.: USAEC, Division of Technical Information, pp. I-191–J-281.

———. 1971. *Environment, power, and society.* New York: John Wiley.

———. 1983. *Systems ecology.* New York: John Wiley.

———. 1988. Self-organization, transformity, and information. *Science* 242:1132–1139.

———. 1996. *Environmental accounting: EMergy and environmental decision making.* New York: John Wiley.

Odum, H. T., and E. C. Odum. 1982. *Energy basis for man and nature,* 2nd ed. New York: McGraw-Hill.

———. 2000. *Modeling for all scales.* San Diego, Calif.: Academic Press.

———. 2001. *A prosperous way down: Principles and policies.* Boulder: University Press of Colorado.

Odum, H. T., and E. P. Odum. 1955. Trophic structure and productivity of a windward coral reef community on Eniwetok Atoll. *Ecological Monographs* 25:291–320.

———. 2000. The energetic basis for valuation of ecosystem services. *Ecosystems* 3:21–23.

Odum, H. T., and R. F. Pigeon, Eds. 1970. *A tropical rain forest: A study of irradiation and ecology at El Verde, Puerto Rico.* Springfield, Va.: National Technical Information Service.

Odum, H. T., and R. C. Pinkerton. 1955. Times speed regulator, the optimum efficiency for maximum output in physical and biological systems. *American Scientist* 43:331–343.

Odum, H. W. 1936. *Southern regions of the United States.* Chapel Hill: University of North Carolina Press.

Odum, H. W., and H. E. Moore. 1938. *American regionalism.* New York: Henry Holt.

Odum, W. E. 1982. Environmental degradation and the tyranny of small decisions. *BioScience* 32:728–729.

———. 1988. Comparative ecology of tidal freshwater and salt marshes. *Annual Review of Ecology and Systematics* 19:137–176.

Odum, W. E., and E. J. Heald. 1972. Trophic analysis of an estuarine mangrove community. *Bulletin of Marine Science* 22:671–738.

———. 1975. The detritus-based food web of an estuarine mangrove community. In *Estuarine research,* G. E. Cronin, Ed. Volume 1. New York: Academic Press, pp. 265–286.

Odum, W. E., and C. C. McIvor. 1990. Mangroves. In *Ecosystems of Florida,* R. J. Myers and J. J. Ewel, Eds. Orlando: University of Central Florida Press, pp. 517–548.

Odum, W. E., E. P. Odum, and H. T. Odum. 1995. Nature's pulsing paradigm. *Estuaries* 18:547–555.

Office of Technology Assessment, United States Congress. 1982. *Global models, world futures, and public policy: A critique.* Washington, D.C.: United States Government Printing Office.

Oksanen, L. 1990. Predation herbivory, and plant strategies along gradients of primary productivity. In *Perspectives on plant competition,* J. B. Grace and D. Tilman, Eds. New York: Academic Press, pp. 445–474.

———. 2001. Logic of experiments in ecology: Is pseudo-replication a pseudo-issue? *Oikos* 94:27–38.

Oliver, C. D. 1981. Forest development in North America following major disturbances. *Forest Ecology and Management* 3:153–168.

Oliver, C. D., and E. P. Stephens. 1977. Reconstruction of a mixed-species forest in central New England. *Ecology* 58:562–572.

Olson, J. S. 1958. Rates of succession and soil changes on southern Lake Michigan sand dunes. *Botanical Gazette* 119:125–170.

O'Neill, R. V., D. L. Deangelis, J. B. Waide, and T. F. H. Allen. 1986. *A hierarchical concept of ecosystems.* Princeton, N.J.: Princeton University Press.

Ophel, I. L. 1963. The fate of radiostrontium in a freshwater community. In *Radioecology*, V. Schultz and W. Klement, Eds. New York: Reinhold, pp. 213–216.

Opie, J. 1993. *Ogallala water for a dry land.* Lincoln: University of Nebraska Press.

Osborn, F. 1948. *Our plundered planet.* Boston: Little, Brown.

Ostfeld, R. S. 1997. The ecology of Lyme-disease risk. *American Scientist* 85:338–346.

Ostfeld, R. S., and C. G. Jones. 1999. Peril in the understory. *Audubon*, July–August, pp. 74–82.

Ostfeld, R. S., C. G. Jones, and J. O. Wolff. 1996. Of mice and mast: Ecological connections in eastern deciduous forests. *BioScience* 46:323–330.

Ostfeld, R. S., R. H. Manson, and C. D. Canham. 1997. Effects of rodents on survival of tree seeds and seedlings invading old fields. *Ecology* 78:1531–1542.

———. 1999. Interactions between meadow voles and white-footed mice at forest-old-field edges: Competition and net effects on tree invasion of old fields. In *Landscape ecology of small mammals*, G. W. Barrett and J. D. Peles, Eds. New York: Springer Verlag, pp. 229–247.

P

Pacala, S. W., and M. J. Crawley. 1992. Herbivores and plant diversity. *American Naturalist* 140: 243–260.

Paine. R. T. 1966. Food web diversity and species diversity. *American Naturalist* 100:65–75.

———. 1974. Intertidal community structure: Experimental studies on the relationship between a dominant competitor and its principal predator. *Oecologia* 15:93–120.

———. 1976. Size-limited predation: An observational and experimental approach with the *Mytilus-Pisaster* interaction. *Ecology* 57:858–873.

———. 1984. Ecological determinism in the competition for space. *Ecology* 65:1339–1348.

Palmer, M. A., C. M. Swan, K. Nelson, P. Silver, and R. Alvestad. 2000. Streambed landscapes: Evidence that stream invertebrates respond to the type and spatial arrangement of patches. *Landscape Ecology* 15:563–576.

Palmgren, P. 1949. Some remarks on the short-term fluctuations in the numbers of northern birds and mammals. *Oikos* 1:114–121.

Palo, R. T., and C. T. Robbins. 1991. *Plant defenses against mammalian herbivory.* Boca Raton, Fla.: CRC Press.

Park, T. 1934. Studies in population physiology: Effect of conditioned flour upon the productivity and population decline of *Tribolium confusum. Journal of Experimental Zoology* 68: 167–182.

———. 1954. Experimental studies of interspecific competition. Volume 2. Temperature, humidity and competition in two species of *Tribolium. Physiological Zoology* 27:177–238.

Patrick, R. 1954. Diatoms as an indication of river change. Proceedings of the Ninth Industrial Waste Conference. *Purdue University Engineering Extension Series* 87:325–330.

Patten, B. C. 1959. An introduction to the cybernetics of the ecosystem trophic-dynamic aspect. *Ecology* 40:221–231.

———. 1966. Systems ecology: A course sequence in mathematical ecology. *BioScience* 16: 593–598.

———, Ed. 1971. *Systems analysis and simulation in ecology.* Volume 1. New York: Academic Press.

———. 1978. Systems approach to the concept of environment. *Ohio Journal of Science* 78: 206–222.

———. 1991. Network ecology: Indirect determination of the life-environment relationship in ecosystems. In *Theoretical studies in ecosystems: The network perspective,* M. Higashi and T. P. Burns, Eds. Cambridge, U.K.: Cambridge University Press, pp. 288–351.

Patten, B. C., and G. T. Auble. 1981. System theory and the ecological niche. *American Naturalist* 117:893–922.

Patten, B. C., and S. E. Jørgensen. 1995. *Complex ecology: The part-whole relation in ecosystems.* Englewood Cliffs, N.J.: Prentice Hall.

Patten, B. C., and E. P. Odum. 1981. The cybernetic nature of ecosystems. *American Naturalist* 118:886–895.

Paul, E. A., and F. E. Clark. 1989. *Soil microbiology and biochemistry.* San Diego, Calif.: Academic Press.

Pavlovic, N. B., and M. L. Bowles. 1996. Rare plant monitoring at Indiana Dunes National Lakeshore. In *Science and ecosystem management in the national parks,* W. L. Halvorson and G. E. Davis, Eds. Tucson: University of Arizona Press, pp. 253–280.

Peakall, D. B. 1967. Pesticide-induced enzyme breakdown of steroids in birds. *Nature* 216: 505–506.

Peakall, D. B., and P. N. Witt. 1976. The energy budget of an orb web-building spider. *Comparative Biochemistry and Physiology* 54A:187–190.

Pearl, R., and S. L. Parker. 1921. Experimental studies on the duration of life: Introductory discussion of the duration of life in *Drosophila*. *American Naturalist* 55:481–509.

Pearl, R., and L. J. Reed. 1930. On the rate of growth of the population of the United States since 1790 and its mathematical representation. *Proceedings of the National Academy of Sciences* 6:275–288.

Pearse, A. S., H. J. Humm, and G. W. Wharton. 1942. Ecology of sand beaches at Beaufort, North Carolina. *Ecological Monographs* 12:136–190.

Pearson, O. P. 1963. History of two local outbreaks of feral house mice. *Ecology* 44:540–549.

Peles, J. D., and G. W. Barrett. 1996. Effects of vegetative cover on the population dynamics of meadow voles. *Journal of Mammalogy* 77:857–869.

Peles, J. D., S. R. Brewer, and G. W. Barrett. 1998. Heavy metal accumulation by old-field plant species during recovery of sludge-treated ecosystems. *American Midland Naturalist* 140: 245–251.

Peles, J. D., M. F. Lucas, and G. W. Barrett. 1995. Population dynamics of agouti and albino meadow voles in high-quality grassland habitats. *Journal of Mammalogy* 76:1013–1019.

Perry, J., and J. G. Perry. 1994. *The nature of Florida.* Gainesville, Fla.: Sandhill Crane Press. Reprint, 1998, Athens: University of Georgia Press.

Peterman, R. M. 1978. The ecological role of mountain pine beetle in lodgepole pine forests, and the insect as a management tool. In *Theory and practice of mountain pine beetle man-*

agement in lodgepole pine forests, Berryman, Stark, and Amman, Eds. Moscow: University of Idaho Press.

Peterson, D. L., and V. T. Parker, Eds. 1998. *Ecological scale: Theory and applications.* New York: Columbia University Press.

Petrusewicz, K., and R. Andrzejewski. 1962. Natural history of a free-living population of house mice, *Mus musculus Linnaeus,* with particular references to groupings within the population. *Ecology and Politics—Series A* 10:85–122.

Phillipson, J. 1966. *Ecological energetics.* London: Edward Arnold.

Pianka, E. R. 1970. On *r*- and *K*-selection. *American Naturalist* 104:592–597.

———. 1978. *Evolutionary ecology,* 2nd ed. New York: Harper and Row.

———. 1988. *Evolutionary biology,* 4th ed. New York: Harper and Row.

———. 2000. *Evolutionary biology,* 6th ed. San Francisco: Benjamin/Cummings.

Picone, C. 2002. Natural systems of soil fertility: The webs beneath our feet. In *The Land Report* 73. Salina, Kans.: The Land Institute, pp. 3–7.

Pielou, E. C. 1966. The measurement of diversity in different types of biological collections. *Journal of Theoretical Biology* 13:131–144.

———. 1974. *Population and community ecology: Principles and methods.* New York: Gordon and Breach.

Pierzynski, G. M., J. T. Sims, and G. F. Vance. 2000. *Soils and environmental quality,* 2nd ed. Boca Raton, Fla.: CRC Press.

Pimentel, D., L. E. Hurd, A. C. Bellotti, M. J. Forster, I. N. Oka, O. D. Sholes, and R. J. Whitman. 1973. Food production and the energy crisis. *Science* 182:443–449.

Pimentel, D., and F. A. Stone. 1968. Evolution and population ecology of parasite-host systems. *Canadian Entomologist* 100:655–662.

Pimm, S. L. 1984. The complexity and stability of ecosystems. *Nature* 307:321–326.

———. 1997. Agriculture—In search of perennial solutions. *Nature* 389:126–127.

Pippenger, N. 1978. Complexity theory. *Scientific American* 238:114–124.

Pitelka, F. A. 1964. The nutrient recovery hypothesis for Arctic microtine cycles. In *Grazing in terrestrial and marine environments,* D. J. Crisp, Ed. Oxford: Blackwell, pp. 55–56.

———. 1973. Cyclic patterns in lemming populations near Barrow, Alaska. In *Alaska Arctic tundra,* M. E. Britton, Ed. Technical Paper 25. Washington, D.C.: Arctic Institute of North America, pp. 119–215.

Platt, R. B. 1965. Ionizing radiation and homeostasis of ecosystems. In *Ecological effects of nuclear war,* Number 917, G. M. Woodwell, Ed. Upton, N.Y.: Brookhaven National Laboratory, pp. 39–60.

Polis, G. A. 1994. Food webs, trophic cascades and community structure. *Australian Journal of Ecology* 19:121–136.

Polis, G. A., and D. R. Strong. 1996. Food web complexity and community dynamics. *American Naturalist* 147:813–846.

Pollard, H. P., and S. Gorenstein. 1980. Agrarian potential, population, and the Tarascan state. *Science* 209:274–277.

Polynov, B. B. 1937. *The cycle of weathering,* A. Muir, Trans. London: T. Murby.

Pomeroy, L. R. 1959. Algal productivity in Georgia salt marshes. *Limnology and Oceanography* 4:386–397.

———. 1960. Residence time of dissolved phosphate in natural waters. *Science* 131:1731–1732.

———, Ed. 1974a. *Cycles of essential elements*. Benchmark Papers in Ecology. Stroudsburg, Pa.: Dowden, Hutchinson and Ross; New York: Academic Press.

———. 1974b. The ocean's food web, a changing paradigm. *BioScience* 24:299–304.

———. 1984. Significance of microorganisms in carbon and energy flow in aquatic ecosystems. In *Current perspectives in microbial ecology*, M. J. Lug and C. A. Reddy, Eds. Washington, D.C.: American Society of Microbiologists, pp. 405–411.

Pomeroy, L. R., H. M. Mathews, and H. S. Min. 1963. Excretion of phosphate and soluble organic phosphorus compounds by zooplankton. *Limnology and Oceanography* 8:50–55.

Popper, K. 1959. *The logic of scientific discovery*. New York: Harper and Row.

———. 1979. *Objective knowledge: An evolutionary approach*. New York: Clarendon Press.

Porcella, D. B., J. W. Huckabee, and B. Wheatley, Eds. 1995. *Mercury as a global pollutant: Proceedings of the third international conference, Whistler, British Columbia, July 10–14, 1994*. Dordrecht; Boston: Kluwer.

Postel, S. L. 1989. *Water for agriculture: Facing the limits*. Worldwatch Paper 93. Washington, D.C.: Worldwatch Institute.

———. 1992. *Last oasis: Facing water scarcity*. New York: W. W. Norton.

———. 1993. The politics of water. *World•Watch* 6:10–18.

———. 1998. Water for food production: Will there be enough in 2025? *BioScience* 48: 629–637.

———. 1999. When the world's wells run dry. *World•Watch* 12:30–38.

Postel, S. L., G. C. Daily, and P. R. Ehrlich. 1996. Human appropriation of renewable freshwater. *Science* 271:785–788.

Potter, V. R. 1988. *Global bioethics: Building on the Leopold legacy*. East Lansing: Michigan State University Press.

Power, M. E. 1992. Habitat heterogeneity and the functional significance of fish in river food webs. *Ecology* 73:1675–1688.

Preston, F. W. 1948. The commonness and rarity of species. *Ecology* 29:254–283.

———. 1962. The canonical distribution of commonness and rarity: Parts 1 and 2. *Ecology* 43:185–215; 410–432.

Price, P. W. 1976. Colonization of crops by arthropods: Nonequilibrium communities in soybean fields. *Environmental Entomology* 5:605–611.

———. 1997. *Insect ecology*. New York: John Wiley.

———. 2003. *Macroevolutionary theory on macroecological patterns*. Cambridge, U.K.: Cambridge University Press.

Prigogine, I. 1962. *Introduction to nonequilibrium statistical mechanics*. New York: Interscience.

Primack, R. B. 2004. *A primer of conservation biology*, 3rd ed. Sunderland, Mass.: Sinauer.

Provine, W. B. 1971. *The origins of theoretical population genetics*. Chicago: University of Chicago Press.

Prugh, T., R. Costanza, J. H. Cumberland, H. Daly, R. Goodland, and R. B. Norgaard. 1995. *Natural capital and human economic survival*. Solomons, Md.: International Society for Ecological Economics Press.

Pulliam, H. R. 1988. Sources, sinks, and population regulation. *American Naturalist* 132: 652–661.

Pyke, G. H., H. R. Pulliam, and E. L. Charnov. 1977. Optimal foraging: A selective review of theory and tests. *Quarterly Review of Biology* 52:137–154.

R

Ramirez, C. C., E. Fuentes-Contreras, L. C. Rodreiguez, and H. M. Niemeyer. 2000. Pseudo-replication and its frequency in olfactometric laboratory studies. *Journal of Chemistry and Ecology* 26:1423–1431.

Randolph, P. A., J. C. Randolph, and C. A. Barlow. 1975. Age-specific energetics of the pea aphid. *Ecology* 56:359–369.

Randolph, P. A., J. C. Randolph, K. Mattingly, and M. M. Foster. 1977. Energy costs of reproduction in the cotton rat, *Sigmodon hispidus. Ecology* 58:31–45.

Rappaport, R. A. 1968. *Pigs for the ancestors: Ritual in the ecology of a New Guinea people.* New Haven, Conn.: Yale University Press.

Rapport, D. J., R. Costanza, and A. J. McMichael. 1998. Assessing ecosystem health. *Trends in Ecology and Evolution* 13:397–402.

Rasmussen, D. I. 1941. Biotic communities of Kaibab Plateau, Arizona. *Ecological Monographs* 11:229–275.

Redfield, A. C. 1958. The biological control of chemical factors in the environment. *American Scientist* 46:205–221.

Redman, C. L., J. M. Grove, and L. H. Kuby. 2004. Integrating social science into the Long-Term Ecological Research (LTER) network: Social dimensions of ecological change and ecological dimensions of social change. *Ecosystems* 7:161–171.

Rees, W., and M. Wackernagel. 1994. Ecological footprints and appropriated carrying capacity: Measuring the natural capital requirements of the human economy. In *Investing in natural capital: The ecological economics approach to sustainibility,* A. M. Jansson, M. Hammer, C. Folke, and R. Costanza, Eds. Washington, D.C.: Island Press.

Reifsnyder, W. E., and H. W. Lull. 1965. *Radiant energy in relation to forests.* Technical Bulletin Number 1344. Washington, D.C.: United States Department of Agriculture, and Forest Service.

Rensberger, B. 1982. Evolution since Darwin. *Science* 82:41–45.

Rhoades, D. F., and R. G. Cates. 1976. Toward a general theory of plant antiherbivore chemistry. In *Biochemical interactions between plants and insects,* J. Wallace and R. Mansell, Eds. Recent Advances in Phytochemistry, Volume 10. New York: Plenum Press, pp. 168–213.

Rhoades, R. E., and V. D. Nazarea. 1999. Local management of biodiversity in traditional agroecosystems. In *Biodiversity in agroecosystems,* W. W. Collins and C. O. Qualset, Eds. Boca Raton, Fla.: CRC Press, pp. 215–236.

Rice, E. L. 1952. Phytosociological analysis of a tallgrass prairie in Marshall County, Oklahoma. *Ecology* 33:112–116.

———. 1974. *Allelopathy.* New York: Academic Press.

Rich, P. H. 1978. Reviewing bioenergetics. *BioScience* 28:80.

Richards, B. N. 1974. *Introduction to the soil ecosystem.* New York: Longman.

Riechert, S. E. 1981. The consequences of being territorial: Spiders, a case study. *American Naturalist* 117:871–892.

Riedel, G. F., S. A. Williams, G. S. Riedel, C. C. Gilmour, and J. G. Sanders. 2000. Temporal and spatial patterns of trace elements in the Patuxent River: A whole watershed approach. *Estuaries* 23:521–535.

Riley, G. A. 1944. The carbon metabolism and photosynthetic efficiency of the Earth. *American Scientist* 32:132–134.

Riley, J., and P. Edwards. 1998. Statistical aspects of aquaculture research: Pond variability and pseudo-replication. *Aquaculture Research* 29:281–288.

Risser, P. G., J. R. Karr, and R. T. T. Forman. 1984. *Landscape ecology: Directions and approaches.* Champaign, Ill.: Natural History Survey, Number 2.

Robertson, G. P., and P. M. Vitousek. 1981. Nitrification potential in primary and secondary succession. *Ecology* 62:376–386.

Rodenhouse, N. L., P. J. Bohlen, and G. W. Barrett. 1997. Effects of habitat shape on the spatial distribution and density of 17-year periodical cicadas, Homoptera Cicadidae. *American Midland Naturalist* 137:124–135.

Roe, E. 1998. *Taking complexity seriously: Policy analysis and triangulation and sustainable development.* Boston: Kluwer.

Romme, W. H., and D. G. Despain. 1989. Historical perspective on the Yellowstone fires of 1988. *BioScience* 39:695–699.

Root, R. B. 1967. The niche exploitation pattern of the blue-gray gnatcatcher. *Ecological Monographs* 37:317–350.

———. 1969. The behavior and reproductive success of the blue-gray gnatcatcher. *Condor* 71:16–31.

———. 1996. Herbivore pressure on goldenrods, *Solidago altissima:* Its variation and cumulative effects. *Ecology* 77:1074–1087.

Rosenberg, D. K., B. R. Noon, and E. C. Meslow. 1997. Biological corridors: Form, function, and efficacy. *BioScience* 47:677–687.

Rosenzweig, M. L. 1968. Net primary production of terrestrial communities: Prediction from climatological data. *American Naturalist* 102:67–74.

Rossell, I. M., C. R. Rossell, K. J. Hining, and R. L. Anderson. 2001. Impacts of dogwood anthracnose, *Discula destructiva* Redlin, on the fruits of flowering dogwood, *Cornus florida* L.: Implications for wildlife. *American Midland Naturalist* 146:379–387.

Royall, R. 1997. *Statistical evidence.* London: Chapman and Hall.

Ryszkowski, L. 2002. *Landscape ecology in agroecosystems management.* Boca Raton, Fla.: CRC Press.

S

Sachs, A. 1995. Humboldt's legacy and the restoration of science. *World•Watch* 8: 29–38.

Salih, A., A. Larkum, G. Cox, M. Kuhl, and O. Hoegh-Guldberg. 2000. Fluorescent pigments in corals are photoprotective. *Nature* 408:850–853.

Salsburg, D. 2001. *The lady tasting tea.* New York: W. H. Freeman.

Salt, G. 1979. A comment on the use of the term emergent properties. *American Midland Naturalist* 113:145–148.

Sanderson, J., and Harris, L. D. 2000. *Landscape ecology: A top-down approach.* Boca Raton, Fla.: Lewis.

Santos, P. F., J. Phillips, and W. G. Whitford. 1981. The role of mites and nematodes in early stages of buried litter decomposition in a desert. *Ecology* 62:664–669.

Sarbu, S. M., T. C. Kane, and B. K. Kinkle. 1996. A chemoautotrophically based cave ecosystem. *Science* 272:1953–1955.

Sarmiento, F. O. 1995. The birthplace of ecology: Tropandean landscapes. *Bulletin of the Ecological Society of America* 76:104–105.

———. 1997. The mountains of Ecuador as a birthplace of ecology and endangered landscape. *Environmental Conservation* 24:3–4.

Sarmiento, J. L., and N. Gruber. 2002. Sinks for anthropogenic carbon. *Physics Today* 55: 30–36.

Schank, J. 2001. Is pseudo-replication a pseudo-problem? *American Zoologist* 41:1577.

Schaus, M. H., and M. J. Vanni. 2000. Effects of gizzard shad on phytoplankton and nutrient dynamics: Role of sediment feeding and fish size. *Ecology* 81:1701–1719.

Scheiner, S. M. 1993. Introduction: Theories, hypotheses, and statistics. In *Design and analysis of ecological experiments,* S. M. Scheiner and J. Gurevitch, Eds. New York: Chapman and Hall, pp. 1–13.

Schimel, D. S. 1995. Terrestrial ecosystems and the carbon cycle. *Global Change Biology* 1: 77–91.

Schimel, D. S., I. G. Enting, M. Heimann, T. M. L. Wigley, D. Raynaud, D. Alves, and U. Siegenthaler. 1995. CO_2 and the carbon cycle. In *Climate change 1994,* J. T. Houghton, L. G. Meira Filho, J. Bruce, H. Lee, B. A. Callander, E. Haites, N. Harris, and K. Maskell, Eds. Cambridge, U.K.: Cambridge University Press, pp. 35–71.

Schindler, D. W. 1977. Evolution of phosphorus limitation in lakes. *Science* 195:260–262.

———. 1990. Experimental perturbations of whole lakes as tests of hypotheses concerning ecosystem structure and function. *Oikos* 57:25–41.

Schlesinger, W. H. 1997. *Biogeochemistry: An analysis of global change.* San Diego, Calif.: Academic Press.

Schoener, T. W. 1971. Theory of feeding strategies. *Annual Review of Ecology and Systematics* 2:369–404.

———. 1983. Field experiments on interspecific competition. *American Naturalist* 122: 240–285.

Schreiber, K. F. 1990. The history of landscape ecology in Europe. In *Changing landscapes: An ecological perspective,* I. S. Zonneveld and R. T. T. Forman, Eds. New York: Springer Verlag, pp. 21–33.

Schultz, A. M. 1964. The nutrient recovery hypothesis for Arctic microtine cycles. Volume 2. Ecosystem variables in relation to Arctic microtine cycles. In *Grazing in terrestrial and marine environments,* D. J. Crisp, Ed. Oxford: Blackwell, pp. 57–68.

———. 1969. A study of an ecosystem: The Arctic tundra. In *The ecosystem concept in natural resource management,* G. Van Dyne, Ed. New York: Academic Press, pp. 77–93.

Schumacher, E. F. 1973. *Small is beautiful: A study of economics as if people mattered.* New York: Harper and Row.

Sears, P. B. 1935. *Deserts on the march.* 2nd ed., 1947, Norman: University of Oklahoma Press.

Seigler, D. S. 1996. Chemistry and mechanisms of allelopathic interactions. *Agronomy Journal* 88:876–885.

Seligson, M. A. 1984. *The gap between rich and poor: Contending perspectives on political economy and development.* Boulder, Colo.: Westview Press.

Selye, Hans. 1955. Stress and disease. *Science* 122:625–631.

Severinghaus, J. P., W. S. Broecher, W. F. Dempster, T. MacCallum, and M. Wahlin. 1994. Oxygen loss in Biosphere-2. *Transactions of the American Geophysical Union* 75(33):25–37.

Shannon, C. 1950. Memory requirements in a telephone exchange. *Bell System Technical Journal* 29:343–347.

Shannon, C. E., and W. Weaver. 1949. *The mathematical theory of communication.* Urbana: University of Illinois Press.

Shantz, H. L. 1917. Plant succession on abandoned roads in eastern Colorado. *Journal of Ecology* 5:19–42.

Shelford, V. E. 1913. *Animal communities in temperate America.* Chicago: University of Chicago Press. Reprint, 1977, New York: Arno Press.

———. 1929. *Laboratory and field ecology: The responses of animals as indicators of correct working methods.* Baltimore, Md.: Williams and Wilkins.

———. 1943. The abundance of the collared lemming in the Churchill area, 1929–1940. *Ecology* 24:472–484.

Shelterbelt Project. 1934. Published statements by numerous separate authors. *Journal of Forestry* 32:952–991.

Shiva, V. 1991. The green revolution in the Punjab. *Ecology* 21:57–60.

Shugart, H. H. 1984. *A theory of forest dynamics: The ecological implications of forest succession models.* New York: Springer Verlag.

Silliman, R. P. 1969. Population models and test populations as research tools. *BioScience* 19: 524–528.

Simberloff, D. S. 2003. How much information on population biology is needed to manage introduced species? *Conservation Biology* 17:83–92.

Simberloff, D. S., and J. Cox. 1987. Consequences and costs of conservation corridors. *Conservation Biology* 1:63–71.

Simberloff, D. S., and E. O. Wilson. 1969. Experimental zoogeography of islands: The colonization of empty islands. *Ecology* 50:278–296.

———. 1970. Experimental zoogeography of islands: A two-year record of colonization. *Ecology* 51:934–937.

Simon, H. A. 1973. The organization of complex systems. In *Hierarchy theory: The challenge of complex systems,* H. H. Pattee, Ed. New York: G. Braziller, pp. 3–27.

Simon, J. L. 1981. *The ultimate resource.* Princeton, N.J.: Princeton University Press.

Simpson, E. H. 1949. Measurement of diversity. *Nature* 163:688.

Slobodkin, L. B. 1954. Population dynamics in *Daphnia obtusa* Kurz. *Ecological Monographs* 24: 69–88.

———. 1960. Ecological energy relationships at the population level. *American Naturalist* 95: 213–236.

———. 1964. Experimental populations of hydrida. *Journal of Animal Ecology* 33:1–244.

———. 1968. How to be a predator. *American Zoologist* 8:43–51.

———. 1980. *Growth and regulation of animal populations,* 2nd ed. New York: Dover.

Smith, B. D. 1998. *The emergence of agriculture.* New York: W. H. Freeman.

Smith, C. R., H. Kukert, R. A. Wheatcroft, P. A. Jumars, and J. W. Deming. 1989. Vent fauna on whale remains. *Nature* 341:27–28.

Smith, F. E. 1969. Today the environment, tomorrow the world. *BioScience* 19:317–320.

———. 1970. Ecological demand and environmental response. *Journal of Forestry* 68: 752–755.

Smith, J. M.. 1964. Group selection and kin selection. *Nature* 201:1145–1147.

Smith, T. B., R. K. Wayne, D. J. Girman, and M. W. Bruford. 1997. A role for ecotones in generating rainforest biodiversity. *Science* 276:1855–1857.

Smolin, L. 1997. *The life of the cosmos.* New York: Oxford University Press.

Snee, R. D. 1990. Statistical thinking and its contribution to total quality. *The American Statistician* 44:116–121.

Snow, C. P. 1963. *The two cultures: A second look.* New York: Cambridge University Press.

Soil Science Society of America (SSSA). 1994. *Defining and assessing soil quality.* Madison, Wisc.: Soil Science Society of America Special Publication 35.

Solbrig, O. T. 1971. The population biology of dandelions. *American Scientist* 59:686–694.

Solomon, M. E. 1949. The natural control of animal populations. *Journal of Animal Ecology* 18:1–32.

———. 1953. Insect population balance and chemical control of pests. Pest outbreaks induced by spraying. *Chemical Industries* 43:1143–1147.

Soon, Y. K., T. E. Bates, and J. R. Moyer. 1980. Land application of chemically treated sewage sludge. Volume 3. Effects on soil and plant heavy metal content. *Journal of Environmental Quality* 9:497–504.

Soule, J. D., and J. K. Piper. 1992. *Farming in nature's image: An ecological approach to agriculture.* Washington, D.C.: Island Press.

Soulé, M. E. 1985. What is conservation biology? A new synthetic discipline addresses the dynamics and problems of perturbed species, communities, and ecosystems. *BioScience* 35: 727–734.

———. 1991. Conservation: Tactics for a constant crisis. *Science* 253:744–750.

Soulé, M. E., and D. Simberloff. 1986. What do genetics and ecology tell us about the design of nature reserves? *Biological Conservation* 35:19–40.

Sowls, L. K. 1960. Results of a banding study of Gambel's quail in southern Arizona. *Journal of Wildlife Management* 24:185–190.

Sparks, R. E., J. C. Nelson, and Y. Yin. 1998. Naturalization of the flood regime in regulated rivers. *BioScience* 48:706–720.

Spiller, D. A., and T. W. Schoener. 1990. A terrestrial field experiment showing the impact of eliminating top predators on foliage damage. *Nature* 347:469–472.

Sprugel, D. G., and F. H. Bormann. 1981. Natural disturbance and the steady state in high altitude balsam fir forests. *Science* 211:390–393.

Spurlock, J. M., and M. Modell. 1978. *Technology requirements and planning criteria for closed life support systems for manned space missions.* Washington, D.C.: Office of the Life Sciences, National Aeronautics and Space Administration.

Stanton, M. L. 2003. *Interacting guilds: Moving beyond the pairwise perspective on mutualisms.* American Society of Naturalists Supplement to *The American Naturalist.* Chicago: University of Chicago Press.

Stearns, S. C. 1976. Life-history tactics: A review of ideas. *Quarterly Review of Biology* 51:3–47.

Steel, R. G. D., and J. H. Torrie. 1980. *Principles and procedures of statistics.* New York: McGraw-Hill.

Steinhart, J. S., and C. E. Steinhart. 1974. Energy use in the U. S. food system. *Science* 184: 307–316.

Stenseth, N. C., and W. Z. Lidicker, Jr., Eds. 1992. *Animal dispersal: Small mammals as a model.* London: Chapman and Hall.

Stephens, D. W., and J. R. Krebs. 1986. *Foraging theory.* Princeton, N.J.: Princeton University Press.

Stephenson, T. A., and A. Stephenson. 1952. Life between tide-marks in North America: Northern Florida and the Carolinas. *Journal of Ecology* 40:1–49.

Sterner, R. W. 1986. Herbivores' direct and indirect effects on algal populations. *Science* 231: 605–607.

Steven, J. E. 1994. Science and religion at work. *BioScience* 44:60–64.

Stevens, M. H. H., and W. P. Carson. 1999. Plant density determines species richness along an experimental productivity gradient. *Ecology* 80:455–465.

Stewart, P. A. 1952. Dispersal, breeding, behavior, and longevity of banded barn owls in North America. *Auk* 69:277–285.

Stickel, L. F. 1950. Populations and home range relationships of the box turtle, *Terrapene c. carolina* Linnaeus. *Ecological Monographs* 20:351–378.

Stiles, E. W. 1980. Patterns of fruit presentation and seed dispersal in bird-disseminated woody plants in the eastern deciduous forest. *American Naturalist* 116:670–688.

Stoddard, D. R. 1965. Geography and the ecological approach. The ecosystem as a geographical principle and method. *Geography* 50:242–251.

Stoddard, H. L. 1936. Relation of burning to timber and wildlife. *Proceedings of the National Association of Wildlife Conference* 1:1–4.

———. 1950. *The bobwhite quail: Its habits, preservation and increase.* New York: Scribner.

Stone, R. 1998. Ecology—Yellowstone rising again from the ashes of devastating fires. *Science* 280:1527–1528.

Strayer, D. L., M. E. Power, W. F. Fagan, S. T. A. Pickett, and J. Belnap. 2003. A classification of ecological boundaries. *BioScience* 53:723–729.

Strong, D. R., Jr., and J. R. Rey. 1982. Testing for MacArthur-Wilson equilibrium with the arthropods of the miniature *Spartina* archipelago at Oyster Bay, Florida. *American Zoologist* 22:355–360.

Strong, D. R., J. H. Lawton, and T. R. E. Southwood. 1984. *Insects on plants: Community patterns and mechanisms.* Cambridge, Mass.: Harvard University Press.

Strong, D. R., D. Simberloff, L. G. Abele, and A. B. Thistle, Eds. 1984. *Ecological communities: Conceptual issues and the evidence.* Princeton, N.J.: Princeton University Press.

Stueck, K. L., and G. W. Barrett. 1978. Effects of resource partitioning on the population dynamics and energy utilization strategies of feral house mice, *Mus musculus* populations under experimental field conditions. *Ecology* 59:539–551.

Sugihara, G., L. Beaver, T. R. E. Southwood, S. Pimm, and R. M. May. 2003. Predicted correspondence between species abundances and dendrograms of niche similarities. *Proceedings of the National Academy of Sciences* 100:4246–4251.

Sukachev, V. N. 1944. On principles of genetic classification in biocenology. *Zurnal Obshcheij Biologij* 5:213–227. F. Raney and R. Daubenmire, Trans. *Ecology* 39:364–376.

Swan, L. W. 1992. The Aeolian biome: Ecosystems of the earth's extremes. *BioScience* 42: 262–270.

Swank, W. T., and D. A. Crossley, Jr., Eds. 1988. *Forest hydrology and ecology at Coweeta.* Ecological Studies, Volume 66. New York: Springer Verlag.

Swank, W. T., J. L. Meyer, and D. A. Crossley, Jr. 2001. Long-term ecological research: Coweeta history and perspectives. In *Holistic science: The evolution of the Georgia Institute of Ecology, 1940–2000,* G. W. Barrett and T. L. Barrett, Eds. Boca Raton, Fla.: CRC Press, pp. 143–163.

T

Taber, R. D., and R. Dasmann. 1957. The dynamics of three natural populations of the deer, *Odocoileus hemionus columbianus*. *Ecology* 38:233–246.

Tangley, L. 1990. Debugging agriculture: Can farming imitate nature and still make a profit? *Earthwatch* 9:20–23.

Tansley, A. G., Sir. 1935. The use and abuse of vegetational concepts and terms. *Ecology* 16: 284–307.

Taub, F. B. 1989. Standardized aquatic microcosm development and testing. In *Aquatic ecotoxicology: Fundamental concepts and methodologies,* Volume 2, A. Boudou and F. Ribeyre, Eds. Boca Raton, Fla.: CRC Press, pp. 47–92.

———. 1993. Standardizing an aquatic microcosm test. In *Progress in standardization of aquatic toxicity tests,* A. M. V. M. Soares and P. Calow, Eds. Boca Raton, Fla.: Lewis, pp. 159–188.

———. 1997. Unique information contributed by multispecies systems: Examples from the standardized aquatic microcosm. *Ecological Applications* 7:1103–1110.

Taub, F. B., A. Howell-Kübler, M. Nelson, and J. Carrasquero. 1998. An ecological life support system for fish for 100-day experiments. *Life Support Biosphere Science* 5:107–116.

Teal, J. M. 1957. Community metabolism in a temperate cold spring. *Ecological Monographs* 27:283–302.

———. 1958. Distribution of fiddler crabs in Georgia salt marshes. *Ecology* 39:185–193.

Tegen, I., A. A. Lacis, and I. Fung. 1996. The influence on climate forcing of mineral aerosols from disturbed soils. *Nature* 380:419–422.

Teubner, V. A., and G. W. Barrett. 1983. Bioenergetics of captive raccoons. *Journal of Wildlife Management* 47:272–274.

Thienemann, A. 1939. Grundzüge einer allgemeinen Oekologie. *Archiv für Hydrobiologie* 35: 267–285.

Thomas, D. J., J. C. Zachas, T. J. Bralower, E. Thomas, and S. Boharty. 2002. Warming the fuel for the fire: Evidence for the thermal dissociation of methane hydrate during the Paleocene-Eocene thermal maximum. *Geology* 30:1067–1070.

Thomas, D. N., and G. S. Dieckmann. 2002. Ocean science—Antarctic Sea ice—a habitat for extremophites. *Science* 295:641–644.

Thompson, J. N., O. J. Reichman, P. J. Morin, G. A. Polis, M. E. Power, R. W. Sterner, C. A. Couch, L. Gough, R. Holt, D. U. Hooper, F. Keesing, C. R. Lovell, B. T. Milne, M. C. Molles, D. W. Roberts, and S. Y. Strauss. 2001. Frontiers of ecology. *BioScience* 51:15–24.

Thorson, G. 1955. Modern aspects of marine level-bottom animal communities. *Journal of Marine Research* 14:387–397.

Tilman, D. 1986. Nitrogen-limited growth in plants from different successional stages. *Ecology* 67:555–563.

———. 1987. Secondary succession and the pattern of plant dominance along experimental nitrogen gradients. *Ecological Monographs* 57:189–214.

———. 1988. *Plant strategies and the dynamics and structure of plant communities.* Princeton, N.J.: Princeton University Press.

———. 1999. Diversity by default. *Science* 283:495–496.

Tilman, D., and J. A. Downing. 1994. Biodiversity and stability in grasslands. *Nature* 367: 363–365.

Tilman, D., S. Naeem, J. Knops, P. Reich, E. Siemann, D. Wedin, M. Ritchie, and J. Lawton. 1997. Biodiversity and ecosystem properties. *Science* 278:1866–1867.

Tilman, D., and D. Wedin. 1991. Dynamics of nitrogen competition between successional grasses. *Ecology* 72:1038–1049.

Tilman, D., D. Wedin, and J. Knops. 1996. Productivity and sustainability influenced by biodiversity in grassland ecosystems. *Nature* 379:718–720.

Todes, D. P. 1989. Kropotkin's theory of mutual aid. In *Darwin without Malthus: The struggle for existence in Russian evolutionary thought.* New York: Oxford University Press, pp. 123–142.

Toynbee, A. 1961. *A study of history.* New York: Oxford University Press.

Transeau, E. N. 1926. The accumulation of energy by plants. *Ohio Journal of Science* 26:1–10.

Treguer, P., and P. Pondaven. 2000. Global change—silica control of carbon dioxide. *Nature* 406:358–359.

Troll, C. 1939. Luftbildplan und ökologische bodenforschung. Berlin: *Zeitschrift der Gesellschaft für Erdkunde,* pp. 241–298.

———. 1968. Landschaftsökologie. In *Pflanzensoziologie und landschaftsökologie,* R. Tüxen, Ed. The Hague: Berichte des Internalen Symposiums der Internationalen Vereinigung für Vegetationskunde, Stolzenau, Weser, 1963, pp. 1–21.

Tuckfield, R. C. 1985. *Geographic variation of song patterns in two desert sparrows.* Ph.D. Dissertation. Bloomington: Indiana University.

———. 2004. Purveying ecological science: Comments on Lele (2002). In *The nature of scientific evidence,* M. Taper and S. Lele, Eds. Chicago: University of Chicago Press, in press.

Tufte, E. R. 1997. *Visual explanations: Images and quantities, evidence and narrative.* Cheshire, Conn.: Graphics Press.

———. 2001. *The visual display of quantitative information,* 2nd ed. Cheshire, Conn.: Graphics Press.

Tukey, J. W. 1977. *Exploratory data analysis.* Reading, Mass.: Addison-Wesley.

Tunnicliffe, V. 1992. Hydrothermal vent communities of the deep sea. *American Scientist* 80: 336–349.

Turchin, P., L. Oksanen, P. Ekerholm, T. Oksanen, and H. Henttonen. 2000. Are lemmings prey or predators? *Nature* 405: 562–565.

Turner, B. L., III, Ed. 1990. *The Earth as transformed by human action.* New York: Cambridge University Press with Clark University.

Turner, M. G., Ed. 1987. *Landscape heterogeneity and disturbance.* New York: Springer Verlag.

———. 1989. Landscape ecology: The effects of pattern on process. *Annual Review of Ecology and Systematics* 20: 171–197.

Turner, M. G., R. H. Gardner, and R. V. O'Neill. 2001. *Landscape ecology in theory and practice: Pattern and process.* New York: Springer Verlag.

U

Ulanowicz, R. E. 1980. A hypothesis on the development of natural communities. *Journal of Theoretical Biology* 85: 223–245.

———. 1997. *Ecology, the ascendent perspective.* New York: Columbia University Press.

United Nations Environmental Program (UNEP). 1995. *Global biodiversity assessment,* V. H. Heywood, Ed. New York: Cambridge University Press.

Urban, D. L., R. V. O'Neill, and H. H. Shugart, Jr. 1987. Landscape ecology. *BioScience* 37: 119–127.

V

Van Andel, T. H. 1985. *New views of an old planet: Continental drift and the history of the Earth.* New York: Cambridge University Press.

———. 1994. *New views of an old planet: A history of global change,* 2nd ed. New York: Cambridge University Press.

Van Dover, C. L. 2002. Community structure of mussel beds at deep-sea hydrothermal vents. *Marine Ecology Progress Series* 230: 137–158.

Van Dover, C. L., C. R. German, K. G. Speer, L. M. Parson, and R. C. Vrijenhoek. 2002. Evolution and biogeography of deep-sea vent and seep invertebrates. *Science* 295: 1253–1257.

Van Dyne, G. M., Ed. 1969. *The ecosystem concept in natural resource management.* New York: Academic Press.

Van Mantgem, P., M. Schwartz, and M. B. Keifer. 2001. Monitoring fire effects for managed burns and wildfires: Coming to terms with pseudo-replication. *Natural Areas Journal* 21: 266–273.

Vanni, M. J., and C. D. Layne. 1997. Nutrient recycling and herbivory as mechanisms in the "top-down" effect of fish on algae in lakes. *Ecology* 78:21–40.

Vanni, M. J., C. D. Layne, and S. E. Arnott. 1997. "Top-down" trophic interactions in lakes: Effects of fish on nutrient dynamics. *Ecology* 78:1–20.

Vannote, R. L., G. W. Minshall, K. W. Cummins, J. R. Sedell, and C. E. Cushing. 1980. The river continuum concept. *Canadian Journal of Fisheries and Aquatic Science* 37:130–137.

Van Valen, L. 1965. Morphological variation and width of ecological niche. *American Naturalist* 99:377–390.

Varley, G. C. 1970. The concept of energy flow applied to a woodland community. In *Quality and quantity of food.* Symposium of the British Ecological Society. Oxford: Blackwell, pp. 389–405.

Verboom, J., A. Schotman, P. Opdam, and A. J. Metz. 1991. European nuthatch metapopulations in a fragmented agricultural landscape. *Oikos* 61:149–161.

Verhulst, P. F. 1838. Notice sur la loi que la population suit dans son accroissement. *Correspondence of Mathematics and Physics* 10:113–121.

Vernadskij, V. I. 1945. The biosphere and the noösphere. *American Scientist* 33:1–12.

———. 1998. *The biosphere,* rev. ed. A. S. McMenamin and D. B. Langmuir, Trans. New York: Copernicus Books.

Verner, T. 1977. On the adaptive significance of territoriality. *American Naturalist* 111: 769–775.

Vicsek, T. 2002. Complexity: The bigger picture. *Nature* 418:131.

Vitousek, P. 1983. Nitrogen turnover in a ragweed-dominated old-field in southern Indiana. *American Midland Naturalist* 110:46–53.

Vitousek, P. M., J. Aber, R. W. Howarth, G. E. Likens, P. A. Matson, D. W. Schindler, W. H. Schlesinger, and G. D. Tilman. 1997. Human alteration of the global nitrogen cycle: Causes and consequences. *Issues in Ecology* 1:1–15.

Vitousek, P. M., P. R. Ehrlich, A. H. Ehrlich, and P. A. Matson. 1986. Human appropriation of the products of photosynthesis. *BioScience* 36:368–373.

Vitousek, P. M., H. A. Mooney, J. Lubchenco, and J. M. Melillo. 1997. Human domination of Earth's ecosystems. *Science* 277:494–499.

Vitousek, P. M., and W. A. Reiners. 1975. Ecosystem succession and nutrient retention: A hypothesis. *BioScience* 25:376–381.

Vogl, R. J. 1980. The ecological factors that produce perturbation dependent ecosystems. In *The recovery process in damaged ecosystems,* J. Cairns, Jr., Ed. Ann Arbor, Mich.: Ann Arbor Science, pp. 63–69.

Vogt, W. 1948. *Road to survival.* New York: W. Sloane.

Vogtsberger, L. M., and G. W. Barrett. 1973. Bioenergetics of captive red foxes. *Journal of Wildlife Management* 37:495–500.

Volkov, I., J. R. Banavar, S. P. Hubbell, and A. Maritan. 2003. Neutral theory and relative species abundance in ecology. *Nature* 424:1035–1037.

Volterra, V. 1926. Variations and fluctuations of the number of individuals in animal species living together. In *Animal ecology,* R. N. Chapman, Ed. New York: McGraw-Hill, pp. 409–448.

Von Damm, K. L. 2001. Lost city found. *Nature* 412:127–128.

W

Wackernagel, M., and W. Rees. 1996. *Our ecological footprint: Reducing human impact on the Earth.* Gabriola Island, B. C.: New Society.

Waddington, C. H. 1975. A catastrophe theory of evolution. In *The evolution of an evolutionist.* Ithaca, New York: Cornell University Press, pp. 253–266.

Walsh, J. J., and K. A. Steidinger. 2001. Saharan dust and Florida red tides: The cyanophyte connection. *Journal of Geophysical Research—Oceans* 106: 11,597–11,612.

Walther, G., E. Post, P. Convey, A. Menzel, C. Parmesan, T. J. C. Beebee, J. Fromentin, O. Hoegh-Guldberg, and F. Bairlein. 2002. Ecological responses to recent climate change. *Nature* 416: 389–395.

Warington, R. 1851. Notice of observation on the adjustment of the relations between animal and vegetable kingdoms, by which the vital functions of both are permanently maintained. *Chemical Society Journal* (U.K.) 3: 52–54.

Watt, K. E. F. 1963. How closely does the model mimic reality? *Memoirs of the Entomological Society of Canada* 31: 109–111.

———. 1966. *Systems analysis in ecology.* New York: Academic Press.

———. 1973. *Principles of environmental science.* New York: McGraw-Hill.

Weaver, J. E., and F. E. Clements. 1929. *Plant ecology.* 2nd ed., 1938, New York: McGraw-Hill.

Wedin, D. A., and D. Tilman. 1996. Influence of nitrogen loading and species composition on the carbon balance of grasslands. *Science* 274: 1720–1723.

Wein, G., S. Dyer, R. C. Tuckfield, and P. Fledderman. 2001. *Cesium-137 in deer on the Savannah River site.* Technical Report WSRC-RP-2001-4211. Aiken, S. C.: Westinghouse Savannah River Company.

Welch, H. 1967. *Energy flow through the major macroscopic components of an aquatic ecosystem.* Ph.D. Dissertation. Athens: University of Georgia.

Wellington, W. G. 1957. Individual differences as a factor in population dynamics: The development of a problem. *Canadian Journal of Zoology* 35: 293–323.

———. 1960. Qualitative changes in natural populations during changes in abundance. *Canadian Journal of Zoology* 38: 289–314.

Werner, E. E., and D. J. Hall. 1974. Optimal foraging and size selection of prey by bluegill sunfish. *Ecology* 55: 1042–1052.

Wesson, R. G. 1991. *Beyond natural selection.* Cambridge: Massachusetts Institute of Technology Press.

West, G. B., J. H. Brown, and B. J. Enquist. 1999. The fourth dimension of life: Fractal geometry and allometric scaling of organisms. *Science* 284: 1677–1679.

Whelan, R. W. 1995. *The ecology of fire.* Cambridge Studies in Ecology. Melbourne: Cambridge University Press.

White, L. 1980. The ecology of our science. *Science* 80: 72–76.

White, R. V. 2002. Earth's biggest "whodunnit": Unravelling the clues in the case of the end-Permian mass extinction. *Philosophical Transactions of the Royal Society of London* A 360(1801): 2963–2985.

Whitehead, F. H. 1957. Productivity in alpine vegetation. *Journal of Animal Ecology* 26: 241.

Whittaker, R. H. 1951. A criticism of the plant association and climatic climax concepts. *Northwest Science* 25: 17–31.

———. 1952. Vegetation of the Great Smoky Mountains. *Ecological Monographs* 26: 1–80.

———. 1960. Vegetation of the Siskiyou Mountains, Oregon and California. *Ecological Monographs* 30: 279–338.

———. 1967. Gradient analysis of vegetation. *Biological Review* 42:207–264.

———. 1975. *Communities and ecosystems,* 2nd ed. New York: Macmillan.

Whittaker, R. H., and P. P. Feeny. 1971. Allelechemics: Chemical interaction between species. *Science* 171:757–770.

Whittaker, R. H., and G. E. Likens, Eds. 1973. The primary production of the biosphere. *Human Ecology* 1:301–369.

Whittaker, R. H., and G. M. Woodwell. 1969. Structure, production, and diversity of the oak-pine forest at Brookhaven, New York. *Journal of Ecology* 57:155–174.

———. 1972. Evolution of natural communities. In *Ecosystem structure and function,* J. A. Wiens, Ed. Corvallis: Oregon State University Press, pp. 137–156.

Wiegert, R. G. 1974. Competition: A theory based on realistic, general equations of population growth. *Science* 185:539–542.

Wiener, N. 1948. *Cybernetics: Or control and communication in the animal and the machine.* New York: The Technology Press, John Wiley.

Wiens, J. A. 1992. What is landscape ecology, really? *Landscape Ecology* 7:149–150.

Wilcox, B. A. 1984. In situ conservation of genetic resources: Determinants of minimum area requirements. In *National parks: Conservation and development,* J. A. Neeley and K. R. Miller, Eds. Washington, D.C.: Smithsonian Institution Press, pp. 639–647.

Wilhm, J. L. 1967. Comparison of some diversity indices applied to populations of benthic macroinvertebrates in a stream receiving organic wastes. *Journal of Water Pollution Control Federation* 39:1673–1683.

Williams, C. M. 1967. Third-generation pesticides. *Scientific American* 217:13–17.

Williamson, M. H. 1981. *Island populations.* Oxford: Oxford University Press.

Wilson, C. L. 1979. Nuclear energy: What went wrong? *Bulletin of Atomic Science* 35: 13–17.

Wilson, D. S. 1975. Evolution on the level of communities. *Science* 192:1358–1360.

———. 1977. Structured genes and the evolution of group-advantageous traits. *American Naturalist* 111:157–185.

———. 1980. *The natural selection of populations and communities.* Menlo Park, Calif., and Reading, Mass.: Benjamin/Cummings.

———. 1986. Adaptive indirect effects. In *Community ecology,* J. Diamond and T. J. Case, Eds. New York: Harper and Row, pp. 437–444.

Wilson, E. O. 1973. Group selection and its significance for ecology. *BioScience* 23:631–638.

———. 1980. Caste and division of labor in leaf-cutter ants, Hymenoptera: Formicidae: *Atta,* I: The overall pattern in *A. sexdens. Behavioral Ecology and Sociobiology* 7:143–156.

———. 1987. The arboreal ant fauna of Peruvian Amazon forests: A first assessment. *Biotropica* 2:245–251.

———. 1988. The current state of biological diversity. In *Biodiversity,* E. O. Wilson, Ed. Washington, D.C.: National Academy Press, pp. 3–18.

———. 1998. *Consilience: The unity of knowledge.* New York: Vintage Books.

———. 1999. *The diversity of life.* New York: W. W. Norton.

———. 2002. *The future of life.* New York: Vintage Books.

Wilson, E. O., and E. O. Willis. 1975. Applied biogeography. In *Ecology and evolution of communities,* M. L. Cody and J. M. Diamond, Eds. Cambridge, Mass.: Harvard University Press, pp. 522–534.

Wilson, J. T., Ed. 1972. *Continents adrift: Readings from Scientific American.* San Francisco: W. H. Freeman.

Wilson, S. D., and D. Tilman. 1991. Components of plant competition along an experimental gradient of nitrogen availability. *Ecology* 72: 1050–1065.

———. 1993. Plant competition and resource availability in response to disturbance and fertilization. *Ecology* 74:599–611.

Winemiller, K. O., and E. R. Pianka. 1990. Organization in natural assemblages of desert lizards and tropical fishes. *Ecological Monographs* 60:27–55.

Winterhalder, K., A. F. Clewell, and J. Aronson. 2004. Values and science in ecological restoration—A response to Davis and Slobodkin. *Restoration Ecology* 12:4–7.

Witherspoon, J. P. 1965. Radiation damage to forest surrounding an unshielded fast reactor. *Health Physics* 11:1637–1642.

———. 1969. Radiosensitivity of forest tree species to acute fast neutron radiation. In *Proceedings of the second national symposium on radioecology,* D. J. Nelson and F. C. Evans, Eds. Springfield, Va.: Clearinghouse of the Federal Science Technical Information Center, pp. 120–126.

Wolfanger, L. A. 1930. *The major soil divisions of the United States: A pedologic-geographic survey.* New York: John Wiley.

Wolff, J. O. 1986. The effects of food on midsummer demography of white-footed mice, *Peromyscus leucopus. Canadian Journal of Zoology* 64:855–858.

Wolfgang, L., W. Sanderson, and S. Scherbov. 2001. The end of world population growth. *Nature* 412:543–545.

Woodruff, L. L. 1912. Observations on the origin and sequence of the protozoan fauna of hay infusions. *Journal of Experimental Zoology* 12:205–264.

Woodwell, G. M. 1962. Effects of ionizing radiation on terrestrial ecosystems. *Science* 138: 572–577.

———, Ed. 1965. *Ecological effects of nuclear war,* Number 917. Upton, N.Y.: Brookhaven National Laboratory.

———. 1967. Toxic substances and ecological cycles. *Scientific American* 216:24–31.

———. 1977. Recycling sewage through plant communities. *American Scientist* 65:556–562.

———. 1992. When succession fails. In *Ecosystem rehabilitation: Preamble to sustainable development,* Volume 1, M. K. Wali, Ed. The Hague: SPB, pp. 27–35.

Woodwell, G. M., and R. H. Whittaker. 1968. Primary production in terrestrial communities. *American Zoologist* 8:19–30.

Woodwell, G. M., C. F. Wurster, and P. A. Isaacson. 1967. DDT residues in an East Coast estuary: A case of biological concentration of a persistent insecticide. *Science* 156:821–824.

Wynne-Edwards, V. C. 1962. *Animal dispersion in relation to social behavior.* New York: Hafner.

———. 1965. Self-regulating systems in populations of animals. *Science* 147:1543–1548.

Z

Zewail, A. H. 2001. Science for the have-nots: Developed and developing nations can build better partnerships. *Nature* 410:741.

Credits

This page constitutes an extension of the copyright page. We have made every effort to trace the ownership of all copyrighted material and to secure permission from copyright holders. In the event of any question arising as to the use of any material, we will be pleased to make the necessary corrections in future printings. Thanks are due to the following for permission to use the material indicated.

Chapter 1. 3: © NASA

Chapter 2. 32: left, Courtesy of Wayne T. Swank **32:** right, Courtesy of the U.S. Forest Service **35:** left, Courtesy of Nicholas Rodenhouse **35:** right, Courtesy of Soil Conservation Service **43:** © Tim McKenna/CORBIS **46:** top, Courtesy of Gary W. Barrett **46:** bottom, Courtesy of Gary W. Barrett **63:** top, Courtesy of J. Whitfield Gibbons **63:** bottom, Courtesy of Gary W. Barrett **66:** top, Courtesy of Space Biosphere Ventures **66:** bottom, Courtesy of Space Biosphere Ventures

Chapter 3. 104: Courtesy of Nicholas Rodenhouse **116:** © Michael and Patricia Fogden/CORBIS **131:** Courtesy of Gary W. Barrett

Chapter 4. 147: © Wally Eberhart/Visuals Unlimited **166:** © Farrell Grehan/CORBIS

Chapter 5. 188: © Niall Benvie/CORBIS **192:** Courtesy of Gary W. Barrett **195:** top, © Raymond Gehman/CORBIS **195:** center, Courtesy of the Joseph W. Jones Ecological Research Center at Ichauway **195:** bottom, Courtesy of the Joseph W. Jones Ecological Research Center at Ichauway **198:** top, Courtesy of Donald W. Kaufman **198:** bottom, Courtesy of Gary W. Barrett **215:** Courtesy of Terry L. Barrett **219:** Courtesy of Gary W. Barrett

Chapter 6. 249: left, © The American Society of Mammalogists **249:** right, © The American Society of Mammalogists **262:** © Tom Tietz/Getty Images **277:** Courtesy of Gary W. Barrett **278:** Courtesy of the Florida Fish and Wildlife Conservation Commission **279:** left, Courtesy of Gary W. Barrett **279:** right, Courtesy of the Florida Fish and Wildlife Conservation Commission

Chapter 7. 294: Courtesy of Terry L. Barrett **299:** © Cleve Hickman, Jr./Visuals Unlimited **300:** Courtesy of Walter Carson **303:** top, Courtesy of C. H. Muller **303:** bottom, Courtesy of C. H. Muller **310:** Courtesy of S. A. Wilde, University of Wisconsin **332:** Courtesy of Steven J. Harper **333:** © Dan Guravich/CORBIS

Chapter 8. 339: left, Courtesy of Gary W. Barrett **339:** right, Courtesy of Gary W. Barrett **343:** © Royalty-Free/CORBIS **344:** Courtesy of D. G. Sprugel and F. H. Bormann **349:** Courtesy of Terry L. Barrett **360:** © The Visual Communications Unit at Rothamsted

Chapter 9. 378: top left, Courtesy of Gary W. Barrett **378:** top right, Courtesy of Gary W. Barrett **379:** Courtesy of the Woodland Cemetery and Arboretum Archives **380:** Courtesy of Gary W. Barrett **381:** left, Courtesy of Gary W. Barrett **381:** right, Courtesy of Gary W. Barrett **382:** top left, © Joe Sohm/Visions Of America, LLC/ PictureQuest **382:** top right, Courtesy of Nicholas Rodenhouse **382:** bottom left, © University Archives, NDSU, Fargo **383:** top left, Courtesy of Gary W. Barrett **383:** right, Courtesy of Nicholas Rodenhouse **383:** bottom left, Courtesy of Gary W. Barrett **385:** top, Courtesy of Jerry O. Wolff, University of Memphis **385:** bottom, Courtesy of Gary W. Barrett, Ecology Research Center, Miami, University of Ohio **400:** left, Courtesy of Gary W. Barrett **400:** right, Courtesy of Terry L. Barrett **402:** Courtesy of Gary W. Barrett **407:** Courtesy of Terry L. Barrett **409:** © The Atlanta Journal-Constitution

Chapter 10. 416: top, Courtesy of Woods Hole Oceanographic Institution and D. M. Owen **416:** bottom, Courtesy of Woods Hole Oceanographic Institution and D. M. Owen **421:** top, © Terry Donnelly/Getty Images **421:** bottom, Courtesy of Lawrence Pomeroy **423:** left, Courtesy of Frank B. Golley **423:** right, © Cousteau Society/Getty Images **425:** bottom left, Courtesy of Gary W. Barrett **425:** bottom right, Courtesy of Gary W. Barrett **430:** top, Courtesy of Eugene P. Odum **430:** bottom, © East Momatiuk/National Geographic/Getty Images **431:** top left, Courtesy of Lawrence Pomeroy **431:** top right, Courtesy of Lawrence Pomeroy **431:** bottom right, Courtesy of Lawrence Pomeroy **437:** top, Courtesy of R. E. Shanks, E. E. Clebsch, and J. Koranda **437:** bottom, Courtesy of R. E. Shanks, E. E. Clebsch, and J. Koranda **440:** left, Courtesy of the U.S. Forest Service **440:** top right, Courtesy of the U.S. Forest Service **440:** bottom right, © Darrell Gulin/CORBIS **441:** left, Courtesy of Gary W. Barrett **441:** top right, Courtesy of Eugene P. Odum **441:** bottom right, Courtesy of Tall Timbers Research Station **444:** Courtesy of Donald W. Kaufman **445:** Courtesy of the Joseph W. Jones Ecological Research Center at Ichauway **446:** top, © Donald I. Ker, Ker and Downey Safaris Limited, Nairobi, East Africa **446:** bottom, Courtesy of Eugene P. Odum **447:** © Rich Reid/National Geographic/Getty Images **449:** top, Courtesy of Eugene P. Odum **449:** bottom, Courtesy of R. H. Chew **451:** Courtesy of Carl F. Jordan **453:** top, © Adalberto Rios Szalay/Sexto Sol/ Getty Images **453:** bottom, © Paul Edmondson/Getty Images **454:** © Chris Hellier/CORBIS **456:** top, Courtesy of Thomas Luhring **456:** bottom, Courtesy of Thomas Luhring

Chapter 11. 471: © Hulton-Deutsch Collection/CORBIS

Index

Page numbers in italics refer to figures and tables.

A

Abiotic substances, 27
Absolute humidity, 206
Academic disciplines, 474, *475*
  publishing, 496–498
Acid rain
  impact, 153
  sulfur dioxide, 153
Acute stress, 219–220
Acyclic, 142
Adaptation
  natural selection, 276
  time needed, 370
Adreno-pituitary feedback hypothesis, 253
Aerobic respiration, 51, 52
Aesthetics, 467
Age distribution, 233–235
Age pyramids, *234*, 235
Aggregation, Allee principle of, 260–261, *262*
Agricultural production, Green Revolution, 88, *90*, 90–91
Agricultural productivity, energy inputs, 91–93
Agriculture
  carrying capacity, 129–130
  hierarchical nature of systems, 396–397, *398*
  underdeveloped countries, 471
  See also Agroecosystems
Agroecosystems, 33
  as climaxes, 359
  as domesticated ecosystems, 405
  genetic engineering, 367–368
  irrigated deserts, 448
  long-term experiments, 359, *360*
  natural ecosystems compared, 457
  phases, 405, *406*, 407–408
  terrace farming, *454*, 454
  traditional practices, 457
Agroforestry, 408, 457
Agrolandscape perspective, 408
A horizon (topsoil), 188
Air pollution
  augmentative synergism, 221
  nitrogen and sulfur cycles, 151, 153
Albinism, 276, 277
Allee principle of aggregation, 260–261, *262*
Alleles, 276
Allelopathic substances, 59
Allelopathy, 59, 296
  examples, 302–303
Allocation of energy. *See* Energy partitioning
Allochthonous inputs, 47
Allogenic forces, long-term evolution, 360
Allogenic succession, 337
  autogenic compared with allogenic influences, 342–345
Allopatric speciation, 364, 365
Alluvial deposits, 189
Alpha diversity, 321–322, 325
Alpha particles, *202*, 202
Alpine tundra, 438
Alternative agriculture, 33, *35*, 37, *406*, 407
Alternative hypothesis, 494–495
Altruism, 289
Altruistic behaviors, 278–279
Amensalism, 283, *284*, *285*, 302
Anaerobic, 29
Anaerobic respiration, 51, 52
Answerable questions, 496
Anthropogenic stress, 219–223
Anthropogenic subclimax, 359
Anthrosol, 191
Apparent plants, *304*, 304
Apportionment, 38
  species diversity, 316–317
Aquatic ecosystem types, 76
Aquifers, 158
  groundwater storage, 209
  Ogallala aquifer, 158, 210
  overdrafts, 210
Artesian wells, 209
Artificial selection, 366–367
Ascendancy, 355
Assimilation efficiencies, 109, *110*
Associations, 432
Asymptotic certainty, 498
Atmosphere
  Earth and Mars/Venus compared, *44*, 44
  evolution, *361*, 361–364
Atmospheric gases, 212–213
Atomic energy, age of, 136
Augmentative synergism, 221
Autochthonous production, 47
Autogenic forces, 360
Autogenic succession, 337
  allogenic compared with autogenic influences, 342–345
  climax, 358

Autogenic succession (*continued*)
ecological succession tabular model, *338*, 339–340
Autotrophic stratum, 21–22, 85
Autotrophic succession, 337
Auxiliary energy flow, 91

B

Background radiation, 203
Barnacle model, 293–294, *294*
Barrett, G. W., 475
Beech Creek Reserve, *410*, 411
Benthic microflora, 422
Benthic zones, 426
Benthos, 28
Beta diversity, 322, 326
Beta particles, *202*, 202
Beyond the Limits (Meadows et al. 1992), 463
B horizon soil, 188–189
Big Government, 474
Biochemical oxygen demand (BOD), 29
Biodiversity
affected by pollution, 318–320
community and landscape levels, 386–389
concern about loss, 324–326
productivity, 100–101, 323–324
stability, 320–322
See also Diversity
Bioenergetics, ecosystem development, 340
Bioethics, *473*, 473
Biofixation, 143–148
Biogeochemical cycles, 141
basic types, 141–143
cycling of carbon, 153–156
cycling of nitrogen, 143–149
cycling of nonessential elements, 168–169
cycling of phosphorus, 149–151
cycling of sulfur, 151–153
global climate change, 175–176
hydrologic cycle, 156–162
nutrient cycling in the Tropics, 170–172
recycling pathways, 172–175
turnover and residence times, 162–163
watershed biogeochemistry, 163–168
Biogeochemistry, 141
Biogeological clock, 361, *362*
Biological attributes, 225
Biological clocks, 185
Biological magnification, 168
radioactive substances, 204
toxic substances, 215–218
Biological oxygen demand (BOD), 54
Biomass, productivity vs., 88
Biomes, 6
concept of, 432
defined, 432
listed, *413*
See also Terrestrial biomes
Biosphere-2, 60, 65, *66*, 67
Biosphere, evolution of, 360–364
Biosystems, *4*, 5
Biotic community, 18
defined, 225
Biotic potential
defined, 239
reproductive potential, 239–240
Birth rates, 8, 229
Black box, 20
Bleaching, 423
BOD. *See* Biochemical oxygen demand; Biological oxygen demand
Bog wetlands, 428–429
Bottom-up controls, *120*, 120–121
Bottom-up hypothesis, 119–120
Box-and-whisker plots, 498, *499*, 501
British thermal unit (BTU), *80*
Brundtland Report, 467–468
BTU. *See* British thermal unit
Budworm cycles, 250–251
Buried seed strategy, 167, 220

C

C_3 pentose phosphate cycle, 48
C_3 plants, 48, *49*, 49–50
C_4 dicarboxylic acid cycle, 48
C_4 plants, 48–50, *49*, *50*
Cadmium (Cd), biological magnification, *218*, 218
Calcium cycle, 164–166, *165*, *166*
quantitative model, 164, *165*
Calculus, 237
Calorie, *81*
Calvin cycle, 48
CAM. *See* Crassulacean acid metabolism
Carbon cycle, 143, 153–156
global carbon cycle, *155*
reservoir sizes and turnover times, *144*
Carbon dioxide, aquatic systems, 212
Carnivores, 28
Carrying capacity, 127–130
concept of, 129–130, 241–246
maximum, 241, *244*, 244
Caves, 456–457
Cesium-137, *205*, 205–206
impact, 169
Chaparral, 446–447
Character displacement, 295
speciation, 365
Chelation, 57
Chemolithotrophs, 50
Chemosynthetic bacteria, 50, 420
Chestnut blight, 298, *299*, 320
China, 477
Chitty-Krebs genetic feedback hypothesis, *253*, 253
Chlorophyll, 101
C horizon soil, 189
Chronic disturbances, 219
Chronic stress, 219, 220
Circadian rhythm, 185
Cities
Allee principle of aggregation, 263
ecological footprint, 74–75
network law, 372
as parasites, 73–74
population growth, 245
urban-industrial technoecosystems, 408–411
Civilizations, 464
Clark, William, 473–474
Climate change. *See* Global climate change
Climate regime, 22
Climatic climax, 356, *357*, 358
vegetation, 433
Climax, 337
community metabolism, 337
concept of, 356–360
Climax community, 356, *357*
Climographs, 211–212
Cloud forests, 452
Club of Rome, 462–463
Clumping, *258*, 258–259
aggregation, 260–261
Coarse-grained habitats, 379
Coarse particulate organic matter (CPOM), 57
river continuum, 159, *160–161*
Coastal Chaparral, 446–447
Coevolution, 286–288, 361

Collective properties, 8
Colluvial deposits, 189
Commensalism, 283, *284*, *285*
as first step, 305–306
Commons, 461–462
Community
coevolution, 286–288
two-species relationship, 283–286
use of term, 5
Community metabolism, measuring, 28–31
Community selection, evolution, 361
Compartment models, *12*, 13
for land use, *369*, 369–370, *372*, 372–373
Compensation depth, 29, 426
Competition
direct interference type, 283, *284*, 289–290, 293
exploitation, 289–290, 293
growth rate, 284–285
habitat selection, *295*, 295–296
interspecific competition and co-existence, 289–296
resource use type, 283, *284*
two-species interactions, *285*
use of term, 290
Competitive exclusion principle, 290, *291*, 293
Complexity theory, 126–127
Computer-simulated models, 11
Concentration factors, radioactive substances, 204–206, *205*, *205*
Confounding, 486
Conservation biology, 458
Conservation easements, 467
Conservation ecology, 458
intent of, 506
two cultures compared, *507*
Conservation Reserve Program (CRP), 192–193
Consumer organisms, 28
Consumers
density and biomass compared, *30*
pond, 28
Continental climates, 211
Continental drift, 364, 418
Continental shelf, 415–418, *419*
Continuum analysis, 386, *387*
Control, 481
Controlled experimental design, 488
Control processes, 9
Conventional agriculture
forest removal compared, 170
tillage, 308–309, *309*
See also Industrial agriculture
Cool deserts, 448–449
Cooperation
ecological assessment, 464
group selection, 288–289
Cooperative breeding, 278–279
Copper, biological magnification, *218*
Coprophagy, 116, 307
detritivores, 57–58
Coral reefs, 422–423, *423*
Core-satellite species hypothesis, 38
Core species, 38
Corpuscular radiation, 202
Corroboration, 492
Cosmic rays, 202–203
Costanza, Robert, 474
Cotton cultivation and pesticides, 223
Countertechnology, 470
Covey, Stephen R., 466
CPOM. *See* Coarse particulate organic matter
Crassulacean acid metabolism (CAM), 50
Crossdisciplinary approaches, 16
Crown fires, 194, *195*, 196
CRP. *See* Conservation Reserve Program
Crude density, 226
Cryptic plants, *304*, 304
Cultivation, 366–367
Cultural eutrophication, 31, 407
Culture of permanence, 475
Cyanobacteria, 48
Cybernetics, 9, 67
elementary principles, 67–68, *68*
science of, 68–71
Cyclic climax, 344
Cyclic hierarchy concept, 472
Cyclic oscillations, *248*, 248–255
Cyclic species, 247
Cyclic succession, 344
Cycling index, 172, 173–175
Cycling pool, 141, 142

D

Darwin, Charles, 276
allopatric speciation, 365, *366*
Mendelian inheritance, 491
publish or "parish," 496–497
Data description, 498
Data display, 408, 509
Day length, 185–187, *186*
DDT, 91
biological magnification, *216*, 216–218, *217*
Deciduous forests, 440–442
Decomposer organisms, 28
Decomposition, 23
carbonaceous matter, *53*
four stages, 55–56
functions it controls, 59
Georgia salt marsh, *55*, 55
global production-decomposition balance, 59–60
human activities' impact, 60
molecular methods of study, 59
nitrogenous organic matter, *53*
overview, 54–59
types of, 51–52
Deductive inferences, 507–508
Deep sea hydrothermal vents, 418, *420*, 420
Deficiency diseases, 213–214
Delayed photosynthesis, 50
Demographic stochasticity, 394
Density dependent, 255–258
Density governing, 248
Density independent, 255–258
Deserts, 448–450
annuals, 187
Desulfovibrio, 52
Deterministic, 492
Detrital systems, 111
food chains, 114, *115*, 116
Detritivores, 28, 57–58
Detritus, 22
Detritus cycle, 173
Detritus food chain, 108
Detritus pathway, 173
Developed country, defined, 97–98
Developmental stages, 337
Dew, 208
Diminishing returns to scale, 126
Disciplinary approaches, 15–16
Disclimax, 359
Diseconomies of scale, 126–127
Dissolved organic matter (DOM), 22
Disturbance corridors, 381, *382*, 383
Diversity
above and below the species level, 322–323
agro-urban landscape, *326*, 326
ecosystem development, 352–354

Diversity (*continued*)
habitat or region, 321–322
See also Biodiversity
Diversity-productivity hypothesis, 323
Diversity-stability hypothesis, 323
Dogwood (*Cornus florida*), 164–166, *166*
DOM (dissolved organic matter), 22
Domesticated landscapes, 404–411
Domestication, 366–367
Domestic ecosystem types, 76
Dominance-diversity curves, 38
ecosystem development, *353*, 353
metacommunity, 393–394, *394*
Dominance-diversity profiles, Shannon-Weaver index, *318*
Dominant age class, 235, *236*
Dominant species, 38, 316
Dominion versus stewardship, 466
Doubling time, 240
Drainage basins, 163
Dual capitalism, 469
Dust Bowl of 1930s, *471*, 471

E

Earth Day, 4
Earthscape, *3*, 3
Earth Summits, 468
EBPM. *See* Ecologically based pest management
Ecological assessment, 464
Ecological density, 226
Ecological economics, 2, 139
Ecological efficiencies, 118–119
ratios, *118*
theory examined, 493–494
Ecological equivalents, 312, 314–315, *315*
Ecological footprint
biogeochemical cycles, 143
concept of, 74–75
Ecologically based pest management (EBPM)
cotton yields, 223
density-dependent controls, 256
Ecological niche, 311–313
Ecological pyramids, *102*, 102–103
Ecological or realized mortality, 229
Ecological or realized natality, 229
Ecological study unit (ESU), 484–486
Ecological succession, 337
aquatic habitats, 349–351
autogenic type, model, *338*, 339–340
comparison of laboratory microcosm and forest, 340–342, *342*
general pattern, *348*
models, 354–355
when succession fails, 371
Ecological system, 18
Ecology
concept of, 2
defined, 3
early history, 2, 3
economics compared, 2
as profession, 496–497
roots, 475, *476*
streams, 506
Economic capital
defined, 10
tragedy of the commons, 462
Economic carrying capacity, 130
Economic defensibility, hypothesis of, 266
Economic development, 468
Third World, 465–466
Economic growth, 468
Economics
concept of, 2
dual capitalism, 469
ecology compared, 2
holistic economists, 469
holoeconomics, 469
nature as value, 2–3
Economic welfare threshold, 130
Economies of scale, 126–127
Ecoregions, 18
defined, 432–433
Ecosphere, 6
Ecosystem cybernetics, 67–71
Ecosystem development, 337–356
bioenergetics, 340
defined, 337
diversity trends, 352–354
energy flow model, 340, *341*
heterotrophic succession, 351–352, *352*
nutrient cycling, *345*, 345–346
production/respiration (*P/R*) model, 340, *341*
relevance to human ecology, 368–373
selection pressure, 352
species replacement, 346–351
systems (cybernetic) model, 340, *341*
theoretical considerations reviewed, 354–355
Ecosystem diversity, 37–41
defined, 37
Ecosystem ecology, 19
Ecosystem health, 45–46
Ecosystem management, 18
Ecosystems
classes of energy source and level, *133*, 133
classification, 75–76
cybernetics, 67–71
defined, 18
ecological footprint, 74–75
ecosystem diversity, 37–41
energy-based classification, 132–135
energy budgets, 56
examples, 26–37
agroecosystems, 32–37
pond and old field, 27–31
watershed concept, 31–32
explanation, 18–21
functional diagram, *20*
Gaia hypothesis, 43–46
gradients and ecotones, 24–26
graphic model, *19*, 20
major types, *75*
microcosms, mesocosms, and macrocosms, 60–67
as open systems, 19
study of, 42–43
technoecosystems, 71–74
trophic structure, 21–24
types listed, *413*
use of term, 5
Ecosystem science, 480
Ecotones, 24–25, 388
defined, 386
increase in density, 389
Ecotopia, 474
Ecotypes, 183–185
Ectomycorrhizae, 308
Ectosymbiotic association, 307
Edaphic climaxes, 356, *357*, 358, *359*
Edge effects, 388–389
defined, 386
diversity, 323
ecotones, 24

Edge habitats, 403
Edges, defined, 386
Edge species, 24
defined, 386
Effective population size, 277
Effect size, 483
Electromagnetic radiation, 202
Embodied energy, 123
Emergence, 355
Emergent property principle, 7–9
eMergy, 121–124
origin of concept, 123
solar vs. fossil fuel, *123*
Emigration, 246
Endemic, 50
Endosymbiotic association, 307
Energetics of scale, 126–127
Energy, 19
defined, 78
degradation of, 78
functional classification, 75
history of civilization, 135–136
money and, 137–139
prosperity and shortages, 474–475
Energy cost, 132
Energy drains, 87, 92
subsidy-stress gradient, 93–94
Energy flow
biogeochemical cycle superimposed, *142*
defined, 82
energy support systems for humans, 137, *138*
money flows, 137, *138*
oak leaf example, *79*
red fox, 106–107, *107*
as suitable index, 103
universal model, 105–107
Y-shaped diagram, 109–110, *110*
Energy optimization, 269
Energy partitioning, 268–270, *270, 273, 274*
food chains and food webs, 108–121
Energy penalty, 132
Energy quality (concentration), 121–124, *122*
Energy subsidy, 91–93
rates of production, 87
Entropy, 78, 80
Environmental chambers, 212
Environmentalism, 477
Environmental resistance, 239
models to explain, 243
Environmental resistance phase, 245
Eolian deposits, 189
Ephemerals, 187
Epifauna, 415
Epilimnion, 426
Erosion, soil, 191–193
Error acceptance criterion, 483
Essential elements, 141
Estuaries and seashores, 420–422, *421*
ESU. *See* Ecological study unit
Ethics, 467, 473
Eureplicates, 483
Eury-, use of, 180
Eurythermal, 180, *181*
Eusociality, 278
Eutrophication, 118, 427
Evaluation goal, 506
Evapotranspiration, 207
Everglades, 32, 150
Evidence, 492–494
defined, 481
hypothesis testing, 494–495
See also Weight of evidence
Evolution
anaerobic world, 54
biosphere, 360–364
defined, 276
effect on atmosphere, *361,* 361–364
Excessive edge, 389
Exchangeable nutrients, 142
Existence energy, 269
Experimental design, 480, 485
Experimental versus observational method, 489–490
Experiment units, 481, 487
Exploitation competition, 289–290, 293
Extrinsic factors, in population fluctuations, 247, 252

F

Fabricated environments, *369,* 373
Facilitation model, 354
Factor compensation, 183–185
Factor interaction, 178
Facultative anaerobes, 48
Facultative cooperation, 283, *284*
Falsification approach, 492–493
Feasibility, 488
Feedback, 68
local feedback, 461
Feedback control systems, 67–71, *68*
Feedback loops, 11, *12*
control mechanisms, 12
Fen wetlands, 428
Feral organisms, 367
Fermentation, 52
Filter effects, 383
Fine-grained habitats, 379
Fine particulate organic matter (FPOM), river continuum, 159, *160–161,* 161
Fire
grasslands, 444
periodicity, 370
prescribed burning, *195,* 196, 197, *198, 445*
wildfires, 194
Fire climaxes, 370
Fire ecology, 194–199
Fire-type ecosystems, 54
First law of thermodynamics, 78, 79
Fishing industry, 302
mariculture, 415
Fish kill, acute stress, *219,* 219–220
Fish parks, 417
Fission products, 205
Fitness, 280
Floaters, 266, *267*
Flooding, 214–215
Flood irrigation, 36
Floodplains, 161
Flood pulse concept, 161
Flow pathways, 11
Fluctuating water-level ecosystems, 370
Fog, 208
Food chains, 108, 111
anaerobic pathway, *112,* 112
branched model, *111,* 111–112
detritus food web, 114, *115,* 116
detritus pathway, *110*
energy flow diagram, *106*
isotopic tracers as aids, 121
length, 117–118
sport-fishing ponds, *113,* 113–114
Food crops, primary production, 96–99
Food quality hypothesis, *253,* 253–254
Food-to-fiber ratio, 88

Food webs, *108,* 108
  bottom-up control, 311
  interaction of bottom-up and top-down control of energy flow, 119–121
  role of heterotrophs, 117–118
  top-down control, 311
Foot-pound, *81*
Forbs, 443
Forcing function, 11
Forest edge, 405
Forested wetlands, *430,* 430–431
Forests
  agroforestry, 408, 457
  clear cuts, 442–443
  fire ecology, 194–199
  harvesting, 442–443
  litter fall, *189*
  northern coniferous forest biomes, 439–440
  old-growth, 442
  soils, *189,* 190
  temperate deciduous forests, 440–442
  tropical, 170–172
    rain forests, 450–454
  tropical scrub or thornwoods, 454
  tropical seasonal forests, 450
  wood as fuel, 99–100
Forest succession, 355
Fossil fuels
  age of, 135–136
  energy content, *81, 82*
Fossil pollen, 328–329, *329*
Fossil record, 327–329
FPOM. *See* Fine particulate organic matter
Frequency of occurrence, 226, 227
Frequency polygon, 259
Freshwater ecosystems, 424–432
  forested wetlands, *430,* 430–431
  freshwater wetlands, 424, 428–430
  lentic ecosystems, 424, *425,* 426–427
  lotic ecosystems, 424, 427–428
  tidal freshwater marshes, *431,* 431–432
Freshwater wetlands, 424, 428–430
  microbial decomposition and recycling, *429*
Fuel-powered ecosystem, *133,* 135
Fuel-powered systems, 133
Functional schemes, ecosystems, 75
Fundamental niche, 313

G

Gaia hypothesis, 43–46
Gallery forests, 452
Gamma diversity, 322, 326
Gamma rays, *202,* 202
  effects studied, 204
Gap phase succession, 343
Gaps, ecological-societal, 465–467, 473–474
Gaseous biogeochemical cycles, 141
Gause principle, 290, *291*
Gcal. *See* Gram-calorie
Generalists, foraging, 271
Gene resources, 452
Genes, 276
Genetic diversity, 316
Genetic drift, 277
Genetic engineering, agroecosystems, 367–368
Genetic feedback, coevolution, 287
Genetic fixation, 183
Genetics. *See* Population genetics
Genotype frequencies, 277
Genotypes, 276–277
Geochemistry, 141
Geographic approach, 413
Gizzard shad, 331–333, *332*
Glacial deposits, 189
Global climate change, 45, 175–176
  average mean temperature, *176*
Global models, 462–465
Global production and decomposition, ecosystems, 46–60
Global sustainability, 467–472
Global warming
  evidence, 480–481
  global climate change, 175–176, *176*
GMOs. *See* Genetic engineering
Gnotobiotic cultures, *61,* 61
GNP. *See* Gross national product
Good model definition, 13
GPP. *See* Gross primary productivity
Gradient analysis, 386, *387*
Gradient analysis approach, 386
Gradients, 24–26
Grain, 379
Gram-calorie (gcal), *81*
Granivorous food chain, 112
Graphics science, 501–503, *502, 503,* 509
Grassland biome, 443–445
  temperate grasslands, 443
  tropical grasslands and savannas, 445–446, *446*
Grassland soils, *189,* 190
Grazing
  secondary compounds, 287
  species diversity, 322
Grazing food chain, 108
Great Smoky Mountains National Park, *455,* 455–456, *456*
Greenhouse effect, 154
Greenhouse gases, 154
  global climate change, 175, 176
Greenhouses, for research, 212
Green Revolution, 88, *90,* 90–91, 469–470
Gross national product (GNP), 10
Gross primary productivity (GPP), 87
  variation by latitude, *90*
Groundwater, 209–211
  contamination, 210–211, 220
  depletion, 209–210
  freshwater ecosystems, 424
  U.S. distribution, *209*
Group attributes, 225
Group behavior, 256
Group selection
  cooperation, 288–289
  evolution, 361
Group survival value, 261–262
Growth curves, 237, *238*
Growth equation models, 284–285
Growth rate curves, *238,* 238
Growth rates, 237
Guilds, 312, 314
Guppy yields, *301,* 301

H

Haber process, 90
Habitat diversity, 316
Habitat loss, *403,* 403
Habitat niche, 312
Habitats, 311, 312
  types listed, *413*
Habitat selection, competition, *295,* 295–296
Hardy-Weinberg equilibrium law, 277
Harvest ratio, 88

Harvest ratios, 367
Heat energy, 79
Herbicides, calcium and nitrogen loss, 166–167
Herbivores, 28
mammalian, 434
Herbivory, 296
insect herbivory, 299–301, *300*
Heterotrophic ecosystems, *73*
Heterotrophic stratum, 22
Heterotrophic succession, 337
ecosystem development, 351–352, *352*
Hierarchy, 4
High tundra, 438
Holdridge life zone system, 432, *434*
Holistic approach, 329–330, *330*
ecological assessment, 464
Holistic economists, 469
Holistic science, 488
Holistic thinking, 8–9
transdisciplinary holism, *15*, 16
Holological approach, 42
Holoplankton, 415
Homeorhesis, *6*, 9, 68
Homeostasis, *6*, 9, 68
Homeostatic mechanisms, 68
Home range, 263–264
Homozygosity, 278
Horsepower (hp), *81*
Hot deserts, 448–449, *449*
Hp. *See* Horsepower
H. T. Odum energy language, 13
ecosystem model using, *14*
symbols used in this book, *14*
Human density, 100
Human ecology, 142
ecosystem development related, 368–373
Human ecosystems, 18
Human-subsidized solar-powered ecosystems, *133*, 134–135
Humic substances, 22, 23
Humidity, 206–207
Humification, 188
Humus (or humic substances), 55–56
chemical structure, 56, *57*
Hydrogen, aquatic systems, 212
Hydrological factors
principal factors, 206–208
temperature and moisture together, 211–212
Hydrologic cycle, 143, 156–162
energetics of, *157*
reservoir sizes and turnover times, *158*
Hydroperiod, 428, 429
Hydrothermal vents, 418, *420*, 420
Hypervolume niche, 312–313
Hypolimnion, 426
Hypothesis testing, and evidence, 494–495
Hypothetico-inductive inference method, 488

I

IBP. *See* International Biological Program
Ice algae, 438
Ice caps, 438–439
Identically distributed variables, 483
Immature soils, 190
Immigration, 246
corridors, 391–392, *392*
Importance percentage value, 228
Importance value, 226, 227
Impoundments, 427
Inbreeding, 277–278
Inclusive fitness, 279
Increasing returns to scale, 126
Independent variables, 483
Index of sustainable economic welfare (ISEW), GNP compared, 130, *131*
Individual organisms, respiration and weight, *125*
Individual organism size, 3/4 power law, 124–126
Inductive inferences, 488, 507
Industrial agriculture, 33, *34*, *35*, 36–37
as domesticated landscape, 405, *406*, 407
Georgia study, 398–399
history in U.S., *36*
history of, 407–408
soil erosion, 192
Industrial fishing, 302
Industrialized agriculture, as techno-ecosystem, 74
Industrial melanism, 366
Industrial pollution, rapid natural selection, 366
Industrial Revolution, 461–462
Industrial societies, anthropogenic stress, 219–223
Infauna, 415
Inference methods and reliability, 486–489
Inference-reliability tradeoff, 486, *487*
Inhibition model, 354
Inorganic substances, 22
Input environments, 19, 21
Input management, *470*, 470–471
Insect herbivory, 299–301, *300*
Insecticides, biological magnification, 216
Instantaneous coefficient of population growth, 239
Instantaneous growth rate, 237, 238
Integrated pest management (IPM)
cotton yields, 223
density-dependent controls, 256
Integrative science, 16, 475
Interaction functions, 11
Interactive approach, 330–331, *331*
Interdisciplinary approaches, 16
Interface fields of study, 10
Interference competition, 283, *284*, 289–290, 293
Intermediate disturbance hypothesis, 397, *399*
International Biological Program (IBP), 435
Interspecific competition, 289–296
Intrinsic factors, in population fluctuations, 247, 252–254
Intrinsic rate of natural increase, 238–240
Ionizing radiation, 201–206
IPM. *See* Integrated pest management
Irrigated deserts, 448
Irruptions, predator pressure, 297
Irruptive populations, 246, *247*, 247–248
ISEW. *See* Index of sustainable economic welfare
Island, defined, 392
Island biogeography theory, 389–392, *390*
neutral theory, 393
Island colonization, 352
Isolation, 264
Isotopic tracers, 121

J

Joule, *81*
J-shaped growth form, *241,* 241–242, *244*
  density independent or extrinsic factors, 256

K

Kcal. *See* Kilocalorie
Keystone species, 324
Kilocalorie (kcal), *81*
Kilowatt-hour (KWh), *81*
Kinetic energy, 79
Kin selection, 278
Knowledge, 495
K-selection, 269, 271–272, *273,* 274–275
KWh. *See* Kilowatt-hour

L

Labile pool, 141, 142
Lacustrine wetlands, [illegible]28
L[illegible] phase, [illegible]44
Lakes. *See* Lentic ecosystems
Landscape, defined, 6, 375
Landscape architecture, 400
Landscape corridors, 381–385
  basic types summarized, 377
  defined, 377
  landscape geometry, *401,* 401–402
  rate of immigration, 391–392, *392*
Landscape corridor width, *402*
Landscape ecology
  defined, 375–377
  history, 376
Landscape elements, 377–385
Landscape geochemistry, 141
Landscape geometry, 399–404
Landscape matrix, 377–378
  landscape geometry, *401,* 401–402
Landscape mosaic, 377
  role of edge habitats, 403
Landscape patches, 377–380, *378, 380, 381*
  experimental patch sizes/shapes, 385, *386, 400*
  landscape geometry, *401,* 401–402
  mesocosms, *63,* 64
  patch connectivity, *402,* 402
  streambed research, 503–504, *504*
Landscape patterns, 376
Landscape processes, 376
Landscape sharpening, 26
Landscape sustainability, concept of, 404
Land use, compartment models, *369,* 369–370, *372,* 372–373
Large-numbers systems, 335
Law of diminishing returns, 126–127
Law of entropy, 78
Lead (Pb), biological magnification, *218,* 218
Leaf-area index, 126
Lemming cycles, 248–250, *250*
Lentic ecosystems, 424, *425,* 426–427
Leopold, Aldo, 467
Levels of organization, 4
Levels-of-organization hierarchy, 4–7, *5*
  landscape ecology related, 375–377
Lichens, 309–310, *310*
Liebig law of the minimum, 178–183
Life, evolution of, anaerobic procaryotes, 54
Life-forms, 433
Life history, traits and tactics, 280–281
Life-support resources, 45
Life tables, mortality, 229–231, *230*
Life zones, 432
Light, as limiting factor, 200–201
Light saturation, 201
Lignin, 55, *56*
Limiting conditions, 178
Limiting factors
  atmospheric gases, 212–213
  concept of, 178–183
  ecophysiology, 199
  groundwater, 208–211
  ionizing radiations, 201–206
  light, 200–201
  macronutrients and micronutrients, 213–214
  temperature, 199–200
  temperature and moisture together, 211–212
  water, 206–208
  wind and flooding, 214–215
Limits of tolerance, 179–180
Limits to Growth, The (Meadows et al. 1972), 462
Limnetic zones, 29, 426
Lincoln index, 227–228
Littoral zones, 426
Local climaxes, 356, *357,* 358
Local feedback, 461
Locust cycles, 251–252
Loess, 189
Lognormal distribution, 395
Long-day plants, 186
Lotic ecosystems, 424, 427–428
Lotka-Volterra equations, 291
Low-input sustainable agriculture, 37, *406,* 407
Low-numbers systems, 335
Low tundra, 438
Lynx-hare oscillations, *248,* 248, *249*

M

Macroconsumers, 22, 23, 28
Macrocosms, 60
Macroevolution, 364
Macroflora, 422
Macronutrients and micronutrients, 213–214
Macrophytes, 27
Mad Max, 474
Main effect, 485
Maintenance energy, 107, 269
Mangroves, 422, *423*
Mariculture, 415
Marine climates, 211
Marine ecosystems, 414–423
  continental shelf, 415–418, *419*
  deep sea hydrothermal vents, 418, *420,* 420
  ocean, 414–415
  upwelling regions, 418
  See also Freshwater ecosystems
Mark-recapture methods, 227–228
Mature-die species, 199
Mature soils, 190
Maximum carrying capacity, 128, 241, *244,* 244
Maximum natality, 229
Maximum sustained yield, 128
Mean, 485, 509
Median, 501, 509
Megacities, 74
Mega-cyclic oscillations, 248–252
Meiofauna, 114, 116
Mercury, impact, 169
Merological approach, 42
Meroplankton, 415

Mesocosms, 60
  Biosphere-2, 65, *66*, 67
  experimental, 62, *63*, 64
  terrestrial, *63*, 63–64
Metabolic alliance, 307
Metabolism, size of individuals, 124–126
Metabolism (CAM), 50
Metacommunity, dominance diversity curves, 393–394, *394*
Metacosms, 485
Metallo-activators, 214
Metapopulations, 267–268, *268*
  immigration and emigration, 94
Methane bacteria, 52
Methane (CH_4), 156
Microconsumers, 22, 23
Microcosms
  aquaria, 62
  defined, 60
  gnotobiotic, *61*, 61
  types of laboratory, 61–62
Microevolution, 364
Micronutrients, 213–214
Middle-numbers systems, 335
Mid-oceanic ridges, 418, *420*
Migration, 246
Mineralization, 56, 188
Minimum known alive (MKA) method, 228
Minimum mortality, 229
Mississippi River, 208
MKA method. *See* Minimum known alive (MKA) method
Models
  computer-simulated, 11
  defined, 10–11
  elements of good definition, 13
  first steps in building, 11–13
  formal and informal, 11
  graphic, 11
Monarch butterfly, *333*, 333, *334*, 335
Money, energy, 137–139
Montane rain forests, 452
Mortality, 229–233
Mountain geoecology, 390
Mountaintop ice caps, 438
Mount Saint Helens, *46*
Multidimensional or hypervolume niche, 312–313
Multidisciplinary approaches, 16
Multiple-use strategies, 369–370
Mutualism, 283, *284*, *285*, 285
  detritus food chains, 307
  diverse organisms, 306
  network mutualism, 311
Mycorrhizae, 307–309, *309*, *310*

N

Natality, 229
Natural capital, 134, 404
  defined, 10
  tragedy of the commons, 462
Naturally subsidized solar-powered ecosystems, *133*, 134
Natural selection
  cooperation and mutualism vs., 305
  defined, 276
  population genetics, 275–280
Nature preserves, 391
Nectar food chain, 112
Negative feedback, 68
  ecological assessment, 464
−3/2 power law, *257*, 257
Nekton, 28
Nested hierarchies, 7
Net community productivity, 87
Net energy, *132*, 132, 269
Net plankton, 414
Net primary productivity (NPP), 87
  food crops, *97*, 97
  plant biomass estimates, *89*
Net radiometer, 86
Network complexity theory, 371–372
Network law, 372
Network mutualism, 311
Neutralism, 283, *284*, *285*
Neutral theory, 392–395
Niche
  spatial representation, *313*
  types of, 312–314
  uses of term, 313
Niche assembly theory, 394
Niche partitioning, 314
Niche width, 314
Nitrogenase, 147
Nitrogen cycle, 143–149
  complexities, *145*, *146*
  detrimental effects of too much nitrogen, 148–149
  reservoir sizes and turnover times, *144*
Nitrogen fixation, 146–148
Non-equilibrium ecosystems, 320
Nonessential elements, 141
  cycling of, 168–169
Nonnested hierarchies, 7
Nonparametric statistics, 259
Nonreducible property, 7
Noösphere, 71
Normality, 483
Northern coniferous forest biomes, 439–440
NPP. *See* Net primary productivity
Null hypotheses, 393, 494
Nutrient cycling, 141, *345*, 345–346
  Tropics, 170–172, 452
Nutrient enrichment
  biodiversity, 101
  See also Eutrophication

O

Obligate anaerobes, 48
Obligate symbiosis, 306. *See also* Mutualism
Observational versus experimental method, 489–490
Ocean, 414–415, *416*
Oceanography, 414
Ocean thermal energy conversion (OTEC), 136–137
Odum, E. C., 474–475
Odum, H. T., 13, 474–475
Ogallala aquifer, 158, 210
  See also Aquifers
Oligotrophic condition, 342, 427
Omnivores, 109
Optimal foraging, 270–271
  optimization cost-benefit models, *271*
Optimization cost-benefit models, *271*
Optimum carrying capacity, 128, 131, 244
Optimum density, 128
Ordination, 386, 387
Organic compounds, 22
Organic detritus, 22
Oscillating steady state, 363
OTEC. *See* Ocean thermal energy conversion
Outliers, 501
Output environments, 19, 21
Outwelling, 94
Overgrazing, 111

Overpopulation, tropical agriculture, 171
Oxygen, aquatic systems, 212

P

Paleoecologists, 327
Paleoecology, 327–329
Palustrine marshes, 429
Palustrine wetlands, 428–429
Panarchy, 472
Parametric statistical methods, 259
Parasite-host relationship, 297, *298*
Parasites, cities as, 73–74
Parasitism, 283, *284, 285,* 285–286
 types of relationships, 296–304
Parasitoid insects, 299
Parasitology, 285–286
Particulate organic matter, 22
Pastoralism, 36
Patch connectivity, *402,* 402
Patch-matrix relationships, 377–378, *378*
Pattern diversity, 316, 322–323
Peaceful coexistence, 305
Peck order, 262–263
Pedosphere, 187
Peer assessment, 506
Percentage growth rate, 237
Periodic flooding, 161
Permafrost, 438
Perturbation dependent, 345
Perturbation ecology, 42
Pesticides, immunities developed, 222–223
Phagotrophs, 22
Phanerozoic eon, 361
Phosphorus cycle, 143, 149–151
 model diagram, *150*
Phosphorus cycle, reservoir sizes and turnover times, *144*
Photic zones, 426
Photochemical smog, 153
 stress on system, 221
Photoperiod, 185–187, *186*
Photosynthesis
 forest systems, 22
 global production, 46–47
 kinds of, and producer organisms, 47–51
 origin, 362–363
Photosynthetic bacteria, 48
Photosynthetic organisms, 420
Phycosphere, 117
Phylogeography, 395
Phytoplankton, 27, 422
Phytotrons, 212
Picoplankton, 414
Pine looper insect, 250, *251*
Pioneer plant species, 337, 346–347
Pioneer stage, 337
Plankton, 414, 415
 freshwater, 426
Planted corridors, *383,* 383–384
Plants
 micronutrients necessary, 214
 patterns of world plant formations, 432, *433*
Plate tectonics, 364
Plotless methods of sampling, 228
Point of inflection, growth rate maximum, 238, *244,* 244
Poisson series, 259, *260,* 260
Pollution, 143
Pollution effects
 diversity, *318,* 318
 evenness, 318, *320*
 richness, 318, *319*
Polyclimax, 356
POM (particulate organic matter), 22
Ponds. *See* Lentic ecosystems
Population
 aggregation, 260–261, *262*
 annual fluctuations, 246–247
 China, 477
 defined, 225, 276
 density, biomass, and trophic relationships, 226–227
 effective population size, 277
 energy partitioning, 268–270, *270,* 273, *274*
 fluctuations and cyclic oscillations, 246–255
 inbreeding, 277–278
 increase (future), 475, 477
 indices of density, 225–226
 life history traits and tactics, 280–281
 metapopulations, 267–268, *268*
 methods for estimating densities, 227–229
 mortality, 229–233
 natality, 229–233
 overshoot of human, 460–461
 patterns of dispersion, 258–260
 population age distribution, 233–235, *236*
 properties, 225–236
 seasonal changes, 246
 statistical methods, 491
 use of term, 5
Population age distribution, 233–235, *236*
Population density, 225–226
Population dispersal, 246
 patterns of dispersion, 258–260
Population dynamics, 237
Population equivalents, 37
Population fluctuations, 246–255
 overview of cycling, *255,* 255
Population genetics, 275–280
 defined, 276
Population growth forms, 241
Population-level data displays, 504
Population limitation hypothesis, 266
Population regulation
 density dependent/independent, 255–258
 multifactorial model, *254,* 254
 quality, 257
Positive feedback, 68
 ecological assessment, 464
Potential energy, 79
 units, 80, *81*
Potter, V. R., 473
Power, 483
Power density, 133
Power or rate units, 80, *81*
Precambrian, *362,* 363
Precautionary principle, 468
Predation, 283, *284, 285*
 interspecific competition, 295
 species diversity, 322
 types of relationships, 296–304
Pre-industrial agriculture, 33, *34,* 36
 as domesticated landscape, 405, *406*
Prescribed burning, *195,* 196, 197, *198, 445*
Primary consumers, 28
 size-metabolism relationships, 126
Primary production
 chlorophyll, 101
 distribution of, 94–96
  vertical distribution and biomass, 95
  world distribution, *96*

human use, 96–100
food, supply, 96–99
as fuel, 99–100
measurement problems, 104–105
Primary productivity, 86
history, 95–96
Primary succession
defined, 337
species replacement, 346
Prisoner's dilemma game model, 288
Producer organisms, 27–28
Producers, 22
density and biomass compared, *30*
pond, 27–28
Production efficiency, 119
Productivity, 86–91
biodiversity, 100–101, 323–324
evapotranspiration, 207
Profundal zones, 29, 426
Protocooperation, 283, *284*, 285
cooperation and aggregation, 306
Pseudo-replication, 483
Publishing scholarly research, 496–498
Pulse stability, 370–371
Pulsing balances, 9, 472
Punctuated evolutionary theory, 364–365
p-value reliance, *497*, 497–498, 509

Q

Quadrat or transect sampling, 228
Qualitative defenses, 304
Quantitative defenses, 304

R

Radial-arm irrigation, *400*, 402–403
Radiation
types of measurement, 203
See also Ionizing radiation
Radiation dose, *203*, 203–204
Radioisotopes, 201
Radiometers, 86
Radionuclides, 201
biological magnification, 216
Rainfall, 206
Rain forests
ecosystems, 171
temperate rain forests, 439, *440*
tropical rain forests, 450–454
tree layers, 451
Rain shadow, 206, *207*
Random distribution, *258*, 258, 259
Randomization, 481
Random variables, 483
Random variation, 483, 492
Rapids zone, 427
Rare species advantage, 395
Rate
basic concepts, 236–238
obtaining, 237
and yield, 87
Rate of production. *See* Productivity
Reciprocal transplants, 183
Recycling, 143
ecological theory of, *174*
paper, *174*, 174–175
Recycling pathways, 172–173
Red tides, 101
nitrogen fixation, 149–150
Reductionist approaches, 8–9, 329–330
disciplinary reductionism, *15*, 15–16
statistics, 488
Redundancy, 324
Refuges, 263
Refuging, 261
Regenerated corridors, 384
Regional climax, 356, *357*, 358
Regionalism, 18
Regions
defined, 6
economics and population, 473–474
Regions of influence, 143
Regular clumped, *258*, 258
Regular or uniform dispersion, *258*, 258, 259
Regulatory factors, conditions of existence, 185–187
Relative abundance, 226, 227
Relative density, species, 228
Relative dominance, species, 228
Relative frequency, species, 228
Relative humidity, 206–207
Relative mobility, 167
Relay faunistics, 346
Relay floristics, 346
Reliability, statistical, 486–489
Remnant corridors, 381, *382*
Removal sampling, 228
Reproductive potential, 239–240
Rescue effect, 392
Research design, 482–483
Reservoir pools, 141, 142
sulfur, 151, *152*
Residence times, 162–163
Residential housing developments, *409*, *410*, 411
Resilience stability, 70, *70*
Resistance stability, *70*, 70
Resource corridors, *383*, 384
Resource quality, food, 116–117
Resprout species, 198–199
Restoration ecology, 342, *343*
conservation biology, 458
key groups, 472
prevention vs., 45
revisited, 471–472
when succession fails, 371
Reward feedback, 69, 117–118
Rhizospheres, 188
Rice paddies, 172, 429, 457
Richness, 37–38
species diversity, 316
River continuum concept, 159, 427
River discontinuum concept, 162
Riverine forests, 452
Riverine wetlands, 428
Rivers. *See* Lotic ecosystems
Robust measures, 501
R-selection, 269, 271–273, 274–275, *275*
attributes, 272
MacArthur's model, *273*
Ruminants, 434

S

Salt-water marshes, *431*, 432
SAM. *See* Standardized aquatic microcosm
Sample, 491
Sample size, 483
Sampling methodologies, 228
Saprophages, 22
density and biomass compared, *30*
Saprotrophs, 22
Satellite species, 38
Savanna
fires, 194, 196–197
tropical grasslands and savannas, 445–446, *446*
Scale
concept of, 396
ecosystems and, 480–482
temporal and spatial, 396–399

Scenarios, 472–477
defined, 472
Scientific hypothesis, 494
Sclerophyllous woodland, 447
Scope of inference, 486
SCS. *See* Soil Conservation Service
Sea level changes, *159,* 159
Seashores and estuaries, 420–422, *421*
Secondary chemistry, 286
Secondary consumers, 28
size-metabolism relationships, 126
Secondary metabolites (secondary compounds), 58–59
Secondary plant compounds, 254
deterring herbivores, 287
Secondary production (SP), 107
Secondary productivities, 87
estimates of global, *89*
Secondary succession
defined, 337
example, 347, *348,* 349
models, 354
Second law of thermodynamics, 78, *79,* 80
Sedimentary biogeochemical cycles, 141
Self-organization, 355–356
Self-thinning, 257
Self-thinning curve, 257
Semi-evergreen seasonal tropical forests, 450, *451*
Seral stages, 337
Sere, 337
secondary plant succession, 347, 349
Serpentine soils, 181
Servomechanisms, 68
Shade-adapted
chlorophyll, 101
defined, 95
Shade adapted, light intensities, 201
Shannon index, 38, *40,* 40, *41*
Shannon-Weaver index
described, 317
dominance diversity profiles, *318*
Shelford law of tolerance, 179
Shifting agriculture, 36, 171
Short-day plants, 186
Sigmoid growth form, *241,* 241–242, *244*
density dependent or intrinsic, 256
Sigmoid growth pattern, 244–245, *245*
Sigmoid population growth, carrying capacity, *128,* 128–129
Significance levels, 486
Silent Spring (R. Carson 1962), 91
Simpson index, 39–40, *40*
Skewed distribution, 497
Small watershed technique, 31
Social hierarchy, 262–263
Social traps, 466
Soil, 187–191
definition, 187
soil displacement, 191–193
soil quality and environmental quality, 193–194
Soil Conservation Service (SCS), 191–192
Soil horizons, 188
Soil profile, 188
Soil quality, 193–194
Soil types
anthrosol, 191
classification, 190
major types, 190, *191*
mature and immature, 190
Solar constant, 83
Solar energy, fossil fuel, energy quality, *123,* 123–124
Solarimeters, 86
Solar-powered systems, 133
Solar radiation, 82–86
average daily, *85,* 85
energy dissipation, *86,* 86
Source reduction, 221
Source-sink concept, metapopulation, 268
Source-sink energetics, 94
South Pacific coral atoll, 43, *44*
SP. *See* Secondary production
Spacecraft, 60
as an ecosystem, 64–65
Spatial and temporal scale, 396–399
Spatial niche, 312
Specialists, foraging, 271
Speciation, 364–366
Species
keystone species, 324
population density, *226,* 226–227
types of interaction between two species, 283–286
Species diversity
additional graphic method, *317*
components, 316
ecosystem development, 353–354
Species replacement, 346–351
population interactions, 339
Species structure indices, 38–39, *39*
Specific growth rate
defined, 237
intrinsic rate of natural increase, 238–240
Sport-fishing ponds, food chains, *113,* 113–114
Springs, 428
S-shaped growth form, *241,* 241–242, *244*
See also Sigmoid growth form
Stability
concept of, 69–70
diversity-stability hypothesis, 323
resistance and resilience, *70,* 70
species diversity related, 320–322
Standardized aquatic microcosm (SAM), 61, *62*
Standing crop present, 88
Star Trek, 474
Statement, use of term, 13
State variables, 11
Statistical inferences, 483
Statistical thinking, 490–492
two ways of discovery, 506–508
Stem-and-leaf displays, 498, *500,* 501–502
Steno-, use of, 180
Stenothermal, 180, *181*
Stewardship versus dominion, 466
Stochastic disturbances, 219
Streams. *See* Lotic ecosystems
Stream spiraling, 162
Stress ecology, 42
Stress tests, 179
Strontium, 169, *205,* 205
Structural features, ecosystems, 75
Subsidy-stress gradient, *93,* 93–94
Sulfur cycle, 143, 151–153, *152*
reservoir sizes and turnover times, *144*
Sun-adapted
chlorophyll, 101
defined, 95
Sun adapted, light intensities, 201
Sunflower, *407,* 407
Surface fires, 194, *195,* 196
Survival rates, 229
Survivorship curves, *231,* 231–233, *232, 233*

Sustainability
concept of, 130–131
defined, 131, 404
global, 467–472
Swamps, 429
Swidden agriculture, 171
Symbiosis, 285
Sympatric speciation, 364, 365
Synergetics, 355
Systematic variation, 492
Systems, defined, 4

T

Taiga, 439
Tall-grass prairies, 443–444, *444*
Technoecosystems, 71–74
human-dominated, 21
natural ecosystems compared, *73*
natural ecosystems related, *72*
parasitic nature, *372*
urban-industrial, 408–411, 458
Technology
countertechnology, 470
paradox in technological development, 469–470
tools for ecology, 9
Temperate deciduous forests, 440–442
Temperate grasslands, 443
Temperate rain forests, 439, *440*
Temperature
interaction with moisture, 211
as limiting factor, 199–200
variability, 200
water's thermal properties, 424, 426
Temporal and spatial scale, 396–399
Terrestrial biomes, 432–457
biomes of North America, *436*
biomes of world, *435*
caves, 456–457
chaparral and sclerophyllous woodland, 446–447
deserts, 448–450
mountains, 454–456
northern coniferous forest biomes, 439–440
polar and high-mountain ice caps, 438–439
semi-evergreen seasonal tropical forests, 450, *451*
temperate deciduous forest, 440–442
temperate grasslands, 443–445
tropical grasslands and savannas, 445–446, *446*
tropical rain forests, 450–454
tropical scrub or thornwoods, 454
tundra (Arctic and Alpine), *437*, 437–438
Terrestrial ecosystems
soil, 187–191
types, 76
Territoriality
defined, 264, 266
home range and, 263–264, *265*, 266–267
population density, 256
Territories
defined, 264
song thrushes, *265*
Tertiary consumers, 28
Theory, knowledge and research design, 482–483
Theory of punctuated equilibria, 364–365
Thermal pollution, 221–222
Thermal radiation, 84–85
Thermocline, 426
Thermodynamics, laws of, 78–82
Thermometers, 200
Thornwoods, 454
3/4 power law, 124–126
Throughflow, 173
Tidal freshwater marshes, *431*, 431–432
Tillage, 308–309, *309*
Tilth, 308
Time element, 87
Tolerance model, 354
Top-down controls, *120*, 120–121
Top-down hypothesis, 119–120
Top-down management, 471
Total counts, 228
Total human ecosystem, 72
Toxic substances, biological magnification, 215–218
Trace elements, 213–214
Tragedy of the commons, 461–462
social traps, 466
Transcending functions, 9
Transdisciplinary holism, *15*, 16
Transformity, 123
Transgenic crops, 367–368
Transpiration efficiency, 208
Treatment, 481
Treatment effects, 485
Treatment levels, 485
Tree farms, 370, 457
Tribolium-Trifolium model, 291–293, *292*
Trophic cascade hypothesis, *120*, 120–121
Trophic levels, 105
energy flow in linear food chain, *106*
food web, 109
Trophic niche, 312
Trophic structure, 21–24
Tropical agriculture, 170–172
Tropical rain forests, 450–454
Tropical scrub, 454
Tropical seasonal forests, 450, *451*
Tuckfield maxim, 482
Tundra, *437*, 437–438
food chains, 112–113
radioactive fallout, 169
Turnover, defined, 30
Turnover rates, 162–163
Turnover times, 162–163
Two-sample alternative hypothesis, 495
Type I error, 495, 506
Type II error, 495, 506
Type III error, 496

U

Ultraviolet radiation, 362
Uncertainty
limits of statistics, 491
peer assessment, 506–507
vanishing, 498
weight of evidence, 507
weight of evidence paradigm, 508–509
Underdeveloped countries
agriculture, 471
defined, 98
Undergrazing, 111
Unequivocal results, 481
Universal energy flow model, 105–107
Universal environmental ethic, 477
Universal model, defined, 105
Unsubsidized solar-powered ecosystems, *133*, 134
Upwelling regions, 418
fishing, 415
major upwellings and currents, *417*
Urban-industrial technoecosystems, 408–411, 458

Urban sprawl, soil erosion, 192
Use of diversity profiles, 40, *41*
Utilization efficiencies, 109, *110*

V

Vagility, 246, 379
Vapor pressure deficit, 207
Variables, 483
Variation, 491–492
  random, 483, 492
  systematic, 492
Vents, deep-sea, 50–51, *51*
Vesicular-arbuscular (VA) mycorrhizae, 308
Visions of the future, 474
Volatile organic matter (VOM), 22–23
Voles
  cycles, 248, *250*
  landscape-based hypotheses, 504, *505*

W

W. *See* Watt
Water
  cycle (*see* Hydrologic cycle)
  desert plants and animals, 450
  principal factors, 206–208
  temperature changes, 424, 426
Watersheds
  biogeochemistry, 163–168
  community types, 387
  defined, 6, 31
  ecological succession, 349–351
  fossil record, 329
Watershed systems, concept, 31–32
Watt (W), *81*
Wave-generated succession, *344*, 344
Wavelengths, 200–201
Weight of evidence
  corroboration, 492–493
  evidence-oriented approaches, 498–505
  statistical thinking, 490
  uncertainty, 507
  See also Evidence
Weight of evidence paradigm, 508–509
Wetlands, 424, 428–431, 464
Wildfires, 194
Wilson, E. O., 475
Wind, 214–215, *215*
Wood, as fuel, 99–100

Y

Yellowstone National Park fires, 197–198
Yields
  fish culture, 114
  and rates, 87
Youth-maturity parallels, 464, 477, *478*

Z

Zinc, biological magnification, *218*
Zonal approach, 386
Zonation, 24, *25–26*, 26
Zooplankton, 28

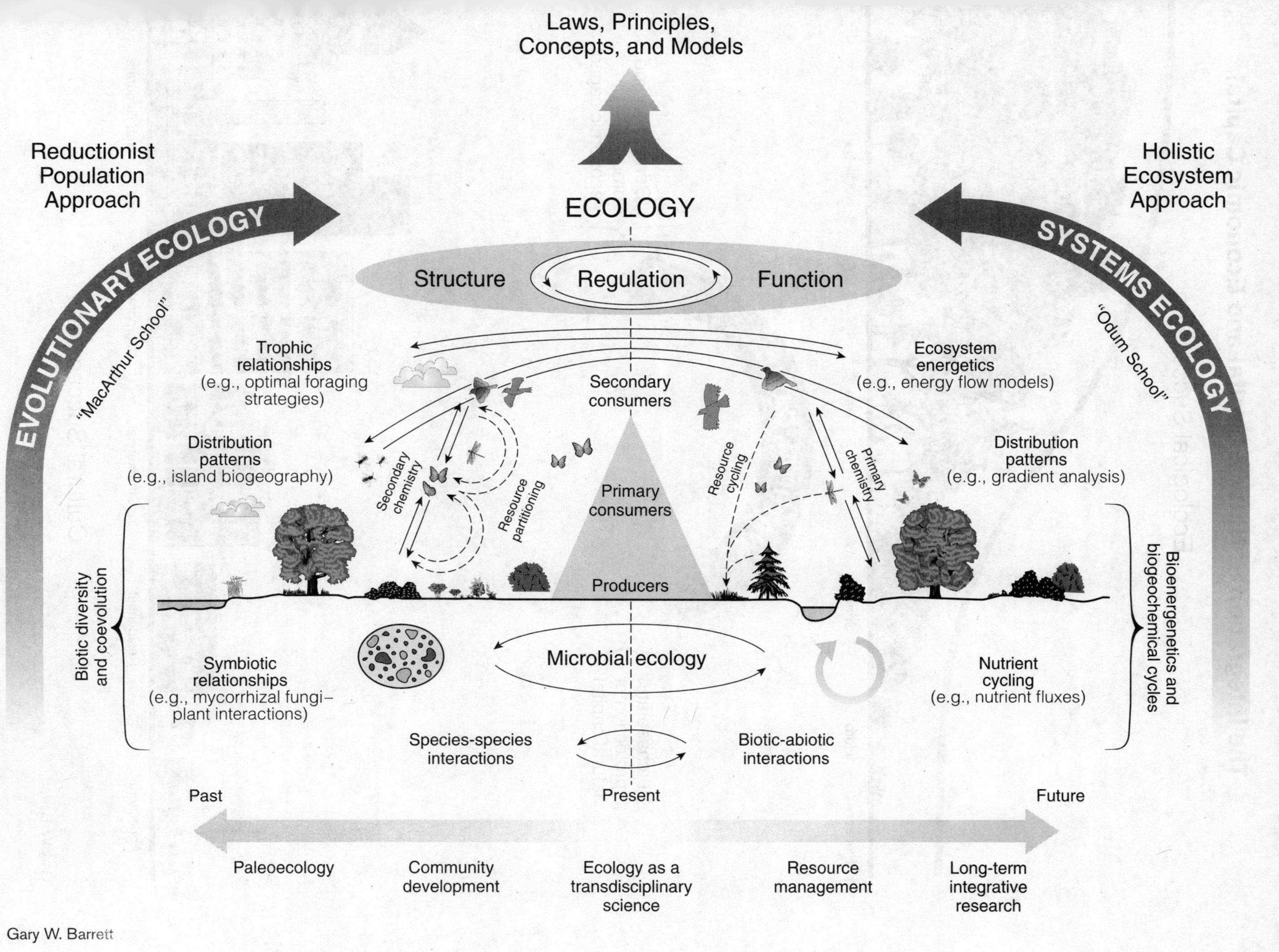
Laws, Principles, Concepts, and Models
Reductionist Population Approach
Holistic Ecosystem Approach
EVOLUTIONARY ECOLOGY
SYSTEMS ECOLOGY
"MacArthur School"
"Odum School"
ECOLOGY
Structure
Regulation
Function
Trophic relationships (e.g., optimal foraging strategies)
Ecosystem energetics (e.g., energy flow models)
Secondary consumers
Distribution patterns (e.g., island biogeography)
Distribution patterns (e.g., gradient analysis)
Secondary chemistry
Resource partitioning
Primary consumers
Resource cycling
Primary chemistry
Producers
Biotic diversity and coevolution
Bioenergenetics and biogeochemical cycles
Microbial ecology
Symbiotic relationships (e.g., mycorrhizal fungi–plant interactions)
Nutrient cycling (e.g., nutrient fluxes)
Species-species interactions
Biotic-abiotic interactions
Past
Present
Future
Paleoecology
Community development
Ecology as a transdisciplinary science
Resource management
Long-term integrative research
Gary W. Barrett

The Integration of Natural Capital and Economic Capital

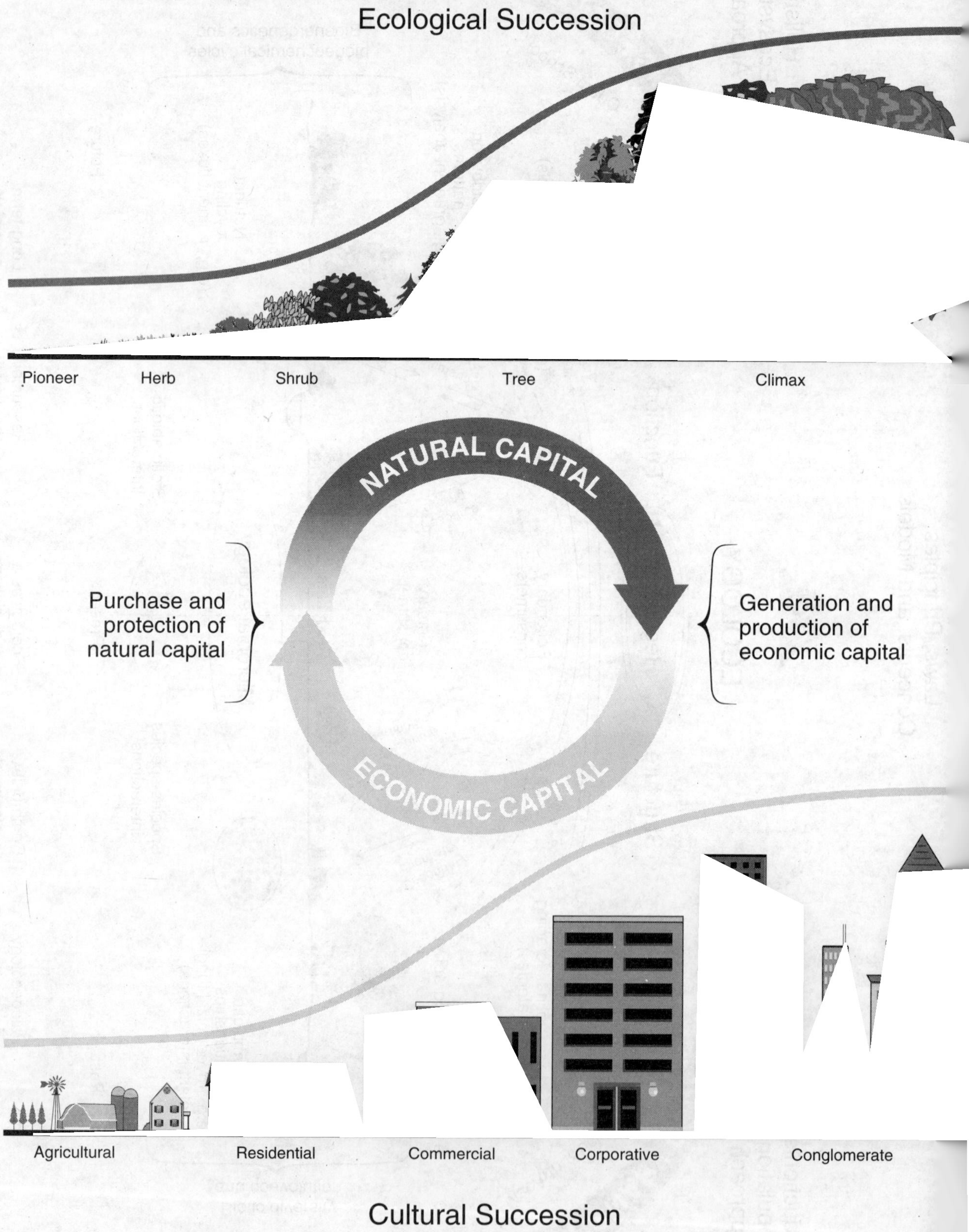

Gary W. Barrett